LASER PULSES – THEORY, TECHNOLOGY, AND APPLICATIONS

Edited by **Igor Peshko**

Laser Pulses – Theory, Technology, and Applications
http://dx.doi.org/10.5772/2624
Edited by Igor Peshko

Contributors

Igor Peshko, E. Nava-Palomares, F. Acosta-Barbosa, S. Camacho-López, M. Fernández-Guasti, Zhiyi Wei, Binbin Zhou, Yongdong Zhang, Yuwan Zou, Xin Zhong, Changwen Xu, Zhiguo Zhang, Kazuyuki Uno, Ricardo Elgul Samad, Leandro Matiolli Machado, Nilson Dias Vieira Junior, Wagner de Rossi, A. Yu. Ivanov, S. V. Vasiliev, Roman V. Dyukin, George A. Martsinovskiy, Olga N. Sergaeva, Galina D. Shandybina, Vera V. Svirina, Eugeny B. Yakovlev, V.I. Mazhukin, Kallepalli Lakshmi Narayana Deepak, Venugopal Rao Soma, Narayana Rao Desai, Zhongyi Guo, Lingling Ran, Yanhua Han, Shiliang Qu, Shutian Liu, Emmanuel d'Humières, Akira Endo, Evgenii Gorokhov, Kseniya Astankova, Alexander Komonov, V. V. Apollonov, Guofeng Zhang, Ruiyun Chen, Yan Gao, Liantuan Xiao, Suotang Jia, Kun Huang, E Wu, Xiaorong Gu, Haifeng Pan, Heping Zeng, J. Degert, S. Vidal, M. Tondusson, C. D'Amico, J. Oberlé, É. Freysz

Published by InTech

Janeza Trdine 9, 51000 Rijeka, Croatia

Publishing Process Manager Oliver Kurelic
Technical Editor InTech Prepress, Novi Sad
Cover InTech Design team

Additional hard copies can be obtained from orders@intechopen.com

Laser Pulses – Theory, Technology, and Applications, Edited by Igor Peshko
p. cm.
ISBN 978-953-51-0796-5

For EU product safety concerns:
IN TECH d.o.o., Prolaz Marije Krucifikse Kozulić 3, 51000 Rijeka, Croatia,
info@intechopen.com or visit our website at intechopen.com.

Meet the editor

Professor Igor Peshko is currently with Wilfrid Laurier University, Department of Physics and Computer Science, Waterloo, Canada. During his career he worked in Universities, National Labs, Government Agencies, and private companies in Ukraine, Poland, Germany, Canada, and USA. His priorities were in lasers physics, applied optics, and sensor areas. He is currently working on general concepts, hardware, and data processing algorithms of multifunctional scientific instruments. Among them: 1) instruments for extra-terrestrial research; 2) chemical identifiers; 3) field-labs for environmental contamination monitoring and for exploration of oil and gas sites; 4) medical devices for pathologies identification; and 5) mobile and stationary sensory robotic terminals for operating within security networks at high-risk zones. He has authored and co-authored more than 120 publications.

Contents

Preface

This book is devoted to some aspects of ultrashort laser pulses generation, characterization, and applications. As of today, hundreds of books discussing these subjects have been published. More and more techniques go to practical use every year. Shorter and shorter pulses are routinely achievable. New spectral ranges, like X-rays, deep UV, middle and far IR, including TeraHertz bands, became a reality.

In modern laser world, the word "Pulse" typically covers pulse durations from microseconds to femtoseconds. In principle, it is possible to generate attosecond pulses (10^{-18} s) by using non-linear processes. Recently, new time ranges were discussed in publications: zeptosecond (10^{-21} s) and yoctoseconds (10^{-24} s). To generate pulses from milliseconds to femtoseconds, hundreds of different laser systems have been developed. They can typically generate pulses of specific durations, which are due to laser principles of operation, specific construction, parameters of gain medium, type of modulator, and so on. Smooth tuning of laser pulse duration continues to be a big problem. The shorter the achieved pulse durations are, the more difficult the problem is to determine how to measure such pulses. The solution is inside the laser pulses. Light itself contains information about Time. Generating "light in time", the pulsed lasers combine these two categories as light is a periodical, cyclic process and can be a measure of time, of length, and of frequency.

During the 50 years of their history, the pulsed lasers passed from seconds to zeptoseconds or about 20 orders of magnitude into the short duration's side. This road was difficult. Every time, starting from huge, complex, ineffective, and very expensive machines, the lasers became elegant and economical; they are being transformed to powerful and smart instruments in science, technology, medicine, and everyday life.

Very popular picosecond and femtosecond pulses typically are achieved by mode-locking technology. The theoretical concept of this is based on the light wave-approximation. However, in many cases the description of the pulse formation and its evolution is better understandable and easily describable in terms of corpuscular approximation. In introductive chapter "Time and Light", the theory of optical noise is applied for explanation of ultrashort pulses phenomena. This approximation can estimate statistical parameters of generation: probability of single-pulse formation (on the round-trip cavity period), depending on active media gain spectrum width, rate of gain increase, cavity length, output mirror reflectance, and other laser practical parameters.

In total, the book consists of five sections housing seventeen chapters. In such complicated and multidisciplinary area as laser pulse generation and optical pulse-matter interaction it is sometimes difficult to specify, which domain of research the chapter belongs to. Conditionally, the chapters have been separated into five interconnected sections:

Introduction: Time and Light – Historical/philosophical/technical overview, optical noise theory of mode-locked lasing, and exotic self-mode-locking technologies;

1. Pulsed World – Characterization of ultrashort pulses, pulsed laser gain media and technologies;
2. Cutting and Shooting – Material processing, refractive index modulation, special structure recordings, interaction of laser radiation with solids;
3. Unusual Applications – Water purification, accelerators, modification of the solids structures;
4. Seeing invisible – Non-linear and single-molecule spectroscopy, THz-technologies.

The scientific editor of this book does not always agree with some concepts, models, and explanations demonstrated in this book. However, the principle of open access publishing is to give possibility for each author to freely demonstrate his/her understanding of phenomena. We know a lot of examples where future Nobel Prize Winners were rejected by solid journals because of negative and non-discussable opinion of the reviewers – the best specialists in specific area. The judge cannot be judged and it is only history that can judge and corrects everyone and everything.

In any case, this book is not a milestone in physics and/or technologies; rather it is like a running train with coaches that will be changed at each new station. The book does not want to go to specific destinations but allows you to sit and think of where your destination could take you. We hope this book will be useful for a wide spectrum of specialists, for professors and students, and for those who are interested in history and in future of the laser technologies.

Igor Peshko
Department of Physics and Computer Science,
Wilfrid Laurier University, Waterloo,
Canada

Section 1

Introduction

Chapter 1

Time and Light

Igor Peshko

Additional information is available at the end of the chapter

http://dx.doi.org/10.5772/54208

1. Introduction

1.1. From seconds to attoseconds

This book is devoted to Laser Pulses. In the modern laser world, the word "Pulse" covers pulse durations from microseconds (free-running laser) to tens of femtoseconds (1fs is 10^{-15} s) (mode-locked laser). It is possible to generate attosecond pulses (1as is 10^{-18} s) by using non-linear processes. Recently, a new time range was discussed in publications: zeptosecond (1zs is 10^{-21} s). To generate pulses from milliseconds to femtoseconds, hundreds of different laser systems have been developed. They can typically generate pulses of specific durations, which are due to laser principles of operation, specific construction, parameters of gain medium, type of modulator, and so on.

In the solid-state free-running laser the parameters of the electromagnetic field interaction process with an inversed population of the gain medium play a dominant role in shaping the laser spikes. These characteristic parameters limit the pulse duration from the long side of the range. A chaotic sequence of such spikes can be as long as the pumping source could effectively excite the gain medium. In some technological applications this can be hundreds of milliseconds envelop. To achieve laser pulse duration of a few seconds, it is necessary to use for modulation the processes with characteristic times of the same order of magnitude. On the short side of achievable durations, another limitation exists. At certain conditions, waves, interacting with solids, can shape a single peak of energy, propagating "alone". In optics, such a single wave is called a "soliton". A tsunami is an example of a mechanical soliton. To propagate in a crystal, an "optical tsunami" can excite the medium and get back the energy at some conditions. In a vacuum, there is no medium that can accumulate energy and support existence of a relatively short, lossless wave. Hence, in a vacuum, a single pulse could be shaped as a wave-package, which is a result of interference of many independent electromagnetic waves, propagating co-axially in the same direction. Since light is an electromagnetic wave repeatable in space and time, a single wave period with minimal length (among all present in the wave set) is the minimal possible duration of an energy pack. Out of the pulse, the negative interference suppresses the energy presence.

Pulse measurement techniques were being developed together with each type of laser and are based on principles depending on the specific range of pulse duration. The shorter the achieved pulse durations, the more difficult the problem of how to measure such pulses. The answer is inside the laser pulses. Light itself contains information about Time.

In summary, we can say that lasers, generating "light in time", combine these two categories as light is a periodical, cyclic process and can be a measure of time and of length. Since this is an introductive chapter, let us consider some historical milestones on the way of Time/Light understanding first.

1.2. In the past

On the way of analysing and understanding Nature, ancient scientists interpreted Light and Time as two absolutely different categories. To this day, it is not clear what Time is - is it a characteristic of processes, or an independently existing parameter? Theoretical physics operates with 4D space: 3 space coordinates and the fourth as time. To measure any distance, a researcher can compare some etalon of length with the object of interest. However, historically, the etalon of length was voluntarily chosen: 1 m is not a "natural" unit of length. At the same time, the light specific wavelength is a natural, repeatable etalon of length. The same is true for time – any process running with macro-objects being involved is just approximately repeatable. Since the speed of light (in vacuum) is the same everywhere, it can be used for measuring time and length: the same number of light waves with specific length corresponds to the same distance in any corner of the Universe – at least, we think so.

If existence of Light is absolutely evident, what supposed to be Time is not very clear until now. A lot of serious discussion and speculation was done but the problem still has to be solved. Time is not a single example of a "questionable" phenomenon. Very often, scientists propose a phantom model that helps make a very helpful device or theoretical approximation. The most popular parameter in optics - "index of refraction" - is probably the most investigated phenomenon that does not exist in Nature at all. This is a very useful model that helps provide theoretical research and find technological solutions in the fields of optics, photonics, and other related areas of activity.

We see or measure light as a result of photon interaction with more "solid" matter: atoms in our eyes or other detectors. In the case of time, the situation is more complicated. Since the early ages of humanity, people observed some regularly repeatable events, like day-night, winter-summer, etc., that, de-facto, were connected with complete or partial appearance and disappearance of light. Probably the first instrument for estimation of day time was the sundial – a vertically installed bar that showed the time period relative to specific points when the position of the sun over the horizon is maximally high and bar shadow is maximally short (midday). However, in cloudy days or at night, the system was useless. The next step in time measuring devices was the sand or water clock (Clepsydra): a spring of water or sands running through a small hole filled or emptied some calibrated tank. This is the most ancient experimental 3D demonstration of the basic formula of time: $T = M/R$ where M is mass and R is rate of mass changes. A very well-known 1D variant of this formula reads: $T = L$ (length)/V (velocity). In case of a "floating clock", M is the total amount of water in the tank and R is the amount of water that drops through the hole during a unit of time. In this case, to build two identical clocks, one needs to make holes with ideally identical cross-sections. So, accuracy of time measurement is linked to accuracy of length measurement.

The next step was a mechanical clock with a pendulum. This device periodically transforms kinetic energy of motion into potential energy of a pendulum in the gravitation field. The electronic clocks use a periodic process of transformation of the electrical energy stored in a capacitor into magnetic energy of a coil. Each repeatable process can be used for measuring time. This could be a planet's rotation around a star, or a planet's rotation around its own axis, or mechanical vibrations – sound, or electromagnetic waves – light.

As of today, the history of pulsed lasers spans about 50 years. Hundreds of books devoted to this subject have been published during this period. More and more techniques go to practical use every year. Shorter and shorter pulses are routinely applicable. New spectral ranges, like deep UV, middle and far IR, including TeraHertz bands became a reality. This book demonstrates new achievements in theory, experiments, and commercial applications of pulsed lasers.

1.3. What is this chapter about?

Until the mid-eighties of the previous century, most of the lasers generating ns- and ps-pulses operated with passive or active Q-modulator. However, it became more and more clear that such non-linear and multi-parametric system as a laser could support self-effecting, self-modulating mode of operation without any external devices or internal elements. Starting from the time of the laser's invention, the theoretical models of laser operation include mode spectrum genesis. This is logically understandable, as the open Fabry-Perot cavity is one of the main components of the laser system. However, the picosecond or, moreover, femtosecond pulse generator has thousands and thousands of modes. Theoretical description of few modes is possible and can be used for analysis of laser operation. However, solving, for example, fifty thousand equations that describe all operating modes is a problem and the results could be very far from the reality.

The author of this chapter considers another approximation. The logic of this is as follows. The laser generation develops from luminescence radiation, which is de-facto, an optical noise. Let us analyze how this noise is being modified propagating through the gain medium after the laser threshold is achieved. At each moment of time, the spectral width and averaged power level determine the statistical properties of radiation. So, depending on laser parameters, the exceeding of the maximal, stochastically appeared optical spike over the next smaller one irradiated during a cavity round-trip period, could be found. At some conditions, the multiple, small-amplitude spikes saturate the amplification of the gain medium and stop the in-cavity power accumulation. Only the highest spikes that are able to saturate the non-linear absorber continue growing.

Around the 1960s, the properties of noise in radio-band were very well investigated on the way of radio rangefinders development. This theory was applied to describe the mode-locking process. Some difference was just in mathematical description: in radio-band the measured value is amplitude (strength) of the electromagnetic wave, and in optics, the square of this value, namely, intensity (power) of light. This approximation can estimate statistical parameters of generation: probability of single-pulse formation on the round-trip cavity period, depending on gain media spectrum width, rate of gain increase, cavity length, output mirror reflectance, and other laser practical parameters.

In this chapter, one can find interesting concepts, theoretical models, and experimentally proven techniques that for different reasons were not revealed in public at the moment they were proposed and demonstrated. Two different laser systems will be discussed: 1) single-frequency laser that, at certain conditions, is capable of slow self-modulation initiated by thermal processes in the cavity; 2) multi-mirror laser that, at certain conditions, is capable of self-mode-locking and generation of ultrashort pulses without any modulator.

This chapter represents philosophy of creation of laser pulses of different gradations:

1. "Slow" pulses of second duration
2. Microsecond pulses in millisecond envelope
3. Nanoseconds: passive and active Q-modulation
4. Picoseconds: passive, active, and natural auto-mode-locking
5. Femtosecond lasers
6. Attosecond and Zeptosecond

For today, hundreds of books describing the laser operational modes, design, constructions, and technologies are available. The condensed description can be found in an open-access encyclopaedia by Dr. Rüdiger Paschotta [1]. In this chapter, we often refer to the encyclopaedia pages, which typically contain a wide list of publications, discussing specific area of research. This chapter focuses mainly on some specific laser regimes and theoretical concepts, which are not considered in traditional books and in on-line encyclopaedia, and can be useful for the design of simple, low power-consumption, and environmentally-stable systems.

2. "Slow" pulses of second's duration

First of all, the initial question should be answered: how to slowly modulate such a fast operating and sensitive system as a laser. In principle, this can be provided by slow "delicate" changes of the gain or losses (or both) in a stationary operated laser with narrow spectral band. Such modulation is very difficult to provide in case of a multi-mode laser because of mode competition process: if output power of one mode is decreased, the others immediately grow up. Hence, first of all, a single-frequency operation has to be provided.

Let us consider an example of specific single-frequency laser [2, 3] that is capable of providing cavity losses periodical self-modulating. This regime has been achieved in the diode-pumped single-frequency Nd:YVO$_4$, Nd:YAG, and Nd:YLF lasers with metallic thin-film (10nm) selector. The main idea of the absorbing thin film selector operation is the following. The metallic film, with thickness Δd significantly smaller than the standing wave period ($\Delta d<\lambda/100$), is placed in the linear cavity. If a thin film plane is adjusted to the node surface area of any mode, the losses for this mode become close to zero and single longitudinal mode starts to operate. A detailed description of the interferometer with an absorbing mirror can be found in [2, 3].

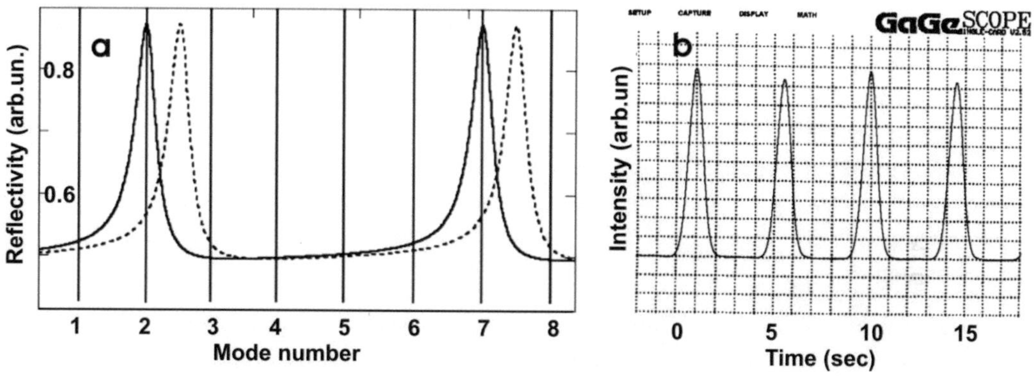

Figure 1. Mode switching mechanism: a) Vertical lines demonstrate laser cavity modes. The laser operates when the interferometer maxima (solid curve) coincide with the laser modes spectral positions, and does not – when the interferometer modes are located between the cavity modes (broken line); b) Oscillogram of the laser output power. One division of time scale is 1 sec. The peak power is approximately 0.5W. The radiation maxima correspond to the spectral positions of the interferometer maxima shown by the solid line in Figure 1a.

The pulses with duration of approximately 1 to 3 s and period of about 3 to 10 s (see Figure 1b), depending on pump power and thermo-optical properties of the cavity and gain crystal, have been observed. The effect has been explained by the thermal changes of the cavity length connected with the difference of the heat generation rate for operating and non-operating laser [4].

The self-pulsation regime of operation can be explained by periodical modulation of losses of a cavity with a thin-film selector caused by the thermally induced changes of optical length of an active medium. Let us assume that the thin-film selector is placed between the two nodes of neighbouring modes and the pump level is slightly below the threshold for both of the modes. At this moment, the laser does not operate; the maximum possible portion of pump power is transformed into heat and the temperature of the crystal increases. Because of the thermal elongation of active medium (and a total cavity), the positions of some nodes move to the absorbing thin-film location and the losses for this mode fall down. The mode switching mechanism is explained in Figure 1. Relative positions of the modes and the absorbing interferometer reflection peaks are shown. The interferometer is formed by a thin-film selector and an output coupler. The solid curve demonstrates resonance position that provides laser operation. The laser starts to operate in the single-frequency regime. At maximum output power, the heat dissipation rate becomes minimal and crystal temperature starts decreasing. Further, the thin-film location "comes out" from the node position and interrupts the laser emission process (dashed curve in Figure 1a). After the laser action break-off, the temperature rises up again and the process repeats.

3. Microsecond pulses in millisecond envelop

The operation of a free-running laser can be described and understood in terms of gain-loss dynamics [5,6]. When a pumping source intensively irradiates the laser medium, the gain grows until it becomes equal to all losses. Starting from this moment, the number of newly "born" (generated) photons is greater than those "dying" (absorbed or scattered). The light intensity grows very fast (avalanche conditions) and the laser starts generating. Very soon, the stimulate emission becomes so strong that the pumping radiation cannot refill the inversion population - in other words, the gain drops down to the level when generation is interrupted. The laser pulse is finished and during the next generation pause the inversion population grows up again until the gain achieves the loss level and the next cycle of generation begins.

When counter-propagating narrow-bandwidth light waves are superimposed, they form a so-called standing-wave interference pattern, the period of which is half the wavelength [7]. This results in so-called spatial and spectral "hole-burning": a) in maxima of standing wave the inversion population is falling down, shaping periodical spatial grating of the gain along the active medium; b) the operating modes are "eating" the gain at their spectral positions and other modes start generating. These have various consequences for the operation of lasers:

1. Difficulties in achieving a single-frequency operation or operation with stable and repeatable parameters when generating in linear cavities;
2. The optical bandwidth and spectral structure of a free-running laser radiation is different when the gain medium is located in different resonator areas;
3. Spatial hole-burning can reduce the laser efficiency, when the excitation in the nodes cannot be utilized.

The ring cavities support running waves and spatial hole-burning can be eliminated.

4. Nanosecond pulses: Passive and active Q-modulation

The Q-switching is a technique which provides generation of laser pulses with extremely high peak power: MegaWatt to GigaWatt [5,8]. Here Q means a quality-factor of the laser cavity. Typical durations of such pulses are between several nanoseconds to few tens of nanoseconds. This effect is achieved by modulating the intracavity losses. Initially, the resonator losses are kept at such a high level that the laser cannot operate at that time. Then, the losses are suddenly reduced by modulator to a small value. So, the laser becomes highly "over-excited" and starts generating just after several flights of light along the cavity. When the intracavity power has reached the value of the gain saturation, the laser radiation intensity drops down to the luminescence level. The resonator losses can be switched in different ways: through active and passive Q-switching.

For active Q switching, the losses are modulated with an active control element, typically either an acousto-optic or electro-optic modulator. Initially, there were also mechanical Q switches such as spinning mirrors or prisms. In any case, shorter cavities and more intensive pumping provide shorter lasing pulses.

For passive Q switching, the losses are automatically modulated with a saturable absorber. This can be a specific dye dissolved in liquid or solid matrix, some doped crystal or glass, and some bulk elements demonstrating the Kerr-type non-linearity. The pulse is formed as soon as the energy stored in the gain medium has reached a level sufficient to keep the absorbing particles in an excited state. The recovery time of a saturable absorber is ideally longer than the pulse duration. At this condition additional unnecessary energy losses would be avoided. However, the absorber should be fast enough to prevent the lasing when the gain recovers after the pulse termination.

5. Mode-locking

Mode-locking is a group of methods for laser generation of ultrashort pulses [9-13]. Typically, the pulse duration is roughly between 50 fs and 50 ps. To provide synchronous (phase-locked) operation of different modes, the laser cavity should contain either an active or a nonlinear passive element.

Figure 2 and Figure 3 demonstrate the difference between mode intensity and phase distributions for: (a) radiation that just is emitted at close to threshold conditions and (b) for already mode-locked radiation at the moment close to gain medium saturation. Simulation was done just for 10 modes to clearly illustrate the difference. In reality the number of operating modes may be tens or hundreds of thousands. The red line in Figure 2 shows the averaged mode intensity distribution. It is the same for any number of modes and any phase distributions. In any case, the initial gain medium luminescent radiation is a stochastic optical noise – a random sequence of occasionally irradiated spikes, which have different spectral content. The passive saturable absorber automatically selects (emphasizes) some spikes with relatively high energy and with relatively short duration. This results in some temporal uncertainty of the generation appearance and in generation of few ultrashort pulses on the cavity round-trip period because the gain medium continuously generates new noise patterns. The temporal distribution of radiation is "breathing" and is not ideally repeatable even on the neighbour round-trip periods of time. Moreover, the general view of time-intensity distribution of generation is unrepeatable from shot to shot – it could be one, or two, or even more high intensity pulses on the round-trip period.

In Figure 4 [14] two oscillograms of the laser output radiation are depicted: (a) an initial stage of generation development and (b) view of the mode-locked generation after non-linear absorber action. The arrows on the oscillogram (a) indicate the positions of strong spikes (peaks are out of the picture) that were transformed later to the single ultrashort pulses. The oscillograms have been acquired with coaxial vacuum photodiode and analogue wide-band oscilloscope.

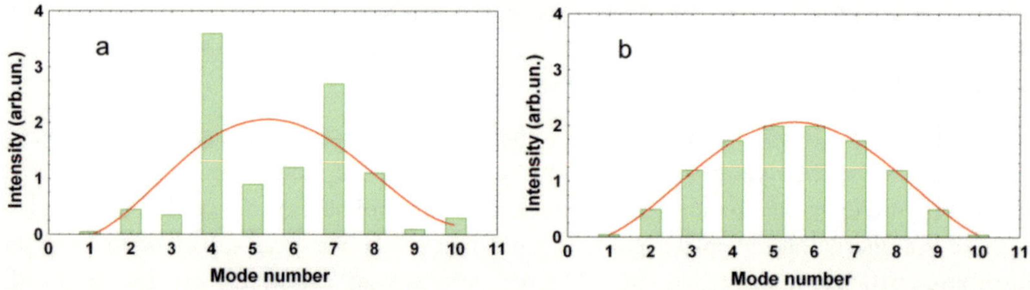

Figure 2. Spectral distribution of intensity of the modes: (a) at the beginning of generation development randomly modulated spectrum of a single noise patter; (b) at moment of pulse train emission. In both cases the red bell-shape line shows the averaged intensity distribution. Distribution (a) is chaotically being changed at each laser shot and during the generation development until achieves the distribution (b). The same is for the phases in Figure 3.

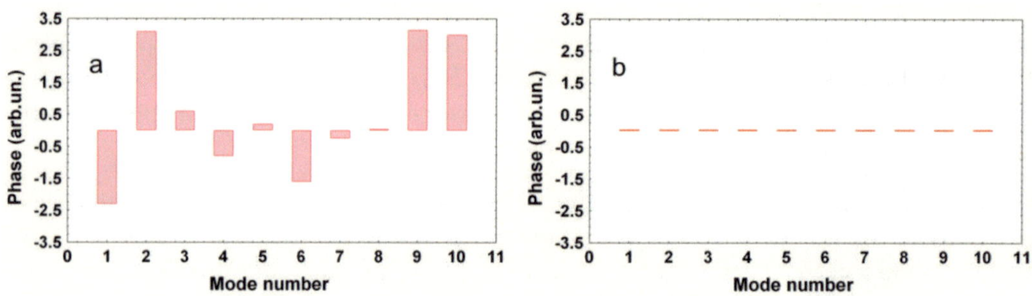

Figure 3. Spectral distribution of the phases of the modes: (a) at the beginning of generation development; (b) at moment of pulse train emission.

Each initial noise spike can be shaped by interference of the modes that belong to different parts of the spectrum and have different phase distribution. During the process of intracavity power increase even the neihbour (in time) spikes coud be modified differently. The ellipses in Figure 4 shows such case when only during two round trip periodes the first spike among three, initially being the smallest, became the highest.

In case of active modulation, the maximum of the modulator transparency very rarely coincides with occasional generated maximal spike of luminescence. However, since any spike is in-time repeatable on the round-trip period and, if coinciding with periodical maximal cavity Q-value, it will grow faster than even bigger spikes irradiated out of high Q-window. After hundreds of round-trips, this initially, maybe minimal pulse achieves the maximal amplitude and is generated as a train of laser pulses. Typically, at active modulation the pulse duration is longer but the jitter of the pulse appearance is much less as compare with passive modulation.

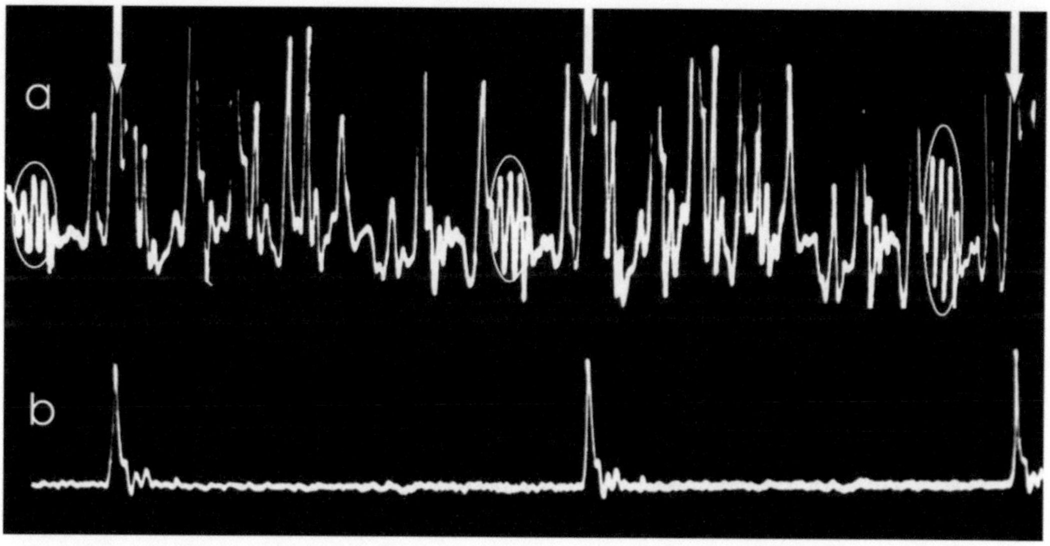

Figure 4. Oscillograms of the laser output radiation: (a) on the initial stage of generation development and (b) after non-linear absorber action. The arrows on the oscillogram (a) indicate the positions of strong spikes (their peaks are out of picture) that were transformed later to the single ultrashort pulses.

6. Propoperties of the optical noise

6.1. Mode formation and radiation dynamics

The phase and amplitude relations between the modes respond to the statistics of a normal (Gaussian) process. At the absence of phase modulation, a minimal possible duration of light random spike is inversely proportional to the total luminescence spectrum width. The amplification band typically has some specific intensity modulation. For relatively narrow spectral bands, the most energetic central modes introduce the main income into the pulse shape formation. At non-stationary pumping, the intensity and spectral width of the gain band is changeable in time that results in changes of random spikes statistics. Moreover, spectral content of the spikes additionally is being deformed because of dispersion, absorption and scattering processes in the cavity elements. As a result, the final time-intensity distribution may be absolutely different from the initial optical random noise distribution that was emitted at threshold conditions.

For the radiation propagating within the linear cavity of length L, the correlation function defers from zero, at least, on the time intervals $T_C = 2L/c$ (for the ring cavity $T_C = L/c$). The existence of signals correlation on T_C intervals means that the operational set of the modes is formed. At this moment, the averaged intensity of circulating in the cavity radiation should be significantly higher than the intensity of firstly irradiated luminescence pattern. After that moment, the temporal intensity distribution does not change significantly after each round-trip through the cavity. However, because of spectral dispersion of an amplification band and different spectral content of each spike, the intensity growth rate for each spike is different. During hundreds of cavity round-trips, the general view of a random noise-like pattern (realization) changes and may be significantly reshaped by the moment of the final ultra-short pulses emission. The shortest spikes shaped during the total period of radiation development may be "late" and not have enough time to be "accelerated" to the maximum intensity. It means that the gain bandwidth may exceed the width, actually used at certain values of pumping, cavity length, material dispersion of the elements, and other laser parameters.

With a Fourier analysis, it is possible to transfer a signal from time domain to frequency and inversely. The mode-locking technology provides zero phase difference between all operational modes. At Gaussian shape of the operational spectral envelope, the pulse intensity has the same shape in time. If one modulates statically the spectrum (by interferometer, for example), the respective modulation appears in temporal intensity distribution. In case of the pulsed pumping, it is possible to modify the gain spectrum by variations of pump rate, cavity length, dispersion of the cavity elements, losses of the cavity to achieve finally different ultrashort pulse durations.

6.2. Statistical properties of optical noise

The theory of electro-magnetic noise was initially developed for the radio band. The main parameter considered in this range is electro-magnetic wave amplitude (*E*). In optics, because of very high optical signals carrying frequency, the intensity (*I*) that is a square of field amplitude (*E*) is considered and measured. Mathematical analysis of the optical stochastic signals with formal exchange E^2 to I has been done in [14]. For detailed description of the mode-locked process, let us repeat here this analysis with taking into account the parameters of the laser cavity. At the beginning, let us rewrite the formulas of random process theory [15] but in terms of intensity, instead of the amplitudes.

The famous Wigner's formula demonstrates the connection between autocorrelation function $k(\tau)$ (time domain) and spectral density $S(w)$ (spectral domain) [16]:

$$k(\tau) = \frac{1}{2\pi} \int_{-\infty}^{\infty} S(w) e^{jw\tau} dw, \tag{1}$$

$$S(w) = \int_{-\infty}^{\infty} k(\tau) e^{-jw\tau} d\tau. \tag{2}$$

The commonly used optical parameter Δf_e is an energetic spectral width of the random process:

$$\Delta f_e = \frac{1}{S_0} \int_{-\infty}^{\infty} S(w) \frac{dw}{2\pi}, \tag{3}$$

here $S_0 = S(w_0)$ is spectral density at maximum of spectral density distribution.

The luminescence field of a solid-state laser has a view of an optical noise with relatively narrow spectral width ($\Delta\nu/\nu \sim 10^{-1} \div 10^{-3}$). During evolution of the laser radiation on initial stage (after the pumping applied), it becomes "more and more harmonical". The spectral width decreases for one to three orders of magnitude because of dispersion of the amplification coefficient [17].

The narrow band process is the stationary random process with a zero average value, the spectral density of which is concentrated near specific frequency f_0:

$$f_0 >> \Delta f_e \tag{4}$$

A narrow-band stationary random process that has symmetrical spectral density with relation to f_0 is named a quasi-harmonic one. Luminescence bands of most laser ions are shaped by a superposition of several spectral lines, even in such homogeneously broaden system as, for example, Nd:YAG.

This results in important conclusions that typically are missed in books and papers:

- Typically, a wide gain band is not symmetrical relatively to the maximum and may have several local maxima. In principle, luminescence radiation of gain crystal is not quasi-harmonic signal. However, it transforms to one because of significant narrowing during generation evolution.
- Zeros of the dispersion curves do not coincide with intensity maxima. Non-monotonic dispersion curve (within the operational spectrum band) stimulates compression or extraction of the equidistant mode spectrum. This means the variations of round-trip time (for different frequencies), and, automatically, the extraction of the single ultrashort pulse duration.

The correlation function of the narrow band random process may be represented as:

$$k(\tau) = \sigma^2 \rho(\tau) \cos\left[w_0 \tau + \gamma(\tau)\right], \tag{5}$$

here σ^2 is a dispersion of the process, $\rho(\tau)$ and $\gamma(\tau)$ are slowly modified functions as compare with $\cos(w_0\tau)$, $\rho(0) = 1$ and $\gamma(0) = 0$.

Correlation function of the wide band process typically is represented as:

$$k(\tau)=\sigma^2\rho(\tau)\ , \tag{6}$$

here $\rho(\tau)$ is normalized correlation function. As it follows from (1) - (6):

$$\sigma^2 = \frac{1}{2\pi}\int_{-\infty}^{\infty} S(w)dw = \bar{I} \tag{7}$$

Evidently, the parameter σ^2 represents total power of radiation and may be used for the normalization of the amplitudes or intensities of the random spike process. From a general view of the formula, describing probability density for the set of random values ξ_n [15], it is easy to get an expression for electromagnetic wave density distribution with substitution $n = 1$.

$$w_1(\xi_1) = \frac{1}{\sigma_1\sqrt{2\pi}} \exp\left[-\frac{(\xi_1 - m_1)^2}{2\sigma_1^2}\right]. \tag{8}$$

For the alternative electromagnetic field, the average value $m = 0$, and the final expression for the function, describing a density of probability of the amplitude A for the output laser radiation with an average intensity value of $\overline{I}$, is:

$$W(A) = \frac{A}{\sqrt{\overline{I}}} \exp(-A^2 / 2\overline{I}) \tag{9}$$

At that, the phase is distributed homogeneously on the interval $-\pi \leq \phi \leq \pi$.

6.3. Statistics of the laser optical noise pattern

From the statistics point of view, the different stages of the generation evolution may be classified as:

1. Luminescent field below the threshold: wideband random optical field;
2. First part of "linear" generation development (mode formation period): summarizing of the wideband and narrowband random signals;
3. Second part of "linear" generation development: quasi harmonic quasi periodical signal formation;
4. Nonlinear stage – gain or losses saturation.

Respectively, the statistical properties of radiation are different on each stage of generation genesis. By manipulating the initial internal radiation and laser external parameters, it is possible to achieve different results in the end of the path.

The sum of harmonic and quasi-harmonic signals [15] may be represented as follows:

$$S(t)+\xi(t) = A_m \cos(w_0t+\varphi_0) + A_c(t) \cos w_0t - A_s(t) \sin w_0t = V(t) \cos[w_0t+\psi(t)], \tag{10}$$

here $A_c(t) = V(t) \cos\psi(t) - A_m \cos\varphi_0$, $A_s(t) = V(t) \sin\psi(t) - A_m \sin\varphi_0$.

A random function $V(t)>0$ is envelope, and a function $\psi(t)$ is a random phase of the sum of harmonic signal and quasi-harmonic noise. From (10) it is easy to achieve:

$$V(t) = \{[A_c(t) + A_m \cos\varphi_0]^2 + [A_s(t) + A_m \sin\varphi_0]^2\}^{1/2} \; V(t)>0, \tag{11}$$

$$\psi(t) = arctg\frac{A_s(t) + A_m \sin\phi_0}{A_c(t) + A_m \cos\phi_0}, -\pi \leq \psi(t) \leq \pi. \tag{12}$$

Envelope $V(t)$ and random phase $\psi(t)$ have one-dimension probability densities:

$$W(\upsilon) = \upsilon \exp\left(-\frac{a^2 + \upsilon^2}{2}\right) I_0(a\upsilon), \;\; \upsilon = V / \sigma, \tag{13}$$

$$W(\psi)=\frac{1}{2\pi}\exp\left(-\frac{a^2}{2}\right)\left\{1+\sqrt{2\pi}a\cos(\psi-\phi_0)\Phi\left[a\cos(\psi-\phi_0)\right]\exp\left[\frac{a^2\cos^2(\psi-\phi_0)}{2}\right]\right\}, \quad (14)$$

here $a=A_m/\sigma$ is signal to noise ratio, $I_0(z)$ is Bessel function of zero's order with imaginary argument.

$$I_0(av)=\frac{1}{2\pi}\int_{-\pi}^{\pi}\exp\left[\pm iav\cos(\psi)\right]d\psi \quad (15)$$

$$\Phi(z)=\frac{1}{2\pi}\int_{-\infty}^{z}\exp\left(-\frac{x^2}{2}\right)dx \text{ is a probability integral} \quad (16)$$

From a general consideration, it is clear that the higher the average radiation power, the higher the number of random spikes amplitude variants that can be realized. Increasing the width of the probability density curve means that the probability of generation of two spikes with the close amplitudes increases versus time. This is an undesirable effect, because the closer the amplitudes are, the more difficult separating them and suppressing lower spikes on the non-linear stage of generation development becomes. To achieve complete mode-locking, it is necessary to maximize the difference between the amplitudes of the random spikes that makes the single pulse separation on the round-trip period of the cavity easier.

In [15], it has been shown that to calculate the number of excesses over certain curve C, it is necessary to estimate the joint density of probability for the envelope and for the derivative of this function. The final formula for the average number of positive excesses of the envelope $V(t)$ of the sum of random (noise) and quasi harmonic processes in unit of time is:

$$N_1^+(C)=\sqrt{\frac{-\rho_0''}{2\pi}}W=\sqrt{\frac{-\rho_0''}{2\pi}}\frac{C}{\sigma}\exp\left[-\frac{a^2}{2}-\frac{C^2}{2\sigma^2}\right]I_0\left(\frac{aC}{\sigma}\right). \quad (17)$$

Respectively, for the quasi-harmonic signal (a=0) the view for (17) may be simplified:

$$n_1^+(C)=\sqrt{\frac{-\rho_0''}{2\pi}}\frac{C}{\sigma}\exp\left[-\frac{C^2}{2\sigma^2}\right]. \quad (18)$$

It is known that the square of the energetic width of the signal spectrum is proportional to the second derivative from the correlation coefficient. For the Gaussian spectral density function, taking into account (6), (7) it is possible to write:

$$\sqrt{\frac{-\rho_0''}{2\pi}}=\Delta f_e. \quad (19)$$

Thus, for quasi-harmonic optical signal that is correlated on time T_C (round-trip period), the average number of the intensity spikes that surely exceed the average radiation intensity level $\overline{I}$ may be found from the following:

$$n^{+} = T_C \Delta f_e W\left(\frac{I}{\sqrt{\bar{I}}}\right) \tag{20}$$

Versus the generation development, the value Δf_e decreases, but $\sqrt{\bar{I}}$ increases. Thus, temporal dependence of n^+ depends on the functions mentioned above and might have complicated non-monotonous behaviour. Formula (20) demonstrates that spectral width and total radiation power determine all properties (statistics) of the laser radiation noise pattern. Hence, calculating these values at any moment of generation development, one can evaluate the statistical parameters of radiation.

6.4. Statistical properties of radiation on the "linear" stage of the generation development

In [18-20], the process of mode-locking in solid-state lasers has been analysed. It was shown that at the end of the linear stage, the noise pattern is a superposition of a great number of patterns emitted after the threshold conditions have been achieved. For increase of the output parameters repeatability, it was proposed to decrease a total number of the operating modes but with keeping the same a total spectrum width. Such mode number thinning out results in increase of repetition rate of the pulses and improves the pulses parameters repeatability. To check these dependencies, a Nd:glass laser with round-trip period in range 20 ps - 20 ns has been built and studied experimentally.

Starting from (9) and (20) with n =1, 2, one can find two maximal amplitudes C_1 i C_2 (the first and the second) probably generated on the laser cavity round-trip period. Typically, for small relative difference $\Delta = (C_1 - C_2)/C_1$ <<1 and for significant number of spikes $q = f_e T_C$ >>1, amplitude difference may be estimated as:

$$\Delta = \ln 2 / 2[1 + \sqrt{2\ln q} / (2\ln q - 1)]\ln q. \tag{21}$$

Spontaneous luminescence of a gain laser medium starts practically at the same moment with pumping. If the gain medium is located in the optical cavity, its emission may be separated on the portions equal to the light round-trip time between the cavity mirrors. Each of these patterns is independent noise realization. The total field at the end of the linear stage is the superposition of a great number of such patterns. Because of the amplification coefficient spectral dispersion, the initial spectral-temporal distribution deforms even on, so called, "linear" stage. It is clear that this classical term cannot be applied to the generation evolution period, which is characterized by signal frequency transformations.

Let suppose that an averaged in time envelope of the spectral intensity distribution is Gaussian: $I(w) = I_0 exp(-\Delta w^2/\delta^2)$. Total power after N round-trips of light is an integral through the spectrum with taking into account the growth of amplification coefficient. After each round-trip, a "fresh" noise pattern is added to the previous amplified radiation pattern. However, for a new noise realization, amplification exceeds losses in wider and wider spectral range. Because of these small differences in starting conditions, the summarized radiation becomes non-Gaussian, even if each realization is Gaussian (but with different spectral width!). The function that describes average intensity change is the following:

$$\sigma_m^{\ 2} = \sigma_0^{\ 2} \exp\{\beta T_p (N-m)(N-m+1)/2\} / \sqrt{1+\alpha_0 (N-m)}, \tag{22}$$

here σ_0^2 is average luminescence power, β is amplification coefficient growth rate; T_C is axial period of the cavity; N is number of round-trips from the moment of threshold completion for the central frequency; m is number of round-trips from the moment of the threshold up to m-th noise realization emission, α_0 is linear initial losses (dimensionless). Spectral width of m-th noise realization changes respectively:

$$\delta_m = \delta_0 \left[1+\alpha_0 (N-m)\right]^{-1/2}, \tag{23}$$

where δ_0 is width of the luminescence of the gain medium. Formula (23) was acquired in assuption $\beta T_C N_{lin} << \alpha_0$, N_{lin} is a number of round-trips during total linear stage. Average intensity of final field accumulated in the cavity in the end of linear stage is as following:

$$\sigma^2 = \sigma_0^{\ 2} \exp\left[\beta T_C \frac{N_{lin}(N_{lin}+1)}{2}\right] \sum_{m=0}^{N_{lin}} \frac{\exp\left[-\beta T_C \frac{m(m+1)}{2}\right]}{\sqrt{1+\alpha_0 (N_{lin}-m)}}. \tag{24}$$

It follows from (24) that the noise realizations with the numbers from $m = 0$ to

$$M = [2/\beta T_C]^{1/2}, \tag{25}$$

introduce the main income into the average intensity. They dictate the spectral parameters of the final field. For the typical laser parameters $\beta = 10^3 s^{-1}$, $T_C = 10^{-8}$s one can estimate $M \approx 450$. Because of (23) the spectrum is not exactly Gaussian. However, difference in first actual M realizations is not significant, because usually $M << N_{lin}$.

Taking into account that on the "linear" stage, until there is no gain medium saturation, each realization develops independently, the final field may be considered as quasi harmonic random process with an average intensity:

$$\sigma^2 = \frac{\sigma_0^{\ 2} M}{\sqrt{1+\alpha_0 N_{lin}}} \exp\left[\beta T_C \frac{N_{lin}(N_{lin}+1)}{2}\right] \tag{26}$$

and the spectrum width

$$\delta = \delta_0 / \sqrt{1+\alpha_0 N_{lin}}. \tag{27}$$

Thus, the field formed up to the "linear" stage end is described by the statistics of quasi-harmonic random process. However, its evolution time is shorter as compared with classical estimations because of storage of energy by adding a significant number of optical noise realizations. It means that the spectral width is higher and the amplitude difference between maximal spike and the second one is less than that predicted in linear model. As a result, the final probability of complete mode-locking is much less than it was estimated before. For different gain media, activator doping concentrations, methods of pumping, and levels of losses into a cavity, the linear stage duration is different. Hence, in certain limits the final spectrum width is not proportional to the starting (luminescent) width when different systems are compared. Though, affecting some parameters of the laser system, it is possible to satisfy requirements, providing the guaranteed achievement of single pulse generation on the cavity round-trip period.

6.5. Spectral-temporal dynamics

The noise pattern spectrum envelope only in average is proportional to the gain profile. In each specific "shot" the number of quanta in each certain mode may significantly differ from the statistically averaged value. Initial mode phases are distributed homogeneously in the range $-\pi \le \varphi \le \pi$. Mode-locking process results in two actions: introducing certain constant phase relations between the modes and in transferring the energy between the randomly modulated modes, so the spectrum envelope becomes smooth without chaotic intensity modulation.

Let us write the modes oscillations in specific point of cavity as the series:

$$\begin{aligned} A^*(w,t) &= A_0\sin(w_0t+\varphi_0)+A_1\sin(w_1t+\varphi_1)+ \\ & A_{-1}\sin(w_{-1}t+\varphi_{-1})+\ldots+A_n\sin(w_nt+\varphi_n)\ +\ A_{-n}\sin(w_{-n}t+\varphi_{-n}) \end{aligned} \tag{28}$$

here $w_{\pm n}=w_0\pm\Delta w\cdot n$, w_0 is the central (maximal) frequency of the operational spectrum, $\Delta w=c/2L$ is an intermode frequency space, n is the mode number starting from the central one.

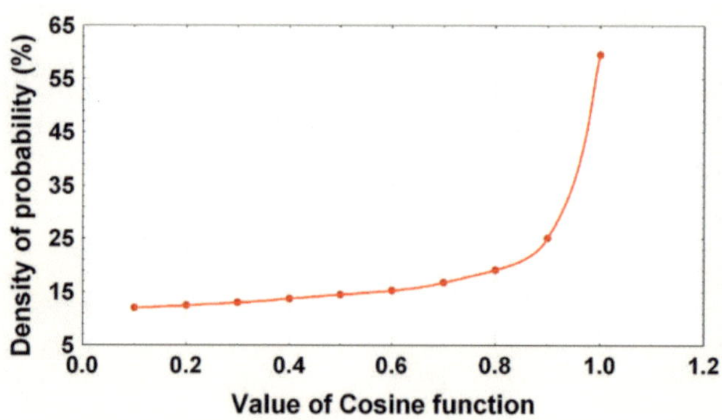

Figure 5. Probability to find a value of $cos(\varphi)$ function in intervals 0 – 0.1, 0.1 – 0.2,...0.9 – 1 at φ homogeneously distributed on the interval $-\pi/2 < \varphi < \pi/2$.

If the initial phase φ_n is not equal to zero, $\sin(w_nt+\varphi_n)$ may be represented as a series with the components: $A_c\sin(w_nt)+ A_s\cos(w_nt)$, where $A_c = \cos\varphi_n$ and $A_s = \sin\varphi_n$ are the random values that changes slowly, as comparing with $\cos(wt)$. Let pay attention to the fact, that even in case when a random phase is distributed homogeneously on the range $-\pi \leq \varphi \leq \pi$, the distribution density of A_c or A_s relate to cosine or sinus functions; in other words, this is not a homogeneously distributed function any more on the range of existence ($-1 \leq A \leq 1$). The probability to find a function on specific interval of meanings is inversely proportional to the rate of function changes, or to the derivative. For the function $A_c = \cos\varphi$ density of probability to find meaning into the interval $1 < A < -1$ is proportional to $\sin\varphi$. Thus, the most probable meaning for A_c is 1 in the interval $-\pi/2 < \varphi < \pi/2$ and -1 in the intervals $\pi/2 < \varphi < 3\pi/2$. For A_s the probable meaning in these intervals is close to zero. Figure 5 shows that in 58% attempts with phase distributed inside the range $-\pi/2 < \varphi < \pi/2$, the meaning of $\cos(\varphi)$ function is inside the range 0.9-1. Respectively, interval 0.8-1 involves about 85% of all noise realization. The joint probability that two neighbour modes are generated with close phase values within this range is 0.72. At small number of operating modes in most number of the laser shots the sum $\sum_n \sin(w_n t + \phi_n)$ may be exchanged to the $\sum_n \sin(w_n t)$.

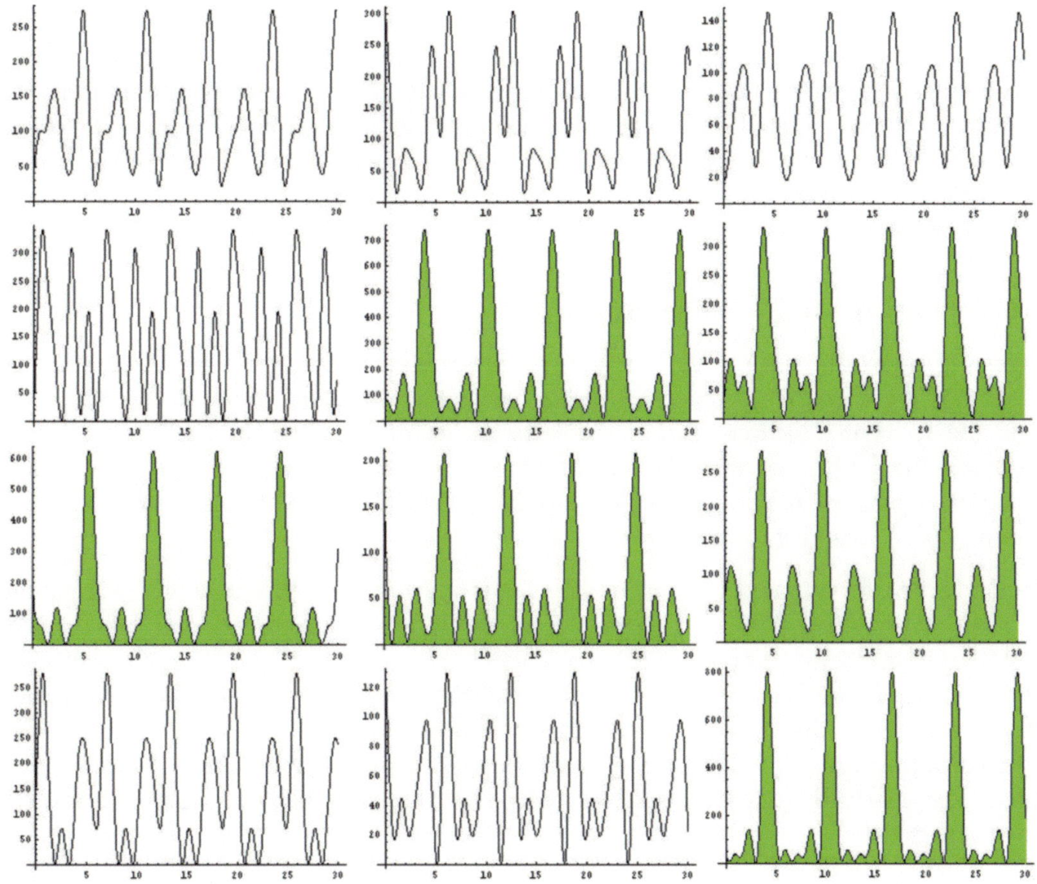

Figure 6. Computer simulation of the five mode interference (natural mode-locking process). Green oscillograms show the cases when the main peak contains above 80% energy of all pulses located on the round-trip period.

Figure 6 demonstrates a computer simulation of natural "mode-locking". The program generates random phases and amplitudes of five modes. Evidently, that half of the oscillograms contains 80% of round-trip period energy in a single pulse. This demonstrates "natural" mode-locking process, achieved without any non-linear elements placed into the cavity. Even with a weak non-linearity, such surely separated spike amplitudes may be easy additionally emphasized. However, there are more modes – less probability to generate "fully synchronized" spectrum. The solution is to keep a wide total spectral width but to decrease a total mode number. This can be done with low quality interferometer that does disturb the equidistant mode spectrum.

Moreover, in a real laser, the mode behaviour is not independent, especially if the amplification band is relatively narrow. Because of inversion hole-burning [7], two neighbour modes have maximal amplification because they are not overlapping in the centre of cavity where typically the gain medium is located. However, the next pair of the modes is overlapping with the first pair in the centre and near the mirrors. For the next 5th and 6th modes the amplification drops down because of spectral envelope's bell-like shape. Because of this phenomenon, practically anytime the laser output from the "pure" cavity (without selection) demonstrates modulation with axial period, which is a beating of two central high power modes. Taking into account the high probability that a "phase coefficients" A_c would be close to 1, one can understand why in big number of laser shots the emission demonstrates periodical structure, even without of any modulators or nonlinear elements.

6.6. Optimal spectral width of the ps-laser

It is well known that the wider the gain spectrum is, the shorter the pulses that can be generated are. However, some experiments have shown that this is not always true. This and the following sections analyse how the index of refraction and gain dispersions influence the duration of the spikes on the "linear" stage of generation evolution. The basic question is: what set of parameters provides the minimal duration of pulses before the non-linearity of the absorber or gain medium starts working.

In principle, the luminescence band of glass lasers has a sufficient width to generate femtosecond pulses. However, some effects result in an increase of the initial spike duration in the process of amplification and generation. First of all, because of the spectral dispersion of the amplification coefficient, the typical spike duration increases approximately 50 times [17]. To avoid this phenomenon, in [21], the interferometer was located in the cavity. It was aligned so that it flattened the maximum of the amplification band. The shortening of the pulses was about 1.5 to 2 times. However, it was much less than expected.

The next reason that makes it difficult to achieve short pulses in solid-state laser hosts is the dispersion of the material refractive index that results in broadening of the ultrafast pulses. At the conditions of low amplification near the threshold, the initial amplitude of the random short and intensive spike decreases during several round-trips through the cavity.

For the initially shorter pulses, the time of spike intensity development up to the absorber saturation energy is longer than for the relatively long pulses. Certainly, longer pulses survive in the process of generation evolution. Let us analyse the random spike spectrum genesis by taking into account amplitude-frequency transformations along the linear stage period.

To simplify the calculations, we assume that the spike spectrum shape is Gaussian, with an energy normalized to one. To study the process, we apply the Fourier transformation and follow all spectrum components developing.

$$I(t) = \frac{\delta}{\sqrt{\pi}}\exp(-\delta^2 t^2) = V(t)V^*(t), \tag{29}$$

Here δ is a luminescence spectrum width (dimension of δ is [s^{-1}]).

$$V(t) = \frac{1}{\sqrt{2\pi}}\int_{-\infty}^{\infty}\{\frac{\exp[-\frac{w-w_0}{\delta}]^2}{\delta\sqrt{\pi}}\exp[\beta T_C\frac{N(N+1)}{2} - N\alpha_0(\frac{w-w_0}{\delta})^2]\}^{1/2}\exp(-i\Delta\Phi)\exp(-iwt)dw \tag{30}$$

The first exponent in (30) describes the amplitude distribution of the spectral components of a pulse with maximum at w_0, the second one shows the amplification depending on the number N of the light field round-trips with the axial period T_C, initial (linear) coefficient of the losses α_0, and amplification coefficient increase rate β. An expression for the phase modification is as following:

$$\Delta\Phi = \frac{2\pi\Delta nL}{\lambda} = \frac{dn}{dw}\left|w - w_0\right|wT_qN, \tag{31}$$

here Δn is refractive index changes (at tuning out of w_0); L is a total length of the gain media passed by the light during the whole linear stage of generation; dn/dw is the refractive index dispersion; T_q is time of a light single pass through the gain medium.

Let us consider, at first, the pulse evolution that is represented by (29) – (31). The final expression for the single pulse intensity is as following:

$$I(t) = \exp[\beta T_C\frac{(N+1)N}{2} - \frac{\Delta(aw_0 + t)^2}{2(\Delta^2 + a^2)}][2\delta\sqrt{\pi(\Delta^2 + a^2}]^{-1}, \tag{32}$$

here the designations are: $\Delta = \frac{1+\alpha_0 N}{2\delta^2}$, $a = \frac{dn}{dw}T_q N$.

From (32), it follows that the final inverse duration is connected with the initial one as:

$$\delta' = \delta\left[1 + \alpha_0 N\right]^{1/2}[(1+\alpha_0 N)^2 + (2\delta^2\frac{dn}{dw}T_q N)^2]^{-1/2}. \tag{33}$$

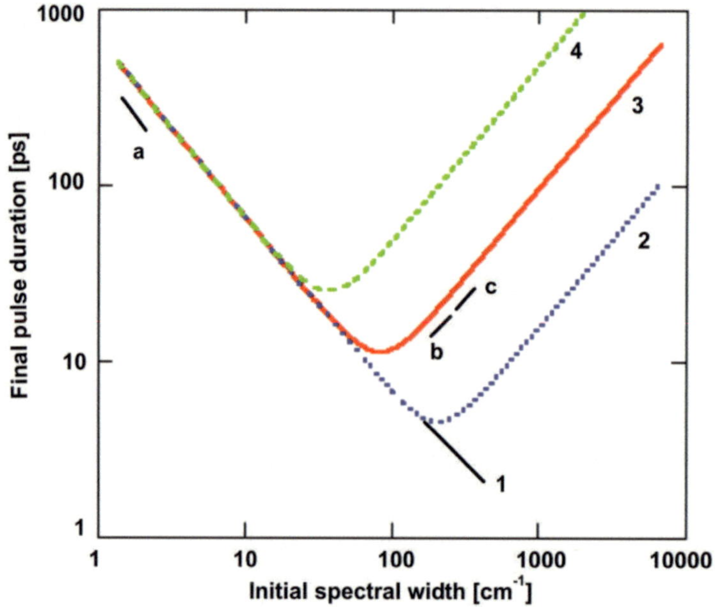

Figure 7. Dependence of the spike duration at the end of linear stage vs the initial noise pattern spectral width for different values of the refractive index dispersion $dn/dw = 0$ (1), $0.13{\cdot}10^{-17}$s (2), $0.8{\cdot}10^{-17}$s (3), $5{\cdot}10^{-17}$s (4).

Figure 7 demonstrates the dependency of the pulse duration at the end of the linear stage Δt on the start spectrum width δ of the random spike at different values of the product $T_q(dn/dw)$. For $(dn/dw) \neq 0$, the curves coincide in the range of small δ meanings. For wide δ, the final Δt depends on the specific value of the product $T_q(dn/dw)$. In Figure 7, the letters mark the spectrum width for (a) Nd:YAG, (b) Nd:phosphate glass, (c) Nd:silicate glass.

At the end of the linear stage of generation, $N\alpha_0 >>1$ and expression for the δ' may be simplified:

$$\delta' = \frac{\delta}{\sqrt{\alpha_0 N}} \text{ for the small } \delta; \tag{34}$$

$$\delta' = \frac{1}{\delta}\sqrt{\frac{\alpha_0}{N}}\left(2T_q\frac{dn}{dw}\right)^{-1} \text{ for the big } \delta. \tag{35}$$

From Figure 7, it is evident that the final spike duration Δt has a minimum. This area determines the certain start spectral width that, at some given set of the parameters (dn/dw, α_0, β, T_C, T_q), can provide the minimal duration of the spikes Δt_{min} at the end of the linear stage. From the (33) the conditions of the Δt_{min} achievement may be deduced:

$$2\delta^2\frac{dn}{dw}T_q = \alpha_0 \tag{36}$$

It follows from (36) that, for example, for the Nd:silicate glass (T_q =10^{-9} s, (dn/dw)= $8{\cdot}10^{-18}$ s) the optimal initial spectrum width is ≈ 70 cm^{-1} but the real one is about 250 cm^{-1}. Thus, to achieve minimal ultrashort pulses in the laser with Nd:silicate glass, it is necessary to narrow the initial luminescence spectrum width 3-4 times. The use of prisms is not appropriate in this case because with spectrum narrowing, it simultaneously results in the increase of the product $T_q(dn/dw)$. As an element with anomalous dispersion, an interferometer may be used. In this case, the etalon automatically plays the role of a gain spectral dispersion flattener.

By decreasing the losses (e.g., by improvement of cavity construction), the value Δt_{min} moves to the smaller spectral width values. If β stays the same, the relative amplification growth is higher in a system where α_0 is less. In this case, the spike spectrum width grows faster and natural spectrum selection process is restrained. If one decreases α_0 at relatively small initial spectrum Δ <30 cm^{-1}, the spike duration decreases because of the process of natural spectrum selection weakening. Inversely, for Δ >60 cm^{-1} the duration increases because of negative influence of the refractive index dispersion.

The starting losses α_0 include linear losses on cavity elements, on mirrors (output), and non-saturated losses in absorber. To achieve Δt_{min}, it is necessary to vary mirror reflection only, because decrease of the absorber density makes the pulse discrimination worse. For garnet type media, the highly reflecting mirrors are preferable; for glasses, it is better to use low reflecting mirrors.

Increase of the product βT_C (with the other parameters staying the same) results in Δt_{min} decrease. This phenomenon is connected with diminishing of the total number of the round-trips during the linear stage of the generation development. Having a low threshold and high amplification increase rate β, the garnets have lower located curves $\Delta t(\delta)$ as compared with glass hosts. This explains why these media, being with about two order of magnitude different luminescence bandwidth, demonstrate ps-pulses with durations just 3-5 times different. The phosphate glasses that have narrower luminescence bandwidth as compared with the silicate ones can provide shorter pulse generation. In all cases, possible variations of the laser parameters may minimize the random spike duration at the end of the linear stage of the generation development.

7. Experimental results

7.1. Spectrum modulation with dynamical interferometer

The mode-locking process supposes introducing of the certain amplitude-phase relations between all generating modes. In known techniques, phase locking is typically provided simultaneously along the whole actual spectral range. However, the condition of simultaneous phase regularization is not obligatory. In principle, the modes may be linked step by step, if a total mode number is relatively small or the time of generation development is relatively long [14]. In this case, the process of mode-locking spreads consequently from small group of modes to the whole spectrum. The simplest way to build such a system is to install into a laser cavity an additional mirror with a piezo-driver. This three-mirror cavity is a dual-interferometer system. Let us place an interferometer with base length l_{0i} into the cavity with an optical length $L_C = kl_{i0}$, where k is integer. A diagram in Figure 8 shows the case where $k = 10$. Because of different cavity and interferometer lengths, there are k cavity modes located between two peaks of the interferometer. Internal mirror should be with low reflectivity and do not disturb significantly the mode spectrum of the main cavity. For two interferometers with bases of different optical length, the velocity of the reflection peak motion in the frequency domain is higher for the interferometer, which has shorter optical base. In principle, it is possible to move the interferometer peak from mode to mode position exactly during the light round-trip time along the cavity. In this case, relatively fast modulation with intermode beating frequency is realized by the relatively slowly moving mirror. This system can be named as an "optical lever" technology.

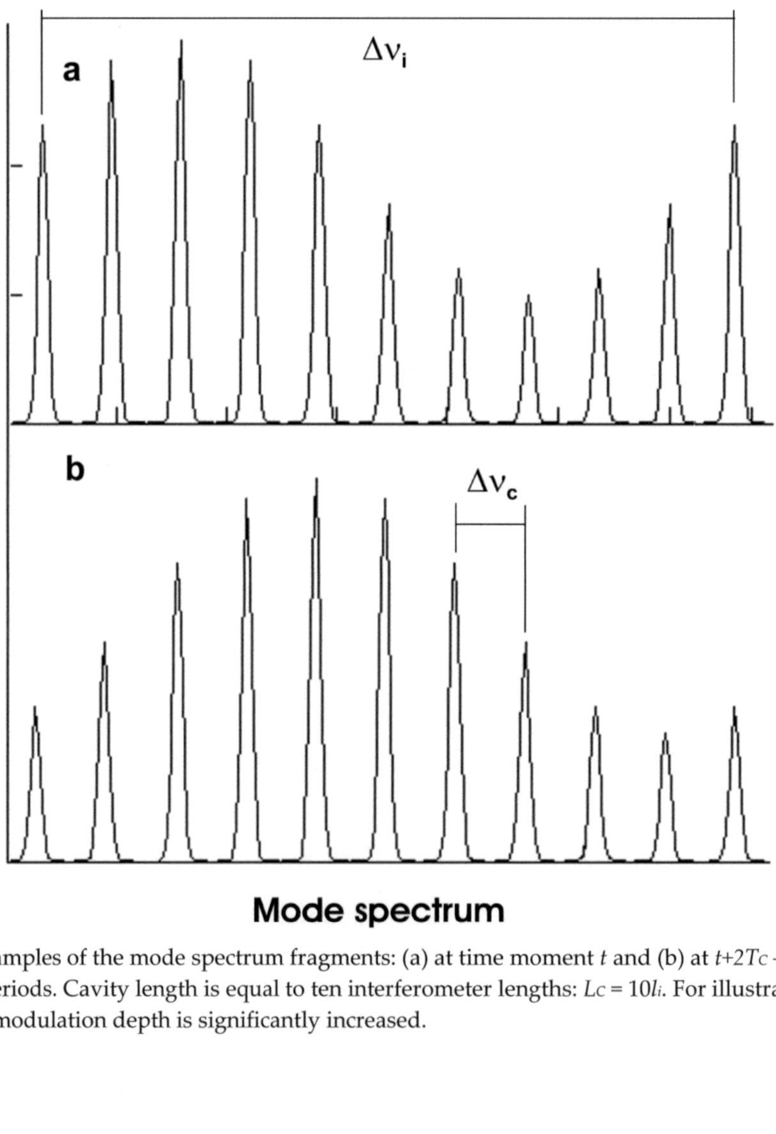

Figure 8. Examples of the mode spectrum fragments: (a) at time moment t and (b) at $t+2T_C$ – after two round-trip periods. Cavity length is equal to ten interferometer lengths: $L_C = 10l_i$. For illustration purpose the modulation depth is significantly increased.

To find the conditions of such mode-locking, let us suppose the optical length of the interferometer l_i changes with constant speed V. In this case, $l_i = l_{0i} + Vt$, here l_{0i} is the start interferometer length. The position of the n-th transparency maximum (of an unmovable interferometer) is described as:

$$\nu_n = 1/\lambda_n = n/2l_i, \text{ here } \nu = 2l_i/\lambda_{n0} \tag{37}$$

Differentiate (37) with respect to time and get n-th interferometer maximum frequency changes rate. With condition $Vt \ll l_{0i}$ (because $Vt \simeq (1-10)\lambda_{n0}$, and $l_{0i} \simeq 10^4\lambda_{n0}$) it is easy to achieve:

$$\frac{d\nu}{dt} = -\frac{nV}{2l_i^2} = -\frac{l_{i0}V}{\lambda_{n0}\left(l_{i0}+Vt\right)^2} \cong -\frac{V}{\lambda_{n0}l_{i0}}. \tag{38}$$

With the moving mirror the new cavity mode has the maximum quality consequently in each new T_C interval. Let us find such a mirror speed that provides the interferometer peak motion from the mode position to the next one for the time of the light round-trip in a linear cavity $T_C = 2L_C/c$:

$$\left(\frac{d\nu_n}{dt}\right)T_C = -\Delta\nu_C. \tag{39}$$

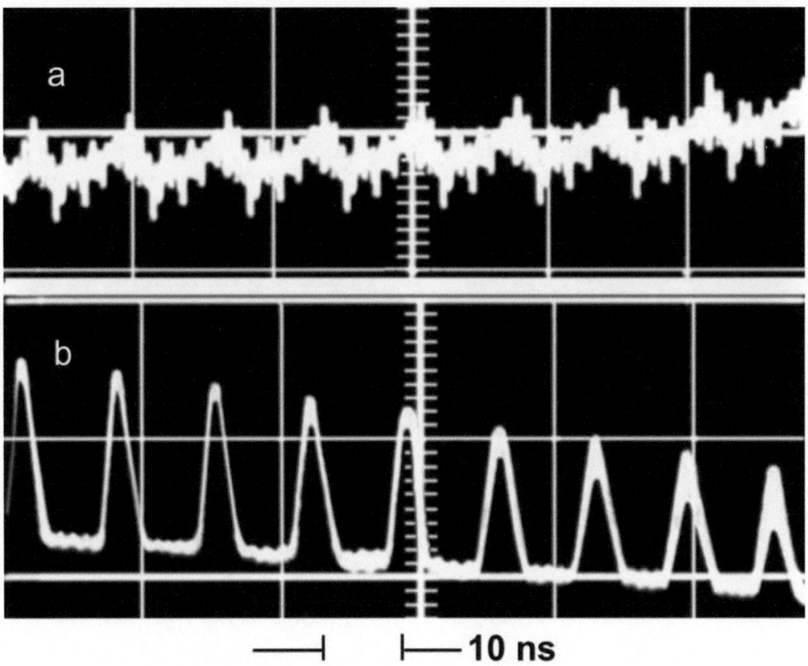

Figure 9. The oscillograms of laser output (a) with unmovable internal mirror and (b) with movable mirror.

After substitution into this equation dv/dt and Tc the value of V can be found. Intermode space $\Delta\nu_i$ for the ring cavity is $\Delta\nu_i = 1/l_i$, and for the linear one – $\Delta\nu_i = 1/2l_i$. Introducing the coefficient $\gamma = 1/2$ for the linear cavity and $\gamma = 1$ for the ring one, finally the formula for the resonant mirror speed is:

$$V = \gamma\lambda_{n0}\Delta\nu_p^{\,2} / \Delta\nu_i. \tag{40}$$

Figure 9 demonstrates the oscillograms of the laser output (a) without interferometer and (b) with scanning interferometer. The interferometer mirror was moved by the piezo-transducer. Cavity length Lc was 200cm, interferometer base l_i was 0.25 cm that provided k = 800. With laser operational wavelength λ = 1.06 microns, the mirror velocity was about V = 5 cm/s. For lower k values the necessary mirror velocity should be higher. Hence, the mechanical motion of the mirror could be hardly achievable. In this case the electro-optical system should be used to control the cavity modes properly. An inclined interferometer limits the total spectrum width and the ultrashort pulses are not achievable with this technique. To solve this problem, the normally installed mirror should be used. To avoid back reflection that results in parasitic selection, a thin film selector should be used as a mirror. The properties of such a system have been described in section 2.

7.2. Laser with anti-mode-locking: "Black" pulses generation

Any periodical signal may be represented as a sum of harmonics with specific amplitude and phase coefficients. The scanning interferometer mirror may be moved with changeable velocity, following the special function. In this case, because of interference of the modes with certain combinations of the phase, the pulse shape may be specially designed [14].

One of the interesting examples of such laser generation is "anti-mode-locking". If the interferometer mirror is moved with the speed as doubled to the resonance one, the zero-phase difference is dictated to each second mode. The neighbour modes are modulated in anti-phase conditions. Figure 10 demonstrates several examples of simulated (left column) and experimentally generated (right column) pulse trains. The cavity length was 525 cm. That corresponds to a 35 ns round-trip time. The interferometer base was about 1.5 cm, with a resonance speed of about 4 cm/s.

Figure 10a – resonance mirror motion; it is classical mode-locking. Figure 10b – average mirror speed is two times higher as compare to the resonance one; motion is with acceleration. Figure 10c – motion with acceleration two times higher as in case (b). Figure 10d – average mirror speed is two times higher compared to the resonance one; even modes are in anti-phase conditions to the odd modes. ("anti-mode-locking"). The picture is inversed to the classical mode-locking (Figure 10a). On the background of the quasi-cw generation there are narrow peaks of radiation absence, or "black pulses".

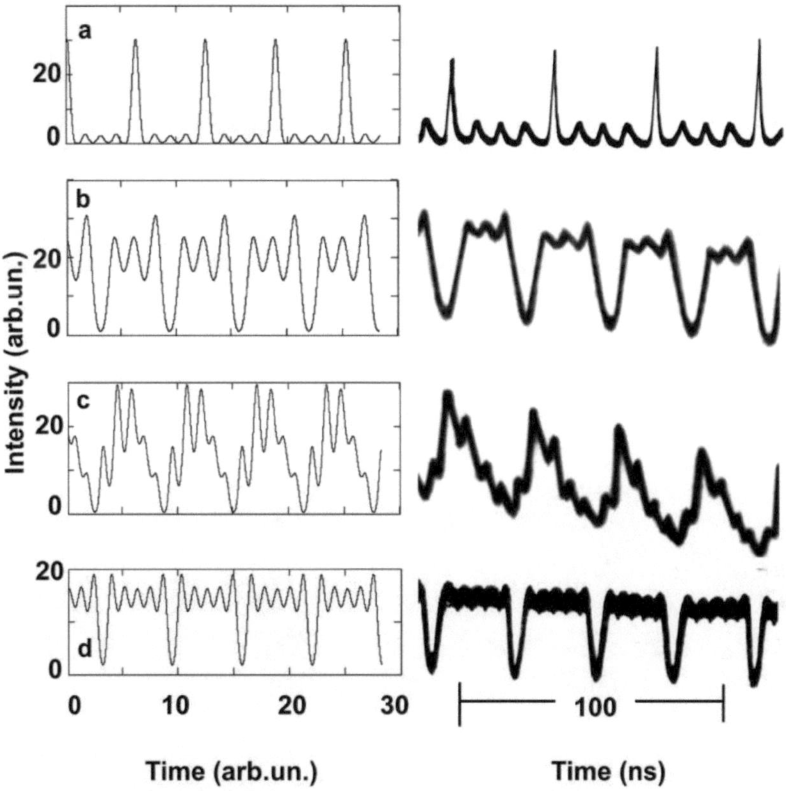

Figure 10. Simulated (left column) and experimental (right column) oscillograms in case of mode-locking (first row), anti-mode-locking (fourth row) and with some phase-shifted modes (second and third row).

7.3. High pulse repetition rates

In various applications such as telecommunications, pulse trains with multi-gigahertz pulse repetition rates are required. The resonators of the bulk lasers are usually too long to achieve such repetition rates with fundamental mode-locking. The high repetition rate pulse trains were obtained with harmonic mode locking [22, 23], when the active modulator worked at frequency multiply integer higher than the inverse cavity round-trip period. Similar research but at passive mode-locking has been provided in [24].

In this section, a 50-GHz repetition rate generation from the Nd:glass laser with 45-MHz cavity is described. The system was of modulator free [14]. This work was completed approximately nine years before the first self-starting Ti:sapphire laser was demonstrated. However, honestly saying, during Nd:glass laser's experiments, the role of the Kerr-type non-linearity was not understood.

The construction of the high repetition rate laser is very simple. The cavity contains just one additional low reflecting mirror that is installed so that the short base interferometer length (*l*) is exactly an integer multiple (*n*) of the full cavity length (*L*) (with heated gain medium): *L*

= nl. At this condition, the ultrashort pulse, running back-and-forth inside the interferometer, after n reflections, exactly coincides with the pulse that was running along the entire cavity. Since the total spectral width is the same, the pulse duration remains the same. The multiplication of ps-pulses results in periodical modulation of spectrum and, respectively, in a decrease of total mode number. At this condition, even without a non-linear absorber but with an aperture installed near the output mirror, the laser generated a train of ultrashort pulses. With a fourth mirror, which shapes a new interferometer with a base length integer to the first interferometer, the pulse repetition rate was increased again. Figure 10 demonstrates oscillograms of the mode-locked generation (a, c) and harmonic mode-locked (b, d). A photograph (e) shows the two-photon luminescence track achieved with a four-mirror cavity without a non-linear absorber. The pulse duration is 11±2 ps and the period of repetition is 20 ps. In all cases, the total cavity round-trip length was 6.6m (22ns).

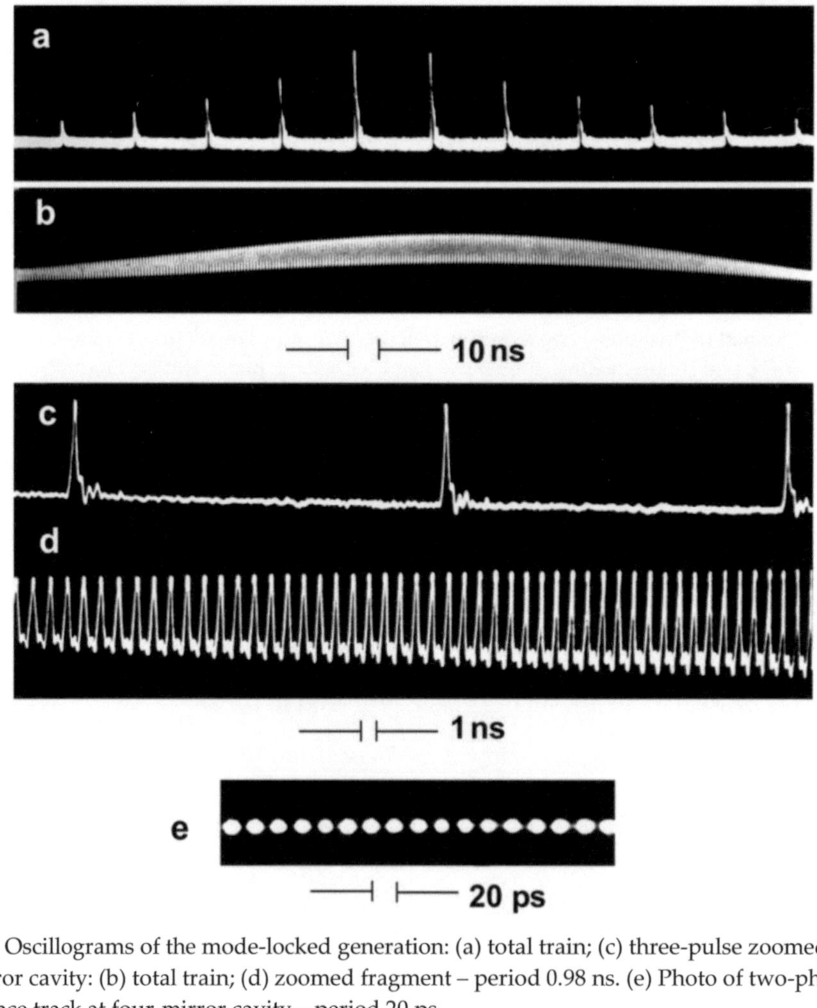

Figure 11. Oscillograms of the mode-locked generation: (a) total train; (c) three-pulse zoomed fragment. Three-mirror cavity: (b) total train; (d) zoomed fragment – period 0.98 ns. (e) Photo of two-photon luminescence track at four-mirror cavity – period 20 ps.

From my personal experiences, the idea to use an interferometer in the mode-locked laser cavity is typically rejected by project or paper reviewers because of supposed strong interferometer dispersion, which affects equidistant mode spectrum and makes short pulse generation impossible. In response, it should be specially emphasized that maximal exceeding of the gain over the losses in an free-running laser could be around 3-4%; hence, to shape the necessary spectral structure, the interferometer should provide about 5% modulation depth, which practically does not shift the modes. Moreover, it is possible to use the interferometer with absorbing mirror (see in section 2), which demonstrates a unique property – it does not reflect light at normal beam falling on the interferometer. In other words, there is no disturbance of the mode spectrum with such an interferometer.

8. Femtosecond pulses

From a theoretical point of view, the basic principles of picosecond and femtosecond pulses generation are the same [9-13]. The difference is in some parameter values. Femtosecond pulses have three orders of magnitude shorter duration than the picosecond ones. Hence, they require much wider operational spectral bands and faster mechanisms of modulation. Because of dispersion, the broadening of fs-pulses happens faster when propagating through the optical elements. Because of very high pulse peak power, the non-linear processes are routine problems for fs-lasers. Practically all types of ps-lasers were repeated in fs-lasers. They are bulk lasers with open cavity, fiber lasers, dye lasers, and semiconductor lasers. Disregarding the construction variety, principles of operation of all these systems are the same, or very similar. In the last five years, fiber lasers were the most dynamically developed segment of the fs-laser. The efforts of researchers are focused on suppression of non-linear effects, on increase of output average power, and on exploitation of new spectral ranges.

9. Attosecond and zeptosecond pulses on the way to yoctoseconds

The attosecond pulse generation and measurement are the hottest subjects of popular and solid scientific journals [25]. With achievement of this duration range, these super-ultra-short laser pulses became an instrument of intra-atomic process investigation. For the first time ever, researchers are capable of seeing how an electron "jumps" between atoms. The peak intensity of the as-pulses approaches and overcomes 10^{18} Wcm^{-2}. This is a powerful tool in studying the ultrafast phenomena, such as the chemical/biological transformations occurring on the femtosecond range of durations. An attosecond control of collective electron motion in plasmas has been provided in [26]. There are two ways to generate an attosecond pulse or pulse train [27]. One is through the nonlinear processes of the superposition of high order harmonics generated in the laser-gas atom interactions. The limitation is that the laser intensity should be low enough to avoid the ionization of atoms. Typically, the efficiency of harmonic emission from the atoms is low. The other way is to generate high order harmonics from the dense surface of plasma created by the high-intensity femtosecond laser. Till now, 12 attoseconds is the world record for shortest controllable time [28].

The femtosecond lasers are more and more used as the sources in high energy atomic and nuclear investigations – see, for example [29]. It was shown that in the scattering of 100-fs laser pulse with an intensity of around 10^{19} W/cm^2 by a counter propagating electron with an initial energy of 10 MeV, a crescent-shaped pulse with pulse duration of 469 as and the photon energy ranging from 230 eV to 2.5 keV is generated in the backward direction. Recently the Internet brought us a new word: Yoctosecond – 10^{-24} seconds. The explanation what this is and where it could be used is very simple and fundamental: this is time taken for a quark to emit a gluon [30].

10. Conclusions and epilogue

1. Theory of random processes, initially developed for the radio-band frequencies, has been applied to describe the properties of the narrow-band optical noise and dynamics of the linear stage of generation evolution.
2. Theoretical analysis of the optical noise properties demonstrated high probability of the "natural mode-locking", when the main part of light energy accumulated in the cavity may be concentrated in a single pulse.
3. It has been shown that for each set of the solid-state laser parameters, there is a specific width of the amplification band that provides minimal duration of random spikes at the end of linear stage of generation. It was shown that two media, having, at least, one order of magnitude difference in luminescence spectrum width, might generate pulses of the same duration.
4. The harmonic mode-locked generation, achieved in a three-mirror cavity without a non-linear absorber, has been demonstrated in 22-ns cavity of Nd:glass laser.
5. A new method for controllable mode-locking with the help of "optical lever" – movable mirror placed into the laser cavity – has been proposed. The anti-mode-locking process with "black pulses" generation has been realized. Generation of rectangular and triangular ns-pulses has been achieved.

During the 50 years of their history, the pulsed lasers passed from seconds to zeptoseconds or about 20 orders of magnitude into the short duration's side. This road was not smooth and easy. Every time, starting from huge, complex, ineffective, and very expensive machines, the lasers became elegant, economical, more powerful and smart instruments in science, technology, medicine, and everyday life. We hope this chapter and this book will be useful for a wide spectrum of specialists, for professors and students, and for those who are interested in history and in future of the laser technologies.

Author details

Igor Peshko
Department of Physics and Computer Science, Wilfrid Laurier University, Canada

Acknowledgement

I very much appreciate my former University of Toronto Master's program student Inderdeep Matharoo who did a great job as a technical editor and as a first reader. At last, this work would be impossible without excellent coffee that my wife Nataliya regularly supplied me. Moreover, only because of her punctuality and accuracy some old photographs with oscillograms used in this chapter survived the journey between several countries and continents where we were lucky to live and work.

11. References

[1] Encyclopedia of Laser Physics and Technology. http://www.rp-photonics.com/encyclopedia.html (accessed 6 May 2012).

[2] Jabczyński J, Firak J, Peshko I. Single-frequency, thin-film tuned, 0.6 W diode-pumped Nd:YVO_4 laser. Applied Optics 1997; 36(12) 2484-2490.

[3] Peshko I, Jabczyński J, Firak J. Tunable Single- and Double-Frequency Diode-Pumped Nd:YAG Laser. IEEE J QE 1997; 33(8) 1417-1423.

[4] Peshko I, Jabczynski J. Thermally induced pulsation in the solid-state single-frequency diode pumped laser. Optica Applicata 1999; XXIX (3) 319-325.

[5] Tang C, Statz H, de Mars G. Spectral output and spiking behavior of solid-state lasers. J. Appl. Phys. 1963; 34 2289-2295.

[6] Zvelto O, Principles of Lasers. Heidelberg: Springer; 2010.

[7] Encyclopedia of Laser Physics and Technology. Spatial Hole Burning. http://www.rp-photonics.com/spatial_hole_burning.html (accessed 6 May 2012).

[8] Encyclopedia of Laser Physics and Technology. Nanosecond Lasers. http://www.rp-photonics.com/nanosecond_lasers.html (accessed 6 May 2012).

[9] Rullière C. Femtosecond Laser Pulses: Principles and Experiments. New York, USA: Springer; 2005.

[10] Diels J.-C, Rudolph W. Ultrashort Laser Pulse Phenomena: Fundamentals, Techniques, and Applications on a Femtosecond Time Scale. Technology & Engineering. Elsevier Science Publishing Co Inc Academic Press Inc; 2006.

[11] Kärtner F.X. Few-Cycle Laser Pulse Generation and its Applications. Berlin: Springer; 2004.

[12] Topics in Applied Physics: Ultrashort Light pulses Picosecond Techniques and Applications Ed.S.L.Shapiro. Berlin, Heidelberg, New York Springer-Verlag 1977.

[13] Encyclopedia of Laser Physics and Technology. Mode locking. http://www.rp-photonics.com/mode_locking.html (accessed 6 May 2012).

[14] Peshko I. Self-effecting processes in the solid-state lasers. Honorary Dr. of Sciences dissertation. Institute of Physics Kiev; 2003.

[15] Tikhonov V I. Excesses of random processes. Moscow: Nauka; 1970. (in Russian).

[16] Wigner distribution function. http://en.wikipedia.org/wiki/Wigner_distribution_function (accessed 6 May 2012).

[17] Sooy W. The natural selection of modes in a passive Q-switched laser. Appl.Phys.Lett. 1965; 2(6) 36-58.

[18] Peshko I, Khiznyak A, Soskin M. Mode-locked laser with controllable parameters. New York: Plenum-press; 1985. (Plenum Press bought the author's rights for initially published brochure: Peshko I, Khiznyak A, Soskin M. Laser of ultrashort pulses with tunable parameters. Kiev: Publishing of Institute of Physics, Academy of Sciences; 1984, #4).

[19] Peshko I, Khizhnyak A. Generation of the pulses of extreme short duration by solid-state lasers. Quantum Electronics 1987; 33, Kiev, Naukova Dumka p.14-20. (in Russian).

[20] Peshko I, Soskin M, Khiznyak A. Generation of the picosecond pulse train with controlable parameters. Quantum Electronics 1982; 9(12) 2391-2398. (in Russian).

[21] Graf F, Low C. Passively mode locked Nd-glass laser with partially suppressed natural mode selection. Opt. Communs. 1983; 47(5) 329-334.

[22] Becker M, Kuizenga D, Siegman A. Harmonic mode locking of the Nd:YAG laser. IEEE J. Quantum Electronics 1972; 8(8) 687-693.

[23] Encyclopedia of Laser Physics and Technology. Harmonic Mode Locking. http://www.rp-photonics.com/harmonic_mode_locking.html (accessed 6 May 2012).

[24] Zhan L. et al. Critical behavior of a passively mode-locked laser: rational harmonic mode locking. Opt. Lett. 2007; 32(16) 2276-xxx

[25] Attosecond pulse generation and detection www.mpq.mpg.de/lpg/research/attoseconds/attosecond.html (accessed 6 May 2012).

[26] Borot A, Malvache A, Chen X, Jullien A, Geindre J.-P, Audebert P, Mourou G, Quéré F, Lopez-Martens R. Attosecond control of collective electron motion in plasmas. Nature Physics 2012; 8 416-421.

[27] Zhu J, Xie X, Sun M, Bi Q, Kang J. A Novel Femtosecond laser System for Attosecond Pulse Generation. Advances in Optical Technologies 2012; Hindawi Publishing Corporation, article ID 908976, doi: 10.1155/2012/908976, 6 pages.

[28] Phys.Org.News. http://www.physorg.com/news192909576.html (accessed 6 May 2012).

[29] Lan P, Lu P, Cao W, Wang X. Attosecond and zeptosecond x-ray pulses via nonlinear Thomson backscattering. Phys. Rev. E 2005; 72, 066501[7 pages].

[30] Femtosecond, Attosecond, and Yoctosecond http://nextbigfuture.com/2009/10/attoseconds-zeptoseconds-and.html (accessed 10 May 2012).

Section 2

Pulsed World

Chapter 2

Femtosecond Laser Cavity Characterization

E. Nava-Palomares, F. Acosta-Barbosa, S. Camacho-López
and M. Fernández-Guasti

Additional information is available at the end of the chapter

http://dx.doi.org/10.5772/50174

1. Introduction

Ultrafast pulses are used now days in physics, engineering, medicine and other research areas such as, materials processing, time resolved spectroscopy, optical coherence tomography, pulse propagation, evolution of fast chemical processes, time resolved interference, etc. It is important to understand the physical processes that govern ultra fast laser pulse generation and shaping. In the present work, we characterize and find optimum operating conditions for a commercial Ti:Sa mode locked laser oscillator that generates pulses around 60 fs duration at 80 MHz repetition rate. Typical output power is 200 mW in mode locked operation.

The sharply peaked time structure at the oscillator output is explained in terms of the superposition of monochromatic waves with integer wavelength multiples of the round trip of the cavity. If a large number of waves with the same absolute phase are added with periodically shifted frequencies, the amplitude envelope becomes a periodically peaked train of temporal pulses. In Ti:Sa oscillators self mode-locking is achieved via Kerr lens mode-locking (KLM). The cavity is tuned so that self focusing of the more intense beam profile is favored.

The ultrafast oscillator is pumped by a Nd:YVO_4 CW laser. This laser is intra-cavity frequency doubled, the power output at 532 nm is 2.07 W. The pump beam is steered via two 45° mirrors into the IR cavity. The beam is focused into the Ti:Sa crystal with a 50 mm focal length lens as shown in figure 1. This beam passes through an IR convex mirror (C1) before reaching the active medium. The Ti:Sa emission resonates in a folded cavity comprising 2 concave mirrors and 6 plane mirrors. The group velocity dispersion (GVD) is compensated with a two prisms setup. A variable slit is placed between the back resonator mirror and the second prism. Figure 1 shows the Quantronix Ti-light laser layout and table 1 lists its main components. The oscillator together with an amplifier and a parametric generator were placed on a very stable passively damped holographic table [1].

In section 2, we revise the mathematical description of ultra short pulses. In particular, Gaussian pulses and hyperbolic secant pulses bringing together results that are scattered here

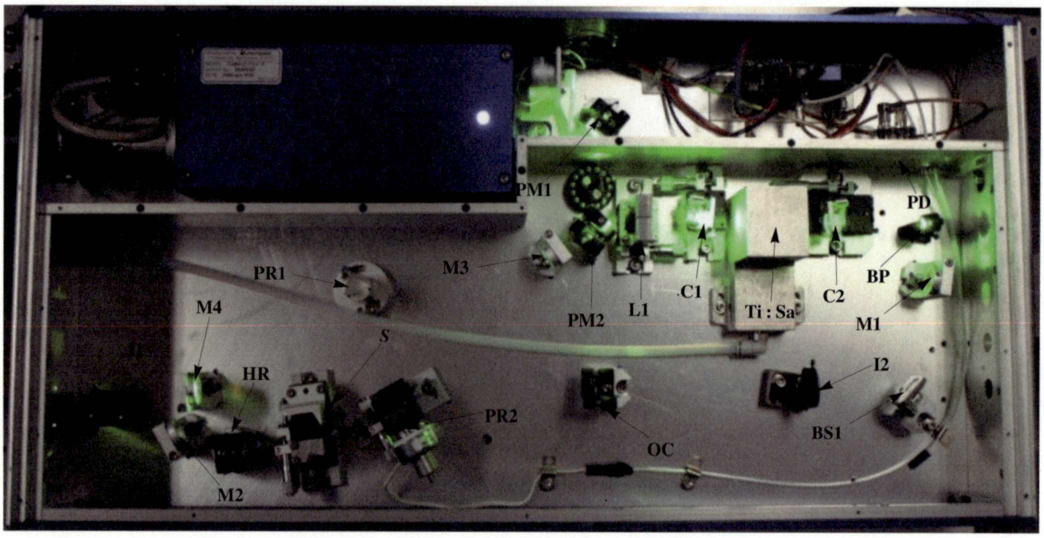

Figure 1. Ti:Sa laser at the Quantum Optics Laboratory, UAM-Iztapalapa

Components	Description
L1	pump beam focusing lens
C1, C2	first and second curved mirrors focused at crystal
M1- M4	IR Cavity folding mirrors
HR	rear highly reflective mirror
OC	output partially reflective mirror
Ti:Sa	Titanium Sapphire crystal
BS1	beam splitter
S	slit
PR1, PR2	first and second GVD compensating prisms
PM1, PM2	mirrors for 532 nm pump, reflection at 45°
I1, I2	iris
BD	beam dump
PD	photo diode

Table 1. Ti:Sa oscillator components.

and there in the literature. In subsection 2.4, the principles of mode-locked laser operation are glossed. Mode-locking is explained in terms of the superposition of monochromatic waves with integer wavelength multiples of the round trip of the cavity. The relevance of a broadband active medium is highlighted. The self mode-locking mechanism is described using the third order Kerr nonlinearity that leads to self focusing of the transverse Gaussian beam profile. The energy content of the output wave is described via a novel approach in subsection 2.5. This formulation avoids the averaging process required by either the Poynting vector or the irradiance functions proposed by Walther, Marchand, Wolf (WMW) and others.

In section 3, power output was measured as a function of mirror C2 position without mode locking. The laser output spectrum was also recorded for different mirror C2 positions within the mode locked regime (ML) with and without CW breakthrough.

In section 4, the pulse repetition rate was carefully measured in order to evaluate the cavity length. The internal photo diode as well as an external fast avalache photo diode were used to measure the pulse train. The difference between the mode locking peak intensity and the continuous CW background light was estimated. These measurements were also performed with much higher resolution using a streak camera.

In section 5, the pulse spectrum was monitored as a function of spectrum intra-cavity cropping. This was achieved through variation of the slit width (S-W) and slit translation (S-T). In section 5.1 the spectrum was evaluated as a function of GVD compensation. For this purpose, the second prism (PR2) position was varied and the spectrum was recorded.

In section 6, a second harmonic generation (SHG) frequency-resolved optical gating (FROG) setup was used to record the temporal features of the pulse: i) the pulse duration, ii) bandwidth, iii) pulse front tilt and iv) spatial chirp. Slit translation, width (S-T) and second prism (PR2) position were adjusted to achieve minimum time bandwidth product (TBP). The pulse measurements were fitted with appropriate curves and correlated to each other. The data obtained from these measurements together with their corresponding spectra from section 5 were fitted to Gaussian and hyperbolic secant time envelopes.

In the last section conclusions are drawn and the main results are abridged.

2. Theory of operation

2.1. Wave description

Ultrafast pulses are usually modeled by a carrier frequency modulated by a temporal envelope. This approach is usually adequate for pulses where the envelope duration is larger than a few oscillation periods. However, for pulses containing only a couple of periods, spatio temporal effects do not permit these type of solutions except for plane waves [2, 3]. It is customary to group the fast oscillating dependence in the phase function whereas the slowly varying features are grouped in a complex amplitude. This approach allows for slowly varying envelope approximations (SVEA) to be readily made [4]. However, a strict amplitude and phase representation requires that both of these functions are real. It is only under these circumstances that appropriate amplitude and phase functions can be used to establish an exact invariant [5].

The representation of a wave using complex variable in its polar representation is

$$\mathbf{E}\left(A,\Phi\right) = \mathbf{A}\exp\left(i\Phi\right), \tag{2.1}$$

where A is the amplitude and Φ is the phase, both of which are real valued functions. To represent a plane wave, the phase is a linear function of time and space, $\Phi = \omega t - \mathbf{k}\cdot\mathbf{r}$, where ω is the angular frequency, t represents time, $\mathbf{k}$ is the wave vector and $\mathbf{r}$ represents position. A plane wave with finite duration, is then described by a linear phase modulated by a time dependent amplitude $A\left(t\right)$

$$E\left(\mathbf{r},t\right) = A\left(t\right)\exp\left[i\left(\omega t - \mathbf{k}\cdot\mathbf{r}\right)\right]. \tag{2.2}$$

For the sake of simplicity, we shall focus on the temporal domain and neglect polarization effects. Pulsed Nd:YAG lasers usually have a Gaussian envelope. This profile is one of the few functions that can be tackled analitically in the time and frequency domains. On the other hand, ultrafast Ti:Sa lasers usually exhibit a profile closer to a hyperbolic secant curve. However, a hyperbolic secant with linear chirp cannot be solved analitically. For this reason, the problem has been tackled with a hyperbolic tangent chirp that is approximately linear in its central region. Analytical results in terms of Gamma functions can then be obtained. These two procedures are described in subsections 2.2 and 2.3.

2.2. Gaussian envelope

A Gaussian envelope has the form $A(t) = \exp(-\Gamma t^2)$, where $\Gamma = 1/\tau_G^2$. So the field (2.2) temporal function takes the form

$$E(t) = A_0 \exp\left(-\frac{t^2}{\tau_G^2}\right) \exp(i\omega_\ell t), \tag{2.3}$$

where ω_ℓ is a constant frequency. Figure 2 shows a plot of a Gaussian pulse. The intensity is defined as $I = |E|^2 = EE^*$, so the intensity of a Gaussian pulse without chirp in the temporal domain is given by the function (2.3) times its complex conjugate

$$I(t) = \frac{1}{2}\varepsilon_0 c\, n E(t) E^*(t) = A_0^2 \exp\left[-\frac{2t^2}{\tau_G^2}\right]. \tag{2.4}$$

Figure 4 shows the intensity in the temporal domain. The full width at half-maximum (FWHM) is obtained by equating at the time when the intensity falls to one half $I = 1/2$,

$$\exp\left[-2\left(\frac{t}{\tau_G}\right)^2\right] = \frac{1}{2},$$

The time half width at half maximum is then

$$\Delta t_{HWHM} = \sqrt{\frac{1}{2}\log(2)}\,\tau_G.$$

In order to get the full width at the half-maximum multiplication by two is required

$$\Delta t_{FWHM} = \sqrt{2\log(2)}\,\tau_G = 1.17741\tau_G. \tag{2.5}$$

To analyze the electric field in the spectral domain, the Fourier transform[1]is evaluated,

$$E(\Omega) = \frac{1}{\sqrt{2\pi}} \int_{-\infty}^{\infty} A_0 \exp\left(-\frac{t^2}{\tau_G^2} + i\omega_\ell t - i\Omega t\right) dt = \frac{A_0 \tau_G}{\sqrt{2}} \exp\left(-\frac{\tau_G^2 (\Omega - \omega_\ell)^2}{4}\right). \tag{2.6}$$

[1] $E(\Omega) = \frac{1}{\sqrt{2\pi}} \int_{-\infty}^{\infty} \mathcal{E}(t) \exp[-i\Omega t]\, dt$

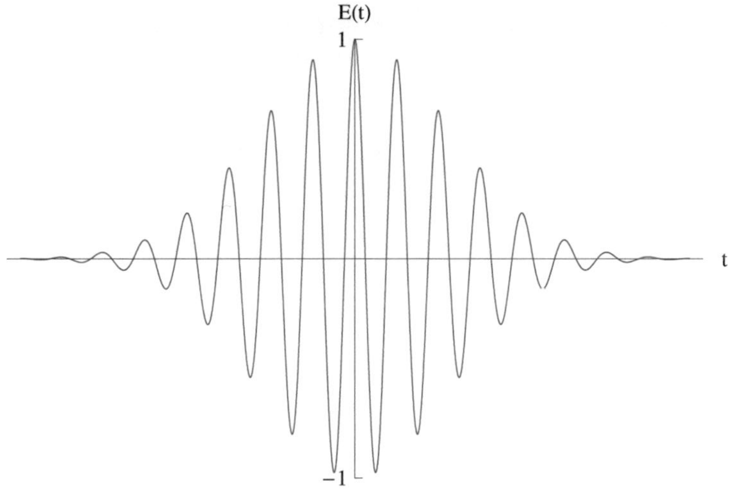

Figure 2. Gaussian pulse.

The intensity in the frequency domain for a Gaussian pulse without chirp is

$$I(\Omega) = \frac{1}{2}\varepsilon_0 c\, n \frac{A_0^2 \tau_G^2}{2} \exp\left[-\frac{\tau_G^2 (\Omega - \omega_\ell)^2}{2}\right]. \tag{2.7}$$

The normalized intensity is obtained upon division by $|E(0)|^2$. The FWHM in the frequency domain is obtained in an analogous fashion as before. The bandwidth measured from the carrier frequency $\triangle\omega_{HWHM} = \triangle\Omega_{HWHM} - \omega_l = \triangle\Omega_{HWHM}$

$$\exp\left[\frac{-\tau_G^2 (\triangle\omega_{HWHM})^2}{2}\right] = \frac{1}{2}.$$

The full width at half maximum is twice the solution to this equation

$$\triangle\omega_{FWHM} = \frac{2\sqrt{2\log(2)}}{\tau_G} = \frac{2.3548}{\tau_G}. \tag{2.8}$$

2.2.1. Gaussian with chirp

Let us now consider a Gaussian pulse with a quadratic phase

$$\Phi(t) = \omega_\ell t - \frac{a}{\tau_G^2} t^2. \tag{2.9}$$

Its instantaneous frequency is given by

$$\omega(t) = \frac{\partial \Phi}{\partial t} = \omega_\ell - \frac{1}{2}\frac{a}{\tau_G^2} t. \tag{2.10}$$

This means that the frequency is time dependent and the pulse is say to be chirped, a is the chirp factor. Figure 3 shows the plot of a chirped Gaussian pulse. With the present sign conventions i.e. eq. (2.10), the chirp factor is positive for down-chirp (frequency decreasing in time) and negative for up-chirp (frequency increasing in time). The electric field for a chirped pulse in the time domain is

$$E(t) = A_0 \exp\left[-(1+ia)\frac{t^2}{\tau_G^2} + i\omega_\ell t\right]. \tag{2.11}$$

The intensity in the time domain of the Gaussian chirped pulse is identical to the Gaussian pulse without chirp (2.4). That is, the time dependent intensity does not depend of the chirp constant.

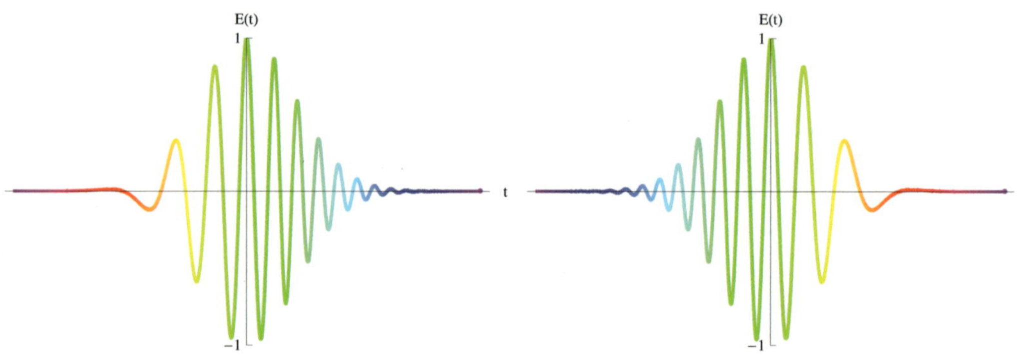

(a) Up-chirp: Negative chirp parameter $a = -3$

(b) Down-chirp: Positive chirp parameter $a = 3$

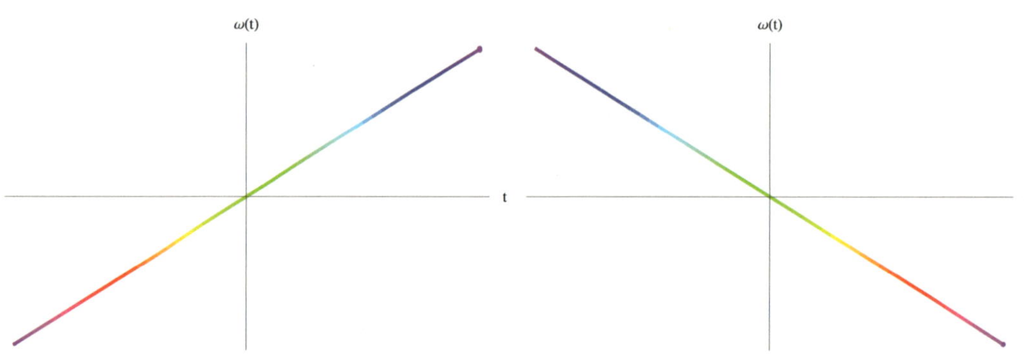

(c) For up-chirp, $a < 0$, the instantaneous frequency has positive slope.

(d) For down-chirp, $a > 0$, the instantaneous frequency has negative slope.

Figure 3. Chirped Gaussian pulse.

The Fourier transform of the quadratic time dependence (2.11) is

$$\tilde{E}(\Omega) = \frac{1}{\sqrt{2\pi}} \int_{-\infty}^{\infty} A_0 \exp\left[-(1+ia)\frac{t^2}{\tau_G^2}\right] \exp(-i\Omega t)\, dt,$$

upon integration

$$\tilde{E}(\Omega) = \frac{A_0}{\sqrt{2}} \frac{\tau_G}{\sqrt{(ia+1)}} \exp\left(\frac{i\Omega\tau_G^2}{4(a-i)}\right). \tag{2.12}$$

This equation can be written as

$$\tilde{E}(\Omega) = E(\Omega) = \frac{A_0}{\sqrt{2}} \frac{\tau_G}{\sqrt[4]{(1+a^2)}} \exp\left[i\phi - \frac{\omega^2\tau_G^2}{4(a^2+1)}\right], \tag{2.13}$$

where $\phi = \left(\frac{1}{2}\arctan(a) + \frac{a\Omega^2\tau_G^2}{4(a^2+1)}\right)$. On the other hand, the intensity in the frequency domain is $I(\Omega) = \varepsilon_0 c\, n\tilde{E}(\Omega)\tilde{E}^*(\Omega)$

$$I(\Omega) = \frac{1}{2}\varepsilon_0 c\, n\varepsilon_0 c\, n \frac{A_0^2}{2} \frac{\tau_G^2}{\sqrt[2]{(1+a^2)}} \exp\left[i\phi - \frac{\Omega^2\tau_G^2}{4(1+a^2)}\right] \exp\left[-i\phi - \frac{\Omega^2\tau_G^2}{4(1+a^2)}\right],$$

$$I(\Omega) = \varepsilon_0 c\, n \frac{A_0^2}{4} \frac{\tau_G^2}{\sqrt[2]{(1+a^2)}} \exp\left[-\frac{\Omega^2\tau_G^2}{2(1+a^2)}\right], \tag{2.14}$$

and the normalized frequency dependent intensity is

$$I(\Omega) = \exp\left[-\frac{\Omega^2\tau_G^2}{2(1+a^2)}\right]. \tag{2.15}$$

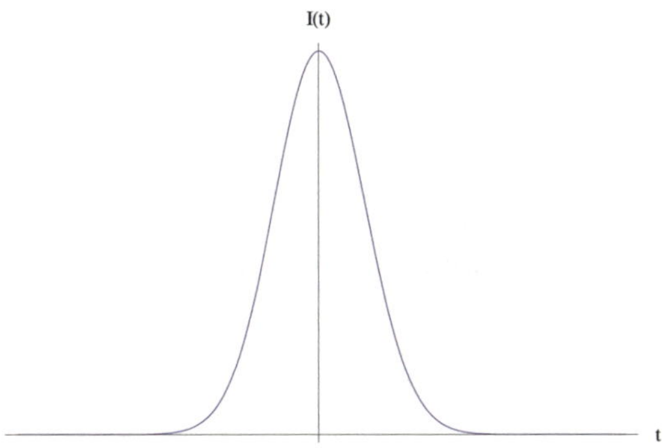

(a) Intensity of a Gaussian pulse with and without chirp in the temporal domain (curves are overlapped).

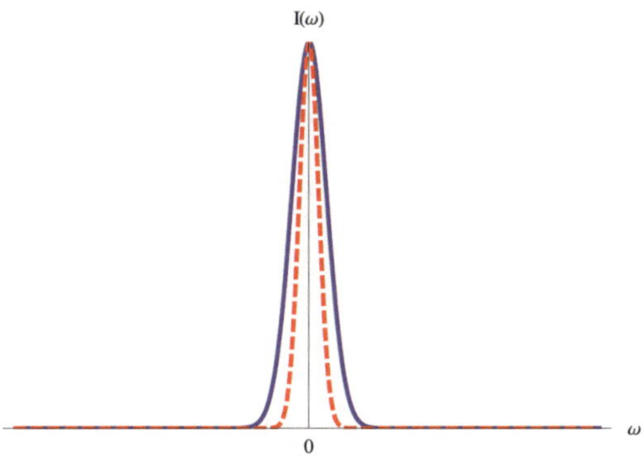

(b) Intensity of a Gaussian pulse with (solid) and without chirp (dashed) in the frequency domain.

Figure 4. Gaussian chirped pulse.

2.2.2. *Time bandwidth product*

The time bandwidth product of a Gaussian pulse without chirp from (2.5) and (2.8) is

$$(\triangle t \triangle\omega)_{FWHM} = \sqrt{2\log(2)}\tau_G \frac{2\sqrt{2\log(2)}}{\tau_G} \approx 2.2226, \tag{2.16}$$

or in terms of the frequency

$$(\triangle t \triangle \nu)_{FWHM} = \frac{2\log(2)}{\pi} \approx 0.4413 \,. \tag{2.17}$$

For a Gaussian chirped pulse, the FWHM in the temporal domain is the same as the Gaussian pulse without chirp. However, the FWHM in the frequency domain for the chirped pulse is obtained from (2.15)

$$\exp\left[-\frac{\Delta\omega^2_{HWHM}\tau_G^2}{2(1+a^2)}\right] = \frac{1}{2}.$$

The frequency FWHM in the frequency domain for a Gaussian chirped pulse is then

$$\triangle\omega_{FWHM} = \frac{\sqrt{8(1+a^2)\log(2)}}{\tau_G}. \tag{2.18}$$

The time bandwidth product from (2.5) and (2.18) is

$$(\triangle t \triangle \omega)_{FWHM} = 4\log(2)\sqrt{1+a^2}, \tag{2.19}$$

or in terms of the linear frequency

$$(\triangle t \triangle \nu)_{FWHM} = \frac{2\log(2)}{\pi}\sqrt{1+a^2}. \tag{2.20}$$

To abridge, a Gaussian pulse with linear frequency chirp produces a Gaussian spectral distribution broadened by $\sqrt{1+a^2}$. The time bandwidth product increases by this same amount since the temporal width is unaltered.

2.3. Hyperbolic secant envelope

2.3.1. *Phase function without chirp*

The electric field with carrier frequency $\frac{d\Phi(t)}{dt} = \omega_\ell$ modulated by a hyperbolic secant envelope $A(t)$ is

$$E(t) = A(t)e^{i\omega_\ell t} = \text{sech}\left(\frac{t}{\tau_s}\right)e^{i\omega_\ell t}.$$

The time dependent intensity is

$$I(t) = \varepsilon_0 c\, n\,|A(t)|^2 = \varepsilon_0 c\, n\, \text{sech}^2\left(\frac{t}{\tau_s}\right).$$

The time $\triangle t_{HWHM}$ where the intensity is at half maximum is

$$\text{sech}^2\left(\frac{\triangle t_{HWHM}}{\tau_s}\right) = \frac{1}{2} \quad \Rightarrow \quad \triangle t_{HWHM} = \text{arcsech}\sqrt{\frac{1}{2}}\tau_s = 0.881374\tau_s.$$

The intensity full width at half maximum is twice this value

$$\triangle t_{FWHM} = 2\text{arcsech}\sqrt{\frac{1}{2}}\tau_s \approx 1.76275\tau_s.$$

The field spectrum is obtained from the Fourier transform of the field

$$E(\Omega) = \mathcal{F}\left\{E(t)\right\} = \sqrt{\frac{\pi}{2}}\tau_s \text{sech}\left(\frac{\pi\tau_s\left(\omega_\ell + \Omega\right)}{2}\right),$$

the frequency dependent intensity is then

$$I(\Omega) = \left|E\left(\Omega\right)\right|^2 = \varepsilon_0 c\, n \frac{\pi}{4}\tau_s^2 \text{sech}^2\left(\frac{\pi\tau_s\left(\omega_\ell + \Omega\right)}{2}\right). \tag{2.21}$$

The maximum intensity is reached when $\Omega = \omega_\ell$, then $I_M = \varepsilon_0 c\, n \frac{\pi}{4}\tau_s^2$. The frequency $\Delta\Omega_{HM} = \omega_\ell + \Omega_{HM}$ where the intensity is at half maximum is then

$$\frac{\pi}{2}\tau_s^2 \text{sech}^2\left(\frac{\pi\tau_s\left(\Delta\Omega_{HM}\right)}{2}\right) = \frac{\pi}{4}\tau_s^2.$$

The frequency FWHM is then

$$\Delta\Omega_{FWHM} = \frac{4}{\pi\tau_s}\text{arcsech}\left(\sqrt{\frac{1}{2}}\right).$$

The Time-bandwidth product is thus

$$\Delta\Omega_{FWHM}\triangle t_{FWHM} = \left(\frac{4}{\pi\tau_s}\text{arcsech}\left(\sqrt{\frac{1}{2}}\right)\right)\left(2\text{arcsech}\sqrt{\frac{1}{2}}\tau_s\right) \approx 1.979.$$

In terms of the linear frequency

$$TBP = \Delta\nu_{FWHM}\triangle t_{FWHM} = \left(\frac{4}{2\pi^2\tau_s}\text{arcsech}\left(\sqrt{\frac{1}{2}}\right)\right)\left(2\text{arcsech}\sqrt{\frac{1}{2}}\tau_s\right) \approx 0.3148\,. \tag{2.22}$$

2.3.2. *Phase function with chirp*

For a pulse with linear chirp $\frac{d\Phi}{dt} = -\frac{2at}{\tau_s^2}$, the phase is quadratic in time $\Phi(t) = \omega_\ell - \frac{at^2}{\tau_s^2}$. The representation of the chirped pulse with hyperbolic secant envelope is then

$$E(t) = A(t)e^{i\Phi} = \text{sech}\left(\frac{t}{\tau_s}\right)\exp\left[i\left(\omega_\ell t + \frac{at^2}{\tau_s^2}\right)\right].$$

The Fourier transform $\mathcal{F}$ of the function with linear chirp cannot be solved analytically, so a different approach is presented. The frequency rate of change is a measure of the frequency deviation as a function of time $\frac{d\Phi(t)}{dt}$. The spectral width of solid state laser systems can be described via laser rate equations [6]. The refractive index variation arising from the rate equation can describe frequency chirping [7]. This model can be used to describe a nonlinear chirp for hyperbolic secant pulses that is mathematically tractable. The phase equation in terms of laser intensity is [8]

$$\frac{d\Phi(t)}{dt} = \frac{a}{2}\frac{1}{I}\frac{dI}{dt},$$

where a is the phase amplitude coupling factor that will correspond to the chirp parameter as we shall presently see. For $I = \varepsilon_0 c\, n\, \mathrm{sech}^2\left(\frac{t}{\tau_s}\right)$,

$$\frac{d\Phi(t)}{dt} = \frac{a}{2}\left(-\frac{2}{\tau_s}\tanh\left(\frac{t}{\tau_s}\right)\right) = -\frac{a}{\tau_s}\tanh\left(\frac{t}{\tau_s}\right).$$

The Gaussian profile with linear chirp and the hyperbolic secant profile with hyperbolic tangent chirp are similar near the central frequency for time $t \ll \tau_s$, $\frac{d\Phi(t)}{dt} = \left(-\frac{a}{\tau_s^2}t\right)$ as can be seen in figure 5.

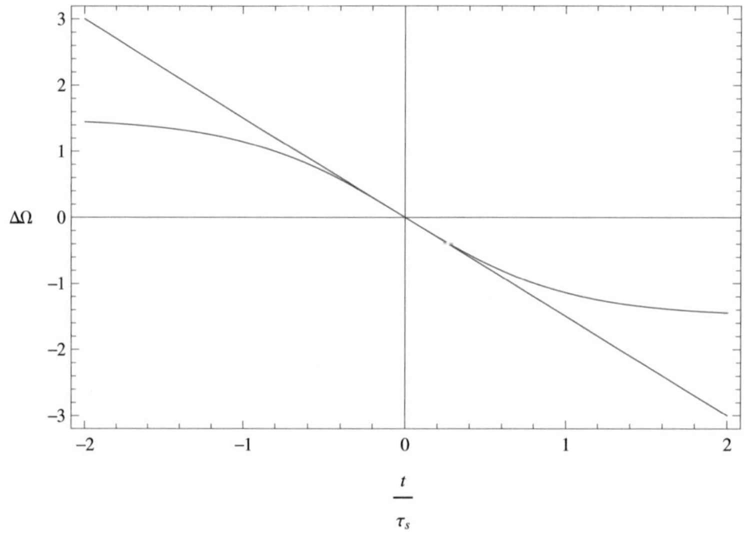

Figure 5. Instantaneous frequency deviation for a hyperbolic tangent chirp and a linear chirp.

The field phase can be obtained by direct integration

$$\frac{d\Phi(t)}{dt} = \frac{a}{2}\frac{1}{I}\frac{dI}{dt} = \frac{a}{2}\frac{d\ln I}{dt} = \frac{d}{dt}\ln I^{\frac{a}{2}}$$

so that $\Phi(t) = \ln I^{\frac{a}{2}}(t)$. Since $I(t) = |E(t)|^2 = \varepsilon_0 c\, n\, \mathrm{sech}^2\left(\frac{t}{\tau_s}\right)$, the phase is then

$$\Phi(t) = \ln\left[\mathrm{sech}^a\left(\frac{t}{\tau_s}\right)\right].$$

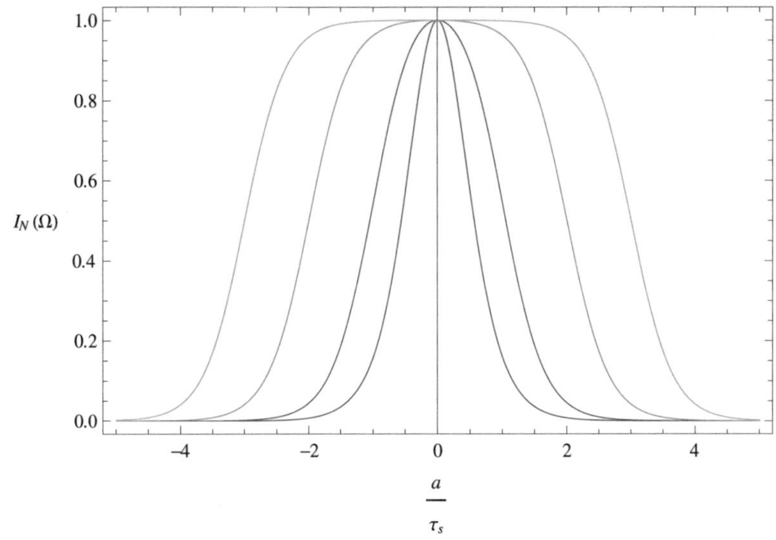

Figure 6. Different profiles as a function of chirp parameter a: 0, 1 ,2, 3 with increasing width and flattened top.

The complete electric field is therefore

$$E(t) = A(t)e^{i\Phi(t)} = \operatorname{sech}\left(\frac{t}{\tau_s}\right)\exp\left[i\left\{\ln\left[\operatorname{sech}^a\left(\frac{t}{\tau_s}\right)\right]\right\}\right],$$

that can be economically written as

$$E(t) = \operatorname{sech}^{1+ia}\left(\frac{t}{\tau_s}\right). \tag{2.23}$$

The electric field for a non-linearly chirped pulse is thus described by a hyperbolic secant function raised to a complex power. The imaginary part corresponds to the chirp parameter. The Fourier transform of this function has an analytical solution in terms of Gamma functions with complex argument:

$$E(\Omega) = \mathcal{F}\{E(t)\} = \frac{2^{ia}\tau_s}{\Gamma(1+ia)}\Gamma\left(\frac{1+ia+i\Omega\tau_s}{2}\right)\Gamma\left(\frac{1+ia-i\Omega\tau_s}{2}\right).$$

The normalized power spectrum can then be written as

$$\frac{|E(\Omega)|^2}{|E(0)|^2} = \frac{\operatorname{sech}\left[\frac{\pi}{2}(\Omega\tau_s+a)\right]\operatorname{sech}\left[\frac{\pi}{2}(\Omega\tau_s-a)\right]}{\operatorname{sech}^2\left(\frac{\pi a}{2}\right)}. \tag{2.24}$$

This function is plotted in figure 6 for various chirp parameters. The envelope flattens on the top as the chirp parameter is increased. Notice that when $a = 0$ (unchirped pulse) we obtain simply the previous $\operatorname{sech}^2\left[\frac{\pi}{2}\Omega\tau_s\right]$ power spectrum.

2.4. Mode-locking mechanism

Phase locked laser operation requires a coherent superposition of waves with a well defined phase relationship. The waves have frequencies separated by the cavity round trip time inverse. Each mode has the form

$$E_j = Ae^{i(\omega_0 t + j\delta\omega\, t)},$$

where $\delta\omega$ is the frequency separation between modes and A is the amplitude of each mode. The modes amplitudes may be considered constant and equal in a first approximation within the gain bandwidth. However, they will obviously fall down at the edges of the medium's bandwidth.

2.4.1. Coherent sum of fields

The mode-locked cavity can be explained in terms of the superposition of monochromatic waves with integer wavelength multiples of the round trip of the cavity. The sum of m_{tot} waves traveling in the same direction (labeled with subindex $\Rightarrow$) with constant amplitude A in complex notation is

$$G_{\Rightarrow} e^{i\gamma_{\Rightarrow}} = \sum_{j=m_i}^{m_f} Ae^{i(\omega t + j\delta\omega\, t)}, \tag{2.25}$$

where $\delta\omega = 2\pi\delta v$, m_i is the initial mode and m_f is the final mode. Waves do not interact directly between them. However, they are actually added in the presence of charges that respond to the superposition of the fields [9]. In this case, it is within the Ti:Sa crystal where the summation is carried out

$$G_{\Rightarrow} e^{i\gamma_{\Rightarrow}} = A \frac{\sin\left(\frac{(m_f - m_i + 1)\delta\omega\, t}{2}\right)}{\sin\left(\frac{\delta\omega\, t}{2}\right)} \exp\left[i\left(\omega t + \frac{\left(m_f + m_i\right)}{2}\delta\omega\, t\right)\right]. \tag{2.26}$$

The amplitude is now time dependent with a periodicity $\frac{\delta\omega\, t}{2} = 2\pi b$ where b is an integer. The amplitude maxima are obtained from L'hopital rule, the magnitude is

$$G_{\Rightarrow max} = A \left. \frac{\cos\left(\frac{m_{tot}\delta\omega\, t}{2}\right)}{\cos\left(\frac{\delta\omega\, t}{2}\right)} \frac{\frac{m_{tot}\delta\omega}{2}}{\frac{\delta\omega}{2}} \right|_{\frac{\delta\omega\, t}{2} = 2\pi} = Am_{tot}, \tag{2.27}$$

where $m_{tot} = m_f - m_i + 1$ is the total number of modes. Therefore, if a large number of waves with the same absolute phase are added with periodically shifted frequencies, the amplitude envelope becomes a periodically peaked train of temporal pulses. For an estimate of the width, consider the first zero of the amplitude function in (2.26), that is $\frac{m_{tot}\delta\omega\triangle t_0}{2} = \pi$. The full width between the first positive and negative zero of the function is

$$\triangle t_{FW0} = \frac{1}{\delta v} \frac{2}{m_{tot}} \tag{2.28}$$

Series expansion of the amplitude function is

$$\frac{\sin\left(\frac{m_{tot}\delta\omega\, t}{2}\right)}{\sin\left(\frac{\delta\omega\, t}{2}\right)} = m_{tot} - \frac{\delta\omega^2}{24}\left(m_{tot}^3 - m_{tot}\right) t^2 + \cdots$$

FWHM estimated from this series for large m_{tot} is

$$\triangle t_{FWHM} = \frac{1}{2\pi\delta\nu}\frac{4\sqrt{3}}{m_{tot}} \tag{2.29}$$

Three mechanisms are invoked in laser cavities in order to amplify the mode locked peaks: active feedback - pumping, intra-cavity saturable absorber and self mode-locking via self focusing. In Ti:Sa oscillators self mode-locking is the mechanism responsible for pulse generation. It is achieved via Kerr lens mode-locking (KLM). Kerr refractive index modulation is a third order nonlinear process where the refractive index is proportional to the intensity of the incoming beam $n = n_0 + n_2 I$. The cavity is tuned so that self focusing of the more intense beam profile is favored. In our setup, the second curved mirror (C2) distance from the Ti:Sa crystal is varied in order to achieve the mode-locking regime. The continuous light (CW) pump is a frequency-doubled Nd:YVO_4 laser at 532 nm with 2.07 W.

2.5. Energy content

In electromagnetic phenomena, Poynting's theorem is considered to represent the field energy conservation equation. However, the conservation equation obtained either from Maxwell's equations or the wave equation is not unique. The choice of Poynting's conservation equation as representing the electromagnetic energy is to some extent arbitrary and not exempt of counterintuitive predictions [10, p.27-6]. The Wigner function wave description [11, p.287] involves the function times its conjugate; Therefore, it also relies on a complex version of Poynting's energy theorem.

Another conservation equation derived from the wave equation is obtained by invoking the two linearly independent solutions [12]. These two wave solutions are referred to as the complementary fields. The complementary fields approach to wave phenomena requires the existence of two fields that establish a dynamical balance between two forms of energy. The energy content of the wave goes from one field to the other and back such that the total energy is constant for an infinite wave-train [13]. The formalism attempts to address two issues: i) a representation that avoids taking time averages and hence is suitable to describe ultrafast phenomena and ii) a description that gives a clearer picture of the apparent redistribution of energy when waves are superimposed.

The complementary fields energy content of an electromagnetic wave with electric field amplitude $A(\mathbf{r},t)$ and phase $\Phi(\mathbf{r},t)$ is equal to

$$\mathcal{E} = \varepsilon A^2(\mathbf{r},t)\frac{\partial\Phi(\mathbf{r},t)}{\partial t}. \tag{2.30}$$

The energy density proposed in an earlier communication [12, eqs.(24)-(25)] has been scaled by the permittivity factor ε to establish the equality. This quantity represents the energy exchange between the two complementary fields. Whereas the flow is equal to

$$\mathbf{S}_{\perp} = \frac{1}{\mu} A^2(\mathbf{r}, t) \nabla\Phi(\mathbf{r}, t). \tag{2.31}$$

The exchange energy flow can be written in terms of the temporal derivative since $\nabla\Phi = \frac{\partial\Phi}{\partial t}\left(\frac{\hat{\mathbf{k}}}{v}\right) = \sqrt{\mu\varepsilon}\frac{\partial\Phi}{\partial t}\hat{\mathbf{k}}$, where $\hat{\mathbf{k}}$ is a unit vector in the direction of propagation. The flow of the exchange energy is then

$$\mathbf{S}_{\perp} = \sqrt{\frac{\varepsilon}{\mu}} A^2(\mathbf{r}, t) \frac{\partial\Phi(\mathbf{r}, t)}{\partial t}\hat{\mathbf{k}}. \tag{2.32}$$

The impedance is often written in optics for non magnetic media as $\sqrt{\frac{\varepsilon}{\mu}} = \varepsilon\frac{c}{n} = \varepsilon_0 c\, n$. For a plane monochromatic wave (labeled with subindex j), the amplitude is constant $A_j(\mathbf{r}, t) = A_j$ and the phase is linear in the time and space variables $\Phi_j(\mathbf{r}, t) = \mathbf{k}_j \cdot \mathbf{r} - \omega_j t$, where the wave vector $\mathbf{k}_j$ and the angular frequency ω_j are constant. The assessed quantities (2.30) and (2.31) are then also constant

$$\mathcal{E} = \varepsilon A_j^2 \omega_j, \qquad \mathbf{S}_{\perp} = \varepsilon_0 c\, n A_j^2 \omega_j \hat{\mathbf{k}}_j. \tag{2.33}$$

Two issues should be highlighted, the frequency dependence of the energy density and the fact that no cycle averages have been performed. The linear frequency dependence means that when summing up over the phase locked monochromatic fields there is a frequency weighing factor. On the other hand, since no cycle averages are performed the energy density gives the instantaneous energy content of the wave-field.

2.5.1. *Superposition of fields*

The energy content of the sum of two or more fields is not equal to the sum of their individual energies but in the case of incoherent superposition. To wit, consider that the fields coexist in a certain region of space-time but there is no medium that allows for the superposition of the fields, that is, the fields do not sum up. The energy content of the jth monochromatic wave from (2.33) is

$$\left|\mathbf{S}_{\perp j}\right| = \varepsilon_0 c\, n\, A^2 (\omega_0 + j\delta\omega),$$

since the modes are equally spaced by $\omega_j = \omega_0 + j\delta\omega$ and $A_j = A$. If the waves are not added, the total energy is equal to the sum of energies of each monochromatic wave is

$$\sum_{j=m_i}^{m_f} \left|\mathbf{S}_{\perp j}\right| = \varepsilon_0 c\, n \sum_{j=m_i}^{m_f} A^2 (\omega + j\delta\omega) = \varepsilon_0 c\, n \left(m_f - m_i + 1\right) A^2 \left(\omega + \left(\frac{m_f + m_i}{2}\right)\delta\omega\right),$$

that can be written as

$$I_{tot} = \left|\mathbf{S}_{\perp tot}\right| = \sum_{j=m_i}^{m_f} \left|\mathbf{S}_{\perp j}\right| = \varepsilon_0 c\, n\, m_{tot} A^2 \overline{\omega} \tag{2.34}$$

where $m_i = 1$, $m_f = m_{tot}$, the initial mode is one and the final mode equals the total number of modes m_{tot} and the mean mode angular frequency is $\overline{\omega} = \omega + \frac{m_{tot}+1}{2}\delta\omega$. The medium bandwidth gain is $\Delta\omega = (m_{tot}+1)\,\delta\omega$. Notice that the total energy flow as well as the energy flow of each mode are time independent.

In the presence of a material medium, the Ti:Sa crystal in the laser oscillator case, the modes interact via the gain medium. The medium thus responds to the superposition of the fields. In this case, it is the energy of the superimposed fields that needs to be evaluated. The energy content of the coherent sum from (2.30) and (2.26) is

$$I_{coh} = |\mathbf{S}_{\perp coh}| = \varepsilon_0 c\, n\, A^2 \left[\frac{\sin\left(\frac{m_{tot}\delta\omega\, t}{2}\right)}{\sin\left(\frac{\delta\omega\, t}{2}\right)}\right]^2 \overline{\omega}.$$

The energy of the sum of waves is now time dependent and sharply peaked at $\delta\omega\, t = 2\pi b$ for integer b. The maximum intensity is

$$I_{max} = |\mathbf{S}_{\perp coh-max}| = \varepsilon_0 c\, n\, A^2 m_{tot}^2 \overline{\omega}. \tag{2.35}$$

However, this result does not necessarily imply that there has been an energy redistribution in time. It does mean that a device or entity capable of responding to the sum of the fields, such as an electric charge, will detect a large amplitude at the peaks. Outside this region, from the comparison of (2.34) and (2.35), it will detect an intensity approximately $1/m_{tot}$ lower.

3. Medium gain and mode locking operation

3.1. Gain without mode-locking.

The second curved C2 mirror was varied without mode locking while maintaining the slit width (S-W) totally opened (7.73 mm). The IR power output was recorded with a thermopile power detector [2]. The center wavelength was measured with two different monochromators. On the one hand, a scanning Czerny Turner monochromator, 0.5 m focal length[3]. On the other hand, a miniature fiber optic Czerny Turner spectrometer, 0.075 m focal length [4] with 2048 pixels detector array. The C2 mirror distance to the crystal is smaller for larger micrometer readings (see figure 1). The integration time in this latter system was 259 ms. The position was scanned from 9.60 mm up to 10.75 mm. From 9.60 mm to 10.30 mm it was varied in 0.10 mm steps. In the interval between 10.30 mm and 10.75 mm the variation was in steps of 0.05 mm.

Figure 7 shows the overlap of two runs. The output power lies around 200 mW between 9.9 and 10.7 mm C2 mirror position. The gain falls off rapidly before or after this range. However, there is a dip at 10.50 mm. This decrease is due to lasing operation at two CW wavelengths as shown in figure 8.

[2] Ophir 3A with Nova 2 meter, 3% accuracy
[3] Pacific MP-1018B
[4] Avantes AvaSpec 2048

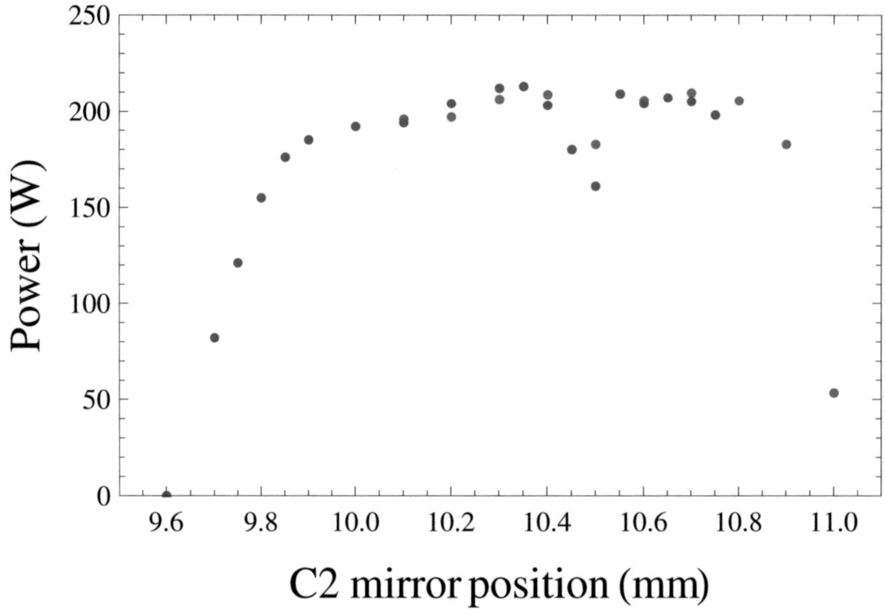

Figure 7. Mirror position vs. output power in non-mode locked operation. Points in pink were recorded directly on the computer whereas the blue ones were taken from the thermopile meter.

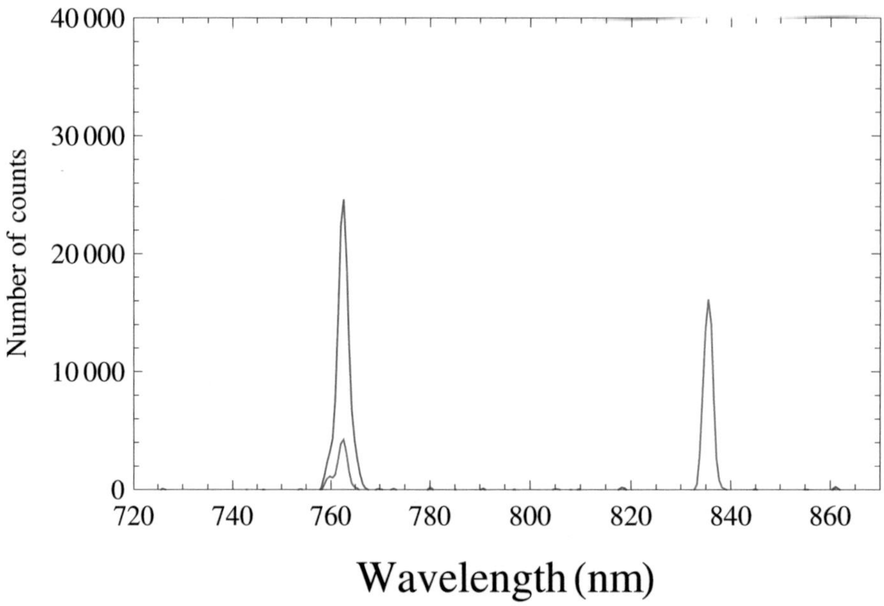

Figure 8. Continuous wave (CW) non-mode locked emission at 9.70 mm (blue) and 10.50 mm (magenta) depending on C2 position. Monochromator integration time is 259 ms.

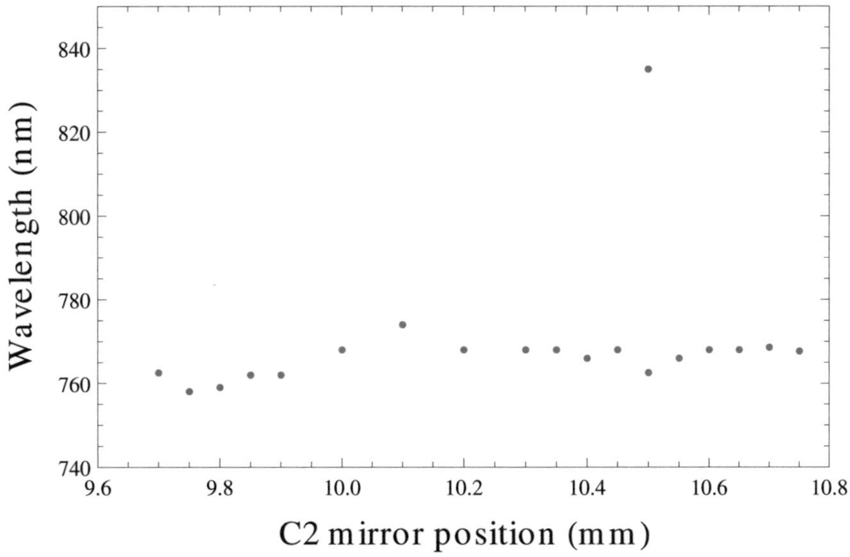

Figure 9. Emission wavelength vs C2 mirror position.

Figure 9 shows the wavelength where the maximum emission took place when C2 mirror was varied. The wavelength remains between 758 nm and 774 nm with the exception of the point at 10.50 mm, that emits at 762.6 nm and 835 nm.

3.2. C2 adjustment in mode-locking operation.

The focusing point within the active medium can be varied with the second curved mirror (C2) in order to obtain mode locking operation [14]. The position when self focussing of the intense beam attains maximum overlap with the pumped region will produce pulsed operation. This position lies on the edge of the gain curve (10.80 mm lies on the right hand border in figure 7). It is necessary to be within the gain curve so that many modes are amplified. However, if some modes experience a very large gain they will prevail over the pulsed many mode superposition. There are two ways to suppress this continuous (CW) light breakthrough: 1) lowering the pump current of the seed laser beam and 2) by moving the second curved mirror C2 [5]. To get the maximum power optimization the second option was chosen. Mode locking operation without CW breakthrough was obtained from 10.81 mm to 10.85 mm micrometer reading as shown in figure 10. Beyond the 10.85 position laser operation was not sustained. The spectrum profile shifts towards longer wavelengths as the beam is focused further into the crystal. From 10.80 mm to 10.81 mm the amplitude decreases due to less overlap between the focused beam and the gain region. Thereafter the amplitude remains constant presumably because self-focusing compensates and focuses the beam within the gain volume. On the other hand, the continuous light CW breakthough vanishes at 10.83 mm position. For C2 greater than 10.83 there is no CW breakthrough because no single mode is amplified above other modes.

[5] Quantronix, Ti-Light Operator's manual.

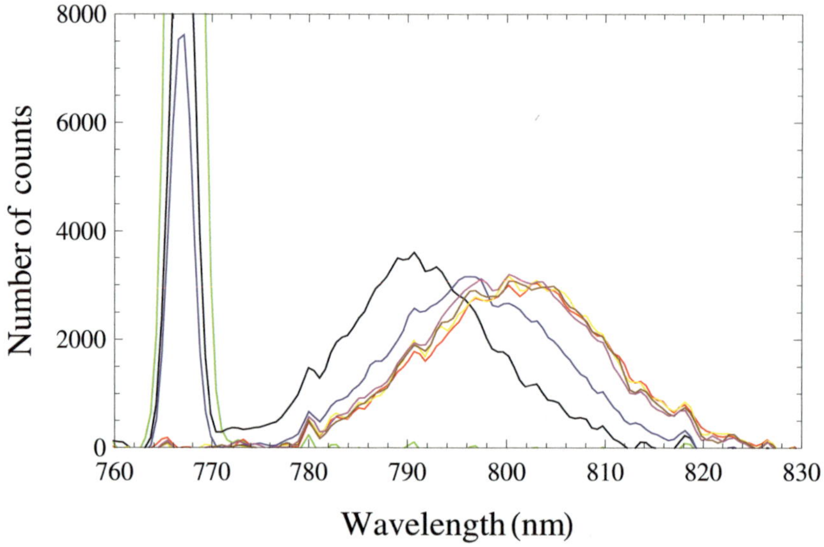

Figure 10. Spectra for different C2 mirror positions. At 10.80 mm two spectra were taken, with (black) and without (green) mode locking . All other spectra were made in mode locked operation; 10.81 mm (blue), 10.82 mm (red), 10.83 (yellow), 10.84 mm (brown) y 10.85 mm (magenta) (Curves from left to right). The continuous sharply peaked contribution vanishes when C2 mirror is located at 10.82 mm and further away. Spectrometer integration time is 259 ms.

4. Pulse train measurement and cavity length

4.1. Repetition rate

The output oscillator pulses were measured with an internal photo diode as well as a fast avalanche external photo diode. The internal photo diode is electronically amplified whereas the external photo diode was not amplified to provide the fastest possible risetime. Both signals gave similar results. The pulse repetition frequency was monitored with a fast oscilloscope[6]. The repetition frequency obtained from the scope trace shown in figure 11 is 77.5 ± 0.1 MHz with a period of $\tau = \frac{1}{\delta\nu} = 12.90 \pm 0.02$ ns which gives a round trip cavity of $\tau c = \frac{c}{\delta\nu} = 12.90 \times 10^{-9} \times 3 \times 10^{8} = 3.86\,\text{m} = 2L$, thus the cavity optical length is $L = 1.928\,\text{m}$. The angular frequency mode separation is then $\delta\omega = 2\pi\delta\nu = 486.9 \pm 0.6$ MHz or in wavelength units @ 800 nm, $\delta\lambda = -\delta\nu\frac{\lambda^2}{c} = -77.5 \times 10^{6}\frac{\left(800\times10^{-9}\right)^2}{3\times10^{8}} = 1.65 \pm 0.02 \times 10^{-13}$m. For a typical bandwidth of 35 nm (see figure 22b) the number of modes is $m_{tot} = \frac{\Delta\omega}{\delta\omega} = \frac{35}{1.65\times10^{-4}} = 2.21 \times 10^{5}$. The coherent superposition of these modes gives a pulse width, according to (2.28)

$$\triangle t_{FW0} = \frac{1}{\delta\nu}\frac{2}{m_{tot}} = 12.9 \times 10^{-9}\frac{2}{2.21 \times 10^{5}} = 116\,\text{fs}.$$

[6] Tektronix 485/R485 analogue oscilloscope with 350 MHz bandwidth or Picoscope 5302 digital scope with 250 MHz bandwidth.

The measured pulse width is in this order of magnitude, although somewhat lower. Recall however, that this estimate is for the first zero rather than width at half maximum. Evaluation from FWHM series estimate (2.29) gives a closer estimate to the values reported in section (6),

$$\Delta t_{FWHM} = \frac{1}{2\pi\delta\nu}\frac{4\sqrt{3}}{m_{tot}} = 12.9 \times 10^{-9}\frac{1.1027}{2.21 \times 10^5} = 64.36 \text{ fs}.$$

Nonetheless, it should be stressed that pulse width is critically dependent on the pulse envelope. The spectra shown in subsequent sections were integrated over 259 ms as mentioned in section 3.1. Since the repetition rate is 77.5 MHz, the spectra show the average of 2.007×10^7 pulses.

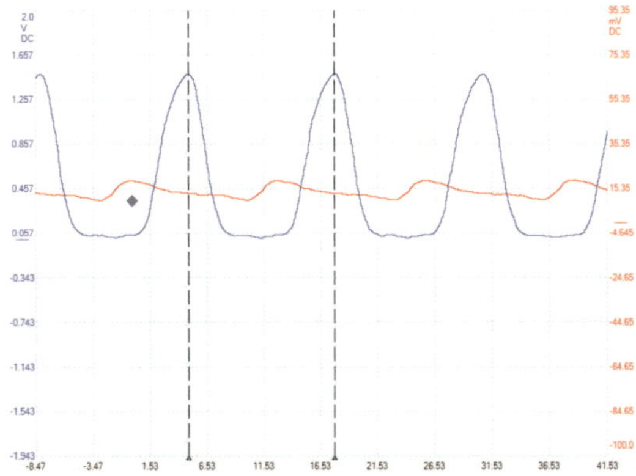

Figure 11. Oscillator output pulses detected with the internal photodiode (blue) and an external avalanche hamamatsu photodiode (red). The horizontal axis represents time in ns and the vertical axis depicts photodiode voltage.

4.2. Mode locking power and CW broadband background

An external fast photo diode was used to evaluate the difference between the mode locking (ML) peak intensity and the background light in between the high intensity peaks. This CW broad spectrum background light should not be confused with the CW narrow bandwidth breakthrough mentioned in subsection 3.2. The broad CW background is always present and is due to the superposition of the cavity modes when they are out of phase. Typical traces are shown in figure 12. The width of the peaks in the order of 1 ns are due to the detection system (photo diode + scope) since the optical pulse duration is four orders of magnitude shorter. The light level stabilizes at $I_{bb} = 10$ a.u. in between pulses. This measurement corresponds to the broad background (bb) emission between pulses in mode-locked operation. There is possibly electrical ringing after some pulses in the blue trace. In all three curves, there are spurious peaks due to analog to digital conversion uncertainty. This assertion is suported by the lack of subsidiary peaks in the streak camera recordings shown in the next section. The average intensity of the narrow bandwidth (nb) in non-mode locked operation is $I_{nb} = 6.3$ a.u. The ratio of these two intensities is $\frac{I_{bb}}{I_{nb}} = 1.59$. This quantity is an estimate of the energy content when a large number of out of phase modes (aprox. 10^5) are present in comparison with flow detected when a few modes are present.

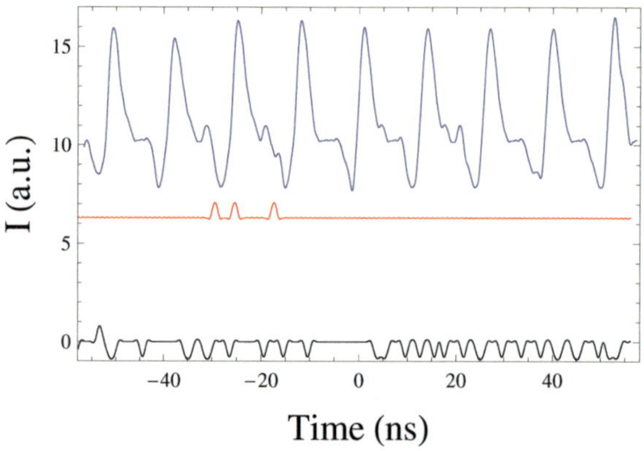

Figure 12. Oscillator pulse train detected with fast avalache photo diode and digital oscilloscope. Mode locked operation (blue trace), non-mode locked (red trace), residual trace with beam blocked in black. Small peaks are most likely due to analog to digital conversion uncertainty.

A streak camera was used to record the oscillator output. Firstly, a fraction of the laser beam output was fed through an optical fiber of 400 μ diameter and two fibers with 5 m length. This way the pulse was spread in time due to dispersion in the multimode fibre to insure that the peak intensity in the streak photocathode was not too high. The streak camera[7] was swept with a 330 ps/mm - 25 ns/mm sweep unit with 100 kHz maximum trigger frequency[8]. This unit was triggered by a pulse generator[9] that was in turn triggered by the laser oscillator electrical signal output. The unit was swept at 6.6, 10, 20, 50 and 100 ns full time bases. A typical intensity plot is shown in figure 13. The time span between pulses is $\tau = 13.12 \pm 0.5$ ns. The error is due to 2.5% non-linearity error in the time sweep. This result is consistent, within error, with the repetition rate obtained from the photo diode measurement.

5. Pulse spectrum and center wavelength

The pulse spectrum can be altered by the position of the second prism or the position and width of the slit placed between this prism and the end mirror.

5.1. Pulse spectrum as a function of prism compensation

In order to keep a narrow pulse, the two prisms are used to compensate the group velocity dispersion (GVD) introduced by the pulse propagation through the titanium sapphire crystal [15, 16]. The second prism (PR2) has a translational motion while the first prism position is fixed. The cavity design would benefit if the first prism (PR1) is also mounted on a traslation stage. The initial operating conditions are shown in table 2.

The spectrum in figure 14 shifts towards longer wavelengths as the second prism (PR2) position decreases. The wing on the right hand side of figure 14 shows that all the profiles

[7] Optronis SC-10
[8] Optronis TSU12-10PG sweep unit
[9] Stanford Research DGC535

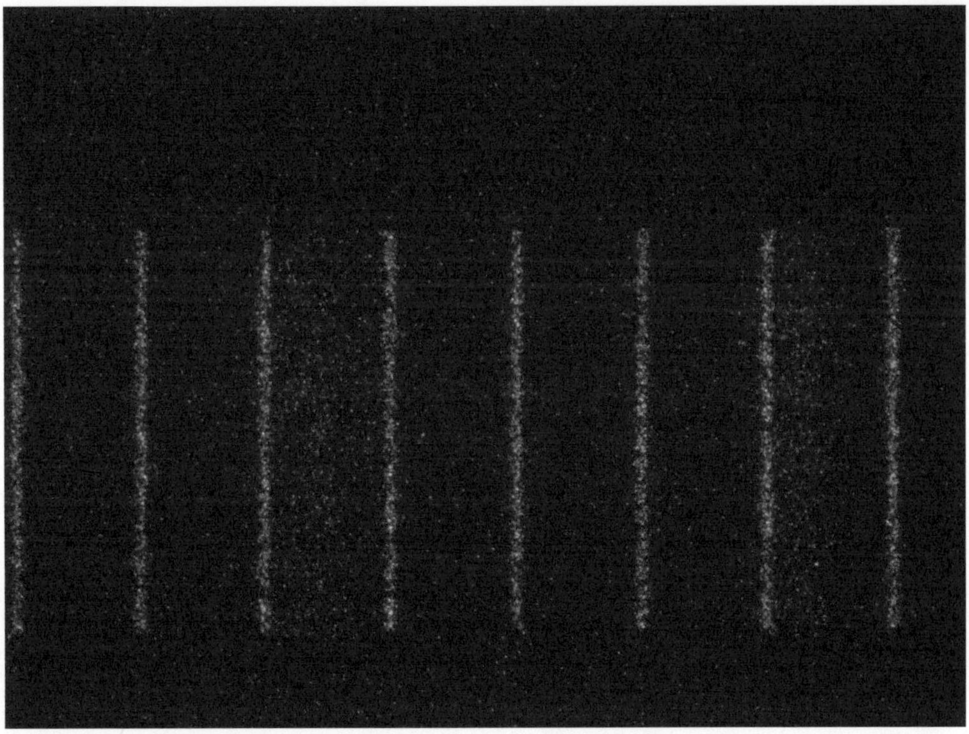

(a) Streak camera image. The vertical direction shows a transverse beam spatial direction and the horizontal direction represents the time axis.

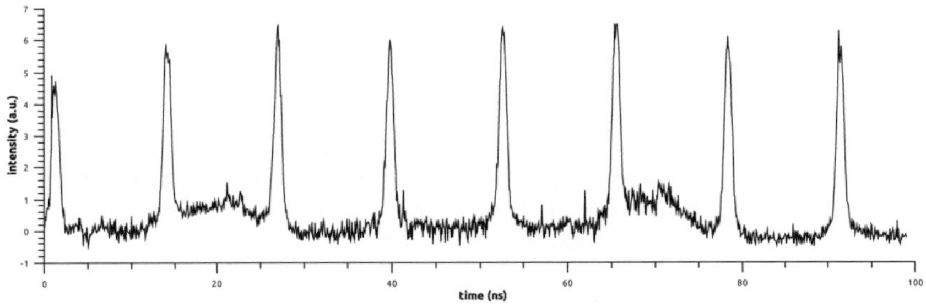

(b) Streak camera intensity time profile.

Figure 13. Streak camera image and graph of Ti:Sa oscillator pulses. The sweep rate is 5 ns/mm, 100 ns full sweep (small background bumps are due to pulses during backtrace).

converge towards 860 nm. However, the wings on the left hand side are displaced as the curve maximum is displaced towards shorter wavelengths. On the other hand, the amplitude increases at longer wavelengths and the bandwidth becomes narrower. In figure 15 a plot of the bandwidth as a function of the prism PR2 position is shown. Another set of spectra under the same operating parameters as shown in table 2 are depicted in figure 16. The same features described in the previous case are observed except for an amplitude decrease rather than increase at longer wavelengths. This discrepancy is possibly due to a slightly different cavity alignment so that shorter or longer wavelengths' gain were favoured in either case.

Element	Position (mm)
Mirror C1 position	3.98
Mirror C2 position	10.85
Lens L1 position	4.12
Prism PR2 initial position	1.30
Slit width (S-W)	7.73
Slit translation (S-T)	3.00

Table 2. Initial conditions when prism PR2 position was varied.

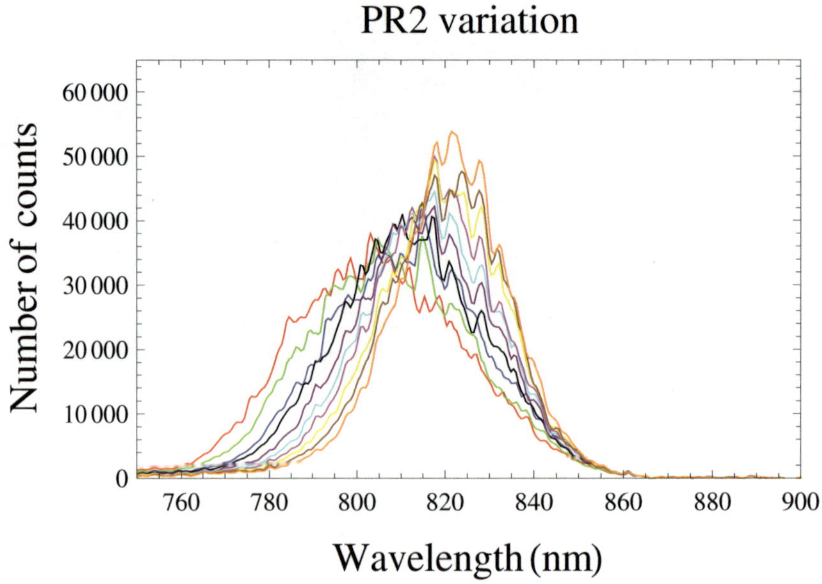

Figure 14. Spectra with mode lock taken at different prism PR2 positions: 0.40 mm(orange), 0.50 mm(brown), 0.60 mm(yellow), 0.70 mm(magenta), 0.80 mm(cyan), 0.90 mm(purple), 1.00 mm(black), 1.10 mm(blue), 1.20 mm(green) y 1.30 mm(red). The PR2 position was varyied by 0.10 mm steps. Spectrum curves shift towards longer wavelengths for larger micrometer readings.

The prism apex moves further into the beam as the micrometer reading increases (see figure 21a). As the beam travels through more prism material, the ML emission shifts towards shorter wavelengths.

5.2. Adjustment of the slit width (S-W)

The slit has two degrees of freedom, translation (S-T) and slit width (S-W) as can be seen in figure 17. Each of these displacements was varied separately.

The slit width micrometer is not calibrated to the real aperture width. The slit aperture was measured with a Vernier and a difference of 0.73 mm was obtained when the slit micrometer was set to zero. Hereafter, this value is added to slit width (S-W) micrometer readings in order to report the actual slit width. It should also be noted that one blade is fixed and only the blade

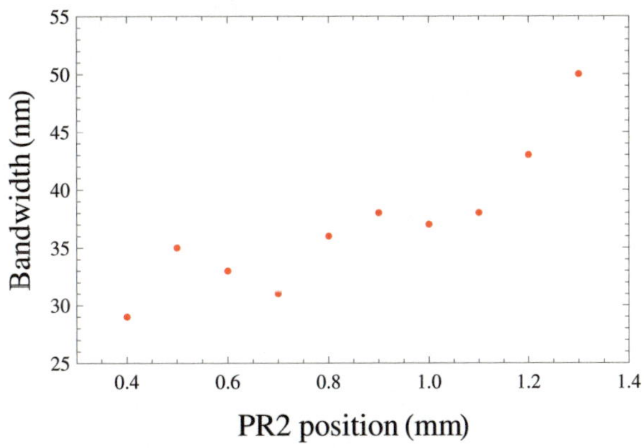

Figure 15. Bandwidth variation as a function of second prism (PR2) position. Micrometer readings between 0.40 and 1.30 mm.

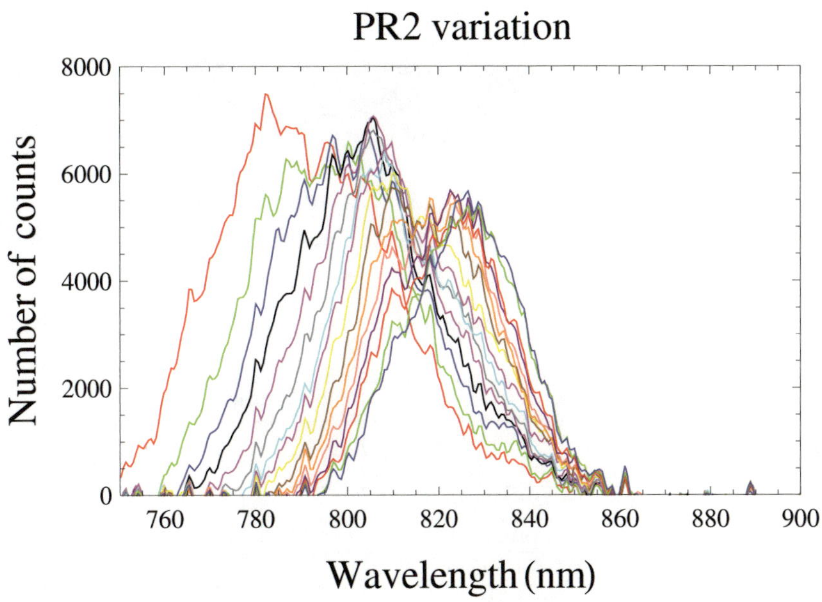

Figure 16. Spectra with mode lock for prism PR2 position between 0.00 and 1.50 mm: 0.00 mm (blue), 0.10 mm (green), 0.20 mm (red), 0.30 mm (purple), 0.40 mm (pink), 0.50 mm (orange), 0.60 mm (brown), 0.70 mm (yellow), 0.80 mm (magenta), 0.90 mm (cyan), 1.00 mm (gray), 1.10 mm (magenta), 1.20 mm (black), 1.30 mm (blue), 1.40 mm (green) and 1.50 mm (red). The position was varied in 0.10 mm steps. Micrometer readings decrease in curves from left to right.

on the blue side of the dispersion actually moves (see figures 21a and 17). The setup would benefit with a slit where both blades are displaced from a central position.

The laser parameters used while the width was changed are abridged in table 3. S-W was set at 3.73 mm and increased in steps of 0.50 mm up to 7.73 mm. Between 3.73mm to 7.73 mm the spectra did not show a noticeable variation because the laser beam is not blocked by the slit. That is, the beam width at the slit plane is smaller than the slit width. This assertion was

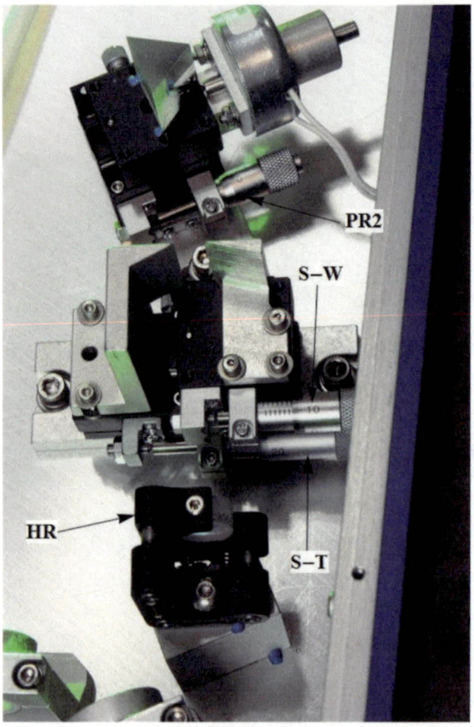

Figure 17. Slit displacement and width: S-T micrometer controls the translation mechanism whereas S-W alters the slit width.

confirmed by direct view with an IR viewer that no appreciable part of the beam was being blocked.

Element	Measurement (mm)
First curved mirror (C1)	3.98
Second curved miror (C2)	10.85
Pump beam focusing lens (L1)	4.12
Second prism position (PR2)	1.10
Initial S-W value	3.23
S-T position	5.50

Table 3. Initial conditions of slit width S-W variation .

The spectra for slit widths ranging from 3.23 to 1.43 mm are shown in figure 18. The S-W was decreased in 0.50 mm steps from 3.23 to 2.23 mm. Thereafter, it was decreased in steps of 0.10 from 1.93 to 1.43 mm. The spectra shifts towards longer wavelengths as the slit is reduced. This is expected because the moving blade crops from the blue side of the dispersion. On the other hand, the bandwidth decreases as the slit is closed although the intensity does not vary in an appreciable way. These features become relevant below the 2 mm width where the beam dispersion width is being considerably clipped. The pulse duration does not necessarily increase as the bandwidth is reduced until the pulse is time-bandwidth limited as we shall see in section 6.3.

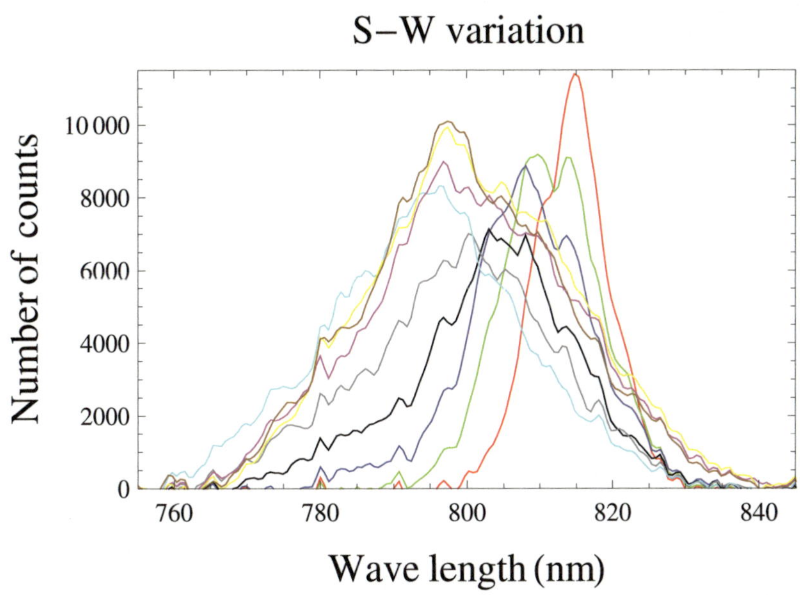

Figure 18. Spectra with mode lock for different slit widths: 1.43 mm (red), 1.53 mm (green), 1.63 mm (blue), 1.73 mm (black), 1.83 mm (gray), 1.93 mm (cyan), 2.23 mm (magenta), 2.73 (yellow) and 3.23 mm (brown). . From 2.23 mm to 3.23 mm the steps were of 0.50 mm. From 1.43 mm to 1.93 mm the S-W the steps were of 0.10 mm. Narrower slit width spectra lie on the right side of the figure.

5.3. Slit translation adjustment

The slit translation was varied from 7.60 mm to 0.40 mm while the slit width (S-W) was totally opened at 7.73 mm. The operating conditions while performing the S-T translation are shown in table 4.

Element.	Measure (mm)
First curved mirror (C1)	3.98
Second curved miror (C2)	10.85
Pump beam focusing lens (L1)	4.12
Second prism position (PR2)	0.5
S-W value	7.73
S-T initial position	7.60

Table 4. Initial conditions of S-T translation.

The spectra are shown in figure 19. ML emission is more intense at shorter wavelengths. The intensity decreases about tenfold for emission centered at 820 nm compared with emission centered at 730 nm. Nonetheless, stable operation is obtained from 730 to 850 nm, that is, in a 110 nm span. The curves suggest that emission can be obtained further to the blue but the micrometer mechanical translation table could not go any further.

In order to analyze these spectra we have separated the contributions in four different groups shown in figure 20. The slit translation (S-T) was decreased in steps of 0.10 mm from 7.60 mm to a 0.40 mm. At 7.60 mm the slit was blocking the redder frecuencies while the blue ones were

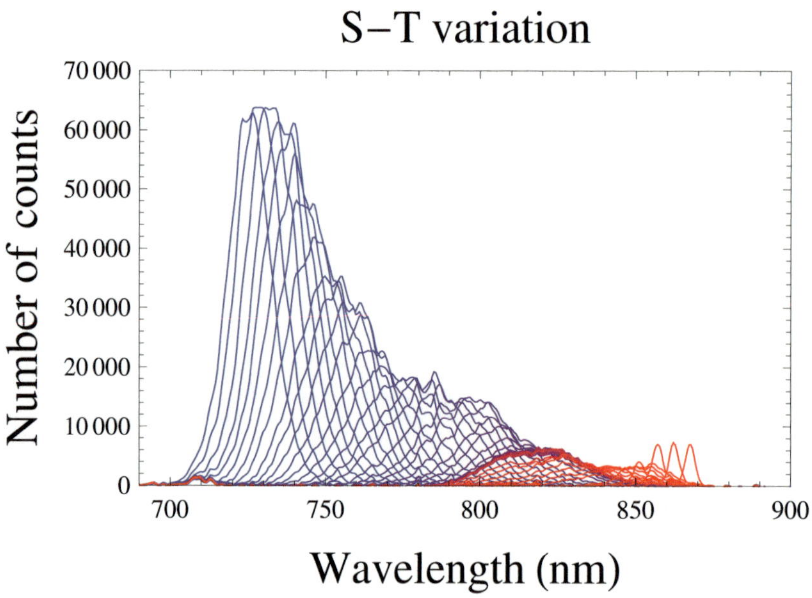

Figure 19. Mode locked spectra as a function of slit translation (S-T) from 7.60 mm to 0.40 mm. The spectrum varies from shorter to longer wavelengths and the intensity decreases for lower micrometer readings.

unblocked. A sketch of the slit together with the beam dispersed light is shown in figure 21a. Blue is dispersed at a larger angle than red in the first prism PR1 shown in photograph 21b. After a two mirror reflection it impringes on the second prism that is inverted with respect to the first one. When the slit was moved towards lower micrometer readings, the longer wavelength frequencies became progressively unveiled. As a result, the center wavelength moved towards longer wavelengths; The bandwidth increased and the spectrum intensity decreased as seen in figure 20a. At 5.20 mm the red side of the spectrum was completely unblocked.

From 5.10 mm to 3.50 mm, , the spectra do not suffer any change and the center wavelength remains constant as well as the bandwidth as shown in figure 20b. The beam is not blocked on neither side by the slit since the beam width is smaller than the slit width. This behaviour continued until the translation micrometer was at 1.60 mm.

Between 1.50 mm to 0.70 mm, the slit started to block the blue side of the spectrum while the red side remained unobstructed. The central emission thus shifts to longer wavelengths as seen in figure 20c. The bandwidth increases while the spectrum intensity continues to decrease.

Finally, from 0.80 mm to 0.40 mm the blue side of the spectrum is further blocked. However, the emission spectra in figure 20d show a different behavior. From left to right, the first two profiles exhibit a bandwidth reduction and an intensity increase on top of a much broader but lower intensity emission. The last three curves exhibit a central wavelength shift but the intensity remains constant while the broader emission is no longer present. This behaviour

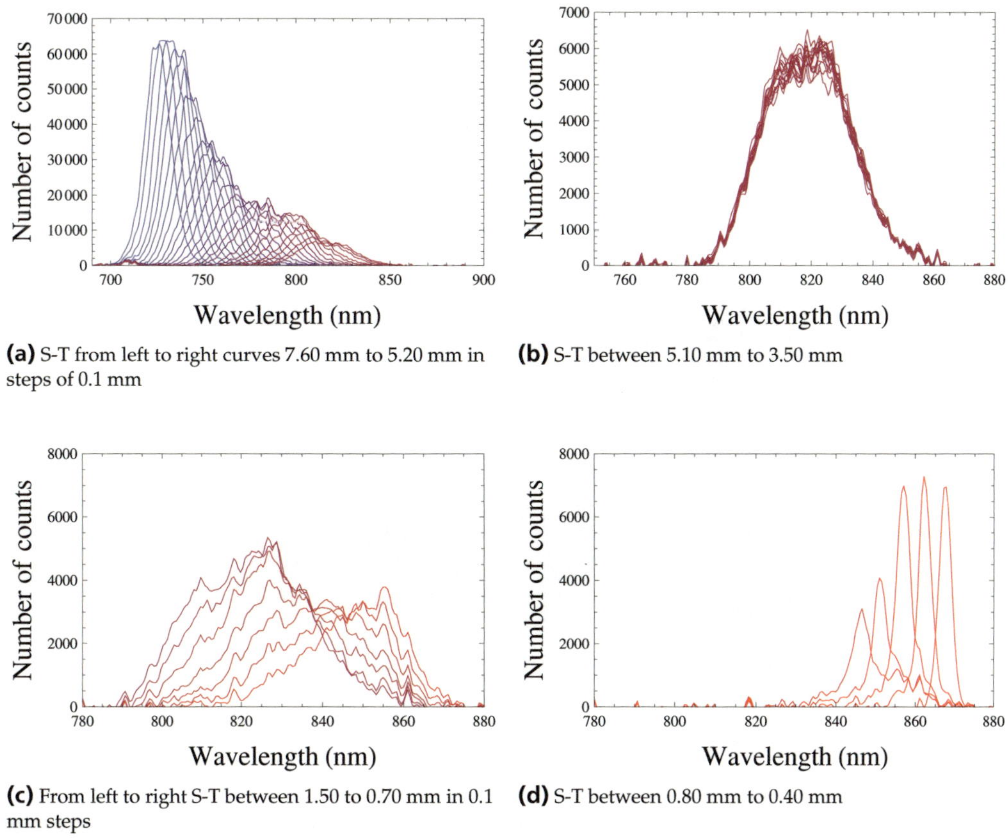

(a) S-T from left to right curves 7.60 mm to 5.20 mm in steps of 0.1 mm

(b) S-T between 5.10 mm to 3.50 mm

(c) From left to right S-T between 1.50 to 0.70 mm in 0.1 mm steps

(d) S-T between 0.80 mm to 0.40 mm

Figure 20. Oscillator pulse train spectra, with mode lock, as a function of slit translation (S-T). 20a) At higher micrometer readings the slit blocked most part of the beam frequencies on the red side. 20b) Does not exhibit any change in the spectrum since the beam is not blocked by the slit. 20c) For lower micrometer readings most part of the beam blue frequencies are blocked. 20d) Bandwiths are narrower when laser is hardly mode locking.

could suggest continuous light (CW) emission. However, even under these circumstances mode locking (ML) was still being held.

6. Pulse duration and time bandwidth product (TBP)

6.1. Frequency-Resolved Optical Gating (FROG)

FROG is a special kind of nonlinear autocorrelator that measures the pulse duration as well as other features of the pulse. This is possible because the autocorrelator signal beam is spectrally resolved. There are different kinds of FROG depending on the nonlinear medium: Polarization-Gate (PG), Self-Diffraction (SD), Transient-Grating (TG), Second-Harmonic-Generation (SHG) and Third-Harmonic-Generation (THG). Second-Harmonic-Generation is the most sensitive because it involves a second order nonlinearity rather than third. However, being an even order nonlinearity it is time

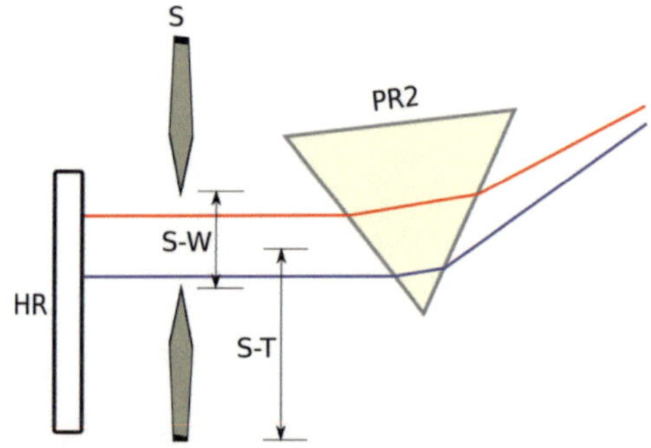

(a) Sketch of the intra-cavity mirror - slit - prism setup.

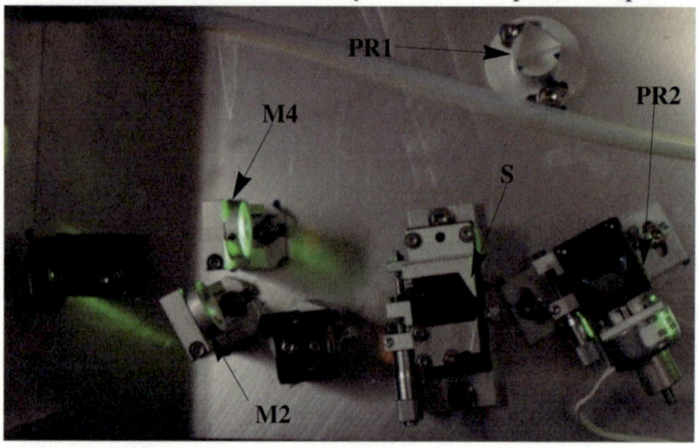

(b) Photograph of the mirror - slit and two prisms setup

Figure 21. Oscillator GVD compensating prisms and spectrum cropping slit. The solenoid that initiates mode locking operation can be seen below the second prism PR2.

symmetric and thus the time direction is ambiguous [17]. In the following experiments we used a simplifed FROG version called grenouille with a SHG crystal, 18 to 180 fs pulse duration measurement capability[10]. Grenouille raw data were processed with Femtosoft QuickFrog software. The pulse retrieval algorithm is reasonably reliable although it sometimes runs wild and it requires resetting. It gives numeric estimates of the pulse duration, TBP, central emission wavelength, pulse front tilt and spatial chirp. However, it does not state a temporal chirp value although the temporal phase curve is presented on screen.

6.2. GVD prism compensation and time bandwidth product (TBP)

The second prism (PR2) position was moved between 1.50 mm and -0.50 mm in steps of 0.10 mm. The initial parameters are summarized in table 5. From the data records, we produced plots of various pulse parameters in figure 22.

The pulse duration is plotted in figure 22a. It decreases as the prism PR2 micrometer reading is reduced. As we mentioned in section 5.1, the beam traverses less prism material as the micrometer reading decreases. The beam is incident on the apex of the first prism. In fact, a little bit of it on one side passes through without hitting the prism. It is expected that the

[10] GRENOUILLE 8-20 from Swamp optics

Element	Measurement(mm)
Mirror C1 position	3.98
Mirror C2 position	10.85
Lens L1 position	4.12
Prism PR2 initial position	1.50
Slit width (S-W)	7.23
Slit translation (S-T)	3.50

Table 5. Initial parameters when prism PR2 was moved.

second prism will produce a good GVD compensation when symetrically placed with respect to the first prism. The micrometer was moved until zero was reached and then moved further down as much as the mount allowed. In the innermost position the minimum time seemed to be reached. If the second prism over-compensates or under-compensates the first prism the pulse duration should increase. A minimum is expected when compensation is optimum. For this reason, we modeled the time duration curve with a quadratic fit

$$57.9953 + 12.8317(0.45 + x)^2. \tag{6.1}$$

The minimum time of the fit is located at 57.99 fs whereas the minimum experimental measurement was 62.4 fs. The bandwidth dependence is shown in figure 22b. It also decreases as the prism PR2 position is decreases. Again, a minimum is expected at optimum compensation. A quadratic fit is phenomenologically assigned

$$22.5591 + 2.79782(1.5 + x)^2. \tag{6.2}$$

While the time minimum seems to be reached at -0.5 mm, it is not clear that the bandwidth minimum has been reached yet. Indeed, the fit in equation (6.2) predicts that the minimum bandwidth is located at -1.5 mm.

The time-bandwidth product is plotted in figure 22c. This value was taken from the Frog retrieval algorithm. It should be proportional to the product of the previous two curves. A quadratic fit gives the following parameters

$$0.65104 + 0.381219(0.6 + x)^2. \tag{6.3}$$

The minimum TBP obtained from the fit is 0.65. However, the minimum TBP achieved with this setup is 0.783. This value is somewhat far from a fourier limited pulse as may be seen from comparison with table (6).

On the other hand the center wavelength increases as the prism position moves out of the beam as seen in figure 22d. A linear fit gives the equation

$$827.012 - 22.6234x. \tag{6.4}$$

This frequency dependence is similar to that observed in section (5.1).

The plot of the pulse front tilt is shown in figure 22e. When the two prism compensate each other transversely they should produce minimum pulse front tilt. However, according to the frog measurements we are far from the minimum position. For this reason, although we expect a quadratic fit with a minimum, a linear fit was sufficient for the actual data

$$-11.8223 + 1.15039x. \tag{6.5}$$

It is not clear whether the pulse front tilt measurement is real or the Grenouille apparatus is slightly miscalibrated. A callibration experiment will be performed to elucidate this point.

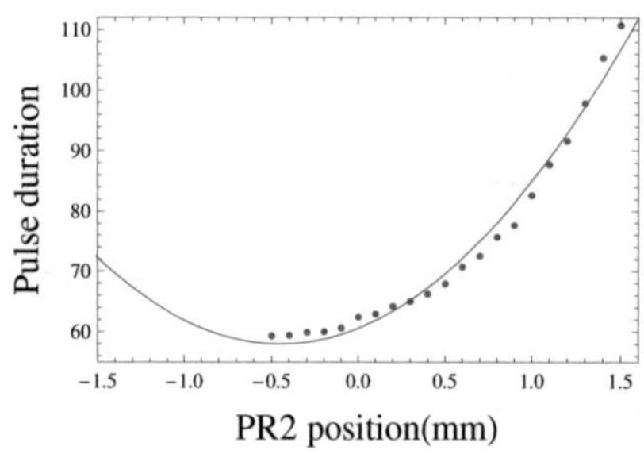

(a) Pulse duration $\triangle t$.

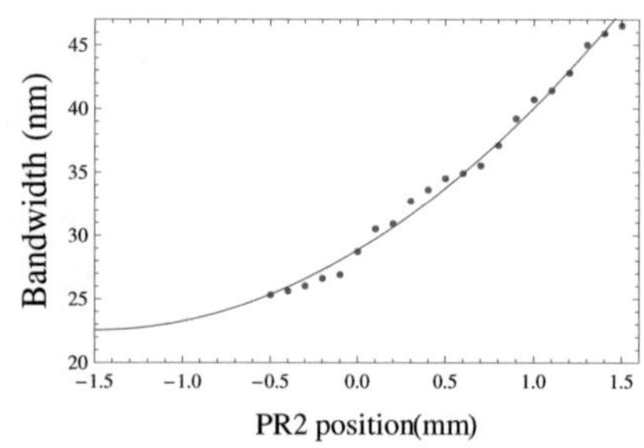

(b) Bandwidth $\triangle\lambda$.

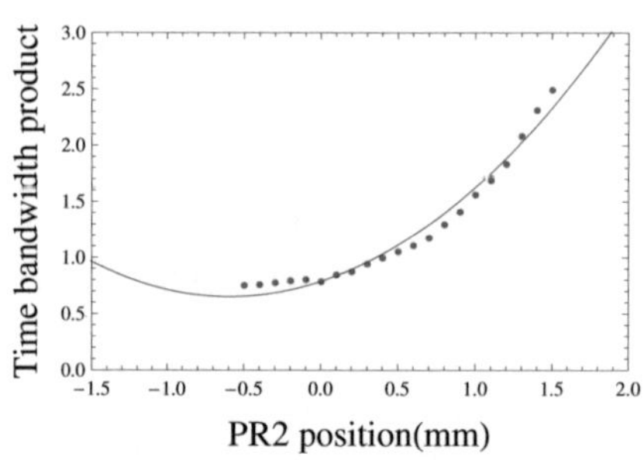

(c) Time bandwidth product (TBP).

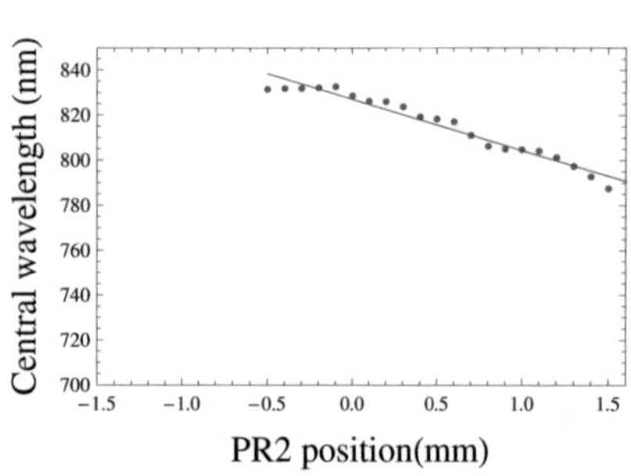

(d) Central emission wavelength λ_c.

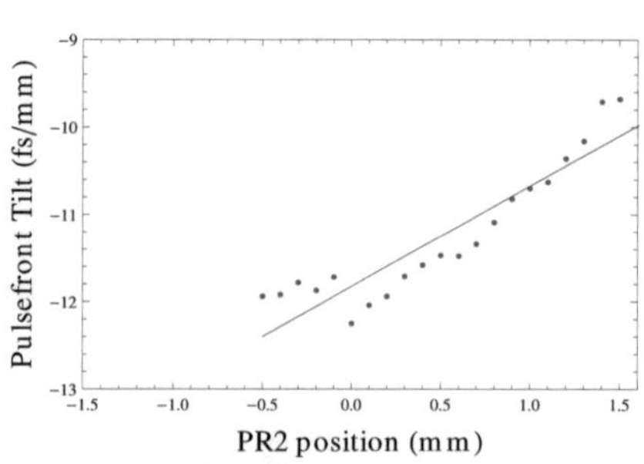

(e) Pulse front tilt (PFT).

Figure 22. $\triangle t$, $\triangle\lambda$, TBP, λ_c, and PFT versus prism PR2 position.

Since we did not achive a Fourier transform pulse, the bandwidth increases while the pulse duration increases too see figures 22a and 22b. Table 6 shows the different Fourier transform limits for different pulse shapes.

Envelope shape	$\varepsilon(t)$	Time bandwidth product
Gaussian function	$\exp\left[-(t/t_0)^2/2\right]$	0.441
Exponential function	$\exp\left[-(t/t_0)/2\right]$	0.140
Hyperbolic secant	$1/\cosh\left[t/t_0\right]$	0.315
Rectangle	-	0.892
Cardinal sine	$\sin^2\left[t/t_0\right]/(t/t_0)^2$	0.336
Lorentzian function	$\left[1+(t/t_0)^2\right]^{-1}$	0.142

Table 6. Fourier transform pulses for different shapes. Data taken from Rulliere et al. [14].

6.3. Spectrum cropping

The slit width (S-W) was varied between 1.43 mm and 7.23 mm. Table 7 shows the initial conditions for this experiment. It was varied in steps of 0.10 mm between 1.43 mm to 1.93 mm. From 1.93 mm to 2.73 mm the slit translation (S-T) was moved to center the laser beam within the aperture. Two different measurements at 2.73 mm on different days give an estimate of the measurement error in the various parameters shown in figure 23. Larger slit widths had little impact on the plotted parameters.

Element	Measurament(mm)
Mirror C1 position	3.98
Mirror C2 position	10.85
Lens L1 position	4.12
Prism PR2 position	1.50
Slit width initial width (S-W)	1.43
Slit translation (S-T)	3.50

Table 7. Initial conditions when S-W was varying

Let us now begin by describing the bandwidth behaviour that should be directly proportional to the slit width. Indeed, from 1.43 to 2.23 the experimental points lie on a straight line. At 2.23 mm the slit no longer crops the beam and thus becomes insensitive to slit width. The maximum bandwidth is around 35 nm consistent with the plateau shown in figure 15 in section 5.1.

The time bandwidth product dependence on slit width is shown in figure 23b. When the slit begins to crop the beam, the TBP decreases until it reaches the Fourier transform limit. A hyperbolic tangent fit of this curve is

$$0.816+\frac{0.793}{2}\tanh\left[\frac{x}{0.259}-4.677\right]. \tag{6.6}$$

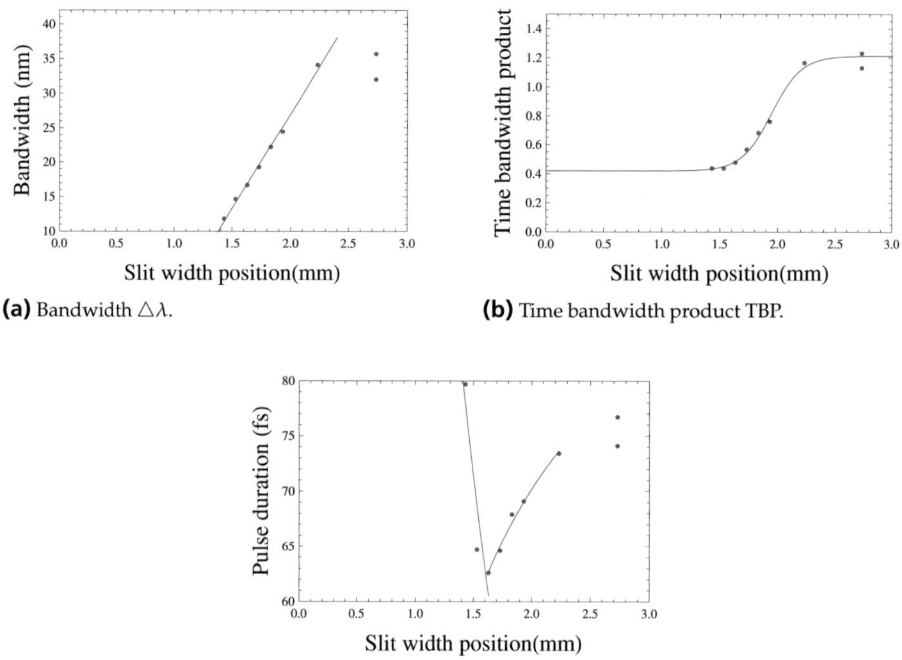

(a) Bandwidth $\triangle\lambda$.

(b) Time bandwidth product TBP.

(c) Pulse duration $\triangle t$.

Figure 23. $\triangle\lambda$, TBP and $\triangle t$ versus slit width.

The minimum asymptotic value is $0.816 - \frac{0.793}{2} = 0.419$. This TBP is below the Gaussian envelope limit as seen from table 6, thus suggesting that the pulse envelope must be closer to the hyperbolic secant proposed for solid state lasers [8].

The time duration of the pulse versus slit width is shown in figure 23c. The experimental points show two rather different behaviors. On the one hand, the pulse duration decreases as the slit width is reduced between 2.73 and 1.63 mm. Beyond this value, the pulse duration increases as the slit width is further reduced.

This apparently wild behaviour of the pulse duration can now be nicely explained in terms of the bandwidth and the TBP. The pulse duration is reduced as frequency components far in the wings are cropped because these frequencies have a large spread in time due to dispersion. For large slit widths, the TBP follows the strict inequality product $\triangle t \triangle \nu > c_B$. Thus both quantities may decrease as long as the equality is not reached. However, when the TBP limit is reached, the pulse is Fourier transformed and it should fulfill the $\triangle t \triangle \nu = c_B$ equality. If $\triangle \nu$ (or equivalently $\triangle\lambda$) is further reduced as it happens beyond the 1.63 mm slit width, the time duration $\triangle t$ must increase so that the equality holds. The pulse duration in the Fourier transform limited region should fulfill an inverse relationship with slit width since

$$\triangle t = \frac{c_B}{\triangle \nu} \propto \frac{1}{\text{s-w}}.$$

The fit gives the following parameters

$$-0.818912 + \frac{55.2703}{x}. \tag{6.7}$$

For the non Fourier limited region the pulse width increases with increasing bandwidth. Since the bandwidth only increases up to the active medium - cavity maximum bandwidth it may also be fitted to an inverse function

$$89.8668 - \frac{24.7288}{x}. \tag{6.8}$$

The asymptotic value for this function is 89.86 fs.

A screen shot of the Fourier limited pulse (TBP=0.437) produced by the grenouille femtosoft software is shown in figure 24. The pulse duration is 64.7 fs and its bandwidth is 14.65 nm. The pulse exhibits a negligible temporal chirp as expected. Nonetheless and contrary to our expectations, the pulse front tilt is reported to be still rather large -10.3 fs/mm.

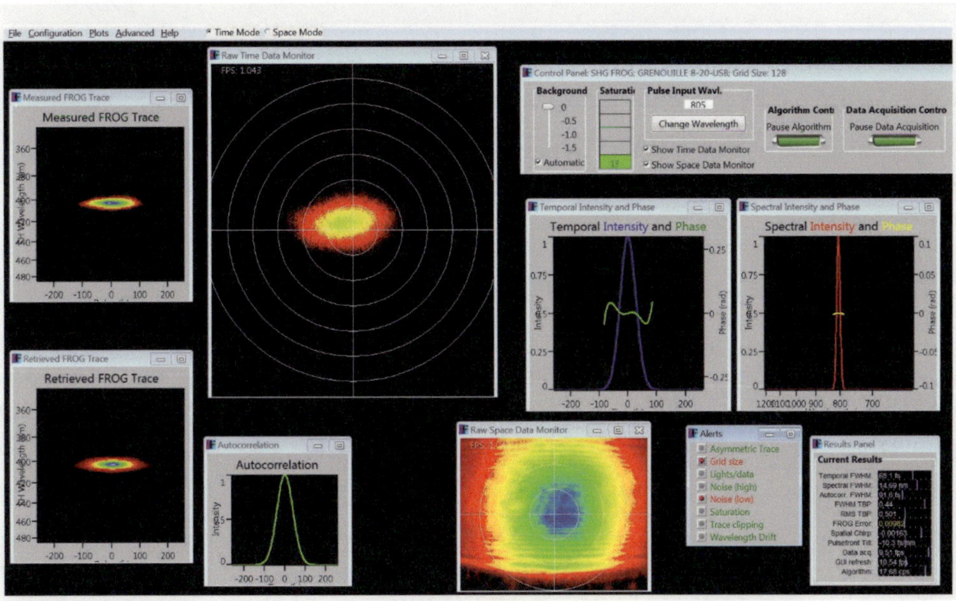

Figure 24. Screen shot for a Fourier limited pulse with slit width 1.53 mm. Grid characteristics 128 points; Delay increment: 4.03 fs; Center Wavelength: 807.3 nm Frog Error was: 0.01039 Temporal FWHM: 64.7 fs Spectral FWHM: 14.65 nm, FWHM Time-Bandwidth Product: 0.437, RMS Time-Bandwidth Product: 0.501.

6.4. Envelope shape

The data obtained from the frequency resolved optical gating measurements as well as the detailed spectra from integrated spectrometer measurements were analyzed in order to obtain the pulse envelope function. Two sets of spectra were analyzed. On the one hand, the spectrum optimized for minimum pulse duration and thereafter minimum TBP while maintaining the pulse duration. On the other hand, typical spectra shown in figure 20b, where the spectrum is rather broad and TBP is large.

6.4.1. Minimum pulse duration and minimum TBP spectrum

A Gaussian function without chirp (2.7) was fitted to the ML spectrum shown in figure 25. The parameters obtained without chirp are: $\tau_G \frac{2\pi c}{\lambda_l^2} 10^{-9} = 0.1086$, center wavelength $\lambda_\ell = 2\pi c/\omega_\ell = 809.628$ nm and the quality of the fit is $R^2 = 0.99770$. Once the time duration $\tau_G = 37.78$ fs was established from the previous fit, the Gaussian function with chirp (2.11) was fitted to the data. The fitted linear chirp parameter a was negligible $a = 4.02016 \times 10^{-8}$. The time duration obtained from the fit is $\triangle t_{FWHM} = \tau_G \cdot 1.17741 = 44.48$ fs. This fourier transform limited calculation underestimates the pulse duration given by the FROG retrieval measurement of 56.4 fs. If the TBP obtained from the FROG measurements is used in order to evaluate the Gaussian enevelope, $\tau_G = 53.43$ fs and the pulse width is $\triangle t_{FWHM} = 62.90$ fs; This time overestimates the FROG value by 11.5 %.

The hyperbolic secant function fit to the same data gave the following parameters: $\tau_s \frac{2\pi c}{\lambda_l^2} 10^{-9} = 0.06678$, center wavelength $\lambda_\ell = 2\pi c/\omega_\ell = 809.572$ nm and the quality of the fit is $R^2 = 0.99705$. Once the time duration τ_S was established from the previous fit, the hyperbolic secant function with chirp was adjusted. The chirp parameter was $a = 0.105855$. This small chirp parameter value improved slightly the curve fit, yielding $R^2 = 0.99790$.

The best fits with the Gaussian and hyperbolic secant envelopes give similar R^2 results. It is therefore not possible to discern whether the envelope follows one or the other profile under these conditions.

6.4.2. Typical spectra with broad spectrum and large TBP

A typical ML spectrum as shown in figure 20b was fitted to a Gaussian envelope function. Since the functional dependence of the pulse width τ_G and the linear chirp parameter a in the exponential argument are similar, no substantial difference is obtained when chirp is added. The Gaussian fit parameters are: $\tau_G \frac{2\pi c}{\lambda_l^2} 10^{-9} = 0.07079$, center wavelength $\lambda_\ell = 2\pi c/\omega_\ell = 817.608$ nm and the quality of the fit is $R^2 = 0.99105$ is rather poor.

The ML spectrum shown in figure 20b was fitted to a hyperbolic secant envelope function with and without chirp. The data are plotted against wavelength rather than frequency; Therefore the scaling $\Omega - \omega_\ell \to \frac{2\pi c}{\lambda^2} (\lambda - \lambda_c)$ is required. The frequency dependent sech^2 intensity function without chirp given by (2.21) has parameters $\tau_s \frac{2\pi c}{\lambda^2} = 0.0357$, centered at $\lambda_\ell = 2\pi c/\omega_\ell = 817.59$ nm and $R^2 = 0.98442$. On the other hand, the normalized intensity function with chirp function (2.24) was fitted with parameters: $\tau_s \frac{2\pi c}{\lambda^2} = 0.0649$, chirp coefficient $a = 1.122$ centered at $\lambda_\ell = 2\pi c/\omega_\ell = 817.56$ nm and $R^2 = 0.99472$. These fits are shown in figure 27. The function with temporal chirp is better adjusted due to the flat top of the central portion of the spectrum. The comparison between these fits is better visualized in a plot of the residuals shown in figure 28.

The hyperbolic secant function with chirp gives the smaller residuals and is thus the best fit. We can conclude that for oscillator pulses with considerable chirp, the pulse envelope is closer to a hyperbolic secant function.

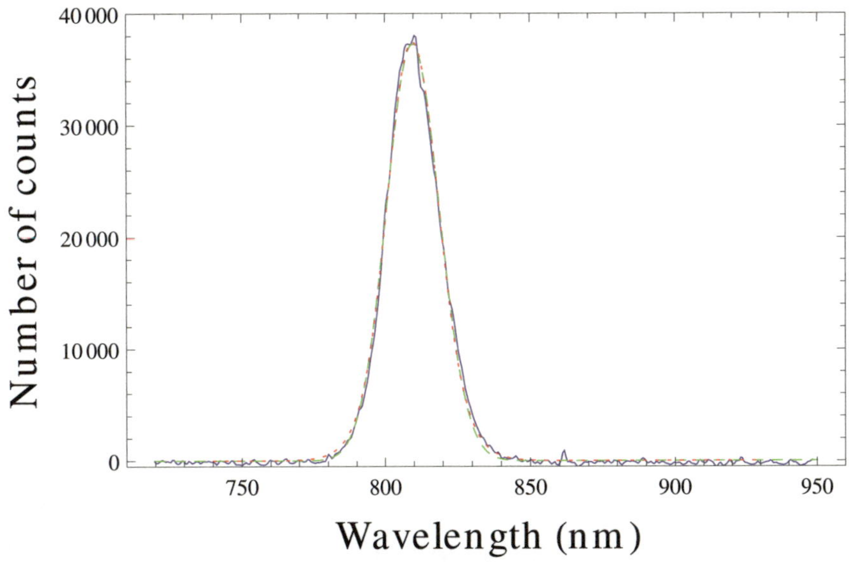

Figure 25. Spectrum of ML femtosecond oscillator taken with miniature spectrometer, 400 ms integration time. The oscillator parameters are optimized for minimum pulse duration (56.4 fs) and minimum TBP (0.624). Fits with Gaussian (dashed green line) and a hyperbolic secant (dotted red line) functions are shown. Fits with or without chirp make little difference.

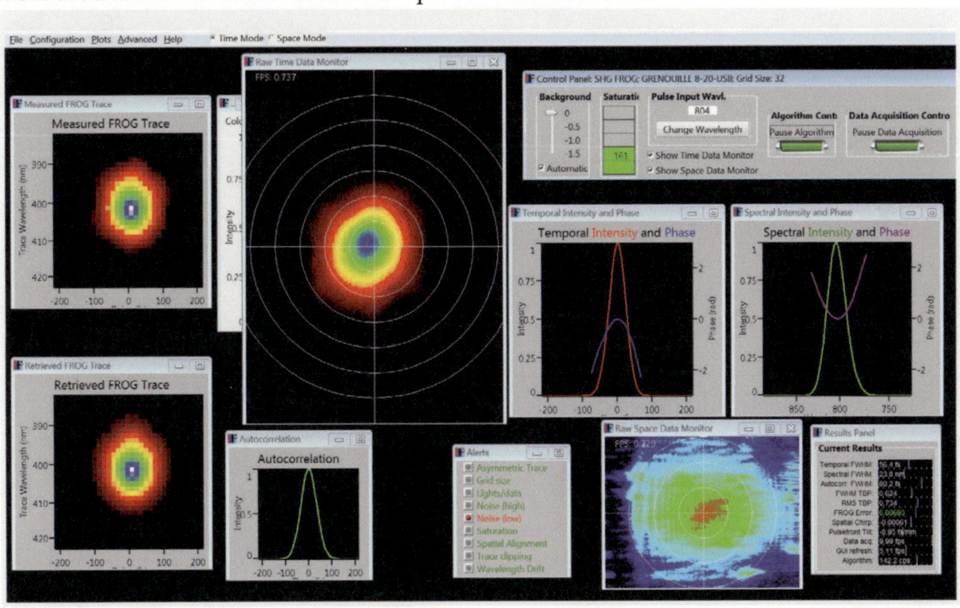

Figure 26. Screen shot for pulse with slit width 1.50 mm and slit translation 5.92mm. Center Wavelength: 809.6 nm Frog Error was: 0.00693 Temporal FWHM: 56.4 fs Spectral FWHM: 23.8 nm, FWHM Time-Bandwidth Product: 0.624.

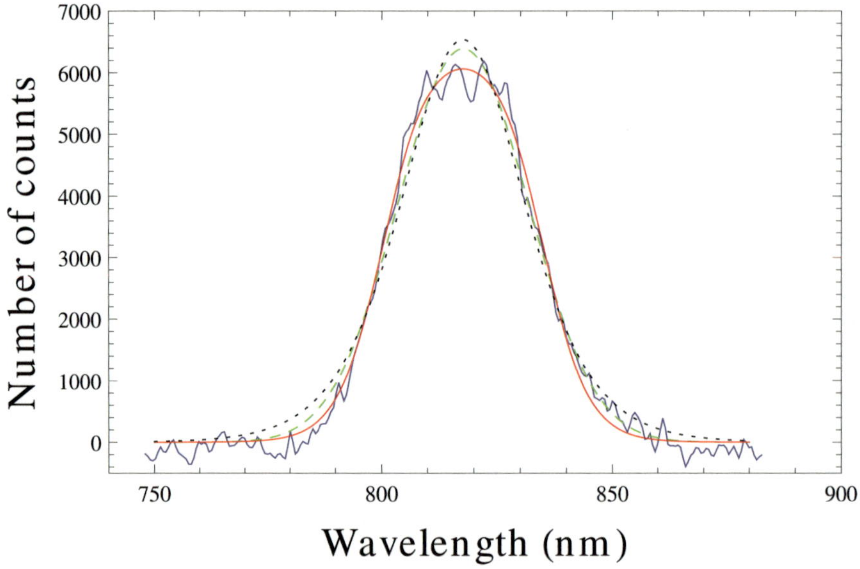

Figure 27. Gaussian (green dotted line) and Hyperbolic secant (HS) envelope fits to a typical spectrum (blue solid line) shown in figure 20b. The black dashed curve is a HS fit without chirp whereas the red continuous curve has a chirp factor of $a = 1.122$.

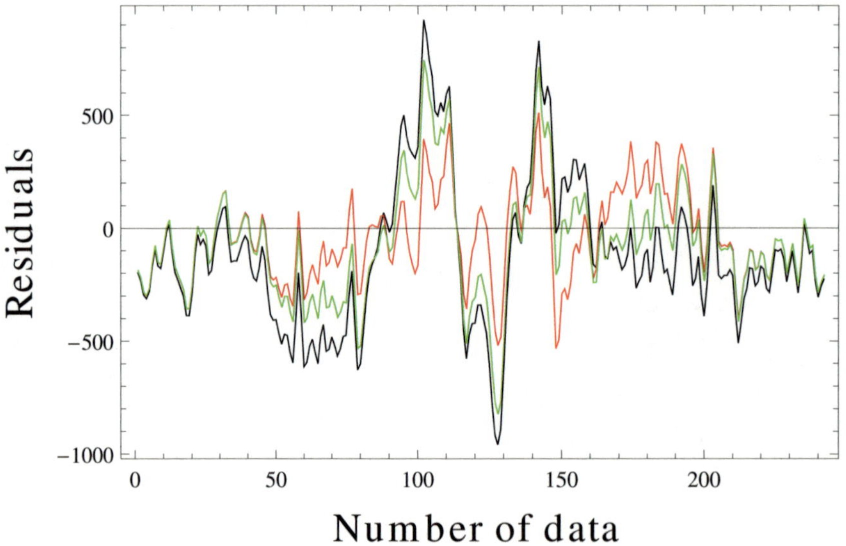

Figure 28. Residuals for the fits shown in figure 27. The Gaussian (green), Hyperbolic secant (HS) with (red) and without (black) chirp show that the best fit corresponds to the HS with chirp.

7. Conclusions

A Ti:Sa femtosecond laser oscillator was optimized for different operation purposes:

- A minimum pulse duration of 56.4 fs was obtained. The prims PR2 and the slit were adjusted to produce the minimum TBP of 0.624 while maintaining the shortest pulse duration.
- A time-bandwidth product of 0.783 with 62.4 fs pulse duration, 28.7 nm bandwidth and central wavelength of 828.6 nm was obtained when the prism PR2 was adjusted and the spectrum was not cropped.
- A Fourier limited pulse with time-bandwidth product of 0.437 with pulse duration between 63 to 79.7 fs, 11.82 to 18 nm bandwidth and central wavelength around 804.1 nm was obtained with appropriate bandwidth cropping with the intra-cavity slit.
- Oscillator emission was tuned from 730 to 850 nm with output power above 200 mW.
- The cavity adjustments would be more flexible if the first prism (PR1) is also mounted on a translation stage. A slit with two movable blades would also be welcome.

The wave description for Gaussian and hyperbolic secant envelopes was presented in detail. Results scattered in the literature were brought together in particular for the hyperbolic profile. The energy content of the wave field has been outlined with a novel continuity equation described in terms of complementary fields. These linearly independent fields do not require a temporal averaging over the pulse cycle. Therefore, they are well suited to describe ultrafast pulses or even few or single cycle pulses.

Acknowledgements

We are grateful to Raúl Rangel CISESE and his collaborators Jacob Licea Rodríguez and Alexandro Ruiz de la Cruz for useful comments and much needed advice.

Author details

E. Nava-Palomares, F. Acosta-Barbosa, S. Camacho-López and M. Fernández-Guasti.
Lab. de Óptica Cuántica, Depto. de Física, Universidad A. Metropolitana - Iztapalapa, 09340 México D.F., Ap. postal. 55-534, Mexico, url: http://luz.izt.uam.mx

8. References

[1] G. Muñoz, C. García, M. Muñoz, and M. Fernández-Guasti. Mesa holográfica pasiva ultra-estable: análisis de vibraciones. In *XXIII reunión anual de óptica*, sept 2010.

[2] Qiang Lin, Jian Zheng, Jianming Dai, I-Chen Ho, and X.-C. Zhang. Intrinsic chirp of single-cycle pulses. *Phys. Rev. A*, 81(4):043821, Apr 2010.

[3] Qiang Lin, Jian Zheng, and Wilhelm Becker. Subcycle Pulsed Focused Vector Beams. *Phys. Rev. Lett.*, 97(25):253902, 2006.

[4] J. C. Diels and W. Rudolph. *Ultrashort Laser Pulse Phenomena: Fundamentals, Techniques, and Applications on a Femtosecond Time Scale, Academic Press, 2nd edition, Elsevier 2006.* Elsevier Science Publishers, 2nd edition edition, 2006.

[5] M. Fernández-Guasti. The Wronskian and the Ermakov - Lewis invariant. *International Mathematical Forum*, 4(16):795 – 804, 2009.

[6] C. H. Henry. Theory of the linewidth of semiconductor lasers. *IEEE J. Quant. Elec.*, 18(2):259–264, 1982.

[7] T. L. Koch and J. E. Bowers. Nature of wavelength chirping in directly modulated semiconductor lasers. *Electr. Lett.*, 20(25/26):1038–1040, Dec 1984.

[8] P. Lazaridis, G.Debarge, and P. Gallion. Time-bandwidth product of a chirped sech2 pulses. *Optics Lett.*, 20(10):1160–1163, 1995.

[9] M. Fernández-Guasti, E. Nava, F. Acosta, and R. Chandrasekar. Physical processes behind a Ti:Sa femtosecond oscillator. In *Optics and Photonics 2011*, volume 8121 of *The nature of light: What are photons? IV*, pages 812118–1–10. SPIE, 2011.

[10] R. P. Feynman, M. Sands, and R. Leighton. *Lectures on physics*, volume II. Addison Weseley, 1972.

[11] M. A. Alonso. Wigner functions in optics: describing beams as ray bundles and pulses as particle ensembles . *Adv. Opt.Phot.*, 3:272–365, 2011.

[12] M. Fernández-Guasti. Complementary fields conservation equation derived from the scalar wave equation. *J. Phys. A: Math. Gen.*, 37:4107–4121, 2004.

[13] M. Fernández-Guasti. The necessity of two fields in wave phenomena. In *Optics and Photonics 2011*, volume 8121 of *The nature of light: What are photons? IV*, pages 81210R–1–12. SPIE, 2011.

[14] Claude Rulliere, editor. *Femtosecond laser pulses*. Advanced texts in physics. Springer, 2nd edition, 2005.

[15] O. E. Martinez, J. P. Gordon, and R. L. Fork. Negative group-velocity dispersion using refraction. *J. Opt. Soc. Am. B*, 1(10):1003–1006, October 1984.

[16] O. E. Martinez. Grating and prism compresors in the case of a finite beam size. *J. Opt. Soc. Am. B*, 3(7):929–934, July 1986.

[17] R. Trebino. *Frequency resolved optical gating*. Kluwer Academic Publ., 2000.

Chapter 3

All Solid-State Passively Mode-Locked Ultrafast Lasers Based on Nd, Yb, and Cr Doped Media

Zhiyi Wei, Binbin Zhou, Yongdong Zhang, Yuwan Zou, Xin Zhong, Changwen Xu and Zhiguo Zhang

Additional information is available at the end of the chapter

http://dx.doi.org/10.5772/48775

1. Introduction

Mode-locking technique is a widely used method for generating ultrashort laser pulses. The mode-locked laser output is a sequence of equally spaced laser pulses. The pulse width is limited by the spectral range of the gain medium and inversely related to the bandwidth of the laser emission. Compared with the active mode-locking technique, passively mode-locked laser with saturable absorber is able to generate much shorter pulse with a simple configuration. In particularly, based on the passive mode-locking mechanism, the Kerr-lens mode locking (KLM) Ti:sapphire laser is recognized as the most important ultrafast laser source. Not only a series of commercial femtosecond lasers with Ti:sapphire crystal were released, but also lead to many innovations in science, such as frequency comb, laser wake field acceleration, attosecond science, laser micro-fabrication *etc*. However, the major drawback of Ti:sapphire laser is its green pump laser source. Currently available green lasers generated by frequency doubled Nd:YAG laser or by Argon laser are relatively bulky and expensive, which limits the practical application of ultrafast Ti:sapphire lasers.

With the remarkable progresses in the semiconductor saturable absorber mirror (SESAM) and laser diode technology, ultrafast laser sources with directly diode-pumped schemes and SESAM mode-locking have attracted more and more attentions because of the compactness and low cost compared to the well developed femtosecond Ti:sapphire laser. Nowadays, intra-cavity SESAM is widely used to start and maintain stable mode locking. Intense studies have been performed on this kind of lasers. Many efforts have been made to find new gain materials for ultrafast laser generations. Our research activity in this field is focused on the development of all solid-state passively mode-locked ultrashort lasers based on a variety of gain media with various wavelength ranges in the near-infrared. Several gain media doped with Nd, Cr or Yb, such as Nd:YVO_4, Nd:$GdVO_4$, Nd:$LuVO_4$, Nd:GSAG,

Cr:forsterite, Cr:YAG, Yb:YAG, Yb:YGG, Yb:GYSO, Yb:LSO etc, have been tested. The results indicate a series of ultrafast laser sources with low cost, compact, simple and robust configuration in the picosecond and femtosecond regimes, which would find a wide range of practical applications.

2. Mode-locking mechanism passively mode-locked solid-state laser with a SESAM

By using an intra-cavity SESAM for passive mode locking, one has to choose the correct parameters of SESAM for a given laser system to get pure CW mode locking [2.1]:

$$\left|\frac{dR}{dI}\right| I < \frac{g_0}{l}\frac{T_R}{\tau_L}. \tag{1}$$

where R is the absorber reflectivity, I is the laser intensity, T_R is the cavity round trip time, g_0 is small signal gain of the laser, l is the total loss coefficient of the laser cavity, τ_L is the upper state lifetime of the laser. That is, the laser tends to operate in Q-switched mode-locking regime with a longer upper state lifetime, larger total loss, or shorter cavity length. In other words, the Q-switching can be suppressed for a large small-signal gain, or small loss, or large saturation intensity, or long cavity. For a fast saturable absorber, the condition is given by [2.1]:

$$\left|\frac{dR}{dE_p}\right| E_p < \frac{g_0}{l}\frac{T_R}{\tau_L}. \tag{2}$$

3. Overview for different solid-state laser materials

Solid-state mode-locked lasers are applied in various fields of physics, engineering, chemistry, biology and medicine *etc*, with application including ultrafast spectroscopy, metrology, superfine material processing and microscopy. To generate ultrafast laser pulses, the laser materials must meet a series of conditions. Firstly, a pump wavelength for which a good pump source is available. Secondly, small quantum defect and high gain of the materials can help get high laser efficiency. Furthermore, it is crucial for laser media to possess a board emission band, which is necessary to generate ultrafast pulses.

We divide materials for generation of ultrafast laser pulses into two types.

The first type is the ones that could only support picosecond laser pulses due to their limited gain bandwidth. Whereas, these laser materials commonly have excellent laser capabilities, such as good thermodynamic property and could be directly pumped by a diode laser with high power. The representative laser media are Nd^{3+} doped materials, which have been applied in diode-pumped energetic and efficient picosecond lasers and amplifiers. A SESAM mode-locked, Nd^{3+}:YAG laser with pulse width of 20 ps and output power of 27 W has been reported [3.1]. Also, an Nd^{3+}:YVO_4 laser with 20 ps of pulse width and 20W of laser power was achieved [3.2]. Malcolm *et al* reported an additive pulse mode-locking (APM) Nd^{3+}:YLF laser, and 1.5 ps pulses were generated [3.3], which is the shortest pulse result by Nd^{3+} doped crystals.

Also, the Nd^{3+} laser on the $4F_{2/3}$– $4I_{9/2}$ transition with emission wavelengths of around 900–950 nm has attracted more and more interest because the quasi three-level pulsed laser could be used for lidar detection of water vapor in atmosphere [3.4]. Up to now, only few works have been reported on passively mode-locked laser on quasi three-level operating of Nd^{3+} doped materials. By Nd:$YAlO_3$ crystal, laser pulses centered at 930 nm with 1.9 ps pulse duration and 410 mW average power has been demonstrated with a Ti:sapphire pump laser[3.5]. With Nd:YVO_4 crystal, 8.8 ps pulses with 87 mW of power [3.6] and 3 ps pulses with 140 mW of power [3.7] centered at 914 nm were achieved, pumped by diode laser and Ti:sapphire laser, respectively.

Other than Nd^{3+} doped crystals, a commonly used host material doped with Nd^{3+} ions is Nd:glass. Different with crystals, glasses are significantly cheaper and have a smoother fluorescence spectrum, which support femtosecond laser pulses generation. As a well known example, laser pulses as short as 60 fs with the output power of 80 mW at the center wavelength of 1053 nm have been achieved with Nd: glass laser [3.8]

The other type of gain media have relatively broad emission spectra, but suffer from one or more unpopular physical properties (Ti:sapphire crystal is the outstanding exception). Yb-doped materials belong to this group. Compared with Nd-doped laser media, Yb-doped materials have attracted even more and increasing interest over the past few decades because of their small quantum defect (resulting a low thermal loading), simple electronic structure (avoiding such unwanted processes as excited-state absorption, up-conversion, and concentration quenching), long fluorescence lifetime (particularly advantageous for Q-switched lasers and high-power ultrashort pulse amplification) and broad absorption and emission bands (compared with Nd^{3+}). These media have been successfully applied in high-power diode-pumped all-solid-state ultrafast laser sources. Until now, extensive mode-locking researches have been reported with various Yb-doped crystals, such as garnet (Yb:YAG [3.9-10]), vanadate (Yb:$GdVO_4$ [3.11]), oxyorthosilicates (Yb:LSO [3.12-13], Yb:YSO [3.13]), tungstates (Yb:KGW [3.14], Yb:KYW [3.15], Yb:KLW [3.16], Yb:NYW [3.17]), borates (Yb:YAB, Yb:LSB, Yb:LYB, Yb:BOYS, Yb:GdCOB, Yb:YCOB)[3.18-23], fluorite (Yb:YLF) [3.24], sesquioxide (Yb:Sc_2O_3, Yb:Lu_2O_3, Yb:$LuScO_3$) [3.25-27], silicate(Yb:SYS) [3.28], niobate(Yb:CN) [3.29]. The short pulse duration of 47 fs has been reported with a Yb^{3+}: $CaGdAlO_4$ crystal [3.30] and the shortest pulse width of 35 fs has been obtained by a Yb:YCOB crystal [3.20].

Transparent ceramics fabricated by the vacuum sintering technique and nanocrystalline technology have the advantages of high doping concentration, easy fabrication of large size samples and high thermal conductivity. This kind of materials has been intensively investigated for the use in ultrashort pulse lasers [3.31-36]. The first reported mode-locked ceramic laser is a Yb:Y_2O_3 laser, which generated 210 mW laser pulses with 450 fs pulse duration and a center wavelength of 1037 nm [3.32]. With the Yb:YAG ceramic, 286 fs pulses with 25 mW output power and 233 fs with 20 mW output power at center wavelength of 1030 nm and 1048.3 nm respectively were achieved [3.35],. Our recent work has boosted the output power from a femtosecond ceramic Yb:YAG laser to 1.9 W [5.19].

To further reduce the thermal effect, especially important for high power laser operation, the thin-disk gain media configuration has been developed. The geometry of the gain media is designed such that the thickness is considerably smaller than the laser beam diameter and the heat generated along with the lasing is extracted dramatically through the cooled end face [3.37-3.40]. Combined with the high absorption of the pump laser and long lifetime of the excited state of the Yb^{3+}-doped active materials, this thin disk laser operation has been a great success in a variety of Yb^{3+}-doped materials [3.41-3.47]. Thin-disk laser based on Yb-doped active materials is a good way to get efficient and high output power laser with excellent beam quality and will find wide applications.

Cr^{4+}:forsterite crystal is an important laser medium to generate ultrashort laser pulses around 1.3 μm. It could be pumped by Nd:YAG laser or Yb-doped fiber laser around 1064 nm. Cr^{4+}:forsterite crystal has wide luminescence spectrum from 1100 nm to 1500 nm [3.48]. Up to now, the shortest pulse duration with Cr^{4+}: forsterite laser is 14 fs [3.49]. Also, another Cr^{4+} doped material (Cr^{4+}:YAG) could be employed as the laser medium to obtain ultrafast laser pulses around 1.4 to 1.6 μm. The absorption spectrum of Cr^{4+}:YAG extends from 950 nm to 1100 nm, so it could share similar pump laser as the Cr^{4+}:forsterite laser. So far, the shortest pulse duration with Cr^{4+}:YAG laser is 20 fs and achieved by Kerr-lens mode locking technology [3.50].

4. Mode-locked Nd-doped lasers at quasi-three levels

4.1. Laser transition at quasi-three levels

Nd^{3+} ions was the first of the trivalent rare earth ions to be used in a laser, and Nd^{3+} doped materials have been remaining the most important laser medium until now. Laser operation has been obtained with this ion incorporated in at least 100 different host materials, and the commonly used host materials are YAG and glass. Take Nd: YAG for example, Figure 1 shows the energy level diagram of Nd^{3+} pumped by 808 nm. The ground state population was pumped to the energy level $^4F_{5/2}$, and then via radiationless transition transfers energy to $^4F_{3/2}$ level. The stimulated emission is commonly obtained at three different groups of transitions centered at 1.34, 1.06, and 0.946 μm, which result from $^4F_{3/2}\rightarrow{}^4I_{13/2}$, $^4F_{3/2}\rightarrow{}^4I_{11/2}$, and $^4F_{3/2}\rightarrow{}^4I_{9/2}$ transitions, respectively.

When radiation at $^4F_{3/2}\rightarrow{}^4I_{13/2}$ and $^4F_{3/2}\rightarrow{}^4I_{11/2}$ transitions, the upper and lower energy levels of pump and laser are separate, this is called four-level operation. Because the lower laser level is well above the ground state $^4I_{9/2}$ and can be quickly depopulated by multi-phonon transitions, there is no appreciable population density ideally in the lower laser level at room temperature. In that way, re-absorption of the laser radiation is effectively avoided and a lower threshold pump power can be achieved.

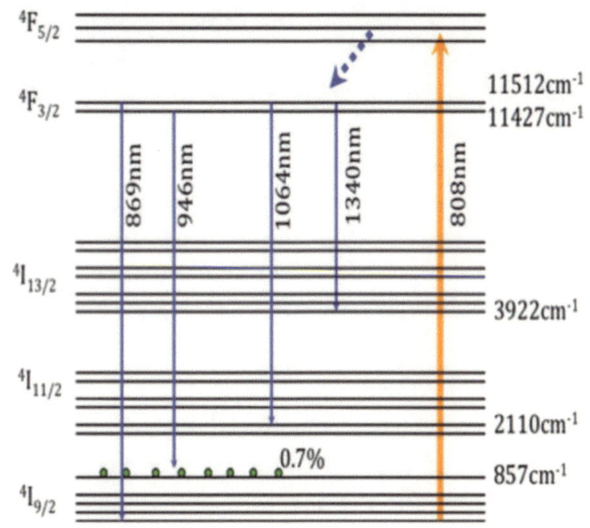

Figure 1. Energy level diagram of Nd:YAG

However, for $^4F_{3/2}\rightarrow{}^4I_{9/2}$ transition, the lower-laser level is the upper sub-level of the $^4I_{9/2}$ multiplet, so there is some population on the lower-laser level at room temperature. This is called quasi-three level operations. For example a thermal population of about 0.7 % in the lower laser level in Nd:YAG at room temperature. This sub-level is thermally coupled with the ground-state and should be efficiently populated with increasing temperature. This would induce a partial re-absorption loss of the laser radiation and cause an increase in the ground-state absorption loss, corresponding to an increasing in passive intra-cavity loss and laser threshold, result in the reducing of the laser output energy. In addition, the laser cross emission section of the quasi-three-level laser is dozens of times smaller than that of four-level system. Therefore, the laser operation with quasi-three levels is more difficult than that one with four levels.

As the typical Nd-doped laser medium, mode-locking researches on Nd:YAG laser for both quasi-three levels and four levels have been extensively reported, therefore, we only introduce our researches on mode-locking with some special Nd-doping media in this chapter.

4.2. Picosecond Nd:GSAG laser at 942nm for three levels operation

During the past few years, the Nd^{3+} laser on the $4F_{2/3}$– $4I_{9/2}$ transition has attracted wide interest because of emission wavelengths of around 900–950 nm. One important application for this quasi three-level pulsed laser is that it could be used for lidar detection of water vapor in atmosphere [3.4] because of characteristic absorption in the 935 nm, 942 nm, and 944 nm wavelength regions. Compared with optical parametric oscillators, generation of above wavelengths by Nd-doped crystal laser is much easier. Also, Frequency doubling of mode-locked quasi-three-level Nd^{3+} lasers generates picosecond blue pulses, which have potential application in many fields, such as life science, holography, and semiconductor inspection [4.1].

In the previous section, we have discussed the difficulty on quasi three-level operation compare with four-level operation. For Nd:GSAG here, a peak emission cross section of 2.7× 10^{-20} cm^2 at 942.7 nm [4.2] and a peak emission cross section of 3.2x10^{-19} cm^2 around 1.06 μm [4.3] were determined by Kallmeyer *et al.* and Brandle *et al.*, respectively. One can see the emission cross section at 942 nm is less than one-tenth the emission cross section at 1.06 μm. This leads to an increased threshold pump power and decreased optical conversion efficiency [4.3–5]. For the first time, we successfully a diode-pumped passively mode-locked Nd:GSAG laser with quasi-three-level operation at the central wavelength of 942.6nm [4.6]. The maximum output power is 510 mW at incident pump power of 16.7 W, and the pulses duration is short as 8.7 ps at a repetition rate of 95.6 MHz.

In the experiment, a standard z-fold resonator with end-pump configuration was employed as shown in Fig. 2. The pump source is a commercial fiber-coupled 808 nm LD with a core diameter of 200 μm, and an NA of 0.22. The crystal used in the experiment has dimensions of 4 mmx4 mm and Nd^{3+} concentration of 1 at. %. The both facets of the crystal have been antireflection (AR) coated at 942 nm, 808 nm and 1061 nm. The crystal was wrapped with indium foil and mounted on a water cooled copper heat sink and the water temperature was maintained at 10°C. Passive mode locking was started by using a SESAM (BATOP GmbH, Germany) with saturable absorptance of 4% at 940 nm, a saturation fluence of 70 μJ/cm^2, and relaxation time of less than 10 ps. The laser spot size on the SESAM is estimated to be 27 μm x20 μm (tangential direction sagittal direction) using the ABCD matrix calculation. The dependence of the total output power on the incident pump power is shown in Fig. 3. Stable CW mode-locked laser can be obtained at the pump power above 11.5 W. When the incident pump power reached 16.7 W, the output power rose to 510 mW, corresponding to a optical efficiency of 3.1%, and the pulse energy inside the cavity and outside the cavity are about 88.3 nJ and 2.65 nJ, respectively. The high threshold pump power and the low efficiency are mainly due to the relatively high total transmission of 6% brought by the folded output coupler.

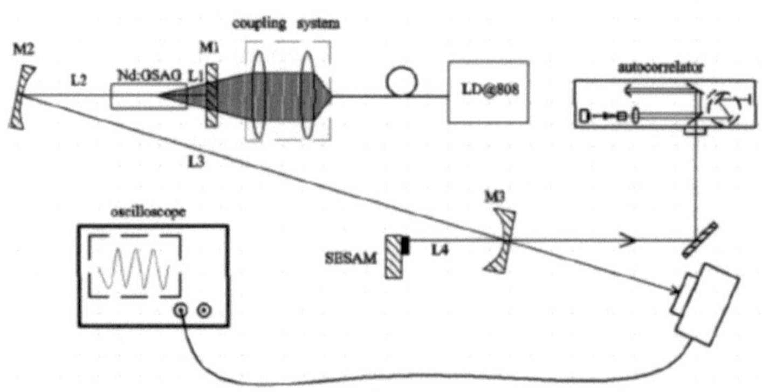

Figure 2. Schematic diagram of the experimental setup. Mirrors: M1: HT@808 nm and 1064 nm, HR@942 nm; M2: HR@942 nm, HT@1064 nm, radius of curvature (RoC)=500nm; M3: T=3%@942 nm, HT@1064 nm, RoC=100 nm.

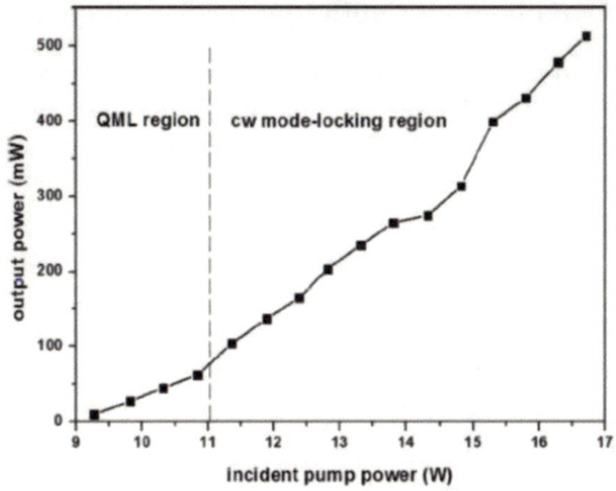

Figure 3. Dependence of the mode-locked laser output power on incident pump power.

The beam quality of the laser was measured with a CCD that could translate along a straight and slick track under incident pump power of 16.7 W. After fitting the measured data, as shown in Fig. 4 (a), we found that M^2 parameters were 1.83 and 1.55 for tangential direction and sagittal direction, respectively. The stable CW mode-locking regime held well for several hours and the repetition rate of the pulses was 95.6 MHz. The pulse duration was measured with an autocorrelator (FR-103MN, Femtochrome Research, Inc.) at the maximum output power, as shown in Fig. 4 (b). The pulse duration is about 8.7 ps assuming a Gaussian shape. Measured by an optical spectrum analyzer (AQ6315A, YOKOGAWA), the spectrum was centered at 942.6 nm which has an FWHM of about 0.65 nm, also shown in Fig. 4 (b). The time–bandwidth product of the pulses is 1.91, which is 4.3 times the transform limit for Gaussian pulses. The primary reason for this is that positive group-velocity dispersion introduced by crystal itself stretched the pulses.

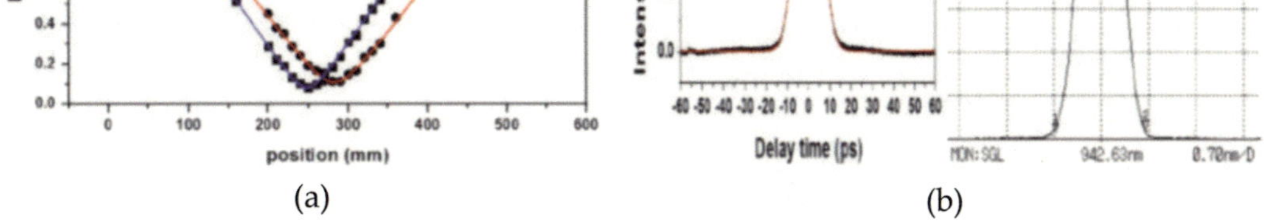

Figure 4. (a) Measured M^2 for tangential direction (Rx) and sagittal direction (Ry); (b) left: Intensity autocorrelation trace of the pulse; right: The laser spectrum of mode-locking operation.

4.3. Picosecond Nd:GGG lasers for both four levels and quasi-three levels

The Nd^{3+} doped gadolinium-gallium-garnet (GGG) crystal, a well known laser gain medium, was firstly reported in 1964 [4.7]. It has many desirable properties as a laser host material, such as excellent thermal conductivity, good mechanic properties, large thermal capacity, high doped concentration, and large size etc, as shown in Tab.1 [4.8-9]. With this laser medium, 30 kW average output power had been demonstrated from a solid-state heat capacity lasers (SSHCL) by Lawrence Livermore National Laboratory in 2004 [4.10]. Although Qin *et al* had obtained mode-locking operation at 1061 nm; the output power is only hundreds of milli-watt [4.11]. In this section, we described a diode pumped mode-locked Nd:GGG laser operation at both four levels and quasi-three levels transition. A laser pulse of 15 ps and 24 ps were generated with an average power of 3.2 W and 320 mW at 1062 nm and 937.5 nm, respectively [4.12-13].

Chemical formula	$Nd:Gd_3Ga_5O_{12}$
Crystal Structure	Cubic
Lattice Constant	a=1.2376 nm
Melting point	1725 °C
Density	7.1 g/cm^3
dn/dT ($\times 10^{-6}$/K)	17
Thermal conductivity	7 W/mK
Heat capacity	0.38×10^3 Ws/(Kg•K)
Thermal expansion coefficient	8×10^{-6}/ °C
Laser Transition	$^4F_{3/2} \rightarrow {}^4I_{11/2}$ 1062 nm $^4F_{3/2} \rightarrow {}^4I_{13/2}$ 1331 nm $^4F_{3/2} \rightarrow {}^4I_{9/2}$ 937.5 nm
Emission cross section at 1061 nm	$2.7\text{-}8.8\times 10^{-19}cm^{-2}$
Fluorescence lifetime	240 μs
Absorption band	808 nm
Photon energy at 1.06 μm	$h\nu = 1.86 \times 10^{-19}$ J
Index of refraction	1.94

Table 1. Properties of Nd:GGG crystal

Laser operation was performed by using a 5-mm-long, antireflection-coated, 0.5 at. % doped Nd:GGG crystal. A high brightness fiber-coupled diode laser, with a core diameter of 200 μm and a numerical aperture of 0.22, was used as pump source. For ease of use, we designed a typical resonator as shown in Fig. 5 for four level running. Passive mode locking was started by a SESAM, which has a saturable absorption of 4% at 1060 nm and a saturation fluence of 70 μJ/cm^2. A plane-wedged mirror with transmission rate of 10 % was used as the output coupler (OC). With this configuration, the laser mode radii in the crystal and on the SESAM were calculated to be 98μm and 105 μm, respectively.

After the optimized alignment, stable mode-locking operation with single mode output was obtained when the incident pump power exceeded 15 W. We measured the intensity autocorrelation trace by using a commercial noncollinear autocorrelator (FR-103MN, Femtochrome Research, Inc.). As shown in Fig. 6 (a), the FWHM width of the autocorrelation trace is about 23 ps. If a $sech^2$-pulse shape is assumed, the mode-locked pulse duration is 15 ps. The central wavelength locates at 1062.5 nm with spectrum bandwidths of 0.25 nm. The time-bandwidth product was calculated to be 0.997, which was about 3 times the transform limit for Gaussian pulses.

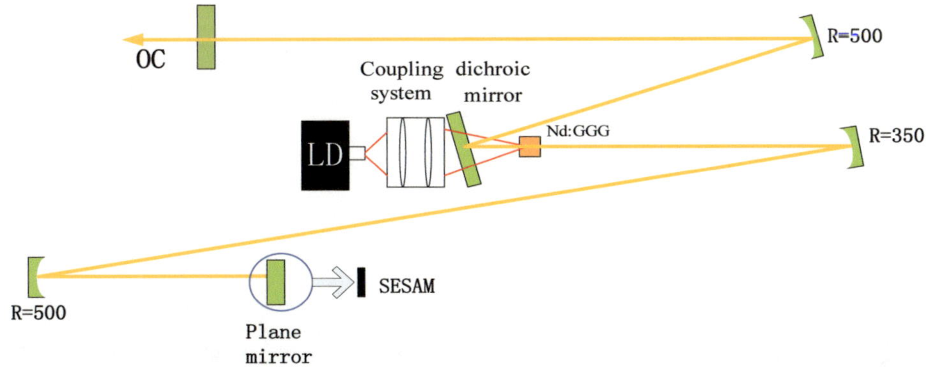

Figure 5. The Experimental layout of passively mode-locked for Nd:GGG laser operating at four level transition.

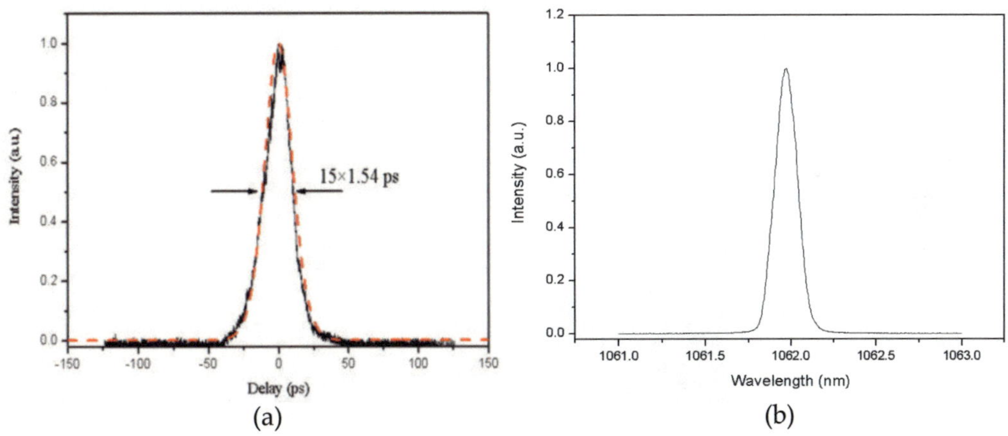

Figure 6. (a) Intensity autocorrelation trace of the pulse; (b) the laser spectrum of mode-locking operation

Mode-locking Nd:GGG laser at quasi-three level transition was realized with the same pumping source and the laser crystal. In order to reduce loss, a simple Z-folded cavity consisted of four mirrors was employed, as shown in Fig.7, the total cavity length was approximately 1.82 m corresponding to a repetition rate of about 82 MHz. To suppress parasitic oscillation at 1.06 μm, most of the cavity mirrors were antireflection coated at this wavelength. A similar SESAM from same company was used to start the mode-locking. For optimization of the cavity alignment, a plane mirror coated for high reflection at 940 nm was used as an end mirror. The waist radii of the laser mode in the crystal and on the SESAM were design to be 80 μm and 20 μm for mode match, respectively. The stable pulses train with the maximum output power of 320 mW has been obtained under the incident pump of 20 W.

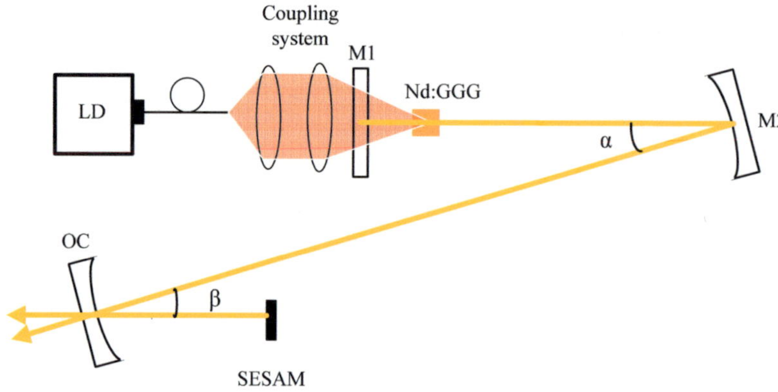

Figure 7. Scheme of Nd:GGG laser mode-locked at quasi-three-level transition.

Intensity autocorrelation trace shown that the FWHM width of the autocorrelation trace is about 39.3 ps (Fig. 8 (a)). If a Gaussian-pulse shape is assumed, the mode-locked pulse duration is 27.8 ps. Fig. 8 (b) depicts the output spectrum at the stable mode locking which was centered at 937.5 nm and had a bandwidth of 0.14 nm. This resulted in a time-bandwidth product of 1.33 which was 3 times the transform limit for Gaussian pulses. The high time-bandwidth product may be resulted from the high group delay dispersion of the laser crystal [4.14].

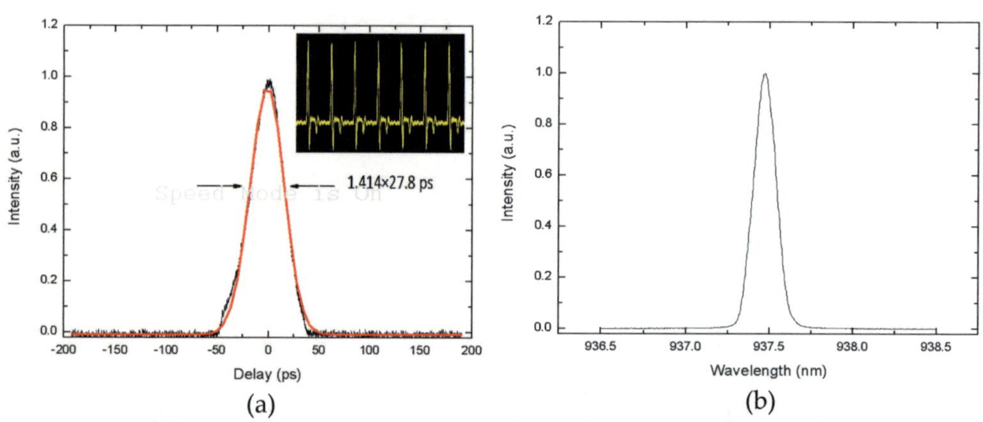

Figure 8. (a) Intensity autocorrelation together with Gaussian fit curve shows the pulse width of 24.7 ps. The inset is oscilloscope traces of the CW mode-locked pulse train. The laser spectrum of mode-locking operation.

4.4. Picosecond $Nd:GdVO_4$ at 912nm for three levels operation

As the most efficient laser host medium, the $Nd:YVO_4$ crystal has been widely used for diode-pumped mode-locking laser research [4.15-20]. Compare to $Nd:YVO_4$ crystal, $Nd:GdVO_4$ possesses many significant advantages such as higher thermal conductivity, larger absorption cross-section and larger stimulated emission cross-section [4.21]. In addition, $Nd:GdVO_4$ emits the shortest wavelength radiation on the $^4F_{3/2}$ - $^4I_{9/2}$ transition for its smallest splitting (409 cm^{-1}) of the lower laser level. However, the quasi-three levels laser is more difficult to operate than the four level laser and would cause the increase in ground-state absorption loss and laser threshold, and then reducing the output energy of the laser.

In the past, much more attention had been paid on passively Q-switched and mode-locked lasers with $Nd:GdVO_4$ crystal operating at 1064 nm and 1342 nm [4.22-26]. However, passively mode-locked $Nd:GdVO_4$ laser at quasi-three-level were less reported. H. W. Yang et al have reported a passively Q-switched $Nd:GdVO_4$ laser at 912 nm with V^{3+}:YAG as the saturable absorber [4.27]. Fei Chen *et al* have obtained passively Q-switched mode-locked $Nd:GdVO_4$ laser at 912 nm [4.28]. In 2006, our group successfully demonstrated the CW passively mode-locked $Nd:GdVO_4$ laser for three level laser operation [4.29]. In this section, we present a stable CW mode-locked $Nd:GdVO_4$ laser at 912 nm with the total output of 128 mw and the pulse width of 6.5 ps [4.30].

We employed a four-mirror folded cavity, shown in Fig. 9. The pump source is a commercially available fiber-coupled diode-laser which could emit a rated maximum power of 30W at 808 nm, with a core diameter of 200 μm and a N.A. of 0.22. The dimensions of the crystal are 3x3x4 mm^3 with a concentration of 0.2 at%. The crystal was coated and wrapped with indium foil and then mounted in a water-cooled copper block. The water temperature was maintained at 10 ℃. The SESAM (Batop GmbH, Germany) used for mode-locking is with a modulation depth of 2% at 912 nm and a saturation fluence of 70 μJ/cm^2. A plane mirror coated HR at 912 nm replaced the SESAM, and then the laser was operating at 912 nm. The maximum output power of 1.45 W was obtained with an incident pump power of 20.3 W. Then we put the SESAM instead of the plane mirror in the cavity. At pump power above 13.4 W, a stable CW mode-locking state was observed. The repetition rate was 178 MHz and the maximum output power was 128 mw at a pump power of 19.7 W, as shown in Fig.10.

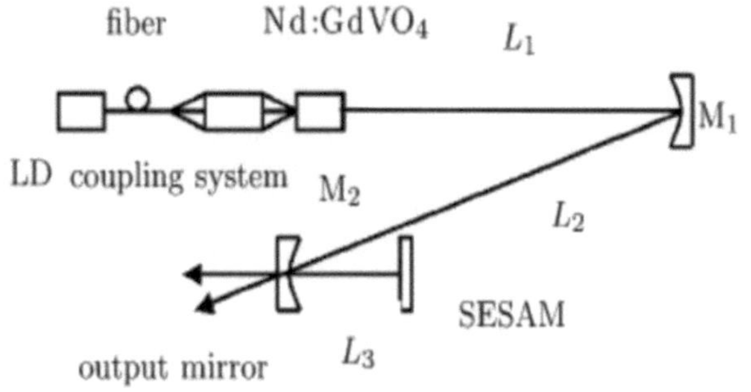

Figure 9. Schematic diagram of the passively mode-locked 912 nm $Nd:GdVO_4$ laser.

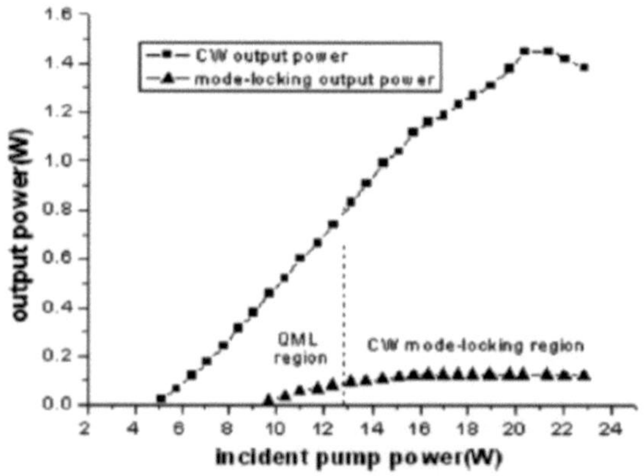

Figure 10. Dependence of the laser CW output and mode-locking output on the incident pump power. QML: Q-switched mode-locking.

The pulse width was measured with a homemade non-collinear second-harmonic-generation autocorrelator. The trace of the SHG autocorrelation is shown in Fig. 11. Assuming a Gaussian pulse profile, we estimated the pulse duration to be approximately 6.5 ps. The corresponding spectrum was measured, it has two peaks at the central wavelengths of 912.3 nm and 912.7 nm, respectively, and the both corresponding bandwidths were 0.25 nm (FWHM).

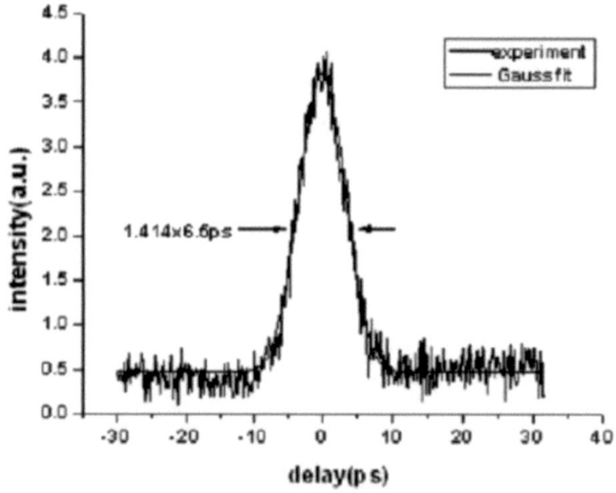

Figure 11. The measured autocorrelation trace of the pulses of the passively mode-locked 912 nm $Nd:GdVO_4$ laser.

4.5. $Nd:LuVO_4$ laser at 916 nm for three levels operation

As in vanadate family, $Nd:LuVO_4$ has larger absorption and emission cross sections than those of $Nd:YVO_4$ and $Nd:GdVO_4$, and thus has attracted much attentions. By employing the $Nd:LuVO_4$ crystal (length of 5.5 mm and doping with 0.1 at.% Nd^{3+} concentration) and using the similar Z-folded cavity design as in passively mode-locked $Nd:GdVO_4$ laser, we obtained a stable passively cw mode-locked $Nd:LuVO_4$ laser at 916nm [4.31]. The pulses train has 6.7 ps of pulse duration (shown in Fig. 12) at repetition rate of 133 MHz. The average output power was 88mW under the pump power of 17.1 W.

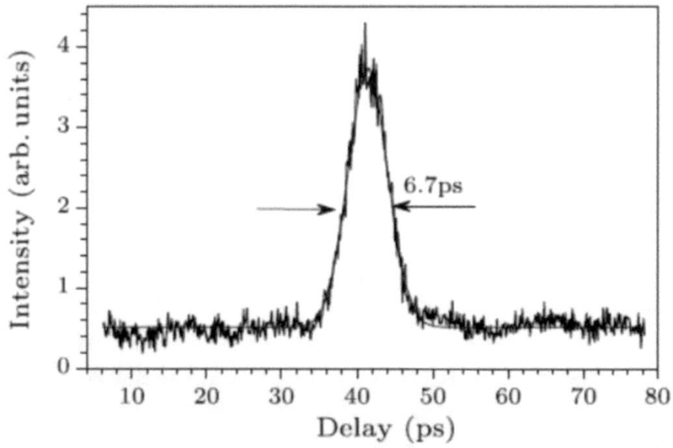

Figure 12. Autocorrelation trace of the mode-locked $Nd:LuVO_4$ laser.

5. Mode-locked femtosecond and picosecond Yb-doped laser

The Yb^{3+}-doped materials have excellent advantages among the available laser hosts, such as high quantum efficiency, absence of excited-state absorption, direct diode laser pumping, as well as large emission bandwidth which can support femtosecond pulses generation, and so on. Up to now, extensive mode-locking research has been reported with various Yb-doped materials. In the following section, we will present some numerical and experimental studies on several passively mode-locked Yb-doped lasers.

5.1. Numerical and experimental investigation of the Yb:YAG laser at a wavelength of 1.05 μm

Among a variety of Yb-doped crystals, the Yb:YAG crystal is one of the most important laser media for several important advantages: excellent thermal-mechanical properties, ease of growth in high-quality crystal, and a high doping concentration without quenching, etc. Remarkable progress has been achieved with ultrashort Yb:YAG lasers because of these favorable properties [5.5, 5.6]. Typically, the Yb:YAG crystal has two main emission wavelengths, 1.03 and 1.05 μm respectively. Ultrafast Yb:YAG laser operating at 1.05 μm has special advantages compared with the one at 1.03 μm. First, the gain around 1.05 μm is flatter and can support femtosecond pulses with broader spectrum and shorter pulse width. Yb:YAG laser pulses as short as 100 fs have been demonstrated at this wavelength. In contrast, limited by the width of the narrow gain peak at 1030 nm, the shortest pulse achieved at this wavelength has the width of 340 fs [5.7], and most femtosecond pulses are longer than 500 fs [5.5, 5.6]. Second, ultrashort pulses at 1.05 μm can be useful in high energy glass laser facilities as the seeding source.

To obtain oscillation at 1.05 μm by the Yb:YAG laser, one must suppress the oscillation at 1030 nm. Some researchers use specially coated mirrors to distinguish these two neighboring wavelengths [5.7,5.8], which, however, inevitably brings additional cavity losses and leads to low laser efficiency. Investigations on the preferred emission wavelength versus the length and the ion concentration of the Yb:YAG crystal will be discussed below, we also described a novel method to obtain efficient 1.05 μm operation based on the Yb:YAG laser[5.9,5.10].

5.1.1. Theoretical investigation

The electronic diagram of the ytterbium ion is shown in Figure 13. It's a typical quasi-three-level system. As the zero phonon line is very narrow and the corresponding absorption cross section is lower, the most efficient pump transition for Yb:YAG is l_1 to u_2 at 940 nm. Emission transitions are from u_1 to l_2 (1024 nm), to l_3 (1030 nm) and to l_4 (1050 nm). For quasi-three-level longitudinally pumped laser system, the thickness of the gain medium is more crucial for the laser oscillation than in a four-level system. In a quasi-three-level system, the terminal level of the laser transition is thermally populated. Thus minimum pump intensity is required for reaching population inversion. As the pump is absorbed when traveling in the gain medium, this minimum intensity is reached after a crystal length which is the so-called optimum length. In the following, the different optimum length at different oscillation wavelength in the Yb:YAG laser will be investigated by the model developed in [5.11] and [5.12].

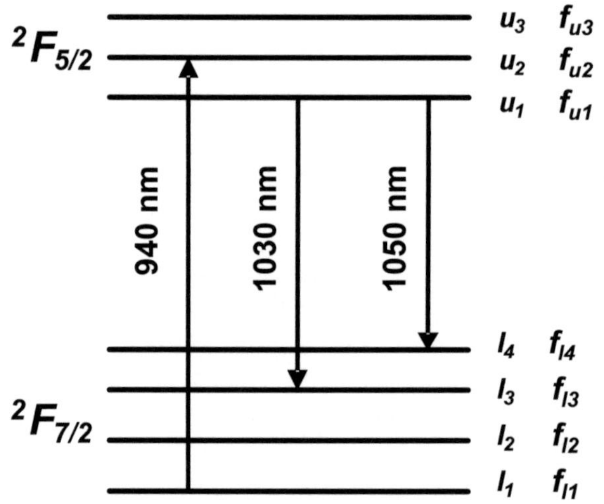

Figure 13. Energy level diagram of Yb^{3+} ion in the Yb:YAG crystal.

The amplification of the laser wave and the absorption of the pump wave are described by equations (5.1a) and (5.1b):

$$\frac{dI_l^{\varepsilon}}{I_l^{\varepsilon}} = \varepsilon g_0 \left\{ X_u - f_l \right\} dz \tag{3a}$$

$$\frac{dI_p^{\varepsilon'}}{I_p^{\varepsilon'}} = -\varepsilon' \alpha_0 \left\{ f_p - X_u \right\} dz \tag{3b}$$

where X_u is the population density of the excited state. $X_u = N_u/N_{Yb}$. N_u is the population of the excited state and N_{Yb} is the Yb^{3+} ions concentration. The linear coefficients of gain g_0 and absorption α_0 are given by equations (4):

$$g_0 = \sigma_l N_{Yb} \left(f_{lk} + f_{u1} \right) \qquad \alpha_0 = \sigma_p N_{Yb} \left(f_{l1} + f_{uj} \right) \tag{4}$$

with $f_l = \dfrac{f_{lk}}{f_{lk} + f_{u1}}$ $\qquad f_p = \dfrac{f_{l1}}{f_{l1} + f_{uj}}$,

where σ_p and σ_l are the absorption and emission cross sections, and f_{jk} are the Boltzmann partition factors of the sublevel k of the manifold j, ε and ε' are ± 1 relative to the direction of propagation of the laser and the pump beams respectively. The dynamic equation for the population reads:

$$\tau_u \frac{dX_u}{dt} = I_p \left(f_p - X_u \right) - X_u - I_l \left(X_u - f_l \right) \tag{5}$$

where τ_u is the excited state life time and I_p and I_l are the pump and laser intensity travelling in both directions normalized to the saturation intensity given by (6):

$$I_{sat}^{l} = \frac{h\nu_l}{\left(f_{lk} + f_{u1} \right) \tau_u \sigma_l} \qquad I_{sat}^{p} = \frac{h\nu_p}{\left(f_{l1} + f_{uj} \right) \tau_u \sigma_p} \tag{6}$$

In CW regime, equation (5) reads: $I_p \left(f_p - X_u \right) - X_u - I_l \left(X_u - f_l \right) = 0$. Then:

$$X_u = \frac{f_p I_p + f_l I_l}{1 + I_p + I_l} \qquad I_i = I_i^{+} + I_i^{-} \tag{7}$$

With regard to equation (3a), the laser beam is reabsorbed when $X_u < f_l$. Then minimum pump intensity required for bleaching the amplifier medium is given by:

$$X_u = f_l \qquad I_p^{\min} = \frac{f_l}{f_p - f_l} \tag{8}$$

For single pass pumping where the pump is travelling following one direction, the pump transmission required for inverting the amplifier medium is given by:

$$I_p(0)\Gamma = I_p^{\min} \qquad \Gamma = \beta = \frac{I_p^{\min}}{I_p(0)} \tag{9}$$

As the pump is absorbed when traveling in the amplifier, this minimum intensity required for bleaching the amplifier medium at the laser wavelength is reached after an amplifier length we call optimum length.

For single pass pumping and CW laser, the output intensity has been derived in [5.12]:

$$I_{out}(L) = (1 - R_s)\sqrt{R_m}\,\frac{g_0\left(\frac{I_p(0)(1-\Gamma)}{\alpha_0} - f_l L\right) + \ell n\sqrt{R_m R_s}}{\left(1 - \sqrt{R_m R_s}\right)\left(\sqrt{R_m} + \sqrt{R_s}\right)} \tag{10}$$

Where R_s and R_m are the reflectivity of the output and rear mirror respectively (assuming all the losses of the cavity are taken into account in the loss of the rear mirror). By (3a) and (3b), it is possible to find a relation connecting the pump and the laser intensities. Then, Γ reads:

$$\Gamma = \sqrt{R_m R_s}^{-\frac{\alpha_0}{g_0}} \exp\left\{-\alpha_0\left(f_p - f_l\right)L\right\} \tag{11}$$

The optimum length for which the laser wave is not reabsorbed leads to maximum output power. Then, the pump transmission must be equal to β leading to:

$$L_{opt} = \frac{-1}{f_p - f_l}\,\ell n\left\{\sqrt{R_m R_s}^{\frac{1}{g_0}}\beta^{\frac{1}{\alpha_0}}\right\} \tag{12}$$

This result can also be obtained by computing the length for which $I_{out}(L)$ is maximum using formula (10). In CW laser operation of a three-level system, inversion density of gain media remains constant above laser threshold. Then, for a given laser medium, pump transmission is fixed and determined by cavity loss level and spectroscopic parameters of gain media. It is worth to mention that L_{opt} as well as β depends on the pump intensity.

	Absorption	Emission	
λ (nm)	940	1030	1050
f_u+f_l	1.04E+00	0.75	7.00E-01
σ (cm²)	7.60E-21	3.30E-20	4.80E-21
τ_u (ms)	0.95	0.95	0.95
f_l	0.838	0.0626	0.0205

Table 2. Values of the crystal parameters

Now the length and the ion concentration of the crystal can be investigated to find out that under what kind of condition the laser oscillation at 1050 nm can be preferred. The emission transitions of Yb: YAG are from upper level u_1 to lower level l_3 (1030 nm) and l_4 (1050 nm). First, as the l_4 energy level is higher than the l_3 one, its thermal population is lower and the f_l parameter at room temperature is 0.0284 against 0.0646. Then, the corresponding optimum crystal length is larger and more pump energy is absorbed. Using (10), the optimum crystal length versus the rear mirror reflectivity for various pump intensities can be calculated. The values of the crystal parameters used for the computation are reported on Table 2. The cross sections are spectroscopic cross sections [5.13].

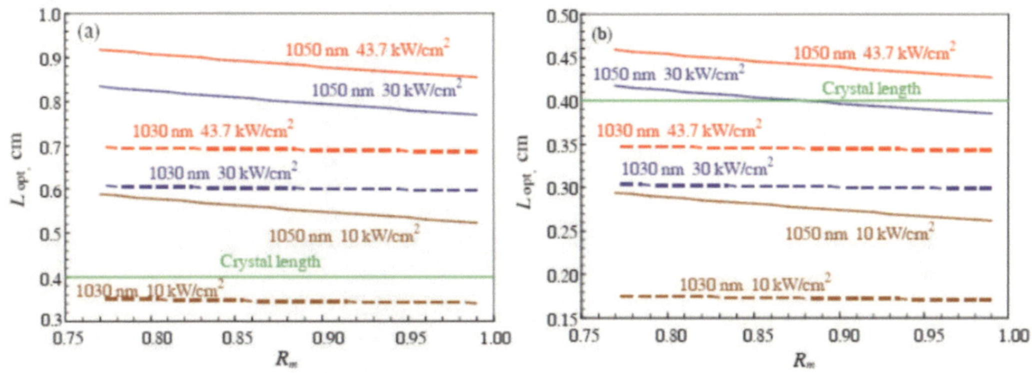

(a) (b)

Figure 14. Optimum length versus reflectivity of the rear mirror for various pump intensities with (a) 5% and (b) 10% doping.

Figure 14 shows the calculated optimum lengths for the two wavelengths of 1030 and 1050 nm and for 5 at. % (a) and 10 at. % (b) doping concentration. It can be noticed that, for a crystal length of 4 mm with 5 at. % doping, the crystal length is much lower than the optimum lengths for 1050 nm oscillation at different pump intensities, but close to the 1030 nm optimum length. As a result, oscillation at 1030 nm is favored. It shows good agreement with the results reported in [5.7]. In that paper, a 5%-doped, 3.5-mm-long Yb:YAG crystal was used, and under the pump intensity of about 36 kW/cm^2, laser oscillation was achieved at the wavelength of 1030 nm. But for the 10 at. % doped crystal, one can notice that the situation is different. The preferred crystal length is much shorter than that of 5 at. % doping crystal. Only when the crystal is very short, oscillation at 1030 nm will be preferred. If the crystal length reaches a proper value, such as 4 mm, the 1030 nm laser is more likely to be re-absorbed and suppressed. This indicates a new way to suppress 1030 nm and get the oscillation at 1050 nm only by choosing the Yb:YAG crystal with proper ion concentration and optimized length.

5.1.2. Mode-locked Yb:YAG laser at the wavelength of 1.05 μm

The experiment is performed with a 10 at. % doped Yb:YAG crystal with the length of 4 mm. A typical Z-shape cavity was used (Figure 15(a)). All the coatings of the intra-cavity mirrors are identical for 1030 and 1050 nm. A CW Ti:sapphire laser at 940 nm with 2 W power was used as the pump. The pump intensity at the front face of the crystal is calculated to be about 43.7 kW/cm^2 at 2 W pump power. The measured absorption of the crystal to the pump light is 91 %, which is in good agreement with the theoretically calculated value of 92.6% from equation (9) by the model.

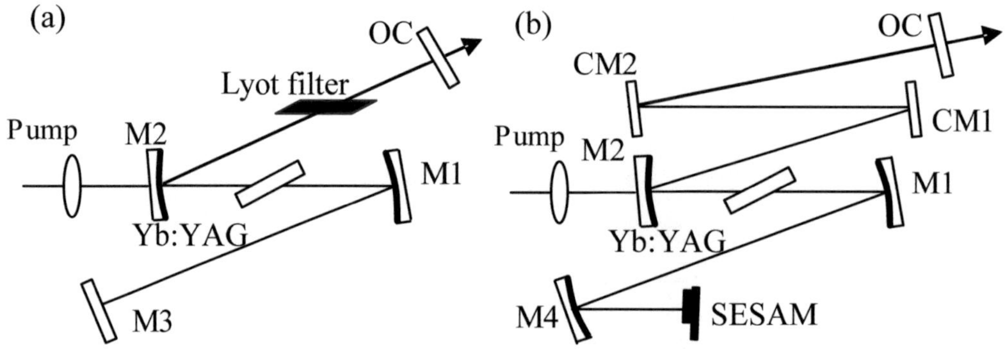

Figure 15. Cavities used to study the (a) CW and (b) mode-locked laser performance of a Yb:YAG laser.

For CW operation, two kinds of output couplers were used, with the transmittivity of 1% and 2.5% respectively. The wavelength of the free-running laser was measured with a scanning spectrometer. When the pump power was increased from the threshold power to 2 W, the emitting wavelength was keeping at 1050 nm. This can be well explained by Figure 14(b). When the pump intensity was increased from zero to 43.7 kW/cm^2, the length of the crystal is near the optimum length of 1050 nm oscillation and much longer than that of 1030 nm oscillation. 1030 nm lasing is more likely to be reabsorbed and oscillation at the wavelength of 1050 nm can be preferred. Under the pump power of 2 W, the maximum output power was as high as 650 mW, leading to a slope efficiency as high as 45.8%.

Based on the CW 1050 nm operation, the passively mode-locking operation of this laser was also investigated. The cavity layout is shown in Figure 15(b). An output coupler with 0.5% transmittivity was used in this setup. Without intracavity dispersion compensation, stable mode-locked pulses in picosecond regime were obtained. A typical intensity autocorrelation trace (obtained by an FR-103MN autocorrelator, Femtochrome Research, Inc.) of the output pulses is shown in Figure 16(a). Assuming a sech^2 pulse shape, one can obtain the FWHM pulse duration of 1.8 ps. A simultaneous measurement of the pulse spectrum is illustrated by the insertion.

A pair of Gires-Tournois interferometer (GTI) mirrors, which introduce group delay dispersion (GDD) of -1200 fs^2 from 1020 to 1080 nm by single pass, were used to obtain femtosecond laser operation. It's enough to compensate the positive GDD caused by the Yb:YAG crystal and the net intracavity GDD remain at a minus value. With this alignment, stable self-starting soliton-like pulses were obtained. Figure 16(b) shows the measured autocorrelation trace and the spectrum of the pulses. The pulse width is 170 fs assuming a sech^2-shape pulse and the FWHM spectral bandwidth reaches 7 nm. The central wavelength redshifted from 1050 to 1053 nm. It's worth to mention that the central wavelength of this femtosecond Yb:YAG laser is exactly the working wavelength of high energy Nd:glass based ultrafast amplifier system. It indicates that the femtosecond Yb:YAG laser has the potential to be an excellent seed source for the above system. Under the full pump power of 2 W, the average power of the femtosecond pulses is 180 mW at a repetition rate of 80 MHz, corresponding to the peak power of 13.3 kW.

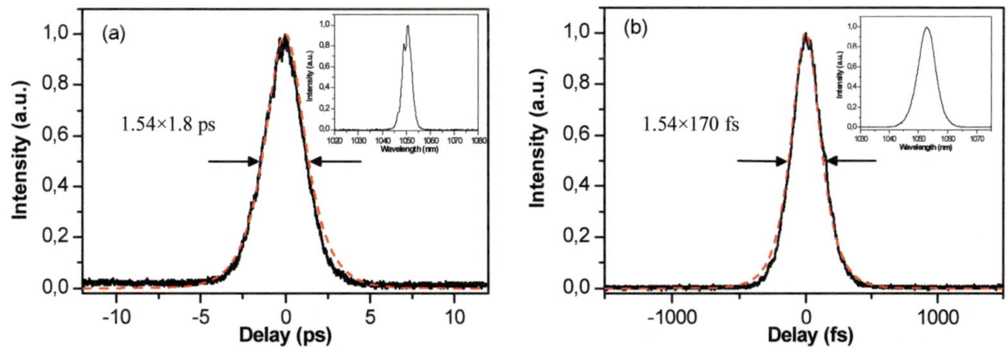

Figure 16. (a) Intensity autocorrelation trace of the picosecond pulses; inset, the corresponding laser spectrum. (b) Intensity autocorrelation trace of the femtosecond pulses; inset, the corresponding laser spectrum.

5.2. Diode-pumped efficient femtosecond ceramic Yb:YAG laser and picosecond Yb:LSO laser

Highly transparent Yb:YAG ceramics for solid-state laser gain medium have been developed in recent years, by the vacuum sintering technique and nanocrystalline technology [5.14]. This Yb:YAG ceramics have such advantages as favorable mechanical properties [5.15], high doping concentration, low cost, and easy fabrication of large-size samples. All of these advantages promise Yb:YAG ceramic extensive potential for the development of high-efficiency and high-power ultrafast laser.

In the first Yb:YAG ceramic laser experiment [5.14], the cw output power was 345 mW with a slope efficiency of 26%. With the improvement on the Yb^{3+} doping concentration, excellent cw laser performances by Yb:YAG ceramics have been demonstrated. Dong et al. realized a 1.73 W cw output power with the absorbed pump power of 2.87 W, corresponding to a slope efficiency as high as 79% [5.16]. A cw output power as high as 6.8 W with the slope efficiency of 72% was further demonstrated by Nakamura et al [5.17]. In the pulsed operation, Yoshioka et al. demonstrated the first mode-locked operation of a Yb:YAG ceramic laser [5.18]. The reported efficiency for the femtosecond operation was, however, relatively low; the output power was 250 mW at the pump power of more than 26 W.

In this section, a high-power and high-efficiency operation of a diode-end-pumped femtosecond Yb:YAG ceramic laser will be described [5.19]. For self-starting femtosecond laser operation, more complicated cavity configuration is needed than for CW laser operation. Lots of additional cavity losses are thereby introduced, such as the insertion losses by the prisms, the non-saturable loss by the SESAM, and the broadband reflection losses by intra-cavity reflectors, etc. To obtain efficient mode-locking operation, several major improvements were made compared with Ref. 5.17. First, other than prisms, highly reflective negative-dispersion mirrors were adopted for dispersion compensation. This avoided the insertion losses by the prisms, which usually caused low efficiency in mode-locked Yb-doped solid-state lasers. Experimental results on this kind of lasers have shown the significant efficiency increase by negative-dispersion mirrors replacing prisms [5.20, 5.21]. Second, a piece of SESAM with the non-saturable loss parameter as low as 0.3% was used to further minimize the intra-cavity losses. Third, the transmission of the output coupler mirror was also optimized for femtosecond operation. Finally, high pumping intensity is much helpful to obtain efficient laser operation. For this purpose, a 7-W high brightness fiber coupled diode laser (Jenoptik laserdiode GmbH) at 968 nm was adopted as the pump. The fiber core diameter is 50 μm, with a numerical aperture of 0.22.

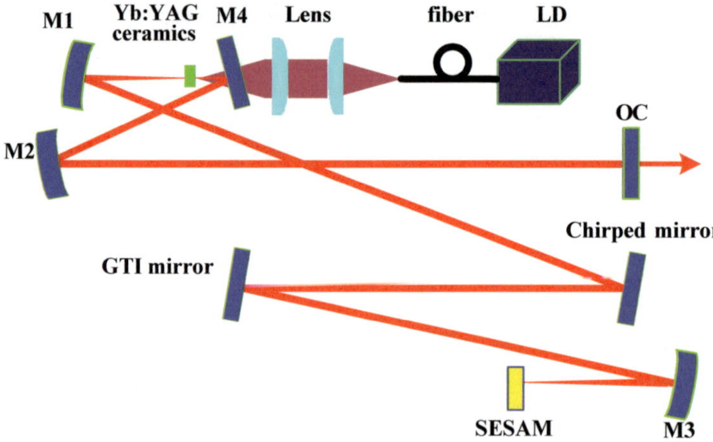

Figure 17. Schematic of the mode-locked Yb:YAG ceramic laser.

A schematic of the laser cavity and pump geometry is shown in Figure 17. The pump laser was reimaged into the Yb:YAG ceramic through a coupling system. The cavity was a standard Z-shape structure, with a piece of high reflective plane mirror folding one arm of the cavity to fit the short focusing length of the pump light. M1 and M3 were curved folding mirrors with ROC of 200 mm, and M2 with ROC of 300 mm. A 2-mm-long 10 at. % doped Yb:YAG ceramic was used for the gain medium. The absorption of the ceramic to the pump laser was around 50% in the experiment due to the lower and narrower absorption peak at 968 nm than at 940 nm. Negative dispersion was introduced by a Gires-Tournois interferometer (G-TI) mirror and a chirped mirror. The G-TI mirror provides second-order dispersion compensation of about -1000 ± 300 fs^2 per rebound in the spectral range from 1030 to 1050 nm and the chirped mirror provides -120 ± 20 fs^2 from 1000 to 1100 nm. The cavity was designed to sustain a fundamental mode with a beam waist of 65 μm × 67 μm in the crystal, and a waist of 58 μm × 65 μm on the SESAM. The total cavity length corresponds to a repetition rate of 64.27 MHz.

Different mirrors were used as output couplers with transmissions of 1%, 2.5%, and 4% from 1020 to 1080nm. Stable soliton mode-locking regime was observed with 1% and 2.5% output couplings. With 4% coupling, the output power under the maximum pump was slightly higher than the output power by the 2.5% coupler, however, cw mode-locking couldn't be realized at this coupling condition. Figure 18 shows the output power as a function of the absorbed pump power for these two output couplers. With the decreasing of mirror reflectivity from 99% to 97.5%, the threshold incident pump power for cw mode-locking operation increased from 1.9 to 2.5 W. The maximum CW mode-locked output power was achieved with the 97.5% reflectivity output coupler. Under the incident pump power of 7 W (the measured absorbed pump power was 3.5 W), stable mode-locked output power of 1.9 W was obtained, corresponding to a slope efficiency of 76% with respect to the absorbed pump power. With the 1% output coupler, the maximum mode-locked power was 1.2 W with the slope efficiency of 44.7%.

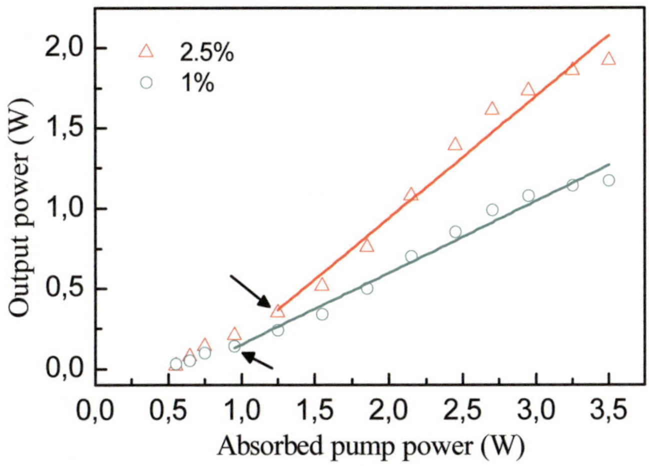

Figure 18. Average output power versus absorbed pump power of the mode-locked Yb:YAG ceramic laser with output couplings of 2.5% (triangles) and 1% (circles). The cw mode-locking thresholds are indicated by arrows.

At the highest output power, the pulse duration was measured to be 418 fs (sech^2 assumption). The FWHM width of the spectrum was 3.4 nm at the central wavelength of 1048 nm (Figure 19 left). The radio-frequency spectrum was also measured by an electrical spectrum analyzer to characterize the performance of femtosecond pulse train at the high power operation. As depicted by Figure 19, the spectrum shows a clean peak at a repetition rate of 64.27 MHz without side peaks, implying that the Q-switching instabilities have been fully depressed. A wider acquisition frequency span (from 0 to 300 MHz with 100 kHz resolution bandwidth) was also performed (inset in Figure 19 right), which was a clear indication for the single-pulse operation in high output power level. The results presented above definitely affirmed that the Yb:YAG ceramic is an excellent laser medium for high-power and high-efficiency diode-pumped ultrafast lasers and amplifiers.

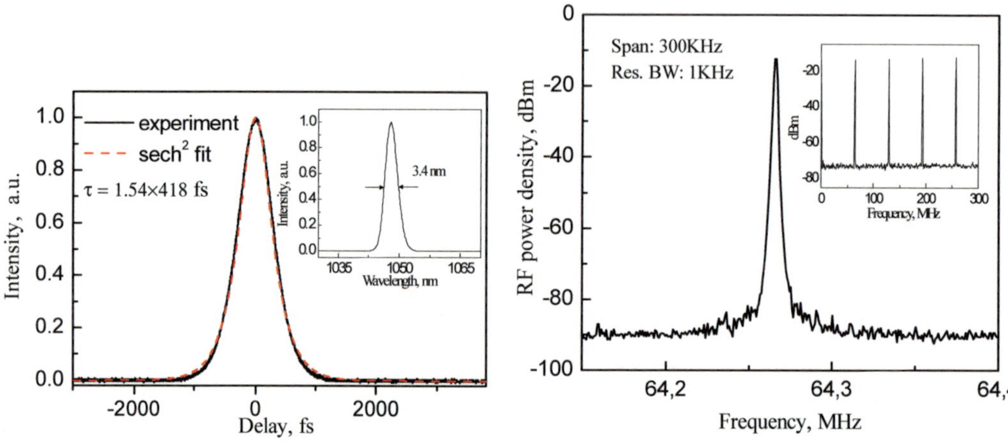

Figure 19. Left, intensity autocorrelation trace of the mode-locking laser pulses with 1.9 W average power; inset, the corresponding spectrum. Right, the rf spectrum at fundamental repetition frequency of the 418 fs pulse train; inset, resolution bandwidth of 100 kHz and span of 300 MHz.

With a similar experimental layout as depicted in Figure 17, a cw mode-locked picosecond Yb:LSO laser was also realized with a 5% doped 3-mm-thick Yb^{3+}:Lu_2SiO_5 crystal as the laser medium [5.22]. At the pump power of 5.4 W from the fiber coupled diode laser, 0.72 W stable mode-locked pulses were obtained with an output coupler of 1% transmittivity. However, the thermal loading prevented higher output power when the pump power increased. The central wavelength of the mode-locked pulses lay on 1058 nm with a FWHM bandwidth of 3.5nm. The pulse duration was 5.1 ps, as shown in Figure 20.

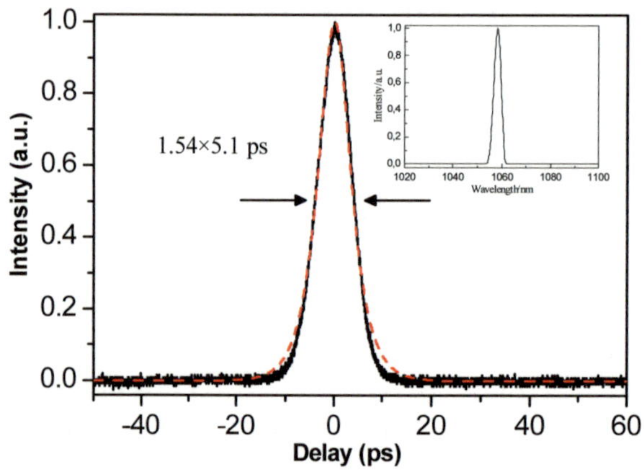

Figure 20. Intensity autocorrelation trace of the mode-locking laser pulses; inset, the corresponding spectrum.

5.3. Femtosecond Yb:GYSO laser

Yb^{3+}-doped $GdYSiO_5$ (Yb:GYSO), a promising ytterbium-doped crystal, has shown several attractive advantages compared to many recent Yb-doped materials. It exhibits $^2F_{7/2}$ ground state splitting as high as 995 cm^{-1}, which makes the population of the transition lower level much less sensitive to the temperature. Yb:GYSO crystal also possesses a comparatively high fluorescence life time of 1.92 ms and good mechanical properties. Du et al. demonstrated the first efficient tunable CW Yb:GYSO laser operation [5.1], and the picosecond Yb:GYSO laser was reported by another group later [5.2]. We describe here the realization of a femtosecond Yb:GYSO laser [5.3]. The setup for the laser is shown in Fig.21. A 3-mm-long, 5%-doped Yb:GYSO crystal is pumped by a CW Ti:sapphire laser at the wavelength of 976 nm. Three pieces of curved folding mirrors with the radius of curvature (ROC) of 100mm (M1-M3) are used to reduce the beam waist inside the active medium and on the SESAM. The SESAM was designed for 0.4% modulation depth and saturation fluence of 120 μJ/cm^2. Its nonsaturable loss was specified to be 0.3%, and the relaxation time less than 500 fs. A pair of chirped mirrors (CM1, CM2) were used in another arm of the cavity for group velocity dispersion compensation. The chirped mirrors were designed with GDD of -120 fs^2 per bounce within the wavelength range from 1000 nm to 1200 nm. Considering the amount of positive GDD introduced by the 3-mm-long 5%-doped Yb:GYSO crystal [5.4], the net intra-cavity GDD was remained at minus value by introducing two bounces onto the chirped mirrors.

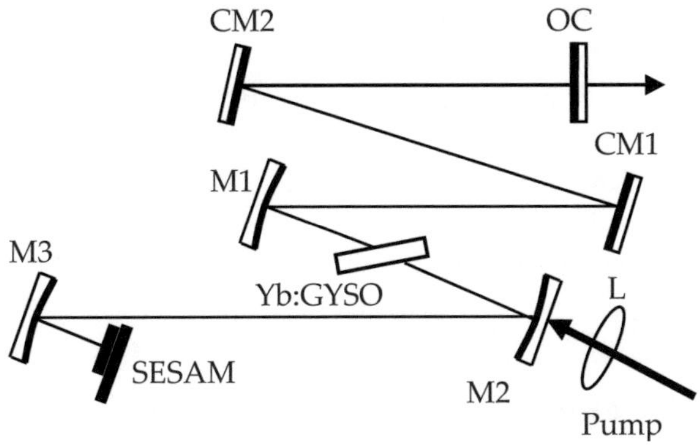

Figure 21. Cavity used to study the femtosecond laser performance of Yb:GYSO. M1, M2, M3, ROC 10 cm; CM1, CM2, flat chirped mirrors; L, lens f=100 mm; OC, 1% output coupler.

With a piece of high reflective mirror in one arm replacing the M3 and SESAM, and removing the chirped mirrors, the CW lasing performance was tested first. At CW lasing, the measured threshold incident pump power was as low as 180 mW, which can be thanks to the large ground state splitting of Yb:GYSO crystal. Output power of 920 mW with diffraction-limited beam pattern was obtained at the wavelength of 1091 nm under 2 W pump power. A tunable range of about 77 nm (from 1033 to 1110nm) was obtained by inserting a Lyot filter in the CW cavity.

After optimization of the cavity alignment shown in Fig. 21, stable and self-starting femtosecond pulses were realized. The autocorrelation trace of the mode-locked pulses is shown in Fig 22(a), the FWHM width of the autocorrelation trace is about 324 fs. If a $sech^2$-pulse shape is assumed, the mode-locked pulse duration is 210 fs.

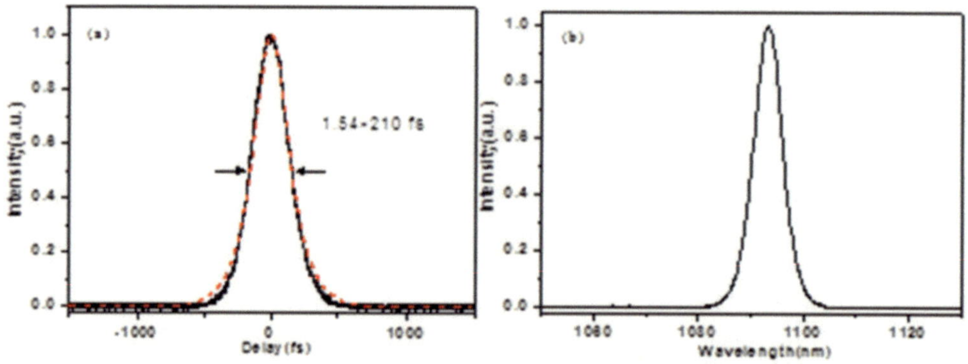

Figure 22. (a) Typical intensity autocorrelation trace of the pulses. The experimental data are shown by the solid curve and the sech2-fitting curve by the dashed curve. (b) The laser spectrum of mode-locking operation.

The spectral width (FWHM) of the pulses was measured as 6.4 nm at the central wavelength of 1093 nm. Fig. 22(b) shows the typical spectrum, compared with the one in CW mode, the central wavelength red shifted slightly from the peak wavelength. The time-bandwidth product is 0.34, which is close to the value of 0.315 for the transform-limited $sech^2$-pulse. This indicates that almost transform-limited pulses were directly obtained from the cavity. However, the gain bandwidth was not fully covered by the obtained spectrum, especially in the shorter wavelength. That means the gain of Yb:GYSO crystal is not fully exploited. Considering the 77 nm tunability achieved in the CW mode, even shorter femtosecond pulses should be possible. Under the full pump power of 2 W from the CW Ti:sapphire laser, the average mode-locking power output from the Yb:GYSO laser was 300 mW at a repetition rate of 80 MHz, corresponding to an energy per pulse of 3.75 nJ and a peak power of 17.9 KW.

A higher mode-locking laser power can be possible by replacing the pump laser by a high power diode laser at 976 nm. In view of the excellent mechanical properties of the Yb:GYSO crystal, a new kind of femtosecond laser with high output power and compactness is very promising.

5.4. Picosecond and femtosecond Yb:YGG laser

Yttrium gallium garnet (YGG) is another garnet crystal. Similar to YAG and GGG, $Y_3Ga_5O_{12}$ (YGG) has many desirable advantages for laser materials—stable, hard, optically isotropic, having good thermal conductivity (9 W/mK), and accepting substitutionally trivalent ions of both rare-earth and iron groups [4.8]. Yb^{3+} doped yttrium gallium garnet (Yb:YGG) was first reported as a scintillator [5.23-24]. The most interesting property is that the bandwidth of its emission spectrum is nearly four times boarder than Yb:YAG 's [5.23]. The high-quality Yb:YGG crystal suitable for laser operation had been grown though optical floating zone method by H. Yu et al for the first time [5.25], and the special thermal properties, including the specific heat, thermal expansion coefficient, thermal diffusion coefficient, and thermal conductivity had been investigated. Tab.3 shows properties comparisons of Yb:YGG, Yb:YAG and Yb:GGG [5.25]. In this section, we reviewed the diode pumped passively mode-locked Yb:YGG laser [5.26-27].

Crystals	Yb:YGG (10at%)	Yb:YAG	Yb:GGG
Symmetry	Cubic	Cubic	Cubic
Thermal expansion coefficient (10^{-6} K^{-1})	3.8	8.18 (10at%)	~7 (Pure)
Specific Heat (J/gK)	0.43	~0.63 (10at%)	~0.37 (Pure)
Thermal diffusion coefficient (mm^2s^{-1})	1.33	1.62 (10at%)	~3 (Pure)
Thermal Conductivity (W/m·K)	3.47	~4.8 (10at%)	7.5 (Pure)
Absorption Cross-sections ($10^{-20}cm^2$)	2.7 (970 nm)	0.77 (941 nm)	0.6 (945 nm)
Emission Cross-sections ($10^{-20}cm^2$)	2.6 (1025 nm)	2.03 (1031 nm)	1.1 (1031 nm)
FWHM at emission peak	22 nm	10 nm	~10 nm
Positive band at effective gain cross-sections (β=0.5)	120 nm	~100 nm	~80 (nm)

Table 3. Comparisons of Yb:YGG, Yb:YAG and Yb:GGG

The Yb:YGG single crystal employed in the experiment was grown by the optical floating zone method, which was fine polished and antireflection coated at a broad spectrum range around 1 μm with cross section of 3 mm×3 mm and length of 3.2 mm. A high brightness fiber-coupled diode laser emitting at 970 nm (Jenoptik, JOLD-7.5-BAFC-105) was used to end-pump the laser medium. The pump laser output from the fiber (with 50 μm core diameter and 0.22 numerical aperture) was coupled into the laser crystal where the laser spot radius was about 30 μm. Fig. 23 (a) is the schematic of the pumping geometry and laser cavity. A Z-fold cavity was employed for mode-locking experiment. M1 was a plane dichroic mirror with high transmission at 970 nm and high reflection at 1020-1100 nm; M2, M3 and M4 were concave mirrors, with radii of curvature of 300 mm, 200 mm and 200 mm, respectively. Passive mode-locking was realized by using a semiconductor saturable absorber mirror (SESAM) (BATOP), which has a saturable absorption of 0.4 % at 1040 nm, a saturation fluence of 120 μJ/cm², relaxation time less than 500 fs, and non-saturable loss of less than 0.3 %. A plane-wedged mirror with transmission rate of 2.5 % was used as the output coupler (OC). According to the ABCD matrix, the cavity was very stable and had large parameter space. After optimization of the cavity alignment, laser pulses as short as 2.1 ps were generated with an average power more than 1 W under incident pumping power of 5.5 W, where multi-pulse will interrupt the stable mode-locking operation when pumping power was higher than 5.5 W.

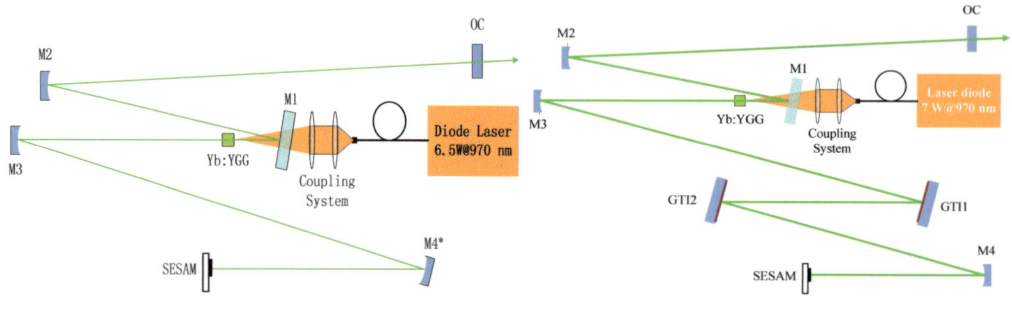

(a) (b)

Figure 23. (a) experiment layout of the picoseconds Yb:YGG laser; (b) experiment layout of the femtoseconds Yb:YGG laser

In order to obtain very short pulses, dispersion conversation is an important issue. A pair of Gires–Tournois interferometer (GTI) mirrors, with group-velocity dispersion of -1400 fs^2 per bounce within the wavelength range from 1025 to 1045 nm, was used for dispersion compensation, as shown in Fig. 23 (b). Limited by the available pump power, the maximum output power was 570 mW under 6.9 W of incident pump power. We measured the intensity autocorrelation trace by using the commercial noncollinear autocorrelator (Femtochrome, FR-103MN). As shown in Fig. 24 (a), the FWHM width of the autocorrelation trace was about 360 fs. If a sech^2-pulse shape was assumed, the mode-locked pulse duration was 245 fs. The Fig. 24 (b) depicts the corresponding spectrum of the stable mode locking, which had a FWHM bandwidth of 5.8 nm at the central wavelength of 1045 nm. The time-bandwidth product was calculated to be 0.39.

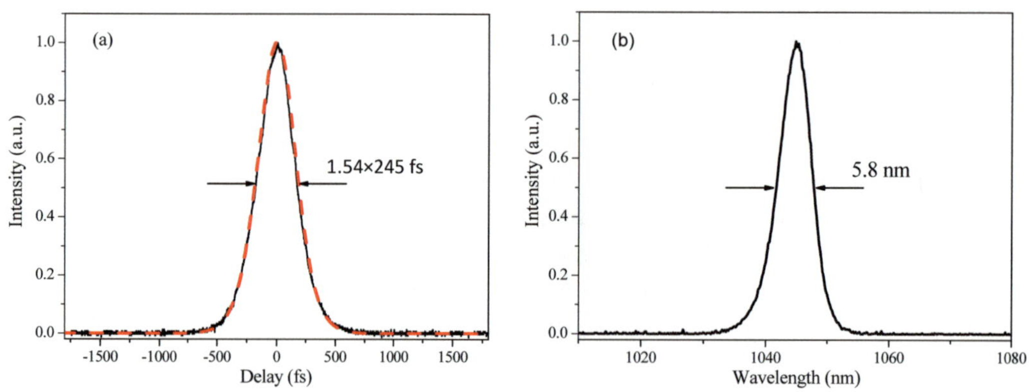

(a) (b)

Figure 24. (a) Intensity autocorrelation trace of the pulse. The experimental data are shown by the solid curve and the sech^2-fitting curve by the dashed curve. (b) The laser spectrum of mode-locking operation.

6. Mode-locked femtosecond Cr^{4+}-doped laser

After several decades of development, there are many kinds of gain media which can operate in near infrared region, but only a few of them can be used in mode-locked lasers. In chapters 4 and 5, Nd-doped and Yb-doped mode-locked lasers have been described. Most Nd-doped lasers work in picosecond region, so do some Yb-doped lasers. Although some Yb-doped lasers can deliver femtosecond pulses, the generated spectral bandwith is relatively narrow and can't support very short pulses.

There are two chromium doped media, namely chromium doped forsterite (Cr^{4+}:Mg_2SiO_4) and chromium doped yttrium aluminium garnet (Cr^{4+}:YAG), are very attractive for femtoseond laser application. They have very broad emission range which can support few tens of femtoseond pulse generation. Combined with the interesting wavelength ranges they cover (1.3 and 1.5 μm, respectively), these lasers have find wide applications in optical communication [6.1], optical coherence tomography [6.2], biophotonics [6.3] and so on. In this section, we will present our work on the development of these two kind of ultrafast lasers.

6.1. Efficient femtoseond Cr:forsterite laser

Forsterite is an anisotropic crystal belonging to the orthorhombic space group. Cr^{4+}:forsterite crystal shows many excellent characteristics as a laser gain medium. Some important parameters are listed in Table 4 and compared with those of Ti:sapphire crystal. Except for low thermal conductivity and relatively small emission cross section, other physical parameters of Cr^{4+}:forsterite crystal are comparable with those of Ti:sapphire crystals. The Cr^{4+}:forsterite crystal can support the generation of very short femtosecond pulses.

Crystal name	Ti:Sapphire	Cr:Forsterite
Density (g/cm^3)	3.98	3.217
Melt point (°C)	2050	1895
Moh's hardness	9	7
Thermal conductivity (W/(m·k))	33	8
Thermal expansion (K^{-1})	5×10^{-6}	9.5×10^{-6}
Tuning range (nm)	660~1050	1130~1367
Upper level lifetime (us)	3.2	2.7
Emission cross section (cm^2)	4×10^{-19}	1.44×10^{-19}
Center wavelength (nm)	795	1244(cw),1235(pulsed)
Shortest pulse width ever achieved (fs)	<5 fs[6.4]	14 fs[6.5]

Table 4. Parameter comparison of Ti:sapphire and Cr:forsterite

Cr^{4+}:forsterite has very wide absorption and emission spectrum, as shown in Figure 25. The use of absorption peaks at 470 nm and 730 nm are typically excluded due to lack of pump sources. The relatively lower but much wider peak near 1 μm is preferred because of the vast availability of the pump laser. Besides, the quantum defect in the crystal will be much smaller when pumped at this wavelength. Although the emission spectrum extends from 600 nm to 1400 nm, wavelengths below 1100 nm overlaps with the absorption spectrum and suffers significant reabsorption, which prevents stimulated emission. Only the part above 1100 nm is suitable for laser operation. Up to now, the shortest pulse duration with Cr^{4+}: forsterite laser of 14 fs was achieved with 80 mW output power [3.49]. We'll present our work on the development of highly efficient self-starting femtosecond Cr:forsterite laser here. With a 7.9 W Yb doped fiber laser as the pump, we obtained stable femtosecond laser pulses with an average power of 760 mW, yielding a pump power slope efficiency of 12.3% [6.6].

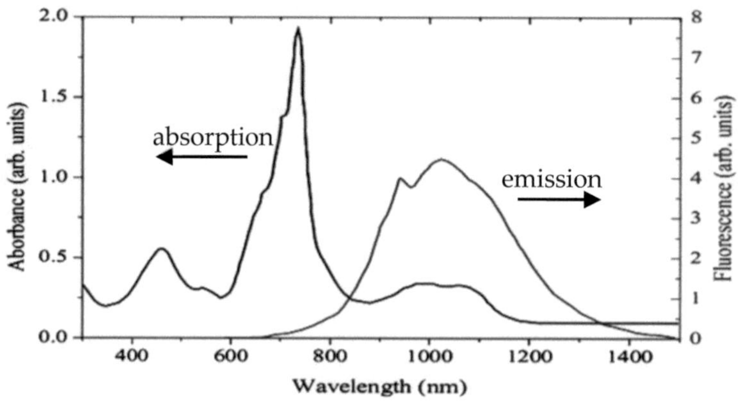

Figure 25. Absorption and emission spectrum of Cr^{4+}:forsterite

The schematic diagram of an efficient Cr^{4+}:forsterite laser is shown in Figure 26 A 1064 nm linearly polarized cw fiber laser is used as the pump (AYDLS-PM-10, Amonics), which deliver 7.9 W average power at maximum. The dimension of Cr^{4+}:forsterite crystal is 4 mm× 2 mm×9 mm and the crystal is cut for propagation of light along the a axis and emitting beam polarization along the c axis (P_{mnb} notation). Both ends are cut at Brewster's angle and polished in order to reduce the reflective loss to the minimum. Due to the low thermal conductivity, the crystal is wrapped in thin indium foil and tightly held in a copper holder. A thermoelectric cooler cools the holder down to 5°C. Although lower temperature can bring higher output power, water condensation on crystal surface will induce unstable operation.

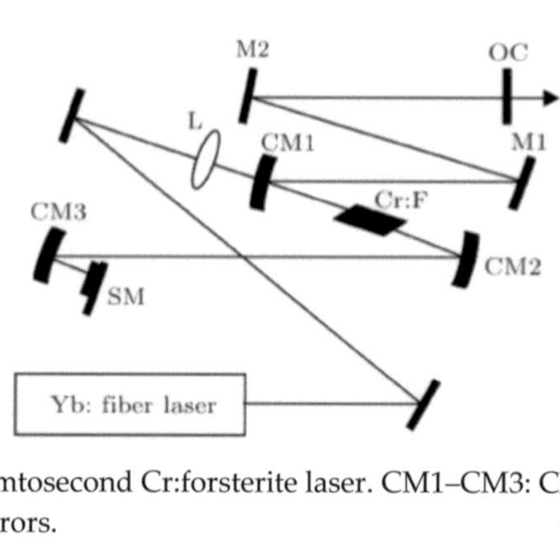

Figure 26. Schematic of the femtosecond Cr:forsterite laser. CM1–CM3: Chirped mirrors, ROC 100 mm; M1 and M2: plane chirped mirrors.

In order to obtain femtosecond pulses, dispersion compensation is important, especially when the used laser crystal is relatively long. By using the formula 13[6.8], the group-delay dispersion of Cr^{4+}:forsterite is calculated and shown in Figure 27.

$$GDD = \frac{1}{2\times 5.17}\left[2\times 84.42 + 6\times 116.70(\omega-\omega_0) + 12\times(-101.21)(\omega-\omega_0)^2 + 20\times 125.08(\omega-\omega_0)^3\right] \quad (13)$$

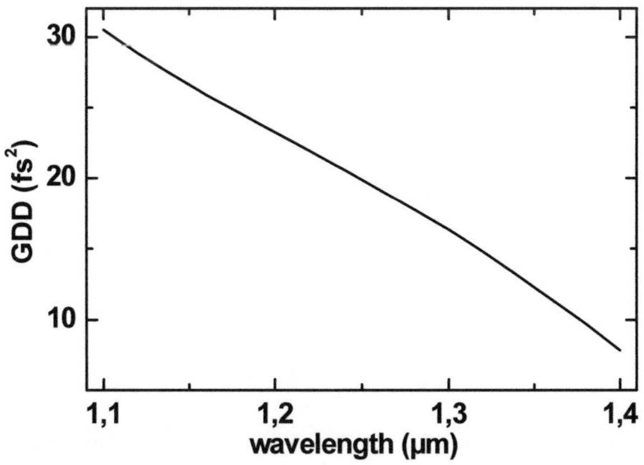

Figure 27. Group-delay dispersion from 1 mm Cr^{4+}:forsterite crystal

The 9 mm long Cr^{4+}:forsterite crystal introduces about 150 fs^2 positive GDD. In order to compensate these dispersion, all the reflective mirrors in cavity are chirped mirrors except the 3% output coupler (OC) and the saturable mirror (SM). Each bounce on CM1, CM2 and CM3 brings in -60±20 fs^2, and -70±20 fs^2 on M1 and M2, respectively. Considering the positive GDD introduced by 1.8-m-long air, the final net intra-cavity GDD after one single trip is about -130 fs^2. Due to the strong oscillations in negative GDD introduced by all the chiped mirrors, a certain amount of negative GDD is necessary to insure the stable mode locking. The semiconductor saturable absorber mirror (SESAM), which is highly reflectively coated from 1240 nm to 1340 nm, has a modulation depth of 2.5% and saturable fluence of 70 μJ/cm^2. In order to reduce the inserting losses, we chose a nonsaturable loss of the SESAM less than 0.5%.

A lens with the focal length of 100 mm is used to focus the pump beam into the crystal. In order to achieve good matching and overlapping between pump and resonant beam, the radius of curvature (ROC) of CM1 and CM2 are 100 mm. Besides, because stable mode locking requires sufficient energy fluence on SESAM, CM3 with ROC of 100 mm is used to focus the resonant beam on it.

The relationship between pump power and output power is shown in Figure 28. The pump power threshold for stable mode locking is 1.8 W, with corresponding output power of 49 mW. The maximum output power of 760 mW is obtained when pump power increases to 7.9 W, which indicates record high slope efficiency as 12.3%. When optimizing the output power of mode-locking operation, we found that a slightly adjustment of the concave mirror

CM2 was necessary when the laser cavity was firstly optimized at other pump powers. We contribute this to the thermal loading effect in the laser crystal. The mode-locking power of the Cr:forsterite laser was almost linear till the pump power up to 7.9 W. Hence we believe that even higher output power can be expected if we use a pump laser with higher power.

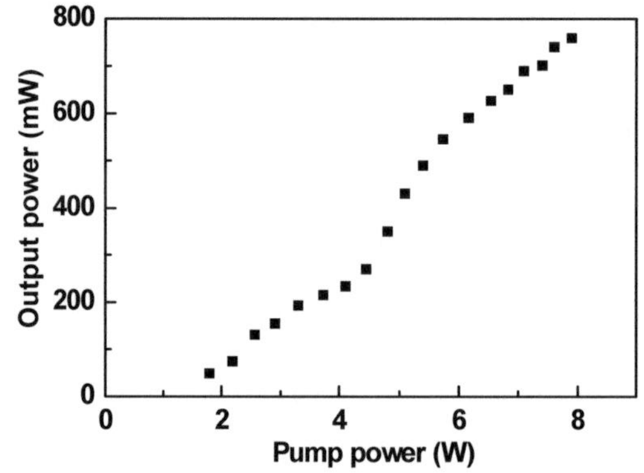

Figure 28. Variation of the mode locking output power as a function of the pump power.

The intensity autocorrelation trace (measured by a FR-103MN autocorrelator, Femtochrome Research, Inc.) and spectrum (measured by AQ-6315A, ANDO) are shown in Figure 29. The autocorrelation trace captured by an oscilloscope shows a FWHM of 71 fs, which indicates the pulse duration of 46 fs by assuming a sech² pulse shape. The spectrum of mode locking operation is centered at 1277 nm, with a width of 45 nm. A time-bandwidth product of 0.38 is obtained, indicating that the output pulses are nearly transform limited.

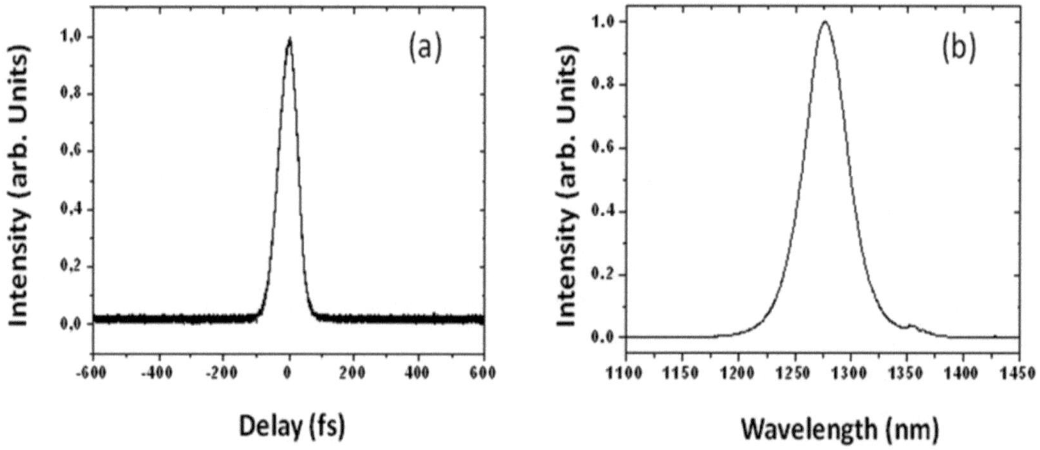

Figure 29. (a) Intensity autocorrelation trace of the output pulses and (b) spectrum of mode locking operation.

The mode-locked Cr^{4+}: forsterite laser can stably work for hours if the environment is dry and stable. The most probable factor which disturbs the mode locking is the water condensation on crystal surfaces. If the environment is quite humid, nitrogen blowing or dehumidifier is required. Based on this efficient femtosecond Cr:forsterite laser, we also realized red light femtosecond laser output with a intra-cavity frequency-doubled Cr:forsterite laser configuration [6.7].

6.2. Self-starting mode-locked femtosecond Cr^{4+}:YAG laser

Chromium doped yttrium aluminium garnet (Cr^{4+}: YAG) is a remarkable gain medium which generates near infrared laser beam. Moreover, it can support femtosecond laser operation for its wide emission spectrum. The first laser operation in Cr^{4+}: YAG was reported in 1988 [6.9]. In 1993, 26 ps pulses were obtained by using active mode locking [6.10]. So far, femtosecond pulses from Cr^{4+}:YAG can be obtained by active mode locking [6.11], kerr-lens mode locking [6.12-13] and SESAM mode locking [6.14-15]. The record of shortest pulses was reported by Ripin *et al.* in 2002 [6.16].

The absorption spectrum of Cr^{4+}: YAG extends from 950 nm to 1100 nm. It's convenient to use all-solid-state Nd:YAG laser or Yb doped fiber laser as pump source. The emission spectrum extends from 1250 nm to 1650 nm, as shown in Figure 30. Although the peak is at 1.39 μm, the most reported output wavelengths are near 1.5 μm.

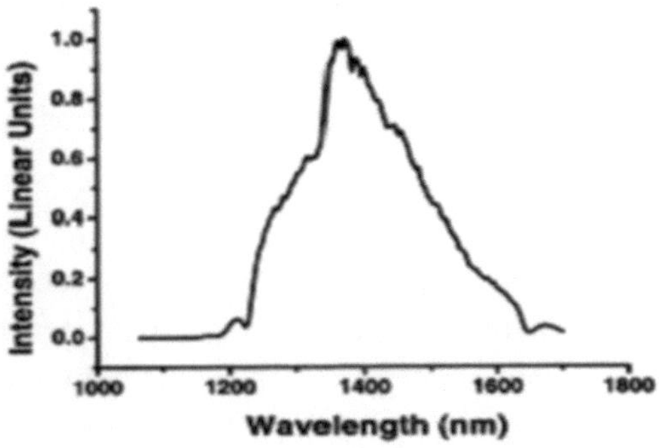

Figure 30. Emission spectrum of Cr^{4+}:YAG

Cr^{4+}:YAG is suitable to generate femtosecond pulses near 1.5 μm for its wide emission spectrum and some other characteristics. There are still some obvious drawbacks though. First, the thermal conductivity is relatively low. The severe thermal lens effect is an annoying problem which causes the unstability of the intra-cavity laser modes. Second, the pump beam and resonant beam both suffer re-absorption effect, which decreases the pump efficiency and increases intra-cavity loss. However, The Cr^{4+}:YAG crystal is still a good gain medium in near infrared range near 1.5 μm. It can support femtosecond pulses generation and has the potential to be used in optical fiber communication, femtobiology, femtochemistry and so on. In this section, a very compact and self-starting femtosecond Cr^{4+}:YAG laser with a pulse width of 65 fs was described [6.17].

The schematic diagram of self-starting mode locking Cr^{4+}:YAG laser is shown in Figure 31.

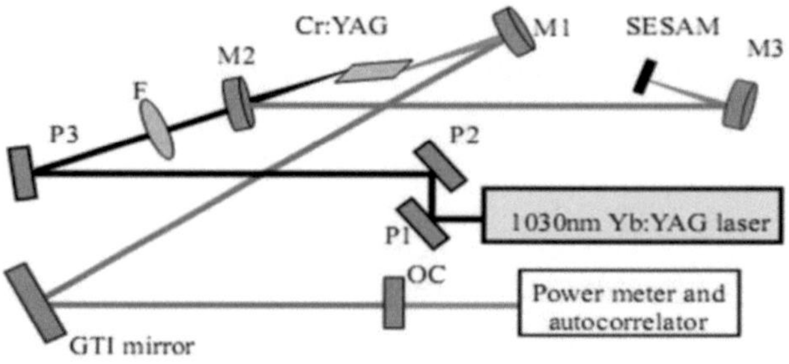

Figure 31. The schematic diagram of the self-starting mode locking Cr^{4+}: YAG laser.

M1, M2 and M3 are concave mirrors with ROC=100 mm, anti-reflective coated (T>98%) at 870~1050 nm and highly-reflective coated (R>99.9%) at 1420~1720 nm. A Gires-Tournois interferometer (GTI) mirror compensates the group-delay dispersion, which can introduce -500±50 fs^2 GDD for each bounce. In order to avoid high intra-cavity loss, an output coupler (OC) with a transmission rate of 1% is used. The pump laser is a commercial Yb:YAG laser which delivers horizontally polarized beam at 1030 nm. The Brewster's angle cut Cr^{4+}:YAG crystal rod is 20 mm long and has a diameter of 5 mm. The laser medium absorbs 90% of the pump energy in this experiment. To reduce the thermal lens effect, the crystal is wrapped in thin indium foil and tightly held in a copper holder. Recirculated cooling water keeps the temperature at 10°C.

For a 20-mm-long Cr^{4+}:YAG, the GDD can be calculated by the following formula [6.12]:

$$GDD = -15296 + 119.83\nu - 0.2054\nu^2 \tag{14}$$

The crystal introduces positive GDD of 226.8 fs^2 at 1500 nm for each single pass. Considering the GDD introduced by air, after compensated by GTI mirror, the net GDD is about -230 fs^2 per single pass in the cavity, as shown in Figure 32. The main drawback by using GTI mirror is that the dispersion compensation isn't precisely adjustable as by using prisms. However, using GTI mirror avoids additional insertion loss and makes the cavity structure more compact and robust.

Under pump power of 9 W, stable mode locked pulses are obtained by fine adjusting the positons of M1, M2, Cr^{4+}:YAG crystal and the SESAM. The highest output power is 95 mW, indicating low slope efficiency. As mentioned before, the re-absorption effect causes large loss for resonant laser. Besides, the OC with low transmission rate also limits the output power. OCs with higher transmission rate are tested, but stable mode locking is unable to achieve.

The mode locking spectra and interferometric autocorrelation trace of output pulses are shown in Figure 33. In Figure a, the solid curve presents the results when the net intra-cavity dispersion is about -230 fs^2. The corresponding spectrum width (FWHM) is 45 nm, centered at 1508 nm. The dashed curve corresponds to the mode-locking results with a net intra-cavity dispersion of -720 fs² by using another mirror to reflect the beam on GTI mirror once more. The spectrum width (FWHM) is 34.5 nm and a small blue shift can be observed. An interferometric autocorrelation trace corresponding to the solid spectrum curve is shown in Figure b. By assuming a sech² pulse shape, the pulse duration is 65 fs, indicating a time-bandwith product of $\Delta\nu\Delta\tau = 0.372$.

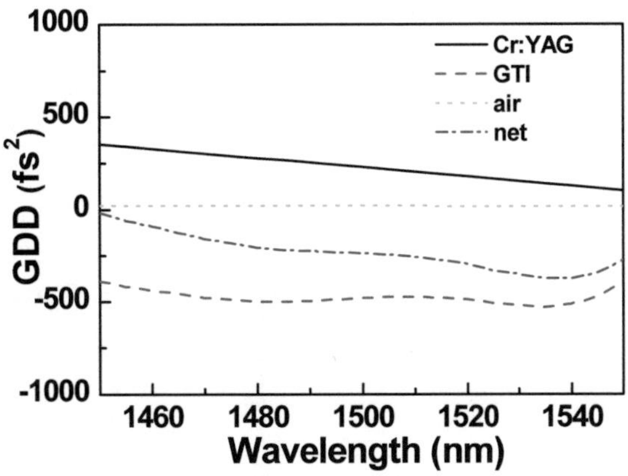

Figure 32. Intra-cavity second-order dispersion contributions.

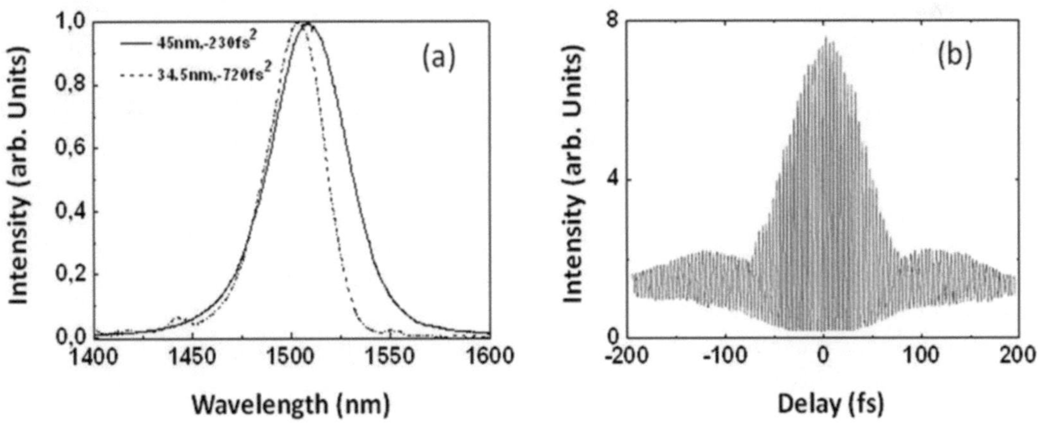

Figure 33. (a) Spectra of mode locking Cr^{4+}:YAG laser, solid curve corresponds to net GDD of -230 fs², dashed curve corresponds to -720 fs²; (b) Interferometric autocorrelation trace of mode locking pulses, the pulse duration is 65 fs.

7. Conclusion

We have described the generation of ultrafast laser pulses with a series of Nd, Yb or Cr-doped gain media. Stable laser pulses were obtained around central wavelengths from 900 nm to 1500 nm and in the picosecond to femtosecond regimes. The pulse duration for Nd-doped mode-locked lasers is in picosecond regime, while for Yb-doped media it can be as short as several tens of femtosecond. The experimental results are summarized in the Table 5. It contains a wide range of very interesting and practically useful all solid-state passively mode-locked ultrafast laser sources in near infrared. Further investigations toward even shorter pulses and higher average power are underway. The direct amplification of all-solid-state ultrafast laser is also in progress toward higher pulse energy.

Gain media	$P_{in}(W)/\lambda_p(nm)$	$P_{out}(mW)$	$\lambda(nm)$	$\Delta\lambda(nm)$	Δt
$Nd:YVO_4$	9W@808	1700	1064		2.3ps
$Nd:GdVO_4$	19.7W@808	128	912		6.5ps
$Nd:LuVO_4$	17W@808	88	916		6.7ps
Nd:GSAG	16.7W@808	510	942.6	0.65	8.7ps
Nd:GGG	24W@808	3200	1062.5	0.25	15ps
	20W@808	320	937.4	0.3	22.7ps
Yb:YAG	2W@940 Ti:S	180	1053	7	170fs
Yb:YAG ceramics	7W@968	1900	1048	3.4	418fs
Yb:GYSO	2W@976 Ti:S	300	1093	6.4	210fs
Yb:LSO	2W@976 Ti:S	70	1047/1066		3.6ps
Yb:YGG	7W@970	570	1045	5.8	245fs
Cr:forsterite	7.9W@1064	760	1277	45	46fs
Cr:YAG	9W@1030	95	1508	45	65fs

Table 5. The experimental results of ultrafast lasers in our lab

Author details

Zhiyi Wei, Binbin Zhou, Yongdong Zhang, Yuwan Zou, Xin Zhong,
Changwen Xu and Zhiguo Zhang
*Beijing National Laboratory for Condensed Matter Physics and Institute of Physics,
Chinese Academy of Sciences, Beijing, China*

Acknowledgments

We thank the helpful technical discussions and support of the laser media by G. L. Bourdet, J.Xu, H.J. Zhang

8. References

[1] U. Keller, K. J. Weingarten, F. X. Kartner, et al, "Semiconductor saturable absorber mirrors (SESAM's) for femtosecond to nanosecond pulse generation in solid-state lasers," IEEE J. Selected Quantum Electron. 2, 435 (1996).

[2] G. J. Spühler, T. Südmeyer, R. Paschotta, M. Moser, K. J. Weingarten and U. Keller, "Passively mode-locked high-power Nd:YAG lasers with multiple laser heads," Appl. Phys. B 71, 19 (2000).

[3] D. Burns, M. Hetterich, A. I. Ferguson, E. Bente, M. D. Dawson, J. I. Davies and S. W. Bland, "High-average-power (>20 W) Nd:YVO_4 lasers mode locked by stain compensated saturable Bragg reflectors," J. Opt. Soc. Am. B 17, 919 (2000).

[4] G. P. A. Malcolm, P. F. Curley and A. I. Ferguson, "Additive pulse modelocking of a diode pumped Nd:YLF laser," Opt. Lett. 15, 1303 (1990).

[5] B. Ileri, C. Czeranowsky, K. Petermann, and G. Huber, "Mixed garnet laser crystals for water vapor detection," in Proceedings of IEEE Conference on Lasers and Electro-Optics Europe, 2005 (IEEE, 2005), p. 10.

[6] T. Kellner, F. Heine, G. Huber, C. Honninger, B. Braun, F. Morier-Genoud, M. Moser, and U. Keller, "Soliton mode-locked Nd:$YAlO_3$ laser at 930 nm," J. Opt. Soc. Am. B 15, 1663 (1998).

[7] A. Schlatter, L. Krainer, M. Golling, and R. Paschotta, D. Ebling, and U. Keller "Passively mode-locked 914-nm Nd:YVO_4 laser," Opt. Lett. 30, 44 (2005).

[8] P. Blandin, F. Druon, F. Balembois, P. Georges, S. Lévêque-Fort and M. P. Fontaine-Aupart, "Diode-pumped passively mode-locked Nd:YVO_4 laser at 914 nm," Opt. Lett. 31, 214 (2006).

[9] J. Aus der Au, D. Kopf, F. Morier-Genoud, M. Moser, and U. Keller, "60-fs pulses from a diode-pumped Nd:glass laser," Opt. Lett. 22, 307 (1997).

[10] C. Hönninger, G. Zhang, U. Keller and A. Giesen, "Femtosecond Yb:YAG laser using semiconductor saturable absorbers," Opt. Lett. 20, 2402 (1995).

[11] M. Weitz, S. Reuter, R. Knappe and R. Wallenstein, "Passive mode-locked 21 W femtosecond Yb:YAG laser with 124 MHz repetition-rate," In Technical Digest of Conference on Lasers and Electro-Optics (optical society of America, Washington, DC, 2004), paper CTucc.

[12] H. Luo, D. Tang, G. Xie, H. Zhang, L. Qin, H. Yu, L. Ng, and L. Qian, "High-power passive mode-locking of a diode pumped Yb:$GdVO_4$ laser," Optics Communications 281, 5382–5384 (2008).

[13] W. Li, Q. Hao, H. Zhai, H. Zeng, W. Lu, G. Zhao, L. Zheng, L. Su, and J. Xu, "Diode-pumped Yb:GSO femtosecond laser," Optics Express 15, 2354–2359 (2007).

[14] F. Thibault, D. Pelenc, F. Druon, Y. Zaouter, M. Jacquemet, and P. Georges, "Efficient diode-pumped Yb^{3+}: Y_2SiO_5 and Yb^{3+}: Lu_2SiO_5 high-power femtosecond laser operation," Opt. Lett. 31, 1555 (2006).

[15] F. Brunner, G. Spühler, J. Au, L. Krainer, F. Morier-Genoud, R. Paschotta, N. Lichtenstein, S. Weiss, C. Harder, A. Lagatsky, and others, "Diode-pumped femtosecond Yb:$KGd(WO_4)_2$ laser with 1.1-W average power," Opt. Lett. 15, 1119–1121 (2000).

[16] F. Brunner, T. Südmeyer, E. Innerhofer, F. Morier-Genoud, R. Paschotta, V. Kisel, V. Shcherbitsky, N. Kuleshov, J. Gao, K. Contag, and others, "240-fs pulses with 22-W average power from a mode-locked thin-disk Yb:KY$(WO_4)_2$ laser," Opt. Lett. 27, 1162–1164 (2002).

[17] U. Griebner, S. Rivier, V. Petrov, M. Zorn, G. Erbert, M. Weyers, X. Mateos, M. Aguiló, J. Massons, and F. Díaz, "Passively mode-locked Yb:KLu$(WO_4)_2$ oscillators," Opt. Express 13, 3465–3470 (2005).

[18] A. Garcia-Cortes, J. M. Cano-Torres, M. Serrano, C. Cascales, C. Zaldo, S. Rivier, X. Mateos, U. Griebner, and V. Petrov, "Spectroscopy and Lasing of Yb-Doped NaY$(WO_4)_2$: Tunable and Femtosecond Mode-Locked Laser Operation," IEEE Journal of Quantum Electronics 43, 758–764 (2007).

[19] F. Druon, S. Chénais, P. Raybaut, F. Balembois, P. Georges, R. Gaum'e, S. Mohr, D. Kopf, and others, "Diode-pumped Yb:$Sr_3Y(BO_3)_3$ femtosecond laser," Opt. Lett. 27, 197–199 (2002).

[20] F. Druon, F. Balembois, P. Georges, A. Brun, A. Courjaud, C. Hönninger, F. Salin, A. Aron, F. Mougel, and others, "Generation of 90-fs pulses from a mode-locked diode-pumped Yb^{3+}:$Ca_4GdO(BO_3)_3$ laser," Opt. Lett. 25, 423–425 (2000).

[21] A. Yoshida, A. Schmidt, V. Petrov, C. Fiebig, G. Erbert, J. Liu, H. Zhang, J. Wang, and U. Griebner, "Diode-pumped mode-locked Yb:YCOB laser generating 35 fs pulses," Opt. Lett. 36, 4425-4427 (2011).

[22] M. Delaigue, V. Jubera, J. Sablayrolles, J. P. Chaminade, A. Garcia, and I. Manek-Hönninger, "Mode-locked and Q-switched laser operation of the Yb-doped $Li_6Y(BO_3)_3$ crystal," Appl. Phys. B 87, 693–696 (2007).

[23] S. Rivier, A. Schmidt, C. Kränkel, R. Peters, K. Petermann, G. Huber, M. Zorn, M. Weyers, A. Klehr, G. Erbert, and others, "Ultrashort pulse Yb:$LaSc_3(BO_3)_4$ mode-locked oscillator," Optics Express 15, 15539–15544 (2007).

[24] M. Lederer, M. Hildebrandt, V. Kolev, B. Luther-Davies, B. Taylor, J. Dawes, P. Dekker, J. Piper, H. Tan, and C. Jagadish, "Passive mode locking of a self-frequency-doubling Yb:$YAl_3(BO_3)_4$ laser," Opt. let. 27, 436–438 (2002).

[25] N. Coluccelli, G. Galzerano, L. Bonelli, A. Di Lieto, M. Tonelli, and P. Laporta, "Diode-pumped passively mode-locked Yb:YLF laser," Optics Express 16, 2922–2927 (2008).

[26] P. Klopp, V. Petrov, U. Griebner, K. Petermann, V. Peters, and G. Erbert, "Highly efficient mode-locked Yb:Sc_2O_3 laser," Opt. let. 29, 391–393 (2004).

[27] U. Griebner, V. Petrov, K. Petermann, and V. Peters, "Passively mode-locked Yb:Lu_2O_3 laser," Opt. Express 12, 3125–3130 (2004).

[28] A. Schmidt, V. Petrov, U. Griebner, R. Peters, K. Petermann, G. Huber, C. Fiebig, K. Paschke, and G. Erbert, "Diode-pumped mode-locked Yb:$LuScO_3$ single crystal laser with 74 fs pulse duration," Opt. let. 35, 511–513 (2010).

[29] F. Druon, F. Balembois, and P. Georges, "Ultra-short-pulsed and highly-efficient diode-pumped Yb:SYS mode-locked oscillators," Opt. Express 12, 5005–5012 (2004).

[30] D. Li, X. Xu, C. Xu, J. Zhang, D. Tang, Y. Cheng, and J. Xu, "Diode-pumped femtosecond Yb: $CaNb_2O_6$ laser," Opt. Lett. 36, 3888–3890 (2011).

[31] Y. Zaouter, J. Didierjean, F. Balembois, G. Lucas Leclin, F. Druon, P. Georges, J. Petit, P. Goldner and B. Viana, "47-fs diode-pumped Yb^{3+}:$CaGdAlO_4$ laser," Opt. Lett. 31,119 (2006).

[32] J. Saikawa, Y. Sato, T. Taira, and A. Ikesue, "Passive mode locking of a mixed garnet Yb: YScAlO ceramic laser," Appl. Phys. Lett. 85, 5845 (2004).
[33] A. Shirakawa, K. Takaichi, H. Yagi, M. Tanisho, J. Bisson, J. Lu, K. Ueda, T. Yanagitani, and A. Kaminskii, "First mode-locked ceramic laser: Femtosecond Yb^{3+}: Y_2O_3 ceramic laser," Laser Phys. 14, 1375–1381 (2004).
[34] M. Tokurakawa, K. Takaichi, A. Shirakawa, K. Ueda, H. Yagi, S. Hosokawa, T. Yanagitani, and A. Kaminskii, "Diode-pumped mode-locked Yb^{3+}:Lu_2O_3 ceramic laser," Opt. Express 14, 12832–12838 (2006).
[35] M. Tokurakawa, K. Takaichi, A. Shirakawa, K. Ueda, H. Yagi, T. Yanagitani, and A. A. Kaminskii, "Diode-pumped 188 fs mode-locked Yb:YO ceramic laser," Appl. Phys. Let. 90, 071101 (2007).
[36] H. Yoshioka, S. Nakamura, T. Ogawa, and S. Wada, "Diode-pumped mode-locked Yb:YAG ceramic laser," Opt. Express 17, 8919–8925 (2009).
[37] H. Yoshioka, S. Nakamura, T. Ogawa, and S. Wada, "Dual-wavelength mode-locked Yb:YAG ceramic laser in single cavity," Opt. Express 18, 1479–1486 (2010).
[38] A. Giesen, H. Hügel, A. Voss, K. Wittig, U. Brauch, and H. Opower, "Scalable concept for diode-pumped high-power solid-state lasers," Appl. Phys. B 58, 365–372 (1994).
[39] A. Giesen and J. Speiser, "Fifteen Years of Work on Thin-Disk Lasers: Results and Scaling Laws," IEEE Journal of Selected Topics in Quantum Electronics 13, 598–609 (2007).
[40] T. Südmeyer, C. Kränkel, C. R. E. Baer, O. H. Heckl, C. J. Saraceno, M. Golling, R. Peters, K. Petermann, G. Huber, and U. Keller, "High-power ultrafast thin disk laser oscillators and their potential for sub-100-femtosecond pulse generation," Appl. Phys. B 97, 281–295 (2009).
[41] C. Kränkel, D. J. H. C. Maas, T. Südmeyer, and U. Keller, "Ultrafast Lasers in Thin-Disk Geometry," High Power Laser Handbook, p. 327, 2011.
[42] R. Peters, C. Kränkel, K. Petermann, and G. Huber, "Power scaling potential of Yb: NGW in thin disk laser configuration," Appl. Phys. B 91, 25–28 (2008).
[43] K. Petermann, D. Fagundes-Peters, J. Johannsen, M. Mond, V. Peters, J. Romero, S. Kutovoi, J. Speiser, and A. Giesen, "Highly Yb-doped oxides for thin-disc lasers," Journal of crystal growth 275, 135–140 (2005).
[44] C. Baer, C. Kränkel, C. Saraceno, O. Heckl, M. Golling, T. Südmeyer, R. Peters, K. Petermann, G. Huber, and U. Keller, "Femtosecond Yb:Lu_2O_3 thin disk laser with 63 W of average power," Opt. lett. 34, 2823–2825 (2009).
[45] C. R. E. Baer, C. Kränkel, C. J. Saraceno, O. H. Heckl, M. Golling, R. Peters, K. Petermann, T. Südmeyer, G. Huber, and U. Keller, "Femtosecond thin-disk laser with 141 W of average power," Opt. Lett. 35, 2302–2304 (2010).
[46] C. Kränkel, J. Johannsen, R. Peters, K. Petermann, and G. Huber, "Continuous-wave high power laser operation and tunability of Yb:$LaSc_3(BO_3)_4$ in thin disk configuration," Appl. Phys. B 87, 217–220 (2007).
[47] O. Heckl, C. Kränkel, C. Baer, C. Saraceno, T. Südmeyer, K. Petermann, G. Huber, and U. Keller, "Continuous-wave and modelocked Yb: YCOB thin disk laser: first demonstration and future prospects," Opt. Express 18, 19201–19208 (2010).

[48] S. Ricaud, A. Jaffres, P. Loiseau, B. Viana, B. Weichelt, M. Abdou-Ahmed, A. Voss, T. Graf, D. Rytz, M. Delaigue, E. Mottay, P. Georges, and F. Druon, "Yb:CaGdAlO$_4$ thin-disk laser," Opt. Lett. 36, 4134–4136 (2011).

[49] N.V. Kuleshov, V.G. Shcherbitsky, V.P. Mikhailov, S. Hartung, T. Danger, S. Kück, K. Petermann, G. Huber, "Excited-state absorption and stimulated emission measurements in Cr^{4+}:forsterite, " J. Lumin. 75, 319 (1997).

[50] C. Chudoba, J. G. Fujimoto, E. P. Ippen, H. A. Haus, U. Morgner, F. X. Kärtner, V. Scheuer, G. Angelow and T. Tschudi, "All-solid-state Cr:forsterite laser generating 14-fs pulses at 1.3 μm," Opt. Lett. 26, 292 (2001).

[51] D. J. Ripin, C. Chudoba, J. T. Gopinath, J. G. Fujimoto, E. P. Ippen, U. Morgner, F. X. Kärtner, V. Scheuer, G. Angelow and T. Tschudi, "Generation of 20-fs pulses by a prismless Cr^{4+}:YAG laser," Opt. Lett. 27, 61 (2002).

[52] http://www.coherent.com.

[53] F. Kallmeyer, M. Dziedzina, X. Wang, H. J. Eichler, C. Czeranowsky, B. Ileri, K. Petermann, and G. Huber, "Nd:GSAG-pulsed laser operation at 943 nm and crystal growth," Appl. Phys. B 89, 305-310 (2007).

[54] C. D. Brandle, Jr. and C. Vanderleeden, "Growth, Optical Properties, and CW Laser Action of Neodymium-Doped Gadolinium Scandium Aluminum Garnet," IEEE J. Quantum Electron. 10, 67 (1974).

[55] S. Wang, X. Wang, F. Kallmeyer, J. Chen, and H. J. Eichler, "Model of pulsed Nd:GSAG laser at 942 nm considering rate equations with cavity structure," Appl. Phys. B 92, 43-48 (2008).

[56] F. Kallmeyer, M. Dziedzina, D. Schmidt, H.-J. Eichler, R. Treichel, and S. Nikolov, "Nd:GSAG laser for water vapor detection by lidar near 942 nm," Proc. SPIE 6451, 64510J (2007).

[57] Changwen Xu, Zhiyi Wei, Yongdong Zhang, Dehua Li, Zhiguo Zhang, X. Wang, S. Wang, H.J.Eichler, Chunyu Zhang and Chunqing Gao, "Diode-pumped passively mode-locked Nd: GSAG laser at 942 nm," Opt. Lett. 34, 2324(2009).

[58] J. E. Geusic, H. M. Marcos, and L. G. Van Uitert, "Laser oscillation in Nd-doped yttrium aluminum, yttrium gallium and gadolinium garnet," Appl. Phys. Lett. 4, 182-184 (1964).

[59] G. F. Albrecht, S. B. Sutton, E. V. George et al, "Solid state heat capacity disk laser ," Laser and Particle Beams 16, 605-625 (1998).

[60] http://www.zlxtech.com.cn/home.asp

[61] M.D. Rotter, C.B. Dane, S. Fochs, K.L. Fortune, R. Merrill, B. Yamamoto, "Solid-state heat-capacity lasers: Good candidates for the marketplace," Photon. Spectra 38, 44-56 (2004).

[62] L. Qin, D. Tang, G. Xie, H. Luo, C. Dong, Z. Jia, H. Yu, and X. Tao, "Diode-end-pumped passively mode-locked Nd:GGG laser with a semiconductor saturable mirror," Opt. Commun. 281, 4762-4764 (2008).

[63] Y. D. Zhang, Z. Y. Wei, C. W. Xu, B. B. Zhou, D. H. Li, Z. G. Zhang, H. H. Jiang, S. T. Yin, Q. L. Zhang, D. L. Sun, "Diode-pumped watt-level mode-locked Nd:GGG laser at 1062 nm," The 7th Asia-Pacitific Laser Symposium, Th-P-78 (2010).

[64] Y. D. Zhang, Z. Y. Wei, C. W. Xu, B. B. Zhou, D. H. Li, Z. G. Zhang, H. H. Jiang, S. T. Yin, Q. L. Zhang, D. L. Sun, "Picoseconds pulse generation with a Nd:GGG laser

operating on quasi-three-level transition," 2009 Lasers & Electro-Optics & the Pacific Rim Conference on Lasers and Electro-Optics, 1-2, 670-671 (2009).

[65] A. Schlatter, L. Krainer, M. Golling, R. Paschotta, D. Ebling and U. Keller, "Passively mode-locked 914-nm Nd:YVO_4 laser," Opt. Lett. 30, 44-46 (2005).

[66] G. J. Spu¨hler, S. Reffert, M. Haiml, M. Moser and U. Keller, "Output-coupling semiconductor saturable absorber mirror," Appl. Phys. Lett. 78, 2733 (2001).

[67] Y. F. Chen, S. W. Tsai, Y. P. Lan, S. C. Wang and K. F. Huang, "Diode-end-pumped passively mode-locked high-power Nd:YVO4 laser with a relaxed saturable Bragg reflector," Opt. Lett. 26, 199-201 (2001)

[68] Ya-Xian Fan, Jing-Liang He, Yong-Gang Wang, Sheng Liu, Hui-Tian Wang and Xiao-Yu Ma, "2-ps passively mode-locked Nd:YVO4 laser using an output-coupling-type semiconductor saturable absorber mirror," Appl. Phys. Lett. 86, 101103 (2005).

[69] Marie-Christine Nadeau, Stéphane Petit, Philippe Balcou, Romain Czarny, Sébastien Montant, and Christophe Simon-Boisson, "Picosecond pulses of variable duration from a high-power passively mode-locked Nd:YVO4 laser free of spatial hole burning," Opt. Lett.35, 1644-1646 (2010).

[70] Y. L. Jia, Z. Y. Wei, J. A. Zheng, W. J. Ling, Y. G. Wang, X. Y. Ma, Z. G. Zhang, "Diode-pumped self-starting mode-locked Nd:YVO4 laser with semiconductor saturable absorber output coupler," Chin. Phys. Lett. 21, 2209 (2004).

[71] Y. L. Jia, W. J. Ling, Z. Y. Wei, Y. G. Wang, X. Y. Ma, "Self-starting passively mode-locking all-solid-state laser with GaAs absorber grown at low temperature," Chin. Phys. Lett. 22, 2575(2005).

[72] A. I. Zagumennyi, V. A. Mikhailov, V. I. Vlasov, A. A. Sirotkin, V. I. Podreshetnikov, Yu. L. Kalachev, Yu. D. Zavartsev, S. A. Kutovoi and I. A. Shcherbakov, "Diode-Pumped Lasers Based on $GdVO_4$ Crystal," Laser Phys. 13, 311-318 (2003).

[73] Jin-Long Xu, Hai-Tao Huang, Jing-Liang He, Jian-Fei Yang, Bai-Tao Zhang, Chun-Hua Zuo, Xiu-Qin Yang and Shuang Zhao, "The characteristics of passively Q-switched and mode-locked 1.06 um Nd:$GdVO_4$ laser with V:YAG saturable absorber," Optical Materials 32, 522 - 525 (2010).

[74] Hou-Ren Chen, Yong-Gang Wang, Chih-Ya Tsai, Kuei-Huei Lin, Teng-Yao Chang, Jau Tang, and Wen-Feng Hsieh, "High-power, passively mode-locked Nd:$GdVO_4$ laser using single-walled carbon nanotubes as saturable absorber," Opt. Lett. 36, 1284-1286 (2011).

[75] Gang Zhang, Shengzhi Zhao, Yufei Li, Guiqiu Li, Dechun Li, Kejian Yang and Kang Cheng, "A dual-loss-modulated Q-switched and mode-locked Nd:$GdVO_4$ laser with AOM and V^{3+}:YAG saturable absorber at 1.34 μm," J. Opt. 13, 035202 (2011).

[76] Kejian Yang, Shengzhi Zhao, Jingliang He, Baitao Zhang, Chunhua Zuo, Guiqiu Li, Dechun Li and Ming Li, "Diode-pumped passively Q-switched and modelocked Nd:$GdVO_4$ laser at 1.34 μm with V:YAG saturable absorber," Optics Express 16, 20176-20185 (2008).

[77] A. K. Zaytseva, C. L. Wangb, C. H. Linc, and C. L. Pana, "Robust Diode-End-Pumped Nd:$GdVO_4$ Laser Passively Mode-Locked with Saturable Output Coupler," Laser Physics 21, 2029 - 2035 (2011).

[78] H. W. Yang, H. T. Huang, J. L. He, S. D. Liu, F. Q. Liu, X. Q. Yang, J. L. Xu, J. F. Yang and B. T. Zhang, "High Repetition Rate Passive Q-Switching of Diode-Pumped

Nd:$GdVO_4$ Laser at 912 nm with V^{3+}:YAG as the Saturable Absorber," Laser Physics 21, 66 – 69 (2011).

[79] Fei Chen, Xin Yu, Xudong Li, Renpeng Yan, Cheng Wang, Deying Chen, Zhonghua Zhang and Junhua Yu, "High power diode-pumped passively Q-switched and mode-locking Nd:GdVO4 laser at 912 nm," Optics Communications 284, 635-639 (2011).

[80] C. Zhang, Z. Y. Wei, L. Zhang, C. Y. Zhang, and Z. G. Zhang, "Passively mode-locked Nd:GdVO4 laser at 912nm," Chinese Physics 15, 2606 (2006).

[81] C.W. Xu, Z.Y. Wei, K.N. He, D.H. Li, Y.D. Zhang, Zhiguo Zhang, "Diode-pumped passively mode-locked Nd:GdVO4 laser at 912 nm," Opt. Commun. 281, 4398 (2008).

[82] He Kun-Na, Wei Zhi-Yi, Xu Chang-Wen, Li De-Hua, Zhang Zhi-Guo, Zhang Huai-Jin, Wang Ji-Yang, Gao Chun-Qing, "Passively Mode-Locked Quasi-Three-Level Nd:LuVO4 Laser with Semiconductor Saturable Absorber Mirror," Chin. Phys. Lett. 25, 4286 (2008).

[83] [5.1] J. Du, X. Liang, Y. Xu, R. Li, Z. Xu, C. Yan, G. Zhao, L. Su, and J. Xu, "Tunable and efficient diode-pumped Yb^{3+}:GYSO laser," Opt. Express 14, 3333 (2006).

[84] W. Li, Q. Hao, L. Ding, G. Zhao, L. Zheng, J. Xu, and H. Zeng, "Continuous-wave and passively mode-locked Yb:GYSO lasers pumped by diode lasers," IEEE J. Quantum Electron. 44, 567 (2008).

[85] B. Zhou, Z. Wei, Y. Zhang, X. Zhong, H. Teng, L. Zheng, L. Su, and J. Xu, "Generation of 210 fs laser pulses at 1093 nm by a self-starting mode-locked Yb:GYSO laser," Opt. Lett. 34, 31 (2009).

[86] W. Yang, J. Li, F. Zhang, Y. Zhang, Z. Zhang, G. Zhao, L. Zheng, J. Xu, and L. Su, "Group delay dispersion measurement of Yb:Gd_2SiO_5, Yb:$GdYSiO_5$ and Yb:$LuYSiO_5$ crystal with white-light interferometry," Opt. Express 15, 8486 (2007).

[87] E. Innerhofer, T. Südmeyer, F. Brunner, R. Paschotta, and U. Keller, "Mode-locked high-power lasers and nonlinear optics-a powerful combination," Laser Phys. Lett. 1, 82 (2004).

[88] S. V. Marchese, C. R. E. Baer, A. G. Engqvist, S. Hashimoto, D. J. H. C. Maas, M. Golling, T. Südmeyer, and U. Keller, "Femtosecond thin disk laser oscillator with pulse energy beyond the 10-microjoule level," Opt. Express 16, 6397 (2008).

[89] C. Hönninger, R. Paschotta, M. Graf, F. Morier-Genoud, G. Zhang, M. Moser, S. Biswal, J. Nees, A. Braun, G. A. Mourou, I. Johannsen, A. Giesen, W. Seeber, and U. Keller, "Ultrafast ytterbium-doped bulk lasers and laser amplifiers," Appl. Phys. B 69, 3 (1999).

[90] S. Uemura, and K. Torizuka, "Center-wavelength-shifted passively mode-locked diode-pumped ytterbium(Yb):yttrium aluminum garnet(YAG) laser," Jpn. J. Appl. Phys. 44, L361 (2005).

[91] B. Zhou, Z. Wei, D. Li, H. Teng, and G. Bourdet, "Numerical and experimental investigation of a continuous-wave and passively mode-locked Yb:YAG laser at a wavelength of 1.05 μm," Appl. Opt. 48, 5978 (2009).

[92] Zhou Bin-Bin, Wei Zhi-Yi, LI De-Hua, Teng Hao, Bourdet G. L, "Generation of 170-fs Laser Pulses at 1053nm by a Passively Mode-Locked Yb:YAG Laser," Chin. Phys. Lett. 26, 054208 (2009).

[93] G. L. Bourdet, "Theoretical investigation of quasi-three-level longitudinally pumped continuous wave lasers," Appl. Opt. 39, 966 (2000).

[94] G. L. Bourdet, and E. Bartnicki, "Generalized formula for continuous-wave end-pumped Yb-doped material amplifier gain and laser output power in various pumping configurations," Appl. Opt. 45, 9203 (2006).

[95] R. J. Beach, "CW theory of quasi-three level end-pumped laser oscillators," Opt. Communications 123, 385 (1995).

[96] K. Takaichi, H. Yagi, J. Lu, A. Shirakawa, K. Ueda, T. Yanagitani, and A. A. Kaminskii, "Yb^{3+}-doped $Y_3Al_5O_{12}$ ceramics – a new solid-state laser material," Phys. Status Solidi A 200, R5 (2003).

[97] A. A. Kaminskii, M. Sh. Akchurin, V. I. Alshits, K. Ueda, K. Takaichi, J. Lu, T. Uematsu, M. Musha, A. Shirakawa, V. Gabler, H. J. Eichler, H. Yagi, T. Yanagitani, S. N. Bagayev, J. Fernandez, and R. Balda, "New results in studying physical properties of nanocrystalline laser ceramics," Kristallografiya 48, 562 (2003).

[98] J. Dong, A. Shirakawa, K. Ueda, H. Yagi, T. Yanagitani, and A. A. Kaminskii, "Efficient Yb^{3+}:$Y_3Al_5O_{12}$ ceramic microchip lasers," Appl. Phys. Lett. 89, 091114 (2006).

[99] S. Nakamura, H. Yoshioka, Y. Matsubara, T. Ogawa, and S. Wada, "Efficient tunable Yb:YAG ceramic laser," Opt. Communications 281, 4411 (2008).

[100] H. Yoshioka, S. Nakamura, T. Ogawa, and S. Wada, "Diode-pumped mode-locked Yb:YAG ceramic laser," Opt. Express 17, 8919 (2009).

[101] B. Zhou, Z. Wei, Y. Zou, Y. Zhang, X. Zhong, G. L. Bourdet, and J. Wang, "High-efficiency diode-pumped femtosecond Yb:YAG ceramic laser," Opt. Lett. 35, 288 (2010).

[102] A. Lucca, G. Debourg, M. Jacquemet, F. Druon, F. Balembois, P. Georges, P. Camy, J. L. Doualan, and R. Moncorgé, "High-power diode-pumped Yb^{3+}:CaF_2 femtosecond laser," Opt. Lett. 29, 2767 (2004).

[103] F. Thibault, D. Pelenc, F. Druon, Y. Zaouter, M. Jacquemet, and P. Georges, "Efficient diode-pumped Yb^{3+}:Y_2SiO_5 and Yb^{3+}:Lu_2SiO_5 high-power femtosecond laser operation," Opt. Lett. 31, 1555 (2006).

[104] Zhou Bin-bin, Zou Yu-wan, Li De-hua, Wei Zhi-yi, Zheng Li-he, Su Liang-bi, Xu Jun, "The experimental study of the continuous-wave mode-locked picosecond Yb:LSO laser," Chinese J. Lasers (in Chinese) 36, 1806 (2009).

[105] I. A. Kamenskikh, N. Guerassimova, C. Dujardin, N. Garnier, G. Ledoux, C. Pedrini, M. Kirm, A. Petrosyan, D. Spassky, "Charge transfer fluorescence and f–f luminescence in ytterbium compounds," Opt. Mater. 24, 267–274 (2003).

[106] S. Heer, M. Wermuth, K. Krämer, and H. U. Güdel, "Sharp ^2E upconversion luminescence of Cr^{3+} in $Y_3Ga_5O_{12}$ codoped with Cr^{3+} and Yb^{3+}," Phys. Rev. B 65, 125112 (2002).

[107] H. Yu, K. Wu, B. Yao, H. Zhang, Z. Wang, J. Wang, Y. Zhang, Z. Wei, Z. Zhang, X. Zhang, and M. Jiang, "Growth and characteristics of Yb doped $Y_3Ga_5O_{12}$ laser crystal," IEEE J. Quantum Electron. 46, 1689-1695 (2010).

[108] Y. Zhang, Z. Wei, B. Zhou, C. Xu, Y. Zou, D. Li, Z. Zhang, H. Zhang, J. Wang, H, Yu, K. Wu, B. Yao and J. Wang, "Diode-pumped passively mode-locked Yb:$Y_3Ga_5O_{12}$ laser," Opt. Lett. 34, 3316-3318 (2009).

[109] Y. D. Zhang, Z. Y. Wei, Z. G. Zhang, D. N. Qian, L. Lv, X. D. Zeng, H. J. Zhang, H. H. Yu, J. Y. Wang, "Diode Pumped Efficient Continuous Wave and Picoseconds Yb:YGG Laser," (In Chinese) Chinese Journal of Lases 38, 0202005 (2011).

[110] S. L. Gilbert, W. C. Swann and T. Dennis, "Wavelength standards for optical communications," Proc. SPIE 4269, 184 (2001).
[111] P. R. Herz, Y. Chen, A. D. Aguirre and et al. "Ultrahigh resolution optical biopsy with endoscopic optical coherence tomography," Opt. Express 12, 3532 (2004).
[112] B. E. Bouma, G. J. Tearney, I. P. Bilinsky and et al. "Self-phase-modulated Kerr-lens mode-locked Cr: forsterite laser source for optical coherence tomography," Opt. Lett. 21, 1839 (1996).
[113] H. M. Crespo, J. R. Birge, E. L. Falcão-Filho and et al. "Nonintrusive phase stablization of sub-two-cycle pulses from a prismless octave-spanning Ti:sapphire laser," Opt. Lett. 33, 833 (2008).
[114] C. Chudoba, J. G. Fujimoto, E. P. Ippen and et al. "All-solid-state Cr:forsterite laser generating 14-fs pulses at 1.3 um," Opt. Lett. 26, 292 (2001).
[115] Zhou Bin-Bin, Zhang Yong-Dong, Zhong Xin, Wei Zhi-Yi, "Highly Efficient Self-Starting Femtosecond Cr:Forsterite Laser," Chin. Phys. Lett. 25, 3679 (2008).
[116] Zhong Xin, Zhou Bin-Bin, Zhan Min-Jie, Wei Zhi-Yi, "Generation of Red Light Femtosecond Pulses from an Intra-Cavity Frequency-Doubled Cr^{4+}:Forsterite Laser," Chin. Phys. Lett. 27, 044204 (2010).
[117] I. Thomann, L. Hollberg, S. A. Diddams and R. Equall, "Chromium-doped forsterite: dispersion measurement with white-light interferometry," Appl. Phys. 42, 1661 (2003).
[118] N. B. Angert, N. I. Borodin, V. M. Garmash and et al. "Lasing due to impurity color centers in yttrium aluminum garnet crystals at wavelength in the range 1.35-1.45 um," Sov. J. Quantum Electronics 18, 73 (1988).
[119] P. M. W. French, N. H. Rizvi, J. R. Taylor and A. V. Shestakov, "Continuous-wave mode-locked Cr^{4+}:YAG laser," Opt. Lett. 18, 39 (1993).
[120] A. Sennaroglu, C. R. Pollock and H. Nathel, "Continuous-wave self-mode-locked operation of femtoseond Cr^{4+}:YAG laser," Opt. Lett. 19, 390 (1994).
[121] Y. Ishida and K. Naganuma, "Characteristics of femtosecond pulses near 1.5 um in a self-mode-locked Cr^{4+}:YAG laser," Opt. Lett. 19, 2003 (1994).
[122] Y. P. Tong, J. M. Sutherland, P. M. W. French and et al. "Self-starting Kerr-lens mode-locked femtosecond Cr^{4+}:YAG and picosecond Pr^{3+}:YLF solid-state lasers," Opt. Lett. 21, 644 (1996).
[123] Z. Zhang, T. Nakagawa, K. Torizuka and et al. "Self-starting mode-locked Cr^{4+}:YAG laser with a low loss broadband semiconductor saturable-absorber mirror," Opt. Lett. 24, 1768 (1999).
[124] S. Naumov, E. Sorokin, V. L. Kalashnikov and et al. "Self-staring five optical cycle pulses generation in Cr^{4+}:YAG laser," Appl. Phys. B 76, 1 (2003).
[125] D. J. Ripin, C. Chudoba, J. T. Gopinath and et al. "Generation of 20-fs pulses by a prismless Cr^{4+}:YAG laser," Opt. Lett. 27, 61 (2002).
[126] Zhou Bin-Bin, Zhang Wei, Zhan Min-Jie, Wei Zhi-Yi, "Self-starting mode-locked Cr^{4+}:YAG laser by Gires-Tournois interferometer mirror for dispersion compensation," Acta Physica Sinica (in Chinese), 57, 1742 (2008).

Chapter 4

Longitudinally Excited CO_2 Laser

Kazuyuki Uno

Additional information is available at the end of the chapter

http://dx.doi.org/10.5772/48525

1. Introduction

In the mid-infrared region (3 – 30 μm), there are only several kinds of commercial laser as shown in Fig. 1. A CO_2 laser is very important because the CO_2 laser emits high output energy and various pulse shapes at the wavelength region between 9.2 μm and 11.4 μm with a high absorptivity in many substances. The first CO_2 laser was developed by C. K. N. Patel in 1964 and had a continuous output power of a few milliwatts (Patel, 1964). In 1968, the pulsed CO_2 laser was developed by A. E. Hill and had a few joules (Hill, 1968). Nowadays, various types of CO_2 lasers have been developed. The characteristics of CO_2 laser pulses are determined by the discharge tube structure (a longitudinal excitation scheme like Fig. 2 (a) or a transversal excitation scheme like Fig. 2 (b)), the discharge types (DC discharge, RF discharge, or pulsed discharge) and the oscillation system (CW oscillation, Q-switched oscillation, or pulsed oscillation).

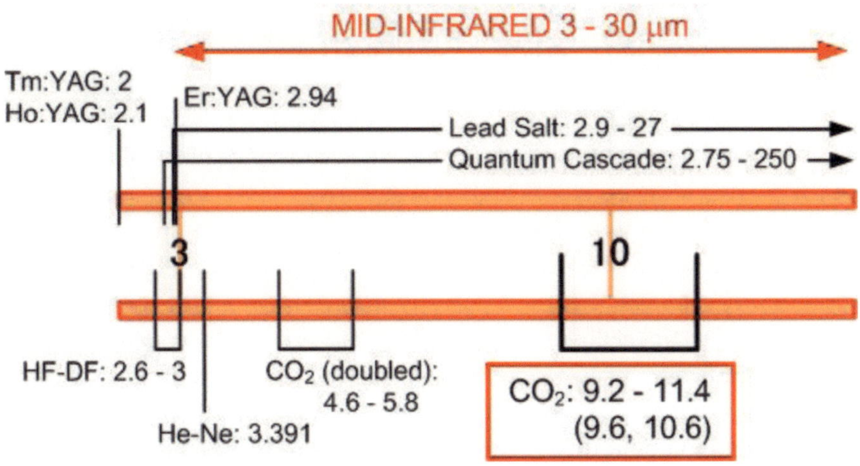

Figure 1. Mid-infrared commercial lasers (Photonics.com). Unit is μm.

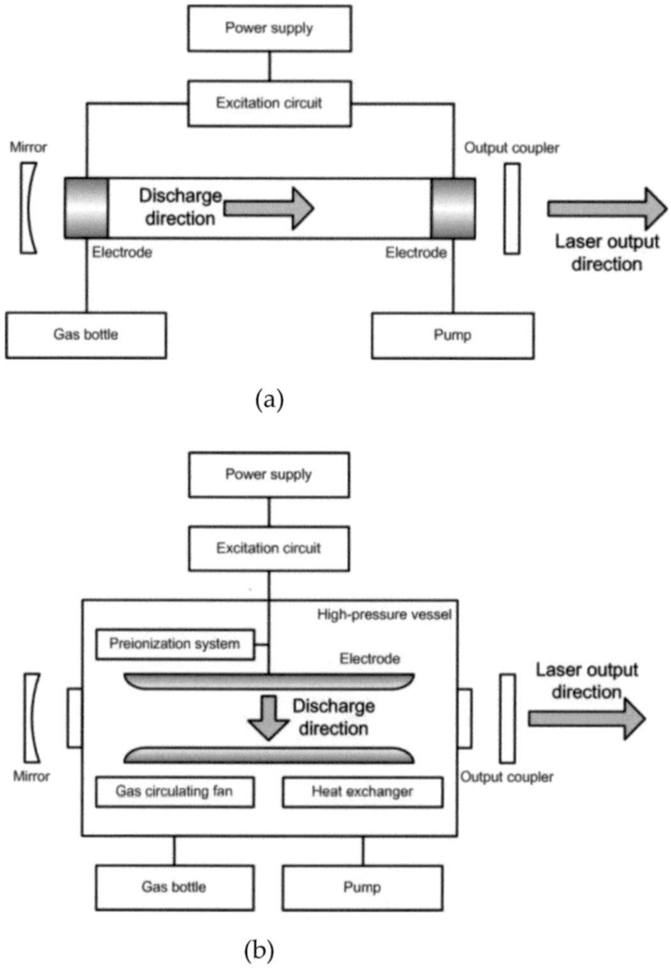

Figure 2. Excitation system. (a) Longitudinal excitation system. (b) Transversal excitation system.

In the CW oscillation and the long pulse oscillation (a pulse width of 10 μs – 10 ms), a low power CO_2 laser (< 10W) is used as a medical laser like tooth and skin treatments. A CO_2 laser emitted 20 – 100 W is used for cutting a nonmetallic substance like a film, cloth, leather, paper, rubber, a plastic, acrylics, etc. A high power laser emitted 100 – 400 W is used for cutting metal, glass, and ceramics. A short pulse CO_2 laser that emits an output energy of a few joules and a pulse width of about 100 ns is used for puncturing and patterning of a nonmetallic substance, removal of surface polymer material which does not give a damage to a metal board, and etc. Specially, a large and high output CO_2 laser called "Lekko VIII" that emitted an output energy of 10 kJ and a pulse width of 1 ns was developed in Institute of Laser Engineering, Osaka University, Japan, 1981 (Yamanaka et al, 1981). Recently, the CO_2 laser attracts attention as a driver laser for a Laser-Produced-Plasma Sn-EUV source (13.5 nm) and a excitation laser for a Terahertz laser (CH_3F, C_2H_2Cl, CH_3OH, D_2O, and etc., 30 – 300 μm) (Ueno et al, 2007; He et al, 2010).

2. Principle of CO_2 laser

The CO_2 laser oscillates in the vibrational level of CO_2 molecular. The CO_2 laser has about 100 oscillation lines at the center of 9.6 μm and 10.6 μm between 9.2 μm and 11.4 μm. Fig. 3 shows the energy diagram of CO_2 laser. Generally, the CO_2 laser is pumped by glow discharge in the low mixed gas (CO_2, N_2, He, and etc.) pressure. The N_2 serves to improve the output energy and the efficiency due to the energy transfer from the vibrational level of N_2 to the upper laser level of CO_2. However, in the short pulse oscillation, N_2 causes a laser pulse tail because of the long lifetime of the vibrational level of N_2. The He serves to make the discharge uniform. The CO_2 molecular is efficiently pumped by very low electron temperature because the energy of the laser upper level 001 is 2349.2 cm^{-1} and the energy required for excitation of CO_2 molecule is about 0.28 eV. However, generally, the CO_2 molecular pumped by the higher electron temperature than 0.28 eV because of an electrical gradient to sustain the discharge. The CO_2 laser oscillates at 10.6 μm band (001 – 100 transition) and 9.6 μm band (001 – 020 transition) and has about 100 oscillation lines because of the P branch and the R branch related to the rotational level. The laser lower levels of 100 and 020 are relaxed by the collision with the CO_2 molecule of ground state. The 010 level serves as a bottleneck. But the thermal relaxation of the 010 level is accelerated by the collision with other molecular, He atom and the laser tube wall. Then, the 010 level returns to the ground state. Therefore, in the efficiently laser oscillation, the N_2 molecular that supplies the upper laser level of CO_2 and media like He that eases the lower laser level of CO_2 are essential. However, in the low-pressure pulse-discharge at the low repetitive operation, He may be unnecessary.

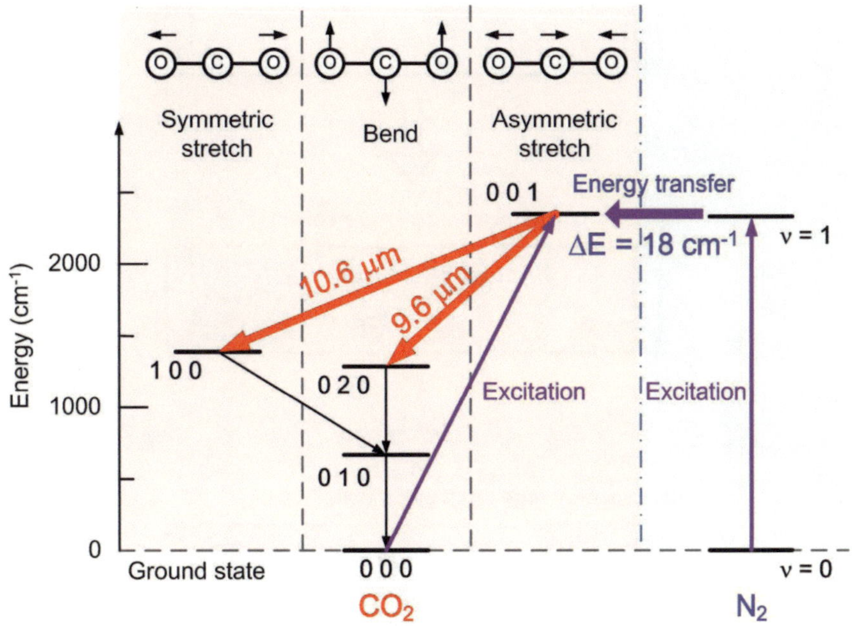

Figure 3. Energy diagram of CO_2 laser.

3. Longitudinal excitation scheme

In the longitudinal excitation scheme, the excitation discharge is in the direction of the laser axis. This system was used in the early stage of development of lasers such He-Ne lasers (633 nm), N_2 lasers (337 nm) and CO_2 laser (9.2 – 11.4 μm). In this system, a dielectric tube with the small inner diameter (1 – 2 cm in CO_2 lasers) and the long length (> 30 cm in CO_2 laser) is used as a discharge tube. Metallic electrodes are attached to each end of the tube. The long discharge length provides a high breakdown voltage (> 20 kV) at a low gas pressure (< 10 kPa). Therefore, this system does not require a high gas-pressure vessel. Moreover, this laser may oscillate at narrow spectral width because the low-pressure operation produces the small pressure broadening. A uniform discharge can be obtained easily because of the fast electron drift velocity (the long mean free path) of the low pressure operation; additionally, the discharge uniformity is not affected by residual charges in the longitudinal excitation because a discharge takes place in a discharge tube with a long length and small inner diameter. In a RF discharge, an electronic trapping occurs. A uniform discharge takes place because electrons vibrate by the high-frequency electric field and are trapped between the discharge gaps. In a pulsed discharge, a uniform discharge takes place by a diffuse streamer discharge is formed by diffusing uniformly and progressing the minute spark discharge with which an avalanche and optical ionization are combined (Fig. 4).

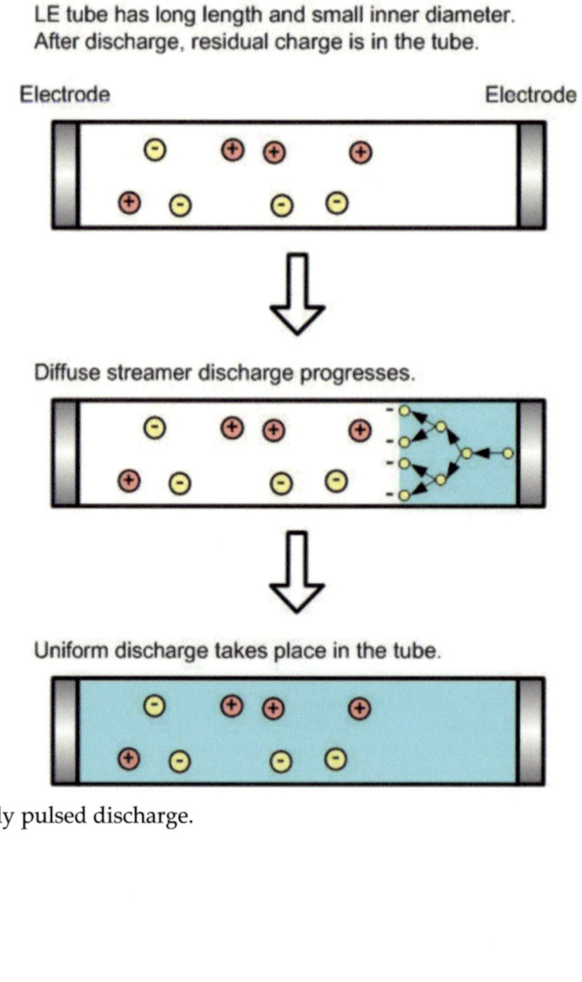

Figure 4. Longitudinally pulsed discharge.

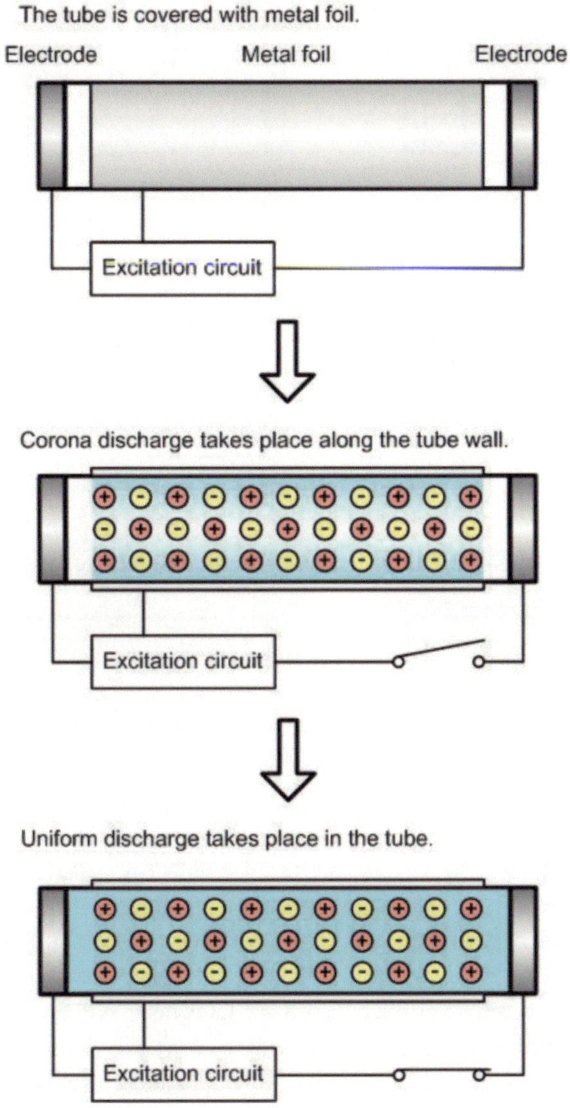

Figure 5. Longitudinally pulsed discharge with strong preionization.

Therefore, the longitudinal excitation scheme does not require a pre-ionization device like UV and X-ray source. On the other hand, when the pre-ionization is required in the high-pressure operation (> 1 atm), the discharge tube is covered with Al foil (Fig. 5, 6) (El-Osealy et al, 2002b; Uno et al, 2009; Uno et al, 2012a). A corona discharge occurs along the tube wall before the main discharge to serve as pre-ionization. Therefore, nothing is put in the discharge tube. In the longitudinal excitation scheme without pre-ionization, a gas lifetime is long. The gas lifetime is a phenomenon in which the laser output energy decreases with operation time in a sealed-off operation. Impurities increases with time in the discharge tube, UV and X-ray for the pre-ionization cannot reach the center part of the discharge space, the uniform discharge cannot take place, the localization of the discharge occurs, and

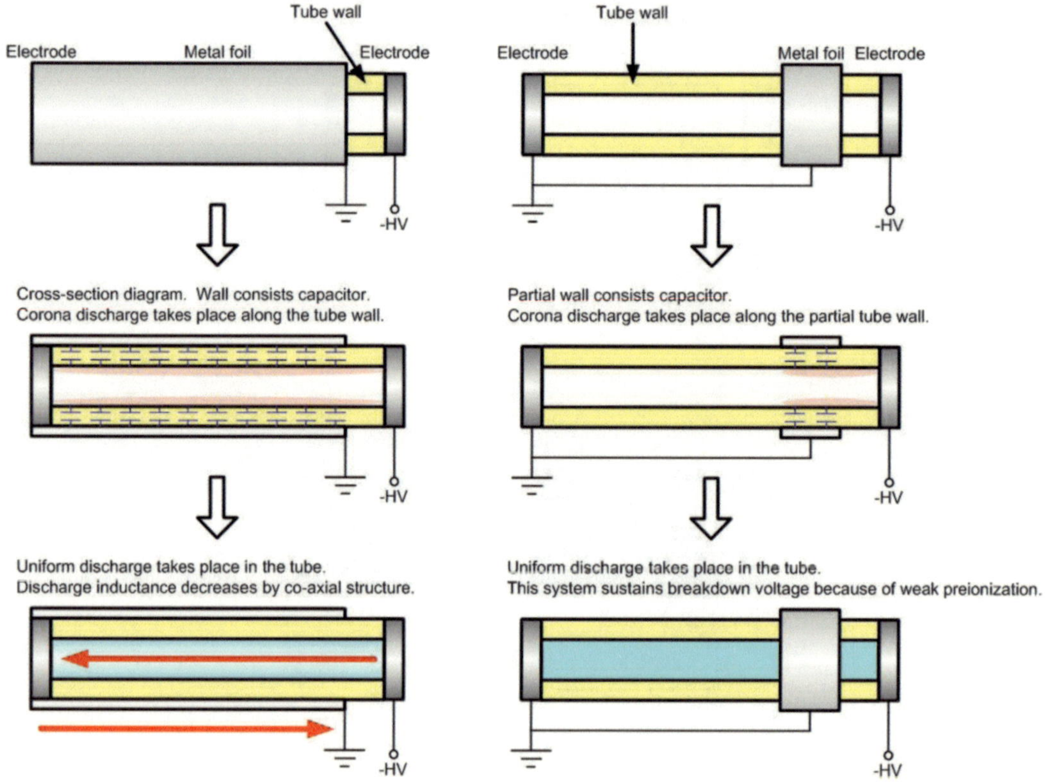

Figure 6. Longitudinally pulsed discharge with simple preionization.

the laser output energy decreases. Therefore, this scheme without the pre-ionization is suitable for the long lifetime operation. Additionally, in this scheme, the perfect hard-sealed discharge tube without a gasket packing like a rubber O-ring is producible because the structure of the discharge tube is simple. Therefore, this scheme is suitable for the long lifetime gas laser. In fact, a low-output sealed-off CO_2 laser has a CW oscillation or a high repetition oscillation without a gas-flowing system and water-cooling system effectively. Additionally, in this scheme, the beam section of the laser is circular because the discharge tube is used a dielectric tube like a glass tube and a ceramics tube. All the optical elements may be circular. This scheme produces laser beam with good beam quality because the length of optical cavity is long enough to the diameter of aperture (the inner diameter of the discharge tube). Therefore, in the longitudinal excitation scheme, a high performance laser can be developed with a simple, portable, and low-cost device.

In the longitudinal excitation scheme, as shown in Table 1, various laser oscillation from mid-infrared to vacuum ultraviolet has been reported. In recent years, especially, new gas lasers, for example, a longitudinally excited CO_2 laser with short laser pulse like TEA and Q-switched CO_2 laser (Uno et al, 2009), a longitudinally excited N_2 laser with the same excitation system as an excimer lamp (Uno et al, 2006), and a longitudinally excited F_2 laser at low-pressure (total pressure of 40 Torr) (El-Osealy et al, 2002b) have been developed.

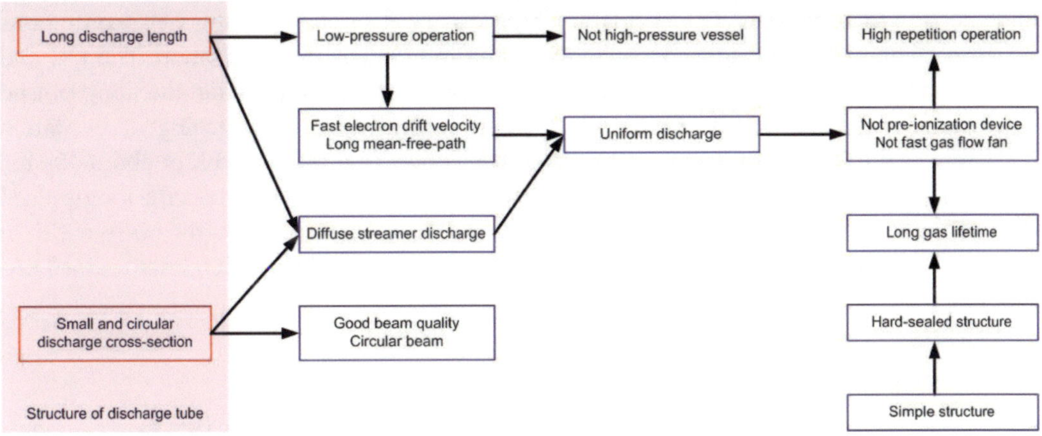

Figure 7. Characteristics of longitudinal excitation system.

Laser	Wavelength		Example of reference
CO_2 laser	Mid-infrared	9.2 – 11.4 μm	Patel, 1964; Hill, 1968; Chung et al, 2002; Uno et al, 2009
Xe laser	Near-infrared	1.73 – 3.51 μm	Komatsu et al, 1991
F laser	Visible	630 – 780 nm	Hocker & Phi, 1976; Uno et al, 2008
He-Ne laser		544, 594, 612, 633 nm	Javan et al, 1961
XeF laser	Ultraviolet	351 nm	Burkhard et al, 1981; Cleeschinsky et al, 1981
N_2 laser		337 nm	El-Osealy et al, 2001; Uno et al, 2006; Uno et al, 2008
XeCl laser		308 nm	Zhou et al, 1983; Furuhashi et al, 1987
KrF laser		248 nm	Newman, 1978; Eichler et al, 1985
ArF laser	Vacuum-ultraviolet	193 nm	Rosa et al, 1986
F_2 laser		157 nm	El-Osealy et al, 2002a El-Osealy et al, 2002b

Table 1. Longitudinally excited gas lasers.

4. Excitation system

The shape of laser pulse depends on the discharge and oscillation types. The discharge type can be divided into the DC (direct current) discharge, the RF (radio frequency) discharge and the pulsed discharge. The oscillation type can be grouped into the CW (continuous wave) oscillation and the pulsed oscillation including Q-switched oscillation. The DC and RF discharge provides the CW oscillation basically and can provide the long-pulsed oscillation (the pulse width from several μs to several ms) by switching of excitation discharge and the short-pulsed (giant-pulsed) oscillation (the pulse width of about 100 ns) by using Q switch. Although it is well known the pulsed discharge provides the long-pulsed oscillation, the pulsed discharge can provide the short-pulsed oscillation (the pulse width of about 100 ns and the pulse tail of several tens μs) like the Q-switched CO_2 laser and a TEA-CO_2 laser.

4.1. DC discharge

Figure 8 shows a general longitudinally excited CO_2 laser pumped by DC discharge. The DC-CO_2 laser commonly is used a high-voltage (a few tens kV) DC power supply and emits the CW power of several tens kW. The output power can control delicately by adjustment of DC voltage.

A discharge tube is made of a heat-resistant hard glass pipe with an inner diameter of several mm and is the co-axial dual structure for cooling by water. Cooling improves the efficiency and output power of the laser oscillation by decreasing the plasma temperature. Additionally, high output power can be provided by a multi-path system folding back the optical path in the resonator and a multi-beam system which uses multi-tubes. High power DC-CO_2 laser emitting CW power of 1 kW using 20 discharge tubes with the inner diameter of 9 mm and the length of 195 cm has been reported (Deshmukh & Rajagopalan, 2003).

A low power CO_2 laser can operate in a sealed-off mode. In a long-life sealed-off CO_2 laser, a gold catalyst is used to regenerate CO_2 from the dissociation products formed during discharge (Tripathi et al, 1994). In the sealed-off CO_2 laser, the laser beam (output power and beam quality) is stable because discharge and the distribution of particles are uniform. In a high power CO_2 laser, a gas-flowing system circulating fresh gas is used for preventing the decrease of the oscillation efficiency by the deterioration of gas. The longitudinal excitation system has only an axially flowing type.

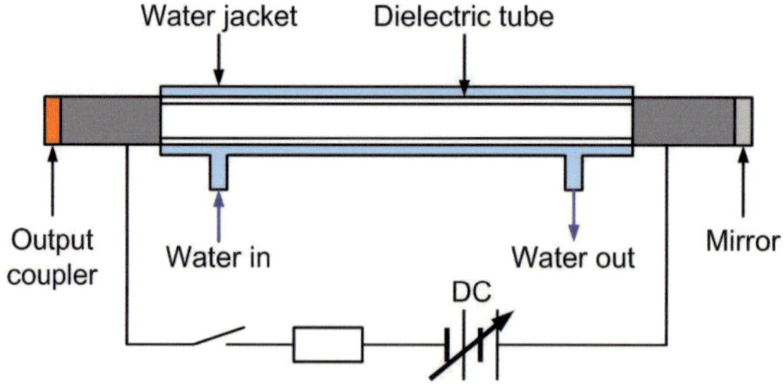

Figure 8. Discharge tube with water jacket.

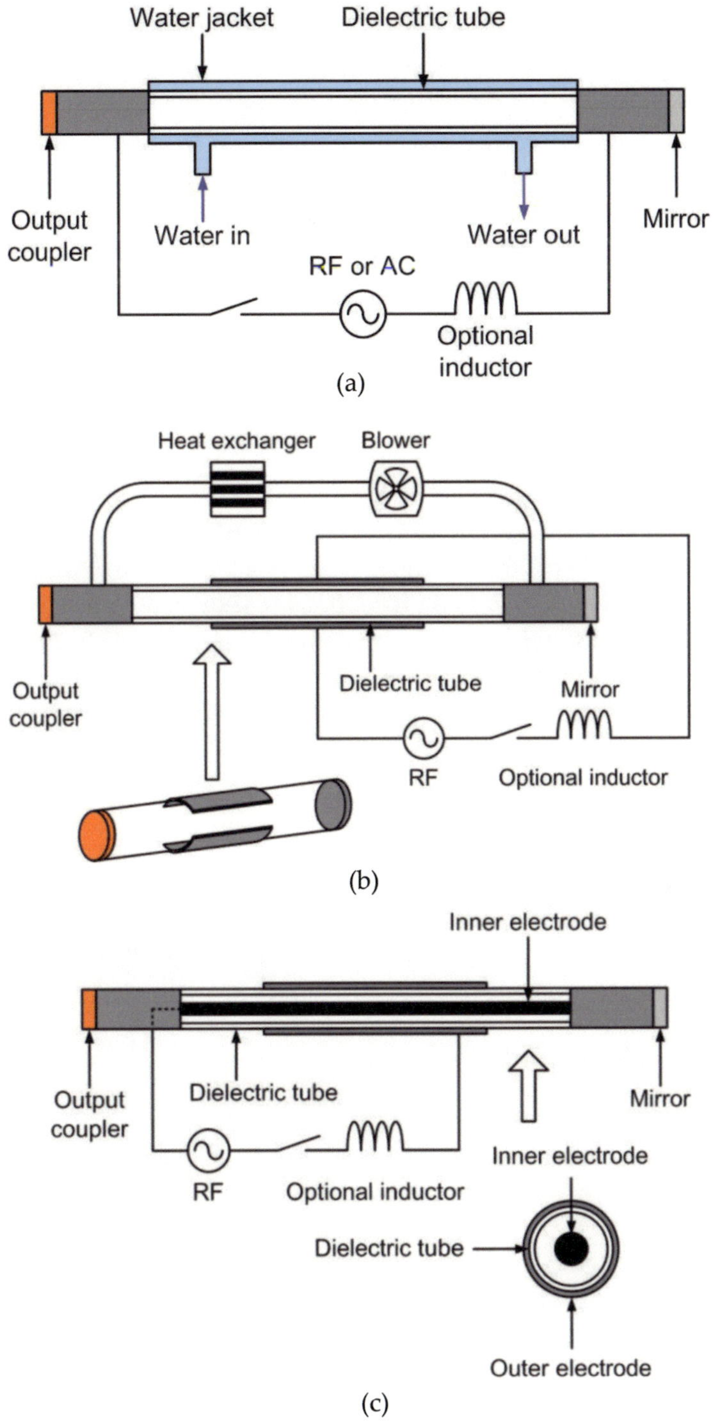

Figure 9. RF discharge tube. (a) Longitudinal excitation (Lee et al, 2000). (b) Transverse excitation (Terai et al, 1993). (c) Coaxial RF-CO_2 laser (Bethel, 1998).

4.2. RF discharge

The first CO_2 laser by C. K. N. Patel used the RF discharge excitation, 27 MC/s (Patel, 1964). In the RF discharge excitation, the medium gas is pumped by a high-frequency voltage (typically 13.56 MHz). In this system, electrodes are able to be put on the outer wall of discharge tube. Thus, the electrodes are not sputtered and there are little degradation of gas and contamination of optical components. A compact high-output CO_2 laser can be developed by high discharge input-energy per unit volume. An electronic trapping provides good pulse characteristics.

However, generally, the RF discharge excitation scheme uses the discharge tube like Fig. 9. Fig. 9 (b) looks like the longitudinal excitation scheme, but is the transversal excitation scheme because the direction of discharge is perpendicular to the direction of a laser axis.

In the low pressure region less than 10 kPa, the laser pumped by RF wave more than 0.1 MHz emits CW and the laser pumped by RF wave less than 10 kHz emits pulse shapes. Generally, the pulse oscillation of RF-CO_2 lasers is realized by switching of excitation discharge and the duty ratio of ON and OFF (Nagai, 2000). Fig. 10 (a) is a example of normal pulses by chopping the CW output. In the axial flowing system, the gas existence time in the discharge tube is long and the rise of gas temperature is large. It is difficult to keep the peak value of a pulse constant for a long time. Therefore, a narrow pulse shape called an enhanced pulse like Fig. 10 (b) or a super pulse like Fig. 10 (c) takes place.

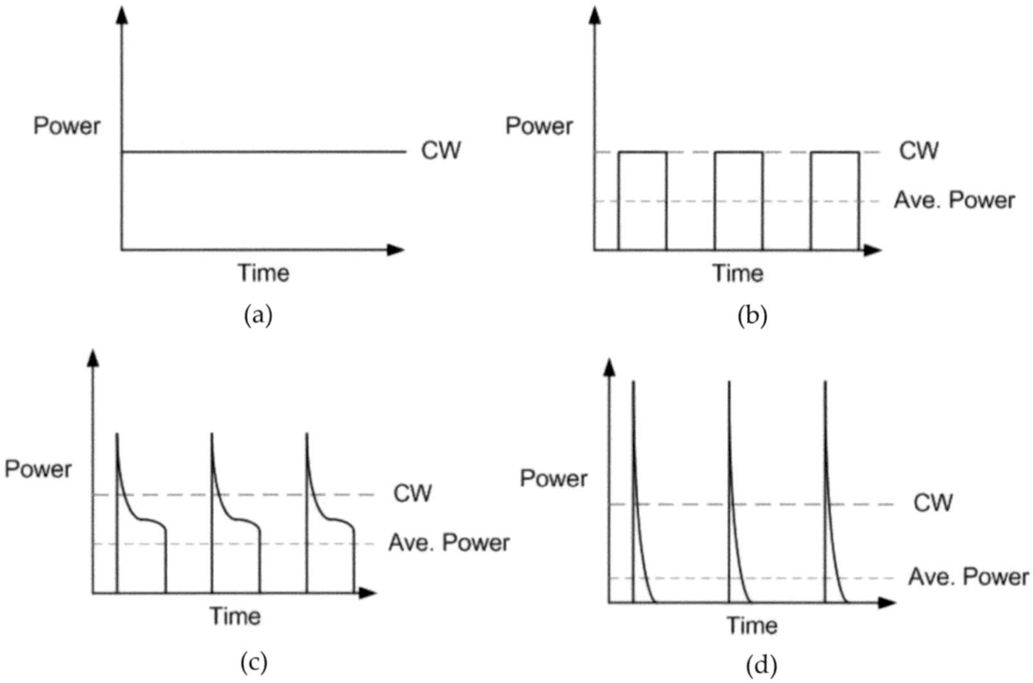

Figure 10. Images of RF-CO_2 laser output (Nagai, 2000). (a) CW. (b) Normal pulses. (c) Enhanced pulses. (d) Super pulses.

In the slab RF-CO_2 laser, commercial sealed-off CO_2 lasers produces the broad output range from CW 20 W to CW 1 kW and the high peak pulse oscillation of 2.5 kW at the pulse width 50 μs at 1 kHz (Coherent, Inc., DIAMOND E-1000). In a compact fast axial flow CO_2 laser, the output power of CW 2.1 kW has been reported (Biswas et al, 2010). Additionally, a low-power CO_2 laser pumped by AC discharge of 60 Hz whose frequency is lower than RF discharge has also been reported (Lee et al, 2000).

4.3. Q-switch

A Q switch is a system to produce a giant pulse with a short pulse and a high peak power. A Q value is a standard of the resonator performance and is a ratio of a loss energy to a storage energy in the cavity. It is hard to carry out a laser oscillation when a Q value is low, and it is easy when a Q value is high. The Q switch rapidly decreases the high resonator losses, that is, rapidly increases the Q value.

A CW-CO_2 laser using a Q switch produces a giant pulse with a pulse width from a few tens ns to a few μs and a peak power of several kW. In a CO_2 laser, the use of a mechanical chopper, rotating mirror (Battou et al, 2008), electrooptical (CdTe) (Tian et al, 2005) or acoustooptical modulators (Xie et al, 2010), or saturable absorber (SF_6) (Soukieh et al, 1999) has been reported.

4.4. Pulsed discharge

A CO_2 laser is easily oscillated by applying a high voltage pulse of about 20 kV to a longitudinal discharge tube filled by CO_2 gas. Fig. 11 shows a longitudinally excited pulsed CO_2 laser with a general and easy excitation circuit (Chung et al, 2002; Loy & Roland, 1977). A voltage pulse of several hundreds V is generated by a power supply and is fed to a step-up transformer. The high-voltage pulse of about 20 kV is directly applied to the discharge tube. The high-voltage pulse has a long rise time of about a few hundred μs because the transformer with a large inductance is directly connected to the discharge tube. When the applied voltage reaches the breakdown threshold, a discharge starts and the laser oscillates. The laser output energy depends on the discharge volume and the input energy. The laser pulse has a long width from several tens μs to several ms because the excitation circuit produces a long and high voltage pulse and the discharge formation time is long. For example, the discharge tube with the length of 80 cm applied the high voltage pulse of 25 kV and produced the output power of 35 W and the width of 3 ms at 60 Hz (Chung et al, 2002).

The longitudinally excited CO_2 laser produces a short laser pulse like TEA-CO_2 laser and Q-switched CO_2 laser by a fast discharge. Fig. 12 shows a CO_2 laser with a capacitor transfer circuit that is used for transversal excited excimer lasers and is known as a fast discharge circuit (Miyazaki et al, 1986). A low-inductance storage capacitor C_s was charged up to DC voltage of about 20 kV. A spark gap was switched by a trigger pulse from a trigger circuit, and the high voltage was transferred to a buffer capacitance C_b. When the voltage reached

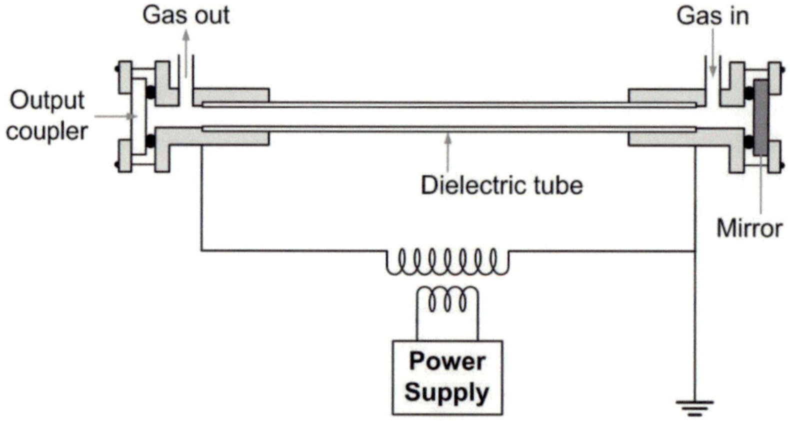

(a)

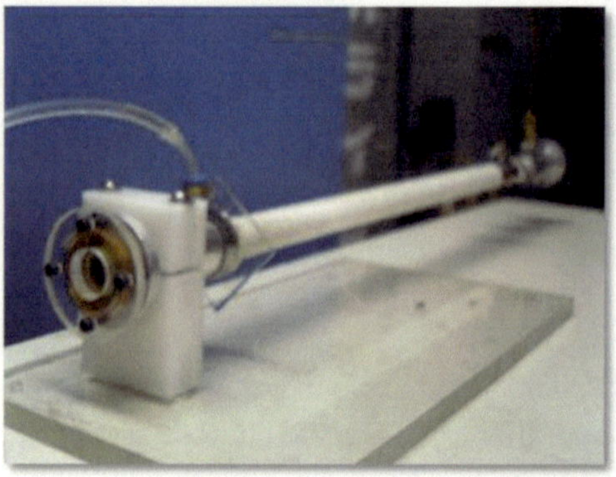

(b)

Figure 11. Image of longitudinally excited long-pulse CO_2 laser. (a) Schematic diagram. (b) Photograph.

the breakdown threshold, a rapid main discharge took place in the discharge tube. The high-voltage pulse of a rise time less than 100 ns is applied to the discharge tube. Additionally, a high-voltage resistor is connected in parallel with the discharge tube. A low resistance acts as a shunt resistance and provides a rapid discharge.

For example, in our study (Uno et al, 2009), the discharge tube was a ceramic pipe made of alumina, with an inner diameter of 13 mm, an outer diameter of 17 mm, and a length of 45 cm. The discharge tube was covered with an Al sheet. Therefore, the laser tube had a coaxial structure for reducing the discharge impedance. An optical cavity was formed by a ZnSe output coupler with a reflectivity of 80% and a high-reflection mirror with a radius of curvature of 20 m. The distance between the output coupler and the mirror (i.e., the cavity length) was 54 cm. In the excitation circuit, the storage capacitance C_s was 5.4 nF and charged up to DC 20 kV, and the buffer capacitance C_b was 2.8 nF. The fall time of the discharge voltage decreased from 32.8 μs (10 MΩ) to 0.97 μs (100 Ω). In the fast discharge, the gain increases gradually from the start of discharge. When the gain becomes enough

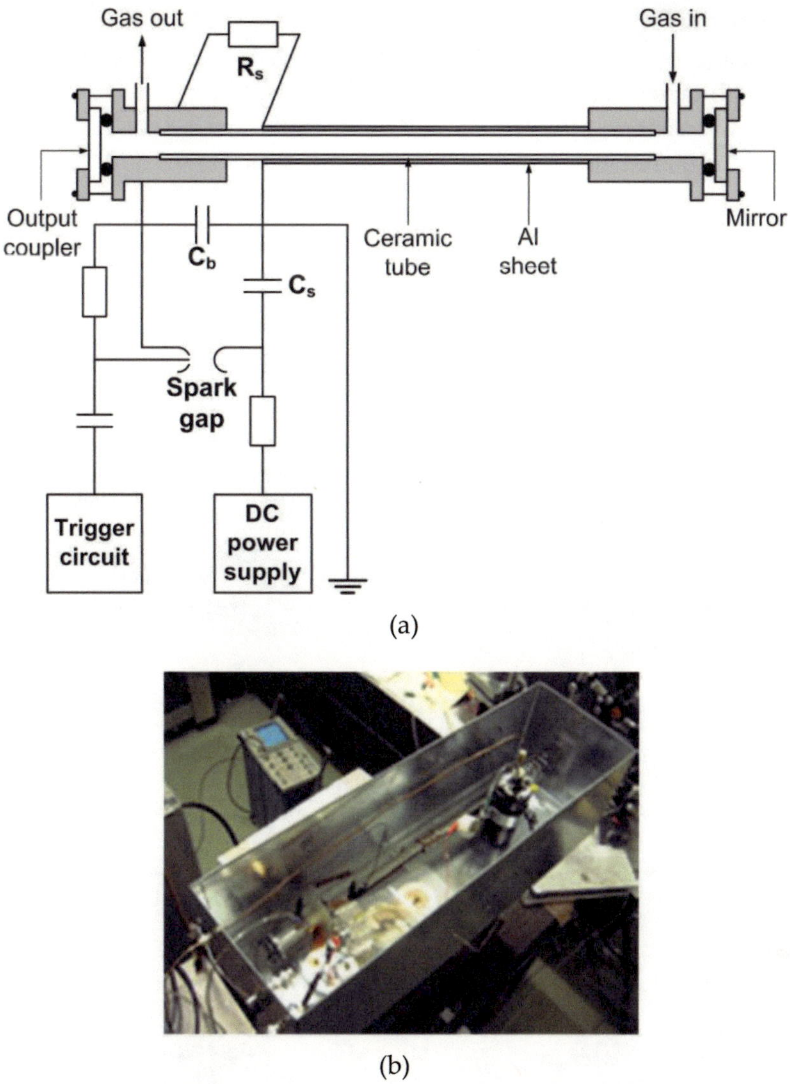

Figure 12. Longitudinally excited short-pulse CO_2 laser with capacitor-transfer circuit (Uno et al, 2009). (a) Schematic diagram. (b) Photograph.

after several μs from the start of discharge, the laser oscillates by a gain Q switch and a spike laser pulse is formed. At that time, the main discharge finishes mostly. Low current after the main discharge and the energy transfer of the upper-level N_2 with the long lifetime causes the laser pulse tail of several tens μs. The pulse tail can be eliminated by eliminating the after current or using of pure CO_2 gas to eliminate the long N_2 tail. Fig. 13 and 14 shows the laser pulse in the discharge formation time of 32.8 μs and 0.97 μs, respectively, at the same discharge tube with the inner diameter of 13 mm and the length of 45 cm and the same gas pressure of 2.9 kPa (CO_2: N_2: He= 1: 1: 2) (Uno et al, 2009). Fig. 13 shows the laser pulse with the spike pulse width of 103.3 ns, the pulse tail length of 61.9 μs, and the output energy of 50 mJ by the discharge formation time of 32.8 μs. The pulse tail length was defined from the end of the spike pulse to the end of the pulse tail. Fig. 14 shows the laser pulse with the spike pulse width of 96.3 ns, the pulse tail length of 17.2 μs, and the output energy of 30 mJ by the discharge formation time of 0.97 μs. These results showed the fast discharge caused the decrease of the pulse tail.

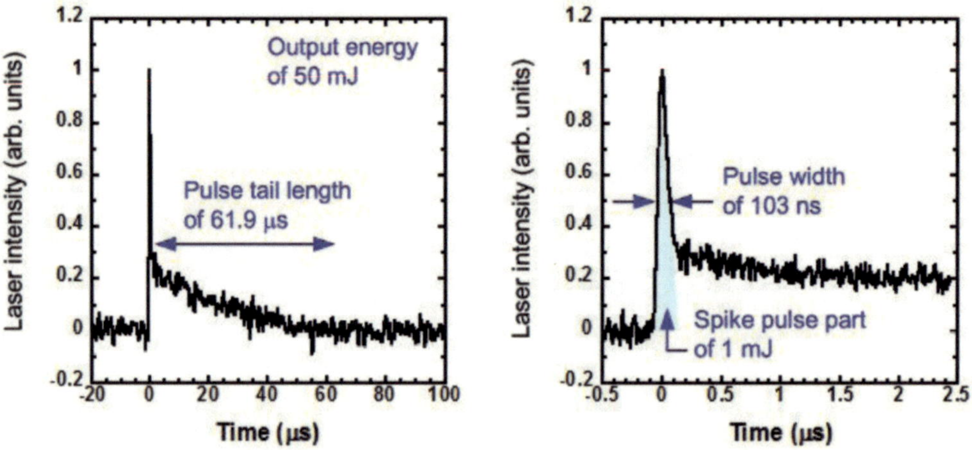

(a) (b)

Figure 13. Laser pulse waveforms at slow discharge (discharge formation time of 32.8 μs and gas pressure of 2.9 kPa (CO_2: N_2: He= 1: 1: 2)) (Uno et al, 2009). Intensity of vertical axis is normalized by peak strength. (a) Overall waveform. (b) Magnified time scale view of spike pulse.

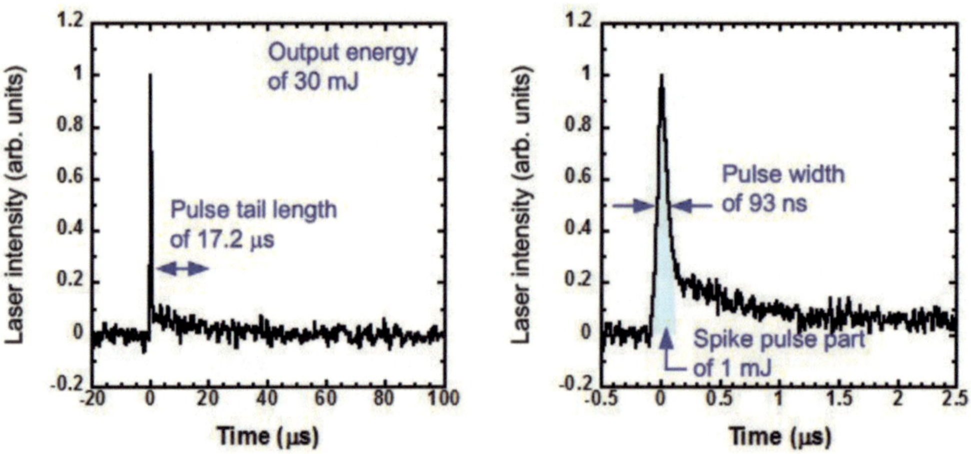

(a) (b)

Figure 14. Laser pulse waveforms at fast discharge (discharge formation time of 0.97 μs and gas pressure of 2.9 kPa (CO_2: N_2: He= 1: 1: 2)) (Uno et al, 2009). Intensity of vertical axis is normalized by peak strength. (a) Overall waveform. (b) Magnified time scale view of spike pulse.

The capacitor-transfer circuit like Fig. 12 developed for the transversely excited excimer laser (Miyazaki et al, 1986). The capacitor-transfer circuit has a buffer capacitance in parallel with the laser tube. The buffer capacitance functions to increase the applied voltage due to the storage capacitance. However, because the transverse excitation scheme has a narrow electrode gap and a small discharge impedance, a buffer capacitance is required to sustain the discharge. On the other hand, the longitudinal excitation scheme has a wide electrode gap and a large discharge impedance and does not require a buffer capacitance to sustain the discharge. Therefore, a circuit without a buffer capacitance, called a direct-drive circuit like Fig. 15, has been developed (Uno et al, 2012a). Fig. 15 shows a longitudinally excited CO_2 laser with a direct-drive circuit (Uno et al, 2012a). A negative several hundreds V pulse was generated by the power supply with a silicon-controlled rectifier and was fed to a

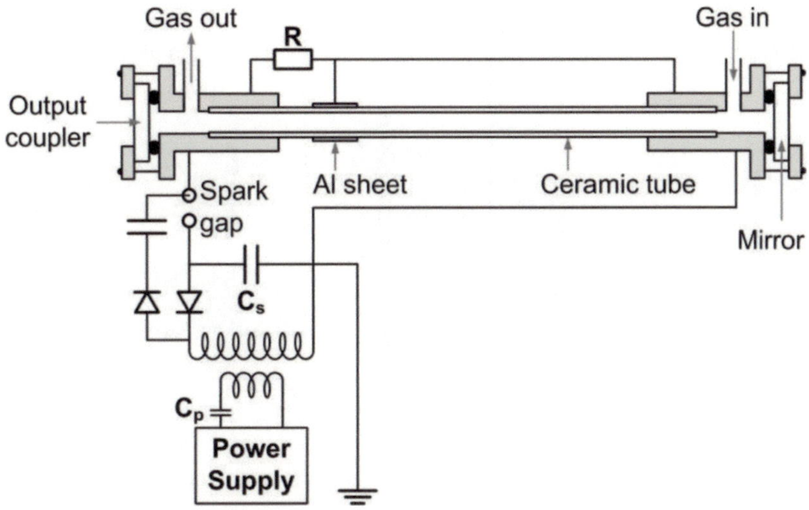

(a)

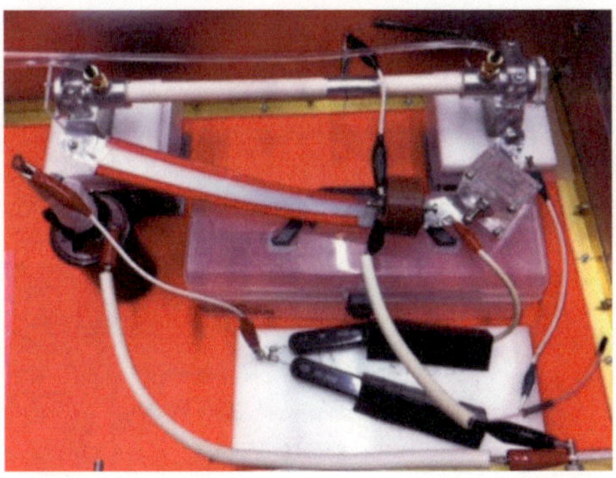

(b)

Figure 15. Longitudinally excited short-pulse CO_2 laser with direct-drive circuit (Uno et al, 2012a). (a) Schematic diagram. (b) Photograph.

transformer, which had a primary capacitance. First, a negative voltage pulse was generated and was used to charge a storage capacitor through a rectifier. Then, a positive voltage pulse generated by the overshoot of the transformer was applied to the trigger electrode of the spark gap through a rectifier and a small capacitor. The storage capacitor was charged, the spark gap was switched, and the high voltage was applied to the laser tube. When the voltage reached the breakdown threshold, a rapid discharge took place in the laser tube. This system produces a short pulse CO_2 laser with a spike pulse.

For example, in our study (Uno et al, 2012a), the laser tube was the almost same as the above longitudinally excited CO_2 laser with the capacitor-transfer circuit. A different point was a partially preionization. Part of the discharge tube was covered with an Al sheet for a high discharge starting voltage. In the excitation circuit, the primary capacitance C_p was 10.2 μF and the storage capacitance was 700 pF.

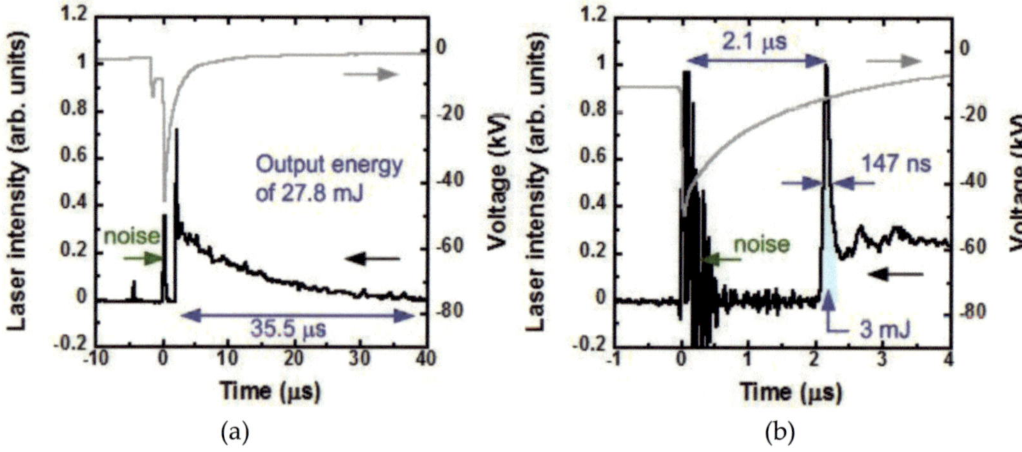

Figure 16. Discharge voltage and CO_2 laser pulse waveforms at 3.4 kPa (mixed gas, $CO_2 : N_2 : He = 1 : 1 : 2$) (Uno et al, 2012a). Black and gray lines represent laser pulse and discharge voltage, respectively. (a) Overall waveform. (b) Magnified time scale view of spike pulse.

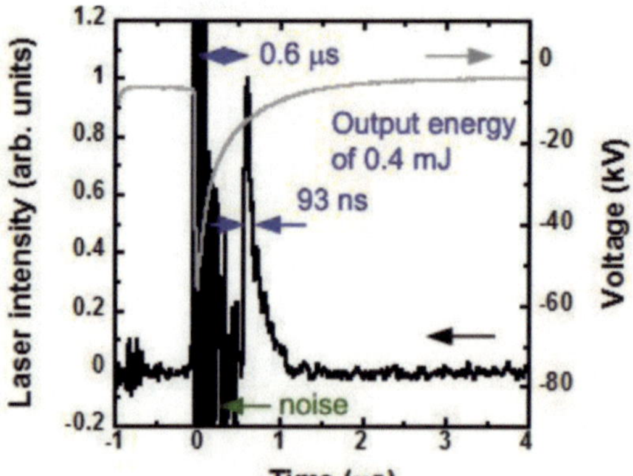

Figure 17. Discharge voltage and CO_2 laser pulse waveforms at 2.2 kPa (pure CO_2) (Uno et al, 2012a). Black and gray lines represent laser pulse and discharge voltage, respectively.

Fig. 16 shows the discharge voltage and laser pulse waveforms at the mixed gas (CO_2: N_2: He= 1: 1: 2) pressure of 3.4 kPa. The discharge starting voltage was –60.2 kV, and the fall time of the discharge was 4.4 μs. The laser oscillation began 2.1 μs after the start of discharge. The laser pulse had a sharp spike, like that from TEA and Q-switched CO_2 lasers. The spike pulse width was 147 ns (FWHM), and the pulse tail length was 35.5 μs. The laser output energy was 27.8 mJ, and the energy of the spike pulse part was estimated to be 3.1 mJ. Fig. 17 shows the discharge voltage and laser pulse waveform at the pure CO_2 gas pressure of 2.2 kPa. The discharge starting voltage was –60.8 kV, and the fall time of the discharge was 1.9 μs. Laser oscillation began 0.6 μs after the start of discharge. The laser output energy was 0.4 mJ. The laser pulse contained a spike pulse only, and the width was 93 ns (FWHM). Pulse-tail-free oscillation was realized by the use of pure CO_2 gas.

Figure 18 shows the simplest short-pulse longitudinally excited CO_2 laser (Uno et al, 2012b). This system has a laser tube, a capacitor, a resister, and a pulse power supply with a low-voltage silicon-controlled rectifier, and a set-up transformer only. The excitation circuit does not have a high-voltage switch; instead, the laser tube plays the role of the switch. Therefore, this system is almost same as the above long-pulse longitudinally excited CO_2 laser. However, the step-up transformer has the fast rise time of about 3 μs. The fast transformer produces the fast discharge which causes the short laser pulse.

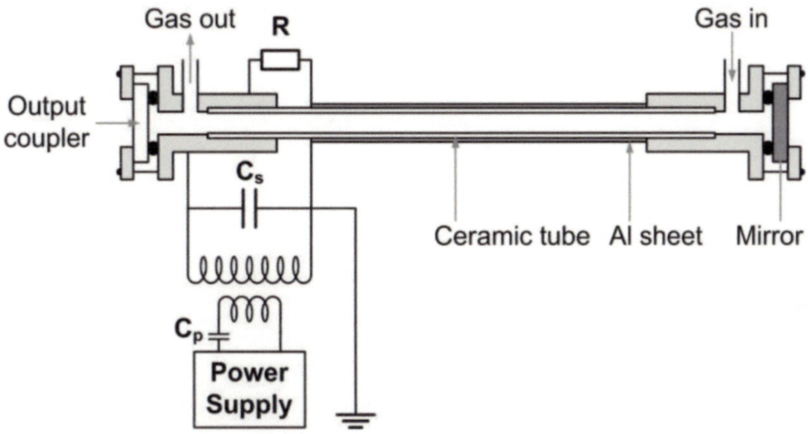

(a)

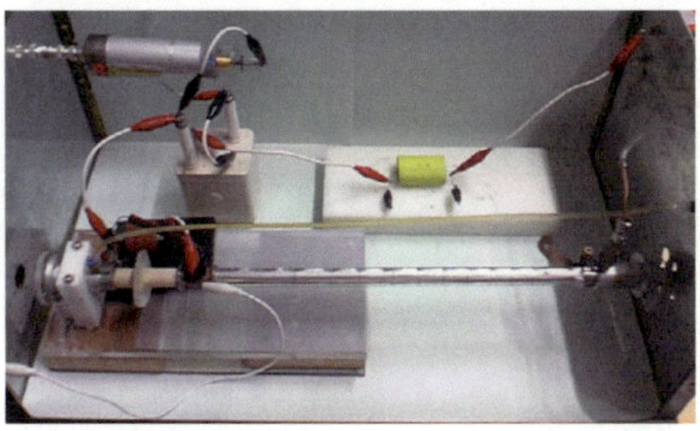

(b)

Figure 18. Short-pulse longitudinally excited CO_2 laser with switchless circuit (Uno et al, 2012b). (a) Schematic diagram. (b) Photograph.

For example, in our study (Uno et al, 2012b), the laser tube was the same as the above longitudinally excited CO_2 laser with the capacitor-transfer circuit (Uno et al, 2009). In the excitation circuit, the primary capacitance C_p was 4.7 μF, the storage capacitance C_s was 700 pF, and the resistor R was 10 MΩ.

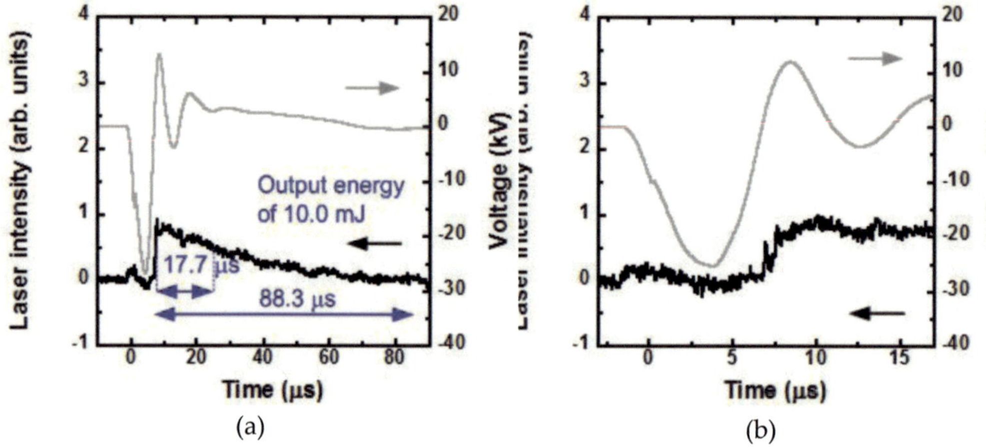

Figure 19. Discharge voltage and CO_2 laser pulse waveforms at 3.8 kPa (CO_2: N_2: He= 1: 1: 2) (Uno et al, 2012b). Black and gray lines represent laser pulse and discharge voltage. (a) Overall waveform. (b) Magnified time scale view of the start of discharge.

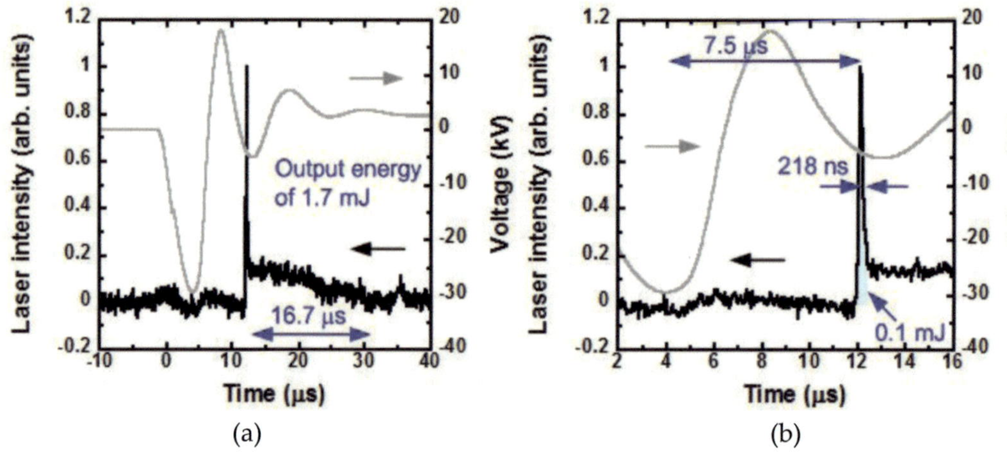

Figure 20. Discharge voltage and CO_2 laser pulse waveforms at 9.0 kPa (CO_2: N_2: He= 1: 1: 2) (Uno et al, 2012b). Black and gray lines represent laser pulse and discharge voltage, respectively. (a) Overall waveform. (b) Magnified time scale view of the start of discharge.

The long pulse oscillation takes place at gas pressure less than 3.8 kPa (CO_2: N_2: He= 1: 1: 2) and the short pulse oscillation takes place at gas pressure more than 4.2 kPa. Fig. 19 shows the discharge voltage and laser pulse waveforms at 3.8 kPa. The discharge voltage reached -25.8 kV at a raise time of 3.4 μs. In the low pressure, the breakdown voltage is low. The discharge starts when the applied voltage reaches the breakdown voltage. The fall time of the main discharge was 2.8 μs. Laser oscillation took place 3.5 μs after the start of discharge. At this time, the discharge continues. The laser pulse did not have a spike pulse and is a conventional long pulse oscillation. The laser output energy was 10.0 mJ. The laser pulse width was the full width at half maximum (FWHM) of 17.7 μs and the pulse length of 88.3 μs. Fig. 20 shows the discharge voltage and laser pulse waveforms at 9.0 kPa. The discharge voltage reached -29 kV at a raise time of 3.1 μs. In the high pressure, the breakdown voltage is high and the applied voltage reaches maximum. The fast discharge is influenced by the impedance matching of the excitation circuit and the discharge space. The fall time of main discharge was 2.2 μs. The laser oscillation took place 7.5 μs after the start of discharge. The laser pulse had a spike pulse width of 218 ns and a pulse tail length of 16.7 μs. The laser output energy was 1.7 mJ. The energy of the spike pulse part was estimated to be 0.1 mJ. It may be necessary to improve the output energy for various supplications.

5. Application

5.1. CW and long pulse

5.1.1. Medical applications

80 wt% of soft biological tissue is water. Evaporation of water is a factor important for soft tissue excision (Awazu, 2008). In laser irradiation, evaporation of water takes place at the critical temperature of 101 K (Auerhammer et al, 1999). The evaporation of water causes the soft tissue excision around the laser irradiation part. The tensile strength of soft tissue is about 10 MPa on skin and several MPa on corneas, blood vessel, and muscle (Duck, 1990). The laser excision of soft tissue is carried out by a high mechanical power which causes a phase transition because 1 atm is about 0.1 MPa.

A CO_2 laser is effectively absorbed in water. In 9.2 – 11.4 μm of the CO_2 laser wavelength band, the absorption coefficient is 10^3 cm^{-1} (Zarrabi & Gross, 2011). Therefore, in the CO_2 laser, the penetration depth to soft biological tissue is short and about 50 μm. The CO_2 laser evaporates only surface of soft tissue. Heat influence of deeper tissue is small and the damage to normal tissue is minimally suppressed. Edema and sharp pain after an operation are little because the heat influence is restrictive. In the CO_2 laser irradiation, shrink, dehydration, and carbonization of surface tissue take place. They cause water evaporation and vaporization of inner tissue. In this process, a protein denaturation layer of residual tissue is certainly generated. The denaturation layer functions as a solidification layer for an arrest of hemorrhage.

At present, a CO_2 laser of about 3 W a pulsed CW of (10 – 1000 ms and a super pulse of 0.1 – 1 ms) is used for the skin treatment such as eliminating a mole, a wart, macule (Gotkin et al, 2009) and wrinkle the aurinasal treatment such as hay fever and allergic rhinitis (Takeno et al, 2009), the eye treatment such as blepharochalasis and droopy eyelid (Rokhsar et al, 2008), and the dental treatment such as section and evaporation of soft tissue, hemostasis, and stomatitis (Zand et al, 2009). The advantages of CO_2 laser surgery are alleviation of bleeding, pain and edema, and shortening of operation and recovery time.

5.1.2. Industrial applications

When laser is irradiated to a material surface, a part is reflected by the reflectance of the material to the laser wavelength at the material surface, and the remainder is absorbed at the material surface. Then, the temperature of the material surface increases. A part of the heat is radiated by the heat radiation rate of the material, and the remainder conducts to inside by the heat conductivity of the material and makes the internal temperature increase. The surface temperature is fixed by the balance between the thermal energy lost by heat conduction and radiation and the thermal energy given by the laser. If the surface temperature does not reach the melting point of the material, a surface treatment takes place. If the surface temperature reaches the melting point, the melting takes place. When the heat input stops by movement or a stop of the laser beam, the material is cooled. Solidification causes welding. Eliminating the molten material (e.g. gas blowing) causes cutting. When the surface temperature exceeds the boiling point of the material by increasing the thermal energy given by the laser beam, a hole called keyhole begins to be formed. The laser reaches the inside of the material through the keyhole. The penetration causes deep melting process called keyhole welding.

The laser cutting is a specialty application of CO_2 lasers. In the laser cutting, a beam control is important. In space, the beam with TEM_{00} mode or the beam with low order mode of axial symmetry like TEM_{00} mode is required. In time, the control of the beam power (W/cm^2) is required. In the laser cutting of metals like steel, a fine cutting without heat influence, strain, and dross is required. It is very important to optimally control heat input by control of the peak power, the repetition rate, the duty ratio, and so on. These factors must be optimally controlled according to the kind of material, the thickness, and the cutting speed. A metal with a few mm of thickness can be cut by the CO_2 laser with the peak power of 100 – 200 W, the repetition rate of about 300 Hz, the cutting speed of about 30 cm/min, and the duty ratio of about 30%. In non-metallic materials such as woods, cloth, paper, ceramics, glass, gum, plastics, polyimide, epoxy, polycarbonate, vinyl and so on, the absorption of CO_2 laser light is very large. Most of these non-metallic materials can be processed (cutting, hole making, marking, etc.) by a low output CW-CO_2 laser emitted less than 300 W. The laser processing is used for the cutting of automobile instrument panels and clothes, the hole making of the material (e.g. ceramics), which the mechanical processing is difficult for, and so on.

5.2. Short pulse

5.2.1. Applications by commercial TEA-CO_2 lasers

An ablation process by irradiation of a short-pulse CO_2 laser is used for hole-making a printed board, marking to a resin board, and cleaning a metal part. In the hole-making of a printed board, improving density of trace is the purpose. Making a via-hole with a diameter of 50 mm or less requires several tens kW of the peak power, a few μs of the pulse width and a few kHz of repetition rate. Thus, TEA-CO_2 lasers are used for the-hole making. The laser marking stamps the model number and name of a product, the date of manufacture, a lot number, etc. In a TEA-CO_2 laser, image transfer with mask is used because the TEA-CO_2

laser produces high output power. CO_2 lasers are used for marking to resin boards, such as glass epoxy. In the laser cleaning, TEA-CO_2 lasers are used for dirt and coat removal on metal material. Dirt and a coat are removed by laser ablation because they tend to absorb the CO_2 laser light than metal. This process does not give the damage by heat to a surface of metal, because the laser light is reflected when the metal material appears. The laser cleaning does not use a medical fluid and blast material, and is a dry process.

5.2.2. *Future applications*

At present, drilling of hard tissue (enamel and dentine) uses an Er:YAG laser with the wavelength of 2.94 μm that is well correspond with the absorption wavelength of water. In 2.94 μm of the Er:YAG laser wavelength, the absorption coefficient is 10^4 cm^{-1} (Zarrabi & Gross, 2011). The laser light is focused on water and micro-explosion of water drills hard tissue. CO_2 laser can also drill hard tissue because the CO_2 laser (9.2 – 11.4 μm) is well absorbed in water. However, a long-pulsed CO_2 laser which is used in the present dental clinic causes carbonization of hard tissue. Fig. 21 (a) and (b) shows the human tooth surface irradiated by a long-pulsed longitudinally excited CO_2 laser with the pulse width of 30 μs and the output energy of 80 mJ. The dentin which contains heat-sensitive protein was carbonized by thermal influence because of the long-pulsed CO_2 laser. On the other hand, Fig. 21 (c) and (d) shows the human tooth surface irradiated by a short-pulsed longitudinally excited CO_2 laser with the spike pulse width of about 100 ns, the pulse tail length of about 60 μs, and the output energy of 80 mJ (Uno et al, 2009). The short-pulsed CO_2 laser resulted in almost no carbonization not only for the enamel but also for the dentine.

Additionally, CO_2 lasers are also absorbed in the human tooth (Heya et al, 2003). The dentine has a weak acid resistance and becomes a saprodontia easily. The dentine is reformed to a hydroxyapatite that has a strong acid resistance by the laser with the wavelength of 8.8 – 10.6 μm (Heya et al, 2003). CO_2 lasers can change the dentine to the hydroxyapatite because the wavelength corresponds to the wavelength of the CO_2 laser. Moreover, the dentin covers dental pulp and is porous. Exposure of the dentine causes dentinal hypersensitivity. The irradiation of the CO_2 laser melts and blocked the dentine surface. Thus, by the CO_2 laser, the medical treatment of dentinal hypersensitivity is also possible. The longitudinally excited CO_2 laser is expected as a complex dental care machine because the laser can produce the long pulse for the soft tissue treatment and the short pulse for the hard tissue treatment with a low-cost and portable device.

Glass marking can be effectively performed by a low-power CO_2 laser because glass absorbs infrared light well. Additionally, the glass marking is energy saving because a CO_2 laser has high electro-optical conversion efficiency. However, a long-pulsed CO_2 laser produces cracks on the surface of glass by heat influence. Fig. 22 (a) shows the glass surface irradiated by a long-pulsed longitudinally excited CO_2 laser with the pulse width of about 30 μs and the output energy of about 45 mJ. The processing part looks white by cracking. On the other hand, a short-pulsed CO_2 laser produces crackless marking. Fig. 22 (b) shows the glass

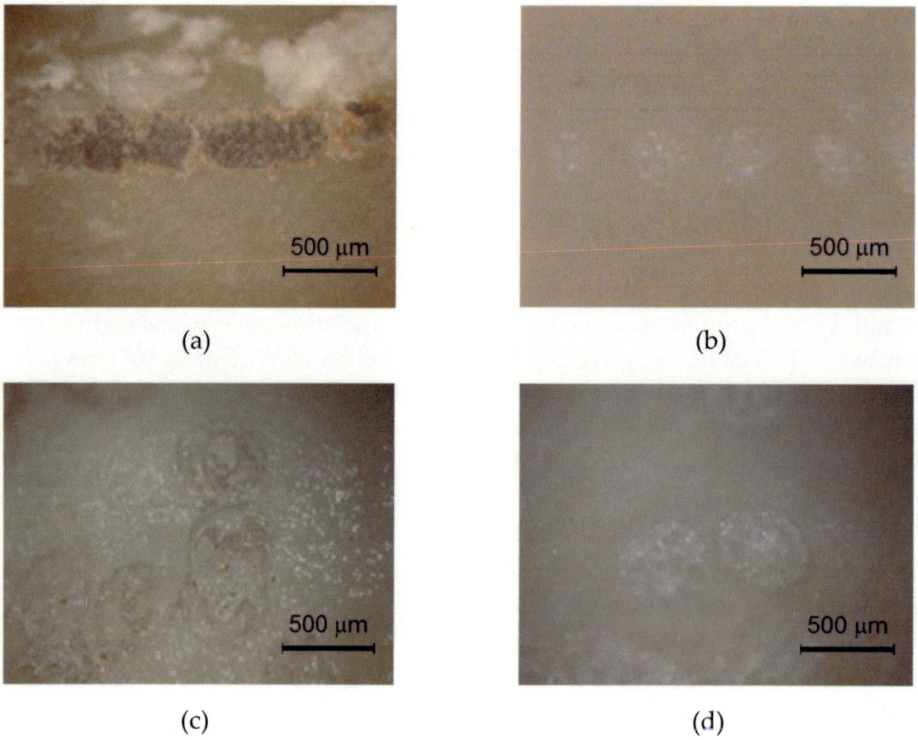

Figure 21. Tooth surface irradiated longitudinally excited CO_2 laser (Uno et al, 2009). (a) Dentine surface by long-pulsed CO_2 laser irradiation. (b) Enamel surface by long-pulsed CO_2 laser irradiation. (c) Dentine surface by short- pulsed CO_2 laser irradiation. (d) Enamel surface by short- pulsed CO_2 laser irradiation.

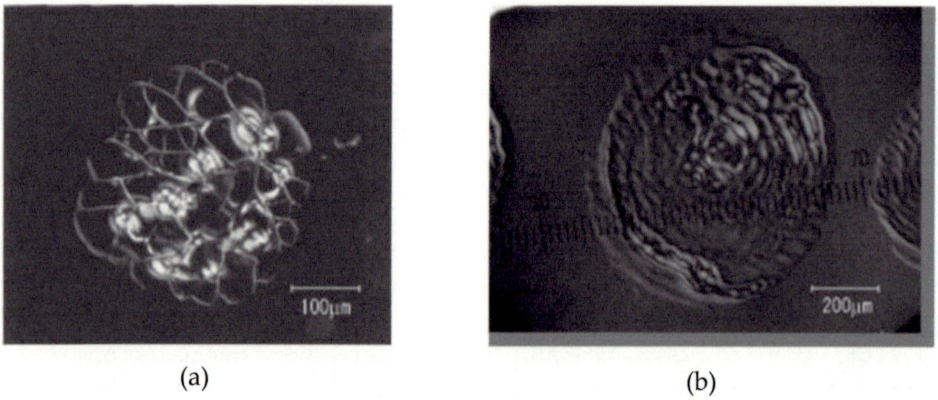

Figure 22. Glass surface irradiated longitudinally excited CO_2 laser (Uno et al, 2009). (a) Long-pulse CO_2 laser irradiation. (b) Short-pulse CO_2 laser irradiation.

surface irradiated by a short-pulsed longitudinally excited CO_2 laser with the spike pulse width of about 100 ns, the pulse tail length of about 60 μs and the output energy of about 45 mJ (Uno et al, 2009). The processing part is not visible with the naked eye because the part does not have a crack. In fact, the short-pulsed CO_2 laser produces stealth marking. However, the irradiation fluence is limited by the heat influence of the pulse tail. Therefore, the development of a short-pulsed CO_2 laser without pulse tail is desired. The short-pulsed longitudinally excited CO_2 laser with a low-cost and portable device will produce a device for product identification and traceability by marking for every process in a factory, and management of a sample by marking in a research institution.

6. Conclusion

In this chapter, the longitudinally excited CO_2 laser has been described. The excitation system and mechanism of the CO_2 laser, the feature of longitudinal excitation system and the excitation discharge, and the application of the CO2 laser have been explained. Especially, in the longitudinally excited CO_2 laser pumped by pulse discharge, new excitation circuits have been introduced and the relation between the discharge and the laser pulse has been explained. The longitudinally excited CO_2 laser can produce CW oscillation, long pulse oscillation, or short pulse oscillation with a compact, simple, and low-cost device. The development of a maintenance-free and warm-up-free device can be expected by the simple structure. Therefore, a user-friendly CO_2 laser which anyone can use easily always anywhere will contribute to various fields.

Author details

Kazuyuki Uno
University of Yamanashi, Japan

7. References

Auerhammer, J. M., Walker, R., Meer, A. F. G. & Jean, B. (1999). Dynamic Behavior of Photoablation Products of Corneal Tissue in the Mid-IR: A Study with FELIX. *Applied Physics B*, Vol. 68, (1999), pp. 119–119.

Awazu, K. (2008). *Infrared Laser for Biomedical Engineering*. in Japanese, Osaka University Press, Osaka, ISBN: 978-4-87259-160-6.

Battou, K., Ameur, K. A. & Ziane, O. (2008). Q-switch of a continuously pumped CO_2 laser with a scanning coupled-cavity Michelson mirror. *Optics Communications*, Vol. 281, (2008), pp. 5234-5238.

Bethel, J. W., Baker, H. J. & Hall, D. R. (1998). A new scalable annular CO_2 laser with high specific output power. *Optics Communications*, Vol. 125, (1998), pp. 352–358.

Biswas, A. K., Bhagata, M. S., Rana, L. B., Verma, A. & Kukreja, L. M. (2010). Indigenous development of a 2 kW RF-excited fast axial flow CO_2 laser. *PRAMANA*, Vol. 75, (2010), pp. 907-913.

Burkhard, P., Gerber, T. & Luthy, W. (1981). XeF excimer laser pumped in a longitudinal low-pressure discharge. *Applied Physics Letters*, Vol. 39, (1981), pp. 19–20.

Chung, H.-J., Lee, D.-H., Hong, J.-H., Joung, J.-H., Sung, Y.-M., Park S.-J. & Kim, H.-J. (2002). A simple pulsed CO_2 laser with long milliseconds pulse duration. *Review of Scientific Instruments*, Vol. 73, (2002), pp. 484-485.

Cleeschinsky, D., Dammasch, D., Eichler, H. J. & Hamisch, J. (1981). XeF-LASER WITH LONGITUDINAL DISCHARGE EXCITATION. *Optics Communications*, Vol. 39, (1981), pp. 79–82.

Deshmukh, S. V. & Rajagopalan, C. (2003). High-power multibeam CO_2 laser for industrial applications. *Optics & Laser Technology*, Vol. 35, (2003), pp. 517-521.

Duck, F. A. (1990). *Physical Properties of Tissue*. Academic Press, London,ISBN: 978-0122228001.

Eichler, H. J., Hamisch, J., Nagel, B. & Schmid, W. (1985). KrF laser with longitudinal discharge excitation. *Applied Physics Letters*, Vol. 46, (1985), pp. 911–913.

El-Osealy, M. A. M., Jitsuno, T., Nakamura, K. & Horiguchi, S. (2002a). Gain characteristics of longitudinally excited F_2 lasers. *Optics Communications*, Vol. 205, (2002), pp. 377–384.

El-Osealy, M. A. M., Jitsuno, T., Nakamura, K., Uchida, Y. & Goto, T. (2002b). Oscillation and gain characteristics of longitudinally excited VUV F_2 laser at 40 Torr total pressure. *Optics Communications*, Vol. 207, (2002), pp. 255–259.

El-Osealy, M. A., Ido, T., Nakamura, K., Jitsuno, T. & Horiguchi, S. (2001). Oscillation and gain characteristics of high power co-axially excited N_2 gas lasers. *Optics Communications*, Vol. 194, (2001), pp. 191–199.

Furuhashi, H., Hiramatsu, M. & Goto, T. (1987). Longitudinal discharge XeCl excimer laser with automatic UV preionization. *Applied Physics Letters*, Vol. 50, (1987), pp. 883–885.

Gotkin, R. H., Sarnoff, D. S., Cannarozzo, G., Sadick, N. S. & Armenakas, M. A. (2009). Ablative Skin Resurfacing With a Novel Microablative CO_2 Laser. *Journal of Drugs in Dermatology*, Vol. 8, (2009), pp. 138–144.

He, Z., Zhang, Y., Zhang, H., Zhang, Q., Liao, J., Zhou, Y., Liu, S. & Luo, X. (2010). Study of Optimal Cavity Parameter in Optically Pumped D_2O Gas Terahertz Laser. *Journal of Infrared, Millimeter, and Terahertz Waves*, Vol. 31, (2010), pp. 551-558.

Heya, M., Sano, S., Takagi, N., Fukami, Y. & Awazu, K. (2003). Wavelength and Average Power Density Dependency of the Surface Modification of Root Dentin Using an MIR-FEL. *Lasers in Surgery and Medicine*, Vol. 32, (2003), pp. 349–358.

Hill, A. E. (1968). MULTIJOULE PULSES FROM CO_2 LASERS. *Applied Physics Letters*, Vol. 12, (1968), pp. 324-327.

Hocker, L. O. & Phi, T. B. (1976). Pressure dependence of the atomic fluorine laser transition intensities. *Applied Physics Letters*, Vol. 29, (1976), pp. 493–494.

Javan, A., Bennett, Jr. W. R. & Herriott, D. R. (1961). Population Inversion and Continuous Optical Maser Oscillation in a Gas Discharge Containing a He-Ne Mixture. *Physical Review Letters*, Vol. 6, (1961), pp. 106–110.

Komatsu, K., Matsui, E., Kannari, F., & Obara, M. (1991). Low Pressure Atomic Xenon Laser Excited by Self-Sustained Longitudinal Discharge. *The Review of Laser Engineering*, in Japanese, Vol. 19, (1991), pp. 490–495.

Lee, D.-H., Chung H.-J. & Kim, H.-J. (2000). Comparison of dc and ac excitation of a sealed CO_2 laser. *Review of Scientific Instruments*, Vol. 71, (2000), pp. 577-578.

Loy, M. M. T. & Roland, P. A. (1977). Simple longitudinally pulsed CO_2 laser and its application in single-mode operation of TEA lasers. *Review of Scientific Instruments*, Vol. 48, (1977), pp. 554-556.

Miyazaki, K., Hasama, T., Yamada, K., Fukatsu, T., Eura, T. & Sato, T. (1986). Efficiency of a capacitor-transfer-type discharge excimer laser with automatic preionization. *Journal of Applied Physics*, Vol. 60, (1986), pp. 2721-2728.

Nagai, H. (2000). *Laser Process Technology*, in Japanese, Optoronics Co. Ltd., Tokyo, ISBN: 978-4-902312-36-2.

Newman, L. A. (1978). XeF^* and KrF^* waveguide lasers excited by a capacitively coupled discharge. *Applied Physics Letters*, Vol. 33, (1978), pp. 501–503.

Patel, C. K. N. (1964). SELECTIVE EXCITATION THROUGH VIBRATINAL ENERGY TRANSFER AND OPTICL MASER ACTION IN N_2-CO_2. *Physical Review Letters*, Vol. 13, (1964), pp. 617-619.

Photonics.com. http://www.photonics.com/LinearCharts/Default.aspx?ChartID=1

Rokhsar, C. K., Ciocon, D. H., Detweiler, S. & Fitzpatrick, R. E. (2008). The Short Pulse Carbon Dioxide Laser Versus the Colorado Needle Tip with Electrocautery for Upper and Lower Eyelid Blepharoplasty. *Lasers in Surgery Medicine*, Vol. 40, (2008), pp. 159–164.

Rosa, J., J. Eichler, H. & Herweg, H. (1986). ArF laser excited in a capacitively coupled discharge tube. *Journal of Applied Physics*, Vol. 54, (1986), pp. 1598–1599.

Soukieh, M., Ghani, B. A. & Hammadi, M. (1999). Mathematical modeling TE CO_2 laser with SF_6 as a saturable absorber. *Optics & Laser Technology*, Vol. 31, (1999), pp. 601-611.

Takeno, S., Hirakawa, K., Ishino, T. & Goh, K. (2009). Surgical treatment of the inferior turbinate for allergic rhinitis: clinical evaluation and therapeutic mechanisms of the different techniques. *Clinical & Experimental Allergy Reviews*, Vol. 9, (2009), pp. 18–23.

Terai, K., Murata, T. & Tamagawa, T. (1993). Characteristics of RF Excited CO_2 Lasers. *The Review of Laser Engineering*, in Japanese, Vol. 21, (1993), pp. 475–484.

Tian, Z., Sun, Z. & Qu, S. (2005). Tunable pulse-width, electro-optically cavity-dumped, rf-excited Z-fold waveguide CO_2 laser. *Review of Scientific Instruments*, Vol. 76, (2005), 083110.

Tripathi, A. K., Gupta, N. M., Chatterjee, U. K. & Bhawalkar, D. D. (1994). Development of a sealed-off cw CO_2 laser using a supported gold catalyst. *Review of Scientific Instruments*, Vol. 65, (1994), pp. 3853-3855.

Ueno, Y., Soumagne, G., Sumitani, A., Endo, A. & Higashiguchi, T. (2007). Enhancement of extreme ultraviolet emission from a CO_2 laser-produced Sn plasma using a cavity target. *Applied Physics Letters*, Vol. 91, (2007), 231501.

Uno, K., Akitsu, T. & Jitsuno, T. (2012a). Longitudinally excited CO_2 laser with short laser pulse using direct-drive circuit. *Journal of Engineering and Technology*, Vol. 2, (2012) pp. 101-106.

Uno, K., Jitsuno, T. & Akitsu, T. (2012b). Simple short-pulse CO_2 laser excited by longitudinal discharge without high-voltage switch. *Journal of Infrared, Millimeter, and Terahertz Waves*, Vol. 33, (2012), pp. 485-490.

Uno, K., Nakamura, K., Goto, T. & Jitsuno, T. (2006). Longitudinally Excited N_2 Laser Pumped by Lamplike Discharge. *Japanese Journal of Applied Physics*, Vol. 45, (2006), pp. 1651-1653.

Uno, K., Nakamura, K., Goto, T. & Jitsuno, T. (2008). Longitudinally excited N_2 lasers without high-voltage switches. *Review of Scientific Instruments*, Vol. 79, (2008), 063107.

Uno, K., Nakamura, K., Goto, T. & Jitsuno, T. (2008). Red-F* Laser and VUV-F_2 Emission Pumped at Low Pressure by Longitudinal, Lamp-Like Discharge. *Plasma and Fusion Research*, Vol. 3, (2008), 037.

Uno, K., Nakamura, K., Goto, T. & Jitsuno, T. (2009). Longitudinally Excited CO_2 Laser with Short Laser Pulse like TEA CO_2 Laser. *Journal of Infrared, Millimeter, and Terahertz Waves*, Vol. 30, (2009), pp. 1123–1130.

Xie, J., Guo, R., Li, D., Zhang, C., Yang, G. & Geng, Y. (2010). Theoreteical calculation and experimental study of acousto-optically Q-switched CO_2 laser. *Optics Express*, Vol. 18, (2010), 12371.

Yamanaka, C., Nakai, S., Matoba, M., Fujita, H., Kawamura, Y., Daido, H., Inoue, M., Fukuyama, F. & Terai K. (1981). The LEKKO VIII CO_2 gas laser system. *IEEE Journal of Quuntum Electronics*, Vol. 17, (1981), pp. 1678-1688.

Zand, N., Ataie-Fashtami, L., Djavid, G. E., Fateh, M., Alinaghizadeh, M.-R., Fatemi, S.-M. & Arbabi-Kalati, F. (2009). Relieving pain in minor aphthous stomatitis by a single session of non-thermal carbon dioxide laser irradiation. *Lasers in Medical Science*, Vol. 24, (2009), pp. 515–520.

Zarrabi, A. & Gross, A. J. (2011). The Evolution of Lasers in Urology : Lasers: A Short History and Simplified Physics. *Therapeutic Advances in Urology*, Vol. 3, (2011), pp. 81–89.

Zhou, Z., Zeng, Y. & Qiu, M. (1983). XeCl excimer laser excited by longitudinal discharge. *Applied Physics Letters*, Vol. 43, (1983), pp. 347–349.

Section 3

Cutting and Shooting

Chapter 5

Ultrashort Laser Pulses Machining

Ricardo Elgul Samad, Leandro Matiolli Machado,
Nilson Dias Vieira Junior and Wagner de Rossi

Additional information is available at the end of the chapter

http://dx.doi.org/10.5772/46235

1. Introduction

Over the past several years, the necessity for micromachining technologies has been growing from scientific research to industry. Segments from medical appliances, microelectronics and the automotive world demand a great variety of applications such as micromotors, microfluidic circuits, MEMS (Micro Electrical Mechanical System), medical devices, electronic tooling, particle filters, micromolds and microvalves, among others [1]. This massive growth in the micromachining segment demands a variety of new micromachining methods.

Techniques to machine the surface of materials are continually being improved. Recently, mechanical micromachining with accurate positioning systems and very small drills were used to trim out materials, covering the range from some millimeters down to a few tens of nanometers [2]. A similar technique that also applies direct micromachining and requires a highly accurate positioning system but no contact, is Electron-Discharge Micromachining (EDM), in which material is removed by an electrode guided along the desired path, very close to the surface of a conducting material submerged in a dielectric fluid; a spark is established between the electrode and the material being processed, removing matter by melting and evaporation [3, 4], with micrometric resolution. Similarly to EDM, the Focused Ion Beam (FIB) technique scans an ion beam, with spot sizes ranging from 10 to 500 nm, over a specimen, etching the material [5]. Additionally to etching, this technique can be used to build new structures by localized deposition using the ion beam-induced decomposition of an organo-metallic gas [6]. Examples of the use of a FIB system are the milling of angled cuts into the suspension ligaments of MEMS accelerometers, and the milling of large depth-to-width aspect ratios tracks to make photonic band gap lattices in $As_{40}S_{60}$ [5, 7]. A disadvantage of this technique is that it does not allow batch production due to the small amount of material removed during machining.

Photolithography, a technique that is available for integrated circuits fabrication since the 1970's, has been improved, and nowadays is used to create structures ranging from micrometers, as in the case of MEMS devices, down to tens of nanometers for integrated circuits [8]. This technique selectively removes material from thin films (nitride/oxide /polysilicon) deposited on silicon substrates, and also from the bulk of the substrate [9]. The process consists in using UV light and optical projection to transfer a pattern from a mask to a light sensitive material (photoresist) film pre-deposited on the substrate surface, selectively curing it. The non-cured photoresist is then removed by a developer, and the substrate exposed portions are etched either by a liquid (wet etching) or by a plasma (dry etching). The repetition of many cycles of photoresist deposition and etch can create complex structures, in a process known as *Surface Micromachining*. Different patterns of polycrystalline silicon can be selectively etched and are used as sacrificial layers to create suspended parts attached to the substrate, building 3D structures on the substrate surface [10]. This process enables the creation of structures such as accelerometers [11] and neural probe arrays [12]. When the etching removes a large quantity of material from the bulk of the substrate, the process is known as *Bulk Micromachining* [9], and is used to create devices such as piezoresistive sensors [13]. Photolithography allows batch production, but it is considered the most indirect method, since material is not removed by a tool.

Although largely used, the described techniques have some disadvantages, such as tool wearing (mechanical micromachining and EDM), high complexity (photolithography) and necessity of processing in vacuum (FIB). Besides, when machining different materials, the etching rates can strongly depend on the material and the tool, and the processing parameters, or even the tool, have to be changed. Additionally, the technique can be restricted to a specific material class, such as metals in the EDM case. Other limitations include the small etching speed in the techniques that require precise positioning (mechanical machining, EDM, FIB), preventing batch processing, and the high complexity of the photolithography, which does not allow fast prototyping. In view of those limitations, the ultrashort pulse laser emerges as a valuable tool for micromachining.

Just a few years after the laser invention, it was already being used as a tool to ablate [14] and machine [15] a wide variety of materials, including metals [16], dielectrics [17], semiconductors [18], composites [19] and biological tissues [20]. At that time, the machining was based on thermal processes arising from the material heating by the laser [21], and the ablation occurred as a consequence of melting and vaporization resulting from phase transitions due to the heat deposited into the material being processed [22, 23].

When machining with laser pulses, the laser beam is focused on the material being processed, the light is absorbed and the material undergoes a physical change around the absorption region. The dimensions of the structures that can be machined depend, on a first approximation, on the focused beam diameter. A diffraction limited TEM_{00} Gaussian beam [24] can be focused to diameters close to the laser wavelength, ranging from a hundred nanometers (Excimer UV lasers) to tens of microns (CO_2 lasers). Nevertheless, even with the smallest attainable laser spot on the material surface, the interaction region follows the heat diffusion volume guided by the process dynamics. For long pulses, which last more than the typical period of lattice vibration (around tens of picoseconds), the dynamics of the laser-matter interaction depends on the laser parameters (wavelength, pulse energy, repetition rate) and on the physical characteristics (absorption, heat capacity, thermal conductivity) of the material being machined. In this long pulse regime, the machining process is a consequence of heating (that results in melting, evaporation, sublimation), depending on the energy absorbed per unit of time and volume, and how it flows inside the material. When the laser radiation is absorbed, the material is heated and thermodynamic processes are responsible for heat diffusion and phase transformations. As the pulse duration shortens, higher intensities are faster achieved, the spatial heat propagation is decreased and phase transformations that lead to material removal occur more efficiently; in this regime, the ablation threshold decreases with the square root of the pulse duration, demonstrating that the ablation is a thermal process [25]. The excess heat that is not used to jettison material flows into the surroundings increasing its temperature, and eventually creating a Heat Affected Zone (HAZ), in which phase transitions modify the material properties, usually in a detrimental way. As a consequence, smaller structures require shorter pulses to be etched, which also produce a reduced HAZ. Nevertheless, although microsecond to sub-nanosecond laser pulses generates intensities high enough to machine metals, semiconductors and even some transparent dielectrics, the heat conduction expands the affected region far beyond the focused beam diameter, creates a large HAZ, and can also generate structures several times bigger than the wavelength used. Collateral effects from heat, like burr, debris and molten material, spread the interaction area limiting the machining precision. Additionally, albeit the high intensities attainable in this temporal regime allow the etching of transparent materials such as sapphire and silica, large amounts of energy are necessary to reach the ablation threshold, resulting in thermal stresses and damages such as chipping and micro-cracking [26, 27]. To handle these problems, different wavelengths are needed for efficient absorption by different materials. For dielectrics, lasers with wavelengths in the UV range are used for better absorption; some glasses have greater absorption in the far infrared [28], hence CO_2 lasers are often employed.

Femtosecond (10^{-15} s) pulses come up as a new possibility for micromachining with some advantages over longer pulses. The machining performed by these ultrashort pulses relies on their very brief duration, which is shorter than the thermal vibration period of the lattice. This small duration minimizes the energy transfer to the material, resulting in an almost nonthermal etching that minimizes the HAZ and preserves the properties of the surrounding material. Also, the ultrashort pulses duration creates very elevated intensities that promote a highly nonlinear interaction with matter, resulting in extremely localized ablations that can etch the material with nanometric precision, allowing the fabrication of very precise, minute structures. Additionally, ultrashort pulses promote a non-selective ablation regarding the material class (dielectric, metallic, etc.), allowing the machining of all kinds of materials with the same laser. The ability of ultrashort pulses to machine any material, even transparent ones, without changing the laser wavelength and the recent availability of relatively cheap, stable, high average power systems that increase the processing speed, are making these pulses very versatile tools to process materials for many high technology applications and devices. Also, in this temporal scale heat diffusion is minimized, and, with the adequate laser parameters, heat effects can be completely eliminated. Furthermore, is worth mentioning that, although it is not strictly a micromachining technique, the creation of micro and nanostructures is possible by the nonlinear interaction of the ultrashort pulses with a liquid solution. In this technique an ultrashort pulses beam is focused, through an optical a microscope optics, into the liquid solution containing a photosensitive polymer [29] or a photon induced reduction composite material [30]; the multiphotonic absorption that happens at the beam focal region promotes the highly localized solidification of the solution, allowing the manufacture of micro and nanometric structures.

The disadvantages in micromachining with femtosecond pulses comes from the high degree of precision demanded, with consequent necessity of tight focusing that imposes very accurate and expensive 3-axes positioning systems. Also, the low average powers that prevented high batch production are being overcome by modern systems with dozens of watts, allowing increased throughputs.

The invention, in the second half of the 1980's, of the Chirped Pulse Amplification Technique [31, 32], the KLM Ti:Sapphire laser [33], and the diode pumped solid state lasers [34] disseminated the use of microjoule-millijoule (μJ-mJ) ultrashort pulses to many laboratories around the world. The widespread availability of these systems in the decades of 1990-2000, resulted in the emergence of many new applications, including the machining by ultrashort pulses. The most used ultrashort pulse lasers nowadays for machining are Ti:Sapphire and Yb:fiber. The Ti:Sapphire systems can deliver pulses as short as a few tens of femtoseconds, centered at 800 nm, with mJ level energies, and repetition rates up to a few kHz; the Ytterbium based lasers generate pulses centered around 1030 nm lasting hundreds of femtoseconds, and, although limited in energy to the μJ range, operate at tens of MHz repetition rates, increasing its machining speed. Recent developments in laser technology design offer high average power systems [35], which in conjunction with beam conforming devices [36] and high speed scanning systems enable the use of ultrashort pulses machining in batch production, without limiting its fast prototyping capabilities.

2. Ablation by ultrashort laser pulses

The typical characteristics of ultrashort pulses for ablation of solids are energies in the range from tens to hundreds of microjoules, pulse widths around 100 fs and focalization to 20 μm radius spot sizes, generating intensities in the range 10^{12}-10^{14} W/cm^2.

When an ultrashort pulse impinges on a solid, the fact that its duration is shorter than the lattice ions vibrational period means that most of the pulse energy will be transferred to the material electrons, heating them. The majority of the electronic thermal energy will be carried away with the ablated material, and the small remaining portion will be added to the energy directly coupled into the ions, heating the lattice, with the possibility of creating a small HAZ. The control of the irradiation conditions can reduce this HAZ to almost zero.

2.1. Ablation physical mechanisms

The underlying mechanism for ultrashort pulses ablation is almost the same for all material classes (metals, semiconductors, dielectrics, polymers, etc.): when the pulse impinges on the material surface, seed electrons are accelerated by the pulse electrical field into a quivering motion, and either generate free electrons in the conduction band by collisions in an exponential avalanche process [37, 38], or are ejected from the surface due to the acquired kinetic energy [39]. The ejection process has a high occurrence probability on the 100 nm surface layer for electrons that acquired a few eV of kinetic energy, and leaves a charge imbalance that can produce a Coulomb explosion [39-41] of the lattice ions after the pulse. Simultaneously to the surface electrons ejection, the avalanche process occurs in deeper layers in the material, increasing the free electrons density and temperature, while the lattice is kept at a lower temperature (two-temperatures model [42, 43]); if the pulse energy and intensity are high enough, the free electrons density reaches a critical value (around 10^{21} cm^{-3} [38, 44]), and then the electrons transfer its energy to the surrounding ions. This relaxation quickly heats the neighboring lattice above its vaporization temperature, creating an unstable phase that undergoes a violent adiabatic expansion (phase explosion) [45-47], which removes material from the surface, carrying most of the thermal energy with it. Also, spallation [47] and fragmentation [45] can occur, although those are not dominant effects. The energy that is not taken away flows to the lattice, heating it. The dominating mechanism will depend on the pulse characteristics (energy, duration) and material.

For metals, the seed electrons are the conduction band free electrons. For the other materials, the pulse leading edge excites electrons from the valence band either by multiphoton ionization [25, 48] or tunneling induced by the laser field [49, 50]. Once the free electrons are present, a metallization occurs, and the electrons heating evolves deterministically in time [38, 50] in almost the same way in all materials [51].

Depending pulse energy, two ablation domains can be identified: the low and high fluence ones. In nonmetals, these regimes are defined by the dominant ablation mechanism: in the low fluence regime the Coulomb explosion predominates [41], while the phase explosion prevails for the high fluence ablation. In metals, Coulomb explosion does not occur because, at the intensities used in machining, the surface charge accumulation is effectively quenched [40] by the electronic mobility, suppressing the positive ions explosion. The electric field needed to promote positive charge accumulation in metals (~10^{10} V/m) demands intensities around 10^{18} W/cm^2 [52], pushing the Coulomb explosion threshold way above the phase explosion one. In metals, the low and high fluence regimes occur when the HAZ is shorter or longer than the optical absorption depth (Beer´s Law [53]), respectively. The regimes threshold can be predicted using the two temperatures model [42, 43] that describes the electronic heating, diffusion and lattice heating.

For all material classes, the main difference between the two ablation regimes is that in the high fluence there will be fusion in the interaction region, and the remaining material can cool down to an amorphous or polycrystalline phase, whose physical characteristics (mechanical, optical, etc.) can differ from the starting material ones [54]. Also, the HAZ will be bigger in the high fluence regime.

Since, by the mechanisms explained, the ultrashort pulses ablation process depends primarily on the electrons answer to the laser field excitation and not on the thermodynamic properties that arise as a consequence of the atomic lattice, the ablation threshold does not show a square root dependence on the pulse duration [25], as in thermal ablation processes; additionally, the ablation by ultrashort pulses has a nonselective character, and the only parameter that has to be known to etch a material is its ablation threshold fluence, F_{th}. As a general rule, since in nonmetals a portion of the pulse energy is used to create free electrons, these material usually present higher ablation thresholds than metals for the same laser conditions. Also, a single material can present two values for the ablation threshold, one for low and other for high fluence, as can be seen in Figure 1, and they have to be known and taken into account when machining the material.

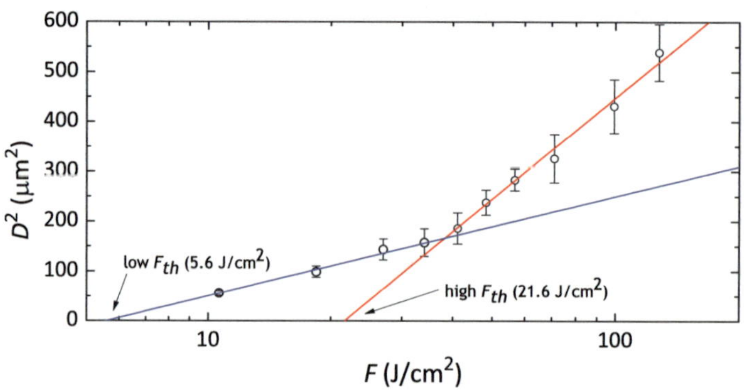

Figure 1. Diameter squared of the region ablated by ultrashort pulses on the surface of Sapphire as a function of the pulse fluence, for single-shot 100 fs pulses [55]. The ablation threshold fluence is given by the fluence at which the ablated diameter is zero, obtained from the fit by equation (1) [56]. The low and high fluence regimes can be clearly identified by two different slopes on the data points.

2.2. Ablation threshold determination

The determination of a given material ablation threshold fluence by ultrashort pulses is usually done using the "zero damage" method, introduced by Liu in 1982 [56]. This method consists in using a TEM_{00} Gaussian beam to ablate the material at various positions in the sample surface, using different pulse energies. Then the diameters of the ablated craters, D, are measured and their square values are plotted as a function of the pulse fluence, F, as show in Figure 1. The data is then fitted by:

$$D^2 = 2w^2 \ln\left(\frac{F}{F_{th}}\right), \tag{1}$$

where w is the beam spot size at the sample surface and F_{th} is the ablation threshold fluence, at which the ablated crater diameter is zero. To execute this measurement, the beam propagation parameters have to be known to calculate its spot size and fluence at the sample surface, and many measurements must be done. Experimentally, this method is demanding because it requires the precise knowledge and stabilization of the beam parameters, a good sample positioning system to maintain the beam spot size constant at the sample surface for all the measurements, and since it can last a long time, up to hours, it is prone to be affected by laser instabilities. Additionally, frequently the use of an electron or atomic force microscope is needed to determine the crater diameter for fluences close to the threshold, at which the damage size is close to zero and is difficult to be measured.

A few years ago we introduced the Diagonal Scan (D-Scan) technique [57, 58], an alternative and simple method to measure the ablation threshold for ultrashort pulses. The method consists in moving a sample longitudinally and transversely (z and y directions in Figure 2a) across the beamwaist of a focused beam, from a position before the beamwaist, where there is no ablation, to a position after the ablation stops. In this way, a symmetrical profile with two lobes, as the one shown in Figure 2b, will be etched on the sample surface. If the etched profile does not present two lobes, the measurement has to be repeated with a higher pulse energy or with a tighter focusing lens [57].

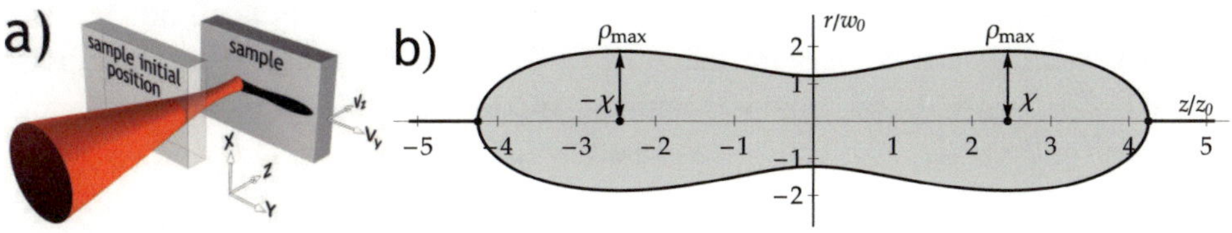

Figure 2. a) Scheme of the D-Scan method. b) profile etched on the sample surface by the diagonal movement across the beam waist position.

It can be shown [57] that the sample ablation threshold fluence is given by:

$$F_{th} = \frac{1}{e\pi}\frac{E_0}{\rho_{\max}^2} \cong 0.117\frac{E_0}{\rho_{\max}^2}, \tag{2}$$

where E_0 is the pulse energy and ρ_{max} is the half maximum transversal dimension of the profile etched.

The determination of the ablation threshold involves knowing the pulse energy and the measurement of the profile maximum transversal dimension (typically a few tens of microns), which can be done in an optical microscope, and the use of equation (2). Compared to the "zero damage" method, the D-Scan is easier to be experimentally performed once it demands only one scan that can be done in a few seconds and a geometrical measurement, replacing a series of measurements that can last a long time, the knowledge of the laser spot size at the sample, and a fitting.

2.3. Ablation by many pulses and incubation effects determination

In a given material, the ultrashort pulses ablation threshold can depend on the presence of defects, dopants, impurities, excitons, etc. [59, 60], which either create intermediate levels in the bandgap or modify the local electronic density. As a consequence, seed and free electrons are created more easily than in the ideal material, and the ablation requires less energy to occur, lowering the ablation threshold value.

The defects can either be intrinsic to the material, or externally originated, such as color centers created by ultrashort pulses [61]. In this case, when etching solids with superimposing pulses, modifications are induced in the portions that are irradiated but not ablated, lowering these regions F_{th} for subsequent pulses due to defects accretion. The following pulses experience a decreasing F_{th} until the defects density saturates and the ablation threshold reaches a constant value. These cumulative phenomena are known as incubation effects [62-64], and the ablation threshold fluence modification caused by them must be taken into account when machining a material.

A few models were proposed by different authors to describe the incubation mechanisms. One of those is the probabilistic defect accumulation model [62] that, although widely used for different material classes [65-67], does not predict the ablation threshold saturation; other is the exponential defect accumulation model, in which the lowering of the ablation threshold increases the defect creation probability for the next pulse, until the defects saturation is reached and a constant value of F_{th} for the superposition of many pulses is established [59, 60, 63]. In this exponential model, the ablation threshold for the superposition of N pulses, $F_{th,N}$, can be described by [63]:

$$F_{th,N} = (F_{th,1} - F_{th,\infty})e^{-k(N-1)} + F_{th,\infty}, \tag{3}$$

where $F_{th,1}$ is the single shot ablation threshold fluence and $F_{th,\infty}$ is the saturated ablation threshold, below which there is no defect accumulation. If $F_{th,1}>F_{th,\infty}$, incubation effects are present and the ablation threshold decreases as more pulses hit the material; if $F_{th,1}<F_{th,\infty}$ laser conditioning occurs, and the material becomes progressively more difficult to ablate as the pulse superposition grows. In equation (3), k is the incubation parameter, which characterizes the strength of defects accumulation, and is equal or greater than zero. If k=0 the ablation threshold is constant and do not depend on the pulses superposition. As k grows, fewer pulses are needed to reach the incubation effects saturation.

To quantify the incubation effects for a given material, its ablation threshold is measured for various pulse superpositions, including the single shot case, and then equation (3) is fitted to the data. Usually this is carried out using the "zero damage" method, which produces a graphic like the one show in Figure 1 to determine F_{th} for each superposition, demanding a lot of experimental time and data analysis.

It is possible to use the D-Scan Method to determine the ablation threshold for the superposition of an arbitrary number of pulses, $F_{th,N}$, and for that the pulse superposition N that etches the profile maximum transversal dimension, $2\rho_{max}$, has to be known. To calculate this superposition, we hypothesize that it is given by the sum of the intensities generated at ρ_{max} by all the pulses that hit the sample during a D-Scan, normalized by the intensity generated by the pulse centered at ◎ (Figure 2b). Under these assumptions, and also considering that the longitudinal and transversal translation speeds (v_z and v_y in Figure 2a) are chosen to etch an elongated profile, it can be shown [68] that the pulses superposition N is given by:

$$N = \vartheta_3(0, e^{-\left(\frac{v_y}{f\,\rho_{max}}\right)^2}), \tag{4}$$

where f is the pulses repetition rate and ϑ_3 is the Jacobi Theta Function of the third kind [69]. It can also be shown [68] that, when the transversal speed is low or the pulses repetition rate is high, situation in which a large value of N is expected, equation (4) can be simplified to:

$$N = \sqrt{\pi}\frac{f\,\rho_{max}}{v_y} \approx 1.8\frac{f\,\rho_{max}}{v_y}, \qquad \text{for} \quad \frac{v_y}{f\,\rho_{max}} \approx 0. \tag{5}$$

Using the previous results, the determination of the ablation threshold for an arbitrary pulses superposition requires performing a D-Scan, then using the experimental values of ρ_{max}, v_y and f in equations (2) and (4) to provide $F_{th,N}$ and N. This is a fast experimental procedure since the D-scan can be done in a few seconds for high values of N (and in a few minutes for small ones), and its repetition for varying values of v_y and f quickly provides the ablation threshold for various superpositions, leading to a prompt determination of the incubation parameter from equation (3).

3. Ultrashort pulses machining main applications

The following sections describe ultrashort pulses machining applications, starting from low energy and fluence ones that produce surface modifications, going to high fluence usages that create 3D structures.

3.1. Low and high fluence regimes

Irradiation of metallic or semiconductor surfaces at fluences near the threshold can results in the formation of the so called "ripples", or "Laser Induced Periodic Structures" (LIPSS), which are regularly aligned, long and periodic structures. Their formation is not yet completely understood and is subject of intense study [70]. Typical periodicity is smaller than the laser wavelength, and for enough number of pulses it can evolve to trenches as deep as 1 μm, meaning an aspect ratio of 10 or more. The spacing and orientation of the ripples seems to depend mainly on the beam properties, as energy, number of shots, polarization and angle of incidence, and then, in principle can be controlled to some extent.

Further surface modifications, however, take place as the number of pulses or their energy increase, and the ripple pattern is broken by the formation of cone-like structures, which present a typical feature size and spacing at the order of few micrometers or even smaller. The shape and size of these structures can be further changed with the increase of fluence and number of superimposed pulses, and the use of an acid atmosphere has also been applied to increase the cones height and sharpness. Figure 3 shows examples of LIPSS and cone like structures produced by low energy ultrashort pulses from a Ti:Sapphire laser.

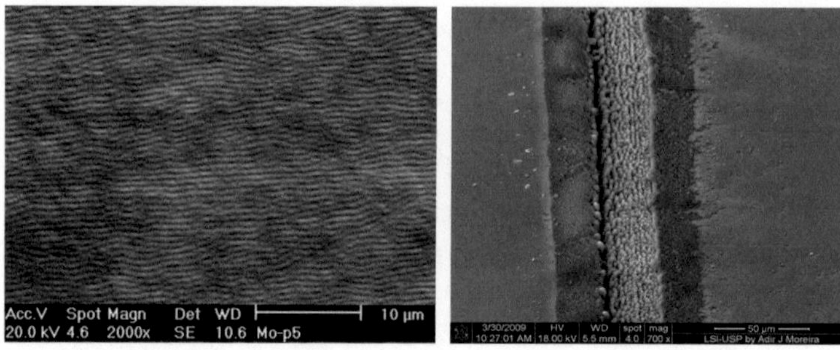

Figure 3. a) Laser Induced Periodic Structures (LIPSS) formed on molybdenum surface. b) Track of cone like structures (white central track) formed on silicon surface.

Up to this point, a minimum amount of material has been ejected from the surface and its topography has not been significantly recessed. To produce substantial depressions it is necessary to further increase the number of shots or the pulse energy. The shape of the created depression resembles the beam intensity profile, and for Gaussian beams the minimum diameter obtained by focusing can be as small as the laser wavelength. If the material and laser conditions allow, heating effects can be neglected, and holes and structures can be directly machined with this size. This is the low fluence ablation regime.

Processing in the low fluence domain can avoid heat accumulation, but the extraction of material is not efficient and only few hundreds of nanometers are etched by each laser pulse. An enormous increase in ejection efficiency is obtained when the fluence is raised to higher values and machining is performed at the high fluence regime. In this regime, heat is accumulated in the affected zone, and several associated phenomena occur. The most important one is the phase explosion, in which the local temperature raises so rapidly that there is not enough time for the material to undergo phase transition. The temperature goes well above the vapor value while the material remains in liquid phase. At this point, a severe explosion happens taking material away from the heated pool.

Machining with high fluence indeed offers high material removal efficiency, but the accumulated heat results in almost the same drawbacks as the ones verified with longer laser pulses that produce thermal ablation. Hence, like in conventional laser processing, this regime leads to debris formation, burr and molten material resolidification on the processed area sidewalls. A HAZ can also appear on the vicinity of the processed volume, causing phase transitions in metals and index of refraction changes in crystals and dielectrics. Due to the very low pulses energy, these effects can extend to a very small region of the produced structure and are irrelevant in many practical cases.

Heating effects also increase the ablated crater size, and the minimum diameter obtained can be far bigger than the focused spot size. On the other hand, low fluence processing can result in affected areas no longer restricted by the diffraction limit. The deterministic character of the laser-material interaction means that ablation will only take place in the portion of the beam cross section in which the intensity is above threshold. Of course this is only possible with a precise control of the laser energy and focus position from shot to shot, and a variation of more than 2% can prevent this possibility.

As already discussed, the complex nature of the ultrashort pulses and matter interaction is responsible for the onset of many different phenomena that depend on the particular characteristics of the laser and the material. The choice of these parameters enables the control of the affected region to produce a rather large amount of diverse structures. Among these possibilities, machining with femtosecond laser pulses can be roughly divided in three great areas: topographic surface modifications, selective ablation of surfaces, and direct writing of microstructures in surfaces and 3D structures in the bulk of transparent materials.

Topographic surface modifications occur when small amounts of material are extracted, resulting in very shallow recesses on the surface, nevertheless affecting its physical and chemical properties. In selective ablation of surfaces, very thin layers of one type of material can be completely etched without affecting the layer bellow it. Direct writing of surface microstructures results in the creation of pre-designed contours and profiles, produced by the relative movement of the laser beam and sample. 3D structures in the bulk of transparent materials are formed when the laser beam passes through the surface of a transparent material and is focused into it, creating complex 3D structures.

3.2. Topographic surface modifications

Nano and micro cone-like structures have been used to trap light and enhance the absorption of reflective surfaces. In particular, this was done by Mei et al. [71] on the surface of silicon with the purpose of using it as an improved sun light absorber. In their experiment, very thin and flexible pieces of silicon were irradiated by femtosecond pulses followed by HF etching. This acid etching enhances the micro spikes sharpness produced by the laser and increases the structure number of reflections, resulting in enlarged light absorption. The irradiation was done in the presence of sulfur hexafluoride (SF_6), which further improved light absorption by creating impurity bands due to sulfur atoms doping during laser irradiation.

Halbwax et al. [72] produced slightly different geometries of such micro structures on n-type silicon doped phosphorus. This was done by varying processing parameters in vacuum. Even using high fluence irradiation, spikes and other structures were formed with size and spacing in the few micrometers range. After boron implantation, these structured surfaces presented an increase of more than 30% in light absorption induced photocurrents, showing its potential for use in photovoltaic cells.

Nano and micro structures have also been used to change physical and chemical properties of titanium and ceramics surfaces in order to improve their biocompatibility for medical implants [73]. This is possible by creating very small structures on their surfaces, enabling, to some extent, the control of cellular growth and adhesion, inhibiting or promoting proliferation of certain types of cells. In this respect, Vorobyev et al. [74], produced several kinds of sub-micrometric structures on pure titanium surfaces by femtosecond laser irradiation. By the use of low and high laser fluences, and varying the number of overlapped laser shots, they produced diverse structures like nanostructured LIPSS, sphere-like nanoprotrusions and columnar microstructures, whose featured size ranged from few dozens of nanometers to hundreds of micrometers.

The same kind of texturing has also been studied to improve the tribological properties of several materials. This has been specially done on thin films of hard materials such as TiN and diamond-like carbon [75], in which nanostructured LIPPS changed the micro and macro frictional properties and hardness of these materials. Another consequence of the formation of micro arrays as cone-like or pillars, is their interaction with water, keenly increasing the contact angle of the liquid with the surface. By controlling the aspect ratio of these arrays, super-hydrophobic surfaces have been produced [76, 77], creating self-cleaning surfaces and which are also believed to take an important role in cell control processes.

As the presence of micro and nano structures on metal surfaces changes its interaction with light, some research has been done to control the surface reflectivity, and hence the perception of its color [78, 79]. The control of shape, size and spacing of gratings on surfaces is a well-known way to control the reflectivity of light; therefore, ripples formed by nano-structured LIPSS obtained by femtosecond laser irradiation are a natural approach to accomplish such a task. These characteristics can be controlled by handling some process parameters, as fluence, polarization direction, angle of incidence and scanning speed. The final shape comprises self-produced micro/nano gratings maintaining an almost fixed pitch and direction, covering a large area. Besides using LIPSS, color in metals has also been done by directly machining small grooves and arrays of micro-holes in their surfaces. Consequently, structuring metal surfaces in scales from nanometers up to hundreds of micrometers allow the control of the optical properties, from UV to terahertz.

3.3. Selective ablation of surfaces

The ability of ultrashort laser pulses to ablate very thin layers of material can be used to selectively remove coatings overlaid on a substrate. Layers as thin as a few hundreds of nanometers can be removed by each laser shot, and the process can be repeated up to the point where all the unwanted material is removed and the substrate is stricken by the laser beam. This process can be a precise and useful tool in many applications [80-84], as cleaning, restoration of art objects, decontamination and decoating of cutting tools.

The differences in physical properties between coating and substrate determine the feasibility, precision and complexity of the process. If the material to be removed absorbs the laser wavelength and presents a lower ablation threshold than the substrate, then the process is relatively simple and there is no need for feedback control. The pulse fluence is adjusted to be above that of the coating and below that of the substrate; an excess of pulses on the substrate surface will not affect it. On the other hand, if the coating damage threshold is close or lower than the substrate one, then a feedback must control the ablation process to prevent damage to the substrate. This can be accomplished by measuring, in real time, the spectrum of the emitted plasma (LIBS [81]) and observing the evolution of characteristics emissions of the coating and the substrate.

3.4. Direct writing microstructures in surfaces

Regardless of the topographic structures formed on a surface hit by femtosecond pulses, they can be used in a bigger scale to process complex structures previously designed, just like in a milling machine. Tiny holes and channels with high aspect ratios can be produced in almost any kind of material, with very high dimensional resolution, no heat effects, and beautiful cosmetic appearance.

Due to the relative low average power of the femtosecond laser systems, large processed areas or high ablated volumes still present a very low productivity, and are only considered when other methods fail. This happens when tiny structures must be produced on sensitive or transparent materials, when heating effects and dross must be avoided, and when sub-micron lateral size and resolution must be attained.

Although some processes use femtosecond laser assistance to produce sub-micron holes [85], these have also been directly drilled on thin films by precisely controlling the fluence just above the threshold [86]. Gaussian beams tightly focused produce a steep spatial intensity (fluence) distribution which precisely determines the diameter within which the fluence is above the threshold of the material. So the intended diameter of a hole can be achieved by using proper laser fluence and few hundreds nanometers holes can be directly drilled in thin films [87, 88].

Alumina is a very sensitive material that frequently presents cracks, delaminations and striations on the laser processed surface. However, some authors [89, 90] have managed to produce high aspect ratio and low tapered holes in alumina wafers suitable for micro-vias used in micro-electronic circuits. In trepanation mode, high fluences can be used in conjunction with other processing parameters in a way that minimizes heat accumulation and avoid recast molten material, cracks and delamination. Drilling in percussion mode produces smaller diameters, but these are dependent on the laser fluence. Unlike with longer pulses, however, the hole sizes almost do not depend on the number of overlapped pulses, which results in an easier and more precise production. Although percussion drilled holes present more tapered profiles, high aspect ratios can be obtained in relatively thick materials; typical diameters of a few dozens of micrometers are easily done on hundreds of micrometers thick alumina plates.

Structures with complex shapes in almost any kind of material can also be done by cutting and sculpting cavities with femtosecond lasers. The exact knowledge of the phenomena occurring during the laser material interaction allows the production of previously designed shapes in the same way as in a milling machining, with very high precision, but in a much smaller scale. Lateral and vertical accuracies depend on the fluence used and on the aspect ratio figure, but sub-micron or a few micrometers resolution is easily attained. Plates with thicknesses in excess of 0.5 mm have been cut to produce microparts and functional elements used, for instance, in MEMS and other microsystems. Carving into bulk or deposited films can also be done to produce trenches, microchannels and molds with applications in microelectronics, microfluidics and MEMS.

Just like in a milling process, machining with ultrashort laser pulses has been done in complete CAD/CAM systems where not only the toolpath has to be controlled, but also the energy and the number of overlapped pulses. Unlike a real milling tool, the size of a damage produced by a fixed laser focused spot increases as the number of shots accumulates (up to a certain point) on the same spot, and the same occurs by the increase in the pulse energy. Hence, the process control software must be fed with the fundamental parameters governing the interaction of the material with ultrashort pulses.

Although scanning mirrors can be used, the most common femtosecond processing systems use stable and precise 2D or 3D translation stages, with a still beam focused by a short focal length lens. Some cases demand vacuum processing, but generally the machining is performed in regular atmosphere with no shrouding gases. The pulse energy is controlled by a combination of a polarization rotator and an analyzer, and the pulses overlapping is calculated by combining the scanning speed with the laser repetition rate and the focused beam diameter.

As isolated components or as parts of a mechanical microsystem, tiny pieces have been machined in a variety of materials and shapes to an increasing number of new applications, where sensors and actuators for MEMS and microchannels for microfluidics are among the most important examples. Silicon, glasses and several kinds of ceramics, polymers and metals have been used to produce structures for these and other applications.

Silicon is one of the most largely studied materials for ultrafast laser machining, and silicon wafers have been cut and etched by laser to produce micro-molds to be used in a large number of micro electromechanical systems and electronic devices [91, 92].

Many kinds of polymers have also been machined by femtosecond lasers, mainly to produce small channels for rapid prototyping of microfluidic devices, and optical gratings. Suriano et al. [93] determined process protocols for laser machining of some polymers, showing the feasibility to obtain good shapes from some kinds of thermoplastic polymeric materials. Despite good geometrical and cosmetic results, some degradation, probably due the use of high fluences, was observed; physicochemical modifications in the processed area caused some darkening which was explained in terms of oxidation, dehydrogenation or thermal depolymerization.

Microstructured optical fibers were transversally machined to provide access to its core with minimum interference in their optical and guiding properties [94]. Micrometric transversal holes machined through polymeric cladding and glass core enable filling the fiber with fluids to take advantage of their optical properties for uses in sensors and in flexible microfluidic devices. Figure 4 shows an example of a square hole machined on the surface of a hollow fiber to allow gas filling to modify its optical properties.

In the work of Alemohammad et al. [95] a series of micro-grooves were transversally and accurately machined on the cladding of a fiber Bragg grating. This structure increases the transmission sensitivity of the fiber to the index of refraction of the external medium, making these engineered fibers a powerful tool for measurements of temperature and liquids concentrations. This is a good example of how controlled such machining can be, since it was precisely done (22 μm wide × 32 μm deep grooves with 50 μm pitch) on a thin polymer curved surface without damaging the core beneath it.

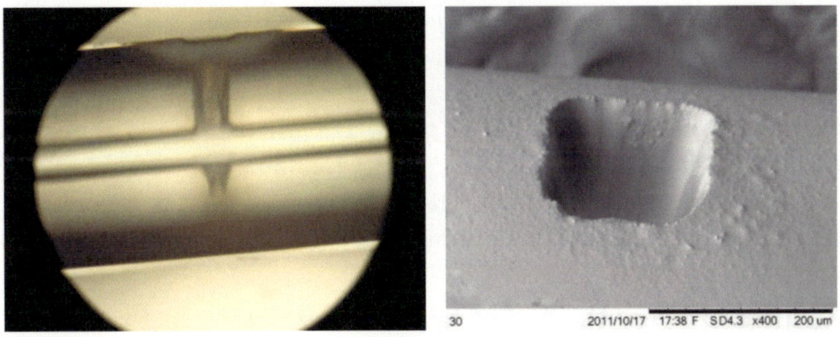

Figure 4. Hole bored in a hollow fiber with ultrashort pulses, using trepanation. a) fiber lateral view optical micrograph, and b) hole entrance view electron micrograph.

Another example of accuracy in laser processing of fibers is the machining of their core end faces [96]. Arrays of sub-micrometric structures boost the signal of a surface enhanced Raman scattering fiber probe, increasing its sensitivity for the detection of many biological and chemical substances. Minuscule cantilevers directly machined in one end of the fiber provided a very sensitive way of movement detection [97].

3.5. Micromachining in metals

Machining metals with ultrashort laser pulses is not as advantageous as machining dielectrics. The presence of free electrons, besides inhibiting Coulomb ablation, widens the heat affected region due to their high mobility. In this sense, the metal thermal properties play a fundamental role in the process characteristics and taking these into account, reproducible, small and accurate structures can still be produced [98].

The control of the pulse duration and wavelength can further improve efficiency and accuracy when machining some metals [99-101]. In a general way, as the pulse shortens (below 1 ps), the finishing improves and the ablation rate increases. Low fluence regime gives the best quality results, but at the same time implies a low ablation rate. Hence, work in this regime is only advantageous with systems offering high repetitions rates at moderate energies [102].

Another approach to increase efficiency is to work in the high fluence regime, where thermal processes, including phase explosion, fully control ablation. In this case, unlike pico or nanosecond pulses, almost all molten material is blown away in the form of nanometric particles, leaving a smooth and clean surface with almost no debris or dross and a negligible HAZ. Indeed, heat can be accumulated also from pulse to pulse as function of the laser repetition rate and thermal properties of the metal. Additionally, for sufficiently high repetition rates, the plume from the previous pulse can also provide shielding that scatters and absorbs pulse energy. Due to this, when processing with high repetition rates, from hundreds of kilohertz to some megahertz, efficiency and accuracy may vary according to the metal and process characteristics.

There are many examples of machining with femtosecond pulses to produce small devices in metals; some of the most common are microcantilevers [103], coronary stents [104] and pieces of NiTi shape memory alloy [105] used in MEMS. Percussion hole drilling in metals [106, 107] is also another application that is quickly growing. Thicknesses up to 0.5 mm and accuracy of few micrometers are reported in many different metals and applications, most of them without shrouding gas or vacuum, using fundamental wavelengths (800 nm or 1030 nm) and pulsewidths from 100 fs to 800 fs. The parts produced frequently do not need re-work or etching for cleaning and better finishing.

3.6. 3D structures in the bulk of transparent materials

The ablation threshold of the bulk of a transparent medium is considerable higher than of its surface (about twice). Even so, damages forming complex 3D structures can be produced in materials transparent to the wavelength used, by tightly focusing the laser beam below their surfaces. The requirement is to ionize and produce plasma in a volume inside the material, without affecting the surface above. To accomplish this, high numerical apertures (N.A.>0.5) objectives must be used to focus the beam under the surface, in a condition where the fluence is below the threshold on the surface, and at the same time overcomes the volume ablation threshold at the focus. In general, for glasses, nonlinear absorptions are induced for intensities in the range of $1\text{-}5\times10^{13}$ W/cm^2, which can be easily reached by any combination between the main processing parameters (pulse energy and duration, and beam diameter at focus).

For a laser beam with good fundamental spatial mode, the diameter of the focused spot can be close to its wavelength (800 nm for Ti:Sapphire laser, 1 μm for Yb:fiber). Practical temporal pulsewidths range from 30fs to 200fs for Ti:Sapphire and from 200 fs to 400 fs for Yb:fiber. Hence, less than 10 nJ are enough to create damages in the bulk of glasses when using Ti:Sapphire lasers, and this figure increases to 30–160 nJ when using Yb:fiber lasers. Under these conditions, induced absorption occurs in a volume of less than 1μm^3, and a single femtosecond pulse is enough to create damage in the bulk of any transparent material [108]. Within this volume, super-hot dense plasma is formed, creating very strong shock waves that destroy any atomic arrangement, resulting in modified structures or even voids.

The use of UV fs pulses enables the creation of even smaller defects into transparent materials, since its wavelength is shorter than near IR. Furthermore, the higher UV photon energy also increases the probability of multiphoton absorption, decreasing the breakdown intensity threshold. Using this approach, Dubov et al. [109] reported the production of 250 nm periodic nanostructures using 30 nJ, 300 fs laser pulses.

For intensities below the threshold, only nondestructive effects can occur, such as changes in the index of refraction of the material and color centers creation [61]. The presence of these defects produces intermediate levels in the band gap of the dielectric, resulting in incubation effects that usually lower the material damage threshold.

Although not a strict micromachining, permanent modification in the optical index of refraction have been extensively used to create light guiding structures in the bulk of transparent media [110, 111]. The relative movement of the laser beam focus inside the material produces a pathway with cross section and index of refraction contrast suitable to guide light as waveguides embedded in the material bulk. The focus movement can be done in the same direction as the beam propagation, or perpendicular to it. In the first case, the modified region presents a cross section with circular geometry, and in the other this geometry is elliptical. In spite of this disadvantage, the transversal geometry is preferred once the sample and waveguide lengths are not limited by the depth of focus as occurs in the longitudinal method.

The refractive index change depends on the laser parameters and material composition, and is induced by many different mechanisms not yet fully understood. The change can be either positive or negative, depending on the material properties, and is caused mainly by thermal effects, defects creation and material densification. High repetition rates (a few MHz) are generally used, causing heat accumulation with consequent melting and fast cooling in the focal volume scanned by the laser. These rapid phase transitions result in local structural modifications, giving rise to changes in the optical index of refraction that can be over 1%.

Many passive and active guiding structures have been incorporated in devices mainly used in communications and biochemical applications. Couplers, splitters and delay lines are some of the more typical, but more sophisticated structures, as interferometers and diffraction gratings, have been also produced.

Focusing the laser beam in the bulk of a transparent material, just above its damage threshold, causes local structural changes that affect its physical and chemical properties. In the case of sapphire and glasses, their chemical resistance to HF acid and KOH are considerable lowered, reaching a ratio of $1{:}10^4$ for sapphire and $1{:}10^2$ for glasses. This phenomenon has been used to produce hollow channels inside these materials when irradiating with femtosecond laser pulses. After laser irradiation, the material with complex modified 3D structures inside, is etched with a solution of HF acid (or KOH) for up to 48 h to remove the modified material, and results in a hollow structure.

Another approach to make 3D hollow structures inside transparent materials is by focusing the laser beam on the rear surface of a glass sample that is in contact with water [112]. Movements of the focused spot in the vertical direction create a microchannel entering the material's bulk while the water removes the debris formed during irradiation. The process is then continued and complex structures can be produced in all directions.

4. Case studies

The following sections present results obtained in our laboratories regarding the determination of ultrashort pulses ablation threshold and incubation parameters in various materials, and the micromachining in the low and high fluence regimes.

4.1. Silicon ablation threshold measurement

The "zero damage" method was used to determine the ablation threshold for crystalline silicon. For this, 30 μJ maximum energy, 100 fs (FWHM) pulses, centered at 800 nm, at 1 kHz maximum repetition rate, generated by a Ti:Sapphire CPA system (Odin, from Quantronix) were focused on the sample surface by a 38 mm lens. The sample was fixed to a computer controlled translation stage. Many measurements were performed to obtain $D^2 \times F_{th}$ data for different pulses superpositions N, as shown in Figure 5a for N=1. In this graph the low and high fluence ablation regimes are clearly seen, and the fits by equation (1) provide the values 0.32 J/cm² and 8.7 J/cm² for the low and high fluence ablation thresholds, respectively. Crystalline silicon shows a curve $D^2 \times F_{th}$ with the same characteristics seen in dielectrics, as exemplified by the sapphire results shown in Figure 1, with an abrupt slope change when thermal effects start to dominate and the material removal rate sharply increases.

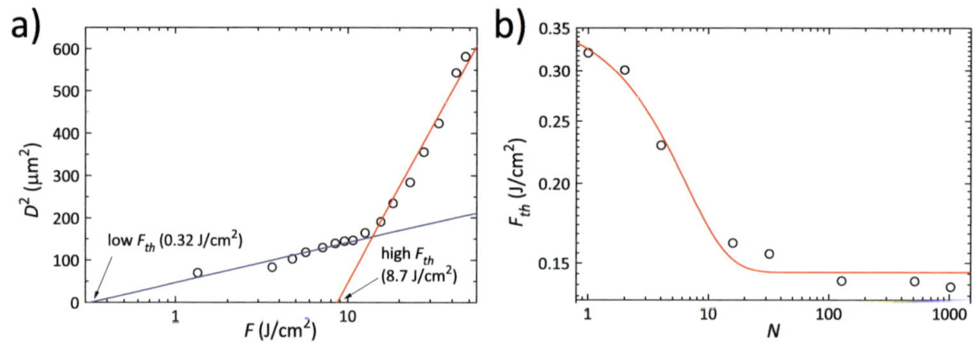

Figure 5. a) Square diameter $D^2 \times F_{th}$ for single pulse ablation (N=1) of silicon; b) Silicon damage threshold as fuction of the pulses superposition.

Figure 5b shows the low fluence ablation threshold measured for different pulses superpositions N, together with a fit by equation (3), which provides $F_{th,1}$=0.32 J/cm², $F_{th,\infty}$=0.15 J/cm² and k=0.22. It can be observed that the defects accumulation is strong in this material, and saturation of the ablation threshold occurs for less than 50 pulses.

4.2. Molybdenum ablation threshold measurement

The "zero damage" method was also used to determine the ablation threshold of metallic molybdenum, repeating the experiment performed with crystalline silicon described in section 4.1. The diameter of the ablated region as a function of the pulse fluence, for the superposition of 4 pulses on the surface of polished molybdenum, is shown in Figure 6a, together with a fit by equation (1), providing F_{th}=0.019 J/cm². This graph shows that the fluence increase does not exhibit an abrupt slope change from the low to the high fluence regime, instead a smooth transition is observed, demonstrating the onset of new processes as the fluence grows.

The ablation threshold dependence on the pulse superposition data, together with a fit by equation (3) is shown on Figure 6b, and, as is the case with dielectrics and semiconductors

(Figure 1 and Figure 5b), it can be seen that the damage threshold decreases as the number of overlapped pulses grows. The values returned by the fit are $F_{th,1}$=0.051 J/cm², $F_{th,\infty}$=0.0136 J/cm² and k=0.84, evidencing that saturation is quickly reached for less than 10 pulses. Nevertheless, more data points around this value would provide a better result.

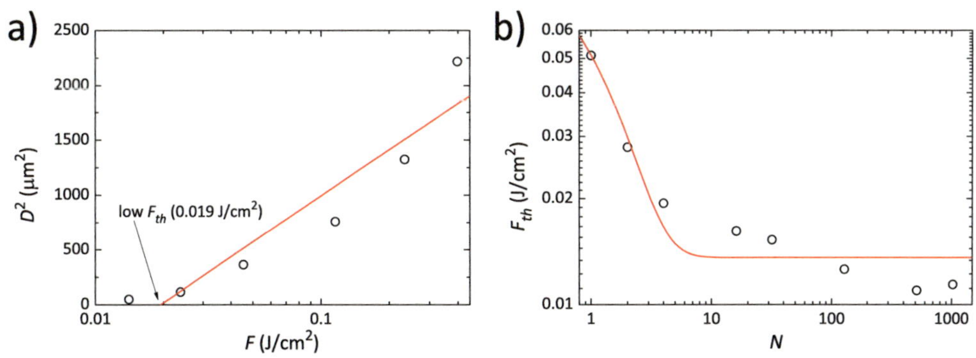

Figure 6. a) Square diameter dependence on the ablation threshold fluence for N=4; b) ablation threshold fluence $F_{th,N}$ as a fuction of the pulses number.

Comparing the single shot fluence ablation thresholds obtained by us for sapphire (5.6 J/cm², Figure 1), silicon (0.32 J/cm², Figure 5a) and molybdenum (0.051 J/cm², Figure 6b), it can be seen that, as the easiness to generate free and seed electrons increase (bandgap decreases), the ablation threshold value drops, as expected by the reasons discussed in section 2.1.

The behavior seen in Figure 6 implies that machining metals with ultrashort pulses is more challenging than machining dielectrics, demanding a better knowledge and control of the process parameters, and sometimes requiring a rework for finishing. The absence of ablation by Coulomb explosion means that always there will be some heat flow to the surrounding area and the conditions of machining must be adjusted to minimize its effects.

4.3. Validation of the D-scan method for ablation threshold and incubation parameter determination

To demonstrate that the D-Scan is a valid method to determine the ablation threshold for an arbitrary pulse superposition, we measured the ablation threshold of BK7 samples by the "zero damage" and D-Scan methods, and compared the results. A Ti:Sapphire CPA system (Femtopower Compact Pro CE-Phase HP/HR, from Femtolasers) was used, continuously generating 100 fs (FWHM) pulses centered at 785 nm with 40 nm of bandwidth (FWHM), at a maximum repetition rate of 4 kHz. The pulses were focused by a 38 mm lens for both methods, and all measurements were performed in air. For the "zero damage" method, the F_{th} was measured for single pulses and for the superposition of 2, 4, 16, 32, 120, 510 and 1020 pulses, and the results, calculated using equation (1), are show as red circles in Figure 7 [68].

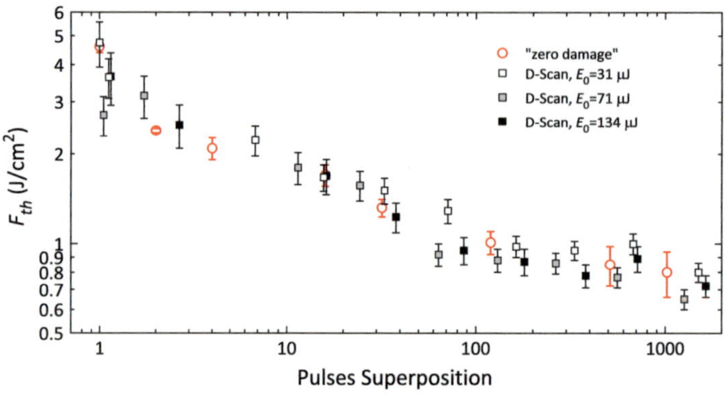

Figure 7. BK7 Ablation threshold measured by the traditional (red circles) and by the D-Scan (squares) method [68]. The good data agreement shows the methods equivalency to determine ablation thresholds in the ultrashort pulses regime.

For the D-Scan measurements, the sample was irradiated by 31, 71 and 134 μJ pulses, at various combinations of repetition rates (50, 100, 500, 1000, 2000 and 4000 Hz) and transversal displacement speeds (6, 12, 25, 50 and 100 mm/min) resulting in superpositions ranging from single pulse to almost 2000 pulses. The Results are show as squares in Figure 7 [68]. This graph shows that the ablation thresholds obtained by both methods agree for superpositions spanning more than 3 orders of magnitude. This implies that the superpositions measured by both methods, although conceptually different (in the "zero damage" the superposition is the number of pulses that completely overlap at the same spot, while for the D-Scan it is given by the sum of the intensity of many pulses spatially apart), are the same, indicating that incubation effects are a linear sum of the intensities that hit a spot. This intensity linear sum effect is also corroborated by the fractionary superpositions that can be observed in Figure 7 D-Scan results for the superposition of less than 10 pulses, which are consistent with the values obtained for integer values of *N*.

4.4. D-scan technique determination of the incubation parameter of optical materials

The D-Scan method was used in our laboratories to quickly determine the incubation parameter for two common optical materials, sapphire and suprasil glass. The same Ti:Sapphire CPA system described in section 4.3 was used, continuously generating 95 μJ, 25 fs (FWHM) pulses. The beam was focused by a 75 mm lens, and the samples were fixed to a 3-axes computer controlled translation stage that executed the diagonal motion. The measurements were done using various combinations of the pulses repetition rate (100, 500, 1000 and 4000 Hz) and transversal displacement speeds (10, 5, 2.5, 1.25, 1, 0.5, 0.25, 0.2, 0.1, 0.05, 0.04, 0.03, 0.02 and 0.01 mm/s), and it took less than 1 hour to etch more than 35 profiles in each sample. The ρ_{max} value for each profile was measured in micrographs taken in and optical microscope, and the ablation threshold and superposition values for each profile were then calculated using equations (2) and (4). The results are exhibited in Figure 8, and show that in about 1 hour of laboratory time, it is possible to perform measurements that produce data spanning more than four orders of magnitude of *N*, revealing the dynamics of the defects accumulation until saturation is reached. One of the strengths of the D-Scan technique, which is fast data acquisition, becomes evident when comparing the graphs exhibited in Figure 8 with the ones shown in Figure 5b and Figure 6b: lots of data points are obtained (in a short time), allowing a detailed analysis of the incubation effects, and deciding which model is better to describe them. This kind of analysis is very arduous to be performed with data gathered with the "zero damage" method, due to its complexity that demands a lot of laboratory and data analysis time.

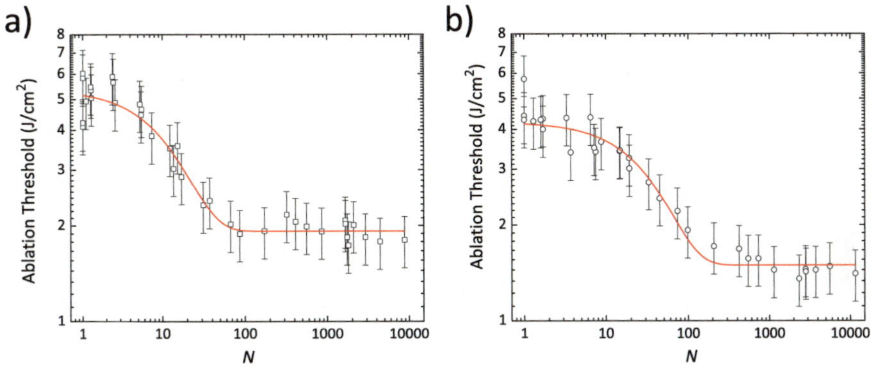

Figure 8. Ablation threshold as a function of the pulse superposition measured using the D-Scan technique for a) suprasil and b) sapphire samples. In each graph, equation (3) is fitted to the data.

To determine the samples ablation thresholds and incubation parameter, equation (3) was fitted to each sample data (red curves on Figure 8), and the parameters obtained are shown in Table 1. The parameters in this table highlight what can be seen in Figure 8: laser created defects have a faster accumulation rate in suprasil (higher *k*), in which saturation of the ablation threshold is reached for less than 100 pulses, in contrast to sapphire where approximately 500 pulses are needed to saturate F_{th}. Nevertheless, saturation fluence is lower in sapphire, indicating that defects electrons are easier excited in this material.

Sample	$F_{th,1}$ (J/cm²)	$F_{th,\infty}$ (J/cm²)	*k*
Suprasil	5.1±0.1	1.92±0.05	0.064±0.008
Sapphire	4.2±0.1	1.49±0.04	0.022±0.003

Table 1. Fit parameters for each sample

4.5. Direct micromachining of microchannels in BK7

In direct machining of cavities, multipulse ablation is often needed to remove the necessary amount of material, and incubation effects occur. These effects must be taken into account when establishing the machining parameters to ensure adequate control of the ablation process. Within this context, this section presents a study on BK7 etching, in which this approach was used and produced microchannels with shapes, dimensions, finishing and HAZ complying with the requirements for microfluidic devices [113].

To evaluate the efficacy of the method, two different strategies were used, and in both approaches, several parallel lines were swept by the focused laser spot to create a channel with the desired width. For each scanning, a different lateral offset distance *d* (Figure 9) was used.

The first strategy, named hard machining, consists in working in the high fluence regime, with a high material removal rate, and low longitudinal displacement speed (large longitudinal pulse superposition *N*). The second approach, dubbed smooth machining, uses a high displacement speed (small *N*) and operates in the low fluence regime, removing small amounts of material per pulse. This condition sweeps the surface with a smaller number of shots when compared to the hard machining, making it necessary to repeat the process many times (many sweeping layers) to reach the desired depth [113].

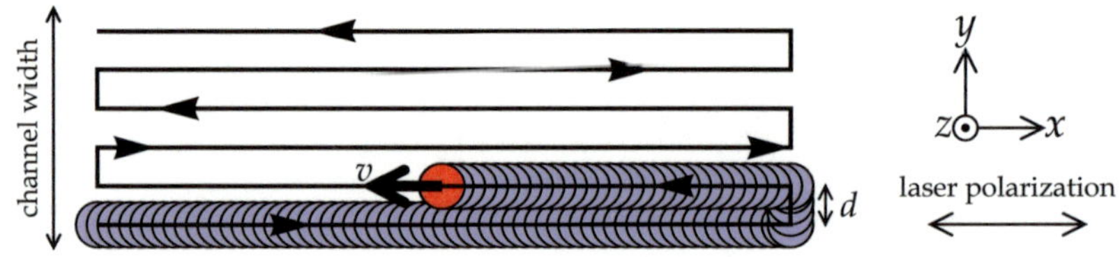

Figure 9. Diagram of the beam path for one etched layer. The beam polarization and scanning direction are parallel to the channel longitudinal axis.

For the hard machining, microchannels were etched with 1 kHz repetition rate, 33 μJ maximum energy pulses, constant scanning speed at 1.6 mm/min chosen to produce a pulses superposition *N*=500, and varying offsets *d* for each different channel. These irradiation conditions lead to asymmetric, shallow and irregular channels, as can be seen on the transversal section OCT images shown in Figure 10. In this image, the ablated regions are clearly seen for each track, together with very deep and tapered modified regions below the ablated surface. These regions, which cannot be seen by optical microscopy, are not related to catastrophic damage, and can only be observed by OCT due to changes in the index of refraction caused by nonlinear effects resulting from the pulses high intensity [114, 115].

The subsurface modified regions are a consequence of incubation effects. As the pulses overlap increases, F_{th} is lowered, and a greater fraction of the pulse energy is used to cause more severe effects on the processed material. The large amount of ablated material also produces a skewed wall for each individual track. As consequence, the sloped walls of the machined track reflect the subsequent pulses to the adjacent track that follows it, making the ablated area erratic and unpredictable [113].

In the smooth ablation strategy, the lateral displacement *d* was changed in each layer to always maintain a fixed overlap of 25% of the created damage diameter, since the size of the ablated area changes due to incubation processes. In each layer, the total width of the machined track takes into account the many sweeps to keep the lateral overlap constant, spreading the amount of pre-damaged material smoothly and homogeneously. When this is not considered, the deposited energy is more quickly absorbed in places where accumulation of pre-damages is greater, resulting in a non-homogeneous ablation.

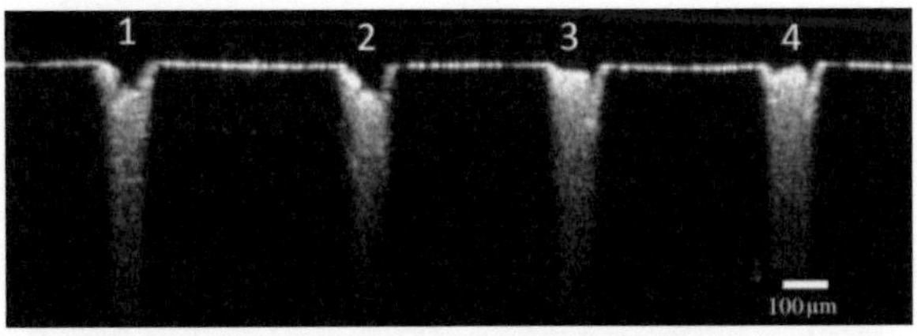

Figure 10. Cross section OCT image of the channels etched by "hard machining". The lateral displacements are d=10, 20, 30 and 40 μm for channels 1 to 4, respectively.

The width D of the ablated track during each sweep depends on the incubation resulting from the *previous* layers etched by N_{n-1} pulses superimposed, and the longitudinal overlap of the *actual* scanning, N_{actual}, where n is the layer number. Within these considerations, the total superposition to calculate D is $N_n=N_{n-1}+N_{actual}$. Thus, D is a function of the beam radius on the machined surface, w, the peak fluence of each pulse, F, and the ablation threshold $F_{th,N}$ for total incubation contribution N, and can be obtained using equation (1) (substituting $F_{th}=F_{th,N}$).

Finally, the value of the longitudinal superposition of the actual scanning, N_{actual}, can be determined by employing equation (4) using the scanning speed v, the repetition rate f and considering that the $\rho_{max}=w$ (beam radius on the sample surface).

A clear improvement in the channels shape occurred when the smooth machining ablation was used, as shown in the OCT image on Figure 11. The channels cross section shows a flat bottom and straight walls, almost perpendicular to the surface. Refractive index modifications below the channels are absent, evidencing that the material properties are preserved in the immediate vicinity of the machined regions.

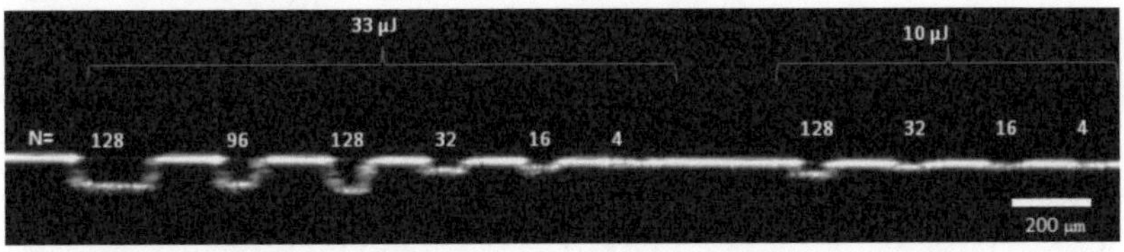

Figure 11. Cross section OCT image of the channels etched by the smoot machining strategy. The total number of overlapped pulses, N, is indicated above each channel. Pulse energies of 10 μJ and 33 μJ were used.

5. Conclusions

In this chapter we presented an overview of micro and nano machining with ultrashort laser pulses. A historical context was provided, the basic physical mechanisms governing the ablation of solids by femtosecond pulses were described, some relevant real-world applications were outlined and experimental results obtained by us were presented. The text references many important and cornerstone works, and the reader who wants to expand his knowledge and have a deeper understanding in this field should start by reading these books, reviews and papers.

The most important parameters to machine a material with ultrashort pulses are its ablation threshold fluence and incubation parameter. Two techniques to determine these parameters were described, the traditional "zero damage" and the D-Scan methods. Experimental results obtained by both techniques were presented, showing that the methods are equivalent, and both can be used to determine these parameters. Nevertheless, we recommend the use of the D-Scan technique because it is experimentally simpler, provides quick results, and the material modifications caused by it are closer to the final machining results once it is executed in a moving sample.

Author details

Ricardo Elgul Samad, Leandro Matiolli Machado,
Nilson Dias Vieira Junior and Wagner de Rossi
Instituto de Pesquisas Energéticas e Nucleares – IPEN-CNEN/SP, Brazil

6. References

[1] Gad-el-Hak M (2006) MEMS : introduction and fundamentals : the MEMS handbook. Boca Raton: CRC Press/Taylor & Francis. p.

[2] Dornfeld D, Min S, Takeuchi Y (2006) Recent Advances in Mechanical Micromachining. CIRP Ann.-Manuf. Techn. 55: 745-768.

[3] Pradhan B, Masanta M, Sarkar B, Bhattacharyya B (2009) Investigation of electro-discharge micro-machining of titanium super alloy. Int. J. Adv. Manuf. Tech. 41: 1094-1106.

[4] Kaminski P C, Capuano M N (2003) Micro hole machining by conventional penetration electrical discharge machine. Int. J. Mach. Tool Manu. 43: 1143-1149.

[5] Langford R M, Dale G, Hopkins P J, Ewen P J S, Petford-Long A K (2002) Focused ion beam micromachining of three-dimensional structures and three-dimensional reconstruction to assess their shape. J. Micromech. Microeng. 12: 111-114.

[6] Malek C K, Hartley F T, Neogi J (2003) Fast prototyping of high-aspect ratio, high-resolution X-ray masks by gas-assisted focused ion beam. Microsyst. Technol. 9: 409-412.

[7] Dale G, Langford R M, Ewen P J S, Reeves C M (2000) Fabrication of photonic band gap structures in $As_{40}S_{60}$ by focused ion beam milling. J. Non-Cryst. Sol. 266–269, Part 2: 913-918.

[8] Chou S Y, Krauss P R, Renstrom P J (1996) Imprint lithography with 25-nanometer resolution. Science 272: 85-87.

[9] Kovacs G T A, Maluf N I, Petersen K E (1998) Bulk micromachining of silicon. Proceedings of the IEEE 86: 1536-1551.

[10] Bustillo J M, Howe R T, Muller R S (1998) Surface micromachining for microelectromechanical systems. Proceedings of the IEEE 86: 1552-1574.

[11] Boser B E, Howe R T (1996) Surface micromachined accelerometers. IEEE J. Solid-St. Circ. 31: 366-375.

[12] Takeuchi S, Suzuki T, Mabuchi K, Fujita H (2004) 3D flexible multichannel neural probe array. J. Micromech. Microeng. 14: 104.

[13] Fuller L F, Sudirgo S (2003) Bulk micromachined pressure sensor In: Proceedings of the 15th Biennial University/Government/Industry Microelectronics Symposium (IEEE, 2003), 317-320

[14] Honig R E, Woolston J R (1963) Laser-Induced Emission of Electrons, Ions, and Neutral Atoms from Solid Surfaces. Appl. Phys. Lett. 2: 138-139.

[15] Boot H A H, Clunie D M, Thorn R S A (1966) Micromachining with a pulsed gas laser. Electron. Lett. 2: 1.

[16] Anisimov S I (1968) Vaporization of Metal Absorbing Laser Radiation. Sov. Phys. JETP-USSR 27: 182-183.

[17] Kocher E, Tschudi L, Steffen J, Herziger G (1972) Dynamics of laser processing in transparent media IEEE J. Quantum Elec. QE 8: 120-&.

[18] Bourg H, Frederick R W (1975) Laser Machining of Silicon. J. Electrochem. Soc. 122: C260-C260.

[19] Longfell J (1971) High Speed Drilling in Alumina Substrates with a CO_2 Laser. Am Ceram Soc Bull 50: 251-&.

[20] McGuff P E, Deterling R A, Gottlieb L S, Bushnell D, Fahimi H D (1964) Surgical Applications of Laser. Ann. Surg. 160: 765-&.

[21] Cohen M I, Unger B A, Milkosky J F (1968) Laser Machining of Thin Films and Integrated Circuits. AT&T Tech. J. 47: 385-405.

[22] Steen W M (2003) Laser material processing. New York: Springer. 408 p.

[23] Ion J C (2005) Laser processing of engineering materials : principles, procedure and industrial application. Oxford: Elsevier/Butterworth-Heinemann. 556 p.

[24] Kogelnik H, Li T (1966) Laser Beams and Resonators. Appl. Opt. 5: 1550-1567.

[25] Stuart B, Feit M, Rubenchik A, Shore B, Perry M (1995) Laser-Induced Damage in Dielectrics with Nanosecond to Subpicosecond Pulses. Phys. Rev. Lett. 74: 2248-2251.

[26] Lan B, Hong M H, Ye K D, Wang Z B, Chong T C (2003) Laser microfabrication of glass substrates by pocket scanning. In: Miyamoto I, Ostendorf A, Sugioka K, Helvajian H,

editors. Fourth International Symposium on Laser Precision Microfabrication. pp. 133-136.

[27] Ben-Yakar A, Byer R L (2002) Femtosecond laser machining of fluidic microchannels for miniaturized bioanalytical systems. In: Sugioka K, Gower M C, Haglund R F, Pique A, Dubowski J J, Hoving W, editors. Photon Processing in Microelectronics and Photonics. pp. 212-217.

[28] Lane D W (1990) The optical properties and laser irradiation of some common glasses. J. Phys. D Appl. Phys. 23: 1727.

[29] Correa D S, Cardoso M R, Tribuzi V, Misoguti L, Mendonca C R (2012) Femtosecond Laser in Polymeric Materials: Microfabrication of Doped Structures and Micromachining. IEEE J. Sel. Top. Quantum Elec. 18: 176-186.

[30] Vora K, Kang S, Shukla S, Mazur E (2012) Fabrication of disconnected three-dimensional silver nanostructures in a polymer matrix. Appl. Phys. Lett. 100: 063120.

[31] Strickland D, Mourou G (1985) Compression of amplified chirped optical pulses. Opt. Commun. 56: 219-221.

[32] Diels J-C, Rudolph W (2006) Ultrashort laser pulse phenomena : fundamentals, techniques, and applications on a femtosecond time scale. Burlington, MA: Academic Press. 652 p.

[33] Sibbett W, Lagatsky A A, Brown C T A (2012) The development and application of femtosecond laser systems. Opt. Expr. 20: 6989-7001.

[34] Keller U (2010) Ultrafast solid-state laser oscillators: a success story for the last 20 years with no end in sight. Appl. Phys. B-Las. Opt. 100: 15-28.

[35] Russbueldt P, Mans T, Weitenberg J, Hoffmann H D, Poprawe R (2010) Compact diode-pumped 1.1 kW Yb:YAG Innoslab femtosecond amplifier. Opt. Lett. 35: 4169 -4171.

[36] Umhofer U, Jäger E, Bischoff C (2011) Refractive and diffractive laser beam shaping optics. Laser Technik Journal 8: 24-27.

[37] Bloembergen N (1974) Laser-induced electric breakdown in solids. IEEE J. Quantum Elec. QE10: 375-386.

[38] Du D, Liu X, Korn G, Squier J, Mourou G (1994) Laser-induced breakdown by impact ionization in SiO_2 with pulse widths from 7 ns to 150 fs. Appl. Phys. Lett. 64: 3071-3073.

[39] Reif J, Costache F (2006) Femtosecond laser interaction with solid surfaces: explosive ablation and self-assembly of ordered nanostructures. In: Rempe G, Scully M O, editors. Advances in Atomic Molecular, and Optical Physics, V. 53. San Diego: Elsevier Academic Press Inc. pp. 227-251.

[40] Stoian R, Rosenfeld A, Ashkenasi D, Hertel I V, Bulgakova N M, Campbell E E B (2002) Surface charging and impulsive ion ejection during ultrashort pulsed laser ablation. Phys. Rev. Lett. 88: 097603.

[41] Bulgakova N M, Stoian R, Rosenfeld A, Hertel I V, Marine W, Campbell E E B (2005) A general continuum approach to describe fast electronic transport in pulsed laser irradiated materials: The problem of Coulomb explosion. Appl. Phys. A-Mat. Sci. Proc. 81: 345-356.

[42] Kanavin A P, Smetanin I V, Isakov V A, Afanasiev Y V, Chichkov B N, Wellegehausen B, Nolte S, Momma C, Tunnermann A (1998) Heat transport in metals irradiated by ultrashort laser pulses. Phys. Rev. B 57: 14698-14703.

[43] Singh N (2004) Relaxation between electrons and surface phonons of a homogeneously photoexcited metal film. Pramana-J. Phys. 63: 1083-1087.

[44] Stuart B C, Feit M D, Herman S, Rubenchik A M, Shore B W, Perry M D (1996) Nanosecond-to-femtosecond laser-induced breakdown in dielectrics. Phys. Rev. B 53: 1749-1761.

[45] Lorazo P, Lewis L J, Meunier M (2003) Short-pulse laser ablation of solids: From phase explosion to fragmentation. Phys. Rev. Lett. 91: 225502.

[46] Perez D, Lewis L J (2003) Molecular-dynamics study of ablation of solids under femtosecond laser pulses. Phys. Rev. B 67: 184102.

[47] Zhigilei L V (2003) Dynamics of the plume formation and parameters of the ejected clusters in short-pulse laser ablation. Appl. Phys. A-Mat. Sci. Proc. 76: 339-350.

[48] Kautek W, Kruger J, Lenzner M, Sartania S, Spielmann C, Krausz F (1996) Laser ablation of dielectrics with pulse durations between 20 fs and 3 ps. Appl. Phys. Lett. 69: 3146-3148.

[49] Keldysh L V (1965) Ionization in the field of a strong electromagnetic wave. Sov. Phys. JETP-USSR 20: 1307-1314.

[50] Joglekar A P, Liu H, Spooner G J, Meyhofer E, Mourou G, Hunt A J (2003) A study of the deterministic character of optical damage by femtosecond laser pulses and applications to nanomachining. Appl. Phys. B-Las. Opt. 77: 25-30.

[51] Gamaly E G, Rode A V, Luther-Davies B, Tikhonchuk V T (2002) Ablation of solids by femtosecond lasers: ablation mechanism and ablation thresholds for metals and dielectrics. Phys. Plasmas 9: 949-957.

[52] Borghesi M, Romagnani L, Schiavi A, Campbell D H, Haines M G, Willi O, Mackinnon A J, Galimberti M, Gizzi L, Clarke R J, Hawkes S (2003) Measurement of highly transient electrical charging following high-intensity laser–solid interaction. Appl. Phys. Lett. 82: 1529.

[53] Born M, Wolf E (1999) Principles of optics : electromagnetic theory of propagation, interference and diffraction of light. New York: Cambridge University Press. 952 p.

[54] Hirayama Y, Obara M (2002) Heat effects of metals ablated with femtosecond laser pulses. Appl. Surf. Sci. 197: 741-745.

[55] Machado L M (to be published in 2012) Micromachining of optical components with femtosecond laser. Ph.D. Thesis (Universidade de São Paulo, São Paulo).

[56] Liu J M (1982) Simple technique for measurements of pulsed Gaussian-beam spot sizes. Opt. Lett. 7: 196-198.

[57] Samad R E, Vieira N D (2006) Geometrical method for determining the surface damage threshold for femtosecond laser pulses. Las. Phys. 16: 336-339.

[58] Samad R E, Baldochi S L, Vieira Jr N D (2008) Diagonal scan measurement of Cr:LiSAF 20 ps ablation threshold. Appl. Opt. 47: 920-924.

[59] Mero M, Clapp B, Jasapara J C, Rudolph W, Ristau D, Starke K, Kruger J, Martin S, Kautek W (2005) On the damage behavior of dielectric films when illuminated with multiple femtosecond laser pulses. Opt. Eng. 44: 051107.

[60] Costache F, Eckert S, Reif J (2008) Near-damage threshold femtosecond laser irradiation of dielectric surfaces: desorbed ion kinetics and defect dynamics. Appl. Phys. A-Mat. Sci. Proc. 92: 897-902.

[61] Courrol L C, Samad R E, Gomes L, Ranieri I M, Baldochi S L, de Freitas A Z, Vieira N D (2004) Color center production by femtosecond pulse laser irradiation in LiF crystals. Opt. Expr. 12: 288-293.

[62] Jee Y, Becker M F, Walser R M (1988) Laser-induced damage on single-crystal metal surfaces. J. Opt. Soc. Am. B 5: 648-659.

[63] Ashkenasi D, Lorenz M, Stoian R, Rosenfeld A (1999) Surface damage threshold and structuring of dielectrics using femtosecond laser pulses: the role of incubation. Appl. Surf. Sci. 150: 101-106.

[64] Martin S, Hertwig A, Lenzner M, Kruger J, Kautek W (2003) Spot-size dependence of the ablation threshold in dielectrics for femtosecond laser pulses. Appl. Phys. A-Mat. Sci. Proc. 77: 883-884.

[65] Lim Y C, Boukany P E, Farson D F, Lee L J (2011) Direct-write femtosecond laser ablation and DNA combing and imprinting for fabrication of a micro/nanofluidic device on an ethylene glycol dimethacrylate polymer. J. Micromech. Microeng. 21: 015012.

[66] Gomez D, Goenaga I (2006) On the incubation effect on two thermoplastics when irradiated with ultrashort laser pulses: Broadening effects when machining microchannels. Appl. Surf. Sci. 253: 2230-2236.

[67] Choi H W, Farson D F, Bovatsek J, Arai A, Ashkenasi D (2007) Direct-write patterning of indium-tin-oxide film by high pulse repetition frequency femtosecond laser ablation. Appl. Opt. 46: 5792-5799.

[68] Machado L M, Samad R E, de Rossi W, Vieira Junior N D (2012) D-Scan measurement of ablation threshold incubation effects for ultrashort laser pulses. Opt. Expr. 20: 4114-4123.

[69] Wolfram Research Inc. (retrieved 2012) Jacobi theta function J_3, http://functions.wolfram.com/EllipticFunctions/EllipticTheta3/06/01/03/.

[70] Reif J, Costache F, Varlamova O, Jia G, Ratzke M (2009) Self-organized regular surface patterning by pulsed laser ablation. Phys. Status Solidi C 6: 681-686.

[71] Mei H, Wang C, Yao J, Chang Y-C, Cheng J, Zhu Y, Yin S, Luo C (2011) Development of novel flexible black silicon. Opt. Commun. 284: 1072-1075.

[72] Halbwax M, Sarnet T, Delaporte P, Sentis A, Etienne H, Torregrosa F, Vervisch V, Perichaud I, Martinuzzi S (2008) Micro and nano-structuration of silicon by femtosecond laser: Application to silicon photovoltaic cells fabrication. Thin Solid Films 516: 6791-6795.

[73] Schlie S, Fadeeva E, Koroleva A, Ovsianikov A, Koch J, Ngezahayo A, Chichkov B N (2011) Laser-based nanoengineering of surface topographies for biomedical applications. Photonic. Nanostruct. 9: 159-162.

[74] Vorobyev A Y, Guo C (2007) Femtosecond laser structuring of titanium implants. Appl. Surf. Sci. 253: 7272-7280.

[75] Yasumaru N, Miyazaki K, Kiuchi J, Sentoku E (2011) Frictional properties of diamond-like carbon glassy carbon and nitrides with femtosecond-laser-induced nanostructure. Diam, Relat. Mater. 20: 542-545.

[76] Tang M, Hong M H, Choo Y S (2008) Hydrophobic Surface Fabrication by Laser Micropatterning. 779 p.

[77] Bizi-Bandoki P, Benayoun S, Valette S, Beaugiraud B, Audouard E (2011) Modifications of roughness and wettability properties of metals induced by femtosecond laser treatment. Appl. Surf. Sci. 257: 5213-5218.

[78] Vorobyev A Y, Guoa C (2008) Colorizing metals with femtosecond laser pulses. Appl. Phys. Lett. 92: 041914.

[79] Ahsan M S, Ahmed F, Kim Y G, Lee M S, Jun M B G (2011) Colorizing stainless steel surface by femtosecond laser induced micro/nano-structures. Appl. Surf. Sci. 257: 7771-7777.

[80] Mateo M P, Ctvrtnickova T, Fernandez E, Ramos J A, Yanez A, Nicolas G (2009) Laser cleaning of varnishes and contaminants on brass. Appl. Surf. Sci. 255: 5579-5583.

[81] Pouli P, Melessanaki K, Giakoumaki A, Argyropoulos V, Anglos D (2005) Measuring the thickness of protective coatings on historic metal objects using nanosecond and femtosecond laser induced breakdown spectroscopy depth profiling. Spectrochim. Acta B 60: 1163-1171.

[82] Dumitru G, Luscher B, Krack M, Bruneau S, Hermann J, Gerbig Y (2005) Laser processing of hardmetals: Physical basics and applications. Int. J. Refract. Met. H. 23: 278-286.

[83] Urech L, Lippert T, Wokaun A, Martin S, Madebach H, Kruger J (2006) Removal of doped poly(methylmetacrylate) from tungsten and titanium substrates by femto- and nanosecond laser cleaning. Appl. Surf. Sci. 252: 4754-4758.

[84] Hermann J, Benfarah M, Coustillier G, Bruneau S, Axente E, Guillemoles J F, Sentis M, Alloncle P, Itina T (2006) Selective ablation of thin films with short and ultrashort laser pulses. Appl. Surf. Sci. 252: 4814-4818.

[85] Atanasov P A, Takada H, Nedyalkov N N, Obara M (2007) Nanohole processing on silicon substrate by femtosecond laser pulse with localized surface plasmon polariton. Appl. Surf. Sci. 253: 8304-8308.

[86] Pronko P P, Dutta S K, Squier J, Rudd J V, Du D, Mourou G (1995) Machining of sub-micron holes using a femtosecond laser at 800 nm. Opt. Commun. 114: 106-110.

[87] Venkatakrishnan K, Tan B, Sivakumar N R (2002) Sub-micron ablation of metallic thin film by femtosecond pulse laser. Opt. Laser Technol. 34: 575-578.

[88] Lim C S, Hong M H, Lin Y, Chen G X, Kumar A S, Rahman M, Tan L S, Fuh J Y, Lim G C (2007) Sub-micron surface patterning by laser irradiation through microlens arrays. J. Mat. Process. Tech. 192: 328-333.

[89] Wang X C, Zheng H Y, Chu P L, Tan J L, Teh K M, Liu T, Ang B C Y, Tay G H (2010) Femtosecond laser drilling of alumina ceramic substrates. Appl. Phys. A-Mat. Sci. Proc. 101: 271-278.

[90] Li C, Lee S, Nikumb S (2009) Femtosecond Laser Drilling of Alumina Wafers. J. Electron. Mater. 38: 2006-2012.

[91] Wang Y, Dai N, Li Y, Wang X, Lu P (2007) Ablation and cutting of silicon wafer and micro-mold fabrication using femtosecond laser pulses. J. Laser Appl. 19: 240-244.

[92] Lee S, Yang D, Nikumb S (2008) Femtosecond laser micromilling of Si wafers. Appl. Surf. Sci. 254: 2996-3005.

[93] Suriano R, Kuznetsov A, Eaton S M, Kiyan R, Cerullo G, Osellame R, Chichkov B N, Levi M, Turri S (2011) Femtosecond laser ablation of polymeric substrates for the fabrication of microfluidic channels. Appl. Surf. Sci. 257: 6243-6250.

[94] van Brakel A, Grivas C, Petrovich M N, Richardson D J (2007) Micro-channels machined in microstructured optical fibers by femtosecond laser. Opt. Expr. 15: 8731-8736.

[95] Alemohammad H, Toyserkani E, Pinkerton A J (2008) Femtosecond laser micromachining of fibre Bragg gratings for simultaneous measurement of temperature and concentration of liquids. J. Phys. D Appl. Phys. 41: 185101.

[96] Lan X, Han Y, Wei T, Zhang Y, Jiang L, Tsai H-L, Xiao H (2009) Surface-enhanced Raman-scattering fiber probe fabricated by femtosecond laser. Opt. Lett. 34: 2285-2287.

[97] Said A A, Dugan M, de Man S, Iannuzzi D (2008) Carving fiber-top cantilevers with femtosecond laser micromachining. J. Micromech. Microeng. 18: 035005.

[98] Nolte S, Momma C, Jacobs H, Tunnermann A, Chichkov B N, Wellegehausen B, Welling H (1997) Ablation of metals by ultrashort laser pulses. J. Opt. Soc. Am. B 14: 2716-2722.

[99] Le Harzic R, Breitling D, Weikert M, Sommer S, Fohl C, Valette S, Donnet C, Audouard E, Dausinger F (2005) Pulse width and energy influence on laser micromachining of metals in a range of 100 fs to 5 ps. Appl. Surf. Sci. 249: 322-331.

[100] Ancona A, Doering S, Jauregui C, Roeser F, Limpert J, Nolte S, Tuennermann A (2009) Femtosecond and picosecond laser drilling of metals at high repetition rates and average powers. Opt. Lett. 34: 3304-3306.

[101] Doering S, Ancona A, Haedrich S, Limpert J, Nolte S, Tuennermann A (2010) Microdrilling of metals using femtosecond laser pulses and high average powers at 515 nm and 1030 nm. Appl. Phys. A-Mat. Sci. Proc. 100: 53-56.

[102] Schille J, Ebert R, Loeschner U, Scully P, Goddard N, Exner H (2010) High repetition rate femto second laser processing of metals. In: Heisterkamp A, Neev J, Nolte S, trebino R P, editors. Frontiers in Ultrafast Optics: Biomedical, Scientific, and Industrial Applications X.

[103] Zhang Q, Guo X P, Dai N L, Lu P X (2009) Corrosion and Fatigue Testing of Microsized 304 Stainless Steel Beams Fabricated by Femtosecond Laser. J. Mater. Sci. Technol. 25: 187-193.

[104] Momma C, Knop U, Nolte S (1999) Laser Cutting of Slotted Tube Coronary Stents – State-of-the-Art and Future Developments. Prog. Biom. Res. 4: 45-51.

[105] Li C D, Nikumb S, Wong F (2006) An optimal process of femtosecond laser cutting of NiTi shape memory alloy for fabrication of miniature devices. Opt. and Las. in Eng. 44: 1078-1087.

[106] Weck A, Crawford T H R, Wilkinson D S, Haugen H K, Preston J S (2008) Laser drilling of high aspect ratio holes in copper with femtosecond, picosecond and nanosecond pulses. Appl. Phys. A-Mat. Sci. Proc. 90: 537-543.

[107] Campbell B R, Palmer J A, Semak V V (2007) Peculiarity of metal drilling with a commercial femtosecond laser. Appl. Surf. Sci. 253: 6334-6338.

[108] Gamaly E G, Juodkazis S, Nishimura K, Misawa H, Luther-Davies B (2006) Laser-matter interaction in the bulk of a transparent solid: Confined microexplosion and void formation. Phys. Rev. B 73: 214101.

[109] Dubov M, Bennion I, Nikogosyan D N, Bolger P, Zayats A V (2008) Point-by-point inscription of 250 nm period structure in bulk fused silica by tightly focused femtosecond UV pulses. J. Opt. A: Pure Appl. Opt. 10: 025305.

[110] Ams M, Dekker P, Marshall G D, Withford M J (2009) Monolithic 100 mW Yb waveguide laser fabricated using the femtosecond-laser direct-write technique. Opt. Lett. 34: 247-249.

[111] Dharmadhikari J A, Dharmadhikari A K, Bhatnagar A, Mallik A, Singh P C, Dhaman R K, Chalapathi K, Mathur D (2011) Writing low-loss waveguides in borosilicate (BK7) glass with a low-repetition-rate femtosecond laser. Opt. Commun. 284: 630-634.

[112] Li C, Chen T, Si J, Chen F, Shi X, Hou X (2009) Fabrication of three-dimensional microchannels inside silicon using a femtosecond laser. J. Micromech. Microeng. 19:

[113] Machado L M, Samad R E, de Freitas A Z, Vieira Jr N D, de Rossi W (2011) Microchannels Direct Machining using the Femtosecond Smooth Ablation Method. Phys. Procedia 12: 67-75.

[114] Taylor R, Hnatovsky C, Simova E, Rayner D, Mehandale M, Bhardwaj V, Corkum P (2003) Ultra-high resolution index of refraction profiles of femtosecond laser modified silica structures. Opt. Expr. 11: 775-781.

[115] Saliminia A, Nguyen N T, Chin S L, Vallee R (2004) The influence of self-focusing and filamentation on refractive index modifications in fused silica using intense femtosecond pulses. Opt. Commun. 241: 529-538.

Chapter 6

Diagnostics of a Crater Growth and Plasma Jet Evolution on Laser Pulse Materials Processing

A. Yu. Ivanov and S. V. Vasiliev

Additional information is available at the end of the chapter

http://dx.doi.org/10.5772/48703

1. Introduction

The scheme of the experimental setup used in the study is presented in Fig. 1. The radiation of the GOR-100M ruby laser *(1)* (λ =0.694 μm) operating in the free oscillation regime (pulse duration τ ~ 1.2 ms, Fig. 2) or rhodamine laser (λ =0.58 μm), pulse duration τ ~ 20 μs) passed through the focusing system *(2)* and was directed through the hole in the electrode *(3)* onto the sample *(4)* that served as the second electrode and was mounted in air at a pressure of 10^5 Pa. The radiation spot diameter with sharp edges on the sample was varied in the course of the experiments from 1 to 2 mm.

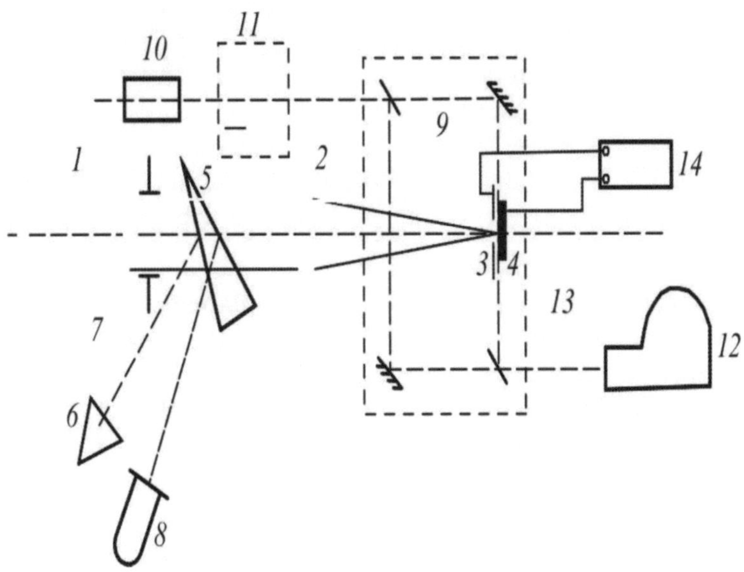

Figure 1. The experimental setup schematic diagram.

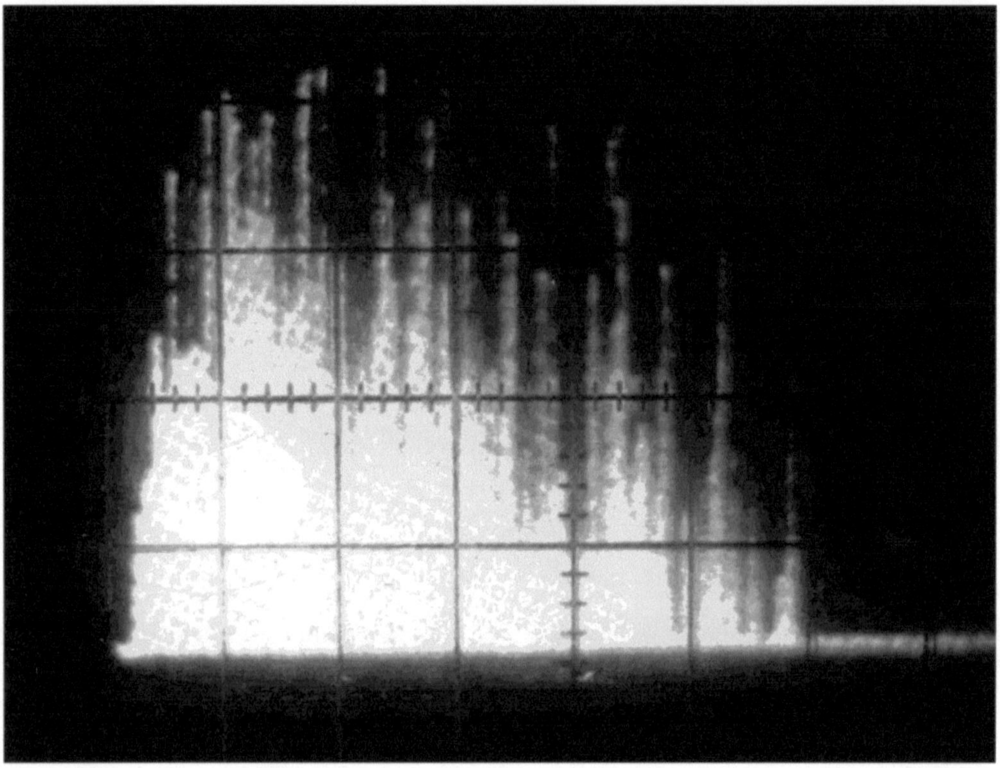

Figure 2. Oscillogram of the radiation pulse from the GOR-100M laser.
The scanning rate is 200 μs/div.

From the front face of the glass wedge *(5)* a part (4%) of laser radiation was directed into the IMO-2N energy meter (6), whose entrance window was located in the focal plane of the lens (7). The energy of the laser pulses varied from 5 to 60 J. The FEK-14 coaxial photodetector *(8),* the signal from which was coupled to the S8-13 oscilloscope, was used to record the temporal shape of the laser pulse. The voltage was applied to the electrodes *(3, 4)* from the source *(14),* built on the basis of the UN 9/27-13 voltage multiplier of the TVS-110 unit. The source allowed the voltage variation within 25 kV and its stabilization in the course of the experiment.

To study the spatial and temporal evolution of the laser plasma plume in the course of laser radiation action on the sample, we used the method of high-speed holographic motion-picture recording. The interelectrode gap was placed in one of the arms of a Mach-Zehnder interferometer (9), which was illuminated with the radiation of the ruby laser *(10)* (λ = 0.694 μm) operating in the free oscillation regime. The pulse duration of the radiation amounted to ~ 400 μs. The transverse mode selection in the probing laser was accomplished using the aperture, placed in the cavity, and the longitudinal mode selection was provided by the Fabry-Perot cavity standard used as the output mirror. The probing radiation after the collimator *(11)* was a parallel light beam with the diameter up to 3 cm, which allowed observation of the steam-plasma cloud development.

The interferometer was attached to the SFR-1 M high-speed recording camera *(12)*, in which the plane of the film was conjugate with the meridian section of the laser beam, acting on the sample, by means of the objective *(13)*. The high-speed camera operated in the time magnifier regime. The described setup allowed recording of time-resolved holograms of the focused image of the laser plasma plume. Separate holographic frames provided temporal resolution no worse than 0.8 μs (the single frame exposure time) and the spatial resolution in the object field ~ 50 μm. The error in the determination of the electron density was ~ 10% and it was governed by the precision with which the shifts of the fringes could be determined in the photographically developed interference patterns.

The diffraction efficiency of the holograms allowed one to reconstruct and record interference and shadow pictures of the studied process under the stationary conditions. The shadow method was most sensitive to grad n, so that the nature of the motion of the front of a shock wave outside the laser plasma and of the motion of the plasma jet could be determined from the reconstructed shadow patterns. This gave information on the motion of the shock front and the laser plasma front generated at the surfaces of metal samples. It was found that the nature of the motion of the shock wave front was practically independent of the target material and was governed primarily by the average power density of the laser radiation.

The reconstructed interference patterns were used to determine the spatial and temporal distributions of the electron density in a laser plasma plume.

The reliability of the results obtained by the method of fast holographic cinematography was checked by determination of the velocity of the front of a luminous plasma jet by a traditional method using slit scans recorded with a second SFR- 1M streak camera.

To study the surface shape of the crater that appears on the plate, we used the fringe projection method, which in the present case appeared to be more efficient than holographic methods of surface relief imaging and the stereophotogrammetric method, since, already at the stage of fringe projecting, it allowed obtaining a picture with controllable sensitivity of measurements and sufficiently good visibility of fringes, controlled visually. The sensitivity of measurements (relative fringe displacement) was set by changing the period of the projected fringes, and the good visibility was provided by changing the angle of illumination of the studied surface till removing the light flares from the crater surface. The present method is thoroughly described and successfully used in [2].

The optical scheme of the apparatus used to visualize the topography of crater is shown on Figure 2. Radiation of helium-neon laser *1* LGN-215 collimated by telescopic system *2* was used to illuminate Mach-Zehnder interferometer *3*. The interference picture was then projected to the sample *4* being studied. During the above procedure a system of dark and light strips was observed on the treated surface, and besides the configuration of strips was connected synonymously with the depth of the crater in the point of interference

$$h = d \cdot \Delta k / (tg\beta + tg\gamma).$$

Here d is a period of interference strips on the flat (nonirradiated) zone of the target, Δk - the displacement of a dark strip, β - an angle between the perpendicular to the surface of the irradiated sample and the projected interference surface, γ - an angle between the normal to the surface of the irradiated sample and the optical axis of the photographic camera.

The sample surface was optically mated with the picture surface 5 with the help of the objective "Helios-44-2". Contour stripes on the picture surface were fixed on the photographic film.

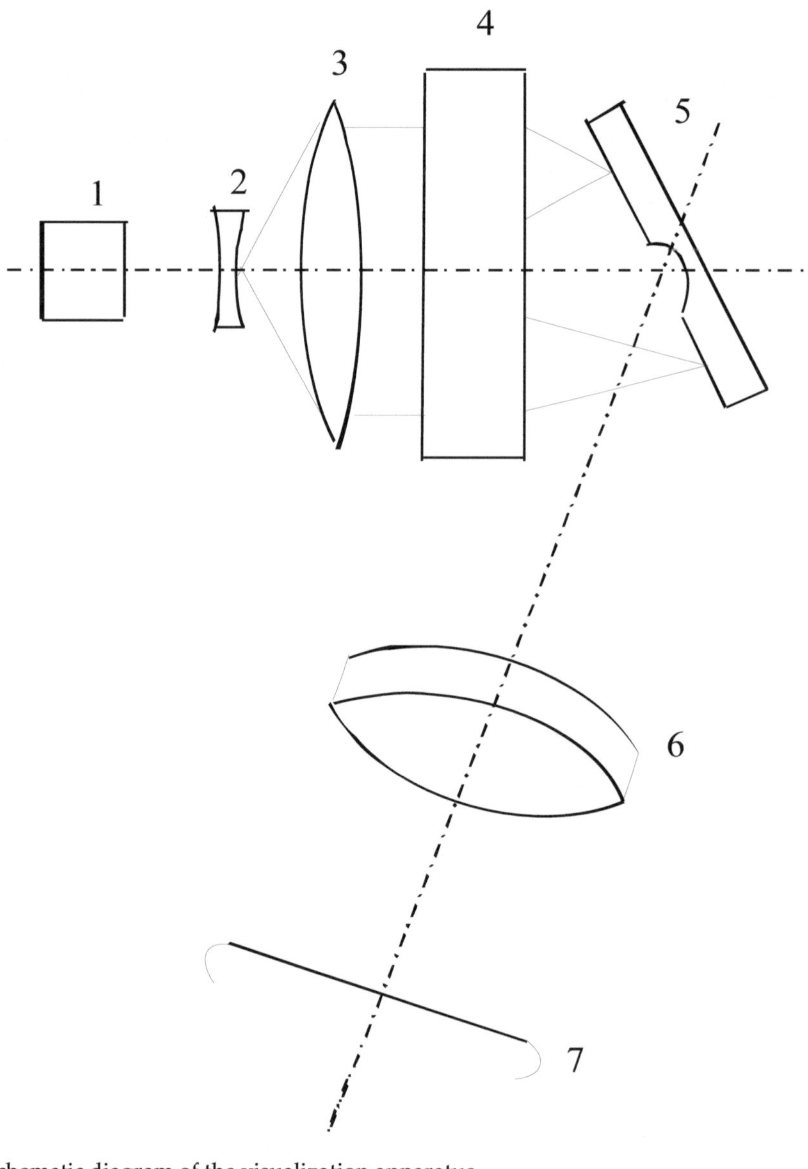

Figure 3. Schematic diagram of the visualization apparatus.

2. Experimental results and discussion

The experimental results have shown that at any polarity of the applied voltage [with positive or negative potential at the irradiated sample with respect to the electrode *(3)* the topography of the crater is practically identical and is determined by the energy distribution over the focusing spot of the laser radiation (Figs 4, 5).

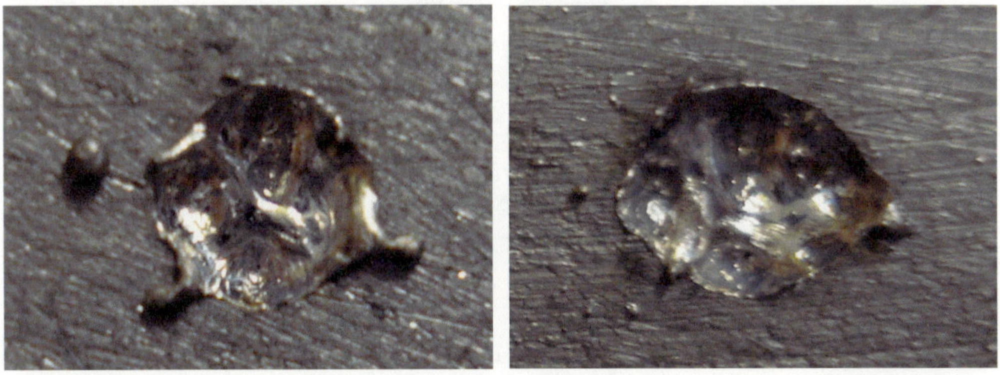

a b

Figure 4. Photographs of the craters obtained under the action of laser pulses on the target in the absence of the external electric field (a) and in the presence of the field (b).

Figures 6 a - c display the interferograms, reconstructed from the holograms recorded at different instants in the course of high-speed holographic motion-picture shooting. The figure clearly illustrates both the initial stage of the laser torch development and the plasma flow around the electrode *(3)* at different directions of the external electric field strength vector.

Figures 6 d - f represent the data on the distribution of concentration of free electrons in the plasma of an evaporated metal at different instants, obtained by processing the interferograms [1]. Although the energy distribution over the laser radiation focusing spot is not uniform, the lines of equal concentration are practically smooth, which is an evidence of relatively uniform ionization of the eroded substance steams. It is essential that, despite a substantial increase in the plasma formation over time, the mean electron concentration in the torch remains practically unchanged and even slightly grows, which may be associated both with a constant increase in the mass of emitted substance and with secondary ionization of the plasma by the laser radiation. Note, that the presence of an external electric field weakly affects the concentration of electrons in the laser plasma plume.

When the interelectrode separation was 2 cm, the maximal transverse size of the steam-plasma cloud at the surface of the electrode *(3)* for negative voltage on the target was 2 cm and in the absence of the external electric field it was *1.5* cm. This may be observed both in the interferograms and by the burn on the polyethylene film protecting the second electrode. As seen from interferograms, after reaching the second electrode in 56, 64, and 72 μs, respectively, the steam-plasma cloud practically does not grow in the transverse dimensions. Probably it is due to the flowing out of the plasma from the interelectrode gap through the hole in the electrode *(3)*, which is used for passing the laser radiation to the target (the hole diameter being 1 cm).

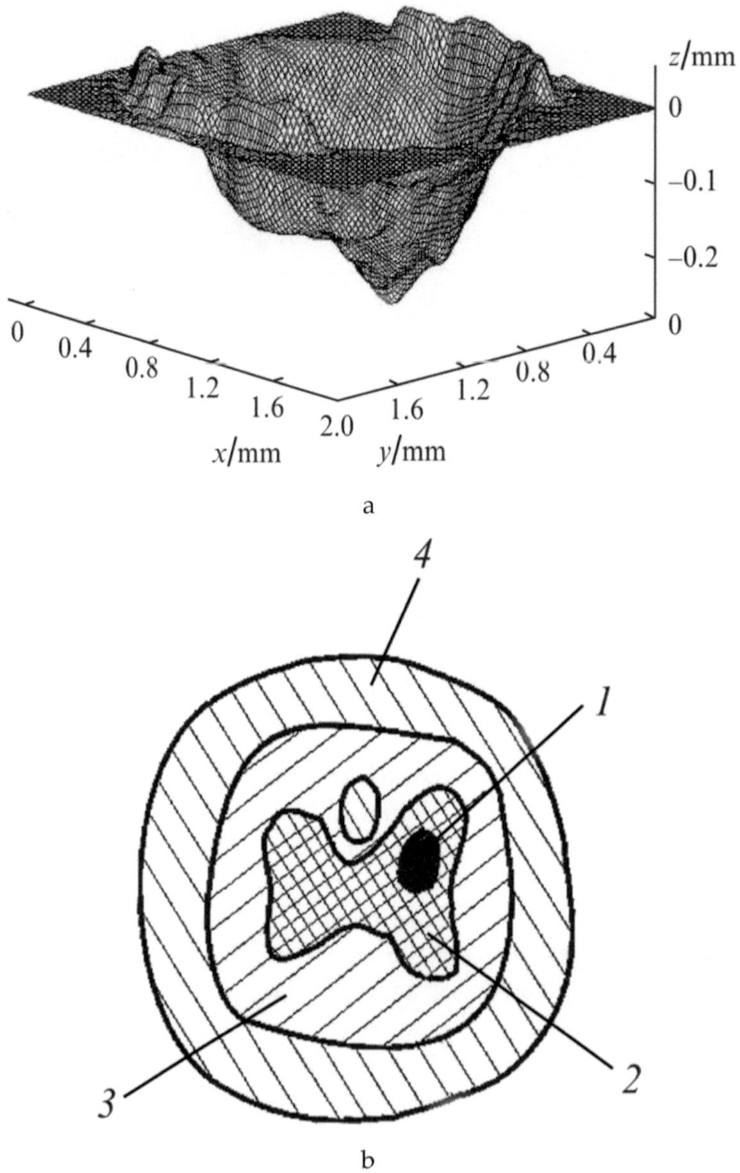

Figure 5. Volume topogram of a crater (a) and the distribution of light energy density over the transverse cross-section of the laser beam (b): 4.5 *(1)*, 3.5 *(2)*, 1.2 *(3)*, and 0.8 J mm^{-2} (4).

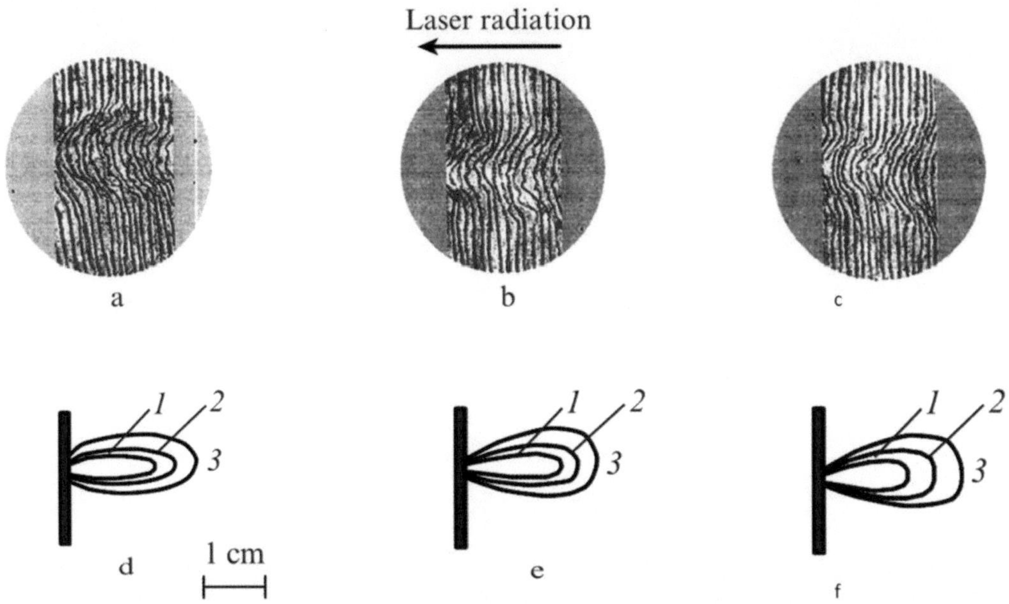

Figure 6. Interferograms of laser plasma torches (a, b, c) and electron concentration isolines in them (d, e, f) at the negative target potential (**b,** e) and at the positive target potential (c, f) at the instants 72 μs after the onset of the laser action; curve *(1)* corresponds to the electron concentration 5x 10^{18}, curve *(2)* to 2.5x 10^{18}, and curve *(3)* to 10^{18} cm^{-3}.

Figure 7 presents the time dependences of the plasma torch front motion velocity at different directions of the external electric field strength vector, calculated by using the information, obtained by analyzing the temporal variation of the interferograms. It is seen that even when the plasma front reaches the electrode (3), its velocity not only does not decrease (which is typical for late stages of the laser plasma torch existence [3]), but even increases; this happens both in the presence of the external electric field of any orientation and in the absence of the field. As already mentioned, this is due to the permanent and significant increase in the mass of the material, carried out under the action of laser radiation on the irradiated sample, as well as to the secondary ionization of plasma by laser radiation.

The maximal expansion velocity of the plasma torch amounted to 350 m/s for the negative voltage at the target, 310 m/s in the absence of the external electric field, and 270 m/s for the positive voltage applied to the target.

Our investigation showed that the time evolution of the leading edge of a luminous plasma moving away from the surface of a sample, deduced from the slit scans, differed from the time evolution of the front of the plasma jet, which was recorded by the shadow method. This allowed us to conclude that the concentration of the heavy particles, responsible for the radiation emitted by the plasma, was low at the front of the laser plasma jet, whereas the electron density was sufficient for reliable determination of the contribution of electrons to refraction in a hologram.

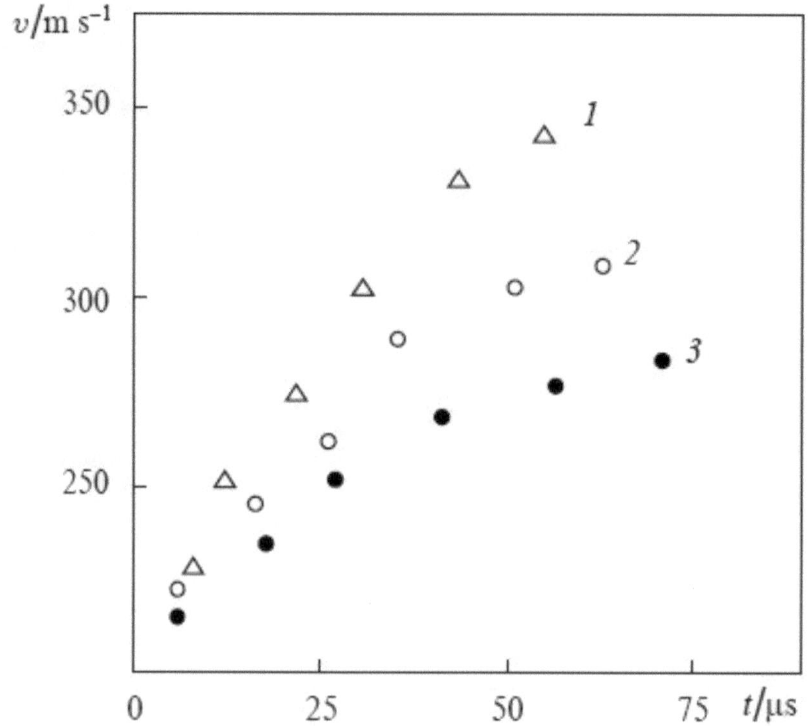

Figure 7. Time dependences of the velocity of the plasma torch front motion at the negative target potential *(1)*, in the absence of the field *(2)*, and at the positive target potential *(3)*.

The distribution of the density of cold air was determined and the electron density distribution was refined by two-wavelength holographic cinematography. We supplemented the system shown schematically in Fig. 1 with a second probe laser and an SFR-1M camera which recorded holograms at the wavelength of the radiation emitted by this laser. The second source of probe radiation was a laser utilizing a rhodamine 6G solution excited by a coaxial flashlamp. The use of a standard power supply system from a GOR-100M laser made it possible to generate output radiation pulses of 30 – 40 μs duration. The line width was reduced employing a plane-parallel Fabry-Perot interferometer. This made it possible to obtain scan holograms of the process at λ_1 = 0.69 μm and λ_2 = 0.58 μm, and to separate the contribution of electrons from that of heavy particles to the refraction of a plasma jet.

This two-wavelength holographic cinematography method was used to determine the radial distributions of the electron density and of the heavy-particle concentration at different moments in time and for different sections of laser plasma near the irradiated surface of a irradiated sample

At distances of 10 - 15 mm from the surface of a sample it was found that heavy particles ("hot" atoms and ions) of metals and molecules of atmospheric gases made only a small contribution to refraction. At large distances (where there were no "heated" luminous particles) the contribution of the cold dense air became significant. This was due to the pushing out of air by a plasma cloud.

When either positive or negative potential is applied to the sample, many small droplets appear on its surface after the laser action (Figures 4 and 8). In particular, at the laser pulse energy 20 J, the diameter of the focusing spot 2 mm, and the electric field strength 10^6 V cm^{-1} we observed ejection of droplets having the mean characteristic size less than 0.1 mm to the distance up to 2 cm from the crater centre. The maximal characteristic size of the droplets was 0.4 mm.

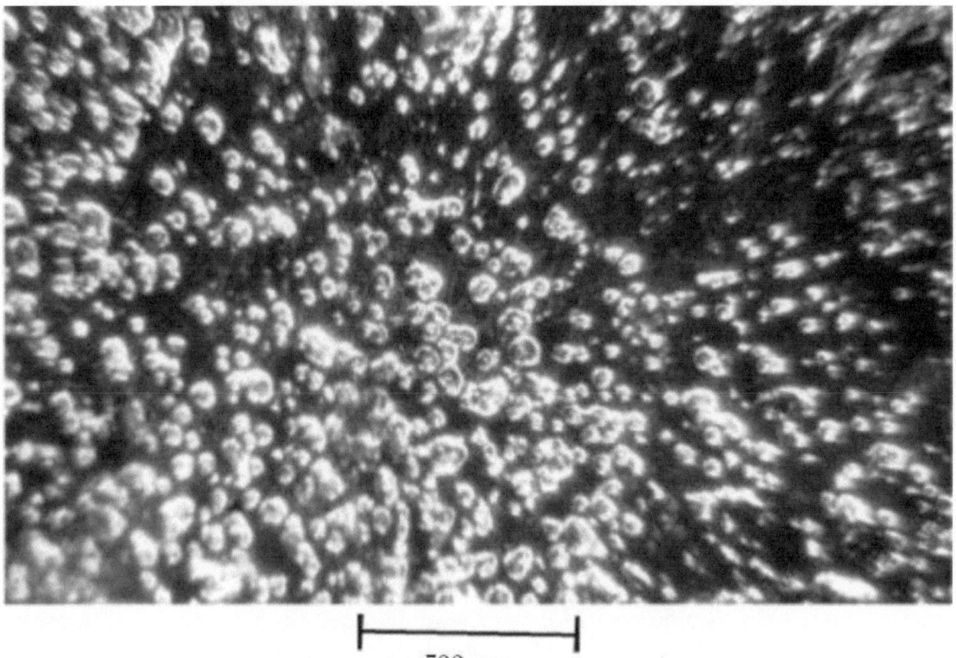

Figure 8. Microscopic surface relief of the crater outer zone photographs.

In the absence of the external electric field the mean size of the droplets was ~ 0.4 mm. The droplets were seen at the distance up to ~ 1 cm from the crater centre.

In accordance with the results presented above, the dynamics of the processes on the surface of a sample, placed in an external electric field with the strength from 0 to 10^6 V m^{-1} and subject to the action of the pulsed laser radiation with the parameters mentioned above, is thought to be the following. The primary plasma formation and the initial stage of the laser torch development, in principle, do not differ from those observed in the absence of the external electric field. The metal is melted and evaporated. As a result of local formation of steam and plasma [4, 5], the erosion torch begins to form with the fine-dispersed liquid-droplets phase. Note, that the bulk evaporation is promoted by the gases, diluted in the metal, and by the spatiotemporal nonuniformity of the laser radiation [4]. At a radiation flux density $10^6 - 10^7$ W

cm^2 the bulk evaporation is typical of all metals used in the experiments [5]. Obviously, the presence of the external electric field affects (increases or decreases depending on the direction of the field strength vector) the velocity of motion of the plasma front and causes some distortion of the plasma cloud shape. It is essential that the mentioned differences (at the considered parameters of laser radiation) are observed only at the initial stage of the laser plume development, because after the steam-plasma cloud reaches the electrode *(3)* an electric breakdown (short-circuit) occurs, and the external field in the interelectrode gap disappears.

Consider now the motion of the molten metal droplets in the steam-plasma cloud. In our opinion, the significant difference in the characteristic size of droplets, observed on the surface of the irradiated sample in the presence of the external electric field (independent of the direction of the field strength vector) and in the absence of the field, is a manifestation of the following mechanism of droplet formation [6]. It is known that at the surface of a liquid (including a liquid metal) the formation of gravity-capillary waves [7] is possible under the action of various perturbations. Undoubtedly, the examples of such perturbations are the spatially nonuniform evaporation of the target material due to nonuniform heating caused by nonuniform energy distribution over the focusing spot [8], the nonuniform primary plasma formation [6, 9, 10] caused by roughness of the irradiated sample surface [8], and, in the first place, the s1op of the molten metal initiated by each spike of laser radiation, acting on the exposed sample [2].

Using the method presented in [11], one can show that at insignificant thickness of the molten metal layer (confirmed by the view of the 'outer' (directed) zone of the crater, particularly, the absence of fillets of significant height at the crater edge) the dispersion equation for the gravity-capillary waves takes the form

$$\omega^2 = \frac{\alpha k^3}{\rho} + gk - \frac{k E_0 E'}{4\pi \rho \xi'}\Big|_{z=0}, \quad (1)$$

where α is the surface tension coefficient of the molten metal; ρ is the metal density; g is the free fall acceleration; k is the magnitude of the wave vector of the gravity-capillary wave; E_0 is the electric field strength at the surface (z = 0) of the molten metal (the z axis is perpendicular to the surface of the irradiated sample, directed towards the laser radiation source and parallel to the vector E_0); $E' = -\partial\varphi'/\partial z$ is the perturbation of the electric field in the space surrounding the molten metal; ξ' is the small displacement of the surface of the liquid in the z axis direction in the gravity-capillary wave.

Because for the uniform field E_0 the potential is $\varphi = -E_0 \cdot z$ (the potential at the metal surface is considered to be zero), the displacement of the mentioned surface by the small quantity ξ' leads to a small distortion of the potential:

$$\varphi'\big|_{z=0} = E_0 \cdot \xi'. \quad (2)$$

It follows from Fig. 3 that the maximal concentration of electrons in the plasma formation does not exceed ~ $.10^{18}$ cm^3, which corresponds to the change in the dielectric constant of the medium ε by approximately 10 $^{-5}$. Therefore, near the metal surface $\varepsilon \cong 1$ and with the boundary condition (2) taken into account

$$\varphi' = E_0 \xi' e^{-kz}.$$

In this case, the dispersion equation for gravity-capillary waves takes the form

$$\omega^2 = \frac{\alpha k^3}{\rho} + gk - \frac{k^2 E_0^2}{4\pi\rho}.$$

Because the frequency of the gravity-capillary waves ω is determined by the temporal characteristics of the abovementioned perturbations and, therefore, does not depend on the strength of the electric field E_0, the growth of the magnitude E_0 (independent of the direction of the vector $\vec{E}_0$) should cause the increase in the magnitude of the wave vector $k = \frac{2\pi}{\Lambda}$ and the decrease in the wavelength Λ of the gravity-capillary wave. If we assume that the droplets are 'torn away' by the plasma flow from the 'tops' of the gravity-capillary wave and, therefore, their characteristic size is proportional to Λ, then it becomes clear why in the presence of the external electric field (of any direction) the observed mean size of the droplets becomes essentially reduced.

The escaped droplets possess the charge of the same sign as the sample. That is why the droplets begin to move with acceleration towards the second electrode. However, since the maximal initial velocity of the outgoing droplets under the analogous conditions [8] is ~ 45 m s^{-1}, i.e., an order of magnitude smaller than the velocity of steam-plasma cloud spreading, the droplets do not reach the electrode (3) before the moment of the breakdown in the interelectrode gap. In what follows (in the absence of the external electric field) the droplets move under the action of the same forces as in [8] and, therefore, in the way, described in [8]. In this case, having acquired at the stage of accelerated motion in the electric field the velocity, exceeding the initial one, the droplets may fly to a greater distance along the surface of the irradiated sample than in the absence of the electric field, which is observed in the experiment. Moreover, having moved to a greater distance from the sample surface and, therefore, being affected by the plasma for longer time before returning to the surface, the droplets may be split into finer parts than in the absence of the external field.

It should be noted that the droplets in the erosion plume may appear not only due to the molten poo1 surface instability, but also due to the condensation of the steams of the erosion products [12, 13]. Moreover, since the droplets produced in the course of condensation of steams may be charged [14], they, similar to those carried out from the molten poo1, in the electric field may be removed from the crater to a greater distance than in the absence of the electric filed. However, this mechanism of plasma formation is dominating under somewhat different conditions of laser radiation acting on the material [12 – 15], namely, at significantly greater mean radiation flux density (~ 10^8 – 10^9 W cm^{-2}) and smaller exposure duration (single pulses of laser radiation were used with the duration ~ 100 – 200 ns and with less smooth temporary shape). In the case of such a regime of laser metal processing one observes the screening of the irradiated sample by the plasma cloud, which is possible only at the concentration of the ablated material steam essentially exceeding 10^{18} cm^{-3} (see Fig. 6). In this case, one observes intense formation of droplets with the dimension ~ 200 nm and smaller, and this process is most active at the late stages of laser radiation action on the material (at decreasing intensity of laser action) [15] and even after its termination [12]. At smaller radiation flux density, characteristic of the experiment considered in the present work (~10^{18} cm^{-3}), droplet condensation from the steam of ablation products is expected to be less intense. Therefore, the essential contribution of the condensation mechanism to the process of formation of large drops (having the size ~ 0.1 – 0.4 mm, see Figures 4 and 8), especially at early stages of the process, i.e., before filling the entire interelectrode gap by the plasma cloud, seems to be hardly probable.

3. Acoustic waves generation

Investigating the acoustic emission let us use the model of a loaded zone radiating waves into elastic medium. Corresponding to this model let us consider the destructed zone as a spherical segment with the curvature radius R, depth d and diameter $2r_1$. z-ax of the coordinate system is directed along the laser beam. It is important that the parameters of the irreversibly deformed zone are changing on the time:

$$R = R(t), \; d = d(t), \; r_1 = r_1(t);$$

here t is the time.

The displacement vector in an elastic zone consists of a longitudinal and a transversal components, $\vec{A} = \vec{A}_l + \vec{A}_t$, and each of them can be described by corresponding wave equation. Because of the presence of the media board direct in the elastic wave generating zone the solution of the wave equations system we shall search as a sum of the volume and the surface components

$$\vec{A}_l = \vec{A}_{lO} + \vec{A}_{l\Pi} = \nabla \psi_O + \nabla \psi_\Pi,$$

$$\vec{A}_t = \vec{A}_{tO} + A_{t\Pi} = \vec{A}_{t0} + rot(\vec{B}).$$

With regard to the symmetry of our problem, $\vec{A}_{t0} = 0$, scalar potential

$$\psi_0(\omega) = -\tilde{A}(\omega) \cdot \frac{\exp(-ik_l r)}{r},$$

$$B_\rho = B_z = 0,$$

$$\psi_\Pi(\omega) = -Z_0(k_R \rho) \cdot (\tilde{B}(\omega) \cdot \exp(-\chi_l z) + \tilde{Q}(\omega) \cdot \exp(\chi_l z)),$$

$$B_\varphi(\omega) = Z_1(k_R \rho) \cdot (\tilde{D}(\omega) \cdot \exp(-\chi_t z) + \tilde{S}(\omega) \cdot \exp(\chi_t z)).$$

Here ω is a frequency, $k_\ell = \omega / c_\ell$, c_ℓ and c_t – longitudinal and transversal sound velocities, $\tilde{A}(\omega)$ – wave amplitude, $k_R = \omega / c_R$, c_R – surface wave velocity, $\chi_i = (k_R^2 - k_i^2)^{1/2}$, $\tilde{B}(\omega)$, $\tilde{D}(\omega)$, $\tilde{Q}(\omega)$, $\tilde{S}(\omega)$, – os cillation amplitudes, $Z_i(x)$ – spherical function.

When $\rho \to 0$ and $z \to \infty$ the result must remain finite, therefore $Z_i(x) = J_i(x)$ (Bessel function), $\tilde{Q}(\omega) = \tilde{S}(\omega) = 0$, and

$$\vec{A}(\omega) = A(\omega)\frac{\vec{r}}{r^3}(1 + ik_l r)\exp(-ik_l(r - R)) + B(\omega)(\vec{\rho}_0 k_R J_1(k_R\rho) + \vec{z}_0 \chi_l J_0(k_R\rho)) \times \\ \times \exp(-\chi_l(z - h)) + D(\omega)(\vec{\rho}_0 \chi_t J_1(k_R\rho) + \vec{z}_0 k_R J_0(k_R\rho))\exp(-\chi_t(z - h)) \quad (3)$$

where $A(\omega) = \tilde{A}(\omega)\cdot\exp(-ik_l R)$, $B(\omega) = \tilde{B}(\omega)\cdot\exp(-\chi_l h)$, $D(\omega) = \tilde{D}(\omega)\cdot\exp(-\chi_t h)$.

Let us consider that on the surfaces $r = R$ and $z = h$ temporal dependence of pressure in the plasma cloud is

$$P/_{r=R} = p(t).$$

On the surface of spherical segment $R = R(t)$

$$\sigma_{rr} = -p(t), \ \sigma_{r\theta} = \sigma_{r\phi} = 0;$$

on the surface $z = h$

$$\sigma_{zz} = \varrho(t), \sigma_{\varrho z} = 0, \ \sigma_{z\phi} = 0.$$

Here σ_{ij} are the components of the stress tensor, r, θ, φ are the coordinates of spherical system.

Out of the spherical segment $R = R(t)$ medium is elastic, and so

$$\left[\lambda\left(\frac{\partial A_r}{\partial r} + 2\frac{A_r}{r} + \frac{1}{r}\frac{\partial A_\theta}{\partial \theta} + \frac{A_\theta}{r}ctg\theta\right) + 2\mu\frac{\partial A_r}{\partial r}\right]_{r=R(t)} = -p(t),$$
$$\left(\frac{\partial A_\rho}{\partial z} + \frac{\partial A_z}{\partial \rho}\right)_{z=h(t),\rho=\rho_1(t)} = 0, \quad (4)$$
$$\left[\lambda\left(\frac{\partial A_\rho}{\partial \rho} + \frac{A_\rho}{\rho} + \frac{\partial A_z}{\partial z}\right) + 2\mu\frac{\partial A_z}{\partial z}\right]_{z=h(t),\rho=\rho_1(t)} = -p(t).$$

Here A_i are the components of the displacement vector $\vec{A}$, λ, μ are the Lame coefficients.

Substituting *Ai* from the equation (3) to the system (4) we can calculate for each temporal moment $A(\omega,R,d,\rho_1)$, $B(\omega,R,d,\rho_1)$, $C(\omega,R,d,\rho_1)$, $\sigma_{zz}(\omega,R,d)$ and

$$\sigma_{zz}(t)=\int_{-\infty}^{+\infty}\sigma_{zz}(\omega,R,d)Exp[i\omega t]d\omega=\int_{-\infty}^{+\infty}\sigma_{zz}(\omega,R(t),d(t))Exp[i\omega t]d\omega=\int_{-\infty}^{+\infty}\sigma_{zz}(\omega,t)Exp[i\omega t]d\omega.$$

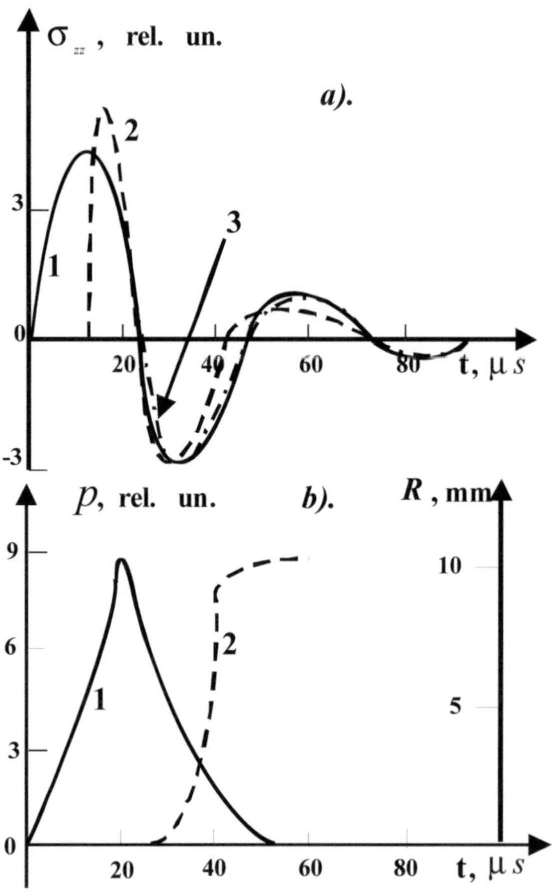

Figure 9. *a).* Temporal dependences of pressure of acoustic wave on the action of laser pulse with duration of 20 μs on cupper sample: 1 – experimental dependence; 2 – dependence calculated without account of crater growth; 3 – dependence calculated with account of crater growth. *b).* 1 – temporal dependences of pressure of plasma cloud on the border of irreversibly deformed zone; 2 – temporal dependences of curvature radius.

The results of the calculations with $R(t)=R_{max}\cdot\exp(\frac{t^2}{\tau_0^2})$ for $t<0$ and $R(t)=R_{max}$ for $t>0$ are presented on the figure 9; $\tau_0=40\ \mu s$.

Evidently that at action on a surface of the copper sample of a rhodamine laser impulse duration ~ 20 μs time of growth of a zone of destruction makes approximately 40 мкс, that will well be coordinated with time of existence of plasma formation at a surface of the target exposed to laser-plasma processing (~ 50 μs). Use of model of the loaded area with the moving borders radiating acoustic waves in the elastic medium allows us to solve the important practical problem – the definition of a law of time growth of a zone of irreversible deformations on a surface of the sample exposed to pulse laser-plasma processing.

4. Laser treating of transparent insulators

Now let us to investigate the dynamics of crater growth and of changes in the density of an inelastically deformed material on the surface of a transparent insulator when it interacts with a millisecond light pulse that has a complex temporal profile. The experimental setup used in the study was similar to presented in Fig. 1. Radiation was provided by a GOR-100M ruby laser operating in the free-running regime. This made it possible to generate pulses of $\tau \sim 1.2$ ms duration with an energy E that could be varied within the limits 5-50 J. The radiation passed through focusing system and was directed to sample. Both single-lens and double-lens focusing systems were used; they formed an image of stop on the surface of the sample. The diameter D of the focusing spot obtained in this way was varied in the course of our experiments from 1 to 2 mm.

Part of the laser radiation was directed by the front face of the glass wedge to an IMO-2N energy meter whose entry pupil was located in the focal plane of lens. The radiation reflected by the rear face of the wedge was directed to FEK-14 coaxial photocell and the photocell signal was applied to the input of an S8-13 oscilloscope, which recorded the temporal profile of the laser pulse.

The sample was placed in a window in one of the arms of a Mach-Zehnder holographic interferometer which was illuminated by radiation from a second ruby laser operating in the free-running regime. The longitudinal modes emitted by this probe laser were selected by a Fabry -Perot etalon, used as the exit mirror, and the transverse modes were selected by a stop placed inside a resonator. The probe radiation was directed to collimator which produced a parallel beam with a diameter of ~ 3 cm. A field of view of this kind was sufficient to study the growth of a crater, the changes in the density of the sample in the inelastically deformed zone, the formation and propagation of elastic waves in the sample, and the processes that occurred in the gas and in the plasma cloud near the sample.

The interferometer was in contact with an SFR-1M high-speed camera. The plane of a photographic film in the camera was made to coincide, with the aid of objective, with the conjugate plane of a meridional cross section of the light beam acting on the sample. The camera was operated in the cine mode.

Fig. 10 shows the holographic interferograms at different moments from the beginning of the interaction of a laser pulse of energy $E \approx 35$ J with a sample made of polymethylmethacrylate (PMMA) on which the pulse formed a focusing spot $D \approx 1$ mm in diameter. The temporal profile of this pulse had the form shown in Fig. 11. We used these interferograms and solved the Abel equation to find the fields representing the distributions of the refractive index $n(z, r, t)$ in space and time, and we then applied the Lorentz - Lorenz expression

$$\rho = \frac{(n^2-1)}{(n^2+2)} \frac{\mu}{R_m}.$$

to find the density fields $\rho\,(z, r, t)$ shown in Fig. 12. Here, μ is the molar mass of the material of the sample; $R_m = 4/3\pi N$, N is the number of particles responsible for refraction in a unit volume. The ratio μ/ρ was found from the parameters of the unperturbed medium.

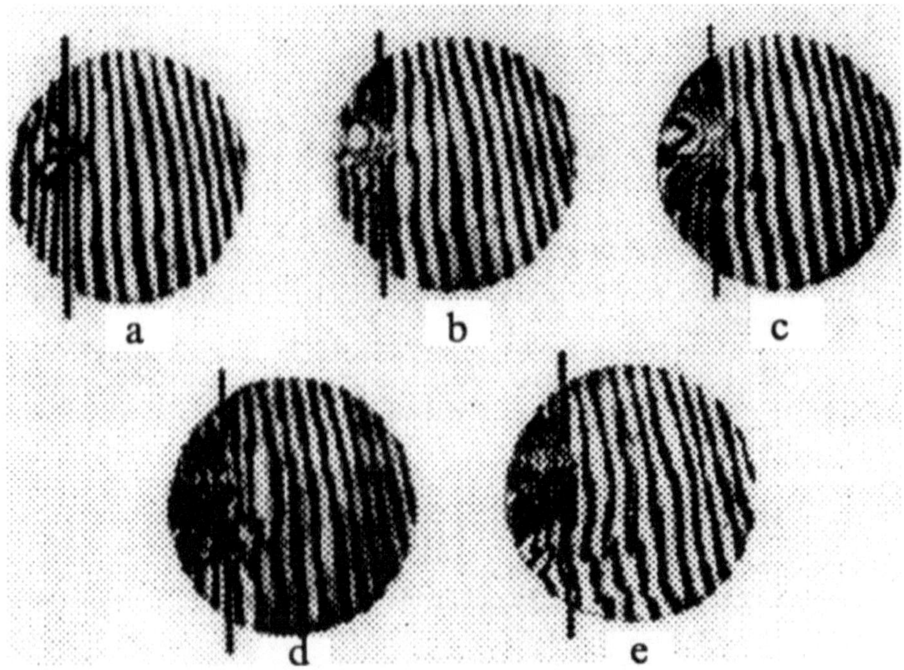

Figure 10. Interferograms recorded at the moments $t = 9.6\ \mu s$ (a), $16.0\ \mu s$ (b), $25.6\ \mu s$ (c), $48.0\mu s$ (d), and $70.4\ \mu s$ (e) relative to the beginning of the interaction.

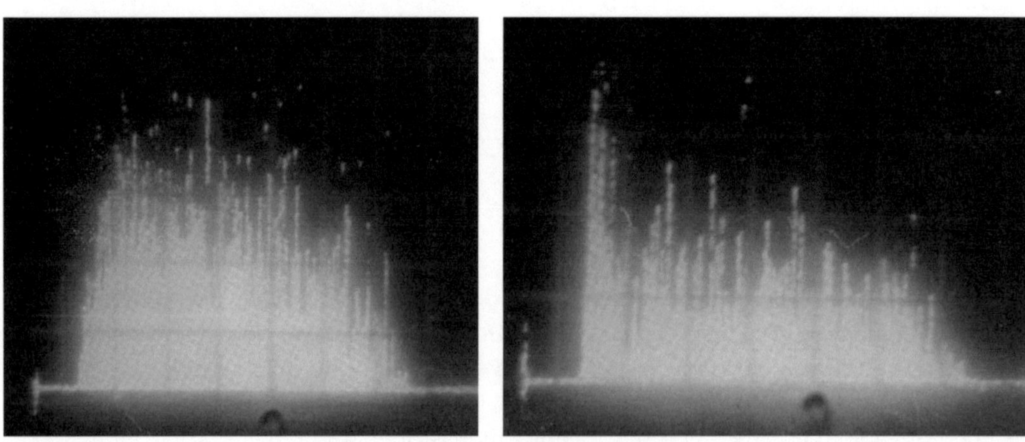

Figure 11. Temporal profile of a laser pulse before and after passing of a sample

Fig. 12 also shows the profiles of a crater formed on the surface of PMMA and recorded at different moments. These profiles yielded the time dependences of the diameter *d*, depth *h*, and volume V of the crater (Fig. 13). Fig. 14 shows the dependences of the final diameter, depth and volume of the crater on the energy of the light pulse interacting with the surface of PMMA.

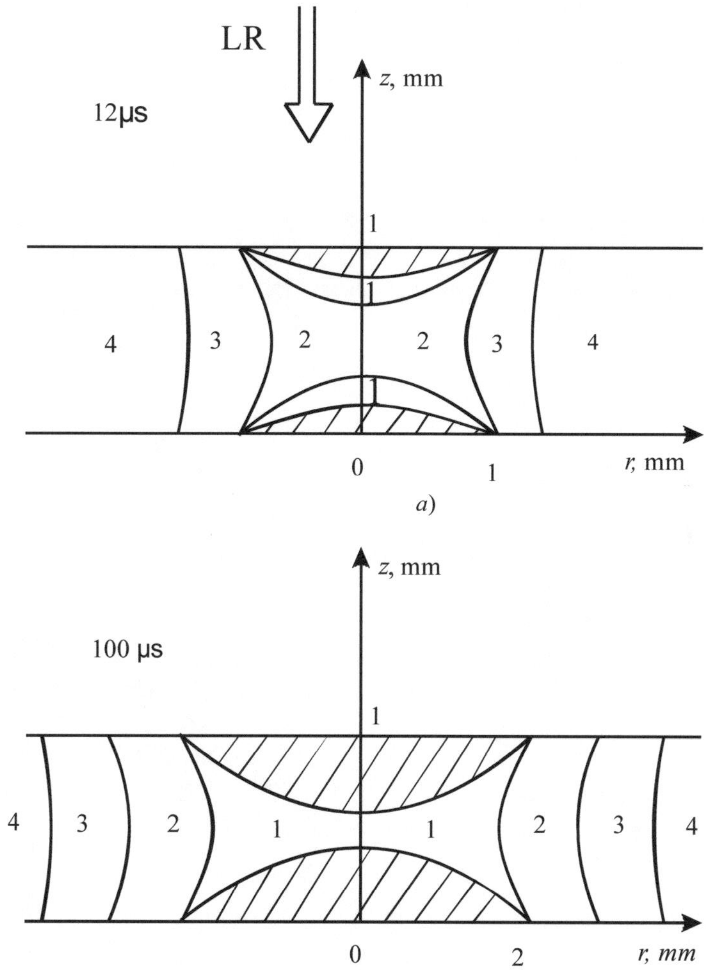

Figure 12. Profile of a crater and of a constant-density field at the moments $t = 12\ \mu s$ (a) and 100 μs relative to the beginning of the interaction.

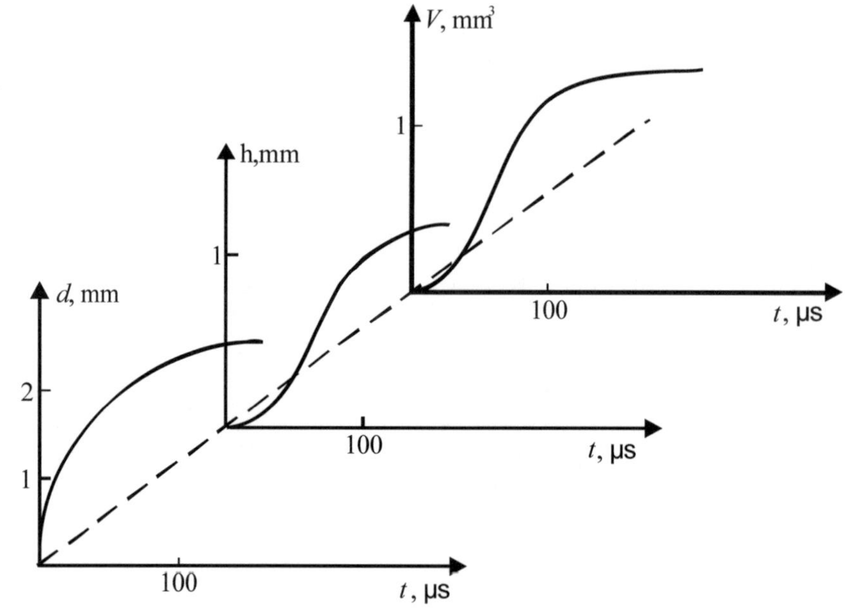

Figure 13. Crater diameter, depth, and volume time dependences.

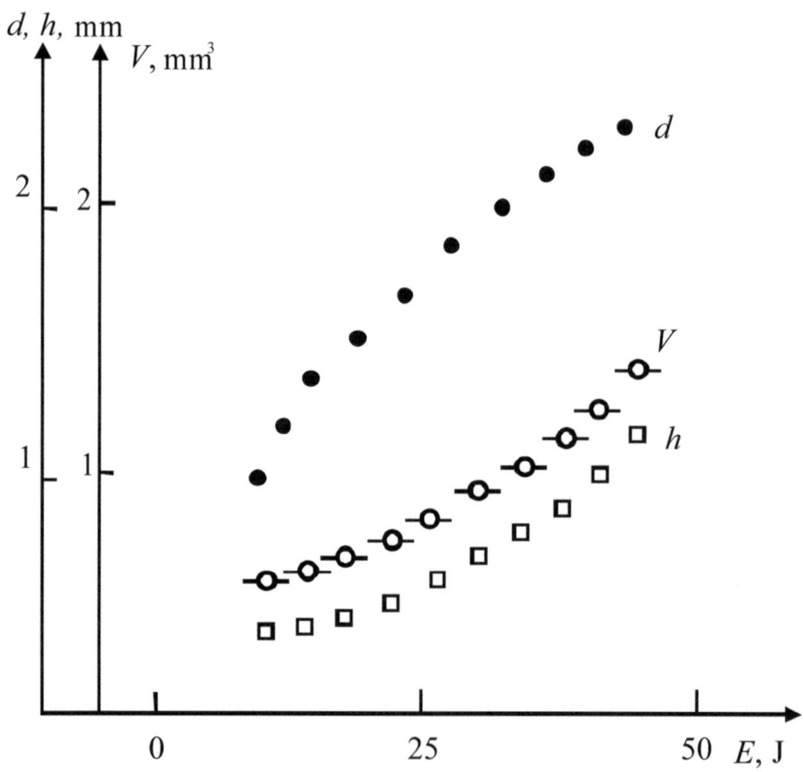

Figure 14. Dependences of the final diameter, depth and volume of a crater on the energy of a laser pulse.

5. Discussion of results

First of all, it is worth noting that the time dependences of the diameter *d*, depth *h*, and volume *V* of the crater (Fig. 13) were smooth and did not react in any way to the separate spikes of the laser pulse profile *E (f)*. The dependences *d(t)*, *h(t)*, and *V(t)* were found to be similar for the same value of

$$E(t) = 2\pi \int_0^t \int_0^{r_0} q(t)\, d\tau\, dr$$

irrespective of the profile *q{t)* (which varied in a random manner from one pulse to the next). The constant-density surfaces changed equally smoothly (Fig. 4).

Obviously, even in the case of a very high power density (~ 10 MW cm^{-2}) the energy of a single spike (~ 0.1 J) was insufficient to stimulate any significant ejection of the mass of the target.

The growth of the crater and the temperature rise in the region adjoining the target surface (it was the temperature rise that altered the density of the target material) were in this case determined by the normal evaporation process, in spite of the appearance of a laser-plasma jet (Fig. 10; the presence of a plasma was detected in an analysis of the field *n (z, r)* by a method similar to that described in [3]). This was deduced from the observation that the mass of the evaporated material *m (I)* was governed entirely by the energy E_t (more exactly, by the absorbed energy which was related to E_t).

It is also worth noting the observation (Fig. 14) that an increase in the energy *E* of a laser pulse first resulted in an increase in the final size of the crater because of the preferential increase in its diameter *d*. When *d* reached ~ 1.5*D*, the three-dimensional crater growth began. In the first stage the volume *V* varied in accordance with the law (Fig. 14)

$$V = D\, exp[C(E\text{-}E_0)] \qquad (5)$$

where $E \approx 7$ J, $C \approx 0.2\ J^{-1}$, and $D = 0.05\ mm^3$, whereas in the second stage it obeyed the law

$$V = V_b + A(E - E_0)^{\alpha}, \qquad (6)$$

where $\alpha \approx 1.5$, $A \approx 2\ mm^3\ J^{-3/2}$, and V_b ~ 0.5 mm^3 (governed by the volume *V* at which the change in the volume became different).

It is important to note that when the energy *E* was sufficient for the observation of both stages, the crater varied with time in exactly the same manner as before: first its diameter *d* increased and the three-dimensional growth was observed only from *d* ~ 1.5 D.

The nature of the crater growth was, in our opinion, governed by the temperature distribution *T (z, r)* in the sample being irradiated (we recall that *T (z, r)* governs the density distribution ρ (z, r); see Fig. 12). In fact, along the crater perimeter the density and temperature gradients can exceed considerably | grad ρ | and | grad T\ on the axis of the system (the z axis), which is to be expected because a transparent sample is exposed to a light beam with sharp boundaries. This sharpness of the boundaries is responsible for a high value of | grad T | at the perimeter of the focusing spot. On the other hand, the transparency of the medium results in a practically homogeneous (and weak) absorption of the energy in layers with different values of z and it is responsible for the low value of | grad T| in this direction.

Since the thermal energy flux is q ~ - grad (T), the peripheral (relative to the axis of the system) part of the sample surface is heated strongly. For this reason the evaporation is strong in the same region. Since the mass of the evaporated material is

$$\Delta m = \rho h \,\Delta S = \Delta g_t \,/\, \beta \sim \Delta ES \,/\, L_b,$$

where β is the heat of evaporation, we find that

$$S \sim \exp E \,/\, (\rho h L_b),$$

in good agreement with equation (5) when h is almost constant (because the component of grad T directed along the z axis is small).

When the boundaries of the evaporation zone spread beyond the focusing spot, the temperature gradient at the perimeter of this spot decreases, resulting finally in the equalization of the gradient over the whole crater profile. The heat flux also becomes equalized in all directions and the evaporation process is then three-dimensional. We are now in the second stage of the process when $\Delta\, d \sim \Delta\, h$.

If we assume that the crater is a spherical segment (in the case of a shallow crater we have h << d and since the crater were shallow in our experiments, this is justified as a first approximation), then

$$\Delta V = 4\pi R^2 \Delta R,$$

where R is the radius of curvature of the crater.

In the case of the processes exhibiting axial symmetry (which was true of the process we investigated) we have $R \sim E^{1/2}$ [16] and therefore $V \sim E^{3/2}$, which is in good agreement with equation (6).

The crater growth practically stopped after $t_{max} \sim 100\ \mu s$ from the beginning of the interaction with the target (we recall that the pulse duration was $\tau \approx 1.2$ ms) and t_{max} depended weakly on E. Obviously, at the power densities of the light flux used in our investigation the energy absorbed at $t > t_{max}$ was no longer sufficient to maintain the vaporization temperature on the surface of the grown crater. The plasma cloud was still maintained near the target surface and, consequently, local (and resulting in a much smaller loss of mass) vaporization still continued.

6. Conclusions

The studies performed have shown that under the action of laser radiation with the mean radiation flux density ~ 10^6 – 10^7 W/cm^2 t the surface of metals in the external electric field with different polarity and the strength up to 10^6 V m^{-1} the characteristic size of the target substance droplets, carried out of the irradiated zone, decreases by several times with increasing external electric field strength. Probably, this is due to a change in the wavelength of the gravity-capillary wave, excited on the molten metal surface. The observed effect offers the possibility to control the size of the metallic droplets in the course of laser deposition of thin films.

At action on a surface of the copper sample of a rhodamine laser impulse duration ~ 20 μs time of growth of a zone of destruction makes approximately 40 мкс, that will well be coordinated with time of existence of plasma formation at a surface of the target exposed to laser-plasma processing (~ 50 μs). Use of model of the loaded area with the moving borders radiating acoustic waves in the elastic medium allows to solve the important practical problem – the definition of a law of time growth of a zone of irreversible deformations on a surface of the sample exposed to pulse laser-plasma processing.

The effective growth of a crater formed as a result of the interaction of a laser pulse of duration of at least ~ 1 ms with the surface of a transparent insulator did not last more than 100 μs. Initially, the area of the crater increased and then, but only after it was twice as large as the focusing spot, the damage region began to grow in three dimensions. The crater growth dynamics was governed primarily by the evaporation mechanism: the individual spikes of a laser pulse interacting with the target did not influence the growth process.

Author details

A. Yu. Ivanov and S. V. Vasiliev
Grodno State University, Grodno, Belarus

7. References

[1] Ivanov A.Yu. Acoustic diagnostics of materials laser treating process. Grodno: GrSU, 2007.

[2] Bosak NA., Vasil'ev S.V., Ivanov A.Yu., Min'ko L.Ya., Nedolugov VI., Chumakov A.N. *Kvantovaya Elektron.*, 27, 69 (1999) *[Quantum Elelctron.*, 29, 69 (1999)].

[3] Barikhin B.A., Ivanov A.Yu., Nedolugov VI. *Kvantovaya Elektron.*, 17, 1477 (1990) *[Soy. J. Quantum Electron.*, 20, 1386 (1990)].

[4] Goncharov V.K., Kontsovoy V.L., Puzyrev MV. *Inzhenerno JIzicheskii zhurnal*, 66, 662 (1994) *[Journal of Engineering Physics and Thermophysics*, 66, 588 (1994)].

[5] Goncharov V.K., Kontsevoy V.L., Puzyrev MV. *Kvantovaya Elektron.*, 22, 249 (1995) *[Quantum Elelctron.*, 25, 232 (1995)].

[6] Zaykin A.E., Levin A.V., Petrov AL., Stranin S.A. *Kvantovaya Elektron.*, 18, 708 (1991) *[Soy. J. Quantum Elelctron.*, 21, 643 (1991)].

[7] Zaykin A.E., Levin A.V., Petrov AL. *Kvantovaya Elektron.*, 21, 486 (1994) *[Quantum Elelctron.*, 24, 449 (1994)].

[8] Vasil'ev S.V., Ivanov A.Yu., Lyalikov AM. *Kvantovaya Elektron.*, 22, 830 (1995) *[Quantum Elelctron.*, 25, 799 (1995)].

[9] Dorofeev IA., Libenson MN. *Opt. Spektrosk.*, 76,73(1994) *[Opt. Spectrosc.*, 76, 66 (1994)].

[10] Chumakov AN. et al. *Kvantovaya Elektron.*, 21, 773 (1994) *[Quantum Electron.*, 24, 718 (1994)].

[11] Rabinovich MI., Trubetskov DI. *Vvedenie v teoriyu kolebanii i voln* (Introduction to the Theory of Oscillations and Waves) (Moscow: Nauka, 1984).

[12] Goncharov V.K., Kozadayev K.V. *Inzhenerno-fizicheskii zhurnal*, 83, 80 (2010) *[Journal of Engineering Physics and Thermophysics*, 83, 90 (2010)].

[13] Chumakov AN., Bereza NA., Hu Dz.D., Bosak NA., Guo Z.H., Hie K.K. *Jnzhenerno-fizicheskiizhurnal*, 84, 524 (2011)*[Journalof Engineering Physics and Thermophysics*, 84, 567 (2011)].

[14] Klimentov SM. et al. *Laser Phys.*, 8 (6), 1(2008).

[15] Goncharov V.K., Kozadayev K.V., Shchegrikovich DV. *Inzhenerno-flzicheskiizhurnal*, 84, 781 (2011) *[Journalof Engineering Physics and Thermophysics*, 84, (2011)].

[16] Ashmarin I.I, Bykovskii Yu A, Gridin V A, et al. *KvantovayaElektron. (Moscow)* 6 1730 (1979) *[Sov. J. Quantum Electron.* 9 1019(1990)].

Chapter 7

Interaction of Femtosecond Laser Pulses with Solids: Electron/Phonon/Plasmon Dynamics

Roman V. Dyukin, George A. Martsinovskiy, Olga N. Sergaeva, Galina D. Shandybina, Vera V. Svirina and Eugeny B. Yakovlev

Additional information is available at the end of the chapter

http://dx.doi.org/10.5772/46237

1. Introduction

Femtosecond lasers bring new opportunities in a variety of technological applications [1] in micro- and nanotechnologies, including electronics, mechanics, medicine and biology. Technologies, based on femtosecond effects, are used, for example, to make light absorbers for solar energy devices [2], for direct fabrication of integrated optical components [3], enhancing performance of photo-electronic devices [4], friction reduction and improvement of mechanical wear resistance [5], surface conditioning of medical implants [6], etc. Further development of the above technologies requires deeper understanding of the physical processes occurring under the ultrashort laser pulse action on different materials.

Changes in the material optical properties under the action of intense radiation represent the key feature of the interaction of laser radiation with condensed media. Dynamics of optical properties of solids under the action of femtosecond laser pulse determines a number of physical effects which are of the great interest for both fundamental science and new applications. In particular, the feedbacks, which are being formed in this case, fundamentally change the properties of condensed matter [7].

During the action of femtosecond pulse on solids the electronic subsystem undergoes intensive photoexcitation while the lattice stays cold. The processes of excitation of the electrons and release of the absorbed energy are spaced in time. High intensity of the laser radiation results in modification of the state of the electron subsystem thus significantly changing the optical properties of the medium [8]. Studies of the femtosecond pulses effects on semiconductors and insulators [9, 10] showed that the concentration of nonequilibrium carriers generated by laser radiation is so high that the surface layer acquires properties of the metals during the pulse.

It was observed experimentally that semiconductors can be disintegrated during the femtosecond laser pulse action [11]. In [11] it was proposed a mechanism of destruction based on the crystal lattice destruction by the electric field resulting from the violation of quasi-neutrality in the irradiated area due to the external electron emission - Coulomb explosion. Conditions for Coulomb explosion occurrence in metals were not found.

The currently used experimental approaches such as femtosecond pump-probe technology [12, 13] and mass-spectroscopy (for example, [14]) provide measurement of integral characteristics, but have limited capability for retrieval of dynamics of the processes. The limitations of the experiment approach are being compensated by extensive use of mathematical modeling, where the fast non-linear processes are simulated in a wide range of the initial data.

Two-temperature model, which is traditionally used to describe the ultrashort laser pulse interaction with matter, has proved its validity in various conditions. The phenomenological two-temperature model of parabolic type was proposed in the 50's of last century by M.I. Kaganov, I.M. Lifshitz, L.V. Tanatarov [15]. It has been used by S.I. Anisimov to describe transient phenomena in a nonequilibrium electron gas and lattice under the submicrosecond laser action [16]. The model represents the primary approach to mathematical description of the nonequilibrium heating of the condensed medium by the action of short- and ultrashort-pulse laser radiation. According to the model it is assumed that the energy absorbed by free electrons increases their temperature, then the interaction of the heated electrons with the lattice results in increasing lattice temperature. Heat transfer takes place through the heat conduction mechanisms.

When analyzing the effect of femtosecond laser pulses on matter one has to consider the following: applicability of two-temperature model to description of the electron temperature, which is determined by the electron equilibrium and applicability of the notion of temperature; and taking into account multi-quantum effects in description of electron emission.

The applicability of a two-temperature model for description of action of femtosecond laser pulse on metals. Femtosecond pulse action on the metal can be generally described as follows. Absorption of photons by free electrons in metals results in increase in the electron kinetic energy and the energy distribution becomes nonequilibrium. This well-known feature determines the behavior of metals in a wide spectral range. One can use the diffusion approximation for obtaining qualitative characterization of photoexcitation of solids during the action of femtosecond pulse. With this approach the distribution of free electrons is described by the integral concentration $n(z, t)$, which varies in time and space (along the axis z, directed into the material) due to photo- and thermionic emission from the surface layers and electron diffusion. It is assumed that the thermalization of the electron gas occurs so fast that the notion of electron temperature can be immediately applied. "Hot" electrons contribute to the photo- and thermal emission: they withdraw a part of the energy stored in the electronic subsystem, thus reducing its temperature and, eventually, the temperature of solid as a whole. At the same time, change in the electrons concentration in the surface region results in a change in the optical characteristics of the material.

It is known that the light is absorbed by the conduction electrons in metals. Depending on the concentration of conduction electrons n and the wavelength of the incident light λ the electrons can be considered as free provided $n{>>}1/\lambda^3$, and the free electron model cannot be applied if $n{<<}1/\lambda^3$. In typical conditions in metals, where $n \approx 10^{22}$ cm^{-3}, and for the light wavelength of $\lambda \approx 1\mu$m the free electron model works well. It is also well known that the electron gas in metals is degenerated at practically all temperature range, and the distribution function of the electron gas just slightly differs from the distribution function at the absolute zero.

The Fermi energy ε_F for metals is very high. For example, for copper ε_F=7.1 eV, for silver ε_F=5.5 eV. For this reason heat effects engage the electrons, whose energy lies in a narrow energy range $\approx 2k_BT$ (k_B - Boltzmann constant) near the Fermi level.

The concentration of electrons, which absorb the incident radiation in metal, can be estimated by using the relation $n=h\nu N/\varepsilon_F$, where ν – the frequency of the incident light. For example, n/N=25% of the conduction electrons are affected in copper under $h\nu$=1.7 eV.

The electron transfer part of excessive energy, which they receive due to light absorption, though their collision to other electrons, ions and lattice defects (dislocations, grain boundaries, etc.). Heating of the metal is determined by efficiency of the collisions, which depends on the particles that exchange with energy. Typically the following relation takes place: $\nu_{ee}>\nu_{ei}>\nu_{ep}$, where ν_{ee} – the frequency of electron - electron collisions, ν_{ei} – the frequency of collisions between electrons and phonons, ν_{ep} – the frequency of electrons collisions with impurities and defects in the metal. The electron-phonon mechanism is quite feasible, since the Fermi energy of electrons in metals is high and is essentially represented by the energy of translational motion of free electrons. The Fermi velocity is $\upsilon_F=(2\varepsilon_F/m_e)^{1/2}=1{,}5\cdot 10^8$ cm/s is also high compared to the speed of sound in metals, that is typically: $\upsilon_F \sim 10^5$ cm/s.

A single collision may not be enough for the electron, which absorbed a photon, to release the excessive energy, i.e. energy relaxation is a multi-stage process of a diffusion character. For this reason, the energy redistribution occurs not only at the skin layer ($\delta_s \sim 10^{-6}$ cm), but in the deeper layer of $l_d=(D/\nu_{ee})^{1/2}=10^{-5}$ cm, where D is the electron diffusion coefficient.

Thus there is a heating of the metal. The heat, which is released in the layer l_D, is further transferred depthward into the material through heat conduction. The characteristic time of the absorbed energy transfer in metal is

$$\tau_{ei} = 1/\nu_{ei} = 10^{-12}\text{-}10^{-11} \text{ s.}$$

The electron gas and the lattice of the metal are two weakly interacting subsystems. Under the conditions:

$$t >> \tau_{ee}, \quad (1)$$

(τ_{ee} – time to establish an equilibrium energy distribution in the electron gas) and

$$\nu_{ii} << \nu_{ei}, \quad (2)$$

(ν_{ii} – frequency of ion-ion collisions) the electron gas and the lattice can be described separately with electron T_e and lattice T_i temperatures. Condition (1) infers rapid redistribution of the absorbed energy between the conduction electrons, and (2) means that the energy transferred to the lattice by electrons, rapidly redistributes between the ions.

Let us excessively consider the relaxation rates ν_{ef}, ν_{ee}, ν_{ei}, ν_{ii}. The relations between these relaxation rates substantially determine the processes in metals during the absorption of radiation.

The frequency of collisions between electrons and photons ν_{ef} is proportional to the power density of laser radiation q absorbed by the metal. It can be estimated using the relation:

$$\nu_{ef} = \frac{\alpha q}{h\nu n}, \tag{3}$$

where α is the absorption coefficient in metals: $\alpha \sim 10^5$ cm^{-1}.

The frequency of electron-electron collisions ν_{ee} in the metal is mainly determined by the number of electrons in the Fermi smearing and is calculated using the relation:

$$\nu_{ee} = \upsilon_F \sigma_{ee} n_e \left(\frac{k_B T_e}{\varepsilon_F} \right)^2, \tag{4}$$

where υ_F – the electron velocity on the Fermi surface ~ 10^8 cm/s; σ_{ee} – electron - electron interaction cross-section, $\sigma_{ee} \approx 5 \cdot 10^{-16}$ cm^2; $k_B T_e$ – Fermi smearing region. Then, at $T_e \sim 10^3$ K, the value of $\nu_{ee} \approx 10^{14}$ s^{-1}, and the time to establish an equilibrium distribution of the electron gas $\tau_{ee} \sim 1/\nu_{ee} \sim 10^{-14}$ s. The rate of energy transfer from the electron gas to the lattice and the lattice temperature is determined by the heat source and heat transfer coefficient of the electrons with the lattice β_{ei}.

The energy, which the lattice obtains from the electron gas per unit volume per unit time is ~ $\beta_{ei}(T_e - T_i)$, $\beta_{ei} \sim 10^{10}$ W/(cm^{3}K).

The frequency of electron-ion relaxation can be expressed in terms of heat transfer coefficient

$$\nu_{ei} = \frac{\beta_{ei}}{\rho_i c_i}, \tag{5}$$

where $\rho_i c_i$ – the volumetric heat capacity of the lattice [10^{-7} W·s/(cm^{3}K)]. Substituting numerical values one obtains ν_{ei}~10^{11} s^{-1} and τ_{ei}~10^{-11} s.

Comparing the expressions (3) and (4), one can show that (1) is always satisfied with the flux densities $q_0 \leq 10^9$ W/cm^2 and the electron gas in a metal is described by temperature T_e. Similarly, the condition (2) is also satisfied, therefore temperature T_i can be introduced to describe the thermal state of the lattice.

Thus, the electron relaxation time is estimated as τ_r~$1/\nu_{ee}$, where ν_{ee} – the frequency of electron-electron collisions. According to the estimates given above, its magnitude is about

10^{-14}–10^{-13} s, which is comparable with the pulse duration (tens of femtoseconds). However, further studies [16-19] have shown that the electron-electron relaxation time can be reduced up to 10^{-16} s at the electron gas temperatures of ~ 100 000 K, achieved by the action of ultrashort laser pulse. In other words, two-temperature model could be applied to analyze the effects of femtosecond laser pulse.

Multiphoton absorption. Although the theory of multiphoton absorption is pretty well developed [17], there is a certain difficulty regarding definition of multi-photon absorption cross-sections of real media, when this theory is used for the analysis of multiphoton absorption during femtosecond laser action.

In terms of quantum mechanics multiphoton process can be represented as a series of successive transitions of an electron to the virtual states [18]. Only the initial and final states are real in this case. The energy conservation law is valid with an accuracy of a natural width of the energy level only for the initial and final states. For virtual states the energy conservation law takes place with the accuracy, that is determined by the energy-time uncertainty relation $\delta E \cdot \delta t \geq \hbar$. At each virtual state the quantum system lives for the time:

$$\delta t \geq \hbar / \delta E. \tag{6}$$

Absorption of another photon makes the system transit to the next state.

It means that the quantum system can potentially absorb a photon of any energy, however the lifetime of the quantum system, absorbing a photon, would differ. If the quantum system absorbs a photon with energy $h\nu = \Delta E_{mn}$, the system transits into a real state, where the lifetime δt is determined by the probability of spontaneous decay of this state. If the quantum system absorbs photon with energy $h\nu \neq \Delta E_{mn}$, the system transits into a virtual state, where the lifetime is determined by the energy-time uncertainty relation: $\delta t \geq \hbar / \delta E$, $\delta E = | h\nu - \Delta E_{mn} |$.

With this approach one can use the following method for estimation of multiphoton absorption cross sections for various medium. One-photon absorption cross section σ_1 for metals can be determined by the known absorption coefficient α and the concentration of free electrons n: $\sigma_1 = \alpha / n$. Estimating the lifetime of an electron on a virtual level with the photon energy of about 1 eV τ_0 ~ 10^{-16} s and assuming the absorption cross section of excited electrons in all virtual levels to be equal to σ_1, one obtains two-photon absorption cross section to be $\sigma_2 = \sigma_1^2 n \tau_0$. Similar consideration provides $\sigma_3 = \sigma_1^3 n \tau_0^2$. Consequently for m+1-photon absorption cross section is $\sigma_{m+1} = \sigma_1^{m+1} n \tau_0^m$. For example, for $\alpha = 10^5 \text{cm}^{-1}$ and $n = 10^{22} \text{cm}^{-3}$ one can obtain $\sigma_1 = 10^{-17} \text{cm}^2$, $\sigma_2 = 10^{-28} \text{cm s}$, $\sigma_3 = 10^{-61} \text{cm}^3 \text{s}^2$.

2. Effect of electron emission on metal heating and destruction by femtosecond laser pulse

Let us analyze the influence of hot electron emission on heating and destruction of metals based on a two-temperature model and using as example the numerical simulation of Coulomb explosion.

In theoretical studies dealing with the different variants of two-temperature model, the most important aspect is to define the nonlinear optical and thermal properties in a wide temperature range, as well as quantitative characteristics of the electron-electron and electron-phonon interaction, controlling the temperature of the electron gas and the energy exchange between electrons and lattice, respectively.

In terms of the numerical analysis classical version of a two-temperature model is a system of interrelated differential equations of heat conduction, the accuracy of whose solution depends strongly on adaptation of the computational grid to the required numerical solution.

Let us consider the two-temperature model of metal heating by a femtosecond laser pulses on metals [19-22]. The model consists of a system of heat-conduction equations for the electrons and the phonons (lattice) subsystems, where thermophysical properties depend on electron temperature T_e and electron concentration n, and the equation describing the temporal evolution of the electron concentration.

$$c_e(T_e,n)\frac{\partial T_e}{\partial t}-\frac{\partial}{\partial z}\left[\lambda_e(T_e,n)\frac{\partial T_e}{\partial z}\right]=-\beta_{ei}(T_e,n)(T_e-T_i)+q_v, \tag{7}$$

$$c_i\frac{\partial T_i}{\partial t}-\frac{\partial}{\partial z}\left[\lambda_i\frac{\partial T_i}{\partial z}\right]=\beta_{ei}(T_e,n)(T_e-T_i), \tag{8}$$

where λ_e, λ_i, c_e, c_i are the electron and lattice heat conductivity and heat capacity, β_{ei} – electron-ion energy transfer coefficient, q_v – the absorbed power density released in the electron subsystem.

Dependence of optical and thermo-physical properties of metal [23] on the electron temperature and concentration is taken into account for electron heat capacity: $c_e=\frac{\pi^2k_b^2n(z,t)T_e}{2\varepsilon_F}$, electron heat conductivity: $\lambda_e=v_e^2\tau_{ee}c_e/3$, electron-ion energy transfer coefficient: $\beta_{ei}=c_e/t_{ei}$, electron velocity: $v_e=\sqrt{3k_bT/m_e}$, the electron mean free path: $l_e=1/(n\sigma\sqrt{2})$. The absorption coefficient of metals depends only on the concentration of free electrons: $\alpha(n)=(\alpha/n_0)n$, n_0 is the initial concentration of free electrons.

The boundary conditions:

$$\lambda_e\frac{\partial T_e}{\partial z}\bigg|_{z=0}=-j_e,\ \lambda_i\frac{\partial T_i}{\partial z}\bigg|_{z=0}=0,\ T_e\big|_{z=\infty}=T_i\big|_{z=\infty}=T_n,$$

where j_e – is the heat flow carried away by the emitted electrons, T_n – the initial temperature. Evolution of the electron density distribution is described by the following diffusion equation:

$$\frac{\partial n(z,t)}{\partial t}=D\frac{\partial^2 n(z,t)}{\partial z^2}, \tag{9}$$

with initial and boundary conditions:

$$D\left.\frac{\partial n}{\partial z}\right|_{z=0} = -F_e, \quad n\big|_{z=0} = n_0, \quad n\big|_{z=\infty} = n_0,$$

where $n_0=10^{22}$ cm^{-3} is the initial concentration of electrons evenly distributed over the volume, F_e is the flow of the electrons resulted from thermionic emission and photo-emission $F_e=F_t+F_{mpho}$.

The thermionic emission is determined by the law of Richardson:

$$F_t = -BT_e^2\big|_{z=0}\exp\left(-\frac{\varphi_e}{k_b T_e\big|_{z=0}}\right)\exp(-z/l_e)/q_e,$$

here B is the Richardson coefficient, ϕ_e is the work function, q_e is the electron charge.

External photoelectric effect implies that the energy of absorbed photons is used by the electrons to overcome the work function, i.e. the minimum energy required for the electron to escape from the surface. The work function in metals is several eV (for silver ϕ_e=4,28 eV). Therefore, for $h\nu$=1.55 eV one should expect the three-photon absorption. For calculating the photoemission let us assume that the free electrons, which are involved into the multiphoton process, reach the surface without energy loss and leave the metal. The emissive layer thickness is limited by the electron mean free path and also depends on the electron concentration and the emission coefficient. The flow of electrons (cm^{-2}/s) caused by photoemission for the m-photon absorption:

$$F_{mpho}\big|_{z=0} = -\int_0^\infty \sigma_m J^m \exp(-z/l_e)dz,$$

J is the absorbed photons flow, $J=q/h\nu$, σ_m is the multi-photon absorption cross section.

The emission of electrons leads to accumulation of positive charge on the metal surface and, therefore, to generation of the electric field. The electric field resulted from breaking of quasi-neutrality of the irradiated area can be calculated from the following equation [24]:

$$\frac{\partial E}{\partial z} = \frac{q_e}{\varepsilon\varepsilon_0}(n_i - n). \qquad (10)$$

This electric field, which is induced by the charge separation, can reach extremely high magnitude and exceed the energy of atomic bonds resulting in Coulomb explosion. To determine the conditions for initiation of the Coulomb explosion the electric field (10) is compared to the threshold magnitude required for removal of an atom from the target. The estimation of the critical electric field [24]:

$$E_{th}\big|_{z=0} = \sqrt{\frac{2\Lambda n}{\varepsilon\varepsilon_0}},$$

where n_0- the concentration of atoms (cm^{-3}), $\Lambda = 2.951$ J/atom - the heat of sublimation, $\varepsilon = 4.9$ - relative permeability of silver, ε_0=8.854·10^{-14} F/cm - dielectric constant.

For the numerical solution of the heat conduction equation and the equation describing the electron density, which are non-stationary partial differential equations, the finite difference method was applied. To calculate the values of the temperature and the electron concentration explicit difference scheme was used. Though explicit scheme provides a relatively high speed of calculation, it has a serious disadvantage related to the need for satisfying the stability conditions, which impose limits on the amount of steps partitions with respect to coordinate and time.

Fig. 1 - 4 show the results for silver for the laser pulse shape $q=q_m\exp(-(t-t_m)^2/t_{m1}^2)$, t_m=100 fs, t_{m1}=50 fs. The calculation was performed until the beginning of the Coulomb explosion, when the electric field resulted from the charge separation exceeds the threshold required for the removal of atoms. For comparison, the simulation was performed without taking the emission into account, but preserving the dependence of material properties on temperature.

Fig. 1 shows that temperature of the electrons and the lattice increase, but does not reach its maximum during the pulse. The difference between the temperature, calculated taking into account the emission and without it at the beginning of Coulomb explosion t=0.15 fs is $\Delta T \approx 290$ K for temperature of the electrons (Fig. 1b) and $\Delta T \approx 60$ K for temperature of the lattice.

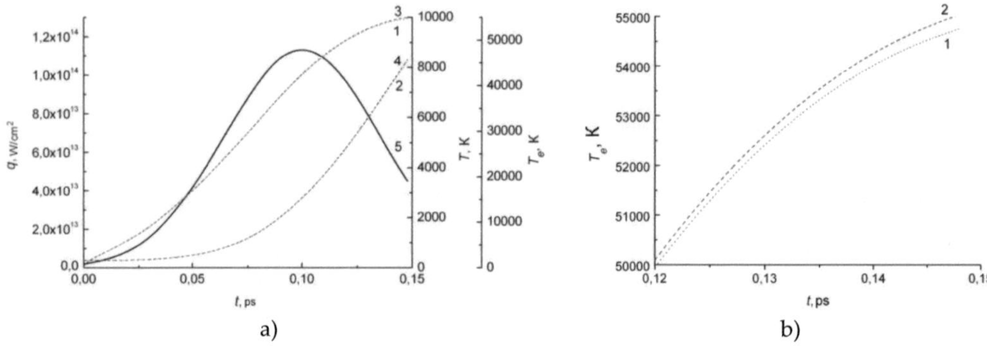

a) b)

Figure 1. a – Transient behavior of electron and lattice temperature: (1) and (2) – with taking into account emission, (3) and (4) – disregarding emission. (5) – laser pulse shape $q=q_m\exp(-(t-t_m)^2/t_{m1}^2)$, t_m=100 fs, t_{m1}=50 fs, maximum power density q_m=5·10^{14} W/cm^2. b – an enlarged part of the temporal dependence of the electron temperature: (1) – with taking into account emission, (2) – disregarding emission

Fig. 2, 3 illustrate temporal behavior of the free electrons concentration, and the number of electrons emitted by photo-and thermionic emission for different pulse shapes. At the initial stage of pulse action photoemission dominates, but thermionic emission increases with the electron temperature rapidly, so thermionic emission begins to prevail over the photoemission. The maximum of the photoemission rate maximum corresponds to the maximum of laser power density. This indicates that the laser pulse shape significantly influences the dynamics of the processes.

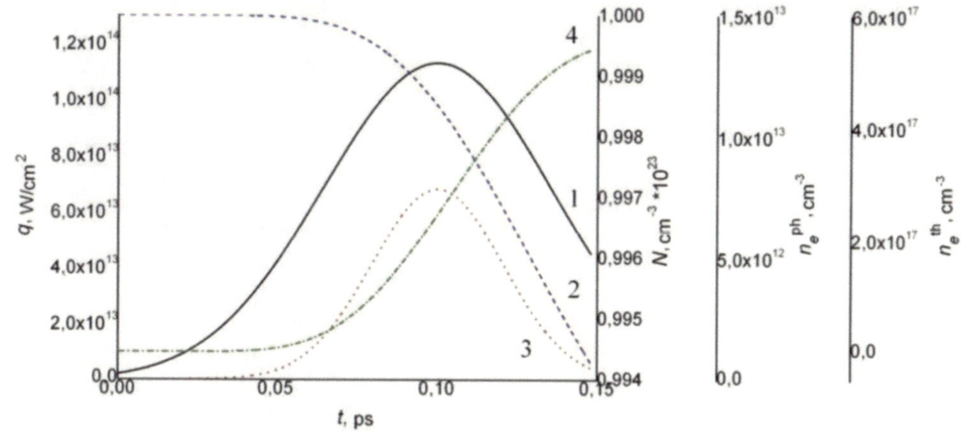

Figure 2. Laser pulse shape (1), the temporal evolution of the net surface electron density N (2) and the density of emitted electrons due to photoemission n_e^{ph} (3) and thermionic emission n_e^{th} (4). Laser pulse shape $q=q_m\exp(-(t-t_m)^2/t_{m1}^2)$, t_m=100 fs, t_{m1}=50 fs

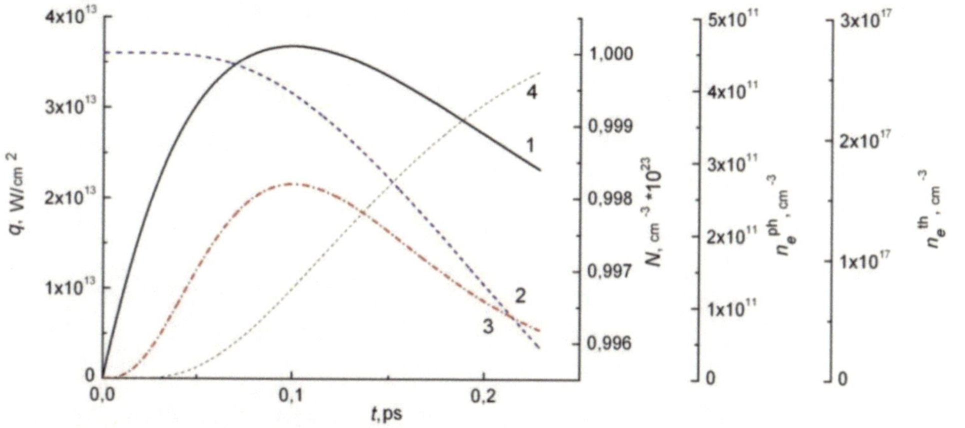

Figure 3. Laser pulse shape (1), the temporal evolution of the net surface electron density N (2) and the density of emitted electrons due to photoemission n_e^{ph} (3) and thermionic emission n_e^{th} (4). Laser pulse shape $q=q_m(t/t_m)\exp(-t/t_m)$, t_m=100 fs

The gradient of the electron density results in an electric field, which grows and reaches the Coulomb explosion threshold (Fig. 4).

Fig. 5 illustrates the nonlinear dependence of the moment when Coulomb explosion starts on the laser power density. It is seen that the dependence is nonlinear, and Coulomb explosion for a pulse duration ~ 100 fs can occur when $q > 10^{15}$ W/cm^2.

We can draw the following conclusions from the results of numerical simulation of the influence of electron emission on heating and destruction of metals irradiated by femtosecond laser pulse.

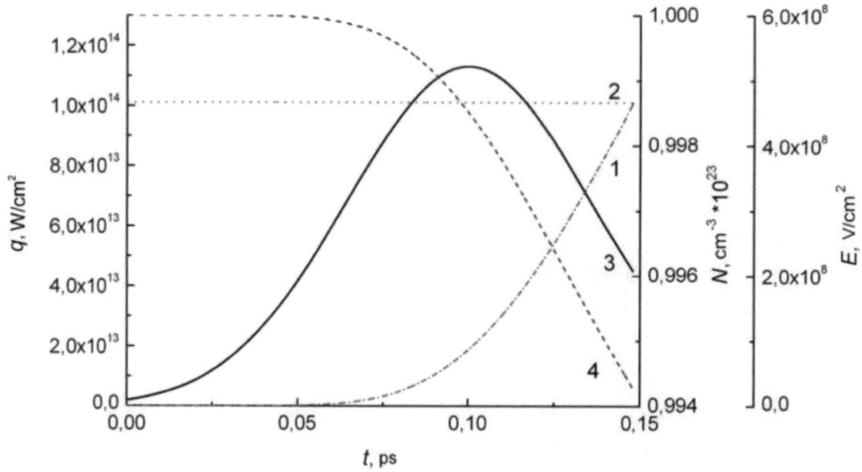

Figure 4. Temporal dependence of the electric field E (1), arising due to emission of electrons and the threshold field value E_{th} (2) corresponding to the beginning of Coulomb explosion. Laser pulse shape $q=q_m\exp(-(t-t_m)^2/t_{m1}^2)$, t_m=100 fs, t_{m1}=50 fs (3) and the temporal evolution of the net surface electron density (4)

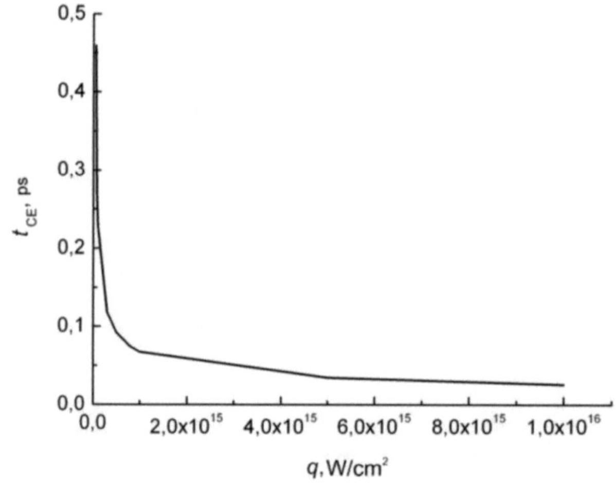

Figure 5. Time of Coulomb explosion onset as function of laser density for laser pulse shape $q=q_m\exp(-(t-t_m)^2/t_{m1}^2)$, t_m=100 fs, t_{m1}=50 fs

Various types of emission have different impacts on the concentration of emitted electrons (see Fig. 2-3): thermionic emission dominates over the photoemission and increases with the electron temperature increasing. The pulse shape significantly affects on the dynamics of the emission processes. However, according to the calculations the effect of emission processes on the electron gas temperature and the lattice temperature (Fig. 1a, b) is negligible. Also, numerical experiment showed that the occurrence of Coulomb explosion (caused by the emission processes during the pulse) in metals requires high-power incident radiation, which is impossible in the real exposure modes.

Emission processes have a significant impact on the processes of heating and destruction of the semiconductors, because the initial concentration of conductivity electrons in semiconductors can be below concentration of free electrons generated by the action of femtosecond laser radiation in contrast to metals. Let us consider the emission impact on the example of a femtosecond microstructuring of the silicon surfaces.

3. Effect of electron emission on changes in optical properties of semiconductors under the femtosecond laser pulse action

The formation of periodic surface structures (PSS) is a perfect evidence of an induced change in the surface optical properties under the femtosecond pulse action on semiconductor and wide-band dielectrics (Fig. 6). Being formed under different conditions PSS exhibit the same formation regularities: the structures orientation depends on the direction of polarization vector of the laser radiation; the structures period depends on the wavelength, incidence angle of the radiation and dielectric permeability of the medium. The observed regularities suggest that the mechanism of PPS formation is determined by electromagnetic field, which is the result of interference of the incident wave and the excited surface electromagnetic waves (SEW).

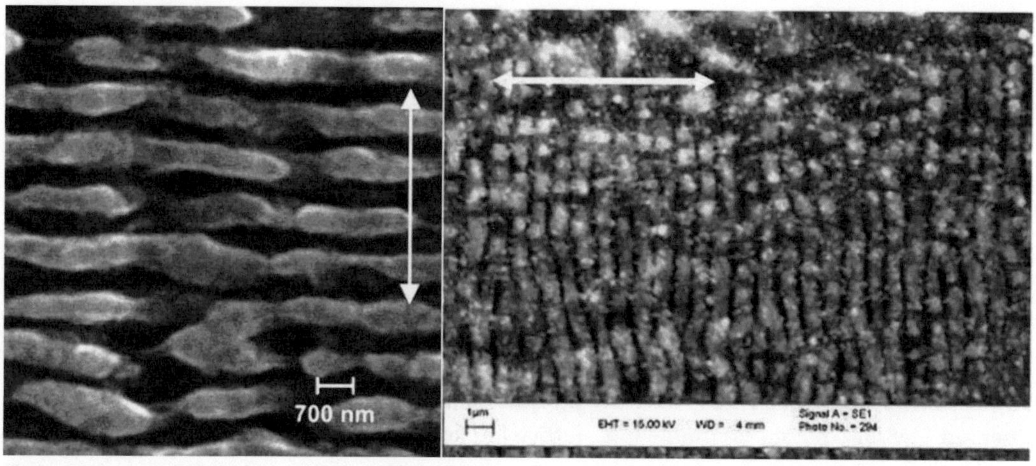

Figure 6. SEM image of the monocrystalline silicon surface irradiated by the laser pulse. Arrow indicates the direction of laser radiation polarization. Left: after irradiation by 1200 pulses, the energy density is ~ 1 J/cm^2. Right: after exposure by 300 pulses, the energy density is ~ 2 J/cm^2

Microstructures shown in Fig. 6, were obtained on silicon irradiated by 80 fs laser pulses at a wavelength of 1.25 μm [25, 26]. Surface structures turned during rotation of the laser radiation polarization vector, and depending on the density of laser flux were oriented either perpendicular (Q_0 ~ 1 J/cm^2) or parallel to the direction of polarization (Q_0 ~ 2 J/cm^2).

It should be noted that low concentration of free electrons in semiconductor in the initial state does not provide the surface optical properties, which are necessary for excitation of the SEW. In case of longer laser pulses (longer than tens of picoseconds) experimentally observed excitation of SEW on semiconductors is related to properties of the melt formed on the surface due to laser heating. This explanation can't be used for femtosecond action because photoexcitation and thermal processes are separated in time and the surface does not melt during the laser pulse. The conditions for excitation of SEW during ultrashort laser pulse result from high concentration of nonequilibrium carriers, which are generated in the semiconductor by the light.

Dynamics of the optical properties change at the surface of semiconductors under femtosecond laser action is related to change of the non-equilibrium carriers plasma frequency. In order to analyze the behavior of the optical properties let us first consider the basic mechanisms of light absorption and recombination of the absorbed energy.

Total absorption coefficient in semiconductors can be considered as a sum of absorption coefficients associated with different mechanisms $\alpha = \sum_{i=1} \alpha_i$. First of all these are fundamental band-to-band absorption of light ($h\nu > E_g$, E_g is the band-gap) and intraband absorption, i.e. absorption by free carriers - electrons and holes (from now on we consider only the electrons for simplicity). Rate of relaxation of the crystal electron system from the excited state to the equilibrium state is determined by the recombination mechanisms with characteristic times $\sim 10^{-12}$-10^{-10}s.

During femtosecond laser pulse the light intensity can achieve very high level without destruction of the matter, therefore initiating multiphoton processes. In the wavelength range from IR to near UV the energy of a single photon is not sufficient for electron transition from the valence band to the conduction band ($h\nu < E_g$). Such transition takes place as a result of simultaneous absorption of several photons. In this case the rate (or probability) of multiphoton ionization is highly dependent on the laser power. The multiphoton ionization rate is proportional to σI^m, where I is the laser radiation intensity, σ_m - m-photon absorption cross-section. The required number of photons is determined by the lowest value of m, satisfying the relation $mh\nu > E_g$.

During femtosecond laser pulse only photoexcitation and fast electronic processes are observed, while recombination and lattice heating can be neglected, because the characteristic times of these processes are much higher than the pulse duration. This is one of the major differences between action of ultrashort and longer laser pulses.

Model. According to the model the dynamics of distribution of nonequilibrium electron concentration $n(z, t)$ during the femtosecond pulse is determined by generation of the nonequilibrium electrons due to two-photon absorption ($h\nu < E_g$) followed by their diffusion and participation in collision processes as described in the diffusion equation (11).

In this consideration nonequilibrium electrons are the electrons that transit from the valence band to the conduction band under photo-excitation, and then contribute to increasing electron gas temperature, emission, and finally recombination after the end of the pulse. Below by "electrons" we will understand "non-equilibrium electrons" and by "electron gas" we will understand "nonequilibrium electrons gas".

$$\frac{\partial n(z,t)}{\partial t} = \alpha_{2phi}J + D\frac{\partial^2 n(z,t)}{\partial z^2} - \frac{n(z,t)}{\tau_e}, \tag{11}$$

where D is the diffusion coefficient of electrons in a solid, τ_e is the time of electron collisions.

Similarly to the above considerations for metals, the losses of electrons at the surface caused by external emission are taken into account in the model as follows:

$$-D\frac{\partial n(z,t)}{\partial z}\Big|_{z=0} = F_{2pho}\Big|_{z=0} + F_t\Big|_{z=0}, \tag{12}$$

$$F_{2pho}\Big|_{z=0} = -\int_0^{\infty} \sigma_2 J^2 \exp(-z/l_e)dz. \tag{13}$$

Expression (13) describes two-photon emission. Single-photon absorption cross-section σ_1 is estimated assuming the maximum absorption coefficient, that corresponds to absorption by free carriers in metal $\sigma_1 \approx \alpha_{max}/n_{max} \sim 10^{-17}$ cm^2 (n_{max} is electron concentration in the metal). Photo-ionization of atoms is considered in terms of electron transition through virtual states. The electron life time (τ_0) at these virtual states is determined basing on the uncertainty relation between energy and time: $\tau_0 \sim 10^{-16}$ s. In this case two-photon cross-section is given by $\sigma_2 \approx \sigma_1^2 n(z,\ t)\tau_0 \sim 10^{-28}$ cm s. Expression (14) describes external thermo-emission (Richardson low).

$$F_t = -BT_e^2\Big|_{z=0} \exp\left(-\frac{\varphi_e}{k_b T_e\big|_{z=0}}\right) \exp(-z/l_e)/q_e \tag{14}$$

Temperature of electrons gas (T_e) is determined by heat conduction equations (15), where heat capacity of electron gas is a function of temperature and concentration.

$$\begin{aligned} \frac{\partial T_e}{\partial t} &= a_e \frac{\partial^2 T_e}{\partial z^2} - \beta_{ei}(T_e - T_i)/c_e + \alpha_e J h\nu / c_e, \\ \frac{\partial T_i}{\partial t} &= a_i \frac{\partial^2 T_i}{\partial z^2} - \beta_{ei}(T_e - T_i)/c_i, \end{aligned} \tag{15}$$

$$c_e = \frac{\pi^2 k_b^2 n(z,t) T_e}{2\varepsilon_F}, \tag{16}$$

where F_{2pho} is determined from (13) for two-photon photoeffect, F_t - from (14).

The Bouguer–Lambert differential law of Eq. (17) determines the intensity distribution $J(z,\ t)$ inside the solid (the z axis is directed depth ward). In this model absorptance of the material (A) is assumed to be constant.

$$\begin{aligned}&\frac{\partial J(z,t)}{\partial z}=-(\alpha_{2phi}+\alpha_e+\alpha_{2pho})J(z,t),\\&J(0,t)=AJ_0(t);J_0(t)=q_0(t)/(h\nu),\\&q_0(t)=q_m(t/t_m)\exp(-t/t_m),\end{aligned}\tag{17}$$

where $AJ_0(t)$ is the density of the absorbed photon flux, with a bell-shaped temporal distribution of intensity of the laser radiation, α_{2phi} is two-photon absorption coefficient of the inner photoeffect, α_{2pho} is two-photon absorption coefficient of the extrinsic photoeffect, and α_e is coefficient for absorption by free electrons, which in turn are defined as

$$\begin{aligned}&\alpha_{2phi}=\sigma_1^2 n_{\max}\tau_\nu J,\\&\alpha_{2pho}=\sigma_1^2 n(z,t)\tau_\nu J,\\&\alpha_e=\sigma_1 n(z,t).\end{aligned}\tag{18}$$

Expressions (11-18) along with initial and boundary conditions at $t = 0$ and $z = \infty$

$$\begin{aligned}&n\big|_{z=\infty}=n\big|_{t=\infty}=0,\\&T_e\big|_{z=\infty}=T_i\big|_{z=\infty}=T_e\big|_{t=0}=T_i\big|_{t=0}=0\end{aligned}\tag{19}$$

allow one to obtain spatial and temporal distribution of the electron concentration in semiconductor $n(z, t)$.

The first qualitative estimates were made for a simplified model, in which the emission mechanism is generalized and the emission flux in the expression (12) is taken into account using the emission factor μ without separation of photo- and thermal emission

$$-D\frac{\partial n(z,t)}{\partial z}\Big|_{z=0}=-\mu n.$$

Fig. 7 shows obtained variation of the electron concentration with depth during the laser pulse action on silicon. The following initial data were used for the calculation: $D = 80$ cm²/s, $\tau_e = 10^{-14}$s, $\mu = 2\cdot 10^8$ cm/s, $Q_0 = 2$ J/cm², pulse duration is 80 fs, wavelength is 1.25 μm. The calculation results indicate that the semiconductor surface acquires properties of a metal during the laser pulse action. The maximum of electron density is located at some distance from the surface. It shifts from the surface into the bulk of material and its value increases during the pulse action. By the middle of the pulse the distribution of electron concentration stabilizes following the shape of the laser pulse.

The above estimates showed that dynamics of the optical properties of semiconductor under action of ultrashort pulse can be described within the same approach used for metals. According to Drude dispersion theory the dielectric permeability of photoexcited semiconductor can be determined by the plasma frequency of the electron gas (ω_p), incident radiation frequency (ω) and the frequency of electron collisions (γ), according to following expressions:

$$\varepsilon = \varepsilon' + i\varepsilon'',$$
$$\mathrm{Re}\,\varepsilon = \varepsilon_n - \frac{\omega_p^2}{\omega^2 + \gamma^2}, \quad (20)$$
$$\mathrm{Im}\,\varepsilon = \frac{\omega_p^2 \gamma}{\omega\left(\omega^2 + \gamma^2\right)},$$

$$\omega_p = \sqrt{\frac{4\pi n(z,t) q_e^2}{m_e}}, \quad (21)$$

where ε_n is the initial value of the semiconductor permittivity, and ε' and ε'' is the real and imaginary parts of the permeability.

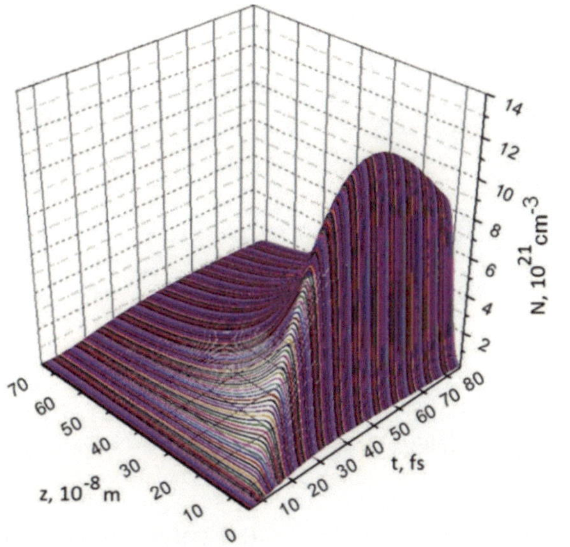

Figure 7. The dynamics of silicon photoexcitation

Let us consider dynamics of the dielectric permeability in the surface layer of semiconductor taking into account the change in the plasma frequency of non-equilibrium carriers by using a mathematical model (11-21). This will help us to identify the role of different emission processes in evolution of the optical properties during a femtosecond pulse.

Numerical simulation was performed for silicon (ε_n = 12, $\gamma = 10^{14}$ s^{-1}) with initial data given above. The calculation results are shown on Fig. 8-10.

When there is no emission surface the real part of dielectric permeability at the surface quickly decreases and becomes negative stabilizing by the end of the first quarter of the pulse (Fig. 10, curve 1). If external photo-emission is taken into account, the character of the dynamics of Reε does not change, but the magnitude at which the permeability stabilizes is increased (Fig. 8, curve 2). Thermionic emission strongly affects the dynamics of the permeability. Influence of the thermionic emission is small in the beginning of the pulse and value of Reε abruptly decreases. Few femtosecond later contribution of the thermionic emission grows and value of Reε, returned to its initial level after several damping oscillations (Fig. 8, curve 3). The observed inertia is typical for the thermionic emission mechanism, and the return of the dielectric permeability to the initial value means that all the "hot" electrons leave the surface as a result of thermal emission. If photo-emission is also taken into account escape of the electrons speeds up (Fig. 8, curve 4).

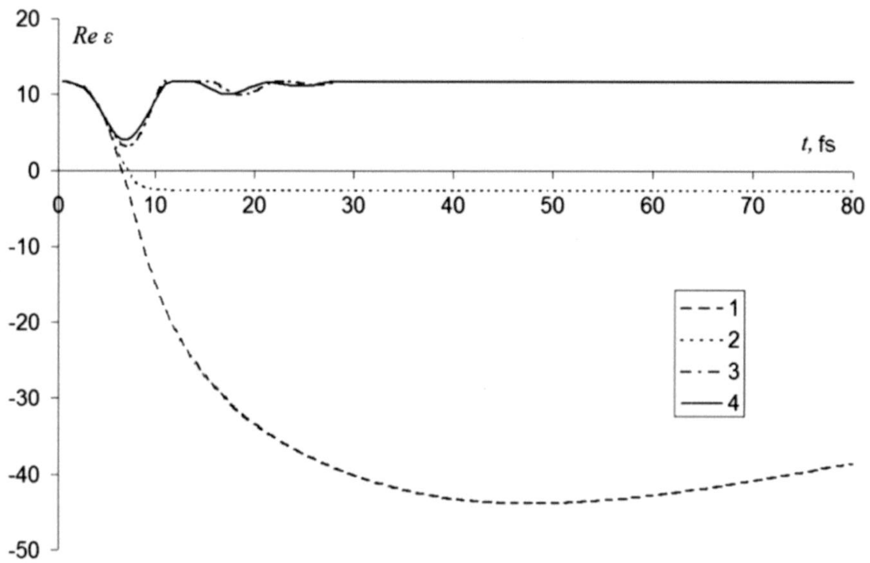

Figure 8. Dynamics of the real part of the permeability at the surface of silicon (Q_0 = 2 J/cm^2). 1 – no external emission, 2 – only photo-emission is taken into account, 3 – only thermionic emission is taken into account, 4 – combined action of photo-emission and thermal emission

Figure 9 shows dynamics of the real part of the permeability at the silicon surface for different values of the light flux density. If only two-photon photo-emission is taken into account, the surface acquires metal-like properties at the first femtoseconds of the pulse, and the permeability remains negative during the entire pulse. Increasing the radiation-flux density reduces the time to reach the steady-state level without changing the character of the dependence (Fig. 9a). When both photo-emission and thermionic emission are taken into account the picture changes (Fig. 9b). If $Q_0 \leq 1$ J/cm^2, Reε smoothly returns to its initial value (curve 1). When the energy density increases the permeability oscillates during transition to its initial value (curves 2 and 3). These oscillations result from dependence of heat capacity of the electron gas on electron concentration. The higher light flux induces higher temperature of the electron gas and contribution of thermionic emission increases. As a result the permeability returns to its initial value faster. At the same time increasing flux density raises the electron concentration and, accordingly, the heat capacity of the electron gas increases thus reducing its temperature and the contribution of the thermal emission. This results in oscillating of dielectric permeability.

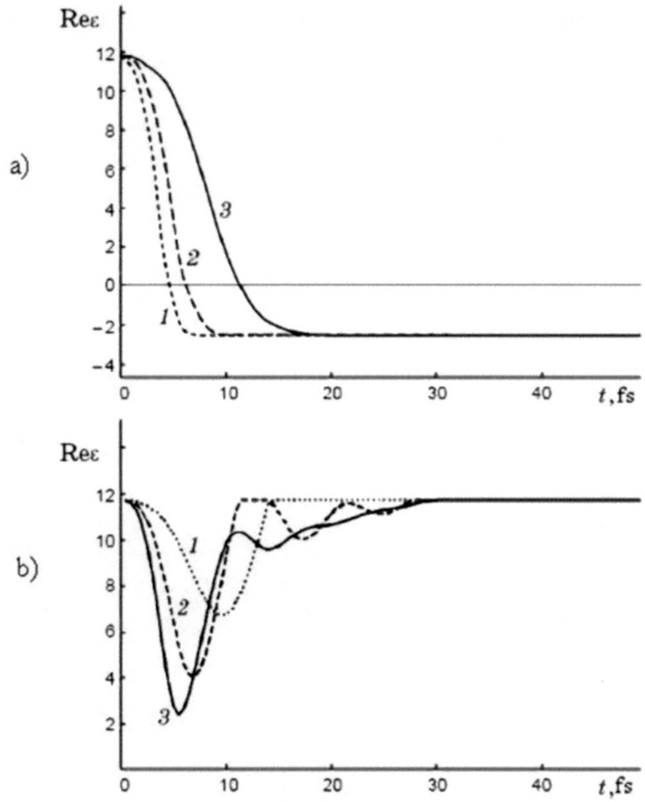

Figure 9. Dynamics of the real part of the permeability on the surface of silicon for various values of the light-flux energy density. (a) Photoemission only, (b) Combined action of photoemission and thermal emission. 1 – Q_0= 1 J/cm², 2 – Q_0 = 2 J/cm², 3 – Q_0 = 3 J/cm²

Figure 10 shows the calculated distribution of permeability Re$\varepsilon(z)$, at the end of the pulse (Q_0 = 2 J/cm²). If there is no emission, a very thin metal-like layer appears on the surface (curve 1). If external photo-emission is taken into account, a metal-like layer (several tens of nanometers thick, curve 2) is formed. The loss of the electrons due to thermionic emission qualitatively changes the permeability distribution depth wards (curve 3). A layer with smaller permeability than its initial value is formed at a distance of tens of nanometers from the surface. Combined action of photo-emission and thermionic emission shifts this layer deeper (curve 4).

Let us consider again the conditions necessary for the SEW excitation in order to compare the results of numerical simulation with the above experimental data on femtosecond silicon microstructuring associated with the SEW excitation.

It is known that the formation of PSS oriented perpendicular to the polarization of laser radiation usually results from excitation of surface plasmon-polaritons under the laser pulse action on the metal. Surface plasmon-polaritons are partially longitudinal electromagnetic waves of the TM-type propagating along the interface between two media with the wave electro-magnetic field being localized near the interface. Excitation of plasmon-polaritons is possible only if one of the media has positive dielectric permeability ($\varepsilon_1>0$), while the real part of dielectric permeability of the other is negative ($Re\varepsilon_2<0$). Also, the condition $|Re\varepsilon_2|>\varepsilon_1$ must be satisfied. A negative dielectric permeability in a metal is determined by a high concentration of free electrons. In the case of the relatively long laser pulse action (tens of picoseconds or more) on the semiconductor the appearance of metal-like optical response is usually associated with the properties of the melt formed on the surface due to laser heating.

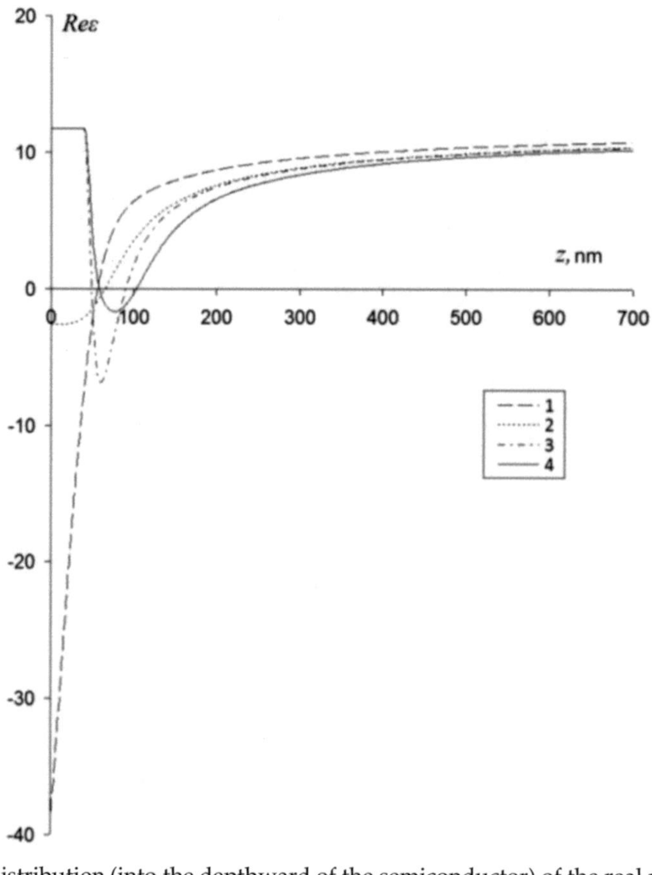

Figure 10. Spatial distribution (into the depthward of the semiconductor) of the real part of the permeability at the end of the laser pulse. 1 – no external emission of electrons, 2 – photo-emission only, 3 – thermionic emission only, 4 – combined action of photo-emission and thermal emission

Semiconductors typically have positive value of their dielectric permeability in the visible and IR range. Under the femtosecond laser pulse action the concentration of nonequilibrium carriers in semiconductors can become so high that dielectric permeability ε would change its sign and creates conditions for excitation of surface plasmon-polaritons. In this case, the formation of PSS perpendicular to the polarization vector is experimentally observed.

The formation of structures parallel to the polarization vector is associated with the excitation of surface waveguide modes (TE-polaritons). It is necessary to create an optically layered structure for excitation of a waveguide mode at the semiconductor surface (Fig. 11).

The refractive index of the waveguide layer in such optically layered structure (n_2) exceeds the refractive index values of adjacent layers (n_1, n_3), ($n_2 > n_1$, $n_2 > n_3$).

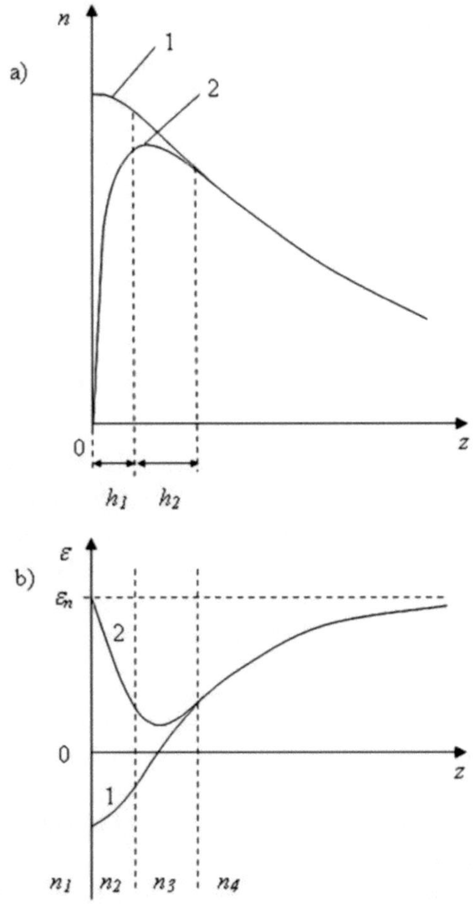

Figure 11. Schematic distribution of the electrons concentration in a semiconductor under the femtosecond laser pulse action (a) and the corresponding distribution of the dielectric permeability (b) for different emissions; n_1, n_2, n_3, n_4 - effective refractive indices of the layers as they are alternating along the coordinate z

In addition, it is necessary to provide a certain minimum thickness of the waveguide layer at a given frequency. If action of laser radiation results in formation of a layer with a refractive index n_3 and closer to surface a layer of thickness h_1 with a refractive index n_2 provided $n_2 > n_3 > n_1$, then [7]

$$h_{\min} \approx \frac{\lambda}{2\pi\sqrt{n_2^2 - n_3^2}} \arccos\sqrt{\frac{n_2^2 - n_3^2}{n_2^2 - n_1^2}}. \tag{22}$$

In particular, for silicon at a wavelength of 1.25 μm the minimum thickness of the waveguide layer is $h_{\min} \sim 70$ nm according to expression (22).

Let us consider the calculated spatial distribution of the silicon dielectric permeability under the femtosecond laser pulse action (see Fig. 10) in terms of the possible conversion of the incident light into surface plasmon-polaritons and waveguide modes. The surface polariton excitation requires transition of the semiconductor surface into a metal-like state, while formation of a dynamic optically-layered structure with a certain minimum thickness of the waveguide layer is necessary for the excitation and propagation of waveguide modes.

As follows from the numerical model (Fig. 10, curves 1-2) in case of relatively low emission a metal-like layer is formed at the surface. Within the layer thickness is about 50-60 nm dielectric permeability becomes negative. This provides the conditions for excitation of surface plasmon-polaritons (TM-type SEW), which is confirmed experimentally by formation of the microstructures perpendicularly to the polarization vector.

If emission rate is high as in case of thermo-emission (Fig. 10, curve 3) an optically layered structure is formed. Although the dielectric permeability does not change its sign, excitation of waveguide modes is possible.

Combination of both photo- and thermal-emission (Fig. 10, curve 4) results in formation of a dielectric layer of thickness ~ 60 nm and a metal-like layer of thickness about 40 nm. The refractive index of this dielectric layer is higher than both the refractive index of the air on one interface, and the refractive index of the metal-like layer on the other interface. Presence of such an optical structure enables excitation of waveguide mode, which results in the formation of periodic relief, which is parallel to the polarization vector on the incident radiation (see Fig. 6, right).

The above consideration showed that multiphoton emission and thermionic emission noticeably vary the optical properties of a semiconductor during a femtosecond pulse action. In particular, layer with different optical properties is formed at the surface and enables excitation of either surface polaritons or waveguide modes in the semiconductors. The considered model allows one to qualitatively and quantitatively interpret available experimental data. This approach allows one to use experimentally observed surface microstructures as relatively simple means of investigation of dynamics of the semiconductor surface properties under femtosecond action.

4. Conclusion

The results of numerical simulation have shown that influence of emission processes on the electron gas temperature and lattice temperature of metals is negligible. Therefore, the emission can be neglected when assessing the parameters of metals processing by femtosecond laser pulse, which simplifies the numerical calculations.

The Coulomb explosion occurrence in metals requires high-power incident radiation, which is impossible for the real exposure conditions.

However, in semiconductors both types of extrinsic emission noticeably change distribution of dielectric permeability near surface providing conditions for excitation of surface polaritons or waveguide modes depending on laser power magnitude.

The proposed method allowed to estimate the cross sections of multiphoton absorption in metals. For example, for metals absorption cross section: $\sigma_1=10^{-17}cm^2$ for one-photon absorption, $\sigma_2=10^{-28}cm\ s$ for two-photon absorption, $\sigma_3=10^{-61}cm^3s^2$ for three-photon absorption.

Author details

Roman V. Dyukin, George A. Martsinovskiy, Olga N. Sergaeva, Galina D. Shandybina,
Vera V. Svirina and Eugeny B. Yakovlev
Department of Laser-Assisted Technologies and Applied Ecology,
National Research University of Information Technologies, Mechanics and Optics,
Saint-Petersburg, Russia

Acknowledgement

This work was supported by grants of RFBR 09-02-00932-a, 09-02-01065-a and State Contract No P1134.

5. References

[1] Korte F., Nolte S., Chichkov B.N., Bauer T., Kamlage G., Wagner T., Fallnich C., Welling H. (1999) Far-Field and Near-Field Material Processing with Femtosecond Laser Pulses. Appl. Phys. A, 69 [Suppl.]: S7–S11.

[2] Vorobyev D.Y., Guo C.L., (2010) Metallic Light Absorbers Produced by Femtosecond Laser Pulses. Advances in Mechanical Engineering, 2010: 1-5.

[3] Chimmalgi A., Grigoropoulos C.P., Komvopoulos K. (2005) Surface Nanostructuring by Nano-Femtosecond Laser Assistant Force Microscopy. App. Phys. 97: 1043191 - 104319112.

[4] Myers R.A., Farrell R., Karger A.M., Carey J.E., Mazur E. (2006) Enhancing Near-Infrared Avalanche Photodiode Performance by Femtosecond Laser Microstructuring. Appl Opt. 45(35): 8825-31.

[5] Etsion I. (2005) State of the Art in Laser Surface Texturing. J. of Tribology, 127(1): 248-253.

[6] Oktem B., Kalaycioglu H., Erdoğan M., Yavaş S., Mukhopadhyay P., Tazebay U. H., Aykaç Y., Eken K., Ilday F. Ö. (2010) Surface Texturing of Dental Implant Surfaces with an Ultrafast Fiber Laser. in Conference on Lasers and Electro-Optics, OSA Technical Digest (CD) (Optical Society of America, 2010), JTuD15.

[7] Libenson M.N. (2007) [Laser-Induced Optical and Thermal Processes in Solids and Their Mutual Influence]. Saint-Petersburg: Nauka. 423 p. (in Russian)

[8] Libenson M.N. (2001) Non-Equilibrium Heating and Cooling of Metals Under Action of Super-Short Laser Pulse. Proc. SPIE. 4423: 1–7.

[9] Carey J.E., Crouch C.H., Mazur E. (2003) Femtosecond-Laser-Assisted Microstructuring of Silicon Surfaces. Optics&Photonics News. 14(2): 32-36.

[10] Shimosuma Y., Kazansky P.G., Qin J.R., Hirao K. (2003) Self-Organized Nanogratings in Glass Irradiated by Ultrashort Light Pulses. Phys. Rev. Lett. 91: 247405-247409.

[11] Kudryashov S.I., Emel'yanov V.I. (2001) Electron gas compression and Coulomb explosion in the surface layer of a conductor heated by femtosecond laser pulse. JETP Lett. 73(12): 666-670.

[12] Evans R., Badger A.D., Fallies F., Mahdieh M., Hall T.A., Audebert P., Geindre J.-P., Gauthier J. C., Mysyrowicz A., Grillon G., Antonetti A. (1996) Time- and Space-Resolved Optical Probing of Femtosecond-Laser-Driven Shock Waves in Aluminum. Phys. Rev. Lett. 77(16): 3359-3362.

[13] Del Fatti N., Arbouet A., Vall′ee F. (2006) Femtosecond Optical Investigation of Electron–Lattice Interactions in an Ensemble and a Single Metal Nanoparticle. Appl. Phys. B. 84: 175-181.

[14] Chen. L.M., Zhang J., Dong Q.L., Teng H., Liang T.J., Zhao L.Z., Wei Z.Y. (2001) Hot Electron Generation Via Vacuum Heating Process in Femtosecond Laser–Solid Interactions. Phys. of Plasmas. 8(6): 2925-2929.

[15] Kaganov M.I., Lifshitz I.M., Tanatarov L.V. (1956) [Relaxation between Electrons and Lattice]. J. Exp. Theor. Phys. 31(2): 232-237. (in Russian)

[16] Anisimov S.I., Kapeliovich B.L., Perelman T.L. (1974) [Electron Emission from Metal Surfaces Exposed to Ultrashort Laser Pulses]. J. Exp. Theor. Phys. 39(2): 375-377. (In Russian)

[17] Anisimov S.I., Benderskii V.A., Farkas G. (1977) Nonlinear Photoelectric Emission from Metals Induced by a Laser Radiation. Soviet Physics Uspekhi. 20(6): 467-488.

[18] Delone N.B. (1989) [Interaction of Laser Beam with Matter. The course of lectures]. Moscow: Nauka. 289 p. (In Russian)

[19] Kato S., Kawakami R., Mima K. (1991) Nonlinear Inverse Bremsstrahlung in Solid-Density Plasmas. Phys. Rev. A. 43(10): 5560-5567.

[20] Wang X.Y., Downer M.C. (1992) Femtosecond Time-Resolved Reflectivity of Hydrodynamically Expanding Metal Surfaces. Optics Letters. 17(20): 1450-1452.

[21] Rethfeld B., Kaiser A., Vicanek M., Simonc G. (2001) Nonequilibrium Electron and Phonon Dynamics in Solids Absorbing a Subpicosecond Laser Pulse. Proc. SPIE. 4423: 250-261.

[22] Rethfeld B., Kaiser A., Vicanek M., Simon G. (2002) Ultrafast Dynamics of Nonequilibrium Electrons in Metals under Femtosecond Laser Irradiation. Phys.l Rev. B. 65: 214303-214313.

[23] Ashcroft N.W., Mermin N.D. (1976) Solid State Physics. New York: Holt, Rinehart, and Winston. 826 p.

[24] Bulgakova N.M., Rosenfeld A., Ehrentraut L., Stoian R., Hertel I.V. (2007) Modeling of Electron Dynamics in Laser-Irradiated Solids: Progress Achieved Through a Continuum Approach and Future Prospects. Proc. SPIE. 6732: 673208-673223.

[25] Zabotnov S.V., Ostapenko L.A., Golovan L.A., Shandybina G.D., Timoshenko V.Yu., Kashkarov P.K. (2005) Third Optical Harmonic Generation of Silicon Surfaces Structured by Femtosecond Laser Pulses. Proc. SPIE. 6161: 0J1-0J5.

[26] Ostapenko I.A., Zabotnov S.V., Shandybina G.D., Golovan L.A., Chervyakov A.V., Ryabchikov Yu.V., Yakovlev V.V., Timoshenko V.Yu., Kashkarov P.K. (2006) Micro- and Nanostructuring of Crystalline Silicon Surface under Femtosecond Laser Pulses. Bulletin of RAS: Physics. 70: 1503–1506.

Chapter 8

Kinetics and Dynamics of Phase Transformations in Metals Under Action of Ultra-Short High-Power Laser Pulses

V.I. Mazhukin

Additional information is available at the end of the chapter

http://dx.doi.org/10.5772/50731

1. Introduction

Action of super-power and ultra-short laser pulses on highly absorbing condensed media is investigated in the past two decades [1] - [5]. The urgency of this problem is primarily determined by a variety of practical applications of pulsed laser irradiation. In a short period of time scientists have mastered such operations as ablation of elemental materials by femtosecond lasers (100 fs) [6], femtosecond nanostructuring [7], generation of metallic nanoparticles and nanostructures by laser ablation of massive targets by 40 fs and 2ps pulses [8], etc.

The use of super-power $G \sim 10^{12} \div 10^{15}$ W/cm^2 and ultra-short $\tau_L \approx 10^{-12} \div 10^{-15}$ s laser pulses for dimensional processing of materials, such as cutting of various materials by pico-femtosecond laser pulses [9], micro-drilling by a femtosecond laser [10], surface etching of metals (Al, Cu, Mo, Ni) and semiconductors (Si) [11], is accompanied by realization of unique physical conditions. In particular, the duration of action becomes comparable with the characteristic times of thermalization and phase transitions in matter. This leads to the need to address complex fundamental problems, including the heating of the material and the kinetics of phase transitions in a strong deviation from local thermodynamic equilibrium.

It should be noted that the majority of laser technologies is associated with the beginning of phase transformations in the material. In particular, the action of pico - femtosecond laser pulses of high intensity on solid targets is one of the ways to create individual particles with unique characteristics, or to form their streams, consisting of a cluster, liquid or solid fragments of the target. The formation of the particle flux in the pulsed laser ablation is observed for a wide range of materials: metals, semiconductors and insulators. The possibility of usage of laser ablation products in practical applications was the impetus for a number of experimental and theoretical studies [12], aimed at studying the conditions and mechanisms of formation of particles of nano - and micro- sizes during the laser exposure.

The accumulation of knowledge of the experimental nature in the first place in this dynamic field leads not only to the variety of practical applications of pulsed laser action, but also to a number of issues of independent physical interest.

The main features of ultrashort action on metals are associated with high speed and voluminous nature of the energy release of the laser pulse. The high rate of heating of a condensed medium is associated with rapid phase transformations of matter, characterized by the transfer of superheated phase boundaries of high-power fluxes of mass and energy. Overheated metastable states at the interface are characterized by temperatures, whose values can be hundreds of degrees higher than the equilibrium values of the melting point or boiling point. Removal of the energy by the flow of matter in conjunction with volume mechanism of energy release of laser radiation contribute to the formation of metastable superheated regions in the volume of solid and liquid phases with near-surface temperature maximum. Calculations [13] showed that the maximum speed of melting front are comparable to the speed of sound ~ (0.5 - 6) km / s, and the phase velocity of the solidification front are (10 - 200) m / s. Accordingly, the maximum superheating / undercooling can reach several thousand / hundred degrees. The achievement of such overheating and overcooling leads to large gradients of the Gibbs energy, which actually determine the driving force for high-speed phase transformations.

Physics of supercooled states in metallic and nonmetallic systems, because of their widespread use in the production of new materials technology, is relatively well-studied [14]. Superheated states received little attention until recently, largely because of the difficulty of experimental investigation and non-obviousness of their application. The situation changed with the advent of femtosecond laser pulses and their application for the production of nanoparticles and nanomaterials [15] - [17].

The purpose of this chapter is the theoretical study of nonequilibrium pulsed laser heating of metals, the kinetics and dynamics of phase transitions in a deep deviation from the local thermodynamic equilibrium.

The main tool for studying the processes initiated by laser pulses of picosecond and femtosecond duration are the methods of mathematical modeling and computational experiment (CE). The possibilities of experimental approaches in this area are very limited due to the large transience of the processes. Computational experiments are preferred in the cases where the natural experiment is not possible, is very difficult or very expensive. The statement of CE is especially convincing in the studies of the kinetics and dynamics of fast processes. For its statement, the computational experiment, which is an important link connecting theory and field experiments, requires the development of appropriate models, determining the properties of all substances studied, the development of computational algorithms and the creation of program codes.

Construction of a theory that covers most of the features of pulsed laser action on materials is very difficult. The methods of mathematical modeling in theoretical constructs have that advantage that you can use the phenomenological and the experimental data.

The description of the kinetics of fast phase transitions of the first kind is carried out in two classes of mathematical models: continuum and atomistic. Continuous models are based on the equations of continuum mechanics, as a rule, are represented as partial differential equations with appropriate boundary conditions and equations of state. Continuum model is used to describe the macro-level processes, heterogeneous kinetics and dynamics of phase transitions of the 1st kind [18]. Atomistic approach is based on a model of molecular dynamics is used to describe the kinetics of homogeneous phase transitions [19].

2. Brief theory of phase transformations of the 1st type

The basis of first-order phase transformations - melting-solidification and evaporation-condensation are two qualitatively different mechanisms: heterogeneous and homogeneous ones. The heterogeneous mechanism is characterized by a sharp interface between the phases (phase front of zero thickness) and determines the dynamics of phase transformations of the 1st type. The homogeneous mechanism of phase transitions that arise usually under the influence of the volume heating or cooling is associated with the processes of volume melting and boiling or spontaneous crystallization and condensation.

Historically, the theory of phase transformations of the 1st type of melting is based on two fundamental sections of classical physics, thermodynamics and kinetics.

2.1. Thermodynamical approach

Thermodynamics is a macroscopic theory [20], which from the energy point of view, considers the properties of macroscopic bodies in equilibrium. This allows to have a great community for the conclusions of thermodynamics. However, thermodynamics does not take into account the internal structure of the considered bodies and some of its conclusions and regulations do not have physical clarity. One way to describe the equilibrium processes in the equilibrium thermodynamics is a theory of thermodynamic potentials.

2.1.1. Phase equilibrium and phase transformations

The properties of a thermodynamic system are determined by thermodynamic parameters. The energy state of a thermodynamic system in equilibrium is uniquely determined by the parameters of the system. There is a unique relationship between the parameters of the system, which is mathematically represented by the state function. From a mathematical point of view, this means that the function has a total differential.

The basis of the method of thermodynamic potentials is just the possibility of introducing the state functions for the equilibrium processes, with total differentials describing the change in the state of a thermodynamic system.

The main identity of thermodynamics of equilibrium processes is usually represented as

$$TdS = d\varepsilon + pdV \text{ or } d\varepsilon = TdS - pdV \tag{1}$$

Depending on the choice of the two independent parameters, one can introduce thermodynamic potentials, which differentiation allows determining the other unknown parameters of the state. In general, a thermodynamic potential may be a function of various parameters. In this notation, the internal energy is given as a function of entropy and volume: $\varepsilon = \varepsilon(S,V)$. It is the function of the state and it has a total differential with respect to its variables. The total differential $d\varepsilon$ can be used to determine the temperature T and pressure p. However, the usage of entropy S and volume V as two independent variables is inconvenient because they are difficult to control in experiment. It is more convenient to use pressure p and temperature T as two independent variables.

After consecutive transformation of internal energy U first into enthalpy $H = H(S,p)$

$$H = \varepsilon + pV, \tag{2}$$

which total differential, taking into account the basic identity of thermodynamics has the form

$$dH = d\varepsilon + pdV + Vdp = TdS + Vdp \tag{3}$$

and then into Gibbs energy $G = G(T,p)$ (Gibbs thermodynamic potential)

$$G = H - TS = \varepsilon + pV - TS \tag{4}$$

one can obtain the total differential of Gibbs energy, which with account of the basic identity of thermodynamics (1), takes the form

$$dG = -SdT + Vdp \tag{5}$$

It is convenient in the fact that the independent variables T and p are easily to modify and to control in experiment.

After reaching the equilibrium state of the system, Gibbs potential takes its minimum value and becomes constant: $dG = 0$. This allows to use the condition of minimum of the Gibbs potential for the description of equilibrium states in which $T = const$ and $p = const$.

Thermodynamic systems in equilibrium state do not necessarily have to be a homogeneous medium. A system in equilibrium may be composed of several phases, different in their physical and chemical properties, separated by the phase boundaries not changing over time.

The multiphase thermodynamic systems are most simply described, components of which are in equilibrium states, and there is no transfer of matter, energy and momentum through interphase boundaries. In this case, such thermodynamic system is in equilibrium and methods of equilibrium thermodynamics apply to describe it.

Given that during phase transformations, each of the phases is a system with variable mass, the notion of chemical potential $\mu(p,T)$ is introduced into the thermodynamic description. It is used to take into account not only exchange of energy but also the exchange of mass (particles). To determine the chemical potential, the term which takes into account the possibility of changing the number of particles in a homogeneous system (the same can be done with the other potentials) is formally introduced into the expression for the thermodynamic Gibbs potential:

$$dG = -SdT + Vdp + \mu dN$$

dN determines the change of the number of particles in the system.

If the thermodynamic potential dG is given as a function of temperature and pressure, then the value of the chemical potential is written as: $\mu = \left(\frac{\partial G}{\partial N}\right)_{p,T}$

The chemical potential can be written through other thermodynamical functions but in this case it will be written in terms of other state parameters:

$$\mu = \left(\frac{\partial G}{\partial N}\right)_{p,T} = \left(\frac{\partial U}{\partial N}\right)_{V,S} = \left(\frac{\partial H}{\partial N}\right)_{p,S} \tag{6}$$

From the relation (6) it follows, that the chemical potential is a physical quantity that is equal to the value of some thermodynamical potential (with constant certain parameters) that is required to add to the system to change its number of particles by unity.

2.1.2. The conditions of equilibrium for two-phase single-component system

If macroscopic transport does not occur through the interphase boundaries, and the phases themselves are in a state of thermodynamic equilibrium, such thermodynamic system, in spite of its heterogeneity, will be in a state of thermodynamic equilibrium.

For the phase equilibrium in the one-component two-phase system, the following three conditions must be fulfilled:

1. the condition of thermal equilibrium, that means the equality of temperatures at both sides of the interphase boundary: $T_1 = T_2 = T_{eq}$,
2. the condition of mechanical equilibrium, consisting of the equality of pressure at the both sides of the interphase boundary: $p_1 = p_2 = p_{eq}$,
3. the condition of the equality of the Gibbs energy per particle, consisting of the requirement of the absence of the macroscopic transfer of molecules (atoms) of this material from one phase to another: $\mu_1(p,T) = \mu_2(p,T)$

In principle, we can use not the Gibbs energy, but any thermodynamic potential, which has a minimum in equilibrium. It is not difficult to show the validity of these conditions. We shall use the total differential for the internal energy, taking into account changes in the number of particles:

$$d\varepsilon = TdS - pdV + \mu dN$$

We write this expression for each phase of the closed-loop system:

$$\begin{aligned} d\varepsilon_1 &= T_1 dS_1 - p_1 dV_1 + \mu_1 dN_1 \\ d\varepsilon_2 &= T_2 dS_2 - p_2 dV_2 + \mu_2 dN_2 \end{aligned} \tag{7}$$

The closedness of the system automatically gives the following equation:

$$d\varepsilon_1 = d\varepsilon_2, \quad dS_1 = dS_2, \quad dV_1 = -dV_2, \quad dN_1 = -dN_2 .$$

Consider an equilibrium two-phase system under some simplifying assumptions. Assume that the phases do not change the volume and do not exchange particles, i.e. $dV_1 = dV_2 = 0$, $dN_1 = dN_2 = 0$. Combining the equations (7) for this case, we obtain the expression: $d\varepsilon_1 + d\varepsilon_2 = dS_1(T_1 - T_2) = 0$, which gives the condition of thermal equilibrium $T_1 = T_2 = T_{eq}$. Assuming the constancy of the entropy and the number of particles in phases, i.e. $dS_1 = dS_2 = 0, dN_1 = dN_2 = 0$ we obtain the condition of mechanical equilibrium: $p_1 = p_2 = p_{eq}$.

Given the conditions of thermal and mechanical equilibrium of (7) we obtain the equality of chemical potentials in the different phases $\mu_1(p,T) = \mu_2(p,T)$. This equation can be solved for the variables T and p and may represent the equilibrium curves of the two phases in the form $p_{eq} = p(T_{eq})$ or $T_{eq} = T(p_{eq})$. If we consider the boundary between liquid and solid, the equilibrium melting curve is obtained $T_m = T(p_m)$.

When describing the interface between liquid and gas, the equilibrium vaporization curve is obtained $p_{sat} = p(T_b)$.

It should be noted that the processes at the interface are static in nature as in the case of equilibrium of different phases and also during phase transitions. There is a constant process of transition of particles from one phase to another at the interface. In equilibrium, these opposing processes compensate each other, and during supply or withdrawal of heat to one of the phases one of these processes begins to dominate leading to a change in the amount of matter in various states of aggregation.

If the components of the thermodynamic system are not in equilibrium with each other, then there are thermodynamic flows through their interface. This will be a process of transformation of matter from one state to another, i.e. phase transformation. Assuming that the occurring processes are quasi-static and the flows are small, one can use the methods of equilibrium thermodynamics to describe such non-equilibrium system. In this case we assume an infinitely small difference between the thermodynamic parameters in different parts of the system.

2.1.3. *The driving force of phase transformations*

In the process of phase transitions of the 1st type, a number of quantities undergo abrupt changes at the interface, so in the following text in the thermodynamic equations, the sign of the differential will be replace by the corresponding value of the difference $dF \approx \Delta F$ for the temperature of the phase transition T_{eq}.

The driving force of phase transitions of the first type is determined by the difference of Gibbs energy (or the magnitude of overheating/overcooling ΔT) for two phases at the interface, defined in (4) and (5) and can be written in two forms

$$\Delta G = (\varepsilon_1 - \varepsilon_2) + p_{eq}(V_1 - V_2) - T_{eq}(S_1 - S_2) = \Delta\varepsilon + p_{eq}\Delta V - T_{eq}\Delta S \tag{8}$$

$$\Delta G = V\Delta p - S\Delta T \tag{9}$$

where $\Delta p = p - p_{eq}$, $\Delta T = T - T_{eq}$, $\Delta\varepsilon = \varepsilon_1 - \varepsilon_2$, $\Delta V = V_1 - V_2$, $\Delta S = S_1 - S_2$

Equilibrium. In equilibrium, $\Delta G = 0$ and the equality (8) takes the form:

$$\Delta\varepsilon + p_{eq}\Delta V - T_{eq}\Delta S = \Delta H - T_{eq}\Delta S = 0$$

where the difference of enthalpy $\Delta H = \Delta\varepsilon + p_{eq}\Delta V$, is known as the equilibrium latent heat of transformation

$$L_{eq} = \Delta H = T_{eq}\Delta S. \tag{10}$$

From equation (9) we can obtain the dependence of the equilibrium pressure P_{eq} on the temperature T_{eq}, that is known as the curve of Clausius-Clapeyron:

$$\frac{dp_{eq}}{dT_{eq}} = \frac{L_{eq}}{T_{eq}}\frac{1}{\Delta V} \tag{11}$$

If one take as ΔV the difference of the volumes of vapor V_{vap} and condensed V_{sol} phases, then since $V_{vap} >> V_{sol}$ for ideal gas $\Delta V = V_{vapor} - V_{solid} \approx \frac{RT}{P}$ one obtain the expression

$$\frac{dp_{eq}}{p_{eq}} = \frac{L_v}{R}\frac{dT_{eq}}{T_{eq}^2}$$

After integration, we find the temperature dependence of the equilibrium vapor pressure that is widely used for many materials:

$$p_{eq} = p_b \exp\left[\frac{L_v}{R}\left(\frac{1}{T_b} - \frac{1}{T_{eq}}\right)\right]$$

The vapor pressure of the material in equilibrium with solid or liquid phase is called the saturated vapor pressure and with the notation $p_{eq} = p_{sat}$, $T_{eq} = T_{sur}$ is usually written as

$$p_{sat} = p_b \exp\left[\frac{L_v}{R}\left(\frac{1}{T_b} - \frac{1}{T_{sur}}\right)\right] \tag{12}$$

where T_{sur} is the temperature of the surface of the condensed phase, p_b, T_b are the equilibrium values of pressure and the boiling point under normal conditions, L_v is the latent heat of evaporation, R is the gas constant.

As it follows from equation (9), at constant pressure, the difference of Gibbs energy ΔG is linearly proportional to the overheating/ overcooling

$$\Delta G = -S \cdot \Delta T = \frac{L_{eq}\Delta T}{T_{eq}} \tag{13}$$

In the future, assuming the difference of energy ΔG to be equal to the rate of the phase transformation, one can obtain, that in the thermodynamic approach, the rate of conversion at constant pressure for small deviations from equilibrium is linearly proportional to the overheating / overcooling ΔT

$$v \approx K \Delta T, \tag{14}$$

where K is the constant of proportionality between the normal speed limits and its overcooling. The constant K does not have any clear physical sense and is chosen experimentally for each material. The dependence (14) by its form coincides with the well known relation for the determination of the linear crystal growth rate obtained on the basis of classical molecular-kinetic models in which the constant of proportionality is called the kinetic coefficient [21]. The main application of the relation (14) and its various modification [22, 23] is found in the description of different processes of melting - solidification. Comparison with experiment showed that the equation (14) gives good agreement mainly at small overcoolings [24, 25]. The kinetic coefficient K is the main parameter, characterizing the mobility of the boundary crystal -melt. Despite the great importance of this characteristic, there are only a few experiments to successfully measure the kinetic coefficient in metals and alloys [26]. The main difficulties of experimental determination are associated with the great complexity of measuring the overcooling at the solidification front. Currently, the main approach to the determination of the quantitative evaluation and qualitative understanding of the physics of the processes behind the coefficient K are the methods of molecular dynamics [27] - [30].

2.2. Kinetic approach

The structure particles of matter are in continuous motion and appear in the main provisions of the molecular-kinetic theory in which all processes are considered at the atomic or molecular level, the particles obey the Boltzmann statistics, and the speed of processes is given by

$$R = R_0 \exp\left(-Q / k_B T\right) \tag{15}$$

where Q is the activation energy, k_B is the Boltzmann's constant, $k_B T$ is the average thermal energy for one atom, R_0 is the pre-exponential factor that affects the process rate. The exponential term $\exp\left(-Q / k_B T\right)$ is known as Boltzmann's factor, that determines the part of atoms or molecules in the system that have the energy above Q at the temperature T.

2.2.1. Melt kinetics

For the first time, the conditions of crystal growth from liquids were formulated by Wilson [31], who suggested that the atoms have to overcome the diffusion barrier in order to make the transition from liquid to solid phase. The rate of accession of atoms to the crystal lattice is expressed by the relation analogous to (15)

$$R_c\left(T\right) = a\nu \exp\left(-Q / k_B T\right) \tag{16}$$

where Q is the activation energy for overcoming the diffusion barrier, ν is the frequency of attempts of transitions, a is the atom diameter.

The rate of accession of atoms to the crystal was estimated in [31] using the diffusion coefficient of liquid $D = \frac{a^2}{6}\nu_D \exp\left(-Q / k_B T\right)$, that allowed to estimate the velocity of the crystallization front as

$$\upsilon_{s\ell} = \frac{6D}{a}\frac{L_m}{k_B T_m}\frac{\Delta T}{T_{s\ell}} \tag{17}$$

where $\Delta T = T_{s\ell} - T_m$, $T_{s\ell}$ is the temperature of the interphase boundary.

Later, Frenkel [32], using the Stokes-Einstein relation $D = \frac{k_B T}{3\pi a \eta}$ between the coefficients of diffusion D and viscosity η expressed the crystal growth rate in terms of viscosity

$$\upsilon_{s\ell} = \frac{2k_B T}{\pi a^2 \eta}\frac{L_m}{RT_m}\frac{\Delta T}{T_{s\ell}} \tag{18}$$

The expressions (17), (18) allowed establishing a linear relationship between the rates of crystal growth and overcooling. Wilson-Frenkel theory [31-33] contributed to a better understanding of the microscopic processes associated with the growth of crystals from the melt. A generalization of the obtained results allowed to obtain that, except for very large deviations from equilibrium, where the homogeneous nucleation mechanism can dominate, the process of melting-solidification proceeds heterogeneously. The heterogeneous nucleation mechanism involves the inclusion of the motion of the liquid–solid interface into the consideration. The velocity of this interface $\upsilon_{s\ell}$, as a function of the deviation from the equilibrium melting temperature T_m is called the response function of the interface and is the main value characterizing the processes of crystallization and melting. As with the similar equation (14) obtained in the thermodynamic approach, the equations of Wilson-Frenkel showed a good agreement with experiments for very small overcooling of the boundary, i.e. for small deviations from equilibrium.

Theoretical studies of the velocity of the interface are based on some modifications and generalizations of Wilson-Frenkel theory. Their meaning is reduced to taking into account several factors, such as the latent heat of melting L_m, interatomic distance a, efficiency coefficient f, showing the proportion of atoms that remain in the solid phase at the border crossing. In the modifications [24,34,35], it was considered that the crystal has always lower enthalpy than the melt. This is the amount of energy needed for atoms of the crystal to make the transition from crystal to melt. Escape rate of atoms of the crystal followed by the addition to the active points of the liquid, contains this energy difference in the form of the Boltzmann factor $\exp\left(-L_m / k_B T_{s\ell}\right)$

$$R_m\left(T_{s\ell}\right) = C_1 \exp\left(-Q / k_B T_{s\ell}\right) \exp\left(-L_m / k_B T_{s\ell}\right) \tag{19}$$

The rate of the reverse flow of atoms into the crystal from the melt depends only on the diffusion process in liquid

$$R_c\left(T_{s\ell}\right) = C_2 \exp\left(-Q / k_B T_{s\ell}\right) \tag{20}$$

where C_1, C_2 are some constants that should be determined.

In the equilibrium point $T_{s\ell} = T_m$, the rates $R_m\left(T_m\right)$ and $R_c\left(T_m\right)$ are equal, which gives $C_1 \exp\left(-Q / k_B T_m\right) \exp\left(-L_m / k_B T_m\right) = C_2 \exp\left(-Q / k_B T_m\right)$, and $C_1 = C_2 \exp\left(L_m / k_B T_m\right)$.

The velocity of the interphase boundary $\upsilon_{s\ell}$ is equal to the difference of the rates $R_c(T)$ и $R_m(T)$

$$\begin{aligned} \upsilon_{s\ell}\left(T_{s\ell}\right) &= R_c\left(T_{s\ell}\right) - R_m\left(T_{s\ell}\right) = C_2 \exp\left(-Q / k_B T_{s\ell}\right) - C_1 \exp\left(-Q / k_B T_{s\ell}\right) \exp\left(-L_m / k_B T_{s\ell}\right) = \\ &= \exp\left(-Q / k_B T_{s\ell}\right)\left[C_2 - C_1 \exp\left(-L_m / k_B T_{s\ell}\right)\right] = C_2 \exp\left(-Q / k_B T_{s\ell}\right)\left[1 - \exp\left(\frac{L_m}{k_B}\left(\frac{1}{T_m} - \frac{1}{T_{s\ell}}\right)\right)\right] = \\ &= C_2 \exp\left(-Q / k_B T_{s\ell}\right)\left[1 - \exp\left(\frac{L_m}{k_B}\left(\frac{T_{s\ell} - T_m}{T_{s\ell}}\right)\right)\right] = C_2 \exp\left(-Q / k_B T_{s\ell}\right)\left[1 - \exp\left(\frac{L_m}{k_B}\left(\frac{\Delta T}{T_{s\ell}}\right)\right)\right] \end{aligned} \tag{21}$$

$\Delta T = T_{s\ell} - T_m$, for overheating $\Delta T > 0$, for overcooling $\Delta T < 0$.

The constant C_2 is associated with other physical constants using the following relation: $C_2 = af\nu$. The studies have shown that the resulting equation (21) well predicts the velocity of the melting-solidification front of silicon [36] in a fairly wide temperature range.

The problems of melting-crystallization of monatomic metals in the modes of rapid heating/ cooling, typical of ultra-short laser irradiation, use a different expression for the velocity $\upsilon_{s\ell}$. It is based on the assumption [37, 38], that crystallization of single-atom metals, (that are characterized by high velocities $\upsilon_{s\ell} \approx 50 \div 100$ m/c [35, 39]) is not diffusion-limited, but is limited only by the collision frequency during transition from liquid to crystal surface. The modification of the equation (21) consists of replacement of the diffusion term with the thermal velocity $\upsilon_T = (3k_B T / m)^{1/2}$. The velocity of interphase boundary $\upsilon_{s\ell}$ is written as:

$$\upsilon_{s\ell}(T_{s\ell}) = \frac{a\,f}{\lambda}(3k_B T_{s\ell} / m)^{1/2}\left[1 - \exp\left(\frac{L_m}{k_B}\left(\frac{\Delta T}{T_{s\ell}}\right)\right)\right] = C \cdot (T_{s\ell})^{1/2}\left[1 - \exp\left(\frac{L_m}{k_B}\left(\frac{\Delta T}{T_{s\ell}}\right)\right)\right] \quad (22)$$

where $C = \frac{a\,f}{\lambda}(3k_B / m)^{1/2}$, λ is the mean free path, m is the atomic mass.

Thus, the kinetic approach allows us to obtain an expression for the response function $\upsilon_{s\ell}(T_{s\ell})$ without fitting coefficients and suitable for a wide range of overheating/overcooling. The kinetic dependence (22) is asymmetric for the processes of melting and solidification. However, large deviations from equilibrium require additional modification of the kinetic dependences (21) (22), because they do not take into account the dynamic effects associated with the occurrence of high pressures generated by the high velocity of propagation of phase fronts.

2.2.2. *Kinetics of vapor*

One of the key processes in the zone of laser irradiation is the transition of condensed matter to the gaseous state. Evaporation process is characterized by high power consumption and large increase in the specific volume of the substance.

Investigation of evaporation process began in the 19th century [40, 41] and continues to this day [42 -]. This fact is defined by practical importance and not fully clarified features of non-equilibrium behavior of matter when it evaporates.

2.2.2.1. *The simplest model of kinetics of evaporation in vacuum*

The thermodynamic relation (9) gives that at a constant temperature the difference between the Gibbs energy between the two phases, one of which is an ideal gas, is linearly proportional to the pressure difference:

$$\Delta G = V\Delta p = k_B T \frac{\Delta p}{p} \quad (23)$$

On the basis of the formula (23), it can be assumed that the rate of phase transformation at constant temperature should be linearly proportional to the pressure difference:

$$\upsilon = \upsilon_0 \frac{\Delta p}{p} \quad (24)$$

The simplest model for the growth kinetics of vapor phase was developed by Hertz [40] and Knudsen [41] about 100 years ago. Its formulation is qualitatively the same as the thermodynamic model (23), (24)

$$j_m = j_m^+ - j_m^- = \frac{1}{\sqrt{2\pi m k_B}}\left(\frac{p_{sat}}{\sqrt{T_{sur}}} - \frac{p}{\sqrt{T}}\right) \tag{25}$$

where j_m is non-equilibrium flow of atoms at the surface of evaporation, j_m^- is the flux of atoms that collide with the surface under the assumption that the adhesion coefficient is 1. Formally, this flow can be determined using the relation connecting the vapor pressure with the equilibrium particle flux directed to the condensed surface: $p = \sqrt{2\pi}\, m \upsilon j_m^-$, where υ is the average velocity in one direction, $\upsilon = \sqrt{\frac{2k_B T}{m}}$, what gives $j_m^- = \frac{p}{\sqrt{2\pi m k_B T}}$. Since the nature of the flux $j_m^{(-)}$ remained undetermined and the values of T and correspondingly p unknown, then during the formulation of the boundary conditions in the problem of evaporation into vacuum it is supposed [20, p.281] to use a single-term version of the Hertz-Knudsen formula to determine the evaporation rate:

$$j_m \approx j_m^+ = \frac{p_{sat}}{\sqrt{2\pi m k_B T_{sur}}}, \tag{26}$$

that takes into account the connection between the pressure of saturated vapor p_{sat} and flux of evaporated atoms at the surface temperature T_{sur}. The equilibrium vapor pressure p_{sat} depends on the temperature T_{sur}, as it is shown in the equation (12).

However, the representation of the process of surface evaporation in the form of a simple model, which does not take into account the reverse influence of evaporated atoms, does not remove the internal contradictions inherent in the model of Hertz - Knudsen. Let us write the expressions for the fluxes of momentum and j_i and energy j_e of the particles, moving away from the evaporation surface

$$\rho < V_z^2 > = j_i^{(+)} + j_i^{(-)} = j_i = \rho_\upsilon R T_\upsilon + \rho_\upsilon u^2$$
$$\rho < V^2 V_z > = j_e^{(+)} + j_e^{(-)} = j_e = \rho_\upsilon u\left(\frac{u^2}{2} + C_p T_\upsilon\right),$$

It is easy to see that these fluxes, which completely describe this one-dimensional flow, are impossible to be characterized by any temperature. If we equate these fluxes to the corresponding thermodynamic expressions containing the velocity, temperature and density and find these values, we will see that the system of equations has two distinct complex solutions, which have no physical meaning.

The ambiguity of the solution for real values of the thermodynamic parameters is associated with the possibility of discontinuous solutions of the type of shock wave. The complexity of the solution in this case is due to the thermodynamic non-equilibrium of the evaporation flow of Hertz-Knudsen, which can not be described in terms of thermodynamic concepts. When taking into account collisions in the non-equilibrium layer, the evaporative flux is thermalized, but the temperature on the outer side of this layer no longer coincides with the surface temperature.

2.2.2.2. Approximation of the Knudsen layer

Intense surface evaporation is essentially non-equilibrium process. In addition to the thermodynamic equilibrium, this process also has a gas-kinetic non-equilibrium in a thin (Knudsen) layer of vapor, directly adjacent to the interface. Gas-kinetic non-equilibrium is due to the flow of material through the phase boundary. The mass flow increases with the growth of the evaporation rate and, consequently, the degree of non-equilibrium of the process increases. From the physical considerations, the maximum velocity of material flow on the outside of the Knudsen layer is limited to the local speed of sound $u \le u_{sound} = (\gamma R T_v)^{1/2}$, or $M = u / u_{sound} \le 1$, where u is the gas-dynamic velocity, M is the Mach number. The maximum deviation from equilibrium is determined by the maximum value of mass flow, which is known to be achieved at $M = 1$.

Under the conditions of phase equilibrium, when the saturated vapor pressure p_{sat} is equal to the external pressure, the flow of vaporized material is balanced by the return flow of particles and the total mass flux through the boundary is zero. The distribution of particle velocities in vapor is in equilibrium and can be described by the Maxwell function with zero average velocity. In cases where the vapor pressure above the surface is less than the saturated vapor pressure, in the system condensed matter-vapor, the directed movement is formed with $u > 0$ and is characterized by non-zero material flow through the phase boundary. The decreasing reverse flow leads to a deviation from the equilibrium in the distribution of the particles. The magnitude of the flux of returning particles decreases with the increase of the rate of evaporation, and the distribution function at the evaporation surface becomes increasingly different from Maxwellian one.

In general, the non-equilibrium distribution function is found by solving the Boltzmann equation in a region with a characteristic size of a few mean free paths. This area is adjacent to the evaporation surface, where the kinetic boundary conditions are set, taking into account the interaction of individual particles with the interface. Similar problem was solved by various methods in many studies, taking into account, in particular, the differences from unity and variability of the coefficient of condensation, which determines the probability of attachment of the particle in its collision with the evaporation surface. (See, for example. [42 - 46]).

Methods of non-equilibrium thermodynamics are used to describe the evaporation process together with other approximate phenomenological approaches [47 -52]. A more general and fundamental approach is to use molecular dynamics method, which was used in [53, 54] to analyze the evaporation process. A recent review on the issue of non-equilibrium boundary conditions at the liquid-vapor boundary is given in [55].

For the equations of continuum mechanics, thin Knudsen layer is a gas-dynamic discontinuity. The knowledge of the relations at this break, connecting the parameters of the condensed medium and the evaporated material, is needed to deal with the full gas-hydrodynamic problem that arises, for example, during the description of laser ablation, taking into account the variability of the Mach number and instability of the evaporation front [56]. The use of kinetic approaches, which explicitly consider the structure of the Knudsen layer, in such cases is difficult because of the emerging problem of significant difference of space-time scales. The solution of these problems is associated with the additional computational difficulties and is not always possible. Therefore, usually another approach is used that allows to determine the matching conditions with certain assumptions about the form of the non-equilibrium distribution function inside the break [57 - 60] without solving the kinetic problem. Approximation of the distribution function in the Knudsen layer was carried out in different models. But to obtain physically reasonable boundary conditions it is necessary to formulate criteria which these conditions must meet. In addition, attention is paid to the peculiarities of the behavior of the fluxes of mass, momentum and energy as the Mach number tends to unity that allows to use the requirement of extremum of total fluxes of mass j_m, momentum j_i and energy j_e as one of the criteria for $M = 1$

"β" - model. In "β" – model [61, 62], a compound distribution function is used to describe one-dimensional non-equilibrium flow of particles on the inner side of the planar Knudsen layer: $f = f^{(+)} + f^{(-)}$. Here, the distribution $f^{(+)}$ for particles, flying out of the surface is given as a Maxwell-shaped function $f^{(+)} = f(\rho_{sat}, T_{sur}, 0)$ with density of saturated vapor ρ_{sat} for the surface temperature T_{sur}. The distribution $f^{(-)}$ characterizes the flow of particles returning to the surface and is supposed to be proportional to a "shifted" Maxwell function $f^{(-)} = f(\rho_v, T_v, u)$ with density ρ_v, temperature T_v and mean velocity u steady flow of vapor on the outside of the Knudsen layer:

$$f^{(+)} = f(\rho_{sat}, T_{sur}, 0), \quad V_z > 0$$

$$f^{(-)} = f(\rho_v, T_v, u) = \beta\left(\frac{m}{2\pi k_B T_v}\right)^{3/2} \exp\left(\frac{V_x^2 + V_y^2 + \left(V_z^2 - u\right)^2}{2 k_B T_v}\right) \quad V_z < 0$$

The flows of mass, momentum and energy $j_k = j_k^{(+)} + j_k^{(-)}, \quad k = m, i, e$ calculated using $f^{(+)}, f^{(-)}$ must be equal to their gas-dynamic values j_k that are determined by function f :

$$\begin{gathered} \rho < V_z >= j_m^{(+)} + j_m^{(-)} = j_m = \rho_v u \\ \rho < V_z^2 >= j_i^{(+)} + j_i^{(-)} = j_i = \rho_v R T_v + \rho_v u^2 \\ \rho < V^2 V_z >= j_e^{(+)} + j_e^{(-)} = j_e = \rho_v u\left(\frac{u^2}{2} + C_p T_v\right), \end{gathered} \tag{27}$$

where $C_p = \frac{\gamma}{\gamma+1} k_B$ is the vapor heat capacity per particle at constant pressure, for single-atom gas $\gamma = \frac{5}{3}$. From the solution of equations (27), it is possible to obtain gas-dynamic conditions on the break, allowing to determine the magnitude of ρ_v, T_v и β in terms of ρ_{sat}, T_{sur} and M. The calculations showed that total fluxes j_m, j_i, j_e have extrema depending on M at $M = 0.88,\ 1.18,\ 1.22$ correspondingly.

Thus, the requirement of the extrema of all flows in the selected point $M = 1$ in the model [61,62] does not met, that can serve as evidence of a bad choice of the distribution function $f^{(-)}$.

"$\varepsilon - \delta$" - model. The drawbacks of the "β" – model can be eliminated by selection of a more general form of the distribution function for the reverse flow of particles $f^{(-)}$ [63]:

$$f^{(-)} = \beta f + \alpha f_1 \quad \alpha + \beta = 1$$

All 3 flows .. will have extrema at $M = 1$, if the function f_1 is set to be equal to $f_1 = f(\rho_1, T_1, u_1)$, where the values of ρ_1, T_1, u_1 are written in terms $\rho_{sat}, T_{sur}, \rho_u, T_v$ using additional fitting parameters ε and δ

$$\begin{gathered} \rho_1, T_1^{1/2} = \varepsilon \rho_{sat} T_{sur}^{1/2} + (1-\varepsilon) \rho_v T_v^{1/2}, \\ \rho_1, T_1^{1/2} = \varepsilon T_{sur}^{1/2} + (1-\varepsilon) T_v^{1/2}, \quad . \\ u_1 = \delta M (\gamma R T_1)^{1/2} = \delta (T_1/T_v)^{1/2} u \end{gathered}$$

For example, at $\varepsilon = 0.70$ and $\delta = 0.32$ all three flows j_k will have extrema at $M = 1$ with values $j_m = 0.853, j_i = 0.557, j_e = 0.892$. It is clear, that this version of selection of fittings coefficients is not the only one.

"α"- model. It is possible to suggest another phenomenological model [63], where strict localization of extrema $j_k(M=1)$ is achieved without usage of fitting coefficients for such distribution function $f^{(-)}$, that do not depend on gas-dynamic values. An example of such function is function

$$f^{(-)} = \alpha^7 f_0 \left(\rho_{sat}, \alpha^2 T_{sur} \right),$$

It takes into account the decrease of temperature $T_\alpha = \alpha^2 T_{sur}$ of the reverse flow of particles as compared to the surface temperature T_{sur}. Due to this change of T_α, the ratio of the normalized fluxes j_e/j_m ceases to be a constant and takes the form $(1-\alpha^8)(1-\alpha^{10})$, which ensures that the correct limiting value is equal to 1.25 in the equilibrium case for $\alpha = 1$. The equation for α, that is obtained from the equity of fluxes (27) has a relatively simple form:

$$\frac{(1-\alpha^8)(1-\alpha^{10})}{(1-\alpha^9)^2}=\frac{\pi}{8}\frac{\gamma^2M^2\left[(\gamma-1)M^2+2\right]}{(\gamma-1)(1+\gamma M^2)^2} \tag{28}$$

The right-hand side of the equation (28) has a maximum at $M=1$, that determines localization of the extrema of α and j_k. The values of $j_k(M=1)$ are equal correspondingly to $j_m=0.85, j_i=0.56, j_e=0.90$.

Modified Crout model. The property of localization of the flows $j_k(M=1)$ also present in the model suggested by D. Crout [57]. It uses non-equilibrium function of distribution of particles, written in analytical form with temperature that is anisotropic by directions: T_L and T_T.

$$f(\rho,T_L,T_T,u,v,w)=\left\{\rho\left[\frac{\pi^3}{h_T m^3}(2k_BT_L)\cdot(2k_BT_T)\right]^{1/2}\cdot\exp\left[-m\left(\frac{(u-u_0)^2}{2k_BT_L}+\frac{v^2}{2k_BT_T}+\frac{w^2}{2k_BT_T}\right)\right]\right\}$$

where T_L is the lateral temperature along x axis, T_T is the transversal temperature along y,z axis; u,v,ω are the components of the velocity vector along corresponding axes x,y,z, u_0 is the drift velocity.

The modification of the Crout model [64] consists of explicit introduction of the Mach number into the main relations that allows to obtain:

$$T_v=\alpha_T(M)T_{sur},\quad \rho_v=\alpha_\rho(M)\rho_{sat}, \tag{29}$$

$$\alpha_T(M)=\frac{2\gamma M^2(m^2+0.5)^2}{(1+\gamma M^2)^2m^2t^2},\quad \alpha_\rho(M)=\frac{1}{\exp(-m^2)+\pi^{1/2}m(1+\mathrm{erf}(m))}\cdot\frac{(1+\gamma M^2)m^2}{\gamma M^2(m^2+0.5)^2},$$

The value of m is determined from non-linear equation

$$F(M)(m^2+0.5)^2-m^2(m^2+a+1.5)=0, \tag{30}$$

where $F(M)=1+\dfrac{3\gamma M^2-1}{(\gamma M^2-1)^2}$,. $a=2t^2-0.5\pi^{1/2}mt-1$, .

$$t=\frac{2m}{\pi^{1/2}}+\frac{1+\mathrm{erf}(m)}{\exp(-m^2)+\pi^{1/2}m(1+\mathrm{erf}(m))},\quad \mathrm{erf}(m)=\frac{2}{\sqrt{\pi}}\int_0^m e^{-y^2}dy.$$

For $M=1$: $T_v=0{,}633T_{sur}$, $\rho_v=0{,}328\rho_{sat}$. $p_v=0.208p_{sat}$,

$M=0$: $T_v=T_{sur}$, $\rho_v=\rho_{sat}$, $p_v=p_{sat}$.

For the numerical solution of the nonlinear equation (30), one can use the Newton's iterative procedure. All fluxes j_m, j_i, j_e have extrema at *M=1*. Calculations using an anisotropic non-equilibrium particle distribution function give the corresponding extreme values of the fluxes: $j_m = 0.84, j_i = 0.55, j_e = 0.88$.

The calculations show that specific choice of the model has relatively little effect on the magnitude of the momentum flux j_2, but significantly affects the flux of mass j_m and energy j_e. Fig.1 shows the dependencies of the ratio of the normalized fluxes $\bar{j}_3 / \bar{j}_1 = j_3 j_1^{(+)} / j_1 j_3^{(+)}$ on M for all discussed models. From the comparison of the extreme values of the fluxes $j_k(M=1)$ and behavior of the curves $j_m(M), j_e(M)$ it follows, that due to bad choice of $f^{(-)}$, the less favorable for the description of evaporation kinetics is "β" – model.

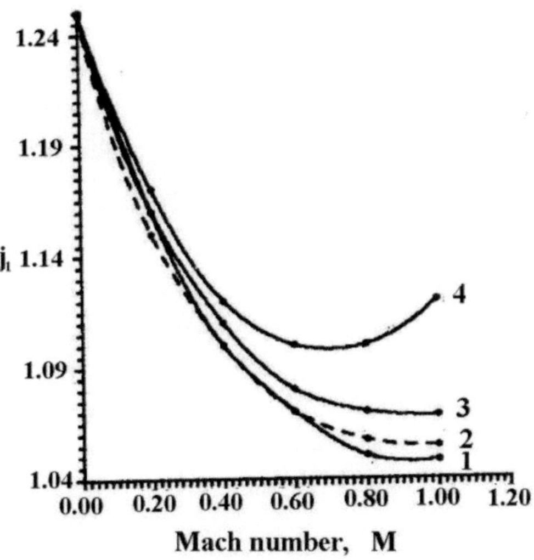

Figure 1. Dependency of the normalized ratio of energy and mass fluxes j_3/j_1 on Mach number for different models: 1 – "ε-δ"-model, 2 – Crout model, 3 - "α"-model, 4 – "β"-model (Knight).

Table 1 for all models shows numerical values of ρ_v and T_v (normalized by ρ_{sat} and T_{sat} correspondingly) depending on the Mach number that changes from zero to unity. Comparative analysis of tabular data, as well as the behavior of the curves j_m, j_e, show a marked difference of the values of ρ_v and T_v, obtained using "β" model, from the values, obtained using other models. The values of ρ_v turned out to be underestimated, and T_v overestimated as compared to their real values at the outer side of the Knuden layer.

	"β"-model (Knight)		"ε-δ"-model		"α"-model		Crout model	
M	T	N	T	N	T	n	T	n
0.0	1.000	1.000	1.000	1.000	1.000	1.000	1.000	1.000
0.1	0.960	0.861	0.958	0.864	0.960	0.869	0.953	0.861
0.2	0.922	0.748	0.916	0.753	0.920	0.758	0.910	0.749
0.3	0.866	0.654	0.876	0.662	0.881	0.666	0.870	0.658

	"β"-model (Knight)		"ε-δ"-model		"α"-model		Crout model	
M	T	N	T	N	T	n	T	n
0.4	0.851	0.576	0.837	0.587	0.844	0.588	0.833	0.582
0.5	0.817	0.511	0.799	0.525	0.808	0.524	0.798	0.519
0.6	0.785	0.457	0.763	0.472	0.773	0.470	0.763	0.466
0.7	0.754	0.410	0.727	0.428	0.740	0.424	0.730	0.421
0.8	0.724	0.371	0.693	0.391	0.705	0.386	0.697	0.384
0.9	0.696	0.337	0.660	0.360	0.672	0.355	0.665	0.352
1.0	0.669	0.308	0.628	0.333	0.640	0.328	0.633	0.326

Table 1. Normalized values ρ_υ and T_υ depending on the Mach number for different models

The modified Crout model, "$\varepsilon-\delta$" – model, "α" – model fulfill the requirement of the extremum of the flows j_k at M=1. The difference of the results that were obtained using these models does not exceed 1.5%. Any of these models can be used to describe the kinetics of the process of non-equilibrium surface evaporation.

2.3. Conclusions

The performed brief analysis of the kinetics of phase transitions is an introduction to the construction of the models that combine mathematical description of kinetics of high-speed phase transformations with dynamics of the macro-processes (heat and mass transfer) under conditions of a strong deviation from local thermodynamic equilibrium that are typical for ultra-short super-power laser action on metals.

3. Thermodynamic and thermo-physical properties of phonon and electron Fermi gas

Determination of physical characteristics of a medium, including equations of state under conditions of local thermodynamic equilibrium can be carried out either experimentally or by means of calculation using distribution functions - a Maxwell-Boltzmann function for ideal gas and ideal plasma and Fermi one for degenerate electron gas, and for the phonon gas – Bose function. In case of violation of the conditions of local thermodynamic equilibrium distribution functions are determined by solving the classical kinetic Boltzmann equation or quantum-kinetic equations. The presence of distribution function is just the required minimum of information that can be used to describe nonequilibrium processes with reasonable accuracy.

The influence of ultrashort high-energy laser on a strongly absorbing media (metals, semiconductors) is in a very short temporal and spatial scales and leads to disturbance of their general local-thermodynamic equilibrium. Irradiated targets in these conditions are presented in the form of two subsystems - electron and phonon each of which is in local thermodynamic equilibrium and are characterized by their temperatures and equations of state. As a consequence, all processes are described in the two-temperature approximation [65], [66]. The target at pico- and femtosecond influence may be heated to very high temperatures and pressures at which the thermal and mechanical properties of matter are not known in general. One of the most important problems for the mathematical modeling is the necessity to determine thermophysical, optical and thermodynamic properties in a wide (tens and hundreds of electronvolts) temperature and frequency ranges for each of the subsystems.

3.1. Electron subsystem

The most important thermophysical and thermodynamic characteristics of the electron Fermi gas within the scope of heat-conducting mechanism of energy transfer are: heat capacity C_e, thermal diffusivity χ_e and thermal conductivity λ_e. For its determination using fundamental physical quantities, which include the electron mean free paths l_{ee}, l_{eph} and the characteristic times (frequency) of interaction for two scattering mechanisms: the electron-electron τ_{ee} and electron-phonon τ_{eph}.

3.1.1. Fermi-Dirac integral and its approximation

For a quantitative description of electrical, thermodynamic and thermophysical properties of degenerate electron gas with distribution function

$$f_e = \frac{1}{\exp\left(\frac{\varepsilon_e - \mu}{k_B T_e}\right) + 1}, \tag{31}$$

widely used Fermi-Dirac functions $F_{k+1/2}(\mu(T_e)/T_e)$ expressed in terms of integrals of the form

$$F_{k+1/2}(\mu(T_e)/T_e) = \int_0^\infty \frac{\left(\frac{\varepsilon_e}{k_B T_e}\right)^{k+1/2} d\left(\frac{\varepsilon_e}{k_B T_e}\right)}{\exp\left(\frac{\varepsilon_e - \mu}{k_B T_e}\right) + 1} = \int_0^\infty \frac{E^{k+1/2} dE}{\exp\left(E - \frac{\mu}{k_B T_e}\right) + 1}, \tag{32}$$

where T_e, ε_e - temperature and energy of electron, $\mu(T_e)$ - chemical potential, $\mu(0) = \varepsilon_F$ - Fermi energy.

In the future, the Fermi integrals will be represented as a function of dimensionless energy of the chemical potential:

$$F_{k+1/2}(\eta) = \int_0^\infty \frac{E^{k+1/2} dE}{\exp(E - \eta) + 1}, \tag{33}$$

where $E = \varepsilon_e / k_B T_e$, $\eta(T_e) = \mu(T_e)/k_B T_e$ - dimensionless energy and chemical potential of electrons.

The integral of form $F_{1/2}(\eta)=\int\limits_0^\infty \frac{E^{1/2}dE}{\exp(E-\eta)+1}$ is used to determine the electron density $N_e=\int\limits_0^\infty g(E)f(E)dE=\int\limits_0^\infty \frac{g(E)dE}{\exp(E-\eta)+1}=\int\limits_0^\infty F(E)dE$,

where $g(E)=\frac{2^{1/2}m^{3/2}}{\pi^2\hbar^3}(k_BT_e)^{3/2}E^{1/2}=g_0\cdot(k_BT_e)^{3/2}E^{1/2}$ is a density of states, $F(E)=g(E)f(E)=g_0(k_BT_e)^{3/2}\frac{E^{1/2}}{\exp(E-\eta)+1}$ is distribution function of the free electron energy, $g_0=\frac{2^{1/2}m^{3/2}}{\pi^2\hbar^3}$. Then

$$N_e=g_0\cdot(k_BT_e)^{3/2}\int\limits_0^\infty \frac{E^{1/2}dE}{\exp(E-\eta)+1}=g_0\frac{2}{3}\varepsilon_F^{3/2}=g_0\cdot(k_BT_e)^{3/2}F_{1/2}(\eta) \tag{34}$$

From the known distribution function of particle energy $F(E)$, using the ratio $<x(E)>=\frac{\int\limits_0^\infty x(E)F(E)dE}{\int\limits_0^\infty F(E)dE}$ the average value of any physical quantity depends on energy can be found. Since the average energy of the electron gas is defined as the ratio of Fermi integrals

$$<\varepsilon_e>=\frac{\int\limits_0^\infty E\cdot F(E)dE}{\int\limits_0^\infty F(E)dE}=(k_BT_e)\frac{\int\limits_0^\infty \frac{E^{3/2}dE}{\exp(E-\eta)+1}}{\int\limits_0^\infty \frac{E^{1/2}dE}{\exp(E-\eta)+1}}=(k_BT_e)\frac{F_{3/2}(\eta)}{F_{1/2}(\eta)}. \tag{35}$$

Similarly, we can determine the other physical quantities of the electron gas.

The chemical potential $\mu(T_e)$ depends on the temperature, so the Fermi integrals can be conveniently represented as a function of dimensionless temperature $\xi=\frac{k_BT_e}{\varepsilon_F}$. In [67] for the integrals of the form (33) has been proposed convenient approximation, which allows to express the integrals $F_{k+1/2}(\xi)$ through the transcendental gamma-functions $\Gamma(k+1/2)$ and the dimensionless temperature ξ:

$$F_{k+1/2}(\xi)=A\xi^{-3/2}\left[1+(B/\xi)^2\right]^{k/2}, \tag{36}$$

where $A=\dfrac{2}{3}\dfrac{\Gamma\left(k+\frac{3}{2}\right)}{\Gamma\left(\frac{3}{2}\right)}$, $B=\left[A\left(k+\frac{3}{2}\right)\right]^{-1/k}$ coefficients expressed in terms of gamma-functions. Equation (36) has correct asymptotics at $\xi\to 0$ and $\xi\to\infty$:

$$F_{k+1/2}(\eta)\approx 1/\xi^{k+3/2}\left(k+3/2\right),\qquad \xi\to 0$$
$$F_{k+1/2}(\eta)\approx\frac{2}{3}\frac{\Gamma\left(k+3/2\right)}{\Gamma\left(3/2\right)}\frac{1}{\xi^{3/2}}\qquad \xi\to\infty\ .$$

The integral of $k=1/2$ order can easily be determined from the expression (34): $F_{1/2}(\xi)=\dfrac{2}{3}\dfrac{1}{\xi^{3/2}}$.

The most frequently used Fermi integrals that are expressed through (36) have the form:

$$F_{-1/2}(\xi)=\frac{4}{3}\frac{\xi^{-1/2}}{\left(\xi^2+\frac{4}{9}\right)^{1/2}},\ F_{3/2}(\xi)=\frac{\left(\xi^2+\left(\frac{2}{5}\right)^2\right)^{1/2}}{\xi^{5/2}},\ F_{5/2}(\xi)=\left[\frac{5}{2}\xi^{-3/2}+\frac{2}{7}\xi^{-7/2}\right]. \qquad (37)$$

The maximum error compared with the exact solution [68] for integrals at $k=-1$ and $k=1$ does not exceed 8%, but increases slightly with increasing of *k*.

Approximation (36) allows to obtain simple analytical expressions for the physical quantities of the electron gas at arbitrary temperatures.

3.1.2. Equations of state

Using the approximating expressions (36) and (37) equations of state for degenerate electron gas can be written as simple analytical expressions at arbitrary temperatures. Since the average electron energy $<\varepsilon_e>$ and its pressure can be represented as

$$<\varepsilon_e>=T_e\frac{F_{3/2}}{F_{1/2}}=\frac{3}{2}\varepsilon_F(\xi^2+0.16)^{1/2}, \qquad (38)$$

$$p=\frac{2}{3}N_e<\varepsilon_e>=\frac{2}{3}N_eT_e\frac{F_{3/2}}{F_{1/2}}=N_e\varepsilon_F(\xi^2+0.16)^{1/2}, \qquad (39)$$

3.1.3. Electron heat capacity $C_e(T_e)$

The expression for the heat capacity of the electron gas $C_e(T_e)=\dfrac{\partial}{\partial T}\left(N_e\langle\varepsilon_e\rangle\right)$ can be obtained from the relations

$$\frac{\partial}{\partial\eta}F_{k+1/2}=(k+1/2)F_{k-1/2},\ \frac{\partial\eta}{\partial T}=\frac{3}{T}\frac{F_{1/2}}{F_{-1/2}},\ C_e(T_e)=\frac{N_e}{F_{1/2}}\left(\frac{5}{2}F_{3/2}-\frac{9}{2}F_{-1/2}\right).$$

Using the approximating expressions (37), heat capacity of electron gas can be represented with an error not exceeding 5% as the following function

$$C_e(T_e)\cong\frac{3}{2}\frac{N_e k_B^2 T_e[K]}{\left[T_e^2+\left(\frac{3\varepsilon_F}{\pi^2}\right)^2\right]^{\frac{1}{2}}},\qquad (40)$$

where $N_e = zN_a$, z - the number of valence electrons, N_a - the concentration of atoms (ions) of lattice.

The resulting expression gives the classical linear dependence of heat capacity of a degenerate electron gas vs. temperature [69] in low temperature region $T_e \ll \varepsilon_F$, $C_e=\frac{\pi^2}{2}\frac{k_B^2 T_e N_e}{\varepsilon_F}$ and constant value at $T_e \gg \varepsilon_F$ equal to heat capacity of gas with Maxwell distribution $C_e = 3/2k_B N_e$. Dependences for copper and aluminum are shown in Fig. 2.

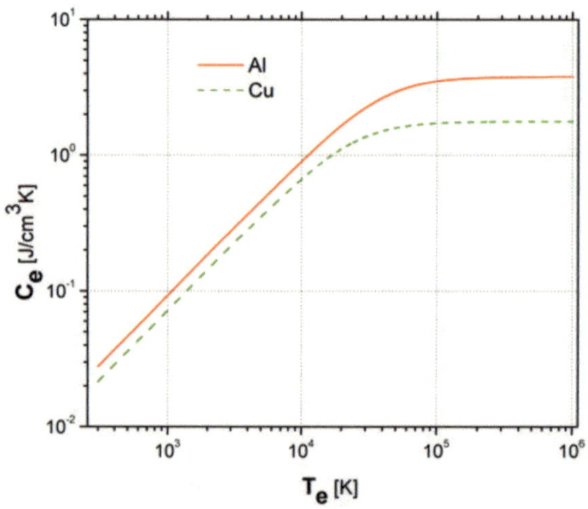

Figure 2. Temperature dependence of electron heat capacity $C_e(T_e)$.

3.1.4. The thermal diffusivity of a degenerate electron gas χ_e.

The thermal diffusivity of electron gas is proportional to the product of the mean free path l_e and average velocity of the electron $<\upsilon_e>$:

$$\chi_e=\frac{1}{3}l_e<\upsilon_e>,\qquad (41)$$

In metals electron mean free path due to several mechanisms: pair of electron-electron collisions, electron-phonon collisions and scattering by plasmons.

Electron-electron collisions dominate at temperatures comparable to the Fermi energy $T_e/\varepsilon_F \approx 1$. Electron-phonon interaction is dominant at low temperatures $T_e/\varepsilon_F << 1$. The interaction associated with the excitation of plasmons occurs at high temperatures, exceeding the plasma frequency energy $T_e > \hbar\omega_r$ ($\hbar\omega_r \cong 10 - 20$ eV). Taking into account high temperature region of occurrence and limitation of experimental data about reducing of the mean free path of electrons due to plasmon excitation (it is known only for some metals), electron scattering by plasmons will not be considered.

3.1.4.1. The electron-electron thermal diffusivity $\chi_{ee}(T_e)$

The mean free path of an electron in pair electron-electron collisions l_{ee} is determined from the known gas-dynamic formula

$$l_{ee} = \frac{1}{N_e \sigma_{ee}}, \tag{42}$$

where σ_{ee} - scattering cross section with energy $\Delta\varepsilon$ transfer for electrons with energies $\varepsilon_1, \varepsilon_2$. The cross section σ_{ee} is expressed through the transport cross section of the collision of two isolated electrons $\sigma_{\ell\ell}^{tr}$ and Fermi integrals $F_{-1/2}, F_{1/2}$. In turn, the transport cross section of collision of two isolated electrons $\sigma_{\ell\ell}^{tr}$ in a field of screened Coulomb potential $U = \frac{e^2}{r}\exp\left(-\frac{r}{d}\right)$ is expressed through the differential scattering cross section $d\sigma$ determined in the Born approximation [70, p.560]. The final cross sections will be written in form

$$\begin{gathered}\sigma_{ee}^{tr} = \frac{2\pi}{9}\left(\frac{4}{9\pi}\right)^{4/3} z^{-4/3}\frac{r^4}{r_B^2}\frac{\left[\ln(t+1) - \frac{t}{(t+1)}\right]}{\left(\xi^2 + 0.16\right)}, \\ \sigma_{ee} = \sigma_{ee}^{tr}\left(\frac{F_{-1/2}}{2F_{1/2}}\right)^2 = \frac{2\pi}{9}\left(\frac{4}{9\pi}\right)^{4/3} z^{-4/3}\frac{r^4}{r_B^2}\frac{\xi^2\left[\ln(t+1) - \frac{t}{(t+1)}\right]}{\left(\xi^2 + \frac{4}{9}\right)\left(\xi^2 + 0.16\right)}\end{gathered}, \tag{43}$$

where $t = 4\langle k\rangle^2 d^2 = \left(\frac{9\pi}{4}\right)^{4/3} z^{1/3}\frac{r_B}{r}\cdot\left[\left(\xi^2 + 0.16\right)\cdot\left(\xi^2 + \frac{4}{9}\right)\right]^{1/2}$, $r_B = \frac{\hbar^2}{me^2}$ = 0.529 10^{-8}cm – Bohr radius, $r = \left(\frac{3}{4\pi N_a}\right)^{1/3}$ - the average distance between atoms, e - the electron charge, d – field acting radius (Debye) $U = \frac{e^2}{r}\exp\left(-\frac{r}{d}\right)$.

At low temperatures $T_e << \varepsilon_F$, the effective cross section σ_{ee} is small and amounts to $\sigma_{ee} = \sigma_{ee}^{tr} \dfrac{\xi^2}{\left(\xi^2 + 4/9\right)} \approx \sigma_{ee}^{tr} \cdot \left(\dfrac{T_e}{\varepsilon_F}\right)^2$. Maximum of cross section is achieved at $T_e \geq \varepsilon_F$, Fig.3, when the degeneracy is passed and the electron-electron collisions with large energy transfer become possible. At very high temperatures $T_e >> \varepsilon_F$ cross section becomes the Coulomb one, Fig.3, and decreases logarithmically. The mean free path of electrons l_{ee} is determined by the formula (42):

$$l_{ee} = \frac{1}{N_e \sigma_{ee}} = \left[N_e \cdot \frac{2\pi}{9}\left(\frac{4}{9\pi}\right)^{4/3} z^{-4/3} \frac{r^4}{r_B^2} \frac{\xi^2\left[\ln(t+1) - \dfrac{t}{(t+1)}\right]}{\left(\xi^2 + \frac{4}{9}\right)\left(\xi^2 + 0.16\right)} \right]^{-1} = \tag{44}$$

$$= \frac{2}{\pi^2}\left(\frac{4}{9\pi}\right)^{1/3} \frac{1}{r^2} \frac{1}{z^{1/3} N_a} \cdot \frac{t^2}{\xi^2} \left[\ln(1+t) - \frac{t}{1+t}\right]^{-1}$$

Calculations indicate that the mean free path for *Al* and *Cu* change in a wide range (~$10^{-2} \div 10^{-7}$) cm and have a minimum at $T_e \approx \varepsilon_F$.

The average thermal velocity of electron $<\upsilon_e>$ is expressed through its average energy $<\varepsilon_e>$:

$$<\upsilon_e> = \left(\frac{2}{m} <\varepsilon_e>\right)^{1/2} = \left(\frac{3\varepsilon_F}{m}\left(\xi^2 + 0.16\right)^{1/2}\right)^{1/2} = \left(\frac{3}{2}\right)^{1/2} \left(\frac{9\pi}{4}\right)^{1/3} \frac{z^{1/3}}{r} \frac{\hbar}{m_e} \left(\xi^2 + 0.16\right)^{1/4}, \tag{45}$$

Taking into account (44) and (45) electronic thermal diffusivity χ_{ee} takes the form:

$$\chi_{ee} = \frac{1}{3} l_{ee} <\upsilon_e> = \frac{1}{3}\frac{2}{\pi^2}\left(\frac{4}{9\pi}\right)^{1/3} \frac{1}{r^2} \frac{1}{z^{1/3} N_a} \cdot \frac{t^2}{\xi^2}\left[\ln(1+t) - \frac{t}{1+t}\right]^{-1} \cdot \left(\frac{3}{2}\right)^{1/2} \left(\frac{9\pi}{4}\right)^{1/3} \cdot$$

$$\cdot \frac{z^{1/3}}{r} \frac{\hbar}{m_e} \left(\xi^2 + 0.16\right)^{1/4} = \left(\frac{2}{3}\right)^{1/2} \frac{4}{3\pi} \frac{\hbar}{m_e} \frac{t^2}{\xi^2} \frac{(\xi^2 + 0.16)^{1/4}}{\left[\ln(1+t) - \dfrac{t}{1+t}\right]} = \chi_0 \cdot \Phi_{ee}(\xi) \tag{46}$$

where $\Phi_{ee}(\xi) = \dfrac{t^2}{\xi^2} \dfrac{(\xi^2 + 0.16)^{1/4}}{\left[\ln(1+t) - \dfrac{t}{1+t}\right]}$ is dimensionless function,

$$\chi_0 = \sqrt{\frac{2}{3}} \frac{4}{3\pi} \frac{\hbar}{m_e} = 0.402, \left[\frac{cm^2}{s}\right].$$

Fig. 4 shows the temperature dependence of $\chi_{ee}(T_e)$ for Al and Cu. Temperature dependence of electron-electron thermal diffusivity $\chi_{ee}(T_e)$ of both metals has a deep minimum at $T_e \approx \varepsilon_F$. Its value reaches ~ 20÷30 [cm²/s], Fig.4. At removal of degeneracy, when $T_e > \varepsilon_F$ the thermal diffusivity increases due to decreasing of the effective cross section σ_{ee}. With further increase of temperature $T_e >> \varepsilon_F$ the thermal diffusivity continues to increase and its dependence $\chi_{ee} \sim \xi^{5/2} / \ln \xi^2$ coincides with temperature dependence of thermal diffusivity of Maxwell electron plasma. At low temperature $T_e << \varepsilon_F$ dependence of thermal diffusivity is inversely proportional to square of temperature $\chi_{ee} \sim T_e^{-2}$. Strong growth of $\chi_{ee}(T_e)$ with decreasing of T_e leads to the fact that χ_{ee} reach $(1-5)\cdot 10^5 \frac{cm^2}{s}$ values which is 3-4 orders higher than the actual electron thermal diffusivity of metals at room temperatures. Thus, the resulting expression (46) is a good approximation only for high temperatures. Under normal conditions ($T_e \sim 300K$) it is necessary to consider the scattering of electrons of metals by phonons to determine thermal diffusivity, this interaction is dominant at low temperatures.

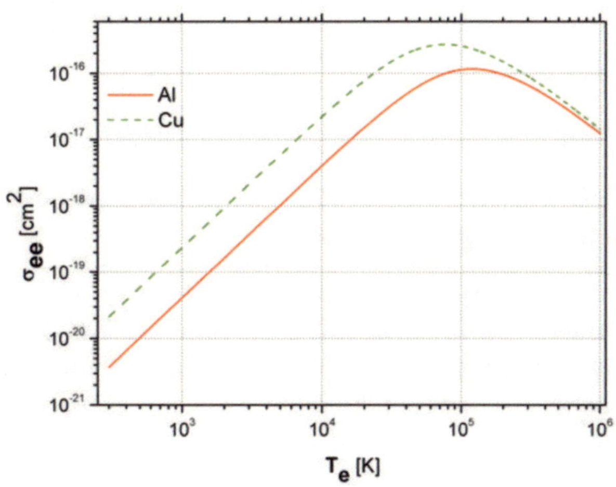

Figure 3. Temperature dependences of electron-electron scattering cross section $\sigma_{ee}(T_e)$.

3.1.4.2. The electron-phonon thermal diffusivity

The mean free path, defined by electron-phonon interaction is described by the assumption of elastic scattering of conduction electrons of metal on lattice oscillations. To determine it is convenient to use the phenomenological approach [70], in which the crystal is considered as an elastic continuum. Lattice oscillations at the same time considered as a wave of elastic deformations. To simplify the density fluctuations are presented as deviations of each atom (ion) from the average, which square of amplitude is directly proportional to temperature T_{ph}. According to the macroscopic theory of elasticity we can obtain an expression for the mean free path, expressed in terms of macroscopic quantities [71] by expressing the force tending to return the atom (ion) to the equilibrium state through the Young's modulus E:

$$\ell_{eph} = \frac{E \cdot r}{N_e k_B T_{ph}} \tag{47}$$

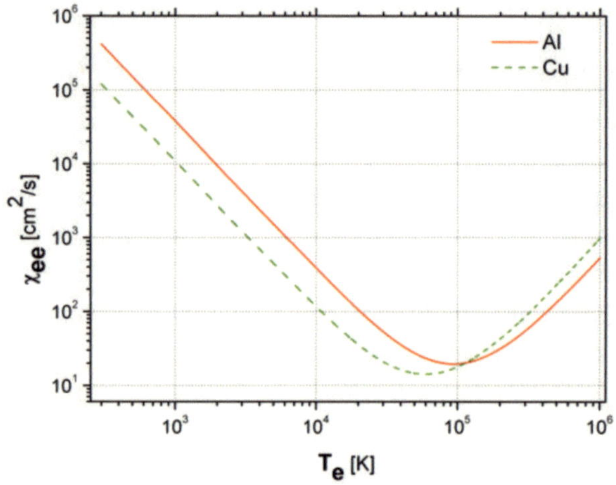

Figure 4. Temperature dependence of electron-electron thermal diffusivity $\chi_{ee}(T_e)$.

From the expression (47) it follows that ℓ_{eph} is inversely proportional to the lattice temperature T_{ph} in electron-phonon interaction.

During melting of a metal the number of collectivized electrons remains practically unchanged. The modulus of elasticity is only one value (excluding the jump in specific volume, which usually does not exceed 10%) that changes. The melting of most metals is accompanied by decrease in elastic modulus by 2-3 times [72]. This decrease causes a corresponding increase in density fluctuation and, consequently, an abrupt decrease in the mean free path ℓ_{eph}:

$$\left(\ell_{eph}\right)_k = \left(\frac{E \cdot r}{z N_a k_B T_{ph}}\right)_k , \tag{48}$$

where $k = s, l$ subscripts denoting membership in the solid and liquid phases, respectively, $r = \left(\frac{3}{4\pi}\frac{1}{N_a}\right)^{1/3} = 0.6204 N_a^{-1/3}$ is a distance between the atoms. From below l_{eph} is limited by Bohr diameter value. The mean free path of electrons taking into account the scattering by phonons , calculated for aluminum and copper from the relation (48) showed that, compared with the mean free path of the electron-electron scattering values for both metals decreased by several orders of magnitude: for the high-temperature region of 1.5 ÷ 2 orders of magnitude, while at low temperatures for 4 ÷ 5 orders of magnitude.

The thermal diffusivity $\chi_{eph}(T_e,T_{ph})$, defined by the mechanism in electron-phonon interaction is calculated by the formula:

$$\left(\chi_{eph}\left(T_e,T_{ph}\right)\right)_k = \left(\frac{1}{3}\,\ell_{eph}\left(T_{ph}\right) < \upsilon_e\left(T_e\right) > \right)_k, \tag{49}$$

Dependences $\chi_{eph}\left(T_e,T_p\right)$ for *Al* and *Cu* calculated from (49) at $T_e = T_{ph}$ are shown in Fig.5. The calculations show that at room temperature values of the electron thermal diffusivity χ_{eph}, for both materials are 3-4 orders of magnitude less than the thermal diffusivity χ_{ee} and reach values of 10^2 cm²/s typical for metals under normal conditions. Under equilibrium conditions with temperature T_{ph} increasing, thermal diffusivity χ_{eph}, undergoing break at the phase transition decreases rapidly to a value of ~ 1 cm²/s at a temperature of ~ 5 000K. Function $\chi_{eph}\left(T_e,T_p\right)$ increase by several orders of magnitude with electron temperature increase.

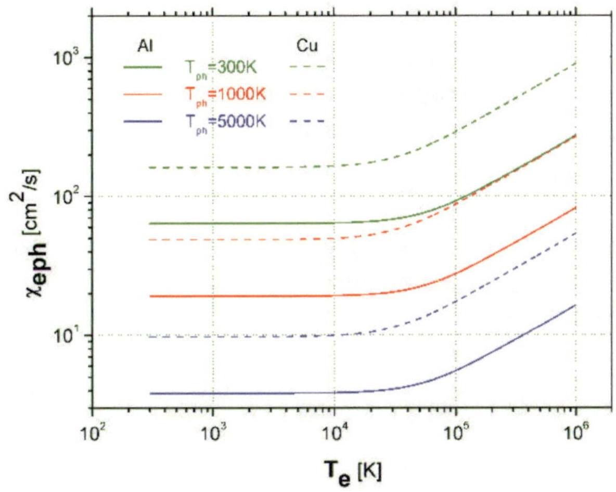

Figure 5. Temperature dependences of thermal diffusivity $\chi_{eph}(T_e,T_{ph})$ for *Al* and Cu with fixed $T = T_e = T_{ph}$.

3.1.4.3. The resulting thermal diffusivity of electron Fermi gas $\chi_e\left(T_e,T_{ph}\right)$

Calculations have shown that taking into account only a pair electron-electron collisions leads to a strong (by several orders of magnitude) overestimation of the thermal diffusivity of the electron gas at low temperatures $T_e << \varepsilon_F$. Accounting for electron-phonon collisions gives a more realistic values of $\chi_e(T_e)$ at low temperatures.

By averaging mean free paths ℓ_{ee}, ℓ_{eph} we obtain an expression for resulting thermal diffusivity, at arbitrary temperature

$$\left(\chi_e\left(T_e,T_{ph}\right)\right)_k=\left(\frac{\chi_{ee}\chi_{eph}}{\chi_{ee}+\chi_{eph}}\right)_k=\left(\frac{1}{3}\ell_e <\upsilon_e>\right)_k=<\upsilon_e>\left(\frac{\ell_{ee}\,\ell_{eph}}{3\left(\ell_{ee}+\ell_{eph}\right)}\right)_{k=s,\ell}, \qquad (50)$$

Fig.6 shows temperature dependences of total electron thermal diffusivity $\chi_e(T_e,T_{ph})$ for *Al* and *Cu* in equilibrium $T_e = T_p$.

3.1.5. The thermal conductivity of the electron gas

According to elementary kinetic theory, the thermal conductivity of the gas is

$$\left(\lambda_e(T_e,T_{ph})\right)_k=\left(\frac{1}{3}C_e\ell_e(T_e,T_{ph})<\upsilon_e>\right)_k=\left(C_e\chi_e(T_e,T_{ph})\right)_k,\left[\frac{\text{W}}{m\cdot\text{K}}\right] \qquad (51)$$

Thus, the thermal conductivity λ_e can be determined through the heat capacity C_e and averaged thermal diffusivity of electron gas $\chi_e\left(T_e,T_{ph}\right)$. The temperature dependences of λ_e for *Al* and *Cu* calculated for the equilibrium case when $T_e=T_{ph}$ are shown in Fig.6. In accordance with the results, the equilibrium electron thermal conductivity $\lambda_e(T_e,T_{ph})$ at temperatures not exceeding the boiling temperature of the equilibrium is practically independent of temperature. In high temperature region T_e>1eV thermal conductivity increases rapidly due to the dominance of electron-electron scattering. It is natural that in this region the thermal conductivity of the electron gas depends on electron density and λ_e increases with increasing of electron concentration. For this reason, electron thermal conductivity of aluminum is higher than the same one of copper at the high-temperature. In low temperature region where electron-phonon interaction is dominated, the ratio is inverse. Because of the greater mean free path ℓ_{eph}, electron thermal conductivity of copper is higher than that of aluminum.

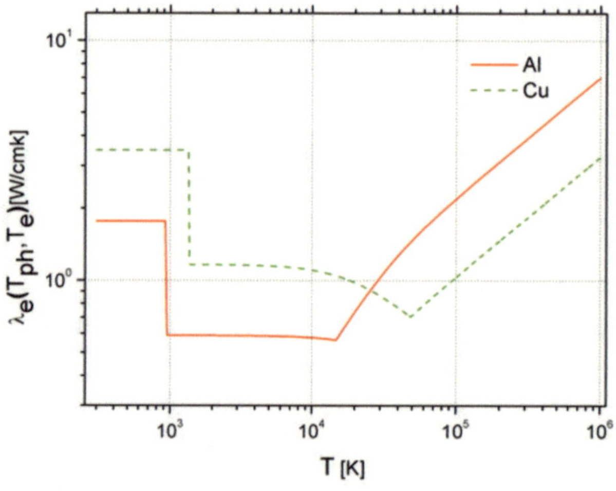

Figure 6. Temperature dependence of total electron thermal conductivity $\lambda_e(T_e,T_{ph})$.

3.2. Phonon gas

3.2.1. Heat capacity of phonons

The main consequence of the existence of lattice oscillations is the possibility of its thermal excitation, which is appeared as a contribution to the heat capacity of solid.

Taking into account the process of melting, heat capacity of phonon gas can be written as [69], [73]:

$$C_{ph,k} = \begin{cases} 3k_B N_{a,k}, & k = s,l, \quad T > T_D \\ \dfrac{12\pi^4}{5} N_a k_B \left(\dfrac{T_{ph}}{T_D}\right)^3, & T < T_D \end{cases}, \left[\frac{J}{m^3 K}\right] \tag{52}$$

T_D is a Debye temperature.

3.2.2. The thermal diffusivity of phonon gas χ_{ph}

Phonons are considered as a gas of particles. From elementary kinetic theory, thermal diffusivity of a gas is given by

$$\chi_{ph} = \frac{1}{3} \ell_{ph} v_{sound} \tag{53}$$

where ℓ_{ph} - phonon mean free path. It is assumed that the phonons move with velocity of sound v_{sound}.

The mean free path of phonons ℓ_{ph} is determined from the description of thermal motion in a solid by means of notions of the phonon gas. The interaction between the phonons can be characterized by some effective cross section which is proportional to the mean square of thermal expansion of the body or the mean square of density fluctuation ρ [69]: $\Delta^2 = N_a k_B T_p \beta$, where β is a coefficient of compressibility. Taking into account $\beta = (\rho v_{sound}^2)^{-1}$:

$$\Delta^2 = N_a k_B T_{ph} \beta = \frac{N_a k_B T_{ph}}{\rho v_{sound}^2} = \frac{k_B T_{ph}}{M v_{sound}^2}$$

Assigning to phonons radius equal to thermal oscillations amplitude, we can taken into account scattering of sound waves and determine the mean free path ℓ_{ph}

$$\ell_{ph} = \frac{r}{\Delta^2 \gamma^2} = \frac{M v_s^2}{k_B T_{ph}} \frac{r}{\gamma^2}, \tag{54}$$

where M - the mass of the atom, γ - Griuneyzen constant.

Accounting for the expressions for the mean free path ℓ_{ph} (54) and the velocity of sound υ_{sound} [73] expressed through the Fermi velocity

$$\upsilon_{sound} = \left(\frac{m}{3M}\right)^{1/2} \upsilon_F = \left(\frac{m}{3M}\right)^{1/2} \left(\frac{9\pi}{4} z\right)^{1/3} \frac{\hbar}{m} \frac{1}{r}, \; M = M_0 A,$$

thermal diffusivity of phonons can be written as:

$$\chi_{ph} = \frac{1}{3} \ell_{ph} \upsilon_s = \frac{M_0 A \upsilon_S^3}{3 k_B T_{ph}} \frac{r}{\gamma^2} = \left(\frac{3}{4\pi}\right)^{1/3} \frac{\pi^2}{(3m)^{3/2}} \frac{\hbar^3}{(M_0 A)^{1/2}} \frac{z}{\gamma^2} \frac{N_a^{2/3}}{k_B T_{ph}}$$

Taking into account the melting process:

$$\left(\chi_{ph}\right)_k = \left(2.831 \times 10^{-13} \frac{z}{A^{1/2} \gamma^2} \frac{N_{a,\kappa}^{2/3}}{T_{ph}[K]}\right)_k \left[\frac{\text{cm}^2}{\text{s}}\right], \; k = s,l \tag{55}$$

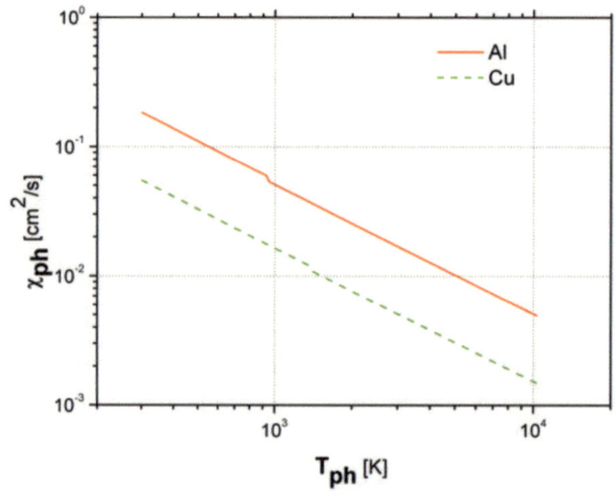

Figure 7. Temperature dependences of phonon thermal diffusivity $\chi_{ph}\left(T_{ph}\right)$.

3.2.3. *Thermal conductivity of phonon gas* λ_{ph}

The thermal conductivity of phonon gas can be expressed through its heat capacity C_{ph} and thermal conductivity. Taking into account expressions for C_{ph} and χ_{ph} in the high-temperature region $T_{ph} > T_D$ thermal conductivity takes the form:

$$\left(\lambda_{ph}\right)_\kappa = C_{ph} \chi_{ph} = \left(\frac{3\pi^2}{(3m)^{3/2}} \frac{\hbar^3}{M_0^{1/2}} \frac{z}{A^{1/2} \gamma^2} \frac{N_a^{5/3}}{T_{ph}[K]}\right)_k = \left(1.17 \times 10^{-35} \frac{z}{A^{1/2} \gamma^2} \frac{N_{a\kappa}^{5/3}}{T_{ph}[K]}\right)_\kappa, \tag{56}$$

Graphic dependences of λ_{ph} for *Al* and *Cu* are represented in Fig. 8.

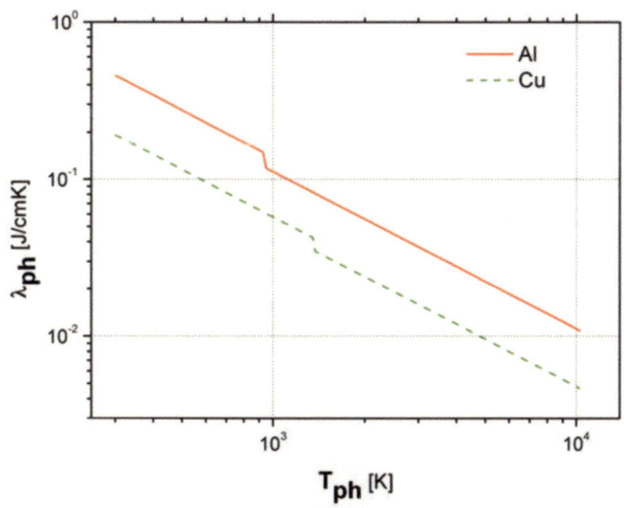

Figure 8. Temperature dependence of the phonon thermal conductivity.

3.2.4. Equilibrium heat capacity and thermal conductivity of metal

The temperature dependences of the total heat capacity and thermal conductivity of a metal can be represented as corresponding sums, which consist of two parts – electron (40), (51) and phonon (52), (55):

$$\begin{gathered} C(T) = C_e(T) + C_{ph}(T), \\ \lambda(T) = \lambda_e(T) + \lambda_{ph}(T) \end{gathered}, \tag{57}$$

Temperature dependences of $\lambda(T)$ and $C(T)$ are shown in Figs 9, 10. For comparison, reference data from [72], [74] - [77] are shown by markers.

According to the equilibrium theory of metals [69], [73] in the temperature range $T_0 < T < T_m$ heat transport by electrons is dominated. As predicted by the Wiedemann-Franz law, equilibrium electron thermal conductivity is practically independent of temperature. Its contribution to the total thermal conductivity of the metal with temperature increasing, is much greater than the contribution of the phonon component $\lambda_e > \lambda_{ph}$. The phonon part of thermal conductivity is inversely proportional to the temperature, and its contribution is noticeable only at low temperatures. As a result, the resulting thermal conductivity of solid-state becomes linearly decreasing temperature dependence. At the phase transition solid - liquid thermal conductivity of both metals decreases abruptly in 2 ÷ 3 times and decreases slowly with increasing temperature.

At very low temperatures, much smaller than Debye temperature $T<T_D$, these relations become invalid, which is associated with the change of interaction mechanisms. Since the

thermal conductivity of metal bodies at low temperatures is determined by the transfer of thermal energy by sound waves, rather than electrons. The heat capacity and the number of phonons in this case are proportional to ~ T^3, and the electrical resistance, defined by the scattering of electrons by phonons is inversely proportional to absolute temperature.

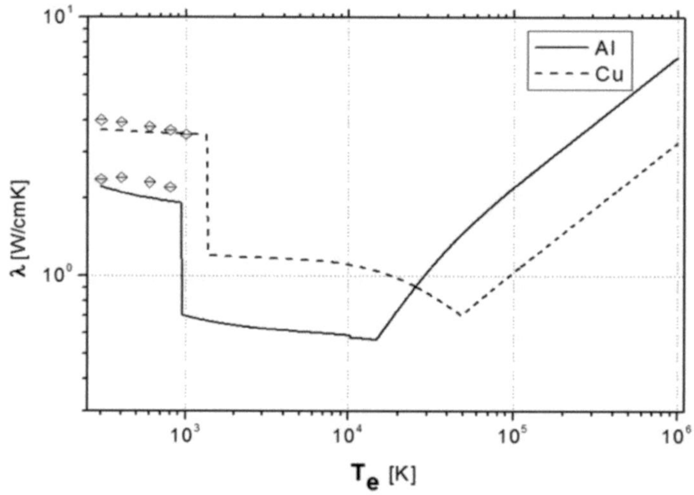

Figure 9. Temperature dependences of total thermal conductivity $\lambda(T)$.

Calculations showed that at temperatures close to the plasma T≥1 eV, the thermal conductivity of metals increases rapidly, and the thermal conductivity of aluminum, due to three times exceeded electron density becomes higher than the thermal conductivity of copper. However, this situation can have meaning only in the case of strongly nonequilibrium with hot electrons and cold lattice. Under equilibrium conditions, the use of calculated data should be restricted to the region of the critical point, in which neighborhood, as known, all physical properties of metals vary sharply.

Comparison of the obtained theoretical data with reference data was carried out in a relatively narrow temperature range, since the physical properties of most metals were measured in the temperature range from several tens to several thousand degrees, and usually not exceeding the boiling temperature of the equilibrium. In the range 300 ÷ 1500 K, the comparison has shown a complete qualitative agreement of results with a good quantitative correlation, Figs.9, 10.

3.3. Electron-phonon interaction

During laser action on metals, entire pulse energy is transmitted directly to the electrons. The result is a highly nonequilibrium region in the solid with hot electrons and cold lattice. Cooling of electron subsystem realized by two mechanisms: energy transfer by electron thermal conductivity, leveling the gradients of electron temperature and electron-phonon interaction, which leads to heating of the lattice.

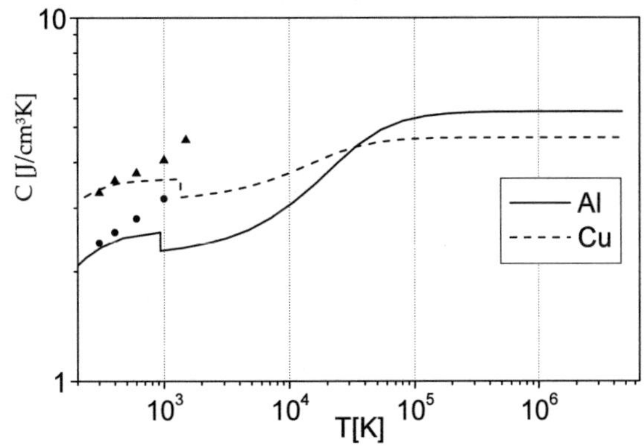

Figure 10. Temperature dependences of total heat capacity $C(T)$.

3.3.1. The average frequency of electron-phonon collisions $\langle \nu_{eph} \rangle$.

Per time unit, the electron undergoes an average of ~ 1/2 τ collisions with the emission of phonons, and roughly the same, but slightly smaller, the number of collisions with the absorption of phonon. Estimate the time of inelastic collisions by using the following equation [70]

$$\frac{1}{2}\tau_{eph} = \frac{1}{N_e \langle \upsilon_e \rangle \sigma_{eph}}.$$

Hence, the average frequency of electron-phonon collisions with energy transfer of electrons to crystal lattice can be represented as

$$\langle \nu_{eph} \rangle = 2N_e < \upsilon_e > \sigma_{eph} = 2zN_a < \upsilon_e > \sigma_{eph} \tag{58}$$

where σ_{eph} is scattering cross section of electrons by phonons.

The cross section for electron-phonon scattering, calculated under the assumption that the product of the wave number of electron and atom displacements from equilibrium [78] has the form

$$\sigma_{eph} = 4\pi \left(\frac{4}{9\pi} \right)^{4/3} \frac{m k_B T_{ph}}{\hbar^2} \left(\frac{r}{r_B} \right)^2 \frac{r^4}{z^{4/3}} \frac{\xi \cdot \left[\ln(t_1+1) - \frac{t_1}{(t_1+1)} \right]}{\left(\xi^2 + 0.16 \right)^{1/2} \left(\xi^2 + \frac{4}{9} \right)^{1/2}} \tag{59}$$

The cross section has a maximum at $T_e \approx \varepsilon_F$. At $T_e << \varepsilon_F$ due to the Pauli principle, the cross section is proportional to temperature T_e. After removing the degeneracy $T_e >> \varepsilon_F$, cross section of electron-phonon collisions decreases with T_e increasing. The linear lattice temperature T_{ph} dependence of the cross section of electron scattering in metals reflects the fact that the effective radius of the atoms can be identified with the amplitude of its thermal fluctuations.

Taking into account expressions for σ_{eph} (59) and $< \upsilon_e >$ (45), the average rate of energy transfer can be written as:

$$\langle v_{eph}\rangle = \left(\frac{3}{2}\right)^{1/2}\frac{8}{3\pi}\left(\frac{r}{r_B}\right)^2\frac{k_B}{\hbar}T_{ph}\Phi_{eph}(\xi)\cdot\left\{\begin{array}{ll}\Phi_{eph}(\xi)=\dfrac{\xi\cdot\left[\ln(t_1+1)-\dfrac{t_1}{(t_1+1)}\right]}{\left(\xi^2+0.16\right)^{1/4}\left(\xi^2+\dfrac{4}{9}\right)^{1/2}} & \text{при } \xi<1 \\ \Phi_{eph}(\xi)=\dfrac{\xi\cdot\left[\ln(t_1+1)-\dfrac{t_1}{(t_1+1)}\right]}{\left(\xi^2+0.16\right)^{3/4}\left(\xi^2+\dfrac{4}{9}\right)^{1/2}} & \text{при } \xi\geq 1\end{array}\right\} \quad (60)$$

The dependences of the mean frequency of transmission electron energy to phonons from the temperature T_e at T_{ph}=300 K and T_{ph} = 1000 for aluminum and copper are shown in Fig.11.

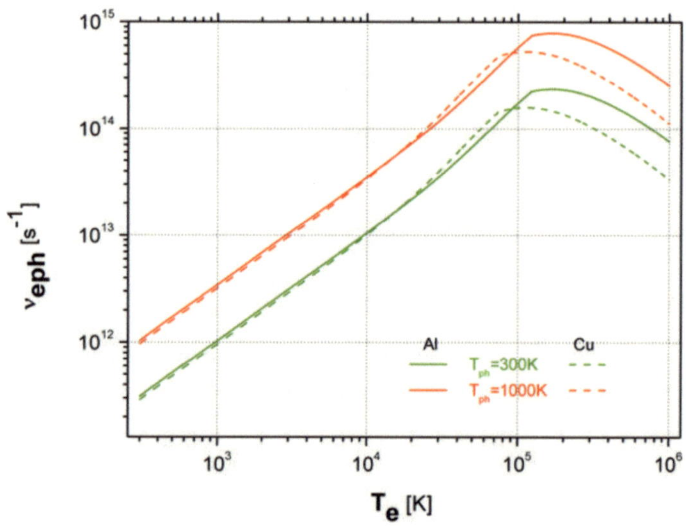

Figure 11. Temperature dependences of frequency of energy transfer of electrons to phonons at 300 K and = 1000 K.

3.3.2. *Energy transfer*

The mechanism of energy transfer to the lattice due to electron-phonon collisions was discussed in several papers [79-81]. The collision of an electron with the lattice phonon is absorbed or is born, for which law of conservation of energy and momentum works:

$$\begin{gathered}\Delta\varepsilon = m\upsilon_1^2/2 - m\upsilon_2^2/2 = \pm\hbar\,\Omega \\ p = 2m\upsilon_F\sin\frac{\varphi}{2} \cong \frac{\hbar\Omega}{\upsilon_{sound}}\end{gathered} \tag{61}$$

where Ω - the phonon frequency, φ - the angle between the initial and final electron momentum $m\upsilon_1$ and $m\upsilon_2$, υ_{sound} - velocity of sound, p - the electron momentum.

From the conservation laws (61) it follows that energy of phonon, excited by an electron with a momentum p equal to $\Delta\varepsilon = \hbar\Omega = 2p\upsilon_{sound}\sin\frac{\varphi}{2}$.

At a spherically symmetric scattering the energy transferred by one electron per unit time, averaged by φ, equals [81]:

$$<\Delta\varepsilon> = \frac{2}{3}\frac{\hbar\Omega_D}{k_B T_{ph}} p\upsilon_{sound}\nu_{eph}\left[1 - \frac{1}{32}\left(\frac{\hbar\,\Omega_D}{p\upsilon_{sound}}\right)^3\right],$$ where $\hbar\Omega_D$ - the Debye energy.

Taking into account velocity of sound υ_{sound}, expressed in terms of the Fermi velocity [73], $<\Delta\varepsilon>$ in its final form is written:

$$\begin{gathered}<\Delta\varepsilon> = \frac{2}{3}p_F\,\upsilon_{sound}^2\frac{\hbar k_D}{k_B T_{ph}}\left[1 - \frac{1}{32}\left(\frac{\hbar k_D \upsilon_{sound}}{p_F\upsilon_{sound}}\right)^3\right]\frac{\nu_{eph}}{k_B T_{ph}} = \\ = \left(\frac{9\pi}{2}\right)^{1/3}\left(\frac{3}{2}\right)^{1/2}\frac{4}{3}\frac{z}{mM}\left(\frac{r}{r_B}\right)^2\frac{\hbar^3}{r^4}\times\left[1 - \left(\frac{1}{2^4 z}\right)\right]\Phi_{eph}(\xi),\end{gathered} \tag{62}$$

3.3.3. *The electron-fonon coupling factor*

The average energy transferred to phonons by electrons per volume unit per time unit is equal to the product of the average energy transferred by a single electron $<\Delta\varepsilon>$ and concentration of electrons, which may participate in the transfer of energy to phonons

$$\frac{d\,\varepsilon}{d\,t} = \begin{cases} <\Delta\varepsilon>\dfrac{k_B\left(T_e - T_{ph}\right)}{\varepsilon_F}N_e & T_e < \varepsilon_F \\ <\Delta\varepsilon> z\left(T_e\right)N_e & T_e \geq \varepsilon_F \end{cases}$$

or

$$\frac{d\,\varepsilon}{d\,t} = <\Delta\varepsilon>\frac{k_B\left(T_e - T_{ph}\right)}{\varepsilon_F}N_e = g(\xi)\left(T_e - T_{ph}\right), \tag{63}$$

where $g(\xi)$ is electron-fonon coupling factor, equal to

$$g(\xi)=\langle\Delta\varepsilon\rangle\frac{k_B N_e}{\varepsilon_F}=\left(\frac{3}{2}\right)^{1/2}\left(\frac{8}{9\pi}\right)^{1/3}\frac{8}{3}\cdot\frac{\hbar k_B}{r_B^2 M_0}\frac{z^2}{A}N_a\left[1-\left(\frac{1}{2^4 z}\right)\right]\Phi_{eph}(\xi),\left[\frac{\mathrm{W}}{m^3 K}\right]. \quad (64)$$

The temperature dependence of the electron-fonon coupling factor $g(T_e)$ is shown in Fig.12. The results calculated for *Al* in [82] are shown to compare by markers.

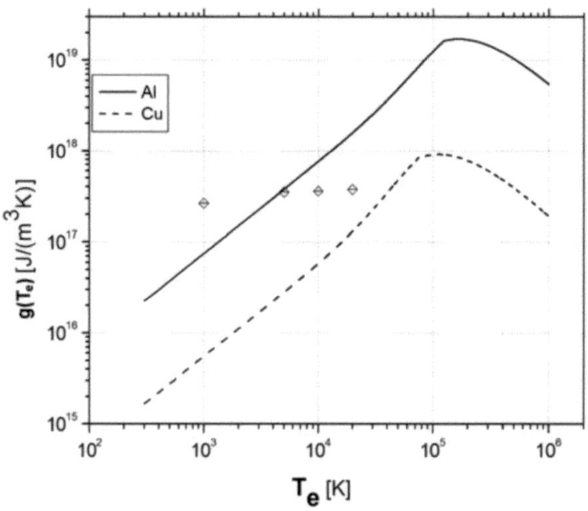

Figure 12. Temperature dependences of electron-fonon coupling factor $g(T_e)$.

3.4. Conclusions

Important thermodynamic and thermophysical properties of a degenerate electron and phonon gas: equation of state, heat capacity, the energy exchange between subsystems in an arbitrary temperature range are determined using fundamental physical quantities - the mean free paths and times (frequency) of the electron - electron and electron-phonon collisions.

Using Fermi integrals technique and its subsequent approximations have provided the temperature dependence of all characteristics in the form of simple analytical expressions.

4. Formulation and solution of the problem of laser action on metals using the method of dynamic adaptation

Action of intense laser radiation on metals may be accompanied by heating, evaporation, plasma formation. The dynamics of the processes in the condensed media and in the flow of vapor and gas environment surrounding the evaporating surface depend on several parameters of the laser pulse: the level of the absorbed intensity, wavelength, duration and spatial and temporal distribution of the energy, as well as optical, thermal and hydrodynamic characteristics of the condensed and gaseous media. It should be noted that the process of interaction is qualitatively different depending on where the target is placed: in the vacuum or the gaseous medium.

As it was already mentioned, the main features of ultra-short super-power action on metals are associated with the high speed and voluminous nature of the laser pulse energy release. The high rate of energy input leads to a strong deviation of the system from the local thermodynamic equilibrium (LTE). The basis of the mathematical description of the pulsed laser heating of metals is a phenomenological two-temperature model (TTM) of parabolic type proposed in the 50s by the authors of [65], [66]. The application of TTM involves the description of the situations with small deviations from LTE. Ultrafast heating of metal targets by super-power laser pulses causes a strong deviation from the LTE and requires appropriate adjustments of the mathematical model. In the first place, the kinetic non-equilibrium phase transitions and powerful dynamic effects associated with the movement of the phase fronts must be taken into account. The consideration of these processes requires the explicit description of the kinetics of the phase transitions and the formulation of the conservation laws at the phase fronts, which are hydrodynamic breaks.

4.1. Mathematical description of processes in the irradiation zone

Laser radiation propagates from the right to the left and strikes the surface of the metal target. Then the radiation is partially absorbed and partially reflected. The absorbed energy is consumed for heating, phase transformations and the generation of shock waves in the solid phase. Fig.13 shows spatial location of phases, moving interphase boundaries $\Gamma_{s\ell}(t)$, $\Gamma_{\ell v}(t)$ and shock wave in solid $\Gamma_{sh,s}(t)$.

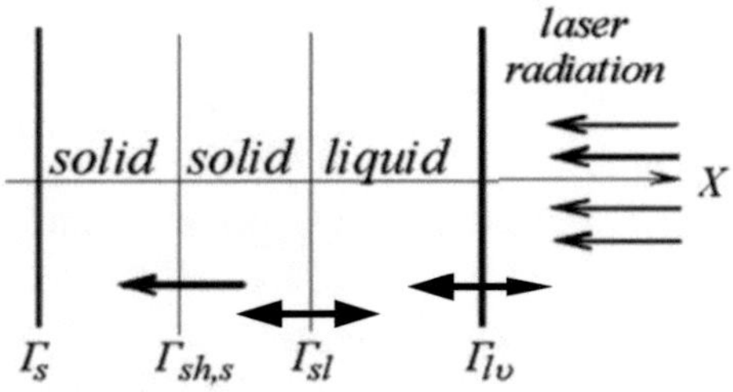

Figure 13. Spatial phase configuration.

Formulation of the problem was carried out under the following assumptions and limitations:

- The mechanisms of volume of melting and evaporation are not included into the consideration.
- The melting and evaporation fronts appear at the irradiated surface during its noticeable overheating $\Delta T_{s\ell} = T_{sur} - T_m$ and $\Delta T_{\ell v} = T_{sur} - T_b(p_{sat})$, the overheated metastable states act sustainably during the consideration.

The mathematical description and modeling of pico- and femtosecond laser action on metal target in vacuum was performed within the framework of a two-temperature and spatially one-dimensional multi-front nonequilibrium hydrodynamic Stefan problem, written for the two phases - solid and liquid.

4.1.1. System of equations

Action of high-power laser pulses on metal targets initiate in them a number of interrelated processes with nonlinear behavior. The description of their behavior is performed using a system of hydrodynamic equations, two energy balance equations for electron and phonon subsystems, and the transport equation of the laser radiation. The system of equations can be written in the domain with three moving boundaries $\Gamma_{sh,s}(t)$, $\Gamma_{s\ell}(t)$, $\Gamma_{kv}(t)$.

$$\left(\begin{array}{l}
\dfrac{\partial\rho}{\partial t}+\dfrac{\partial(\rho u)}{\partial x}=0 \\
\dfrac{\partial(\rho u)}{\partial t}+\dfrac{\partial(\rho u^2)}{\partial x}+\dfrac{\partial p}{\partial x}=0 \\
\dfrac{\partial(\rho_e\varepsilon_e)}{\partial t}+\dfrac{\partial(\rho_e u\varepsilon_e)}{\partial x}=-\left(p\dfrac{\partial u}{\partial x}+\dfrac{\partial W_e}{\partial x}+g(T_e)(T_e-T_{ph})+\dfrac{\partial G}{\partial x}\right) \\
\dfrac{\partial(\rho\varepsilon_{ph})}{\partial t}+\dfrac{\partial(\rho u\varepsilon_{ph})}{\partial x}=-\left(p\dfrac{\partial u}{\partial x}+\dfrac{\partial W_{ph}}{\partial x}-g(T_e)(T_e-T_{ph})\right) \\
\dfrac{\partial G}{\partial x}+\alpha(T_e)G=0, \qquad \rho_e=z\dfrac{m}{M}\rho, \\
P=P_e(\rho_e,T_e)+P_{ph}(\rho,T_{ph}),\;\; \varepsilon_e=C_e(T_e)T_e,\;\; \varepsilon_{ph}=C_{ph}(T_{ph})T_{ph}
\end{array}\right)_{k=s,\ell}$$

$$t>0,\ \Gamma_s<x<\Gamma_{sh,s}(t)\cup\Gamma_{sh,s}(t)<x<\Gamma_{s\ell}(t)\cup\Gamma_{s\ell}(t)<x<\Gamma_{\ell v}(t) \quad , \tag{65}$$

$$W_e=-\lambda(T_e,T_{ph})\frac{\partial T_e}{\partial x},\; W_{ph}=-\lambda(T_{ph})\frac{\partial T_{ph}}{\partial x},\; P(\rho,T)=P(\rho_e,T_e)+P(\rho,T_{ph}),$$

the designations: ρ,u,ε,T,p are the density, gas-dynamic velocity, internal energy, temperature and pressure correspondingly, $\alpha(T_e)$, $R(T_e)$ are the coefficient of volume absorption of the laser radiation and surface reflectivity, G is the density of the laser radiation, $C_e(T_e)$, $C_{ph}(T_{ph})$, $\lambda_e(T_e, T_{ph})$, $\lambda_{ph}(T_{ph})$ are electron and phonon heat capacity and heat conductivity, $g(T_e)$ electron-phonon coupling factor. Indexes s,ℓ,v, designate solid, liquid, vapor phases, e, ph designate electron and phonon components.

4.1.2. Edge conditions

We use edge conditions in the form of initial conditions at $t=0$ and boundary conditions at $x=\Gamma_s$, $\Gamma_{sh,s}(t)$, $\Gamma_{s\ell}(t)$, $\Gamma_{\ell v}(t)$.

Initial conditions

$$t=0:\;\; u(0,x)=0,\;\; p=0,\;\; \rho=\rho_0,\;\; T_e=T_{ph}=T_0=293K \tag{66}$$

Boundary conditions

$x=\Gamma_s$: the condition of zero mass and heat flow is used as boundary conditions at the left (fixed) boundary:

$$x=\Gamma_s:\ \rho_s u_s=0,\quad W_T=0; \tag{67}$$

$x=\Gamma_{s\ell}(t)$: the model of surface melting-crystallization is used as boundary conditions at the moving interphase boundary $\Gamma_{s\ell}(t)$.

The model of surface melting-crystallization is a non-equilibrium kinetic version of the Stefan problem [18], [83,84], formulated for the conditions of significant deviation from LTE and consisting of three conservation laws: mass, momentum and energy, written in the stationary (laboratory) coordinate system:

$$x=\Gamma_{s\ell}(t):\ \rho_s(u_s-\upsilon_{s\ell})=\rho_\ell(u_\ell-\upsilon_{s\ell}) \tag{68}$$

$$P_s+\rho_s(u_s-\upsilon_{s\ell})^2=P_\ell+\rho_\ell(u_\ell-\upsilon_{s\ell})^2, \tag{69}$$

$$\left(\lambda_{ph}\frac{\partial T_{ph}}{\partial x}\right)_s-\left(\lambda_{ph}\frac{\partial T_{ph}}{\partial x}\right)_\ell=\rho_s L_m^{ne}\upsilon_{s\ell}, \tag{70}$$

These conservation laws are accompanied by the kinetic dependence of the interphase front velocity $\upsilon_{s\ell}(\Delta T_{s\ell})$ on the overheating of the solid surface (22).

$$\upsilon_{s\ell}(T_{s\ell})=\frac{a\,f}{\lambda}(3k_B T_{s\ell}/m)^{1/2}\left[1-\exp\left(\frac{L_m}{k_B}\left(\frac{\Delta T}{T_{s\ell}}\right)\right)\right], \tag{71}$$

Additional account of hydrodynamic effects was carried out using the expressions for the curve of equilibrium melting $T_m(P_s)$, temperature dependence of the equilibrium latent melting heat $L_m^{eq}(T_m(P_s))$ and non-equilibrium latent melting heat L_m^{ne}

$$L_m^{ne}=L_m^{eq}(T_m(P_s))+\Delta C_{ps\ell}\Delta T_{s\ell}+\frac{\rho_s+\rho_\ell}{\rho_s-\rho_\ell}\frac{(u_s-u_\ell)^2}{2},\quad T_m=T_m(P_s)=T_{m,0}+kP_s,$$

where $\Delta C_{ps\ell}=(C_{p\ell}-C_{ps})$, $\Delta T_{s\ell}=(T_{s\ell}-T_m(P_s))$.

The electron component is assumed continuous with respect to the electron density N_e and temperature T_e during transition through the phase front:

$$\left(\lambda_e\frac{\partial T_e}{\partial x}\right)_s=\left(\lambda_e\frac{\partial T_e}{\partial x}\right)_\ell,\qquad T_{e,s}=T_{e,\ell}. \tag{72}$$

$x=\Gamma_{k\upsilon}(t)$: the model of surface evaporation is used as boundary conditions at the moving interphase boundary $\Gamma_{k\upsilon}(t)$. The process of surface evaporation within the approximation of the Knudsen layer is described by three conservation laws and three additional parameters at the outer side of the Knudsen layer (temperature T_υ, density ρ_υ and velocity u_υ). In the general case, two of these parameters (usually T_υ and ρ_υ) are determined using certain approximation relations, e.g. (30), while the third one (usually the Mach number $M = u / u_c$) is found from the solution of the gas-dynamic equations.

Three conservation laws are written in the stationary (laboratory) coordinate system and have the form:

$$x = \Gamma_{kv}(t) : \rho_k\left(u_k - v_{kv}\right) = \rho_v(u_v - v_{kv}), \tag{73}$$

$$p_k + \rho_k\left(u_k - v_{kv}\right)^2 = p_v + \rho_v\left(u_v - v_{kv}\right)^2, \tag{74}$$

$$\left(\lambda_{ph}\frac{\partial T_{ph}}{\partial x}\right)_k - \left(\lambda_v\frac{\partial T_v}{\partial x}\right)_v = \rho_k\left(u_k - v_{kv}\right)L_v^{ne}, \tag{75}$$

$$\rho_v = \alpha_\rho(M)\rho_{sat}, T_v = \alpha_T(M)T_{sur}, \tag{76}$$

$$p_{sat}(T_{sur}) = p_b \exp\left[\frac{L_v^{ne}}{RT_b}\left(\frac{\Delta T_{sur}}{T_{sur}}\right)\right], \rho_{sat} = \frac{p_{sat}(T_{sur})}{RT_{sur}}, \Delta T_{sur} = T_{sur} - T_b\left(p_{sat}(T_{sur})\right)$$

$$L_v^{ne} = L_v^{eq}\left(T_{sur}\right) + C_{pv}\left(T_v - T_{sur}\right) + \frac{\rho_k + \rho_v}{\rho_k - \rho_v}\frac{\left(u_k - u_v\right)^2}{2},$$

The boundary conditions for the electron component and the laser radiation are written as:

$$-\lambda_e\frac{\partial T_e}{\partial x} = \sigma T_e^4, \tag{77}$$

$$G(t) = \left(1 - R_k\left(T_e\right)\right) \cdot G_0 \exp\left(-\left(\frac{t}{\tau_L}\right)^2\right). \tag{78}$$

σ is the Stefan-Boltzmann constant.

$x = \Gamma_{sh,s}(t)$: three conservation laws are written at the moving front of the shock wave, so-called Rankine-Hugoniot relations, that have the following form in the laboratory coordinate system:

$$j_{sh,s}^{m} = \rho_1\left(u_1 - v_{sh,s}\right) = \rho_0\left(u_0 - v_{sh,s}\right), \tag{79}$$

$$j_{sh,s}^{i} = p_1 + \rho_1\left(u_1 - v_{sh,s}\right)^2 = p_0 + \rho_0\left(u_0 - v_{sh,s}\right)^2, \tag{80}$$

$$j_{sh,s}^{e} = -W_{T,1} + j_{sh,s}^{m}\left[\varepsilon_1 + \frac{\left(u_1 - v_{sh,s}\right)^2}{2}\right] = -W_{T,0} + j_{sh,s}^{m}\left[\varepsilon_1 + \frac{\left(u_0 - v_{sh,s}\right)^2}{2}\right] \tag{81}$$

The index "0" signifies the values at the background side, "1" relates to the shock wave side.

4.2. Numerical algorithm and finite difference schemes

Mathematical feature of Stefan problems is the lack of explicit expressions for the interconnected quantities at the interphase boundaries: temperature $T_{s\ell}$, T_{sur} and velocities of the phase fronts $\upsilon_{s\ell}$ and $\upsilon_{k\upsilon}$. So the problem (65) – (81) will be nonlinear even for constant values of thermo-physical and optical properties. For low velocities of phase transformations ($\upsilon_{s\ell}$, $\upsilon_{k\upsilon} << \upsilon_{sound}$), the processes behave in a quasi-equilibrium way, and their description can be performed within the framework of equilibrium models. For example, in the problem of melting-crystallization, a phenomenological condition of the temperature equity $T_{s\ell} = T_s = T_\ell = T_m$ is used instead of the kinetic condition (71), and the differential Stefan condition is completely omitted. The influence of the phase transition is taken into account using a singular adding of the latent heat L_m^{eq} to the heat capacity (equation of state) in the point of the phase transition, so-called enthalpy method [85 - 87]. To solve the problems in this statement, the methods of the type of "pass-through" or uniform schemes have found wide application [85], where the velocity and location of the phase transition are not determined explicitly.

Fast phase transitions ($\upsilon_{s\ell}$, $\upsilon_{k\upsilon} << \upsilon_{sound}$), that are typical for powerful pulsed laser action, occur under the conditions of high non-equilibrium caused by the powerful flow of material [88] over the interphase boundary. Using the smoothing procedure of the enthalpy reduces the class of solutions of the phase transformations problems in the material; in particular, it excludes from consideration the effects of overheating and overcooling of the condensed matter [89, 90]. High-speed phase transitions require explicit tracking of the phase boundaries [91]. Typically, the velocity of phase boundaries is determined numerically. In computational respect, the presence of moving boundaries leads to a significant complication of the numerical solution [92].

4.2.1. Method of dynamic adaptation

Finite-difference method of dynamic adaptation [93], [94],was used to numerically solve the discussed problem. This method allows performing explicit tracking of any number of interphase boundaries and shock waves [95].

The method of dynamic adaptation is based on the procedure of transition to an arbitrary nonstationary coordinate system. Its usage allows to formulate the problem of grid generation and adaptation on a differential level, i.e. part of differential equations in the obtained mathematical model describes physical processes, and other part describes the behavior of the grid nodes [96].

4.3. Arbitrary nonstationary coordinate system

The transition to an arbitrary nonstationary coordinate system is performed using an automatic coordinate transformation using the sought solution. Formally, it is possible to use a variable change of the common form $x=f(q,\tau)$, $t=\tau$ to perform a transition from the physical space $\Omega_{x,t}$ with Euler variables (x,t) to some computational space with arbitrary nonstationary coordinate system $\Omega_{q,\tau}$ and variables (q,τ). It is assumed that this transformation corresponds to a univalent nondegenerate reverse transformation $q=\phi(x,t), \quad \tau=t$. During the transition between coordinate systems, the partial derivatives of the dependent variables are connected using the following expressions:

$$\frac{\partial}{\partial t}=\frac{\partial}{\partial \tau}+Q\frac{\partial}{\partial q},\frac{\partial}{\partial x}=\frac{1}{\partial \psi}\frac{\partial}{\partial q},\frac{\partial^2}{\partial x^2}=\frac{1}{\psi}\frac{\partial}{\partial q}\frac{1}{\psi}\frac{\partial}{\partial q}, \tag{82}$$

where $\psi=\partial x/\partial q$ is the Jacobian of the reverse transformation, Q is the transformation function that should be determined.

In the arbitrary coordinate system, moving with velocity Q, where $\frac{\partial x}{\partial \tau}=-Q$, or $\frac{\partial \psi}{\partial \tau}=-\frac{\partial Q}{\partial q}$, the differential model (65) in the new variables (q,τ) takes the form:

$$\frac{\partial \psi}{\partial \tau}=-\frac{\partial Q}{\partial q},\psi=\frac{\partial x}{\partial q} \tag{83}$$

$$\frac{\partial}{\partial \tau}(\psi\rho)+\frac{\partial}{\partial q}\left(\rho\left(u+Q\right)\right)=0 \tag{84}$$

$$\frac{\partial}{\partial \tau}(\psi\rho u)+\frac{\partial}{\partial q}\left(p+\rho u\left(u+Q\right)\right)=0 \tag{85}$$

$$\frac{\partial}{\partial \tau}(\psi\rho_e\varepsilon_e)+\frac{\partial}{\partial q}\left(\rho_e\varepsilon_e\left(u+Q\right)\right)=-\left(p\frac{\partial u}{\partial q}+\frac{\partial W_e}{\partial q}+g(T_e)(T_e-T_{ph})+\frac{\partial G}{\partial q}\right) \tag{86}$$

$$\frac{\partial\left(\psi\rho\varepsilon_{ph}\right)}{\partial \tau}+\frac{\partial}{\partial q}\left(\rho\varepsilon_{ph}\left(u+Q\right)\right)=-\left(p\frac{\partial u}{\partial q}+\frac{\partial W_{ph}}{\partial q}-g(T_e)(T_e-T_{ph})\right) \tag{87}$$

$$\frac{\partial G}{\partial q}+\psi\alpha(T_e)G=0 \tag{88}$$

$$p=p_e\left(\rho_e,T_e\right)+p_{ph}\left(\rho,T_{ph}\right),\quad \varepsilon_e=C_e(T_e)T_e,\quad \varepsilon_{ph}=C_{ph}(T_{ph})T_{ph},\quad W=-\frac{\lambda(T)}{\psi}\frac{\partial T}{\partial q},$$

Thus, usage of an arbitrary nonstationary coordinate system is accompanied by the transformation of the initial differential model (65) into the extended model (83) – (88), in which the equation (83) is the equation of the inverse transform. Equations (84) – (88) describes the physical processes. The type, characteristics and kind of boundary conditions for the inverse transformation equations depend on the type conversion function Q. Because of this equation, the initial (66) and boundary conditions (67) – (81) are changed accordingly.

Thus, the unknown functions in the computational space are not only the functions of the physical fields but also the coordinates of the grid nodes. The equation of the reverse transformation is used after the determination of the function Q to construct the adaptive grid in the physical space. Its differential analogue describes the dynamics of the grid nodes, and the function Q performs controlled movement of the nodes in agreement with the dynamics of the sought solution. In the computational space, the nodes of the grid and all discontinuous and interphase boundaries are steady. The value of the function Q at the boundaries of the domain are determined from the boundary conditions.

Selection of the function Q. The function Q that is in agreement with the sought solution can be determined from the quasi-steady principle [96], which states that we should search for such nonstationary coordinate system where all processes occur in a steady way. Since the energy balance equation includes all hydrodynamic variables, to determine the function Q, we can use only two equations (86), (87), assuming that $\frac{\partial \varepsilon_e}{\partial \tau} = \frac{\partial \varepsilon_{ph}}{\partial \tau} = 0$.Then the function Q will have the form

$$Q = -u - \frac{2p\frac{\partial u}{\partial q} + \left(+ \frac{\partial W_e}{\partial q} + \frac{\partial W_{ph}}{\partial q} + \frac{\partial G}{\partial q} \right)}{\left(\rho_e \frac{\partial \varepsilon_e}{\partial q} + \rho \frac{\partial \varepsilon_{ph}}{\partial q} + re \right)} \tag{89}$$

where $W_e = -\frac{\lambda_e}{\Psi}\frac{\partial T_e}{\partial q}$, $W_{ph} = -\frac{\lambda_{ph}}{\Psi}\frac{\partial T_{ph}}{\partial q}$.

Differential schemes

The differential model (83) – (88) was approximated by a family of conservative differential schemes obtained using integration-interpolation method [97]. Computational grids $\left(\omega_m^j\right)_s$ and $\left(\omega_m^j\right)_\ell$ are introduced in the computational space $\Omega_{q,\tau}$ in each subdomain with integer q_i and half-integer $q_{i+1/2}$ nodes for the spatial variable q and step $\Delta\tau^j$ for the variable τ.

$$\left(\omega_m^j\right)_k = \left\{ \begin{matrix} \left(q_m,\tau^j\right), \left(q_{m+1/2},\tau^j\right); & q_{m+1} = q_m + h, \quad q_{m+1/2} = q_m + 0.5h, \\ \tau^{j+1} = \tau^j + \Delta\tau^j, & i = 0,..,N-1, \quad j = 0,1,... \end{matrix} \right\}_{k=s,\ell}$$

The flow variables $W_{e,}$, $W_{ph,}$, G, and functions u,Q,x, correspond to the integer nodes during approximation, while functions ε_e, ε_{ph}, T_e, T_{ph}, ρ, ψ correspond to half-integers ones $\left(q_{m+1/2},\tau^j\right)$.

The family of conservative differential schemes has the form:

$$
\begin{aligned}
&\frac{\psi^{j+1}_{m-1/2}-\psi^{j}_{m-1/2}}{\Delta\tau^{j}}=\frac{Q^{\sigma_1}_{m}-Q^{\sigma_1}_{m-1}}{h}\\
&\frac{\psi^{j+1}_{m-1/2}\rho^{j+1}_{m-1/2}-\psi^{j}_{m-1/2}\rho^{j}_{m-1/2}}{\Delta\tau^{j}}=\frac{\rho^{\sigma_2}_{m}\left(Q^{\sigma_1}_{m}-u^{\sigma_3}_{m}\right)-\rho^{\sigma_2}_{m-1}\left(Q^{\sigma_1}_{m-1}-u^{\sigma_3}_{m-1}\right)}{h}\\
&\frac{u^{j+1}_{m}\left(\psi^{j+1}_{m-1/2}\rho^{j+1}_{m-1/2}+\psi^{j+1}_{m+1/2}\rho^{j+1}_{m+1/2}\right)-u^{j}_{m}\left(\psi^{j}_{m-1/2}\rho^{j}_{m-1/2}+\psi^{j}_{m+1/2}\rho^{j}_{m+1/2}\right)}{2\Delta\tau^{j}}=\frac{p^{\sigma_4}_{m-1/2}-p^{\sigma_4}_{m+1/2}}{h}+\\
&+\frac{\rho^{\sigma_2}_{m}\left(Q^{\sigma_1}_{m}-u^{\sigma_3}_{m}\right)u^{\sigma_3}_{m+1}+\rho^{\sigma_2}_{m+1}\left(Q^{\sigma_1}_{m+1}-u^{\sigma_3}_{m}\right)u^{\sigma_3}_{m}}{2h}-\frac{\rho^{\sigma_2}_{m}\left(Q^{\sigma_1}_{m}-u^{\sigma_3}_{m}\right)u^{\sigma_3}_{m-1}+\rho^{\sigma_2}_{m-1}\left(Q^{\sigma_1}_{m-1}-u^{\sigma_3}_{m-1}\right)u^{\sigma_3}_{m}}{2h}
\end{aligned}
\tag{90}
$$

$$
\begin{aligned}
&\frac{\left(\psi^{j+1}_{m-1/2}\rho^{j+1}_{e,m-1/2}\varepsilon^{j+1}_{e,m-1/2}\right)-\left(\psi^{j}_{m-1/2}\rho^{j}_{e,m-1/2}\varepsilon^{j}_{e,m-1/2}\right)}{\Delta\tau^{j}}=-\left\{\frac{\left[\rho^{\sigma_2}_{e,m}\varepsilon^{\sigma_5}_{e,m}\left(Q^{\sigma_1}_{m}-u^{\sigma_3}_{m}\right)-\rho^{\sigma_2}_{e,m-1}\varepsilon^{\sigma_5}_{e,m-1}\left(Q^{\sigma_1}_{m-1}-u^{\sigma_3}_{m-1}\right)\right]}{h}+\right.\\
&+\frac{p^{\sigma_4}_{m-1/2}\left(u^{\sigma_3}_{m-1}-u^{\sigma_3}_{m}\right)}{h}+\frac{2\left[\lambda^{\sigma_6}_{m}\dfrac{T^{\sigma_7}_{m+1/2}-T^{\sigma_7}_{m-1/2}}{\psi^{\sigma_8}_{m+1/2}+\psi^{\sigma_8}_{m-1/2}}-\lambda^{\sigma_6}_{m-1}\dfrac{T^{\sigma_7}_{m-1/2}-T^{\sigma_7}_{m-3/2}}{\psi^{\sigma_8}_{m-1/2}+\psi^{\sigma_8}_{m-3/2}}\right]}{h}+\left[g(T_e)^{\sigma_9}(T_e-T_{ph})^{\sigma_{10}}\right]_{m-1/2}+\\
&\left.+\frac{\left(G^{\sigma_{11}}_{m}-G^{\sigma_{11}}_{m-1}\right)}{h}+\frac{\left(W^{\sigma_{12}}_{e,m}-W^{\sigma_{12}}_{e,m-1}\right)}{h}\right\},
\end{aligned}
$$

$$
\begin{aligned}
&\frac{\left(\psi^{j+1}_{m-1/2}\rho^{j+1}_{m-1/2}\varepsilon^{j+1}_{ph,m-1/2}\right)-\left(\psi^{j}_{m-1/2}\rho^{j}_{m-1/2}\varepsilon^{j}_{ph,m-1/2}\right)}{\Delta\tau^{j}}=\\
&-\left\{\frac{\left[\rho^{\sigma_2}_{m}\varepsilon^{\sigma_{13}}_{ph,m}\left(Q^{\sigma_1}_{m}-u^{\sigma_3}_{m}\right)-\rho^{\sigma_2}_{m-1}\varepsilon^{\sigma_{13}}_{ph,m-1}\left(Q^{\sigma_1}_{m-1}-u^{\sigma_3}_{m-1}\right)\right]}{h}+\frac{p^{\sigma_4}_{m-1/2}\left(u^{\sigma_3}_{m-1}-u^{\sigma_3}_{m}\right)}{h}+\right.\\
&+\frac{2\left[\lambda^{\sigma_{14}}_{ph,m}\dfrac{T^{\sigma_{15}}_{ph,m+1/2}-T^{\sigma_{15}}_{ph,m-1/2}}{\psi^{\sigma_8}_{m+1/2}+\psi^{\sigma_8}_{m-1/2}}-\lambda^{\sigma_{14}}_{ph,m-1}\dfrac{T^{\sigma_{15}}_{ph,m-1/2}-T^{\sigma_{15}}_{ph,m-3/2}}{\psi^{\sigma_8}_{m-1/2}+\psi^{\sigma_8}_{m-3/2}}\right]}{h}+\\
&\left.+\left[g(T_e)^{\sigma_9}(T_e-T_{ph})^{\sigma_{10}}\right]_{m-1/2}+\frac{\left(W^{\sigma_{16}}_{e,m}-W^{\sigma_{16}}_{e,m-1}\right)}{h}\right\},
\end{aligned}
$$

Here $f^{\sigma_r}=\sigma_r f^{j+1}+\left(1-\sigma_r\right)f^{j}$, and $\sigma_r=\sigma_1,\sigma_2,\ldots$ □are the weight coefficients, determining the degree of implicitness of the differential schemes.

To solve the obtained system of nonlinear partial-differential equations we use a computational algorithm with enclosed iterative cycles consisting of one external and two internal cycles [98]. Each of the internal cycles uses the Newton iterative procedure.

5. Modeling results

We consider two modes of pulse with pico - and femtosecond laser fluence with a wavelength $\lambda_L = 0.8$ μm on two metal target of aluminum (Al) and copper (Cu). The influence on each of the targets was carried out with the same energy density F, respectively $F = 0.7\,J\,cm^{-2}$ for Al and $F = 2.0\,J\,cm^{-2}$ for Cu.

Temperature dependences of the reflectivity $R(T_e)$ and volumetric absorption coefficient $\alpha(T_e)$ were calculated according to [99, 100]. Fig. 14 shows the time profiles of the incident and the absorbed laser pulse with a Gaussian profile $\tau_L = 10^{-12}$ s and $\tau_L = 10^{-15}$ s.

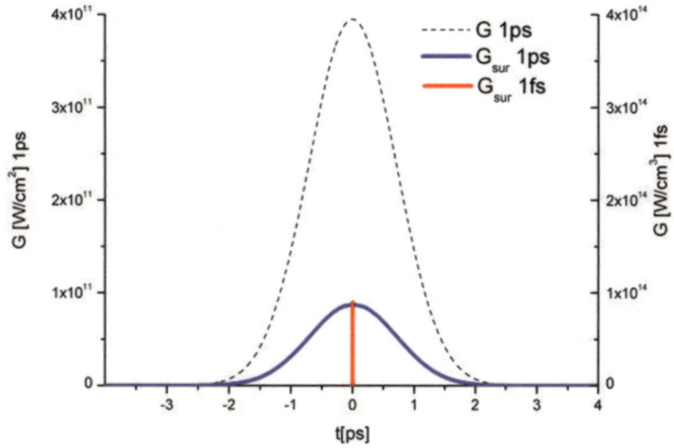

Figure 14. Time profile of pico- and femtosecond pulses on the surface of Al.

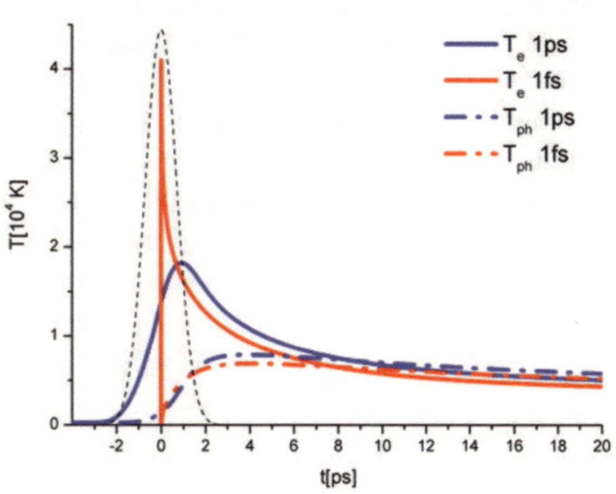

Figure 15. Time dependence of the surface temperature of Al.

The incident laser radiation is absorbed in the electron component, causing it to warm up fast. The main feature of the nonequilibrium heating is that it occurs under conditions of intense competition between the rapid release of the laser energy and the factors limiting the heat - strongly varying electron thermal conductivity coefficient $\lambda_e(T_e, T_{ph})$ and electron-fonon coupling factor $g(T_e)$.

At picosecond influence $\tau_L = 10^{-12}$, the maximum of electron temperature of the surface $T_{e,\max} \approx 1.8 \cdot 10^4 K$ is achieved at the descending branch of the laser pulse at the moment $t \cong \tau_L$, Fig. 15, and the values of the coefficients vary in the range $\lambda_{e,\max} / \lambda_{Al}^{eq} \approx 10$ and $g_{\max} / g_{Al}^{eq} \approx 70$. The maximum value of the lattice temperature is achieved after the pulse $t \approx 5\tau_L$ and amounts $T_{ph,\max} \approx 7.5 \cdot 10^3 K$. The maximum gap between the temperatures reaches the value $\Delta T = T_e - T_{ph} \approx 10^4 K$. Temperature equalization $T_e \approx T_{ph}$ occurs within a time ~ 10 ps.

Femtosecond influence $\tau_L = 10^{-15} s$ differs from picosecond that the same energy is released during the 3 orders of magnitude less than that causes more of a deviation from local thermodynamic equilibrium. The values of the coefficients $\lambda_e(T_e, T_{ph})$ and $g(T_e)$ limiting the heating of electronic components, and also increase $\lambda_{e,\max} / \lambda_{Al}^{eq} \approx 27$ and $g_{\max} / g_{Al}^{eq} \approx 200$, but not as much as $G(t)$. Time profile of the electron temperature of the surface $T_{e,surf}(t)$ significantly shifted relative to laser pulse time profile $G(t)$. The maximum value of the electron temperature is achieved at the end of the pulse $t \cong 2\tau_L$ and is $T_{e,\max} \approx 4.1 \cdot 10^4 K$, Fig. 15. A second difference of a femtosecond influence, despite the rapid energy exchange between the subsystems is a slow heating of the lattice, which differs little from the effects of heating at picosecond influence. The maximum value of the lattice temperature is achieved at the time $t \approx 5$ ps and is $T_{ph,\max} \approx 7.2 \cdot 10^3 K$. The maximum gap between the temperatures reaches a value $\Delta T = T_e - T_{ph} \approx 3.3 \cdot 10^4 K$. Temperature equalization $T_e \approx T_{ph}$ occurs within the time ~ 9 ps.

The high heating rate determines high rate of phase transformations. Fig. 16 shows the time dependences of the melting front propagation velocity $\upsilon_{sl}(t)$ for two modes of influence. Since the rate of heating of the lattice by pico - and femtosecond pulses differ slightly, then the rate υ_{sl} is almost the same. The only difference is in the initial stage of the process. At the picosecond influence, the birth of the melting front starts at the front of the laser pulse $t \approx -0.1 ps$, while at the femtosecond with a considerable delay after the pulse, $t \approx 30 fs$. The maximum values of $\upsilon_{sl\,\max}(t) \approx 3.75$ km/s are achieved in the time interval $t \approx 2 \div 2.5 ps$. Such a high rate of propagation of the front leads to a pressure jump at the melting surface of the solid phase $p_s = 0.16 Mbar$ for $\tau_L = 10^{-12}$ s ($p_s = 0.14 Mbar$ for $\tau_L = 10^{-15}$ s) and the formation of a shock wave moving ahead of the front of melting, Fig. 17.

Appearance of high dynamic pressure causes an increase in the equilibrium melting temperature, the curve $T_m(p_s)$ shown in Fig. 18. Its maximum value reaches $T_m(p_s) \cong 1900 K$. Due to the dynamic pressure of about one order of magnitude as compared to the equilibrium value, increases the value of non-equilibrium heat of melting $L_m^{ne}(p_s)$.

Comparison of the $T_{s\ell}(t)$ and $T_m(p_s(t))$ curves indicates a significant overheating of surface of the phase boundary $T_{s\ell}(t) > T_m(p_s(t))$. The degree of nonequilibrium of melting process is conveniently characterized by the response function $\Delta T_{s\ell}(\upsilon_{sl})$, which is shown in Fig.19.

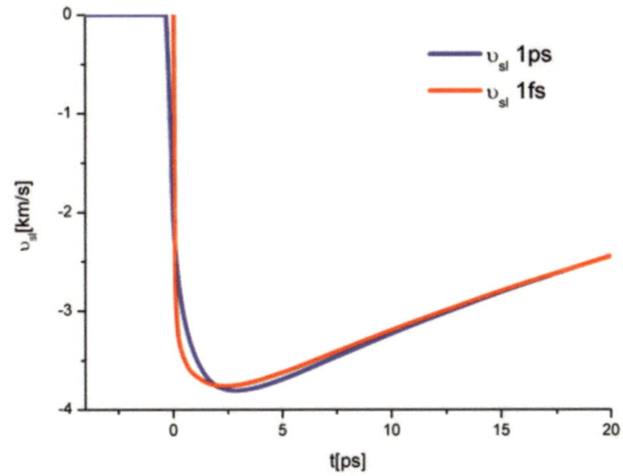

Figure 16. The time profile of the velocity of melting of aluminum.

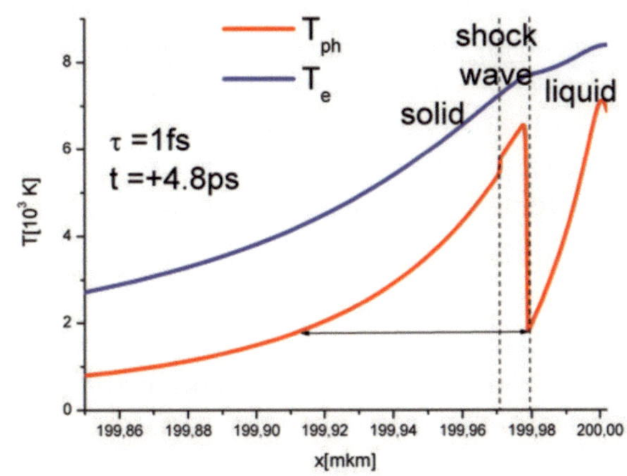

Figure 17. The spatial profiles of temperature of aluminum.

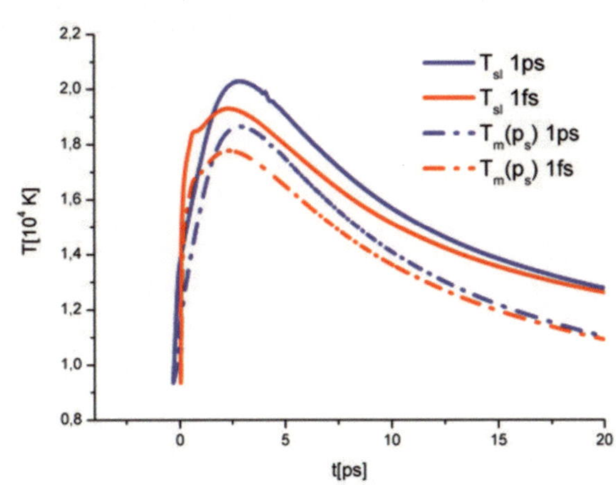

Figure 18. The time profiles of the equilibrium melting temperature and temperature of the melting front of Al.

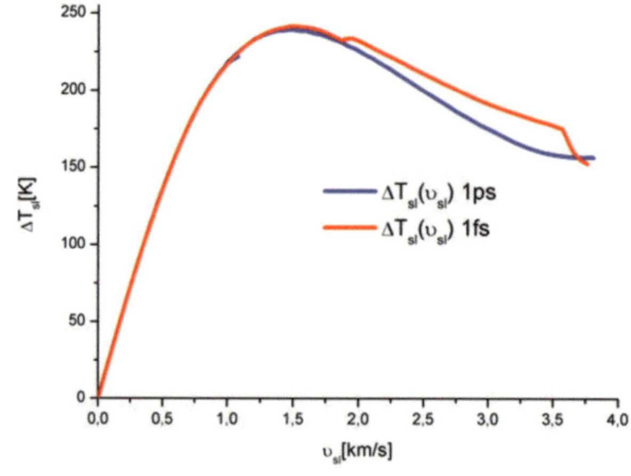

Figure 19. Response functions $\Delta T_{s\ell}(\upsilon_{sl})$ for Al.

At low overheating (supercooling), the function $\Delta T_{s\ell}$ increases linearly with growth υ_{sl}. In the range of υ_{sl} ~ 1 km/s, at which the shock wave is formed in the solid phase, $\Delta T_{s\ell}$ function has a smooth maximum indicating the change of the mechanism of phase transformation. The thermal conductivity mechanism is dominant at low velocities $\upsilon_{sl} \leq 1 km/s$ will be displaced by hydrodynamic at large values $\upsilon_{sl} \leq \upsilon_{sound}$.

Fig. 20 shows fragment of dependence of $\upsilon_{sl}(t) > 0$ which characterizes the hardening process which takes place in the nanosecond time range. Maximum velocity of the solidification front does not exceed 45 m/s, respectively, the maximum undercooling is $10 \div 15K$. High velocity of propagation of melting fronts are associated with strong cross-flows of matter and energy across the phase boundary.

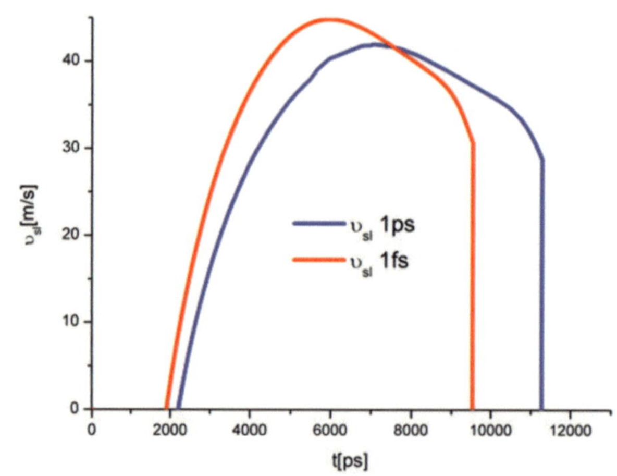

Figure 20. The time profiles of the velocity of solidification of aluminum.

In aggregate with volumetric nature of the energy transfer from the electron gas to the lattice it leads to the formation of highly superheated metastable state, characterized by the appearance of the surface temperature maximum $\Delta T_{ph,\max} = T_{ph} - T_m(p_s) \approx 4.5 \cdot 10^3 K$ in the solid phase, Fig. 17. In the liquid phase is also formed near-surface temperature maximum, Fig.20, but since the maximum velocity of evaporation front $\upsilon_{\ell\upsilon,\max} \approx 40\, m/s$ is two orders of magnitude lower than the velocity υ_{sl}, magnitude and its role are insignificant.

Action of ultrashort laser pulses on metal target with different thermo-physical properties – copper – has several differences and many in common with the action of the aluminum target. The main difference of the non-equilibrium heating and non-equilibrium phase transformation in the condensed media is in the first place connected with the rate of energy exchange between the electron and phonon subsystems. The coefficient of energy exchange $g(T_e)$ in copper is by about 1.5 orders lower than that of aluminum, so that during laser energy release, the action of one of the main factors that limit the heating of the electron subsystem turns out to be significantly lower. As a result, the maximum values of the electron temperature of the surface are several times higher than that for aluminum: $T_{e,\max} \approx 4. \cdot 10^4 K$ for picosecond pulse $T_{e,\max} \approx 1.2 \cdot 10^5 K$ for femtosecond pulse, fig. 21.

Heating and melting of the lattice take place in the picosecond range $t_m \approx +0.1$ *ps*, fig.22. For copper, the rate of energy transfer to the lattice is significantly lower, so that the velocity of the melting front turns out to be several times lower than that for aluminum $\upsilon_{s\ell,\max}(t) \approx 1.3$ km/s, fig.23. The maximum value of the pressure in the solid phase is also lower, $p_{s,\max} = 0.14\, Mbar$. Despite the fact that the velocity $\upsilon_{s\ell}$ for copper is significantly lower than that for Al, the value of overheating of the interphase surface turns out to be significantly higher due to decrease of the hydrodynamic effects: $\Delta T_{s\ell,\max} \approx 350 K$, fig. 24. The overheating of the sub-surface region of the solid phase reaches the value of $4 \cdot 10^3 K$.

The solidification of the liquid phase takes place in the nanosecond rage – the same as for aluminum. The lifetime of the melt reaches the value of $15 \div 20$ ns, fig.25, the maximum velocity of the crystallization front - $\upsilon_{s\ell} \approx 35$ m/s. The maximum value of the overcooling of the front ~ *22 K*. The overcooling of the volume of the solid reaches the value of *800 K*.

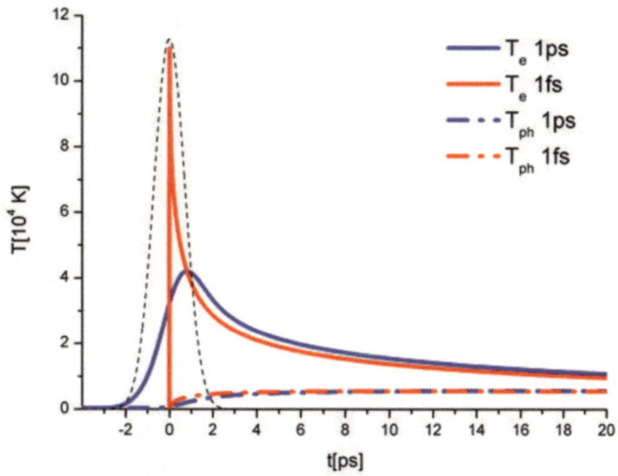

Figure 21. Time dependence of the surface temperature of Cu.

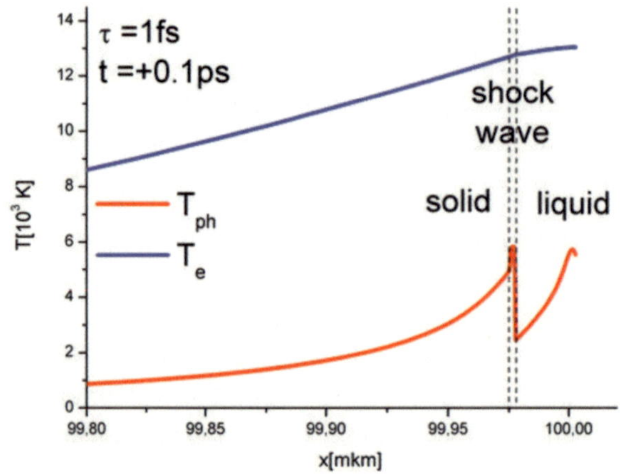

Figure 22. The spatial profiles of temperature of copper.

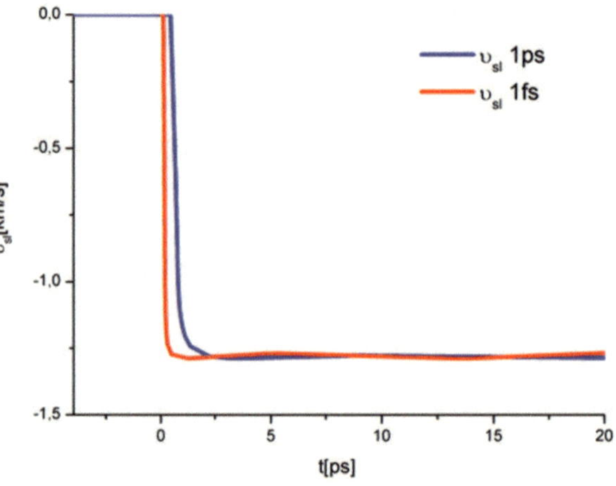

Figure 23. The time profile of the velocity of melting of copper.

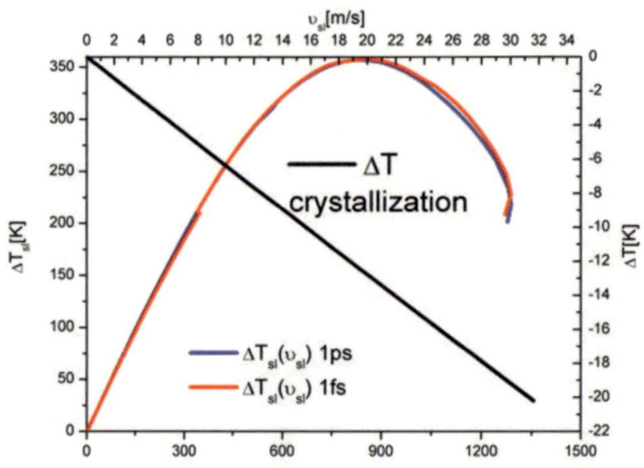

Figure 24. Response functions $\Delta T_{s\ell}\left(\upsilon_{sl}\right)$ for Cu.

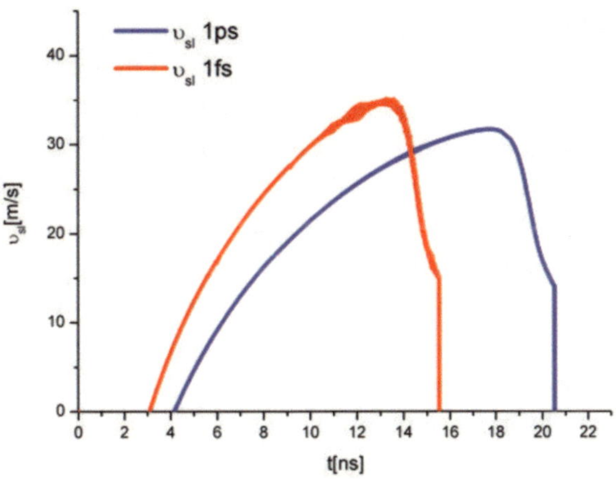

Figure 25. The time profiles of the velocity of solidification of copper.

Modern experimental methods of the investigation of the processes of melting-crystallization of the metal systems determine the velocities of the solid-melt front to be within the range of $10 \div 100$ m/s with corresponding overcooling of liquid within the range of $10 \div 400 K$. Thus, the data from the modeling of the metastable overcooled states in the liquid phase of metals is in relatively good agreement with the experimental measurements. It is not possible to perform the same comparison with the overcooled states in the solid phase because of the lack of reliable experimental data which is caused by the complexity of statement of such experiments.

6. Conclusion

The detailed analysis of the physical processes of the non-equilibrium heating and the fast phase transformations in the metal systems under the action of ultra-short laser pulses was performed in this chapter. A mathematical model was built based on the brief analysis of the thermodynamical and kinetic approaches for the description of the phase transformations of the first order. This model is a non-equilibrium hydrodynamic version of the Stefan problem and describes the kinetics of high-speed heterogeneous phase transformations: melting, crystallization, evaporation. The model is supplemented by the explicit description of the moving phase fronts for correct reflection of the dynamics of the phase transitions.

The methods of mathematical modeling were used to investigate the regimes of pico- and femtosecond action on metals - Al, Cu. The kinetics and dynamics of phase transformations are analyzed. The results of the modeling confirmed the skewness of the high-speed processes of melting – solidification. In particular, the maximum velocity of melting reaches the value of several kilometers per second, while the maximum velocity of solidification does not exceed hundred of meters per second. The temperature dependence $\upsilon_{s\ell}(T_{s\ell})$ does not have a sharp bend in the point of change of the direction of the phase transition.

Special attention is paid to the overheated and overcooled metastable states that are caused by the fast phase transitions. The deviation from the local thermodynamical equilibrium is noticed both in the bulk and at the interphase boundary. The response function $T_{s\ell}(\upsilon_{s\ell})$ has a linearly increasing character for low values of velocity $\upsilon_{s\ell}$ in the range of $[0 \div \pm 50]\,m/s$. As the velocity $\upsilon_{s\ell}$ rises more than $\upsilon_{s\ell} > 100 m/s$, a maximum is formed in the response function for both metals. The appearance of the maximum means that the dominating mechanism of the phase transition changes: the mechanism of heat conductivity is changed by the hydrodynamic one. The maximum value of overheating of the interphase boundary reaches the value of 200 ÷ 400 K, the one for overcooling is by one order lower than that value. The overheating and overcooling of the bulk are significantly higher - $4\cdot 10^3 \div 5\cdot 10^3 K$ for overheating and $6\cdot 10^2 \div 8\cdot 10^2 K$ for overcooling. The results for different materials are compared between each other and with experimental data.

These results do not pretend to be an exhaustive description of the variety of fast phase transitions, because they used the assumption of the thermodynamic stability of overheated / overcooled metastable states during all times considered. Refusal of such assumptions will require further research to include the kinetics of homogeneous phase transformations into the consideration.

Author details

V.I. Mazhukin
Keldysh Institute of Applied Mathematics of RAS, Moscow, Russia

Acknowledgement

This work was supported by the RFBR (projects nos. 10-07-00246-a, 12-07- 00436-a, 11-01-12086-ofi-m).

7. References

[1] Fujimoto J.G., Liu J.M., Ippen E.P., Bloembergen N (1984) Femtosecond Laser Interaction with Metallic Tungsten and Nonequilibrium Electron and Lattice Temperatures. Physical Review Letters. 53: 19/1837 – 1840.
[2] Chichkov B.N., Momma C., Nolte S., von Alvensleben F, Tunnermann A. (1996) Femtosecond, picosecond and nanosecond laser ablation of solids. Appl. Phys. A 63: 109–115.
[3] Shirk M.D., Molian P.A. (1998) A review of ultrashort pulsed laser ablation of materials. Journal of Laser Applications. 10: 1/18-28.
[4] Von der Linde D., Sokolovski-Tinten K. (2000) The physical mechanisms of short-pulse laser ablation. Appl. Surf. Sci. 154-155: 1-10.

[5] (2007) Part 2: Ultrafast Interactions. In: Claude R. Phipps, editor. Optical Sciences. Springer Series in Laser ablation and its applications. Santa Fe, New Mexico: Springer. 129: 121-280.

[6] Gonzales P., Bernath R., Duncan J., Olmstead T., Richardson M. (2004) Femtosecond ablation scaling for different materials. Proc. SPIE. 5458: 265-272.

[7] Chichkov B.N., Korte F., Koch J., Nolte S., Ostendorf A. (2002) Femtosecond laser ablation and nanostructuring. Proc. SPIE. 4760: 19-27.

[8] Hamad S., Krishna Podagatlapalli G., Sreedhar S., Tewari S. P., Rao S. V. (2012) Femtosecond and picosecond ablation of aluminum for synthesis of nanoparticles and nanostructures and their optical characterization. Proc. SPIE 8245: 82450L.

[9] Ostendorf A., Kamlage G., Klug U., Korte F., Chichkov B. N. (2005) Femtosecond versus picosecond laser ablation. Proc. SPIE 5713:1- 8.

[10] Kamlage G., Bauer T., Ostendorf A., Chichkov B.N. (2003) Deep drilling of metals by femtosecond laser pulses. Applied Physics A: Materials Science & Processing. 77: 307-310.

[11] Lopez J., Kling R., Torres R., Lidolff A., Delaigue M., Ricaud S., Hönninger C., Mottay E. (2012) Comparison of picosecond and femtosecond laser ablation for surface engraving of metals and semiconductors. Proc. SPIE 8243: 82430O.

[12] Bulgakov A.V., Bulgakova N. M. Burakov I.M. (2009) Nanosized Material Synthesis by Action of High – Power Energy Fluxes on Matter. Novosibirsk: Institute Thermophysics SB RAS. 461 p.

[13] Mazhukin V.I., Lobok M.G., Chichkov B.N. (2009) Modeling of fast phase transitions dynamics in metal target irradiated by pico and femto second pulsed laser. Applied Surface Science. 255: 5112-5115.

[14] Herlach D.M., Galenko P.K., Holland-Moriz D. (2005) Metastable materials from undercooled metallic melts. Amsterdam: Elsevier. 485 p.

[15] Eliezer S., Eliaz N., Grossman E., Fisher D., Gouzman I., Henis Z., Pecke S., Horovitz Y., Fraenkel M., Maman S., Lereah Y. (2004) Synthesis of nanoparticles with femtosecond laser pulses. Phys. Rev. B 69: 144119/1-6.

[16] Mene′ndez-Manjo′n A., Barcikowski S., Shafeev G.A., Mazhukin V.I., Chichkov B.N. (2010) Influence of beam intensity profile on the aerodynamic particle size distributions generated by femtosecond laser ablation. Laser and Particle Beams. 28: 45–52.

[17] Stratakis E., Barberoglou M., Fotakis C., Viau G., Garcia C., Shafeev G. A. (2009) Generation of Al nanoparticles via ablation of bulk Al in liquids with short laser pulses. Opt. Express. 17: 12650-12659.

[18] Mazhukin V.I., Samarskii A.A. (1994) Mathematical Modeling in the Technology of Laser Treatments of Materials. Review. Surveys on Mathematics for Industry. 4: 2/85-149.

[19] Ivanov D.S., Zhigilei L.V. (2003) Combined atomistic-continuum modeling of short-pulse laser melting and disintegration of metal films. Phys. Rev. B. 68: 064114-1 – 064114-22.

[20] . Landau L. D., Lifshic E. M. (1976) Teoreticheskaja fizika, t. V. Statisticheskaja fizika. Chast' 1. M.: Nauka. 616 p.

[21] Christian J. W. (1965) The theory of transformations in metals and alloys: an advanced textbook in physical metallurgy. Oxford: Pergamon Press. 975 p.

[22] Kirkpatrick R. J. (1975) Crystal Growth from the Melt. Review. American Mineralogist A: 60: 798-614.

[23] Asta M., Beckermann C., Karma A., Kurz W., Napolitano R., Plapp M., Purdy G., Rappaz M., Trivedi R. (2009) Solidification microstructures and solid-state parallels: Recent developments, future directions. Acta Materialia: 57: 146/941–971.

[24] Jackson K. A. (2004) Kinetic Processes:Crystal Growth, Diffusion, and Phase Transitions in Materials. Weinheim: Wiley-VCH Verlag GmbH & Co. KGaA. 409 p.

[25] Hoyt J. J., Asta M. (2002) Atomistic computation of liquid diffusivity, solid-liquid interfacial free energy, and kinetic coefficient in Au and Ag. Phys. Rev. B. 65: 21/214106-1 - 214106-11.

[26] Rodway G.H., Hunt J.D. (1991) Thermoelectric investigation of solidification of lead I. Pure lead. Journal of Crystal Growth. 112: 2–3/554–562.

[27] Hoyt J.J., Asta M., Karma A. (2002) Atomistic Simulation Methods for Computing the Kinetic Coefficient in Solid-Liquid Systems. Interface Science. 10: 2-3/181-189.

[28] Buta D., Asta M., Hoyt J. J. (2007) Kinetic coefficient of steps at the Si(111) crystal-melt interface from molecular dynamics simulations. Journal of Chemical Physics. 127: 7/074703 – 074713.

[29] Amini M., Laird B. B. (2006) Kinetic Coefficient for Hard-Sphere Crystal Growth from the Melt. Physical Review Letters. 97: 216102-1 - 216102-4.

[30] Monk J., Yang Y., Mendelev M. I., Asta M., Hoyt J. J., Sun D. Y. (2010) Determination of the crystal-melt interface kinetic coefficient from molecular dynamics simulations. Modelling and Simulation in Materials Science and Engineering. 18: 015004-1 - 015004-18.

[31] Wilson H.A. (1900) On the velocity of solidification and viscosity of supercooled liquids. Philos. Mag. 50: 238-250.

[32] Frenkel Ja. I. (1932) Note on the relation between the speed of crystallization and viscosity. Phys. Z. Sowjet Union. 1: 498 – 499.

[33] Frenkel' Ja.I. (1975) Kineticheskaja teorija zhidkostei. L.: Nauka. 592 p.

[34] Jackson K.A., Chalmers B. (1956) Kinetics of solidification. Can. J. Phys. 34: 473 – 490.

[35] Chalmers B. (1964) Principles of Solidification., New York: John Wiley &Sons. 129 p.

[36] Klude M. D., Ray J.R. (1989) Velocity versus Temperature Relation for Solidification and Melting of Silicon: A Molecular-Dynamics Study. Phys. Rev. B. 39: 3/1738-1746.

[37] Tymczak C.J. Ray J.R. (1990) Interface response function for a model of sodium: A molecular dynamics study, J. Chem. Phys. 92: 12/7520 – 7530.

[38] Tymczak C.J. Ray J.R. (1990), Asymmetric crystallization and melting kinetics in sodium: A molecular-dynamics study. Phys. Rev. Lett. Vol. 64, Issue 11, pp. 1278–1281.

[39] MacDonald C. A., Malvezzi A. M., Spaepen F. (1989) Picosecond time-resolved measurements of crystallization in noble metals. J. Appl. Phys. 65: 1/129 -137.

[40] Hertz H. (1882) Uber die Verdunstung der Flussigkeiten, insbesondere des Quecksilbers, im luftleeren Raume. Ann. Phys. und Chemie. 17: 177-190.

[41] Knudsen M. (1915) Die Maximale Verdampfungs-geschwindigkeit Quecksilbers. Ann. Phys. und Chemie. 47: 697-705.

[42] Siewert C.E., Thomas J.R. (1981) Strong evaporation into a half space. J. of Applied Mathem. and Physics(ZAMP). 32: 421 – 433.

[43] Siewert C.E. (2003) Heat transfer and evaporation/condensation problems based on the linearized Boltzmann equation. European Journal of Mechanics B/Fluids. 22: 391–408.

[44] Frezzotti A. (2007) A numerical investigation of the steady evaporation of a polyatomic gas. European Journal of Mechanics B/Fluids. 26: 93–104.

[45] Titarev V. A., Shakhov E. M. (2001) Teplootdacha i isparenie s ploskoi poverkhnosti v poluprostranstvo pri vnezapnom povyshenii temperatury tela. Mekhanika zhidkosti i gaza. 1/141 - 153.

[46] Abramov A. A., Butkovsky A. V. (2009) Effect of the Mode of Gas-Surface Interaction on Intensive Monatomic Gas Evaporation. Fluid Dynamics. 44: 1/80–87.

[47] Bond M., Struchtrup H. (2004) Mean evaporation and condensation coefficients based on energy dependent condensation probability. Phys. Rev. E. 70: 061605-1 – 20.

[48] Ford I J, Lee T-L. (2001) Entropy production and destruction in models of material evaporation. J. Phys. D: Appl. Phys. 34: 413–417.

[49] Rose J. W. (2000) Accurate approximate equations for intensive sub-sonic evaporation. Int. Journal of Heat and Mass Transfer.. 43: 3869 - 3875.

[50] Bedeaux D., Kjelstrup S., Rubi J. M. (2003) Nonequilibrium translational effects in evaporation and condensation. J. Chemical Physics. 119: 17/9163 – 9170.

[51] Frezzotti A., Gibelli L., Lorenzani S. (2005) Mean field kinetic theory description of evaporation of a fluid into vacuum. Phys. Fluids. 17: 012102-1 - 012102-12.

[52] Rahimi P., Ward Ch. A. (2005) Kinetics of Evaporation: Statistical Rate Theory Approach Int.J. of Thermodynamics. 8: 1/1-14.

[53] Meland R., Frezzotti A., Ytrehus T., Hafskjold B. (2004) Nonequilibrium molecular-dynamics simulation of net evaporation and net condensation, and evaluation of the gas-kinetic boundary condition at the interphase. Phys. Fluids. 16: 2/223 – 243.

[54] Ishiyama T., Yano T., Fujikawa S. (2005) Kinetic boundary condition at a vapor–liquid interface. Phys. Rev. Lett. 95: 0847504.

[55] Frezzotti A. (2011) Boundary conditions at the vapor-liquid interface. Phys. Fluids. 23: 030609-(1 -9).

[56] Kartashov I.N., Samokhin A.A., Smurov I.Yu. (2005) Boundary conditions and evaporation front instabilities. J. Phys. D: Appl. Phys. 38: 3703-3714.

[57] Crout D. (1936) An application of kinetic theory to the problem of evaporation and sublimation of monatomic gases. J. Math. Physics. 15: 1-54.

[58] Labuntsov D.A., Kryukov A.P. (1979) Analysis of Intensive Evaporation and Condensation. Inter. J. Heat and Mass Transfer. 22: 7/989–1002.

[59] Mazhukin V.I., Samokhin A.A. (1987) Matematicheskoe modelirovanie fazovykh perekhodov i obrazovanija plazmy pri deistvii lazernogo izluchenija na pogloshhajushhie sredy. In: Samarskii A.A., Kurdjumov S.P., Mazhukin V.I., editors. Nelineinye differencial'nye uravnenija matematicheskoi fiziki. Moskva: Nauka. pp. 191-244.

[60] Samokhin A.A. (1990) Effect of laser radiation on absorbing condensed matter. In: Prokhorov A.M. Series editor. Fedorov V.B. editor. Proceedings of the Institute of General Physics Academy of Sciences of the USSR. New-York: Nova Science Publishers. 13: 203 p.

[61] Anisimov S.I. (1968) Ob isparenii metalla, pogloshhajushhego lazernoe izluchenie. ZhEhTF. 54: 1/339-342.

[62] Khight C.J. (1979) Theoretical Modeling of Rapid Surfact Vaporation with Back Pressure. AIAA J. 17: 5/81-86.

[63] Mazhukin V.I., Prudkovskii P.A., Samokhin A.A. (1993) O gazodinamicheskikh granichnykh uslovijakh na fronte isparenija. Matematicheskoe modelirovanie. 5: 6/3-9.

[64] Mazhukin V.I., Samokhin A.A. (2012) Boundary conditions for gas-dynamical modeling of evaporation processes. Mathematica Montisnigri. XXIV: 8-17.

[65] Lifshic I.M., Kaganov M.I., Tanatarov L.V. (1956) Relaksacija mezhdu ehlektronami i kristallicheskojj reshetkojj. ZhEhTF. 31: 2(8)/232-237.

[66] Lifshic I.M., Kaganov M.I., Tanatarov L.V. (1959) K teorii radiacionnykh izmenenijj v metallakh. Atomnaja ehnergija. 6/391-402.

[67] Martynenko Ju.V., Javlinskii Ju.N. (1983) Okhlazhdenie ehlektronnogo gaza metalla pri vysokojj temperature. DAN SSSR. 270: 1/88-91.

[68] Beer A.C., Chase M.N., Choquard P.F. (1955) Extension of McDougall-Stoner tables of the Fermi-Dirac functions. Helvetica Physica Acta. XXVII: 5/529-542.

[69] Kittel C. (1978) Introduction to Solid State Physics, 6h Edition. New York: J. Wiley & Sons. 789 p.

[70] Landau L.D., Lifshic E.M. (1963) Teoreticheskaja fizika. Kvantovaja mekhanika. Nereljativistskaja teorija. T.III. Moskva: Fizmatgiz. 702 p.

[71] Frenkel' Ja.I. (1958) Vvedenie v teoriju metallov. Moskva: Fizmatlit. 368 p.

[72] Touloukian Y. S., Powell R. W., Ho C. Y. and Klemens P. G (1970) Thermophysical Properties of Matter Volume 1: Thermal Conductivity: Metallic Elements and Alloys, New York, Washington: IFI/Plenum Data Corp.

[73] Ziman J.M (1974) Principles of the theory of solid. Gambridge: University Press. 472 p.

[74] Cheynet, B., J.D. Dubois, M. Milesi (1996) Données thermodynamiques des éléments chimiques. In : Techniques de L'Ingenieur, Materiaux metalliques. No. M153. Forum M64. pp. 1-22

[75] Aluminum: Properties and physical metallurgy (1984) In: John E. Hatch editor. American society for metals. Ohio: Metals Park. 421 p.

[76] Fizicheskie velichiny. Spravochnik (1991) Grigor'eva I.S., Mejjlikhova E.Z. editors. Moskva: Ehnergoatomizdat. 1232 p.

[77] Iida T., Guthrie R.I.L. (1988) The Physical Properties of Liquid Metals. Oxford: Clarendon Press. 288 p.

[78] Pines D (1965) Ehlementarnye vozbuzhdenija v tverdykh telakh. Moskva: Mir. 383 p.

[79] Qiu T.Q., Tien C.L (1994) Femtosecond Laser Heating of Multi-Layer Metals. I. Analysis. Int. J. Heat Mass Transfer. 37: 17/2789-2797.

[80] Ginzburg V.L., Shabanskii V.L (1955) Kineticheskaja temperatura ehlektronov v metallakh i anomal'naja ehlektronnaja ehmissija. Dokl. AN SSSR. 100: 3/445 - 448.

[81] Martynenko Ju.V., Javlinskii Ju.N (1987) Vozbuzhdenie ehlektronov metalla oskolkom delenija. Atomnaja ehnergija. 62: 2/80-83.

[82] Lin Z., Zhigilei L.V., Celli V (2008) Electron-phonon coupling and electron heat capacity of metals under conditions of strong electron-phonon nonequilibrium. Phys. Rev. B. 77: 7/075133-1 - 075133-17.

[83] Mazhukin V.I., Mazhukin A.V., Demin M.M., Shapranov A.V. (2011) Ehffekty neravnovesnosti pri vozdejjstvii impul'snogo lazernogo izluchenija na metally. Opticheskiy zhurnal. 78: 8/29-37.

[84] Mazhukin V.I., Mazhukin A.V., Demin M.M., Shapranov A.V. (2011) Mathematical modeling of short and ultrashort laser action on metals. In: Sudarshan T.S., Beyer Eckhard, Berger Lutz-Michael editors. Surface Modification Technologies (STM 24). Dresden: Fraunhofer Institute for Material and Beam Technology IWS. 24: 201 -208.

[85] Samarskii A.A., Moiseenko B.D (1965) Ehkonomichnaja skhema skvoznogo scheta dlja mnogomernojj zadachi Stefana. ZhVM i MF. 5: 5/816-827.

[86] Meyer G.H (1973) Multidimensional Stefan Problems. SIAM J. Numer. Anal. 10: 3/522-538.

[87] White B.R.E (1983) A modified finite difference scheme for the Stefan problem. Mathem. of Comput. 41: 164/337-347.

[88] Klimontovich Ju.L (1982) Statisticheskaja fizika. Moskva: Nauka. 608 p.

[89] Mazhukin V.I., Smurov I., Flamant G (1995) Overheated Metastable States in Pulsed Laser Action on Ceramics. J. Applied Physics. 78: 2/1259-1270.

[90] Mazhukin V.I., Smurov I., Dupuy C., Jeandel D (1994) Simulation of Laser Induced Melting and Evaporation Processes in Superconducting. J. Numerical Heat Transfer Part A. 26: 587-600.

[91] V.I.Mazhukin, M.G.Lobok, B.N.Chichkov (2009) Modeling of fast phase transitions dynamics in metal target irradiated by pico and femto second pulsed laser. Applied Surface Science. 255/5112-5115.

[92] Mazhukin V.I., Samarskii A.A., Chujjko M.M (1999) Metod dinamicheskojj adaptacii dlja chislennogo reshenija nestacionarnykh mnogomernykh zadach Stefana. Doklady RAN. 368: 3/307 - 310.

[93] Mazhukin A.V., Mazhukin V.I (2007) Dinamicheskaja adaptacija v parabolicheskikh uravnenijakh. Zhurnal vychislitel'nojj matematiki i matematicheskojj fiziki. 47: 11/1911 – 1934. Mazhukin A.V., Mazhukin V.I (2007) Dynamic Adaptation for Parabolic Equations. Computational Mathematics and Mathematical Physics. 47: 11/1833 – 1855.

[94] Breslavsky P.V., Mazhukin V.I (2008) Metod dinamicheskojj adaptacii v zadachakh gazovojj dinamiki s nelinejjnojj teploprovodnost'ju. Zhurnal vychislitel'nojj matematiki i matematicheskojj fiziki. 48: 11/2067–2080. Breslavskii P. V. and Mazhukin V. I (2008) Dynamic Adaptation Method in Gasdynamic Simulations with Nonlinear Heat Conduction. Computational Mathematics and Mathematical Physics. 48: 11/2102–2115.

[95] Koroleva O.N., Mazhukin V.I (2006) Matematicheskoe modelirovanie lazernogo plavlenija i isparenija mnogoslojjnykh materialov. Zhurnal vychislitel'nojj matematiki i matematicheskojj fiziki. 46: 5/910 – 925. Koroleva O.N., Mazhukin V.I (2006)

Mathematical Simulation of Laser Induced Melting and Evaporation of Multilayer Materials. Computational Mathematics and Mathematical Physics. 46: 5/848 – 862.

[96] Mazhukin V.I., Samarskii A.A., Kastel'janos Oplando, Shapranov A.V (1993) Metod dinamicheskojj adaptacii dlja nestacionarnykh zadach's bol'shimi gradientami. Matematicheskoe modelipovanie. 5: 4/32-56.

[97] Samarskii A.A (1989) Teorija raznostnykh skhem. Moskva: Nauka. 616 p.

[98] Samarskii A.A., Popov Ju.P (1992) Raznostnye metody reshenija zadach gazovojj dinamiki. Moskva: Nauka. 421pp.

[99] Mazhukin V.I., Mazhukin A.V., Koroleva O.N (2009) Optical properties of electron Fermi-gas of metals at arbitrary temperature and frequency. Laser Physics. 19: 5/1179 – 1186.

[100] Mazhukin A.V., Koroleva O.N (2010) Raschjot opticheskikh kharakteristik aljuminija. Matematicheskoe modelirovanie. 22: 5/15-28.

Chapter 9

Direct Writing in Polymers with Femtosecond Laser Pulses: Physics and Applications

Kallepalli Lakshmi Narayana Deepak, Venugopal Rao Soma
and Narayana Rao Desai

Additional information is available at the end of the chapter

http://dx.doi.org/10.5772/45854

1. Introduction

Nonlinear optical phenomena in the optical spectral range followed by the invention of laser in early 1960's directed the generation of optical pulses using Q switching and mode locking techniques. Ultrafast lasers with extremely short pulse duration (<100 fs) opened a new avenue towards fabrication of integrated photonic and signal processing devices in a variety of transparent materials. A new approach for the local modification of transparent materials through nonlinear optical processes has been investigated due to extraordinarily high peak intensities of short pulses. A variety of materials including metals, dielectrics, polymers, and semiconductors have been successfully processed by the use of fs pulses [1-14]. Bulk refractive index change in transparent materials is found to be useful in applications of waveguides. Various applications resulting from fs laser writing of different materials, especially in polymers, have been successfully demonstrated in the fields of micro-fluidics, bio-photonics, and photonics etc. [15-32]. The minimal damage arising from the generation of stress waves, thermal conduction, or melting has proved to be one of the main responsible mechanisms for various applications demonstrated using fs laser micromachining. In the present chapter we discuss the formation of free radicals and defects which are responsible for emission in polymer systems. The impact of fs lasers pulses causing minimal damage can be utilized to fabricate emissive micro-craters, especially in polymers. These emissive micro-craters find prospective applications in memory based devices.

2. Experimental

Herein we present detailed micro-structure fabrication procedures and spectroscopic investigations of those structures in four different polymers (a) Poly Methyl Methacrylate

(PMMA) (b) Poly Di Methyl Siloxane (PDMS) (c) Polystyrene (PS) and (d) Poly Vinyl Alcohol (PVA). The spectroscopic investigations were carried out in both bulk and thin films of polymers. We used a Ti: Sapphire laser delivering 100 fs pulses at 800 nm with an energy of 1 mJ per pulse at 1 kHz repetition rate. The input energy was varied using half wave plate and polarizer combination. Three Newport stages with 15 nm resolution were arranged three dimensionally and the translation was controlled using computer controlled program. We used 40X and 20X microscope objectives with 0.65 and 0.4 numerical apertures, respectively, to focus the laser beam in our experiments. CCD camera was used while Z stage of the micro-fabrication set up was adjusted for fabricating structures either on the surface or inside the bulk. Figure 1 shows schematic diagram of the micro-fabrication setup. As shown in figure 1 the laser pulses were passed through half wave plate (HWP) and polarizer (BP) so that energy could be varied using this combination. Apertures were introduced in middle to align the beam. M1-M3 are the mirrors used to make the beam incident vertically onto the three dimensionally arranged Newport stages. We fabricated several micro-structures, micro-craters, surface and subsurface diffraction gratings, surface grids, microfluidic channels and arbitrary shapes in these polymers. Since these polymers are transparent and have large band gap, we calculated the Keldysh parameter for these polymers to assess the dominant ionization mechanism. Tunneling ionization was found to be the main responsible mechanism for all the investigated polymers. The spectroscopic investigations were carried out in both bulk and thin films of polymers. Bulk PMMA and PS were purchased from Goodfellow, USA and UK. PVA thin films were prepared by dissolving 8.56 grams of PVA beads in 100 ml of water and thin films were prepared using spin coating technique. Thin films of PS are made by preparing the PS solution first. Solution of PS was prepared by mixing 1 gram of polystyrene beads (ACROS) in 8 ml toluene and stirred for 48 hours for complete miscibility. We prepared thin films of PS on a glass plate by spin coating the solution of PS. All these polymers were cut into 1 cm × 1 cm square pieces, polished using different grades of polishing sheets and sonicated in distilled water before micro fabrication experiments were carried out.

The energy of an 800 nm photon corresponds to 1.55 eV while the optical band gap of pure PMMA being 4.58 eV implies that the nonlinear process involving at least three photons is responsible for structural modification at the focal volume [33]. Schaffer et al. [34] have shown that there are three possible mechanisms viz. tunneling, intermediate, and multi-photon ionizations that take place when transparent material interacts with femtosecond pulses. The Keldysh parameter which tells us the mechanism that is dominant is defined as $\gamma = (\omega/e)\,(m\times c\times n\times \varepsilon_o \times E_g/I)$ where 'ω' is the laser frequency, 'I' is the laser peak intensity at the focus, 'm' and 'e' are the reduced mass and charge of the electron, respectively, and 'c' the velocity of light, n is the refractive index of the material, E_g is the band-gap of the material and ε_0 is the permittivity of free space. For our studies that Keldysh parameter was <1.5 illustrating tunneling as responsible mechanism for structures written using 40X and 20X microscopic objectives. Figure 2 shows the plot obtained for Keldysh parameter with different energies ranging from 150 μJ to 30 nJ (equivalent to 83 to 0.017 PW/cm^2). Results obtained with different polymers using fs laser direct writing are explained in detail in references 35-41.

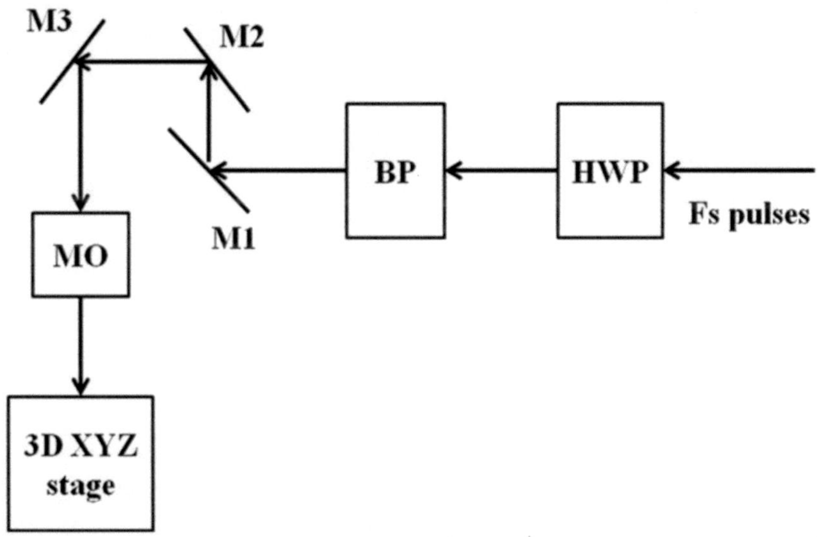

Figure 1. Experimental setup for microfabrication

Our initial studies suggested an increase in structure width with number of scans, energy, and focusing conditions. Figure 3(a) shows some of the Field Emission Scanning Electron Microscope (FESEM) images of the fabricated microstructures on PVA thin film (60-80 μm thickness). We clearly observed the formation of a trough in the central portion of each structure. It is evident as central portion of the incident Gaussian pulse has more intensity (thereby affecting the polymer more) resulting in void formation. This phenomenon was observed for energies ranging from 100 μJ to 10 μJ. Figure 3(b) shows evidently the formation of trough in the central portion of the microstructure fabricated at 100 μJ energy with 1 mm/s speed. For structures fabricated at low energies (typically less than 10 μJ) formation of trough was not observed. Figure 3(c) shows a plot of width of microstructure as well as trough obtained at different energies for different sets. Formation of micro-craters was seen at 1 μJ energy and with 1 mm/s speed. Structures fabricated with 635 and 564 nJ obviously demonstrated formation of micro-craters. Figure 4(a) shows the structure fabricated at 1 μJ energy, while 4(b) and 4(c) show the FESEM images of the microstructures fabricated at 635 and 564 nJ energies with 1 mm/s speed. In our experiments with different polymers we observed that the effect of fs pulses on these polymers, leading to the formation of defects resulting in emission, formation of paramagnetic centers, and broadening of Raman vibrational modes in the fs laser modified regions. Figure 5 shows the part of shape 8 fabricated in PMMA. Since the image was big only a part of 8 shape, which is curved, is shown. We captured the emission from the fs laser modified region using a confocal microscope. We could clearly observe pseudo-green color from the fs laser modified regions which is an indication of emission. Similar results were obtained in case of other polymers also, details of which are summarized in references 35-41. By combining the formation of micro-craters (which took place at high scan speed >3 mm/s and low energies of ~nJ) and the emission coming from the modified regions we establish the possibility of using the fluorescent micro-craters towards memory based applications.

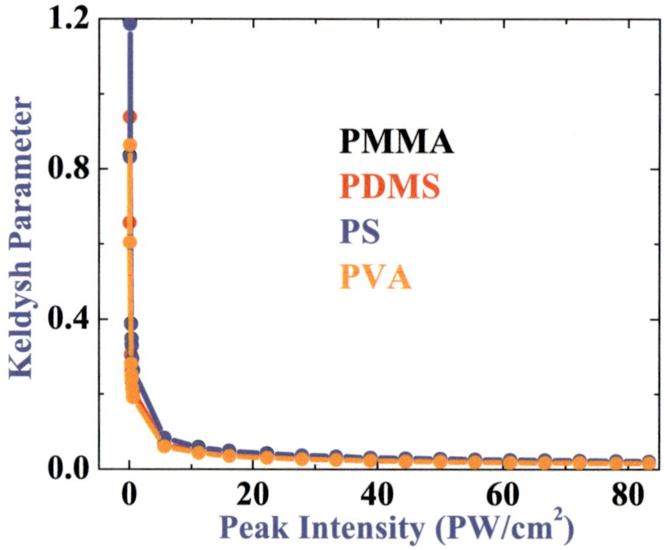

Figure 2. Keldysh parameter of investigated polymers with different peak intensities.

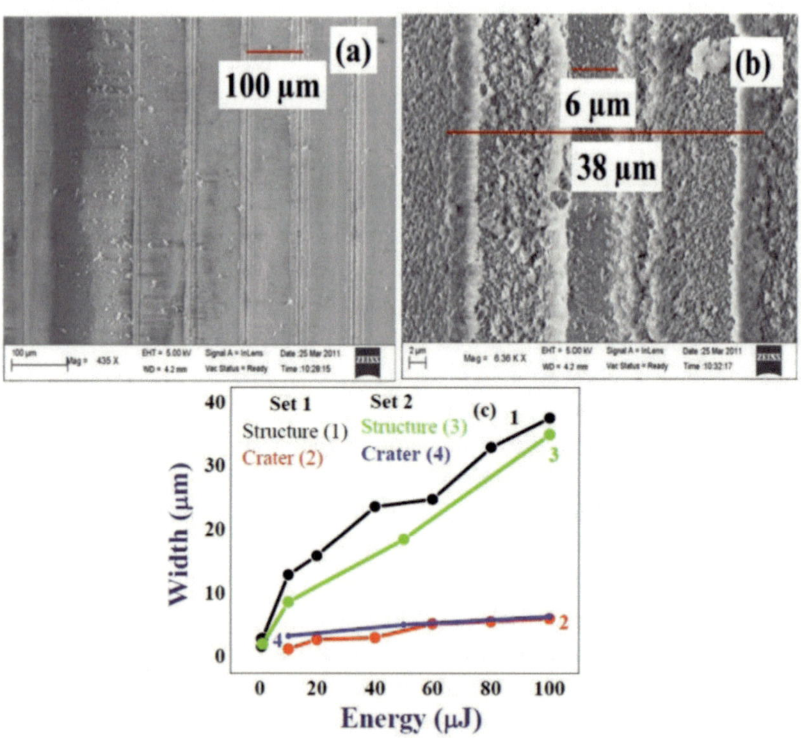

Figure 3. (a) FESEM images of structures fabricated. Structures from right to left were fabricated with energies from 100-10 µJ in steps of 20 µJ. (b) Microstructure fabricated at 100 µJ energy, 1 mm/s speed. (c) Plot of structure and trough widths with energy.

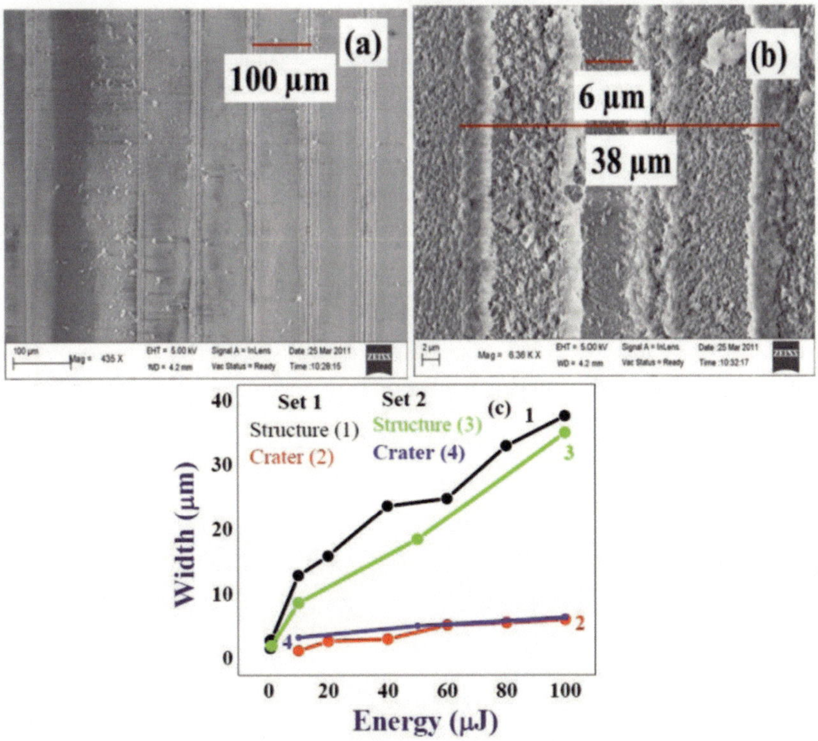

Figure 4. FESEM images microstructure fabricated at 1 mm/s speed, with energy of (a) 1 µJ, (b) 635 nJ, and (c) 564 nJ.

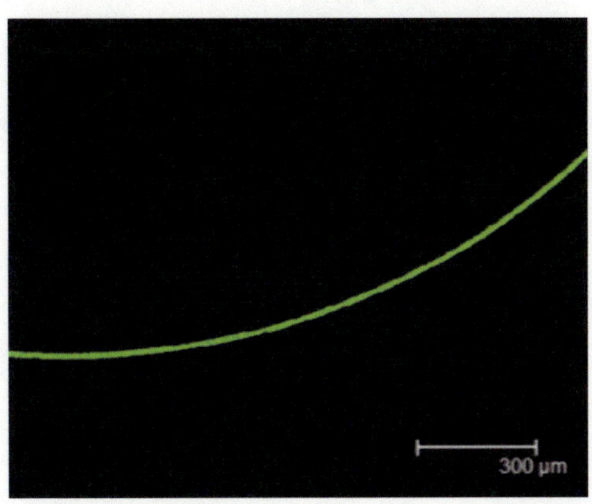

Figure 5. Part of the shape 8 fabricated in PMMA. Pseudo green color shows emission from the modified region excited at 488 nm. Scale bar is 300 µm.

Structures were also fabricated at different energies (1 µJ, 635 nJ, 168 nJ, and 33 nJ) and different scanning speeds to study the formation of micro-craters in PS. Figure 6(a) shows the confocal microscope image of a buried single microstructure in PS fabricated at 1 µJ energy and 1 mm/s speed. For the same energy we observed the appearance of craters when the scanning speed was increased to 2 mm/s. This is depicted clearly in figure 6(b), where the formation of micro-craters was just onset. The process of formation of micro-craters is evident from the edges of figure 6(b). Figures 6(c) and 6(d) illustrate the confocal microscope images of the craters formed at still higher scanning speeds of 3 and 4 mm/s, respectively. From these data we confirmed that by increasing the scanning speed one can reduce the pulse to pulse overlap and the micro-craters can be obtained similar to observations of periodic refractive index modulation observed earlier in glasses. The same trend was observed even at other energies also. However, we found that the formation of craters at higher scanning speeds and lower energies exhibited better profiles. This is obvious because the intensity of heat waves generated at each position where the pulse impinges depends on the pulse energy.

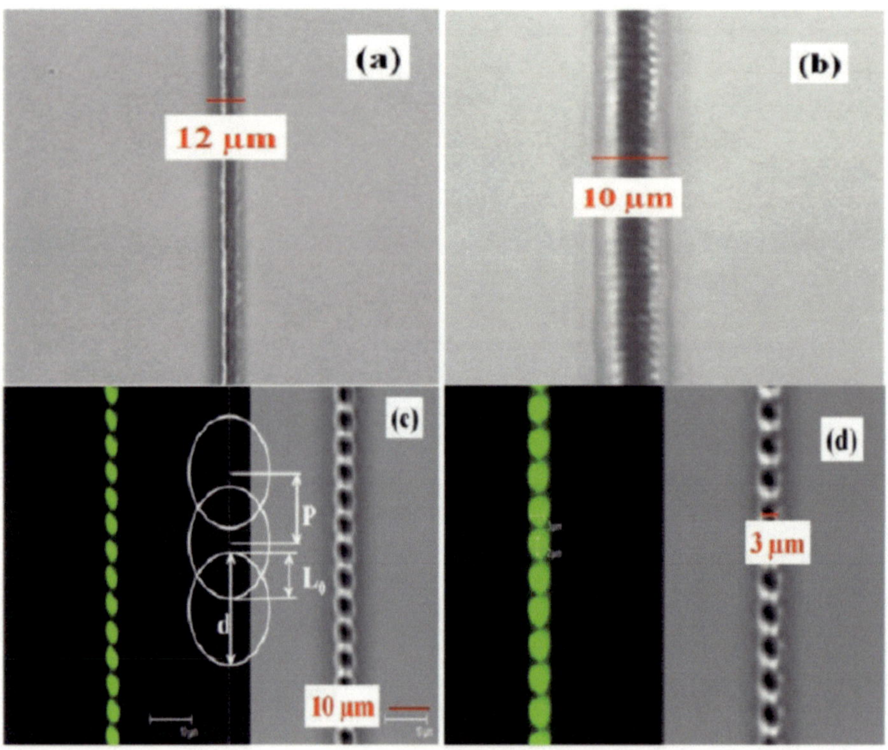

Figure 6. (a) Confocal microscope image of a buried single micro-structure in PS fabricated at 1 µJ energy and 1 mm/s speed. Structure width is ~12 µm. (b) Beginning of formation of micro- craters at the same energy when scanning speed is increased to 2 mm/s. Structure width 10 µm. (c) Scanning speed increased to 3 mm/s and the inset shows model used to fabricate micro- craters (d) Confocal microscope image of the craters formed at still higher scanning speed of 4 mm/s. Pseudo green color indicates emission from the modified micro-craters when excited at 488 nm wavelength. Crater size is 3 µm.

S. No.	Energy (nJ)	Speed (mm/s)	Observation (Line/Crater)	Structure/Crater Size (μm)
1	1000	1	Line	12
		1.5	Line	10
		2	Line	10
		2.5	Crater formation started	6
		3	Crater	6
		3.5	Crater	4
		4	Crater	3
		4.5	Crater	4
		5	Crater	3
2	635	1	Line	14
		1.5	Line	12
		2	Crater formation started	10
		2.5	Crater	3
		3	Crater	3
		3.5	Crater	3
		4	Crater	3
		4.5	Crater	3
		5	Crater	3
3	168	1	Line	4
		1.5	Crater formation started	3
		2	Crater	2
		2.5	Crater	2
		3	Crater	2
		3.5	Crater	2
		4	Crater	2
		4.5	Crater	2
		5	Crater	2
4	33	1	Line	3
		1.5	Crater formation started	3
		2	Crater	2
		2.5	Crater	2
		3	Crater	2
		3.5	Crater	2
		4	Crater	2
		4.5	Crater	2
		5	Crater	2

Table 1. Details of micro crater structures fabricated at different energies and scanning speeds.

Therefore, at higher energies these intense heat waves travel and overlap with the waves generated at other positions. This leads to the formation of a line instead of a crater. This aspect is highlighted in table 1. As energy is reduced, micro-crater formation occurred at lower scan speeds, indicating the fact that the effect of pulse to pulse overlap controlled by scan speed and the phenomena of generation/propagation of heat waves at each position are important factors in the fabrication of micro-craters. Inset of figure 6(c) shows a model that represents the formation of micro-craters. The pitch '**p**' which is the distance from center to center can be varied by setting the scanning speed according to the relation **p=s/f** where '**s**' is scanning speed and '**f**' is laser repetition frequency, which is 1 kHz in our case. L_0 is the overlap region between two consecutive focused spots which is (d-p) where d is the diameter of the spot. The spot size is calculated using the relation D=1.22 λ/NA where D is the diameter of the focused spot, λ is the wavelength and NA is the numerical aperture of the microscope objective used. We had used a 40X microscope objective with NA of 0.65 in our experiments. The estimated spot size was ~ 1.55 μm. In our experiments on micro-crater fabrication we observed clear micro-crater formation from 3 mm/s speed onward with different energies as tabulated in table 1. Hence the minimum pitch (**p**) can be taken as 3 μm.

As the radius of the focused spot was ~0.75 μm, the micro craters start appearing for speeds beyond 3 mm/s. However, we could obtain minimum crater size of ~2 μm in our experiments owing to the size of the focused spot and continuous translation. The fabrication of these micro craters is useful in the areas of memory storage devices and photonic crystals.

For better understanding about the emission originating from modified regions, we fabricated microstructures over a large area to record the emission and excitation spectra using conventional fluorescence spectrometer. To accomplish this, an array of closely spaced lines were fabricated in PMMA (energy 10 μJ, speed 1mm/s, period 30 μm), PDMS (50 μJ, 1mm/s, 10 μm), PS (1 μJ, 0.5 mm/s, 30 μm), and PVA (10 μJ, 1mm/s, 20 ◉m) using a 40X objective. All the four polymers are transparent to visible light before fs laser irradiation and hence do not show any emission when excited with visible light. However, due to multi-photon absorption overcoming the large band gaps of these polymers, they get modified and optical centers are produced that exhibit emission when excited at different excitation wavelengths. We observed this kind of emission when the fs laser fabricated channels were irradiated with visible wavelengths such as 458 nm, 488 nm, 514 nm wavelengths. The observation of emission and change in emission peak with the excitation wavelength were observed and could be interpreted in the context of formation of myriad optical centers or defects. The increase in emission intensity with irradiation dose in bulk and thin films of polymers are due to the formation of more number of optical centers with irradiation dose.

Nurmukhametov et al. [42] have exposed PS films and solutions with UV laser light beam (λ = 248 nm) and observed changes in absorption and luminescent properties. They observed formation of optical centers with absorption band from 280 nm to 460 nm with fluorescence band from 330 nm to 520 nm. They established different optical centers being responsible for such a behavior and hypothesized their results more close to the spectrum of diphenylbutadiene (DPBD). They figured out three optical centers responsible for the behavior of emission namely trans-stilbene, DPBD and diphenylhexatriene (DPHT). The absorption and emission spectra due to the fs laser irradiation in the present study match nearly with their reported work indicating the creation of similar optical centers. Figures 7(a) – 7(c) show the emission peaks are different for different excitation wavelengths. This emission was recorded using a confocal microscope. We feel that the modification of PS was due to 800 nm (infrared) laser through multi-photon absorption process. Consequently, fs laser irradiation results in bond scission followed by the formation of large number of defects that act as optical centers for emission. This resulted in shift in emission peak with excitation wavelength. The observed spectral peaks are mentioned in table 2. Observed spectral peaks from fluorescence spectrometer and the confocal microscope are also compared in the table.

S. No.	Excitation Wavelength (nm)	Emission peak (nm) observed in confocal microscope	Emission peak (nm) observed in fluorescence spectrometer
1	458	509, 544	505,536
2	488	535	550
3	514	565	563
4	400	Excitation source was not available	454, 462, 476

Table 2. Comparison of emission data obtained in irradiated PS (PSG1).

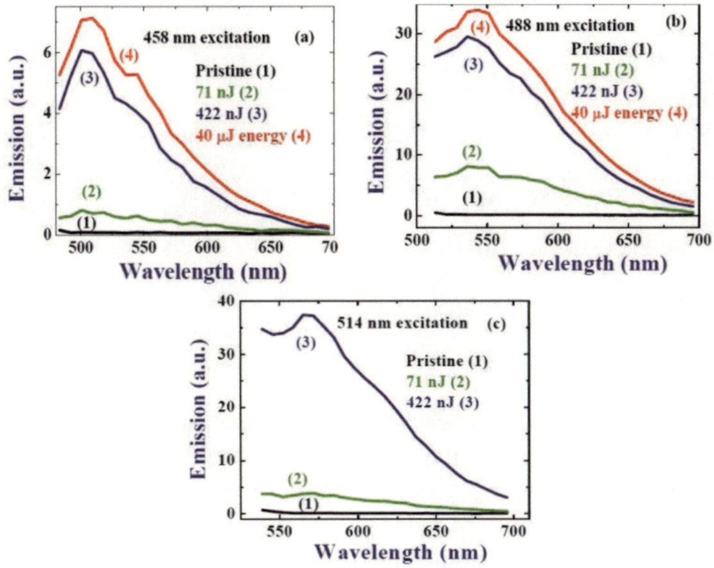

Figure 7. (a) Plots of emission when structures fabricated at 40 µJ, 422 nJ, 63 nJ and pristine region of PS excited at 458 nm; (b) Plots of emission when structures fabricated at 40 µJ, 422 nJ, 63 nJ and pristine region of PS excited at 488 nm; (c) Plots of emission when structures fabricated at 422 nJ, 63 nJ and pristine region of PS excited at 514 nm.

The effect of change in emission peak with excitation wavelength was observed in all polymers and is attributed to the Red Edge Effect (REE). The change in emission peak with excitation wavelength in case of PS is shown in figure 8(a). The excitation spectra collected at different monitoring wavelengths looked alike which is depicted in figure 8(b). Inset of figure 8(b) shows the transmission spectra of un-irradiated and irradiated PS. Varieties of fluoropores in different media with frozen or relatively slow structural dynamics from vitrified and highly viscous solutions to polymer matrices have shown similar effects. These phenomena did not account for concepts of independence of emission energy on excitation energy within the absorption band (Vavilov's law) and the occurrence of emission irrespective of excitation band, always from the lowest electronic and vibrational state of same multiplicity called Kasha's Rule [43-46]. These effects originate not from the violation of fundamental principles but from their operation in specific conditions when the ensemble of excited molecules is distributed in interaction energy with molecules in their surroundings. In a condensed medium, this distribution always exists at the time of excitation, but its display in a variety of spectroscopic phenomena depends on how fast the transitions are between the species forming this ensemble of states. Thus, REE is a site-selective effect which allows probing of the dynamics of redistribution of fluoropores between different environments.

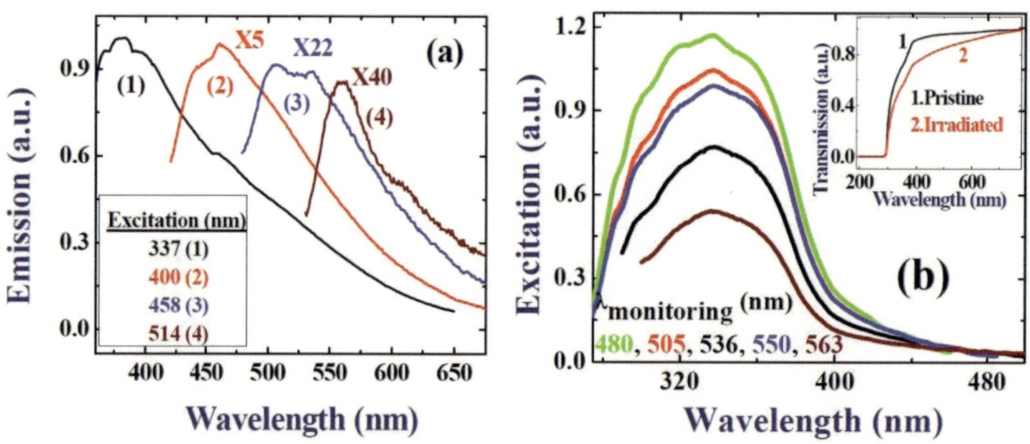

Figure 8. (a) Emission from PSG1 grating excited at different wavelengths. Emissions with longer wavelength excitations are enhanced for clarity (b) Excitation spectra of irradiated PS at different monitoring wavelengths. Inset of (b) shows the transmission spectra of unirradiated and irradiated PS.

The absorption characteristics of these optical centers thus formed showed maximum absorption at same wavelength. As the defect centers are distributed over wide range of energies, the emission spectrum too gets distributed over the entire fluorescence spectrum contribution coming from all types of defects. This results in the spectra being the same in the excitation spectra recorded by monitoring different emission wavelengths. However the emission, when excited at long wavelength side, due to localization of energy we observe different emission spectra. Figure 9(a) shows the plot of excitation wavelength with the observed emission wavelength for all polymers which is an indication of REE effect. Since these polymers contain active functional groups such as aldehyde, ketone etc., the absorption maximum within these functional groups can also lead to the observed emission. In the case of PMMA, the maximum excitation near 370 nm can be ascribed to n →π* transition. In case of PS due to the presence of ring, it could be due to π→π* transition. After these functional groups are excited they can get de-excited to any of the electronic states by emission process. Figure 9(b) shows a schematic diagram for the functional groups within these polymers undergoing different transitions. We also investigated the role of possible vibrational levels in emission spectra we recorded. For this, we excited all irradiated polymers from 250 nm-560 nm in the range of 20 nm and collected emission spectra. Interestingly, some of the emission peaks coincided with Raman peaks of these polymers. Plots 10 (a)-(b) show the emission spectra recorded for fs laser irradiated PMMA at different excitation wavelengths. The energy difference between the excitation and the emission peak is calculated in the wave numbers using the relation $\Delta E/hc\ (cm^{-1}) = 1/\lambda_{ext} - 1/\lambda_{em}$, where λ_{ext} and λ_{em} are excitation and emission wavelengths respectively. For PMMA, the emission peaks due to excitation wavelengths at 458, 488, 530, and 560 nm nearly match the Raman mode of characteristic peak at 1736 cm^{-1} of γ (C=O) of (C-COO) mode [47]. Excitations at 380, and 400 nm nearly match with Raman mode of 3454 cm^{-1} which is $2\gamma_2$ overtone of 1730 cm^{-1}. Emission peak at 543 nm matched with another Raman peak of combination band involving γ (C = C) and γ (C-COO) of PMMA. The spectra therefore contain the peaks due to transition from the excited state to the vibrational states. The shift in the emission peaks with excitation wavelength is shown in figure 10 (b). This shift in the peak of the emission with excitation could be due to the excitation of different localized states created during laser irradiation. Table 3 shows excitation, emission peaks and the calculated energy difference between the excitation and the emission peaks.

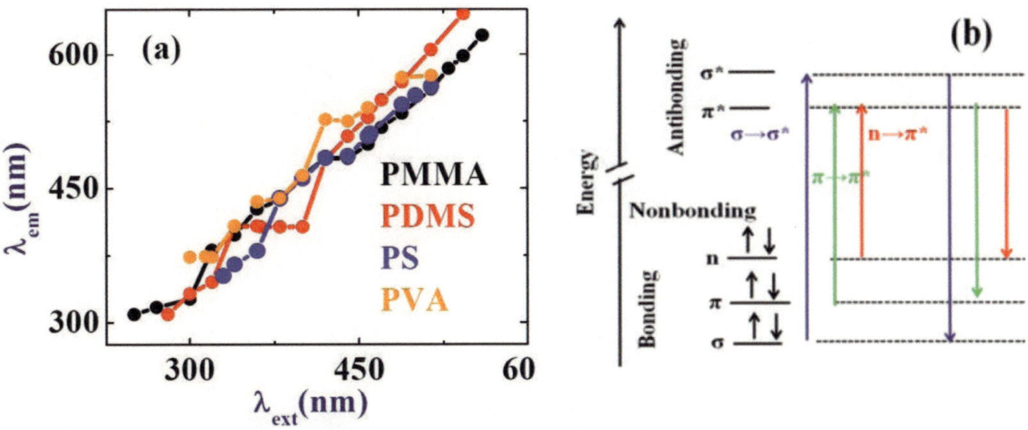

Figure 9. (a) REE in investigated polymers (b) Energy level diagram.

Different research groups have worked on electron spin resonance (ESR) of polymers [48-50]. However, reports on ESR analysis of fs laser irradiated polymers are sparse as the field has gained momentum recently and many theoretical and experimental results need to be explored further. In our endeavor towards understanding ESR of fs irradiated polymers, we fabricated two dimensional grids to increase effective area of irradiation for ESR analysis. We observed peroxide type free radicals when these polymers were treated with fs laser. Pure polymers such as PMMA are not paramagnetic substances and hence do not contain any paramagnetic centers. So, there were no peaks observed in ESR spectrum. When polymers PMMA, PDMS, and PS are treated with fs laser, they showed ESR signal which is an indication of existence of peroxide type free radicals. Figure 11 shows ESR signal in PS. In case of PS, the radicals are alkoxy radicals as reported in literature [51]. Further, confocal micro-Raman studies were carried out for fabricated structures in the channels. Formation of defects such as optical centers and free radicals led to broadening of Raman peaks and reduction in Raman intensity due to high intense shock waves formed at the center of the Gaussian pulse. In order to study the local effect, a microstructure was fabricated on a PVA thin film at two different energies (635 nJ and 10 μJ) with speed of 1 mm/s. Figure 12 shows the Raman spectra collected in the middle and end regions of the structures fabricated. Raman spectra recorded for higher energies showed the broadening of Raman modes along with decrease in Raman intensity compared with structures fabricated at low energies and pristine regions of the polymer from which we could conclude the formation of defects. We observed similar effects even in other polymers also.

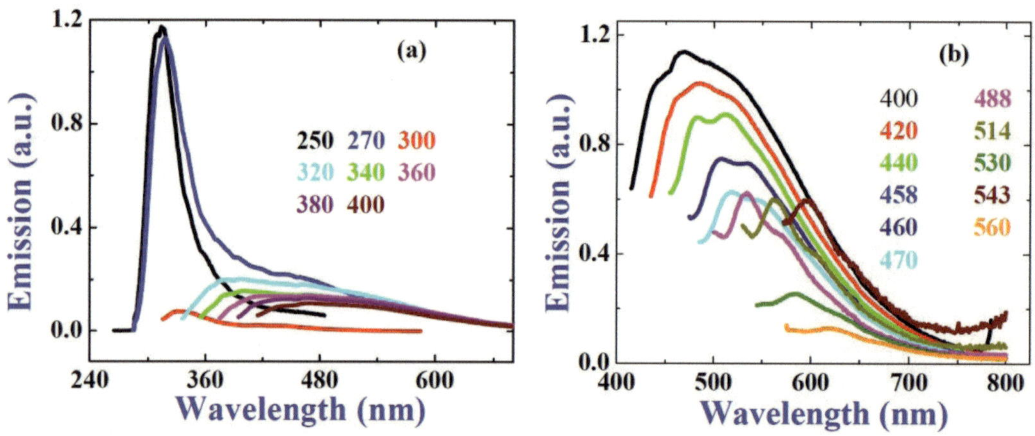

Figure 10. Plot of emission spectra of (a) PMMA at excitation wavelengths from 250-400 nm (b) PMMA at excitation wavelengths from 400-560 nm

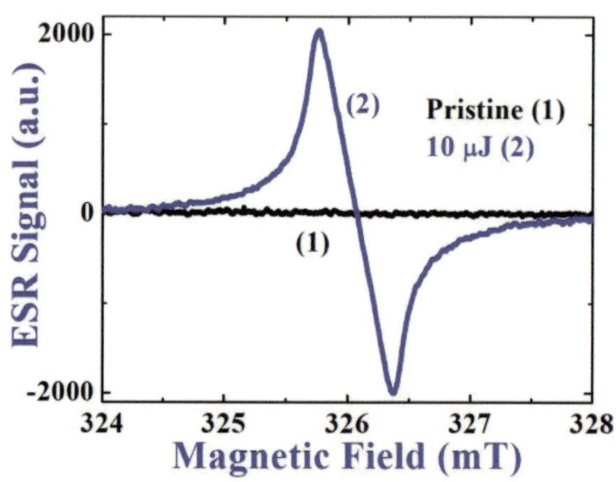

Figure 11. ESR signal observed in irradiated PS.

S. No.	λ_{ext} (nm)	λ_{em} (nm)	$\Delta E/hc$ (cm^{-1}) = λ_{ext}^{-1} - λ_{em}^{-1}
1	250	309	7638
		313	8051
2	270	317	5491
3	300	326	2658
4	320	381	5003
		398	6214
		461	9558
5	340	398	4286
		464	7860
6	360	427	4359
		465	6272
7	380	437	3432
		467	4903
8	400	464	3448
9	420	484	3148
10	440	483	2023
		511	3158
11	458	499	1794
		529	2930
12	460	507	2015
		535	3048
13	470	518	1972
		551	3128
14	488	534	1765
		568	2886
15	514	561	1630
16	530	584	1745
17	543	598	1694
18	560	621	1754

Table 3. Excitation, emission, and Raman shift observed in fs laser irradiated PMMA.

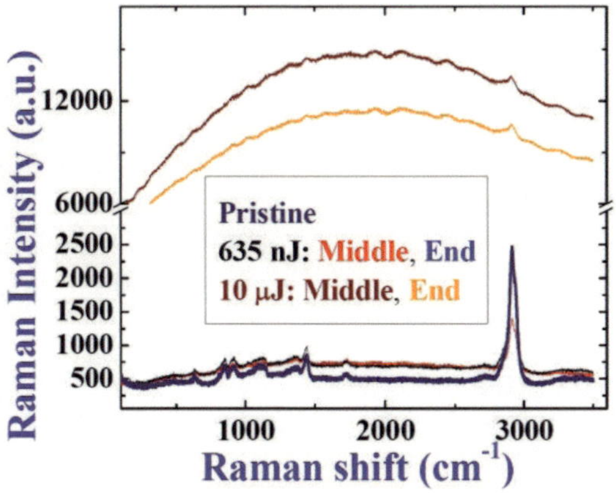

Figure 12. Confocal Raman spectra for structures fabricated on PVA thin film.

3. Conclusions

We presented our results on the physical and spectroscopic investigations of fs laser irradiated polymers. The unusual behavior of emission observed in case of these polymers is attributed to the red edge effect. The transitions involved with in the functional groups such as $n \rightarrow \pi^*$, and $\pi \rightarrow \pi^*$ are responsible for the emission observed. The role of coincidence of Raman signals with emission observed is illustrated. The presence of paramagnetic centers is confirmed through ESR studies. Both the defects which are responsible for emission and ESR signal are further confirmed through confocal micro Raman studies by observing the broadening of Raman modes.

Author details

Kallepalli Lakshmi Narayana Deepak
School of Physics, University of Hyderabad, Gachibowli, Hyderabad, India
Laboratoire LP3, UMR 6182, CNRS-Universite Aix-Marseille, Pole Scientifique et Technologique de Luminy, Case 917, Marseille, France

Soma Venugopal Rao
Advanced Center for Research in High Energy Materials (ACRHEM), University of Hyderabad, Gachibowli, Hyderabad, India

Desai Narayana Rao*
School of Physics, University of Hyderabad, Gachibowli, Hyderabad, India

Acknowledgement

We are extremely grateful to DRDO, India for all the financial support during the tenure of this project. K. L. N. Deepak acknowledges CSIR for financial support to carry out his doctoral work. DNR and SVR acknowledge Department of Science and Technology (DST), India for financial support through a project **SR/S2/LOP-11/2005**.

4. References

[1] Gattass R R, Mazur E (2008) Femtosecond laser micromachining in transparent materials. Nat. Phot. 2: 219-225.

[2] Juodkazis S, Mizeikis V, Misawa H (2009) Three-dimensional Microfabrication of materials by femtosecond lasers for photonics applications. J. Appl. Phys. 106: 051101.

[3] Nolte S, Will M, Burghoff J, Tuennermann A (2003) Femtosecond waveguide writing: a new avenue to three-dimensional integrated optics. Appl. Phys. A 77: 109-111.

* Corresponding Author

[4] Ams M, Marshall G D, Dekker P, Dubov, Mezentsev V K, Bennion I, Withford M J (2008) Investigation of ultrafast laser–photonic material interactions: challenges for directly written glass photonics. IEEE J. Sel. Top. Quant. Electron., 14(5): 1370-1381.

[5] Qiu J, Miura K, Hirao K (2008) Femtosecond laser-induced micro features in glasses and their applications. Journal of Non-Crystalline Solids. 354: 1100–1111.

[6] Della Valle G, Osellame R, Laporta P (2009) Micromachining of photonic devices by femtosecond laser pulses. J. Opt. A: Pure Appl. Opt. 11: 013001.

[7] Qiu J, Miura K, Hirao K (2008) Femtosecond laser-induced micro features in glasses and their applications," Journal of Non-Crystalline Solids. 354, 1100–1111, 2008.

[8] Della Valle G, Osellame R, Laporta P (2009) Micromachining of photonic devices by femtosecond laser pulses. J. Opt. A: Pure Appl. Opt. 11: 013001-.

[9] Taccheo S, Della Valle G, Osellame R, Cerullo G, Chiodo N, Laporta P, Svelto O, Killi A, Morgner U, Lederer M, Kopf D (2004) Er : Yb-doped waveguide laser fabricated by femtosecond laser pulses. Opt. Lett. 29: 2626-2628.

[10] Florea C, Winick K A (2003) Fabrication and characterization of photonic devices directly written in glass using femtosecond laser pulses. Journal of Lightwave Technology 21: 246-253.

[11] Minoshima K, Kowalevicz A M, Ippen E P, Fujimoto J G (2002) Fabrication of coupled mode photonic devices in glass by nonlinear femtosecond laser materials processing. Optics Express 10: 645-652.

[12] Streltsov A M, Borrelli N F (2001) Fabrication and analysis of a directional coupler written in glass by nanojoule femtosecond laser pulses. Optics Letters 26: 42-43.

[13] Davis K M, Miura K, Sugimoto N, Hirao K (1996) Writing Waveguides in Glass with a Femtosecond Laser. Optics Letters 21: 1729-1731.

[14] Glezer E N, Milosavljevic M, Huang L, Finlay R J, Her T H, Callan J P, Mazur E (1996) Three-dimensional optical storage inside transparent materials. Opt. Lett. 21(24): 2023-2025.

[15] Watanabe W, Sowa S, Tamaki T, Itoh K, Nishii J (2006) Three-Dimensional Waveguides Fabricated in Poly(methyl methacrylate) by a Femtosecond Laser. Jap. J. Appl. Phys., Part 2, 45: L675-L767.

[16] Zoubir A, Lopez C, Richardson M, Richardson K (2004) Femtosecond laser fabrication of tubular waveguides in poly(methyl methacrylate). Opt. Lett. 29(16): 1840-1842.

[17] Sowa S, Tamaki T, Itoh K, Nishii J, Watanabe W (2006) Three-dimensional waveguides fabricated in poly(methyl methacrylate) by a femtosecond laser. Jap. J. Appl. Phys., Part 2: Letters 45: 29-32, L765-L767.

[18] Wang K, Klimov D, Kolber Z (2007) Waveguide fabrication in PMMA using a modified cavity femtosecond oscillator. Proc. SPIE 6766: 67660Q.

[19] Watanabe W, Sowa S, Tamaki T, Itoh K, Nishii J(2006) Three-Dimensional Waveguides Fabricated in Poly(methyl methacrylate) by a Femtosecond Laser. Jap. J. Appl. Phys., Part 2, 45: L675-L767.

[20] Zoubir A, Lopez C, Richardson M, Richardson K (2004) Femtosecond laser fabrication of tubular waveguides in poly (methyl methacrylate). Opt. Lett. 29(16): 1840-1842.

[21] Sowa S, Tamaki T, Itoh K, Nishii J, Watanabe W (2006) Three-dimensional waveguides fabricated in poly(methyl methacrylate) by a femtosecond laser. Jap. J. Appl. Phys., Part 2: Letters 45: 29-32, L765-L767.

[22] Cumpston B H, Ananthavel S P, Barlow S, Dyer D L, Ehrlich J E, Erskine L L, Heikal A A, Kuebler S M, Lee I Y S, McCord-Maughon D, Qin J Q, Rockel H, Rumi M, Wu X L, Marder S R, Perry J W (1999) Two-photon polymerization initiators for three dimensional optical data storage and microfabrication. Nature 398: 51-54.

[23] Nie Z, Lee H, Yoo H, Lee Y, Kim Y, Lim K S, Lee M (2009) Multilayered optical bit memory with a high signal-to-noise ratio in fluorescent polymethylmethacrylate. Appl. Phys. Lett. 94: 111912.

[24] Farson D F, Choi H W, Lu C, Lee L J (2006) Femtosecond laser bulk micromachining of microfluidic channels in poly (methyl methacrylate). J. Laser Appl. 18(3): 210-215.

[25] Day D, Gu M (2005) Microchannel fabrication in PMMA based on localized heating by nanojoule high repetition rate femtosecond pulses. Opt. Exp. 13(16): 5939-5946.

[26] Ding L, Blackwell R I, Kunzler J F, Knox W H (2008) Femtosecond laser micromachining of waveguides in silicone-based hydrogel polymers. Appl. Opt. 47(17): 3100-3108.

[27] Si J, Meng Z, Kanehira S, Qiu J, Hua B, Hirao K (2004) Multiphoton-induced periodic microstructures inside bulk azodye-doped polymers by multibeam laser interference. Chem. Phys. Lett. 399: 276-279.

[28] Si J H, Qiu J R, Zhai J F, Shen Y Q, Hirao K (2002) Photoinduced permanent gratings inside bulk azodye-doped polymers by the coherent field of a femtosecond laser. Appl. Phys. Lett. 80(3): 359-361.

[29] Kim T N, Campbell K, Groisman A, Kleinfeld D, Schaffer C B (2005) Femtosecond laser-drilled capillary integrated into a microfluidic device. Appl. Phys. Lett. 86: 201106.

[30] Mendonca C R, Cerami L R, Shih, Tilghman R W, Baldacchini T, Mazur E (2008) Femtosecond laser waveguide micromachining of PMMA films with azoaromatic chromophores. Opt. Exp. 16(1): 200-206.

[31] Zhou G, Ventura M J, Vanner M R, Gu M (2004) Use of ultrafast-laser-driven microexplosion for fabricating three-dimensional void-based diamond-lattice photonic crystals in a solid polymer material. Opt. Lett. 29: 2240-2242.

[32] Nolte S, Will M, Burghoff J, Tuennermann A (2003) Femtosecond waveguide writing: a new avenue to three-dimensional integrated optics. Appl. Phys. A 77: 109-111.

[33] Baum A, Scully P J, Perrie W, Jones D, Issac R, Jaroszynski D A (2008) Pulse-duration dependency of femtosecond laser refractive index modification in poly (methyl methacrylate). Opt. Lett. 33(7): 651-653.

[34] Schaffer C B, Brodeur A, Mazur E (2001) Laser-induced breakdown and damage in bulk transparent materials induced by tightly focused femtosecond laser pulses. Meas. Sci. Technol. 12: 1784–1794.

[35] Deepak K L N, Narayana Rao D, Venugopal Rao S (2009) Fabrication and Optical Characterization of microstructures in PMMA and PDMS using femtosecond pulses for photonic and micro fluidic applications. Appl. Opt. 49 (13): 2475.
[36] Deepak K L N, Venugopal Rao S, Narayana Rao D (2010) Femtosecond laser-fabricated microstructures in bulk poly(methylmethacrylate) and poly(dimethylsiloxane) at 800 nm towards lab-on-a-chip applications. Pramana 75(6): 1221-1232.
[37] Deepak K L N, Kuladeep R, Venugopal Rao S, Narayana Rao D (2011) Luminescent microstructures in bulk and thin films of PMMA, PDMS, PVA, and PS fabricated using femtosecond direct writing technique. Chem. Phys. Lett. 503: 57-60.
[38] Deepak K L N, Kuladeep R, Narayana Rao D (2011) Emission properties of femtosecond (fs) laser fabricated microstructures in Polystyrene (PS). Opt. Commun. 284: 3070-3073.
[39] Deepak K L N, Kuladeep R, Praveen Kumar V, Venugopal Rao S, Narayana Rao D (2011) Spectroscopic investigations of femtosecond laser irradiated Polystyrene and fabrication of microstructures. Opt. Commun. 284: 3074-3078.
[40] Deepak K L N, Kuladeep R, Venugopal Rao S, Narayana Rao D (2012) Studies on Defect Formation in Femtosecond Laser Irradiated PMMA and PDMS," Radiation Effects and Defects in Solids 67: 88-101.
[41] Deepak K L N, Venugopal Rao S, Narayana Rao D (2011) Effect of heat treatment to efficient buried diffraction gratings in Polystyrene. Appl. Surf. Sci. 257: 9299-9305.
[42] Nurmukhametov R N, Volkova L V, Kabanov S P (2006) Fluorescence and absorption of polystyrene exposed to UV laser radiation. J. Appl. Spectr. 73: 55-60.
[43] Galley W C, Purkey R M, Role of Heterogeneity of the Solvation Site in Electronic Spectra in Solution, Proc. Natl Acad. Sci. USA 1970; 67, 1116-1121.
[44] Rubinov A N, Tomin VI (1970) Bathochromic luminescence of organic dyes, Opt. Spektr. 29: 1082-1086.
[45] Terenin A N, *Fotonika molekul krasitelei,* Photonics of Dye Molecules (1967) Nauka:Leningrad, 616.
[46] Birks JB. Photophysics of Aromatic molecules. Wiley-Inter-science: London 1970.
[47] Thomas K J, Sheeba M, Nampoori V P N, Vallabhan C P G, and Radhakrishnan P (2008) Raman spectra of polymethyl methacrylate optical fibres excited by a 532 nm diode pumped solid state laser.J. Opt. A: Pure Appl. Opt. 10: 055303-055307.
[48] Velter-Stefanescu M, Duliua O G, Preda N (2005) On the relaxation mechanisms of some radiation induced free radicals in polymers. J. Optoelectron. Advanced Mat. 7(2): 985–989.
[49] Kaptan H Y, Tatar L (1997) An Electron Spin Resonance study of mechanical fracture of poly (methyl methacrylate). J. Appl. Polym. Sci. 65: 1161-1167.
[50] Abdelaziz M (2008) Electron spin resonance and optical studies of poly (methylmethacrylate) doped with CuCl2. J. Appl. Polym. Sci. 108: 1013-1020.

[51] Ohnishi S, Tanei T, Nitta I (1962) ESR Study of Free Radicals Produced by Irradiation in Benzene and Its Derivatives. J. Chem. Phys. 37: 2402-2407.

Chapter 10

Holographic Fabrication of Periodic Microstructures by Interfered Femtosecond Laser Pulses

Zhongyi Guo, Lingling Ran, Yanhua Han, Shiliang Qu and Shutian Liu

Additional information is available at the end of the chapter

http://dx.doi.org/10.5772/46136

1. Introduction

In recent years, with the developments of the laser technology, femtosecond laser technology is becoming more and more consummate as a novel technology. Femtosecond laser pulse is also becoming into a powerful tool for microfabrication and micro-machining of various multi-functional structures in dielectric materials through multi-photon absorption because of its high-quality and damage-free processing. Up to now, many high-quality material processing techniques have been achieved by using femtosecond laser pulses with the methods of directly writing [1-10] and holographic fabrication [11-22], such as waveguide [1], special diffractive optical elements (DOE) [4-10], micro-gratings [11-15], and photonic crystals [16-20]. Because multiphoton nonlinear effects play a major role in this process, the resulting change in refractive index or cavity formation can be highly localized only in the focal volume where the fluence is above a certain material dependent threshold, which makes it possible to micro-fabricate devices inside the bulk of transparent materials. These structures were usually fabricated with a focused beam and written dot by dot by translation of the sample with respect to the focal point.

Compared to directly writing technology by femtosecond laser pulses, holographic lithography has been considered as a more effective method for fabricating periodic structures because it can be controlled easily by the number of the beam, angles between every two beams, energies of the laser beams, and most importantly, only one pulse needed for holographic fabrication. Especially, holographic lithography is considered to be the most effective method for the fabrication of the photonic crystals. And Cai *et al.* [21-22] have proven that all fourteen Bravais lattices can be formed by interference of four noncoplanar beams (IFNB), and have quantitatively obtained the solutions including the required wavelength and the four wave vectors for each special lattice. The relative simulated results are shown in Fig. 1.

There are also many groups who have attempted to fabricate the periodic structures in photosensitive transparent materials or on the surface of the silica glass and the metal film by interfered multiple femtosecond laser pulses. Especially, it is very easy to fabricate the periodic structures inside of the photosensitive transparent materials with the aid of diffractive beam-splitter (DBS) [18-20]. However it is limited by the angles between two beams and the energy of the pulses. It is therefore difficult to fabricate microstructure with smaller period, especially in the materials with big band gap, such as silica glass. So there are also some researching groups [13, 16-17] focusing on the realization of periodic structures on the surface of the silica glass by a single shot of two or three femtosecond laser pulses originating from one pulse by the beam splitters.

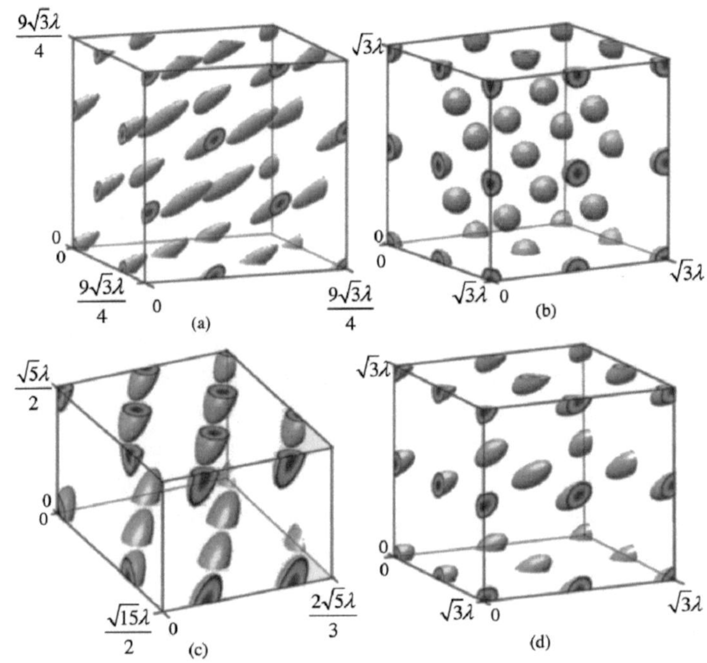

Figure 1. Computer simulations of four Bravais lattices formed by interference of four noncoplanar beams: (a) face-centered cubic lattice, (b) body-centered cubic lattice, (c) hexagonal lattice ($c = 3a/2$), and (d) simple cubic lattice.

In this chapter, we have reviewed the progress of fabrication of the periodic structures on the surface or inside of the transparent silica glass by a single shot of several (two, three, four) femtosecond laser pulses. When a single shot of two pulses interfered with each other, there will be one dimensional grating structures being formed inside of the photosensitive transparent materials or on the surface of the materials. When a single shot of three coplanar pulses interfered with each other, the one-dimensional M-shape grating can be formed on the surface of the silica glass. However, when a single shot of noncoplanar three or four pulses interfered with each other, two-dimensional periodic microstructure can be obtained, which distributed as a hexagonal lattice or tetragonal lattice.

2. Holographic fabrication by a single shot of two femtosecond pulses

2.1. The formation of the common grating and the extraordinary grating on the surface of the silica glass

The experimental setup for the holographic fabrication of microgratings by a single shot of two interfered femtosecond laser pulses is depicted in Fig. 2 (a). A regeneratively amplified Ti:sapphire laser (Coherent. Co.) with a central wavelength of 800 nm, pulse duration of 120 fs, and pulse repetition of 1–1000 Hz was used. A single femtosecond laser pulse with a beam diameter of 6 mm could be selected and split into two pulses that were then redirected at controllable incident angles on the surface of the fused silica glass (K9 glass) which is transparent for the laser with a wavelength of 800 nm. These split pulses generating from the single pulse were focused on the glass surface by two lenses with focal lengths of 20 cm to give a spot size of ~50 μm at the focal plane. Attenuators could be used to obtain a proper energy of the pulses for controlling the results of the holographic fabrication. The two pulses could be adjusted both spatially and temporarily by the optical delay device perfectly. The angle between two pulse and the pulse energies could be controlled easily.

When we set the angle between the split beams as 40° and the pulse energy as 45 μJ, the fabricated grating is shown in the Fig. 2 (b-e) by optical microscopy and atomic force microscopy (AFM) respectively. From the Fig. 2 (d), we can obtain the period of the fabricated grating is about 1.06 μm which agree well with the calculated result according to the formula $d=\lambda/[2\sin(\theta/2)]$ in which θ is colliding angle (θ=40° in this experiment) between two incident beams and λ is the incident laser wavelength of 800 nm. The periods of the micrograting encoded by holographic technology should be controlled easily by varying the angle. Analogously, the depth of the fabricated grating could also be tuned by controlling the pulse energy.

However, when we set the angle θ as 20° and the pulse energy as 65 μJ, the fabricated grating is shown in the Fig. 3. Not only did we get the ordinary grating whose periods accorded with the theoretic equation $d=\lambda/[2\sin(\theta/2)]$, but also obtained the extraordinary grating [15, 23] whose period is a half of the ordinary grating. The extraordinary grating formed at the middle of each bulge of the ordinary grating as shown in Fig. 3 (a-b), so the period of the extraordinary grating is a half of the ordinary grating's observed easily from Fig. 3(c) which also shows that the depth of the extraordinary grating is a half of the ordinary grating's nearly.

In the experiments, with the decrease of the incident energy in the same angle between two beams, the modulation depths of the extraordinary grating are decreasing gradually. With the increase of the angle between two beams with the same energy, the modulation depths of the extraordinary grating are also decreasing gradually. At last the modulation will be vanishing from the central part to the edge of the ordinary grating gradually.

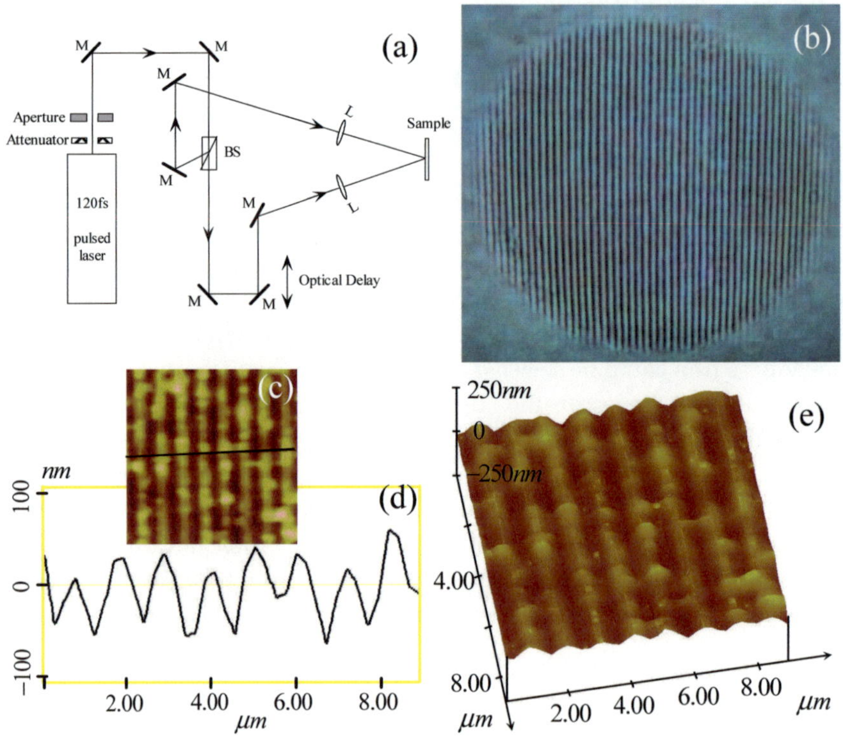

Figure 2. The experimental setup for the holographic fabrication of microgratings, (b) the fabricated grating observed by optical microscopy, (c) top-viewing AFM image of the fabricated grating, (d) cross-section view in the direction of black line shown in (c), (e) three dimentional view of the fabricated structure.

The formation of the modulation grating could be attributed to the higher-order modulation arising from second-harmonic generation (SHG) when the femtosecond laser pulse was incident to the surface of silica glass. As a rule, because of the inversion symmetry of the silica glass, there should be no second order nonlinearity in the silicon glass. However, a layer of plasma could be formed on the surface of the glass when the pulse incident to the sample because of the ultra high electrical field of the femtosecond laser in a time given by the duration of the laser pulse[24,25]. Therefore, when the femtosecond laser pulse with high intensity is incident on this thin plasma layer, there will be the higher harmonic generation because of the electrons quivering nonlinearly, and the SHG can reach to 2% of the fundamental laser [26]. The second-harmonic radiation is emitted in the direction of the reflected fundamental laser direction as depicted in Fig. 2 (d) [24]. So the period of the modulation grating is a half of the common grating because of the same of the angle between the two beams but a half of the wavelength. Although the intensity of the SHG is much smaller than the fundamental laser, the modulated depth of the silica glass can reach to a correspondingly large depth because it is much easier for the silica glass to be ionized by a mechanism of single or two photons absorption than multiphoton absorption.

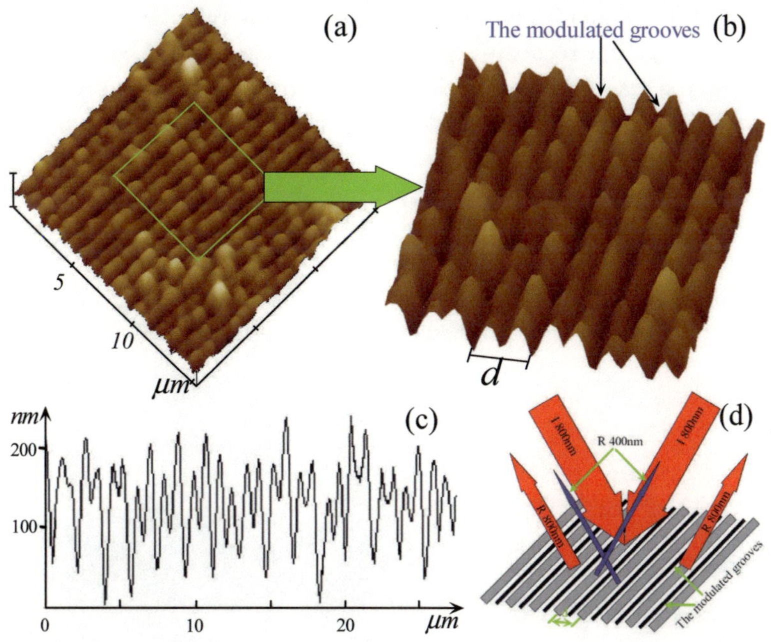

Figure 3. AFM photos of the resulted micrograting encoded with an energy of ~65 μJ for each interfered beam and a colliding angle of ~20°. (a) Image of the central portion of the grating, (b) an enlarged version for the chosen part in (a), and *d* is the period of the common grating. (c) cross-section view of the fabricated grating. (d) schematic illustration for the formation mechanism of the modulating grating.

2.2. Fabrication of multiple layers of grating

Li *et al.* [11] have reported that the multiple layers of grating can be fabricated inside soda–lime glass by a single shot of two interfered femtosecond laser pulses at different depths of the sample. A top view of experimental setup is shown in Fig. 4 (a), which is similar to the Fig. 2 (a). The two beams, with diameters of 5 mm, were focused by two lenses with focal length of 150 mm to give a spot size of 30 30μm at the focal plane.

The multiple gratings were written inside of the soda–lime glass one after another by focusing the beam 1 and beam 2 into special position inside of the glass sample with the recording plane. Without loss of generality, three layers of micrograting could be recorded at depths of 200, 400, and 600 μm, respectively. The grating at depth of 600μm should be encoded firstly. Then the sample was translated along the *z* direction to a depth of 400μm and the second grating could be recorded. After that, the third grating was fabricated after moving the sample to a depth of 200μm lastly.

Because the micrograting was formed inside of the samples around the focal point without damage to the surface or other parts of the sample, multiple layers of grating can be recorded successfully. And the images of the fabricated multiple layers of microgratings could be read out by beam 2 and taken over three different recording planes as depicted in

Fig. 4 (b) and (c). The schematics on the right illustrate the relative positions of the sample, three layers of grating, the recording plane, and the objective. The experimental results demonstrated that the readout image of the first layer (not shown here) and the second layer (as depicted in Fig. 4 (b)) of the recorded grating consisted of well-defined straight bright lines alternating with black lines. However, for the readout of the third layer of the micrograting, some of the straight lines became curved as shown in Fig. 4 (c). This aberration may be caused by the wave-front distortion of the incident beams due to the accumulation of nonlinear effects when the focused high peak-power pulse propagated through the sample. The longer the optical path inside the sample was, the more severe the wave-front distortion and the more obvious the resulting grating aberration.

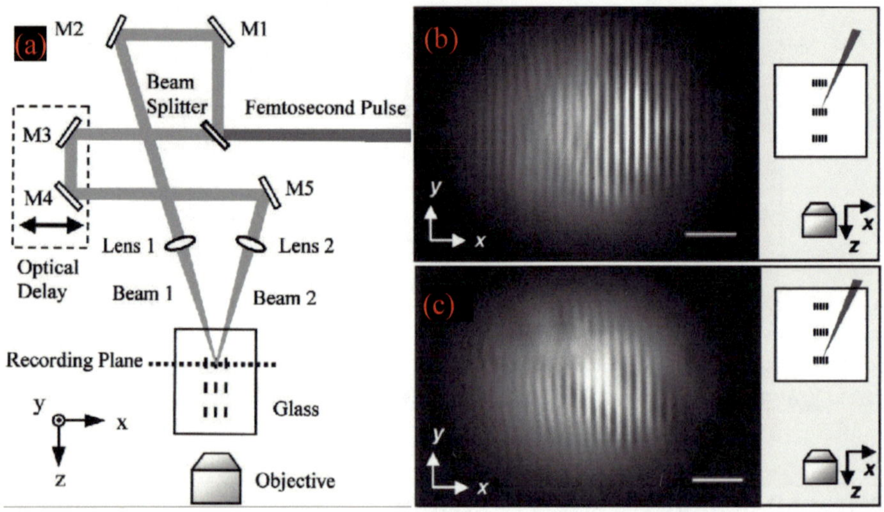

Figure 4. (a) Top view of the experimental setup for the formation of multiple gratings inside glass. The recording plane is in the $x - y$ plane and through the overlapping focal points of the two incident beams: beam 1 and beam 2. (b)(c). Readout images of the second and the third layers of grating inside a slide glass plate by beam 2. The images were taken through a 100X optical objective that was focused onto the recording planes at depths of 400μm (b) and 600μm (c) respectively. The schematics on the right indicate the relative positions of the sample, grating layers, the readout beam, and the 100X optical objective. Scale bar: 5μm.

2.3. Fabrication of micrograting inside the glass doped Au nanoparticles

Noble metal nanoparticle-contained glasses exhibit large third-order nonlinear susceptibility and ultrafast nonlinear response due to the local field effect near surface Plasmon resonance and quantum size effect [27, 28]. In recent years, many studies have been carried out on the fabrication of nanoparticle-doped glasses [29-31]. Shiliang Qu *et al.* realized the formation of micrograting constituted with Au nanoparticles in Au_2O-doped glasses induced by two interfered femtosecond laser pulses followed by successive heat treatment [12].

A typical silicate glass is composed of $70SiO_2.20Na_2O.10CaO$ doped with $0.1Au_2O$ (mol%). Reagent grade SiO_2, $CaCO_3$, Na_2CO_3, and $AuCl_3.HCl.4H_2O$ were used as starting materials.

An approximately 40g batch was mixed and placed into a platinum crucible. Melting was carried out in an electric furnace at 1550 °C for 1 h. The glass sample was obtained by quenching the melt to room temperature. The sample thus obtained was transparent and colorless. The annealed sample was cut and polished, and then subjected to successive experiments for femtosecond laser.

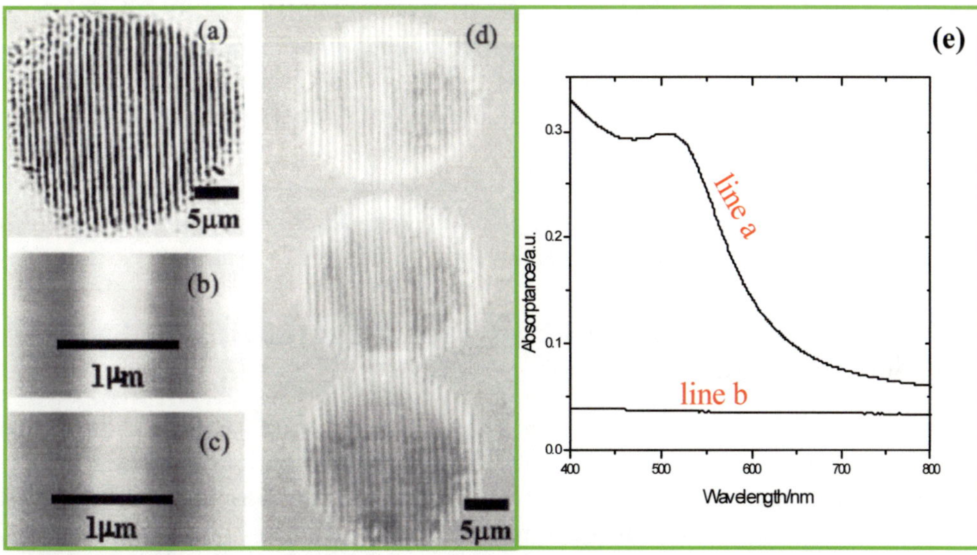

Figure 5. Optical microscopic photos of Au nanoparticles precipitation in periodic arrays in silicate glass (microgratings), taken by a 100X transilluminated optical microscope. (a) Energy is 30 μJ per pulse. (b) Magnified view of (a). (c) Energy is 38μJ per pulse. (d) Part of a group of formed microgratings array inside of the sample. (e) Absorption spectra of the glass samples after holographic irradiation by femtosecond laser pulses with (line a) and without (line b) heat treatment.

The used laser system and the experimental setup are the same as the used setup as shown in Fig. 2 (a). The two incident beams were first focused onto the front surface of Au_2O-doped glass to optimize the incident pulse energy. In the case of sufficiently high energy, the two coherent beams can induce periodic ablation, forming a grating in the glass. Herein, we reduced the incident pulses' energy to a certain lower level at which the two coherent beams cannot directly induce periodic ablation on the surface of the glass. Then the sample was moved 50μm to the lens in the direction of angular bisector of two incident light paths and made the laser pulses be focused inside the glass. After irradiation by a single shot of two interfered pulses, the micrograting can be recorded inside the glass, which can not be observed immediately because there is no obvious changes in the focusing place, however, after heat treating the sample at 550 °C for 1 h, the formed micrograting can be observed because of the Au nanoparticle precipitation. Such grating was constituted by the laser-heating induced Au nanoparticle precipitation in the Au_2O-doped glass.

In Qu's experiments, this lower pulse energy was selected to be 30 and 38 μJ for comparison, and the colliding angle θ between the two incident beams was fixed at 45°. The fabricated micrograting is shown in Fig.5 (a-d). The period d of the obtained gratings was about ~1μm which was agreeing well with the value calculated from the colliding angle θ and the laser wavelength λ according to the formula $d=\lambda/[2\sin(\theta/2)]$.

The absorption spectrum of the grating was measured by a spectrophotometer (JASCOV-570) as shown in Fig. 5 (e) (line a). Apparently, a weak peak occurs around 508 nm, which is induced by the surface plasmon resonance of Au nanoparticles in the glass. The Au nanoparticles with 3 nm average size in the glass were observed in the grid of the fabricated micrograting from a transmitted electronic microscopy (TEM). However, if the glass sample was irradiated only by the interfered pulses but not heat treated, neither could absorption peak in the range of 500–600 nm be observed in the absorption spectrum as shown in Fig. 5 (e) (line b), nor could Au nanoparticles be observed by the TEM. This indicates that the Au nanoparticles can be precipitated in the periodic one-dimensional arrays in the glass through the irradiation of two coherent beams with the aid of heat treatment.

2.4. Fabrication of multiple gratings by a single shot of two interfered femtosecond laser pulses

The interference of two laser beams can create microstructures with one-dimensional periodic patterns in certain materials due to the periodic modulation of the laser intensity with a period scale of the order of the laser wavelength. As stated as above, there has been many groups focusing the fabrication of microgratings in glasses [11, 12], thin films [32], polymers[33], and inorganic–organic hybrid materials [34] by use of two interfered femtosecond laser pulses in a single shot. However, there is a disadvantage for micrograting fabrication with this method: Only a single micrograting can be formed for one pulse. Shiliang Qu *et al.* have reported holographic fabrication of multi-microgratings on silicate glass by a single shot of two interfered femtosecond laser beams with the aid of a mask [14].

In the experimental setup as shown in Fig. 6 (a) which is similar with the Fig. 2 (a) and Fig. 4 (a), a regeneratively amplified Ti:sapphire laser (Spectra-Physics) with a wavelength of 800 nm, pulse duration of 120 fs, and pulse repetition of 1–1000 Hz was used. A single laser pulse with a beam diameter of 8 mm was selected and split into two beams that were then redirected at approximately equal incident angles on a silicate glass surface. The two beams were focused on the glass surface by two lenses with focal lengths of 10 cm (L1) and 20 cm (L2) to yield spot sizes of 40μm and 80μm, respectively.

The colliding angle θ between the two beams was fixed at 40°. After the optical paths were adjusted to realize perfect overlap of the two beams both spatially and temporally, the surface of the glass was adjusted to be approximately normal to the perpendicular bisector of the two incident beams, so that the glass surface became the laser interfering plane. A mask used for laser beam modulation was placed in the optical path in which lens L1 is located. The mask consists of three equilaterally and triangularly arrayed apertures, whose diameters are 2.5 mm and the spaces between them are 3.5 mm as shown in Fig. 6(b).

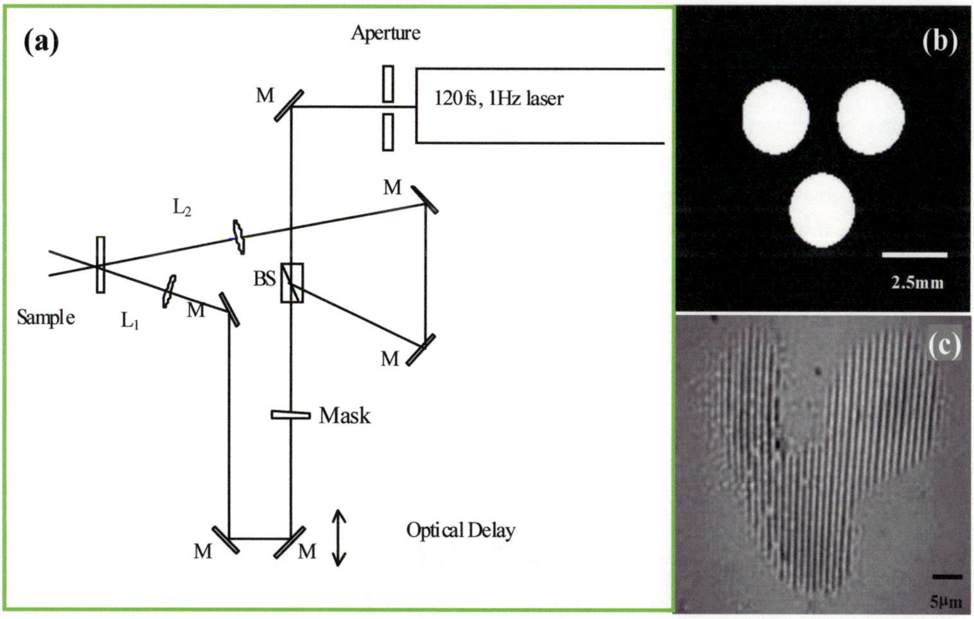

Figure 6. Experimental setup for the one-off writing of multi-microgratings by a single shot of two femtosecond laser pulses. M's, mirrors; BS, beam splitter. (b) Data mask for multimicrograting formation. (c) Optical microscopic observation of a multi-micrograting formed on silicate glass with a period of 1.1μm ($E_{1,2}$~70 μJ, θ~40°).

After irradiation by one single shot of two interfered pulses, the multi-micrograting comprises three micrograting as shown in Fig. 6(c), which has a high fidelity to the configuration of the mask used. The multi-micrograting was formed through periodic ablation resulting from the interference of the reference laser beam with the three beams caused by the mask. And the period of the formed micrograting is also agree well with the theoretical expectation of the common grating $d=\lambda/[2\sin(\theta/2)]$. This means that a multi-micrograting comprising even more micrograting and configurations can also be one-off written by changing only the mask structure. However, multi-micrograting formation can be realized only when the two interfered beams are overlapped at an appropriate position out of their focus on the front surface of the glass.

3. Holographic fabrication by a single shot of multiple femtosecond pulses

3.1. The formation of the M-shaped micrograting by three coplanar interfered beams

As stated above, two interfered femtosecond pulses' interference can induce one and two-dimensional periodic structures by single and double-exposure techniques, respectively [11–15, 40]. However, usually the second pulse could not overlap completely with the microstructure formed by the first pulse in the double-exposure technique due to the rather small size of the focal spot. Here, we will show that the fabrication of M-shape gratings with controllable modulation depth could be realized by adding the third beam into a two-beam interference system. The experimental results show that the depth ratio between neighbor grooves can be conveniently controlled by changing the pulse energy of the third beam. Morphology characterizations of as-fabricated periodic M-shape gratings with a period of 2.6 μm are presented by optical microscopy and atomic force microscopy (AFM). A theoretical simulation has also been done for explaining the concrete experimental results, which shows a higher fidelity to the experiment.

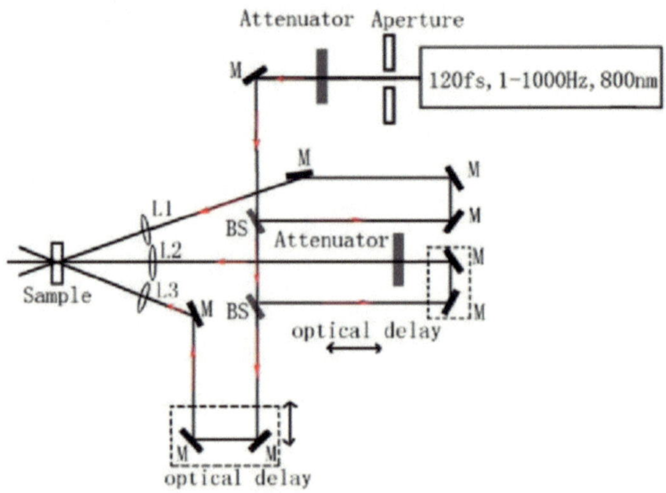

Figure 7. Experimental setup of three-beam interference optical system (AT and AP stand for attenuator and aperture, respectively).

The schematic experimental setup for fabricating M-shape grating is shown in Fig.7. The used ultrafast pulses (with a pulse width of 120 fs, central wavelength of 800 nm, and repetition rate of 1–1000 Hz) are produced by a Ti: sapphire regenerative amplified laser system (Coherent Inc). A laser pulse with a diameter of ~5 mm is split into three beams by two beam-splitters. Then three beams are focused by three lenses (L1; L2, and L3), each with a focal length of 150 mm, to give a spot size of ~50 μm at the focal plane. The collision angles between L1–L2, L2–L3, and L1–L3 are θ, θ, and 2θ (θ = 18° and 30°), respectively. Two time-delay devices are employed for time adjustment of the beams collision. A BBO frequency-double crystal is located at the focal spot to check time superposition by SHG. When three collinear blue points appear on the white screen located on the back of frequency-double crystal, it indicates that three beams have been superposed on each other both spatially and temporarily. Two attenuators are used to adjust the energy of femtosecond laser pulse. The sample silica glass is located in the direction perpendicular to beam L2.

The optical microscope images of the gratings formed on the silica glass at different pulse energies of L2 (θ = 18°, the pulse energies of L1 and L3 are both 50 μJ) are shown in Fig. 8 (a) and Fig. 8 (b). The neighbor grooves with different widths alternatively can be observed in the grating as shown in Fig. 8 (a). However, in Fig. 8 (b), the neighbor grooves of the as-formed grating have the same widths. Fig. 8 (c) shows the AFM characteristic results of as-formed grating (in Fig. 8 (a)). Fig. 8 (c) is a top-view AFM image of the formed micrograting, where the neighbouring grooves with different widths alternately appear more clearly. Fig. 8 (d) displays the three dimensional morphology, which clearly shows that the M-shape grating is formed by the alternate appearing of two kinds of grooves: the deeper grooves and the shallower grooves. Fig. 8 (e) indicates a cross-section scanning picture along the horizontal line 'L' shown in Fig. 8 (c).

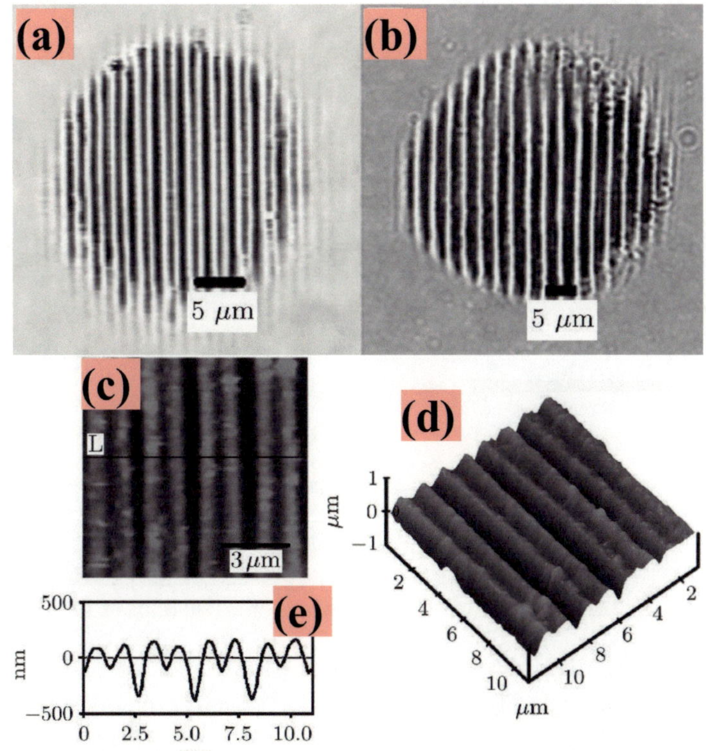

Figure 8. Optical microscope images of as-formed gratings by coplanar three-beam interference with different pulse energies of L2, i.e. (a) 50 μJ and (b) 100 μJ. AFM images of the M-shape grating formed on silica glass by three coplanar interfering beams, each with equal energy of 50 μJ, the collision angles of three beams are 18°, 18°, and 36°, respectively. Panel (c) is for planar image, panel (d) for three-dimensional image, and panel (e) for cross-section pattern along line L.

It is obviously shown that the structure is formed by periodically arranging the M-shape units with a size of about 2.6 μm. The modulation depths of the deeper grooves and shallower grooves of the M-shape grating are ~500 nm and ~240 nm, respectively. The deeper grooves and the shallower grooves of the M-shape grating each have a period of ~2.6 μm revealed in Fig.8 (c). This result accords well with the value calculated from the universally known period formula $d = \lambda/\sin\theta$ (in our case λ = 800 nm, θ = 18°, and d = 2.59 μm).

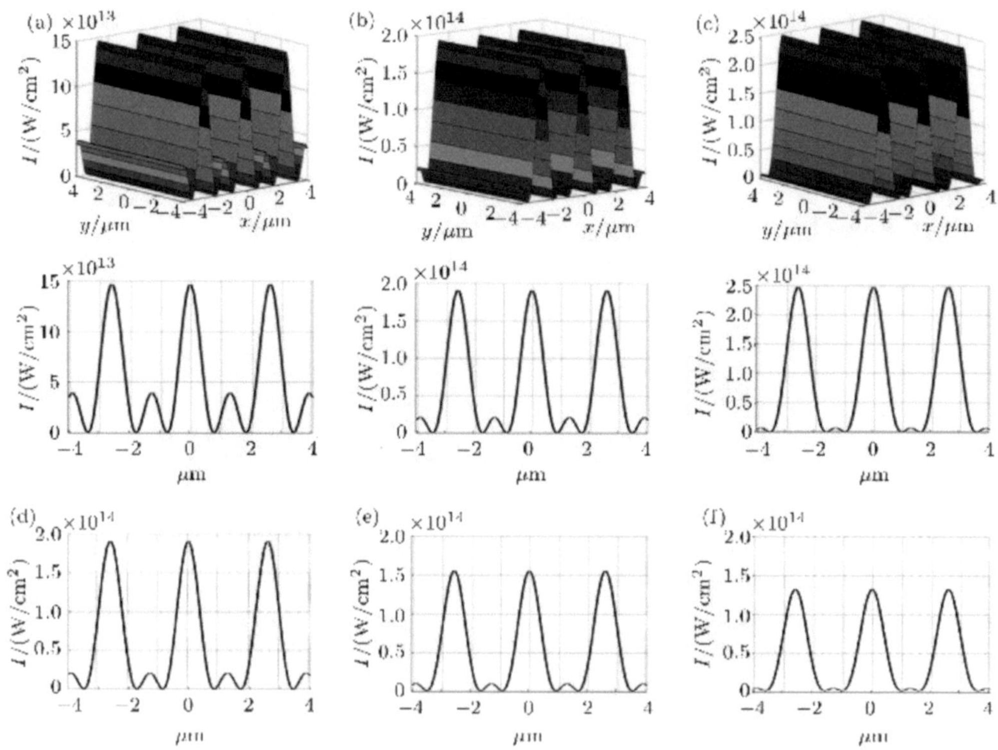

Figure 9. Intensity distributions simulated by using three interfering coplanar beams with different pulse energies of L_2 ((a), (b), and (c)) and L_3 ((d), (e), and (f)), where $k = 2\pi/\lambda$, $\lambda = 800$ nm, $\theta = 18°$ and different values of I_{01}, I_{02} and I_{03}, i.e. (a) $I_{01} = 2.12$, $I_{02} = 0.85$, and $I_{03} = 2.12$, (b) $I_{01} = 2.12$, $I_{02} = 2.12$, and $I_{03} = 2.12$, (c) $I_{01} = 2.12$, $I_{02} = 4.24$, and $I_{03} = 2.12$, (d) $I_{01} = 2.12$, $I_{02} = 2.12$, and $I_{03} = 2.12$, (e) $I_{01} = 2.12$, $I_{02} = 2.12$, and $I_{03} = 1.06$, and (f) $I_{01} = 2.12$, $I_{02} = 2.12$, and $I_{03} = 0.53$, which are all in units of 10^{13} W/cm^2.

The simulated intensity distributions in the interfering region formed by three coplanar interfering beams are shown in Figs. 9 (a), (b) and (c), which correspond to the middle beam (L2) pulse energies of 20 μJ, 50 μJ and 100 μJ, respectively. In Figs. 9 (a), (b) and (c), the upper images show the three-dimensional patterns and the nether curves display the two-dimensional cross-section patterns. From the cross-section images of Figs. 9 (a-c), we can see clearly that the periodic intensity patterns are formed by the periodic arrangement of inversely M-shape structures, which is attributed to stronger intensity peaks and weaker ones arrayed with the same period of about 2.6 μm alternately and periodically. These results are consistent with the AFM results as shown in Fig. 8. We notice that with the energy of the middle pulses increasing, the stronger intensity peaks in the interfering region increase, while the weaker intensity peaks decrease. It indicates that the intensity ratio of the stronger peak and the weaker one increases. If laser pulses with such a periodic inversely M-shape intensity distribution irradiate the materials, the stronger intensity peaks will lead to the formation of deeper grooves, and the weaker intensity peaks will induce the formation of shallower ones. As a result, the M-shape surface structures will be resultantly formed. This fabricating technology for the formation of M-shape gratings provides a fabricating method for special gratings with special use in industrial applications, such as the calibration for 3D reconstruction in computer vision application. And the fabricated M-shape grating can also be used in microfluidic chip devices as transport channels with different flowing speeds.

3.2. The formation of the triangular lattice by three noncoplanar interfered beams

When a single shot of three coplanar pulses interfered with each other, the M-shaped gratings could be fabricated as above. However, when a single shot of three non-coplanar pulses interfered with each other as depicted in Fig. 10 (a) by adjusting two time-delay for obtaining perfect overlap of the three pulses both spatially and temporarily, the two-dimensional periodic microstructure have been obtained, which distributed as a hexagonal lattice as shown in Fig. 10 (b) and agreed well with the simulated results [see Fig. 10 (c)] [16, 17]. In experiments, the geometric angles were kept as $\theta=30^0$ and $\varphi=35^0$ respectively as shown in Fig. 10 (a). When we set the pulse energy as 75 μJ, we can obtain two-dimensional periodic hexagonal lattice of microholes [see Fig. 10 (d-f)], however, when we set the pulse energy as 30 μJ, the two dimensional periodic microstructures present doughnut orbicular platform [see Fig. 10 (g-i)]. The period in the direction of line "*L*" is nearly 2.358 μm according to Fig. 10 (f) and (i) which agrees well to the calculated result [16].

The different microstructures in our experiments were attributed to the formation of plasma and molten liquid at different pulse energy levels. Generally it is hard to form plasma and molten liquid on the surface of silica glass by a laser with a wavelength of 800 nm because of the bigger band gap. However, the intensity at the focal point of the femtosecond laser beam where three beams interfere together could reach to 100 TW/cm^2 nearly. Such a high-energy influence within the focal volume ionized the silica glass quickly through the combined action of the avalanche and multiphoton processes [17, 35]. A layer of plasma formed on the surface of the silica glass at the time of the laser pulse duration. While the intensity decreased to be lower than a certain value, which can be called the ionized threshold as shown in Fig. 11 (a), the plasma vanished and the molten liquid of the material appeared. With the decrease of the intensity to be lower than a certain value, which could be called the molten threshold [see Fig. 11 (a)], the molten liquid of the material disappeared, so there would be a doughnut molten liquid formed in every enhanced spot of the interfered field [see Fig. 11 (b)]

When the pulse energy was set as 75 μJ, a layer of modulated plasma was formed by the interfered field with hexagonal lattice depicted in Fig. 10 (c) after the anterior part of the pulse was incident to the surface of the silica glass. Higher intensity induced a relatively larger plasma area in every enhanced spot of the interfered field. The subsequent posterior part of the pulse removed the plasma very swiftly because of the high light pressure originating from the reflection. Therefore, the microholes formed on the surface of the silica glass, as depicted in Fig. 10 (d-f). When the pulse energy decreased to 30 μJ, there was just a layer of the plasma with smaller areas than that of 75 μJ in the center of every enhanced spot in the interfered field after the anterior part of the pulse was incident to the surface. The

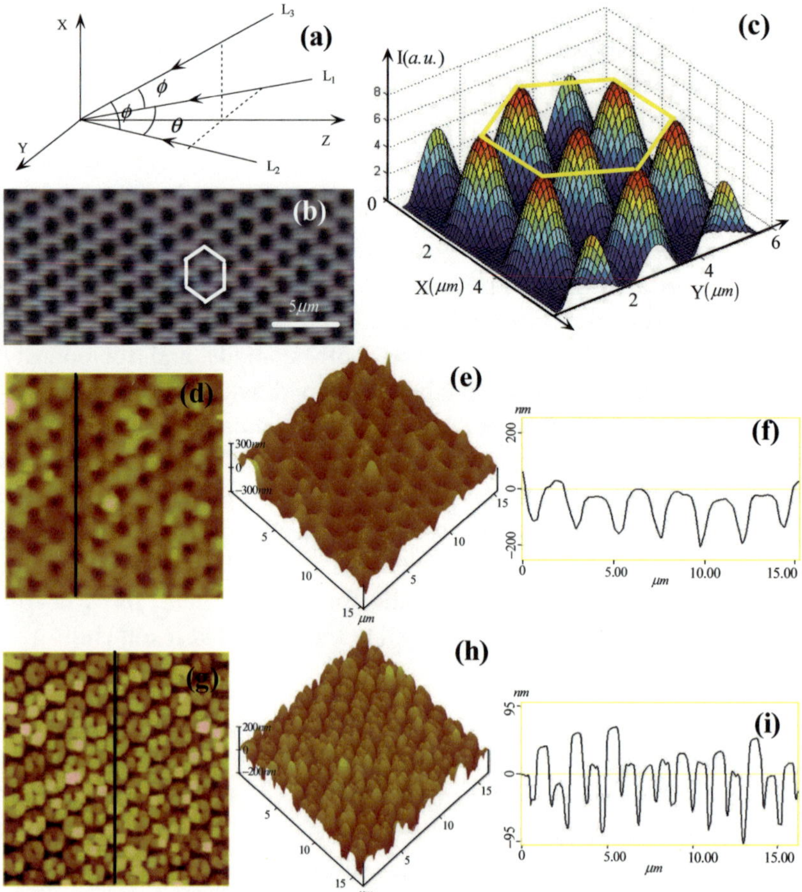

Figure 10. (a) Geometric sketch of the three non-coplanar interfered beams, and the sample is laid in the x–y plane, (b) Photo of the periodic microstructure fabricated in silica glass, taken by an optical microscopy, (c) Calculated intensity distribution of the interference by three non-coplanar beams. The induced periodic microstructure depicted by an AFM: (d) Energy is 75 μJ per pulse, (g) Energy is 30 μJ per pulse, (f) (i) Three-dimensional image of the microstructure of (d) and (g) respectively, (e) (h) The cross section of the microstructure in the direction of black line showed in (d) and (g) respectively

subsequent posterior part of the pulse also removed the central plasma very swiftly by means of light pressure, so there is a tiny hole formed in the center of the enhanced spot. However, in the region of the molten liquid, there were two distinct interaction components because of Marangoni effect, thermocapillary and chemicapillary, which resulted from the thermal potential of a temperature gradient and the chemical potential of a compositional gradient, respectively [17, 36-37]. The thermocapillary force moved the molten material outward from the center, while the chemicapillary force moved the molten material toward the center [38-39] as depicted in Fig. 10 (b). When the chemicapillary force dominated, a platform formed in the center of the spot. The combined action of the light pressure to the plasma and the chemicapillary force to the molten liquid induced a periodic orbicular platform on the surface of the silica glass as depicted in Fig. 10 (g-i).

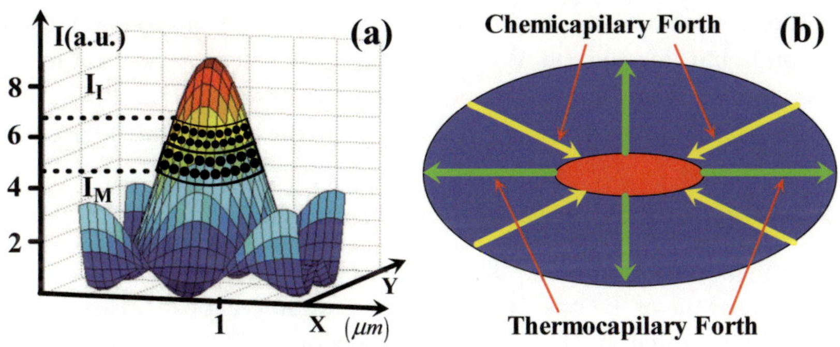

Figure 11. (a) The intensity distribution of the enhanced spot in the interfered field by three non-coplanar beams, I_I and I_M represent the densities of the ionized and molten thresholds respectively, (b) schematic diagram explanation for the mechanism of formed different structures owing to the formation of the plasma and Marangoni effect.

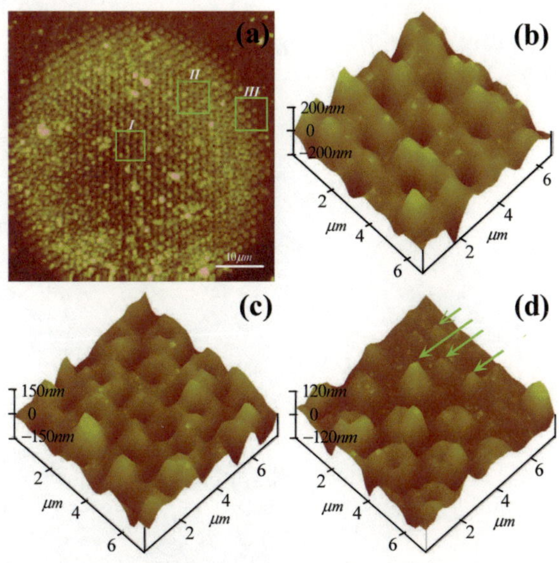

Figure 12. (a) AFM photo of the periodic structure induced by a single shot of three pulses with pulse energies of 50 μJ. Three-dimensional analyzed photos of the three selected parts I, II, and III in (a) are shown in (b)–(d) respectively.

In order to verify our explanation on the different induced microstructures, we set the pulses energy as 50 μJ, and the resulted microstructure is shown in Fig. 12 (a). Because of the Gaussian type intensity distribution of the pulse, three regions are selected from the central part (I), outer part (II) and the edge part (III) of the microstructure as depicted in Fig. 12 (a). The corresponding three dimensional images of the selected areas are shown in Fig. 12 [(b)-(d)], respectively. For the region I, the intensity is high enough to make the light pressure dominate, therefore the periodic microholes formed [see Fig. 12 (b)]. In contrast, in the region II, where the intensity is comparatively low, the chemicapillary force and the light pressure dominate, so that the microstructure present a orbicular platform [see Fig. 12 (c)].

However, in the selected area III, the intensity is very low and there is no plasma layer but just a layer of liquid formed on the surface of the silica glass. In this case, if the chemicapillary force dominates, the microcones can be observed. As shown in Fig. 12 (d), just several microcones (indicated by arrows) formed in the edge of the interfered districts.

3.3. The formation of the tetragonal lattice by four noncoplanar interfered beams

The interference of two beams creates a one-dimensional (1D) periodic pattern. By increasing the number of beams [16-20] or the double-exposure techniques [40], in principle, two-dimensional (2D) and three-dimensional (3D) periodic patterns can be designed. Although, as stated above, the interference of the three beams [13, 16-17] can create a 1D [13] or 2D [16, 17] periodic pattern on the surface of the materials of inside the transparent materials, the complicated optical setup is required for the interference of multiple laser beams, and its precise adjustment is difficult. Kondo *et al*. [18, 19] and Si *et al*. [20] have reported the fabrication of the 2D tetragonal lattice by four noncoplanar interfered pulses originating from the single pulse with the aid of the special DBS (diffractive beam splitter).

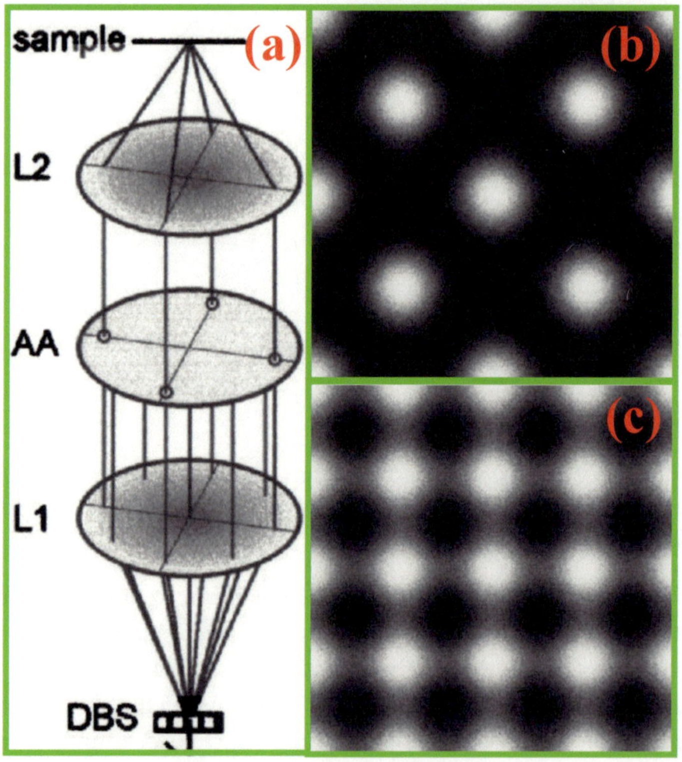

Figure 13. (a) Optical setup. DBS: diffractive beam splitter, L1 and L2: lenses,AA: aperture array. The inset shows the absorption spectrum of 4-mm-thick SU-8 film spin-coated on a coverglass. (b) Calculated intensity distribution by the interference of four beams which have same phase, (c) phase of one beam is shifted by π.

The optical setup used for the present experiments is shown in Fig. 13(a). Briefly, a DBS (G1023A or G1025A; MEMS Optical Inc.) divides the input laser beam into several beams, and the beams are collected on the sample by two lenses. Temporal overlap of the divided pulses is achieved without adjusting the optical path lengths. Each beam was made to be parallel or slightly focused by the adjustment of the distance between the two lenses. Slight focusing increased laser power density and helped to make the MPA efficient. The beams meant to form interference were selected by an aperture array, which is placed between the two lenses. Negative photoresist SU-8 (Microlithography Chemical Corp.) was used as an initial material for the fabrication. The absorption spectrum of SU-8 indicates that one-photon absorption is negligible at an 800-nm wavelength. Consequently, it is expected that photopolymerization, if occurring, is due to a multiphoton reaction. The layer of SU-8 was spin-coated on a coverglass plate having a thickness of about 4 μm, and prebaked before exposure to fs pulses. The interference angles θ_{air} (the angle between the main optical axis and the other beams in air) applied in the experiments were measured to be 33.6°, 21.9°, and 10.8°.

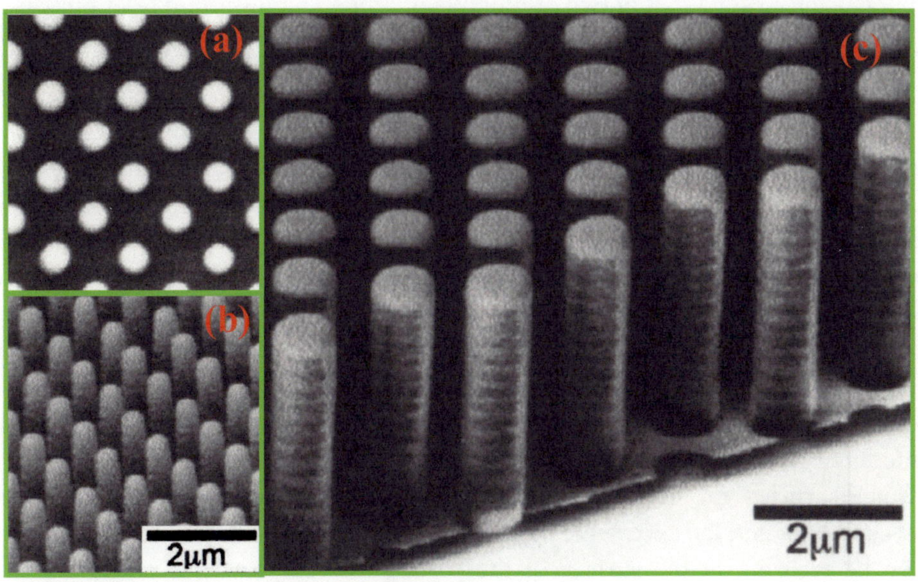

Figure 14. (a)(b) Top view and oblique view SEM images of the structure fabricated by the four-beam interference of fs pulses, (c) Close-up view of rods in the structure fabricated by four-beam interference of fs pulses with an interference angle of 21.9°.

Femtosecond pulses from a Ti:sapphire regenerative amplifier (wavelength of 800 nm, pulse duration of 150 fs, repetition rate of 1 kHz) were used for experiments. By using this method, periodic structures were fabricated. Periodic structures fabricated with an interference angle of 33.6° are presented in Fig. 14. In this figure, (a) and (b) show SEM images of the same sample from different perspectives. The oblique view shown in (b) clearly demonstrates that the periodic structure consists of high-aspect-ratio rods. The height and diameter of the rods are about 4 μm and 0.6 μm, respectively; thus, an aspect ratio of about 7 was achieved. It should be noted that the height of the obtained structure is not limited by the coherent length of the pulse. A higher structure could be obtained if the stiffness of the material allowed the rods to withstand capillary forces during the development procedure.

Figure 14 (c) provides a close-up view of the rods fabricated with an interference angle at 21.9°. To obtain this image, the coverglass containing the fabricated structures was deliberately broken and a small fragment was observed. As seen, the rods are slightly bellows-shaped. Distinct ring-like features repeat periodically along each rod with a period about 0.3μm. This could be attributed as the result of interference between the incident and reflected beams at the resist–coverglass interface.

4. Conclusion

In conclusion, we have reviewed the fabrications of the one-dimensional and two-dimensional periodic microstructures on the surface or inside of materials by multiphoton absorption using a single shot of two or multiple interfered femtosecond laser pulses.

Firstly, we have introduced the fabrication of the microgratings by a single shot of two interfered femtosecond laser pulses. When two interfered pulses overlapped on the surface of the silica glass, not only did we get the ordinary grating whose periods accorded with the theoretic equation $d=\lambda/[2\sin(\theta/2)]$, but also obtained the extraordinary grating whose period is a half of the ordinary grating. The formation of the modulation grating could be attributed to the higher-order modulation arising from second-harmonic generation (SHG) when the femtosecond laser pulse was incident to the surface of silica glass. The multiple layers of the microgratings have been fabricated in the different depths of the silica glass sample by a single shot of two interfered femtosecond laser pulses, and the results show that the fabricated gratings can be read out by one of the recorded beams very easily. The noble metal nanoparticles consisted microgratings have also been realized in silicate glasses by two interfered femtosecond laser pulses with the aid of heat treatment because the noble metal nanoparticles in silica glass can be precipitated after the irradiation of the femtosecond laser and the successive heat treatment. At the same time, the multiple gratings can also be realized on the surface of the silica glass samples by a single shot of two interfered femtosecond laser pulses with the aid of a mask, which is very significative for enhancing the processing efficiency of the fabricated microgratings.

Secondly, we have also introduced the fabrication of the 1-D or 2-D periodic microstructures by a single shot of multiple interfered femtosecond laser pulses. When a single shot of three coplanar interfered femtosecond laser pulses interfered with each other on the surface of the silica glass, M-shaped gratings can be formed and the morphologies of the M-shaped gratings can also be modulated by tuning the incident pulse energy. However, when a single shot of noncoplanar three or four pulses interfered with each other, two-dimensional periodic microstructure can be obtained, which distributed as a hexagonal lattice. Different morphologies of the induced structures such as microvoid, orbicular platform and nanotip, could be formed with the changes of the incident pulse energy. The fabrication of the 2D tetragonal lattice have also been fabricated in Negative photoresist SU-8 by four noncoplanar interfered pulses originating from the single pulse with the aid of the special DBS (diffractive beam splitter). Although the special DBS provided an easy way for realizing multiple beams interference, it also had some shortages, such as difficulty for fabricating microstructure with a smaller period, especially in the materials with big bandgap, such as silica glass.

Author details

Zhongyi Guo*
School of Computer and Information, Hefei University of Technology, Hefei, China
Department of Optoelectronic Science, Harbin Institute of Technology at Weihai, Weihai, China
Department of Physics, Harbin Institute of Technology, Harbin, China

Lingling Ran
Department of Optoelectronic Science, Harbin Institute of Technology at Weihai, Weihai, China
College of Electronic Engineering, Heilongjiang University, Harbin, China

Yanhua Han and Shiliang Qu
Department of Optoelectronic Science, Harbin Institute of Technology at Weihai, Weihai, China

Shutian Liu
Department of Physics, Harbin Institute of Technology, Harbin, China

Acknowledgement

This work was supported by the National Science Foundation of China (NSFC: 10904027; 61108018), the China Postdoctoral Science Foundation (AUGA41001348) and the Heilongjiang Province Postdoctoral Science Foundation (AUGA1100074), and development program for outstanding young teachers in Harbin Institute of Technology, HITQNJS. 2009. 033.

5. References

[1] C. Schaffer, A. Brodeur, J. García, and E. Mazur, "Micromachining bulk glass by use of femtosecond laser pulses with nanojoule energy," Opt. Lett. 26, 93-95 (2001)

[2] J. Koch, F. Korte, T. Bauer, C. Fallnich, A. Ostendorf, B. Chichkov. Nanotexturing of Gold Films by Femtosecond Laser-induced Melt Dynamics. Appl. Phys. A. 2005, 81:325–328.

[3] S. Juodkazis, H. Misawa, T. Hashimoto, E. Gamaly, and B. Luther-Davies, Laser-induced microexplosion confined in a bulk of silica: Formation of nanovoids, *Appl. Phys. Lett.* 88 (2006), 201909

* Corresponding Author

[4] Y. Li, Y. Dou, R. An, H. Yang, and Q. Gong,, "Permanent computer-generated holograms embedded in glass by femtosecond laser pulses," *Opt. Express,* 13 (2005) 2433-2438.

[5] Q. Zhao, J. Qiu, X. Jiang, E. Dai, C. Zhou, and C. Zhu, "Direct writing computer-generated holograms on metal film by an infrared femtosecond laser," *Opt. Express,* 13 (2005) 2089-2092.

[6] Z. Guo, S. Qu, Z. Sun, and S. Liu, "Superposition of orbit angular momentum of photons by combined computer-generated hologram fabricated in silica glass with femtosecond laser pulses", *Chin. Phys. B.* 17 (2008) 4199.

[7] Z. Guo, S. Qu, S. Liu, Generating optical vortex with computer-generated hologram fabricated inside glass by femtosecond laser pulses, Optics Communications, 273, 2007, 286-289.

[8] Z. Guo, H. Wang, Z. Liu, S. Qu, J. Dai, and S. Liu, Realization of holographic storage on metal film by femtosecond laser pulses micromachining, Journal of Nonlinear Optical Physics and Materials, 18(4), 2009, 617-623.

[9] Z. Guo, W. Ding, S. Qu, J. Dai, and S. Liu, Self-assembled volume grating in silica glass induced by a tightly focused femtosecond laser pulses, Journal of Nonlinear Optical Physics and Materials, 18(4), 2009, 625-632.

[10] K. Zhou, Z. Guo, W. Ding, and S. Liu, Analysis on volume grating induced by femotosecond laser pulses, Opt. Express 18(13), 2010, 13640-13646.

[11] Y. Li, W. Watanabe, K. Yamada, T. Shinagawa, K. Itoh, J. Nishii, and Y. Jiang "Holographic fabrication of multiple layers of grating inside soda–lime glass with femtosecond laser pulses". *Appl.Phys.Lett,* 80 (2002) 1508.

[12] S. Qu, J. Qiu, C. Zhao, X. Jiang, H. Zeng, C. Zhu, and K. Hirao, "Metal nanoparticle precipitation in periodic arrays in Au_2O-doped glass by two interfered femtosecond laser pulses," *Appl. Phys. Lett.* 84, 2046-2048, (2005).

[13] Y. Han, S. Qu, Q. Wang, Z. Guo, and X. Chen, Controllable grating fabrication by three interfering replicas of single femtosecond laser pulse, Chinese physics B, 18, 5331, (2009).

[14] S. Qu, C. Zhao, Q. Zhao, J. Qiu, C. Zhu, and K. Hirao, One-off writing of multimicrogratings on glass by two interfered femtosecond laser pulses, Opt. Lett. 29, 2058 (2004).

[15] Z. Guo, S. Qu, L. Ran, and S. Liu, Modulation grating achieved by two interfered femtosecond laser pulses on the surface of the silica glass, Appl. Surf. Sci., 253, 8581, (2007).

[16] Z. Guo, S. Qu, Y. Han, and S. Liu, Multi-photon fabrication of two-dimensional periodic structure by three interfered femtosecond laser pulses on the surface of the silica glass, Optics Communications, 280, 23-26, (2007)

[17] Z. Guo, L. Ran, Y. Han, S. Qu, and S. Liu, "The formation of novel two-dimensional periodic microstructures by a single shot of three interfered femtosecond laser pulses on the surface of the silica glass", *Opt. Lett.* 33 (2008) 2383.

[18] T. Kondo, S. Matsuo, S. Juodkazis, and H. Misawa, "Femtosecond laser interference technique with diffractive beam splitter for fabrication of three-dimensional photonic crystals" *Appl.Phys.Lett.* 79 (2001) 725-727.

[19] T. Kondo, S. Matsuo, S. Juodkazis, V. Mizeikis, and H. Misawa, "Multiphoton fabrication of periodic structures by multibeam interference of femtosecond pulses," *Appl.Phys.Lett.* 82 (2003) 2758-2760.

[20] J. Si, Z. Meng, S. Kanehira, J. Qiu, B. Hua and K. Hirao, "Multiphoton-induced periodic microstructures inside bulk azodye-doped polymers by multibeam laser interference", *Chem. Phys. Lett.* 399 (2004) 276-279

[21] L. Z. Cai, X. L. Yang, and Y. R. Wang, "All fourteen Bravais lattices can be formed by interference of four noncoplanar beams," Opt. Lett. 27, 900-902 (2002)

[22] L. Z. Cai, X. L. Yang, and Y. R. Wang, "Formation of three-dimensional periodic microstructures by interference of four noncoplanar beams," J. Opt. Soc. Am. A 19, 2238-2244 (2002)

[23] Zhongyi Guo, Shiliang Qu, Shutian Liu, and Jung-Ho Lee, Periodic microstructures induced by interfered femtosecond laser pulses, Proc. of SPIE Vol. 7657 76570K-1, 2010.

[24] D. Von der Linde, H. Schulz, T. Engers and H. Schuler, Second harmonic generation in plasmas produced by intense femtosecond laser pulses, *IEEE J. Quantum Electron.* 28, 2388, (1992).

[25] Y. Shen, The Principles of Nonlinear Optics, Wiley, New York (1984) 553.

[26] U. Österberg and W. Margulis, "Dye laser pumped by Nd:YAG laser pulses frequency doubled in a glass optical fiber," Opt. Lett. 11, 516-518 (1986).

[27] S. Qu, C. Zhao, X. Jiang, G. Fang, Y. Gao, H. Zeng, Y. Song, J. Qiu, C. Zhu, and K. Hirao, Optical nonlinearities of space selectively precipitated Au nanoparticles inside glasses, Chem. Phys. Lett. 368, 352 (2003).

[28] H. B. Liao, R. F. Xiao, J. S. Fu, H. Wang, K. S. Wong, and G. K. L. Wong, "Origin of third-order optical nonlinearity in Au:SiO_2 composite films on femtosecond and picosecond time scales," Opt. Lett. 23, 388-390 (1998)

[29] J. Qiu, M. Shirai, T. Nakaya, J. Si, X. Jiang, C. Zhu, Space-selective precipitation of metal nanoparticles inside glasses, Appl. Phys. Lett. 81, 3040 (2002).

[30] James R. Adleman, Helge A. Eggert, Karsten Buse, and Demetri Psaltis, "Holographic grating formation in a colloidal suspension of silver nanoparticles," Opt. Lett. 31, 447-449 (2006).

[31] E. Valentin, H. Bernas, C. Ricolleau, and F. Creuzet, Ion Beam "Photography": Decoupling Nucleation and Growth of Metal Clusters in Glass, Phys. Rev. Lett. 86, 99 (2001).

[32] Y. Nakata, T. Okada, and M. Maeda, Fabrication of dot matrix, comb, and nanowire structures using laser ablation by interfered femtosecond laser beams, Appl. Phys. Lett. 81, 4239 (2002).

[33] Yan Li, Kazuhiro Yamada, Tomohiko Ishizuka, Wataru Watanabe, Kazuyoshi Itoh, and Zhongxiang Zhou, "Single femtosecond pulse holography using polymethyl methacrylate," Opt. Express 10, 1173-1178 (2002)

[34] G. Qian, J. Guo, M. Wang, J. Si, J. Qiu, K. Hirao, Holographic volume gratings in bulk perylene-orange-doped hybrid inorganic-organic materials by the coherent field of a femtosecond laser, Appl. Phys. Lett., 83 (2003), p. 2327

[35] S. Juodkazis, H. Misawa, T. Hashimoto, E. Gamaly, and B. Davies, Laser-induced microexplosion confined in a bulk of silica: Formation of nanovoids, Appl. Phys. Lett. 88, 201909 (2006).

[36] Y. Lu, S. Theppakuttai, and S. Chen, Marangoni effect in nanosphere-enhanced laser nanopatterning of silicon, Appl. Phys. Lett. 82, 4143 (2003).

[37] J. Eizenkop, I. Avrutsky, G. Auner, D. Georgiev, and V. Chaudhary, Single pulse excimer laser nanostructuring of thin silicon films: Nanosharp cones formation and a heat transfer problem, J. Appl. Phys. 101, 094301 (2007).

[38] S. Huang, Z. Sun, B. Luk'yanchuk, M. Hong, and L. Shi, Nanobump arrays fabricated by laser irradiation of polystyrene particle layers on silicon, Appl. Phys. Lett. 86, 161911 (2005).

[39] R. Piparia, E. Rothe, and R. Baird, Nanobumps on silicon created with polystyrene spheres and 248 or 308 nm laser pulses, Appl. Phys. Lett. 89, 223113 (2006).

[40] K. Kawamura, N. Sarukura, M. Hirano, N. Ito, and H. Hosono, Periodic nanostructure array in crossed holographic gratings on silica glass by two interfered infrared-femtosecond laser pulses, Appl. Phys. Lett. 79, 1228 (2001).

Section 4

Unusual Applications

Chapter 11

Ion Acceleration by High Intensity Short Pulse Lasers

Emmanuel d'Humières

Additional information is available at the end of the chapter

http://dx.doi.org/10.5772/46137

1. Introduction

1.1. Context

Lasers can now deliver short and very intense pulses capable of producing hot and dense plasmas. The diffusion of such lasers has allowed studying the behaviour of ionized media in novel ways. Ion acceleration through laser plasma interaction is a recent research subject and its fast-paced progresses highlight the mastering of high power laser systems and the better understanding of the physics of high intensity laser plasma interaction. Short and intense laser pulses have also permitted the development of areas of research like harmonic generation, hard X-ray radiation sources, energetic particles sources and inertial confinement fusion. These researches have lead to numerous applications in transdisciplinary domains: material science, development of X-UV lasers, surface treatments, chemistry and biology are just a few examples. Large national programs are focused on these topics like the National Ignition Facility in California, the Laser Mégajoule and the Apollon laser in France.

These topics use plasmas where kinetic and collective effects are dominant with numerous non linear phenomena. Impressive progresses are being accomplished in the modelling of these phenomena and in their optimization thanks to Particle-In-Cell (PIC) codes. This is due to the fact that the mechanisms at play are complex and difficult to model analytically, or using a fluid model. PIC codes used to describe plasmas solve the coupled system of Maxwell's equations (for electromagnetic fields) and Vlasov equations (for each particle specie). The PIC method consists in solving Vlasov equation for each particle specie by the characteristics method for macro-particles, each representing a given amount of real particles but that conserves the mass and the charge of the particles of the specie that is studied. Several research areas of high intensity laser plasma interaction require PIC codes capable of modelling atomic physics processes (ionization and collisions for instance) and energy loss processes through radiation of energetic charged particles in intense electromagnetic fields.

1.2. Ion acceleration in laser-plasma interaction

Particle accelerators play an important role in numerous scientific programs. Thanks to their versatility, they have found numerous applications in science and medical fields. Plasmas can endure very large acceleration gradients, making them very interesting as accelerating media. Particle acceleration by interaction of a high intensity laser with a gas or a solid is therefore attracting a growing interest. More and more laboratories in the world possess their own laser system and the results obtained by these laboratories, both theoretical and experimental, are very promising. It is nowadays possible to accelerate protons at energies of several tens of MeV [Snavely et al., 2000], and electrons at energies higher than a GeV [Leemans et al., 2006]. Moreover, the observation of intense energetic ion beams [M. Borghesi et al., 2006] represents one of the most interesting recent developments in "High Field Science", made possible by the advent of ultra-short and ultra-intense lasers [Mourou et al., 2006].

High power laser pulses open the possibility of ion acceleration to several tens MeV for protons, and higher than a GeV for Carbon ions on millimeter scale distances. These beams therefore open original perspectives, as new exploration tools of fundamental physics, or as new ways to increase the discovery potential of existing accelerators. Possible applications of these beams are numerous: ion fast ignition for inertial confinement fusion (in which a "spark", a beam of energetic ions, is used to ignite a target and reduces the total energy cost), high resolution radiography of plasmas, hadrontherapy, radio-isotopes production, laboratory astrophysics, etc… Indeed, the properties, not explored until now, in terms of current, duration (of the order of a few ps), laminarity (low divergence) and compactness – the acceleration process takes place on a distance of the order of ten microns – show that these beams are complementary, and even superior on some aspects to conventional accelerators. Finally, the constant development of high intensity lasers, on the duration, energy, repetition rates, and compactness aspects, makes the applicability perspectives of these intense sources extremely promising.

The maximum energy, as well as the properties of these ion/proton beams go well beyond the limits explored during the first experiments performed at the end of the seventies that observed ion beams from laser-matter interaction. In these first experiments, targets were irradiated with CO_2 (λ=10,6 μm) lasers with pulse durations < 1 ns and peak intensities $I\lambda^2$ of the order of 10^{17}-10^{18} W.cm^{-2}.μm^{-2} [Begay et al., 1982]. The ions were emitted from the surface of the target irradiated by the laser, with a large angular divergence and an energy of a few 100 keV/nucleon. The protons originated from contaminants on the target surface. These experiments, repeated using Nd:Glass lasers at λ=1 μm produced similar results, with a higher ion energy due to the increased $I\lambda^2$ [Fews et al., 1994; Clark et al., 2000b] . Simulations using the parameters of these experiments showed that the ion acceleration process could in this case be described as a plasma expansion in vacuum mechanism, first considered isothermal [Gitomer et al., 1986; Sack et al., 1987], or adiabatic.

The discovery of energetic and collimated ion beams was made thanks to a recent fortuitous observation. The related experimental results were published in 2000 and obtained on the

LLNL PetaWatt system in the USA. The scientists conducting the experiments wanted to study the electron beams generated during the interaction of this laser with solid targets. From these first experiments using short pulses (ps) [Clark et al., 2000a; Snavely et al., 2000; Maksimchuk et al., 2000], it therefore appeared that these ion beams presented extremely interesting characteristics. Since, many experiments helped to understand the physics of laser ion acceleration with solid targets, and showed that this source possesses unique characteristics in comparison to beams produced using conventional accelerators:

- Excellent transverse laminarity (more than 100 times better than conventional accelerators [Borghesi et al., 2004; Cowan et al., 2004]).
- Extremely small equivalent source size (μm), characteristic equivalent to the laminarity that allows to obtain an excellent spatial resolution for radiography applications [Borghesi et al., 2002; Mackinnon et al., 2004].
- Very short duration of the source (~ps), allowing to obtain an excellent temporal resolution for pump/probe applications.
- Large energy spectrum (which can be modulated [Hegelich et al., 2006; Schwoerer et al., 2006; Toncian et al., 2006], as can be modulated the divergence [Patel et al., 2004; Toncian et al., 2006]), allowing to obtain in a single radiography, through the different associated times of flight, a movie of the probed phenomena in a single shot.
- Very high current at the source ~kA.
- Intrisic compactness of the acceleration, from 0 to a few tens MeV in a few ten microns (the compactness of the whole "accelerator" is therefore linked to the one of the laser).

It was demonstrated by studying the acceleration physics that the energetic ions observed were accelerated electrostatically from the non-irradiated surface, by fast electrons generated in the interaction zone between the ultra-intense laser and the target, and propagated through the target, as originally predicted [Hatchett et al., 2000; Wilks et al., 2001]. It was also demonstrated that the divergence parameters or the ion energy distribution could be changed using micro-setups triggered by ultra-short lasers [Toncian et al., 2006].

These studies on the understanding of the properties of laser accelerated ion beams have opened original application perspectives in various domains, for fundamental research (study of warm dense matter) and the development of diagnostics for ICF (Inertial Confinement Fusion), as well as for possible medical applications, extremely motivating even if their effective realization is still uncertain (protontherapy for instance):

- Plasma diagnostics for inertial fusion with lasers (analysis of rapidly evolving electric and magnetic fields), analysis of dense objects (static and dynamic like in the case of the compression of a fusion capsule) with resolutions ~ps and ~μm, or for the ignition [Roth et al., 2001; Atzeni et al., 2002; Temporal et al., 2002; Temporal, 2006; Key et al., 2006] of fusion reaction for ICF [Tabak et al., 1994].
- Production of large scale hot and dense plasmas [Patel et al., 2004; Antici et al., 2006], an important aspect for the study of equation of states.
- Injection in conventional accelerators [Krushelnick et al., 2000a; Cowan et al., 2002, Antici et al. 2008b] and evaluation of the potential of laser-accelerated protons for

medical applications (protontherapy [Bulanov et al., 2002a; Bulanov et al., 2002b; Fourkal et al., 2002; Malka et al., 2004] or production of radio-isotopes [Santala et al., 2001; Fritzler et al., 2003; Ledingham et al., 2004; Lefebvre et al. 2006]).

Recently (in 2011), two teams of experimentalists from the Los Alamos National Laboratory beat the maximum laser accelerated ion energy world records one after the other, obtaining 80 MeV and then 120 MeV for protons and Carbon ions with energies higher than a GeV with a 130 TW laser system and an intensity of a few 10^{20} W/cm^2. The main laser ion acceleration mechanisms are now known and understood, the acceleration by hot electrons (or Target Normal Sheath Acceleration, TNSA), and the acceleration by radiation pressure, but new regimes are explored and an important theoretical and experimental effort is ongoing to further our understanding of laser ion acceleration with various types of targets, to understand its limits in the high laser intensity and in the high laser energy regimes, and to get closer to some applications (Fast Ignition, hadrontherapy…).

Section 2 of this Chapter will present laser ion acceleration with flat dense targets, both through plasma expansion and through the laser radiation pressure with some details on the particular case of circularly polarized laser pulses. The transparency regime is also discussed. Section 3 is devoted to laser ion acceleration using low density targets, either gas jets, exploded foils or foams. Section 4 is devoted to laser ion acceleration using structured targets like cones or reduced mass targets to improve specific characteristics. Section 5 is devoted to the ultra high laser intensity and ultra high laser energy regimes that are starting to be accessible with recent laser systems and that will be further explored in the near future with laser systems under construction. Section 6 is devoted to the presentation of several applications of laser ion acceleration and Section 7 presents the conclusions and perspectives of this Chapter.

2. Laser ion acceleration with flat dense targets

Laser ion acceleration using flat solid foils has been the subject of numerous studies in the last ten years, experimental, theoretical and numerical. This Section presents some important results obtained in these studies. First, results on the TNSA regime are presented with the main scaling laws for the maximum proton energy in this regime. Laser ion acceleration through the laser radiation pressure is then introduced with its extension to circularly polarized laser pulses. Both recent theoretical results and first experimental demonstrations are discussed. The special case of the transparency regime is then presented allowing a coupling of the TNSA and of the radiation pressure regime.

When a high intensity laser (>10^{18} W/cm^2) with a short pulse duration (30 fs – 10 ps) irradiates a target, this pulse produces, at the effective critical density (~10^{21} cm^{-3}), where the laser pulse can not penetrate the plasma, a population of "fast" electrons through different mechanisms (acceleration by the ponderomotive potential [Wilks et al., 1992], resonant absorption [Estabrook et al. 1978], the Brunel effect [Brunel, 1987], mechanisms that are not the primary goal of this Chapter). The dominant mechanism accelerating the electrons depends on the gradient on the front surface and on the laser incidence angle [Lefebvre et al., 1997]; moreover, depending on the interaction conditions, the electrons are accelerated, either in the laser direction, or in the target perpendicular direction [Santala et al., 2000].

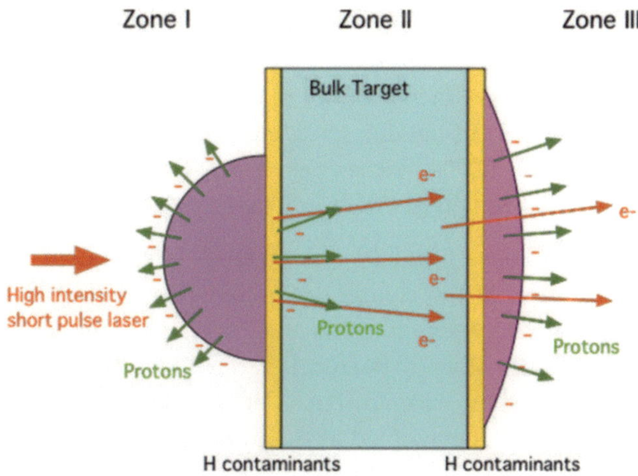

Figure 1. Illustration of the various mechanisms leading to high energy ion acceleration by a short and high power laser interacting on a solid target.

In zones I and III (Figure 1), the mechanism is a plasma expansion, with at its front a Debye sheath of hot electrons. In zone III, where there is no initial gradient (in zone I, it is possible only by using very high temporal contrast lasers, which has only been possible recently [Ceccotti et al., 2007]), there is an important difference with zone I: initially, the situation is the one of a virtual cathode where electrons partially extend in vacuum whereas atoms remain structured on the target surface. This is the origin of the high quality of the beam. Following this initial phase, the electrons steadily transfer their energy to the ions and the charge separation field associated to the electron sheath gradually decreases until the acceleration stops, when the electrons are at the same velocity as the ions. This results in the existence of a "cutoff" energy, i.e. a maximum energy of the ion beam, already observed in the seventies. In zone II, protons are accelerated by the strong electrostatic field set up when the laser pulse pushes electrons inside the target.

Except the energy reached by the ions, the main difference between the accelerations in zone I and II on one hand and III on the other hand is that one develops from a pre-existing plasma while the other one originates from an initially cold surface, unperturbed. Therefore, acceleration in zone I or II produces a beam largely divergent (2π sr) and turbulent (the structure of the ion front is thermal, not ordered) while acceleration in zone III produces a beam with a limited divergence (determined by the spatial structure of the hot electron sheath), and especially, extremely laminar thanks to the structure of the back surface (if it has a good surface quality) [Fuchs et al., 2005; Fuchs et al., 2007a].

2.1. The target normal sheath acceleration

The production of energetic electrons when a high intensity laser interacts with a solid foil has been explained above. These electrons are going to form a hot electron cloud around the target leading to a plasma expansion on both sides of the target [Wilks et al., 2001]. The resulting electrostatic field will then accelerate ions located on both sides of the target to high energies. The amplitude of this field (~TV/m) depends on the hot electron temperature and density. Protons can also be accelerated if they are present in the target or on its surface as contaminants. The basics of the theory of hot plasma expansion [Mora, 2003; Mora, 2005] will be presented and their use in obtaining scaling laws for the maximum proton energy will be explained [Fuchs et al., 2006a; Fuchs et al., 2007a, Robson et al., 2007]. This regime is the so-called "Target Normal Sheath Acceleration" regime (TNSA).

The space charge fields in zones I and III are evidently determined by the local electron density and by the extension of the charge separation [Hatchett et al., 2000; Wilks et al., 2001]. The separation length between electrons and ions is given either by the Debye length (λ_D~1 µm for 1 MeV electrons), either by the local gradient length (notably for the part on the front of the target) if it is greater than λ_D [Fuchs et al., 2007b]. As in most cases the laser pulse has a pedestal, a plasma is generated at the front surface and therefore the field is in general lower in zone I than in zone III, which confirms the experimental measurements: the energy of ions accelerated towards the laser (in zone I) is smaller than the energy of the ions accelerated in the same direction as the laser propagation direction (in zone II or III) [Ledingham et al., 2004; Yang et al., 2004]. Thanks to the density and temperature of electrons in zone III, and to the value of λ_D, the field in zone III is initially of the order of a few TV/m.

Acceleration in zone III is the most suited to produce a high quality beam, laminar, that is adapted to imaging applications. Because of the high energy required for electrons to reach zone III, this mechanism was only made possible by the advent of high power laser with short pulse durations. This mechanism also produces, with nowadays lasers, more energetic ions than the acceleration in zone II. The respective energy that can be reached by the accelerated protons in the three zones is illustrated in [J. Fuchs et al., 2007a]. Acceleration in zone II produces less energetic protons for different laser/target conditions as confirmed by PIC simulations performed in a wide range of parameters (for laser intensities going from 10^{17} to 5×10^{19} W/cm^2, and for laser pulse duration ranging from 10 fs to 500 fs) [Sentoku et al., 2003; Murakami et al., 2001; Pommier et al., 2003; Pukhov, 2001]. It is also noteworthy that the acceleration in zone II produces a beam with a lower quality [Fuchs et al., 2005] due to the stochastic nature of laser-plasma interaction (in the preplasma) which does not produce a charge separation field structure at the critical density interface as ordered as the one existing at the back (non perturbed) surface of the target (zone III). The collimation and low emittance of rear side accelerated ions are considerable assets for numerous applications.

Nevertheless, we can note that the use of high temporal contrast lasers will erase the differences between the acceleration in zones I and III: in these conditions, there will be no preplasma, and therefore the spatial quality of the ion beams produced in zones I and III will be similar [Ceccotti et al., 2007]. If the target is, in addition, very thin, the electron populations will be similar on both sides of the target and the ion energy will therefore also be very close. On the contrary, for thick targets, the electrons will have a lower density when arriving at the back surface of the target (zone III), and the resulting accelerating field as well as the maximum ion energy will thus be lower in this zone than in zone I.

An acceleration model has been developed to obtain a predictive model of maximum ion energy useful for experimentalists [Fuchs et al., 2006; d'Humieres et al, 2006; Fuchs et al., 2007a]. This model uses theoretical results on the isothermal expansion of a hot plasma [Mora, 2003] coupled to a model of characteristic ion acceleration time. By studying the dependence of this characteristic time on various parameters, it was possible to extend the validity of this model to very short pulse durations.

This study has also allowed to link the characteristics of the TNSA accelerated ion beams to the characteristics of the energetic electron population in order to develop a new diagnostic of this electron population for laser-plasma interaction experiments [Antici et al., 2008a].

To obtain an analytical estimate of the maximum energy that can be gained by protons accelerated forward from the back surface of the target (zone III) [Fuchs et al., 2006; d'Humieres et al, 2006; Fuchs et al., 2007a] it is possible to use the approach chosen by Mora [2003], which was based on an approach initiated in the seventies to treat the expansion of a plasma in vacuum, and adapted to the case of a hot electron burst initiating the expansion. In the case of a self-similar isothermal expansion [Gurevitch et al., 1966], a simple evaluation of the accelerated ions maximum energy is:

$$E_{max/back}=2ZT_{hot}*[\ln(t_p+(t_p^2+1)^{1/2})]^2$$

Where T_{hot} is the electron temperature in the sheath in zone III and $t_p=\omega_{pi}*t_{acc}/(2*\exp[1])^{1/2}$ with t_{acc} the acceleration time and $\omega_{pi}=\sqrt{\frac{n_{e0}Ze^2}{M_i\varepsilon_0}}$ the ion plasma frequency (n_{e0} is the initial electron sheath density at the back of the target and M_i the ion mass). As already mentioned, in most experiments, a plasma is located on the front surface during the target irradiation by the main pulse. In this case, as already showed in experiments, T_{hot} can be well approximated using the ponderomotive potential $m_ec^2(\gamma-1)$ [Malka et al., 1996; Popescu et al., 2005]. In reality the expansion is far from being isothermal, and is actually adiabatic [P. Mora, 2005]. Starting from a hot electron cloud, there is a progressive energy transfer through the electrostatic fields to the ions. Nevertheless, such an adiabatic model cannot be solved analytically, only numerically.

The ambition of this model was to see if the isothermal model could be used, having the advantage of being simple and easy to use analytically, by compensating with a finite (determined) acceleration time the fact that the hot electron population is maintained at a constant temperature [Fuchs et al., 2006]. To confront this approach with experimental results and see how interesting it is, it is important to determine the missing ingredients, namely n_{e0} and t_{acc}.

To determine n_{e0}, the quantity of hot electrons accelerated inside the target by the laser has been evaluated in a simplified way, with an energy balance. This quantity can be evaluated as $N_e=f\ E/T_{hot}$, where f is the laser energy fraction absorbed in the plasma and converted in hot electrons. This fraction f was measured by using a buried layer inside the targets and by measuring the X-ray radiation induced in these layers by the hot electrons. This way, it is possible to determine the total energy in the hot electrons and therefore f. It depends of the incident laser intensity: $f = 1.2\times10^{-15}I^{0.74}$ (where I is in W/cm^2 units) with a maximum f of 0.5. We can now assume that these electrons are, longitudinally, confined and transversally diluted in the target [Kaluza et al., 2004] until they are spread on the electron sheath area A_{sheath} at the back of the target. This would give $n_{e0}=N_e/(d\times A_{sheath})$ where $A_{sheath}=\pi(r_0+d\times\tan\theta)^2$, d is the initial target thickness and r_0 is the radius of the zone from which the electrons are accelerated at the critical interface, radius that we assume to be equal to the focal spot radius (r_0=FWHM). A_{sheath} also depends on the hot electron divergence half-angle in the target θ. Several experimental and theoretical studies show that θ is between 15° and 40° [Fuchs et al., 2003; Stephens et al., 2004; Adam et al., 2006], increasing with laser energy [Lancaster et al., 2007; d'Humières et al., 2006].

In the model, θ=25° has been chosen in agreement with what the LULI team had deduced from astigmatic proton emission measures [J. Fuchs et al., 2003]. This model is supported by the fact that:

- The values of the size of the electron sheath at the back of the target obtained with this model are in good agreement with direct measures that can be obtained of the size of the proton source.
- The values of electron density at the back of the target as well as the ones of the zone on which these electrons spread are also in good agreement with direct measures of these parameters performed using an interferometry diagnostic with a probe beam reflected on the back of the target that was adapted to this problem [Antici et al., 2008a].

To determine the accelerating time t_{acc}, different approaches were tested by comparing them not only to data obtained during experiments for laser intensities ranging from 2×10^{18} to 6×10^{19} W/cm^2, and for laser pulse durations between 150 fs to 10 ps [Fuchs et al., 2006], but also to data obtained with shorter pulse durations taken from various articles. A priori, one could think that t_{acc}, which represents physically the time during which in the model the electron temperature is maintained to T_{hot} must be proportional to the laser pulse duration. Nevertheless, for very short pulse durations, it is important to take into account a minimum time for the energy transfer between electrons and ions. Therefore, for very short pulse durations, the acceleration time is not proportional to the laser pulse duration anymore but needs to tend towards a constant value. It is also necessary to include in the model the fact that this acceleration time needs to depend on the laser intensity. Indeed, for low laser intensities, this acceleration time needs to be increased because the expansion is slower. In the end, the function, which is most adapted to the different experimental measures, is [J. Fuchs et al., 2007a]: $t_{acc}=\alpha^*(\tau_L+t_{min})$ where τ_L is the laser pulse duration, t_{min}=60 fs and where α goes linearly from 3 for an intensity of 2×10^{18} W/cm^2 to 1.3 at 10^{19} W/cm^2 and is constant at 1.3 for higher intensities. The figures in [Fuchs et al., 2006; d'Humières et al., 2006], as well as Figure 2 show a comparison between the calculations of the maximum proton energy $E_{max/back}$ and experimental results obtained in a wide range of laser and target parameters (in particular for short pulse durations, and/or low intensities that were not explored experimentally at LULI). This model was also validated using 2D PIC simulations [d'Humières et al., 2006].

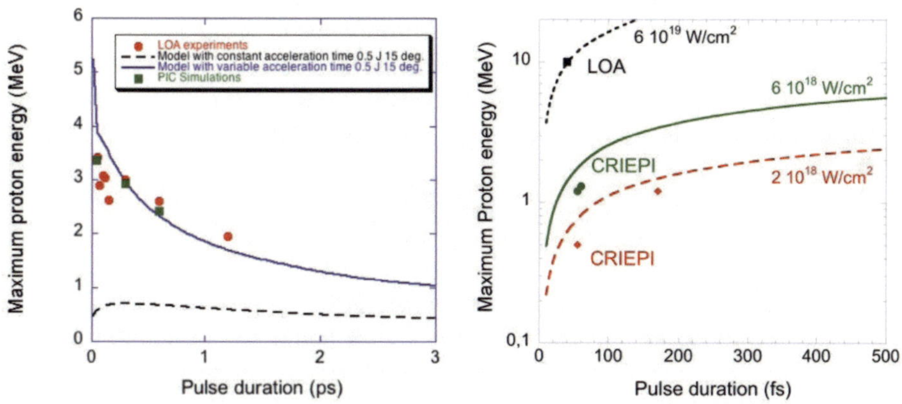

Figure 2. (Left) Data (points), simulations (squares) and model (lines) giving the maximum proton energy as a function of the pulse duration for very short pulse durations. In the model, a focal spot of 6 μm FWHM is used as well as a target thickness of 5 μm. The different lines correspond to different intensities for which the experiments were performed. (Right) Same graph but for various laser systems (points) and the model (lines). Here a thickness of 5 μm was used in the model as most of the reported experiments used a similar thickness.

It is nevertheless important to keep in mind that the acceleration time t_{acc} is not the « true » acceleration time, it is the typical time during which the electron population remains at a « high » temperature. In reality, the hot electron temperature decreases progressively during the energy transfer between electrons and ions.

To conclude this Section, the analytical work on the maximum proton energy produced with the accelerating mechanism in zone III has allowed to obtain an evaluation tool that is able to correctly reproduce the existing data obtained at LULI and other laser systems, and to extrapolate it to other laser parameters (it is important to keep in mind that this model can only be entirely reliable in the parameter range in which it has been tested).

Effects of a small scale gradient at the back of the target on laser ion acceleration in this regime has also been investigated [Fuchs et al., 2007b], and the physics of long pulse interaction and its effects on the creation of the accelerating electron sheath is now being intensely studied.

2.2. Laser ion acceleration through the laser radiation pressure

Laser ion acceleration through the laser radiation pressure has first been studied theoretically and using Particle-In-Cell simulations by Wilks et al. [1992] and Denavit et al. [1992]. It was first referred as the hole boring regime and can lead to the development of strong electrostatic shocks inside the target [Silva et al., 2004]. Electrons are pushed by the laser ponderomotive force at the front of the target and lead to a strong charge separation setting up a high amplitude electrostatic field. This field will accelerate front surface ions to high energies during the duration of the laser pulse. This regime is also referred to as Radiation Pressure Acceleration (RPA) or the laser piston regime.

It is possible to estimate analytically the maximum proton energy of forward accelerated protons from the front surface. Indeed, the maximum velocity of the accelerated protons in zone II is given by two times the receding plasma surface velocity [Pukhov, 2001; Denavit, 1992]. In the case of a total laser reflection at the critical interface (i.e. when the laser piston efficiency is maximum), this receding velocity is given by [Wilks et al., 1992; Wilks, 1993] :

$$\frac{u_s}{c} = \sqrt{\frac{1}{2}\frac{n_c}{n_e}\frac{Zm_e}{M_i}a_0^2}$$

Where n_c is the effective critical density, where the laser wave reflection takes place, i.e. $n_e=\gamma n_c$ [Fuchs et al., 1999] at normal incidence with γ being the Lorentz factor given by $\gamma=(1+a_0^2)^{1/2}$, with $a_0^2=(p_{osc}/mc)^2=I\lambda^2/(1.37\times10^{18})$, I being the laser intensity (in units of W/cm^2) and λ is the laser wavelength in microns.

Therefore:

$$E_{max/front} = 2*M_i*u_S^2=Zm_ec^2a_0^2/\gamma$$

One can note that this evaluation is consistent with the values measured specifically for protons produced in zone II [Kaluza et al., 2004] that can indeed be measured using two different techniques:

- With neutron spectroscopy: by using a CD_2 target and by measuring the emitted neutrons [Youssef et al., 2006; Habara et al., 2004].
- With nuclear activation [Nemoto et al., 2001]: a thin layer containing Deuterium is placed at the front of Al targets. Experimentalists were able to quantitatively determine the D^+ spectrum emitted in zone II for which the maximum cutoff energy compared favourably with the formula above [Fuchs et al., 2005; Fuchs et al., 2007a].

The analytical work on the maximum proton energy produced with the accelerating mechanism in zone II has allowed showing that the acceleration mechanism in zone III produces, in almost every case, the maximum proton energies [Fuchs et al., 2007a].

This conclusion can be obtained using four different means:

- First, experimentally, it was possible to demonstrate that acceleration in zone III produces the forward accelerated protons with the maximum energy.

- Then, the same conclusion related to the fact that acceleration in zone III is dominant can be reached by observing the agreement between the experimental measures and the predictions of the acceleration model in zone III (see above).
- The models described in this Section and in Section 2.1 are also in good agreement with 2D PIC simulations over a wide range of target and laser parameters.
- Finally, when comparing systematically the maximum proton energies obtained by the two mechanisms, one also notes the dominance of acceleration in zone III to generate the most energetic protons. This comparison was done for various pulse durations, target thicknesses and laser intensities [Fuchs et al., 2007a]. The dominance of zone III is systematic, except for ultra-short pulse durations (< 10 fs) for which the laser energy is too low and the number of electrons accelerated towards the back surface is therefore low, thus reducing the amplitude of the accelerating field at the back of the target.

With this study, the controversy on the origin of the most energetic protons was concluded.

More recently, this regime has been revisited in the case of a circularly polarized laser pulse. In this case, a strong electrostatic field is also created at the front of the target as electrons are pushed forward but they are not heated. They therefore create a thin electron layer travelling inside the target. The hot electron temperature and density are therefore decreased, also decreasing the efficiency of the TNSA mechanism. The thin accelerated electron layer will drive a strong charge separation that can lead to the acceleration of quasi-mono-energetic ion beams. Recent numerical and theoretical studies [Klimo et al., 2008; Robinson et al., 2008; Macchi et al., 2009; Grech et al., 2011; Schlegel et al., 2009; Tamburini et al., 2010], as well as recent experimental studies [Henig et al., 2009] have explored this regime. It can also be explored using higher wavelength with CO_2 lasers interacting on gas jets [Palmer et al., 2011; Haberberger et al., 2011].

2.3. Laser ion acceleration in the transparency regime

A new regime of laser ion acceleration which relies on the interaction of high contrast laser pulses with ultra thin foils has been validated using simulations [Dong et al. 2003; d'Humières et al., 2005; Yin et al., 2006; Albright et al., 2007] and experiments [P. Antici et al., 2007; D. Neely et al., 2006]. In this regime, the laser is able to go through the target during the duration of its interaction. Both the TNSA mechanism and the radiation pressure acceleration can play an important role in this case and laser energy absorption is increased as volume absorption occurs.

It has been shown that proton energy could be increased by reducing the solid target thickness [Mackinnon et al, 2002; Kaluza et al., 2004]. For extremely thin targets where relativistic transparency for the laser pulse occurs, theoretical studies [Dong et al., 2003; d'Humières et al., 2005; Esirkepov et al., 2006; Yin et al., 2006] have suggested that proton acceleration could be even more efficient than using the standard regime of acceleration from solids [Allen et al., 2004; Fuchs et al., 2005]. Antici et al. [2007] report on some of the first experiments performed to explore the regime of ion acceleration using ultra-thin targets interacting with an ultra-high temporal contrast 320 fs duration laser pulse. Proton beams accelerated to a maximum energy of ~7.3 MeV from targets as thin as 30 nm thick and for a 10^{18} W.cm^{-2} laser intensity were observed. Neely et al. have also reported [2006] results using similar targets in ultra-high contrast conditions. Here, using a much lower on-target laser intensity and a longer pulse (10^{18} W.cm^{-2} and 320 fs) than Neely et al. (10^{19} W.cm^{-2} and 33 fs), ~2 times higher proton energies were obtained. It was also shown that this acceleration regime produces nearly a tenfold increase of proton maximum energy as compared to the standard regime using thicker targets at the same laser intensity. PIC simulations in close agreement with the measurements suggest that enhanced target electron heating as compared to thicker targets results in the observed proton energy increase. In the simulations much higher proton energy increase is predicted for similar target thickness and somewhat higher laser intensities (3-6 10^{19} W.cm^{-2}).

As in the rear-surface acceleration process with thick targets, ion acceleration in the relativistic transparency regime involving ultrathin targets is observed in simulations to be lead by the fast electrons accelerated by the laser from the front side into the target, except that a higher number of electrons are heated to higher temperatures as the result of the interaction of the laser pulse with the whole volume of the ultra-thin target. Emerging from the target, these electrons form a dense electron plasma sheath at the vacuum interface that accelerates surface ions [Hatchett et al., 2000], mostly hydrogen-rich contaminants [Gitomer et al., 1986].

In the standard regime of thick targets, the proton energy increase for decreasing target thickness is due to a reduced spreading of the fast electrons within the target as discussed in Section 2.1. For the relativistic transparency regime, an optimum target thickness is predicted to be reached when the laser absorption in the target equals the laser transmission [d'Humières et al., 2005], in practice it is in the range of tens to hundreds of nm. At the peak intensity levels (> 10^{18} W/cm^2) required to accelerate ions to large energies, the pedestal pre-leading the main pulse, or ASE (laser amplified spontaneous emission), is generally high enough to ablate such thickness of target material. More importantly, it also induces rear-surface target heating prior to the arrival of the main pulse for thin targets (still thicker than the ablated thickness), reducing the ion energy. This has limited the minimum target thickness to few microns in typical high-energy ion acceleration experiments. This is why laser pulses need to be contrast enhanced in order to explore this relativistic transparency regime. The ultrahigh contrast can be achieved using several methods like self-induced plasma shuttering from solids [Doumy et al., 2004], or frequency doubling of the laser wavelength. Both induce similar reduction in laser energy but the latter has the further disadvantage of decreasing the $I\lambda^2$ factor.

A maximum proton energy of ~7.3 MeV was recorded during these experiments using two plasma mirrors to be able to use thinner targets. The targets were 30 nm thick SiN planar membranes uncoated or coated with a variable thickness of Al. Without plasma mirrors, the laser interaction with ultra-thin targets yielded no result (probably because the target material was evaporated before the main pulse) and the laser interaction with 5 μm thick targets also yielded a maximum energy below the detection threshold, i.e. <0.9 MeV. At similar laser intensities, the proton energy increase going from thick target with a standard contrast to ultra-thin targets with ultra-high contrast was thus higher than a factor 8. Equivalently, to reach similar proton energies in the standard regime of thicker target having cold rear-surfaces, much higher laser intensities are needed, e.g. 2 10^{19} W.cm^{-2} for 25 μm thick targets [Fuchs et al., 2006]. Laser to proton energy conversion, calculated by considering protons from 1.8 MeV (resp. 4 MeV) to the maximum energy is 4 % (resp. 1.1 %) for 30 nm thick target, 3.7 % (resp. 0.9 %) for 80 nm thick target, and 2 % (resp. 0.4 %) for 100 nm thick target. Note that when the same maximum proton energy is reached in the regime of thicker targets at much higher laser intensity, the laser to proton energy conversion is much lower, i.e. ~0.1 % for protons from 4 MeV to the maximum energy. The angular distribution of the proton beams is also observed to be irregular, on the contrary to proton beams accelerated at similar maximum energies from thick metal targets which display a very smooth profile. This could be due to the fact that the focal spot in presence of two plasma mirrors is not optimized.

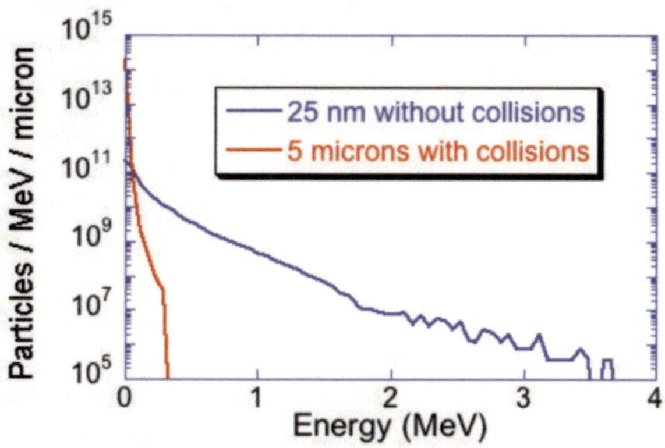

Figure 3. Electron spectrum obtained from the 2D PIC simulations for a 25 nm thick target and a 5 microns target, a 350 fs pulse duration, and a 1.38×10^{18} W/cm^2 laser intensity.

Simulations performed using the 2D PIC code PICLS [Sentoku and Kemp, 2008] of the interaction of a prepulse-free laser pulse with a thin dense target allow to get an insight on the interaction conditions. The incident laser pulse has a 1 μm wavelength, a 350 fs pulse duration, an intensity of 1.38×10^{18} W/cm^2, and a transverse spot size on target of 6 μm FWHM. The temporal and spatial (transverse) profiles are Gaussian and the laser electric field is in the simulation plane. The laser interacts at normal incidence with the 25 nm thick, 100 n_c density target, initially fully ionized, that consists of a layer of Al surrounded by two layers of hydrogen.

The maximum proton energy is 7.33 MeV, as observed 2.8 ps after the beginning of the simulation, in close agreement with the observation. Performing a simulation with the same parameters but adding the treatment of collisions yields a very close maximum proton energy of 7.28 MeV, as can be expected since in such conditions where the target is heated efficiently, the jxB absorption significantly dominates collisional absorption. In both simulations, we observe negligible transmission. The consistency between experiment and simulations support the fact that in the experiment the main laser pulse could interact with a solid density target.

To understand the specific mechanism of proton energy increase obtained using ultrathin targets, an additional simulation using a thicker, i.e. 5 μm thick, target still at 100 n_c, with collisions was performed. These simulations show a higher heating for ultrathin targets compared to the thicker target. We have to note that the laser absorption is comparable for the two targets (~23 %). It is the partition of the absorbed laser energy into the cold and hot electron populations that differs significantly between the two target thicknesses, as shown in Figure 3. In the thicker target, a high number of electrons remain cold and the maximum electron energy is low. In the ultrathin target, the hot electrons spread around the whole target and the maximum electron momentum is high. This is due to the target heating induced by the hot electrons, which are not damped by the cold background as in the thicker target, leading to a density lowering of the target, and an enhanced absorption in hot electrons for subsequent parts of the laser pulse.

Interaction with ultrathin targets leads to a very efficient electron heating, and high proton energies. Going towards longer pulse durations also favors absorption in hot electrons in ultra-thin targets since the trailing part of the pulse can take advantage of the target heating induced by the leading part and of the increased absorption. This explains the increased proton energy observed in these experiments using 320 fs duration pulses as compared to experiments performed with 30 fs duration pulses [Neely et al., 2006], although the intensity used here is ten times lower.

The experiments and simulations presented here show the potential of laser-acceleration of protons from ultra thin targets using ultra-high contrast pulses. In the present experiment, exploring of this regime has been limited to the threshold of the relativistic regime. Both the low efficiency of the consecutive plasma mirrors and the low final focal spot quality are strong limitations. In future studies, this could be improved by adjusting the fluence on the mirrors. It has been indeed demonstrated that 50 % efficiency with good focusing quality could be obtained from two successive mirrors [Wittmann et al., 2006]. This would allow achieving intensities above 10^{19} W/cm^2, thus working with an optimum configuration for the same target thicknesses. In this case, as shown in [Antici et al., 2007], 2D simulations show that very high proton energies (> 100 MeV) should be obtained. Around 120 MeV protons and GeV Carbon ions were recently measured in this regime at the Los Alamos National Laboratory [Jung, 2012]., breaking the existing world records, and plasma mirrors are now routinely used in laser ion acceleration experiments with ultrathin targets.

3. Laser ion acceleration using low density targets

In this Section experimental and theoretical studies on laser ion acceleration with low density targets are presented. This regime offers a promising alternative to the accelerating regimes described in Section 2. The basics of this regime are described and some recent experimental and theoretical results are also discussed.

As discussed in sections 1 and 2, intense research is being conducted on sources of laser-accelerated ions and their applications, e.g. radiography and production of Warm Dense Matter (see review articles [M. Borghesi et al., 2006; J. Fuchs et al., 2009]). This is motivated by the exceptional properties that have been demonstrated for proton beams accelerated from planar targets (see Section 2), such as high brightness, high spectral cut-off, high directionality and laminarity, and short duration (~ps at the source). In most experiments, solid density targets, typically gold or aluminum, are used and the most energetic ions are accelerated from the rear side by a strong electrostatic field created by fast electrons (through the so-called TNSA mechanism). This section presents a different setup aiming at studying the possibility of enhancing the efficiency and ease of laser acceleration of protons and ions compared to what has been achieved up to now using standard TNSA.

Indeed, it is not only important for applications of such ion beams that their parameters (maximum energy and total number) are enhanced, but also that they are easily usable. Regarding the second point, there are currently some issues/limitations linked with using solid targets: 1) targets need to be aligned precisely for each shot, 2) laser temporal contrast needs to be controlled, 3) debris are produced, 4) repetition rate is limited. These limitations would all be greatly alleviated if one could use gas jets as the laser interaction medium. Following preliminary theoretical [Esirkepov et al., 1999; Yamagiwa et al., 1999; Sentoku et al., 2000] and experimental [Krushelnick et al., 1999; Sarkisov et al., 1999] studies on ion laser acceleration using underdense targets, ion acceleration from low density targets was recently the subject of growing attention [Willingale et al., 2006; Yogo et al., 2008; Antici et al., 2009; Willingale et al., 2009]. However, the efficiency of the process was not shown to be high. For example in [Willingale et al., 2006], the maximum energy of the ions produced this way was only half of the one that could be obtained using solid targets.

There are several interpretations about the (best) production of the proton beam in gaseous conditions. One hypothesis is that the protons are produced by electrostatic acceleration at the target–vacuum boundary like in TNSA, although this was subject of debate [Willingale et al., 2007]. For example, acceleration of energetic ions observed by Willingale et al. had been explained in terms of strong inductive electric fields due to magnetic fields variations on a steep density gradient [Bulanov et al., 2007]. Through simulations, it has been found that shock acceleration could be used in such low-density medium to accelerate ions not only very efficiently, but also with a comparable number as TNSA [d'Humières et al., 2010a]. The basic idea behind it is that lower density targets can improve the absorption of the laser energy, hence both the laser-to-ion conversion and the ion energies can be enhanced. However, to reach such optimum working point, precise interaction conditions need to be met. With present-day laser parameters, it requires rather thin gas jets (of the order of 100 microns), which are not readily available. So, in parallel of working to produce such gas jet, it was proposed to work on demonstrating such optimum of ion acceleration exploiting shock acceleration in lower-than-solid density targets [d'Humières et al., 2010b]. For this, it is possible to use, as a substitute for optimized gas jet, targets that have been exploded by a secondary laser prior to the arrival of the main beam. The goal is not only to demonstrate the existence of optimized ion acceleration beyond TNSA, but also to shed light about these processes.

Test experiments have recently been performed at LULI (France) demonstrating the effectiveness of ion acceleration in the regime which had been explored through simulation. However the laser intensity and energy during that experiment were somewhat limited (~J) [Gauthier et al., 2012]. As PIC simulations (see below) show that this acceleration process is more efficient with higher intensities and longer pulse durations, doing experiments on that topic using higher energy lasers would be very beneficial as it would allow to effectively test the increased efficiency of shock acceleration in the domain where it is seen by simulations to be best: with long pulses and high laser energy. This, coupled to progresses on gas jets, would strongly enhance the practicality and possibilities for ion probing or warm dense matter generation on future high-power and high rep-rate laser facilities.

As found out by 1D and 2D Particle-In-Cell simulations performed with the PICLS code with respect to previous experiments using exploded foils having low-density [Antici et al., 2009], the ion acceleration processes in these plasmas depend strongly on the characteristics of the density gradient. For sharp and small density gradients, the most energetic protons are accelerated at the back surface, similarly as in TNSA. The acceleration mechanism is then similar to the one observed with solid foils. For intermediate density gradients, the most energetic protons are accelerated by a collisionless shock mechanism in the decreasing density ramp at the target exit. A two-step acceleration process takes place: first, ions are accelerated in volume by electric fields generated by hot electrons, second, the ion energy is boosted in a strong electrostatic shock, as shown in [E. d'Humières et al., 2010a]. If the density profile becomes too long, the shock cannot be triggered and ions are accelerated by normal expansion in the density gradient which is a low efficiency mechanism [Grismayer and Mora, 2006].

Collisionless shocks have already been studied in decreasing density gradients for spherical plasmas [Peano et al., 2007] and for plasmas located at the back of a solid foil irradiated by a laser [Tikhonchuk et al., 2005]. In [d'Humières et al., 2010b], and [E. d'Humières et al., 2011a; E. d'Humières et al., 2011b], the shock regime and the two step process were studied in detail. The first step, the launch of a fast ion wave, requires a hot electron population and a descending density profile and occurs in a zone of high amplitude magnetic fields. This first step has already been described theoretically by Grismayer and Mora [2006] but since the authors used a Lagragian code, they could not go beyond this first step. The second step, the development of a strong electrostatic shock which boosts the energy of the ions, happens when the ion bunch resulting from the first step enters a low density plasma region where the magnetic field has strongly decreased. The first 3D simulation of this regime were recently performed [d'Humières et al., 2011a] and confirmed the 2D simulation results. Using 1D PIC simulations, it was also confirmed that this regime does not require a high magnetic field to launch a strong collisionless shock. A Vlasov-Poisson code is presently used to study in more details the processes described and to obtain scaling laws of the maximum proton energy and accelerated proton numbers with laser and target parameters [d'Humières et al., 2011b]. The number of accelerated protons is high, similar to what is observed with TNSA, and could be higher as it is possible to accelerate ions in a thicker region of plasma even if the density is low in this region, whereas TNSA is limited to a small region inside the back of a target where the strong electrostatic field can be felt. As exemplified by Figure 4, 2D PIC simulations show that to reach this high efficiency regime, a high laser energy is needed for low density plasmas with a high thickness by density product. The fact that it is difficult nowadays to produce these types of plasmas except from exploded foils naturally gives an advantage to large-scale facilities with several tens of Joules in the main pulse and several tens of Joules in the exploding pulse.

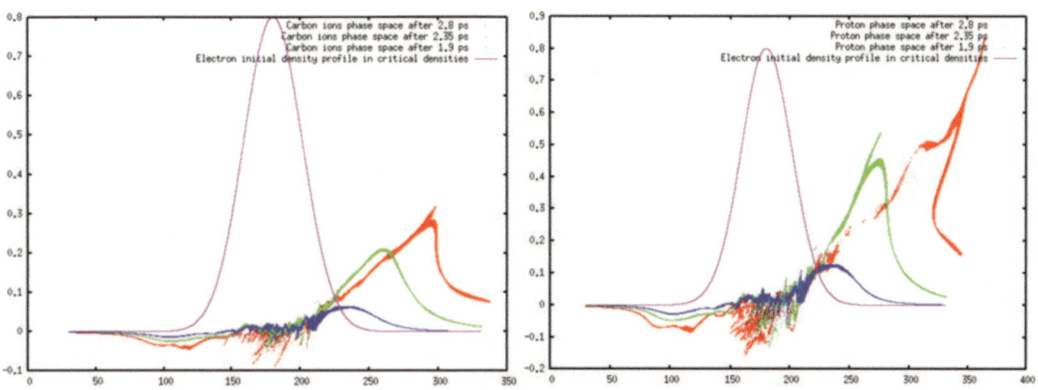

Figure 4. (right) Proton phase space at three different times (1.9 ps after the beginning of the simulation for the blue dots, 2.35 ps for the green dots and 2.8 ps for the red dots) and initial electron density profile (purple line, units of critical density). The laser is injected from the left. The target is composed of Carbon ions, protons and electrons and its FWHM is 50 microns. Laser intensity is $5x10^{20}$ W/cm^2, pulse duration is 700 fs FWHM and the focal spot width is 6 microns. Maximum proton energy at t=2.8 ps is 281 MeV. It reaches 296 MeV after 3 ps. The development of the ion wave is clearly visible at t=1.9 ps and the shock is launched just before t=2.35 ps and has fully developed at t=2.8 ps. (left) Carbon ions phase space at three different times (1.9 ps after the beginning of the simulation for the blue dots, 2.35 ps for the green dots and 2.8 ps for the red dots) and initial electron density profile (purple line, units of critical density). The Carbon ion wave is clearly visible at t= 1.9 ps and 2.35 ps and the shock on Carbon ions starts to develop just before t=2.8 ps. This is later than for the shock on protons. This is normal as the Carbon ion wave takes more time to develop than the proton wave. Maximum Carbon ion energy is already 496 MeV at t= 2.8 ps. It reaches 889 MeV after 4 ps. In both cases, the x-axis is in microns and the y-axis for the phase spaces represents beta*gamma where beta is the ion velocity divided by the velocity of light and gamma is the proton Lorentz factor.

As illustrated in Figure 4, preliminary simulations using the TITAN (LLNL) laser parameters were performed. These show the extremely interesting potential of this regime since in the simulations 296 MeV protons and ~900 MeV C ions are accelerated. Other simulations using TITAN parameters but exploring different target thickness of plasmas (still peaking at 0.8 n_c) have been performed. For all cases, the production of energetic particles is linked to strong electrostatic shocks. More simulations and experiments are needed to fully grasp the potential of this regime and to optimize it.

4. Optimization of laser ion acceleration with various types of targets

In this Section, several innovative studies on the optimization of the characteristics of laser accelerated ion beams are presented. New targets were recently tested to change the interaction conditions, the electron sheath characteristics or the ion distribution. These studies have allowed to find new ways to increase the maximum ion energy, as well as new ways to decrease the beam divergence and energy spread. Several of these targets or setups are described here. Other targets have recently been investigated to improve the coupling of the laser energy with the target electrons but are not described in this Section. Klimo et al. [2011] have for instance tested nano-structured targets to enhance the coupling between the laser and energetic electrons.

4.1. Laser ion acceleration using micro-cones

Studies of the interaction of energetic laser pulses with micro-cones using PIC simulations have been performed in collaboration with experimentalists from Los Alamos National Laboratory and the University of Nevada, Reno. The importance of the geometry of the cone on the increase of the laser energy absorption and of the hot electron temperature was highlighted. It was also shown that these targets are very sensitive to the laser alignment and contrast. Coupling these cones with flat micro-disks, it is possible to strongly increase the maximum proton energy [Flippo et al., 2008] while also increasing the obtained ion beam divergence [Renard Le Galloudec et al., 2010].

The basics of laser ion acceleration with micro-cones are detailed in this Section as well as recent theoretical, experimental and numerical results. The world record in maximum laser accelerated protons energy was broken in 2009 using these types of targets [Gaillard et al., 2011]. A proton beam with energies as high as 69 MeV was obtained using 80 J of laser energy whereas the previous record had remained at 58 MeV [Snavely et al., 2000] for almost 10 years and needed 500 J.

Cone targets are of interest for their potential to increase the hot electron temperature and population density [Sentoku et al., 2003a; Kodama et al., 2004], which are the main contributors to the efficacy of the TNSA mechanism. Sentoku *et al.* [2003a] showed that sharp tip cones can effectively increase the number of electrons available for laser heating while guiding the laser light along the cone wall surface toward the cone apex. This action tremendously increases the interaction area of the laser, producing more electrons and concentrating the laser field at the cone neck near the flat-top surface. UNR physicists in combination with the nano-fabrication group, NanoLabz, were able to design a relatively inexpensive process for mass producing a new type of cone target (Figure 5.a). A process with obvious advantages over the hand-assembled ICF targets to date. Early results have shown these cones can produce excellent ion beams when properly aligned.

Pioneer experiments performed on the Titan laser (LANL) showed that the flat-top cone produces more protons at higher energies than the flat foil, in this case 13 times more above 10 MeV [Flippo et al., 2008]. The total energy present in the beam from the Au flat-foil is measured to be 0.4% of the incident laser energy. When this is done for the flat-top cone target it is measured to be 1.9% of the incident laser energy. This represents a nearly 5 fold increase in the conversion efficiency over the Au flat foil targets and a 3.4 fold increase in the total amount of protons, with nearly 13 times the number above 10 MeV. The proton beam observed from the cone target contains two orders of magnitude more protons than previous experiments at similar intensities [Maksimchuk et al., 2004] and is more than two orders of magnitude more efficient (laser energy conversion to protons) than previously published work for a similar laser intensity [Fuchs et al., 2005] and energy [Fuchs et al., 2005; Borghesi et al., 2006]. The beam also has a 3 to 5 times higher maximum proton energy than previously reported for a similar intensity from flat foils [Fuchs et al., 2006; Robson et al., 2007], at least 1.5 times that of the Trident flat foil, and potentially more than 2 times that of the flat foil as simulations indicate.

The cone targets have been modelled with the particle in cell code PICLS. One simulation was performed exactly matching the experiment at a laser wavelength of 1 μm with an intensity of 10^{19} W/cm^2, a pulse duration of 600 fs and a focal spot of 12 μm. The cone inner neck diameter was 10 μm, and the flat-top thickness was 13 μm with a diameter of 90 μm. Maximum proton energy reached almost 40 MeV (Figure 5.b). Several other simulations where performed as part of a parameter study at a laser wavelength of 1 μm and an intensity of 10^{19} W/cm^2, but with a pulse duration of 350 fs and a FWHM spot of 6 μm to keep the calculation time to a minimum. The various simulations are summarized in [Flippo et al., 2008]. The cone parameters have been varied to highlight the main advantages and drawbacks of the flat-top micro-cones and to understand the driving factors behind the observed enhancement. They reveal that the cone leads to significantly improved laser energy absorption and conversion to hot electrons when well aligned, resulting in higher hot electron temperatures and densities. These increases due to the flat-top cones, which are a function of the cone neck-diameter and opening angle [Nakatsutsumi et al., 2007], provide a potential advantage for laser-ion acceleration as these are the main parameters governing the TNSA accelerated ions [Wilks et al., 2001].

At these laser parameters (1 μm laser wavelength, 350 fs pulse, 6 μm spot size, and 10^{19} W/cm^2 intensity), the flat-foils have been observed experimentally to produce ~3-5 MeV [Fuchs et al., 2006a] for a 20 μm thick target. The simulated flat-foil, which is 10 μm thick, results in a proton maximum energy of 8.6 MeV. This higher energy is consistent with the fact that the thinner simulated target follows the experimental trend of increasing proton energy for a thinner target at the same laser parameters [Mackinnon et al., 2002; Fuchs et al., 2006a]. The smooth cone with a 10 μm inner diameter neck and 10 μm thick walls (for a total neck outer diameter of 30 μm) has the highest performance, yielding a maximum proton energy of ~26 MeV, with a 90 μm top and a top-to-neck ratio of 3. Comparatively, a cone with a 20 μm inner diameter (total of 40 μm) with a top to neck ratio of 2.25 yields 10 MeV, a 62% decrease. Here, when the neck size is increased by a factor of two the electron temperature, the maximum electron energy, and the proton maximum energy are all decreased by about the same factor. Combining information from [Flippo et al., 2008], one can deduce that the neck diameter has a stronger influence on the maximum proton energy and electron temperature than the flat-top diameter. However, it is evident that both dimensions play a role; and thus, their ratio is still a good figure of merit.

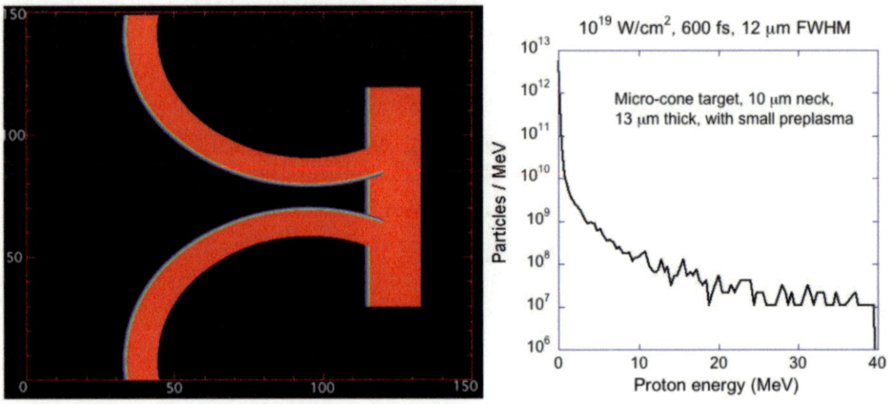

Figure 5. (a) Illustration of the shape of the flat-top cone geometry. Axis are in microns (from 0 to 150 microns). (b) Simulated proton energy spectrum.

One can conclude that for the TNSA mechanism to work efficiently, a smooth surface on which a dense, hot sheath can form is essential. A sharp-tip cone would not be suited for this application, and results have not shown promise for such cones to be used as proton beam generators. One also needs to transport the many hot electrons efficiently to the flat-top surface from wherever the hot electrons are generated. This last point can be described as the hot electron population transport characteristic of the cones. The transverse pointing accuracy can have a major effect on this transport. The farther the laser is focused away from the cone axis, the longer and more complex the path is to the flat-top for the hot electrons. On the way, these electrons can be scattered and lost to other areas of the cone or slowed down by the transfer of energy to the ions present on the rear surface of the cone side-walls. The overall hot electron population and temperature is shown to be only slightly affected by the increase in transverse offset distance; whereas the maximum proton energy is largely perturbed, decreasing by a factor of eight. This is caused by the electrons inability to be efficiently transported to the flat-top, and instead spread out over the relatively large surface area of the cone's side-walls [Flippo et al., 2008].

Another issue affecting the hot electron population, temperature, and transport is the preplasma filling in the inner cone neck. As long as the laser is able to propagate through the preplasma without significant filamentation, a higher density preplasma can enhance the coupling to the hot electrons, which can effectively lead to an increase in the maximum ion energy. However, if the laser is strongly affected by a denser preplasma and is instead reflected, dispersed, or entirely absorbed far from the flat-top, then any hot electrons generated must travel a longer distance to arrive at the flat-top. As in the case of the transverse pointing accuracy, the hot electrons will be scattered or decreased in energy traversing the longer path, or completely lost to forming the cone side-wall sheath, never making it to the flat-top. As the density of the plasma inside the cone neck increases the distance between the interaction zone and the flat-top increases. When the level of preplasma density inside the cone neck is increased from $0.001n_c$ to $2n_c$ on the laser axis (increasing exponentially to $10n_c$ at the cone wall interface), the laser is not able to reach the flat-top. Although the laser absorption has been increased to 83.2%, the maximum proton energy is decreased to 18.8 MeV due to the effect of the longer electron transport distance to the flat-top. Nevertheless, the simulated maximum proton energy is still more than 2 times higher than that from the simulated flat-foil target, an attribute of the enhanced laser absorption and hot electron temperature.

To gauge the effect of the longitudinal focusing, two simulations were performed such that the laser was focused both on the flat-top as well as 80 microns prior to the flat-top (the Rayleigh length is 100 microns in these conditions). The laser absorption, hot electron temperature and maximum proton energy are only slightly changed. This is consistent with the observation that the curved cone geometry dilutes the longitudinal pointing accuracy differences as seen in [Sentoku et al., 2003a]. The main reason for the increase in the maximum proton energy comes from the cone guiding the laser light allowing more energy transfer to electrons as well as a neck diameter being of the right dimensions so as to form a thick, underdense, preplasma for efficient hot-electron generation and a short undisturbed transport to the flat-top.

It has therefore been shown that laser-ion acceleration with flat-top cones is a novel and an efficient method to obtain high quality energetic ion beams. PIC simulations have also shown that the maximum energy of the accelerated ions is proportional to the hot electron temperature and density, and is inversely proportional to the hot electron transport distance. The cone wall smoothness is an important factor to optimize the maximum ion energy, while the longitudinal pointing accuracy has only a small influence on the final proton energy. As long as the laser axis is aligned with the cone axis and as long as the preplasma is not sufficiently dense to affect laser propagation toward the cone tip, the maximum ion energy depends only on the hot electron population characteristics, which is influenced by the cone geometry and not necessarily by the flat-top size. If the preplasma level is high or if the laser is not well aligned, even the higher electron temperatures and densities will not be sufficient to overcome the effect of the elongated electron transport path and associated scattering to the cone side-walls. The long transport results in proton spray emission from the cone side-walls and poor ion beam quality and efficiency.

New target concepts along with new ideas to achieve ignition of fusion targets with laser and particle beams are presently of high interest and have a wide range of applications in the field of high energy density physics. Studies of the cone target along with other shapes have paved the way to enable a better understanding of the cone physics allowing to develop the cone target presented in the rest of this section along with its relevance for an array of applications [Renard-Le Galloudec et al., 2010].

For such a target to give its full potential, because of the physics occurring in a cone, some criteria need to be met (the same as the ones above). The cone target needs to be precisely aligned. The laser enters the cone and starts hitting the faces when its diameter is about 3 to 4 times the inside tip size [Sentoku et al., 2004; Nakamura et al., 2007; Renard-Le Galloudec et al., 2008]. Under low preplasma conditions [Sentoku et al., 2004; Renard-Le Galloudec et al., 2008], so as to not destroy the conical shape the laser interacts with, the cone microfocuses the laser light into the tip [Sentoku et al., 2004]. At the same time, the laser interacts with the faces of the cone, creates electrons and guides them along the faces to the tip where the electron beam gets out [Renard-Le Galloudec et al., 2009]. This increases dramatically the electron density in the tip, enables a higher conversion efficiency of laser light into very energetic or hot electrons [Sentoku et al., 2004; Nakamura et al., 2007; Nakamura et al., 2004; Nakatsutsumi et al, 2007], and thus enhances both electrons and protons [Chen et al., 2005]. Note here that the cone, not the laser, defines the beam diameter [Renard-Le Galloudec et al., 2008]. A smaller cone angle produces more energetic electrons compared to a more open cone [Nakamura et al., 2007; Chen et al., 2005]. In addition, cones show an increased absorption of the laser light compared to flat targets [Nakamura et al., 2007], which makes them more efficient. More complex cone-based geometries have also been studied [Flippo et al., 2008] and also show an increased efficiency compared to flat targets.

In [Renard-Le Galloudec et al., 2010], both the inside and outside tip of the cone are slightly curved. Shaping the back of flat targets has been demonstrated to focus protons beams [Wilks et al., 2001; Ruhl et al., 2001; Roth et al., 2002; Patel et al., 2003; Snavely et al., 2007], it is however the first time that this concept is adapted to a cone geometry in order to use cones as an essential element of the particle beam production and reap the benefits of the increased efficiency of its shape. It does more than a standard flat or curved target. It adds three essential aspects. The first aspect is that making use of the cone faces by allowing the laser to spread on them greatly reduces the amount of preplasma filling the cone, thus enabling an efficient use of the cone shape. It also uses the faces to create the electrons and guide them to the tip. Several articles have showed the imprint of the laser pattern on flat targets into the particle beam [Fuchs et al., 2003]. As the laser bounces several times on the faces on its way to the tip, its imprint disappears. It creates, at the tip, a laser imprint free area of high energy density, enabling more uniform beams. Also, if best focus is positioned toward the entrance of the cone, then all of the laser light available gets in regardless of the fnumber of the focusing optic compared to the cone angle. The second aspect is the fact that the cone, then, not the laser, defines the particle beam [Renard-Le Galloudec et al., 2008]. The laser is clearly not directly driving the characteristics of the beam produced. The third aspect is the ability to control the divergence of the output beam. The tip of the cone is slightly curved in this case. This results in a modification of the divergence of the output particle beam by effectively modifying the accelerating sheath shape, and can be tuned by adjusting the amount of curvature. This efficiently produces a beam with extremely relevant characteristics to various applications. With high-energy high-repetition rate lasers as well as targets that are on the verge of cost effective mass production, cost effective compact applications can be readily envisioned.

Because the new target shape proposed in [Renard-Le Galloudec et al., 2010] has not been fabricated yet, the 2D Particle-In-Cell (PIC) code PICLS was used to run simulations and assess the electromagnetic fields structures and proton beam characteristics in comparison with flat targets. Several intensities were run to span the range available to short pulse lasers. The inner and outer tip diameters are respectively 10 and 30 μm. They are both curved. The target itself is 10μm thick. The incident laser pulse (1 μm, 40 fs, 21 μm FWHM transverse spot size at 3×10^{18} W/cm^2) has a Gaussian temporal and transverse spatial profile. The laser interacts with the target at normal incidence, with its polarization in the simulation plane. The initial target density is $40n_c$ and remains higher than the relativistic critical density a_0n_c, where a_0 is the normalized laser amplitude and n_c is the critical density ($n_c = 1.1 \times 10^{21}/\lambda$ (μm)^{2}cm^{-3}, λ is the laser wavelength). The plasma is composed of Al ions, protons and electrons.

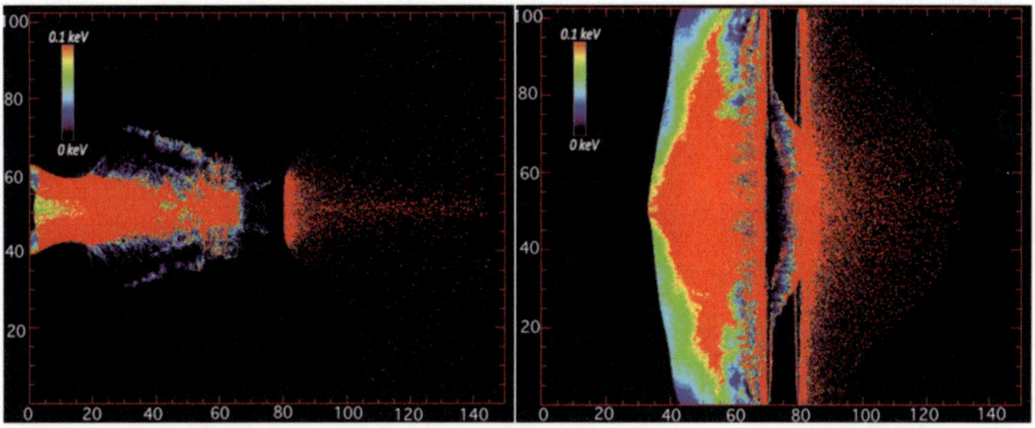

Figure 6. Proton energy density for the cone target at 3.10^{20}W/cm^2 (6.a) and for the flat target for the same laser intensity (6.b). The pulse is injected to the left. Axis are in microns from 0 to 150 for the x-axis and 0 to 102 for the y-axis.

Figure 6.a represents the 2D proton energy density for a 10μm thick curved-tip cone in a high intensity case at 3.10^{20}W/cm^2. Figure 6.b represents the same 2D proton energy density for a 10μm flat target at the same intensity. We clearly see that the protons are a lot more confined in the cone than in the flat target where they tend to diffuse laterally. The particles emitted from the cone are much more collinear to the laser axis compared to the flat target where they expand perpendicularly to the sheath.

In both cases the average divergence is small, especially for the high-energy protons. While the cone target does not seem to do better for these, it does control the divergence much better than the flat target over a wider range of energies. The curvature also allows to focus the most energetic protons in a specific location, and thus to deposit through the ions a higher energy in a smaller volume than in the case of a flat target, which is of special interest to isochoric heating.

The results in [Renard Le Galloudec et al., 2010] confirm that the cone is a much more efficient structure over a range of intensities (from 3×10^{18} W/cm^2 to 3×10^{20} W/cm^2). As the intensity increases, the maximum proton energy increases in general regardless of the target but the cone target clearly shows higher maximum proton energy than the flat target for all intensities. That difference increases with intensity. Especially evident at 3.10^{20}W/cm^2, both electrons (fig. 7.b) and protons [Renard Le Galloudec et al., 2010] are accelerated to higher energies in a higher number for the cone target. Enhanced laser interaction results in much higher maximum proton energies at high intensities (from 54.4 MeV to 98.7 MeV). Laser absorption is greatly increased in cone targets. In the high intensity case, laser intensity reaches a maximum of 2.4 10^{21} W/cm^2 in the tip of the cone (6.10^{20} W/cm^2 for the flat target), highlighting the microfocusing effect of the cone. In the low intensity case, the large preplasma present in both cases tends to give similar laser parameters evolutions, similar electric fields and a moderate increase of maximum proton energy (see Figure 7.a). In the high intensity case, the laser intensity and the longitudinal electric field reach significantly larger values in the cone leading to an important increase in maximum proton energy.

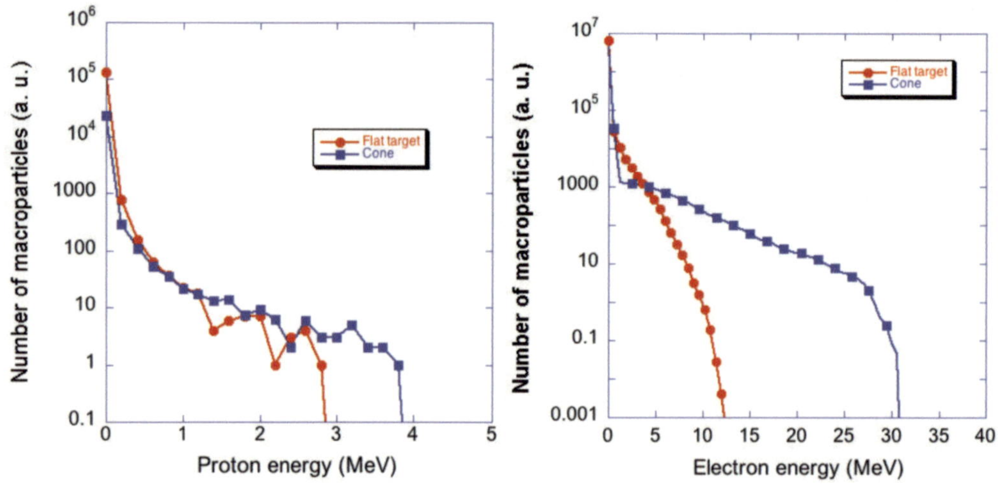

Figure 7. Proton energy spectrum (7.a) for both cone and flat target at 1.98ps for the low intensity case, and electron energy spectrum (7.b) for both cone and flat target at 924fs in the high intensity case.

It is therefore shown that this new conical target shape has the potential to produce proton beam of a higher maximum energy, a higher intensity and a lower divergence. Because of the appropriate use of the cone structure itself, by using the faces leading to the tip, it is also shown that the target itself defines the proton beams characteristics, nor the laser imprint or the focal spot size have an impact on these characteristics. The contrast of the laser can also be mitigated and finally the fnumber of the focusing optic is superseded by that of the cone target itself. All these parameters increase the potential for various groups to join in the research endeavour and pursue exciting new applications.

4.2. Laser ion acceleration using reduced-mass targets

The coupling between high intensity laser pulses with solid foils having a limited transverse extension (~ few tens of μm) has been studied by diagnosing electrons and protons produced during this interaction [Buffechoux et al. 2010]. It was observed that by reducing the area of the target surfaces, it is possible to reflect electrons near the edges of the target during and just after the laser irradiation. This transverse refluxing, which does not occur with usual large planar targets, can maintain a hotter, denser and more homogeneous accelerating electron sheath. As a consequence, when this transverse refluxing takes place during the ion acceleration duration, the maximum ion energy and the conversion of laser energy in the energy of high energy ions are strongly increased. The ion beam divergence is also reduced thanks to the more homogeneous electron sheath. These results will be detailed in this Section. Since, new experiments performed in 2011 using such targets allowed to obtain proton energies of around 80 MeV [Schollmeier et al., 2011].

The dynamics of MeV electrons generated in solids by ultra-intense lasers plays a crucial role in many applications such as electron-driven fast ignition [Tabak et al., 2006] or the production of secondary sources, e.g. X-rays [Quéré et al., 2006], positrons [Chen et al., 2009] or ions [Fuchs et al., 2009], all with important scientific or societal perspectives.

Understanding the hot electrons dynamics requires considering several aspects of their transport through the target. This Section reports an investigation of the influence of the target lateral dimensions on the dynamics of hot electrons and associated energetic proton production [Buffechoux et al., 2010]. It was the first to identify the important role played by the lateral electron recirculation in small targets. As observed in the simulations discussed here, electrons that are injected into the target center are seen to spread along the target surface with a velocity v^{t}_{hot} ~0.7c and reflected at the target edges. They therefore transit from the center to the edges and back in a time $\tau_t = D_s/v^{t}_{hot}$ where D_s is the target transverse diameter. When τ_t is of the order of the laser pulse duration τ_L, the hot electrons, refluxing from the edges, are confined during or shortly after τ_L. This leads to a time-averaged denser, hotter and more homogeneous electron population. This has been observed through the use of a combination of hot electrons and accelerated proton diagnostics.

The effective (time-averaged) hot electron temperature (T_{hot}) as a function of the target surface area was analyzed [Buffechoux et al., 2010]. It was observed that both N_{hot} and T_{hot} increase for targets having surface areas <3-4×10^4 μm^2, corresponding to target transverse diameters D_s<170-200 μm. For larger targets, both T_{hot} and N_{hot} remain, on the contrary, constant. With regards to T_{hot}, the data are consistent with simulations of sharp-interface plasmas irradiated at 45° with S polarization [Lefebvre et al., 1997]. With regards to N_{hot}, the data are consistent with measurements of hot electrons density (n_{hot}) and sheath surface [Antici et al., 2008a] which, combined, yield comparable number of electrons contained in the sheath. All these results are consistent with previous complementary measurements that showed an increase of bulk target heating when reducing the target surface area [Nakatsutsumi et al., 2008; Perez et al., 2010].

It was also observed that the hot electron sheath becomes more uniform when the target surface area is reduced. This result is obtained by analyzing the angular proton beam profile as recorded on the RCF film. It is shown that the beam is more collimated when the target surface area is reduced. This suggests a flatter electron sheath along the target rear side for smaller targets, which is consistent with the picture of geometrically confined electrons. The measurement of the thermal emission from the target rear side further confirms this. As a result of the observed increase of N_{hot} and T_{hot} within the electron sheath, when reducing the target surface area, a clear improvement of the proton beam characteristics was observed (maximum proton energy and laser-to-proton conversion efficiency).

Two-dimensional (2D) particle-in-cell (PIC) simulations of laser target interactions were performed to help identify that lateral refluxing of the hot electrons is the key process leading to hotter, denser and more homogeneous electron sheaths when reducing the target surface area. The mechanism of refluxing, as observed in the simulations, is as follows: because of their high velocity, electrons trajectories can be considered as ballistic. As electrons enter the target with an angle close to the laser incident angle (45°), their average transverse (with respect to the target normal) velocity in the target is $v^t_{hot} \approx c \times \cos(45) = 0.7c$. After several reflections off the sheath fields on the front and back surfaces with that same angle, the electrons will have travelled transversely to the edge of the target to be again reflected back (note that here there is no distinction between electrons turning around the target and electrons reflected back).

Since the electric field accelerating the ions is proportional to $(n_{hot}T_{hot})^{1/2}$ [Fuchs et al., 2009; Mora, 2003], proton acceleration is expected to be enhanced for the small foil compared to the medium foil, and indeed the simulation results agree well with the variation of the experimental proton cutoff energies. The enhancement in ion acceleration can take place only if the electrons can come back to the target center within the ion acceleration time τ_{acc}. From this, an experimental proton acceleration time can be deduced $\tau_{acc} \approx 800\text{-}950$ fs. This is consistent with theoretical estimates. Also in good agreement with the experimental result, simulation results show that refluxing produces in the simulations a more uniform sheath in small targets. The constrained lateral dimension forces the hot electrons to recirculate in the small foil, thus homogenizing the sheath.

Reducing the surface area of solid targets therefore leads to an increase in the effective hot electron number and mean energy due to the lateral electron recirculation. In particular, this effect enhances the properties of laser-accelerated ions, in terms of energy, flux and collimation, as necessary for progress towards applications. Use of this simple mechanism is of interest for future experiments at higher intensity laser facilities, but will impose high temporal contrast laser pulses to avoid preplasma leakage to the target rear-surface. In addition, progress in target fabrication will offer targets that are not only of reduced lateral size, but also thinner as it is now found favorable for ion acceleration (see Section 2.3).

4.3. Focalization of laser accelerated ions using curved targets

The dynamics of the focusing of laser accelerated ions using curved solid targets has been intensively studied in the last years [Patel et al., 2004; Kar et al., 2011; Chen et al., 2012]. The ability to tightly focus, i.e. over tens of microns, dense (> 10^{10} particles), short (with duration ~ps) bunches of positively charged particles is far beyond the possibility of present particle accelerators. The increase of the particle density of these beams would significantly improve the efficiency and prospects of a number of important applications. High energy density proton beams would allow, for example, igniting pre-assembled Inertial Confinement Fusion (ICF) targets, also known as proton Fast Ignition (PFI) [M. Roth et al., 2001], or enable ultrafast heating, above keV, of dense materials to explore their properties [Mancic et al., 2010; Pelka et al., 2010; Carrié et al., 2011]. High energy density positron beams would allow enhancing the rate of antimatter creation in the laboratory [Andresen et al., 2007], opening up investigations of many fundamental laws of nature.

These positively charged beams, broadband or monoenergetic, can nowadays be produced with a high particle number and over a short duration using ultra-intense, short pulse lasers interacting with solids targets [Wilks et al., 2001; Chen et al., 2010] as described in Section 2. When using flat targets as sources, the ion beam is divergent, 0-25 degrees depending on proton energy [Fuchs et al., 2003; Roth et al., 2005; Toncian et al., 2006], since the expanding sheath field front on the target rear-side is Gaussian in shape [Romagnani et al, 2005; Antici et al., 2008]. Such divergence can however be compensated by curving the back surface of the target, so that the accelerated proton beam will converge [Patel et al., 2003; Kar et al., 2011; Offermann et al., 2011]. This points to the necessity of understanding and optimizing the dynamics of the focusing sheath in order to achieve tight focusing, as necessary for the above-mentioned applications.

A study of the ion beam focusing dynamics through temporally and spatially resolved measurements, with picosecond and micrometer resolution, of the shape of the sheath field from a curved target, irradiated by a high-intensity short pulse laser has recently been performed [Chen et al., 2012]. Experiments performed at LULI show that the major part of the energy carried by ions converge at the center of cylindrical targets in a spot having a diameter of 30 microns, which can be beneficial for applications requiring high ion energy densities. The location of the focus is a function of proton energy, although most of the protons focus at the geometric target center. It was also shown experimentally and using 2D PIC simulations that the exact location of laser illumination on the curved target does not adversely affect the ability to focus the sheath-accelerated ion beam, although it modifies the directionality of the ion beam. Despite these advantages, an important filamentation was observed during the focusing, which limits the energy deposition precision in the ion converging zone. This effect is important for the use of such a setup in order to obtain high ion energy densities. It is the case in the fast ignition concept with ions which requires to have a high ion concentration zone with a diameter of 10 to 20 microns. Simulations have shown that at higher laser intensities, closer to the ones expected to be used for ion fast ignition, ion focusing was improved.

4.4. Acceleration of quasi-mono-energetic ions beams using the TNSA mechanism

For many potential applications of laser accelerated ions, controlling the characteristics of the produced ion beam is essential. Several techniques have recently been tested to obtain quasi-mono-energetic ion beams using the TNSA mechanism. Using either a secondary laser to control the thickness of the layer of ions accelerated to high energies [Hegelish et al., 2004], or small plastic dots to reduce the transverse extension of the accelerated proton bunch [Schoewer et al., 2004], these techniques try to constrain the acceleration zone to avoid acceleration gradients and therefore limit the ion energy spread. Another possibility to control the energy spread as well as the beam divergence has recently been developed. This technique consists in using a tuneable plasma microlens that allows focusing the ion beam and selecting specific energies. The control of the characteristics of proton beams using a solid cylinder irradiated by a secondary laser has therefore been studied [Toncian et al., 2006; d'Humières et al., 2007] using simulations and experiments performed at LULI (France). This Section explains these various techniques and discusses their limits (Figure 8).

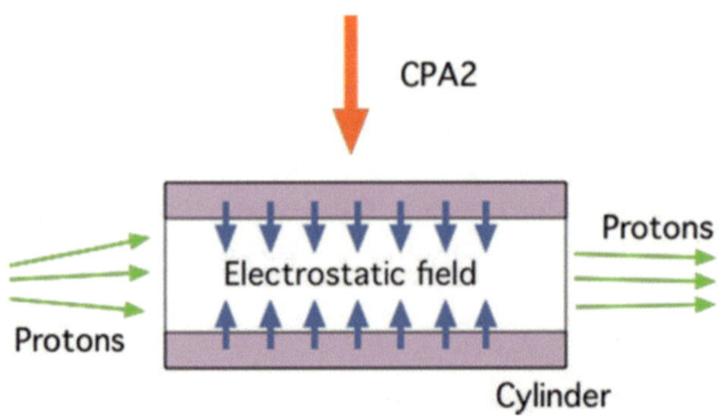

Figure 8. Schematic of the micro-lens setup focalizing a proton beam propagating on its axis. The proton beam is accelerated from a planar metallic foil by a first laser pulse. The proton beam is focalized using the hollow cylinder irradiated on the side by the laser pulse CPA2.

The focalization is effective only for a range of energies: as the cylinder is located at a certain distance from the source, and as the source has a large spectrum, protons with different energies have different times of flight and therefore reach the cylinder at different times. If the high energy protons cross the cylinder before it is irradiated by the secondary laser generating the focalizing fields, they do not undergo any focalizing effect (it is the case of the picture on the left in Figure 9). Protons with a slightly lower energy, synchronized with the focalizing fields will undergo the maximum effect. This is illustrated by the simulations performed in Figure 9.

The focalization symmetry is also observed in PIC simulations [d'Humières et al., 2006] and comes from the fact that fast electrons, even if they are produced in one region of the cylinder wall, propagate faster on the whole diameter than it takes them time to expand in vacuum towards the interior of the cylinder. Therefore, when the plasma expansion into vacuum phase, at the origin of the focalizing radial electric fields, starts, the radial distribution of these fields is symmetric.

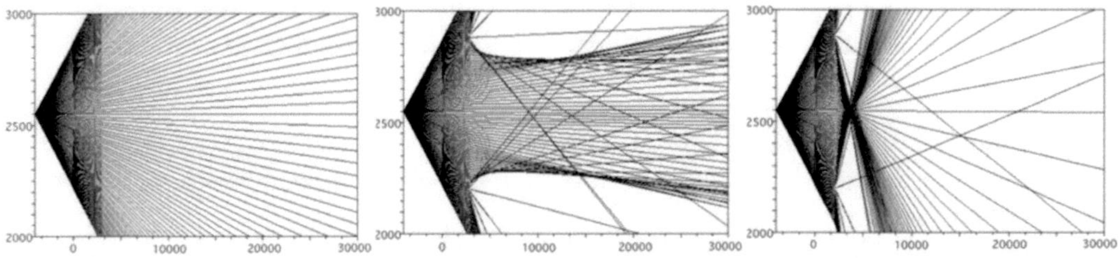

Figure 9. Trajectories of 100 protons at (a) 7.6 MeV; they exit the cylinder (located at x=0) before it is irradiated and their divergence is not affected. The two axis are in microns but at different scales. (b) Protons at 6.25 MeV; they are close to the exit of the cylinder when it is irradiated, sustain small fields and are therefore well collimated. (c) Protons at 4.9 MeV; they are at the middle of the cylinder when it is irradiated, they therefore sustain fields much stronger than the ones on b). They are therefore focalized at a short distance (5 mm) from the cylinder and then strongly diverge.

It appears that this micro-lens (its diameter is typically of the order of a mm, as well as its length) has the advantage, compared to curved targets, to allow a focalization which is not limited to distances of a few mm. It also allows to move away the source from the focalization region, which can be interesting for instance in the perspective of fast ignition using proton beams: increasing the distance between the two would allow to avoid damaging the source during the implosion of the main target. The focalization distance can indeed be tuned simply by varying the intensity of the beam generating the focalizing fields.

The fact that the focalization distance is variable as a function of energy can be exploited to select in energy in a controllable manner part of the spectrum, which is initially large. It is then enough to position a small opening at a calculated distance from the cylinder. A small number of high energy protons (crossing the cylinder before the focalization) will go through the opening. The protons with a focalizing distance corresponding to the distance between the opening and the cylinder will all be transmitted leading to a peak in the spectrum while the protons with lower energies will be focalized before the opening and will therefore reach it unfocused, and a small number of them will cross the opening. The position and the width of the peak can be easily adjusted by tuning the intensity and the delay between the pulse triggering the micro-lens and the primary pulse generating the source, as well as by tuning the distance between the micro-lens and the source.

The use of the micro-lens is not limited to protons. Therefore, the micro-lens presents the advantage, in comparison with previous solutions, of being able to achieve focalizing as well as energy selection in a single step. Moreover the process is simple: the pulse triggering the micro-lens requires just 10% of the main pulse, the micro-lens is cheap (it is just a section of what is used for medical syringes), it does not require any particular target engineering, and finally it is easy to change the desired parameters (focalization distance, size of the focalization point, energy of the protons to be focalized, peak to be selected in energy, width of this peak) by adjusting the laser parameters (intensity, delay between the two pulses) and the cylinder-source distance.

This development, achieving compact collimation of energetic ions, is therefore an extremely important step towards applications of laser accelerated ion sources. These advantages compared to existing (conventional) setups can be summarized as:

- Compactness: mm instead of dm.
- Capability to focus beams of several A instead of several mA; the last two points are the results of the fact that it is a plasma setup, therefore able to sustain much higher electric fields than conventional setups.
- Selective collimation in a few ps.
- Capability to focus in a few mm, cm or m multi-MeV beams in place of a minimum of a few meters for conventional setups or not more than mm for curved targets.

5. Laser ion acceleration limits in the ultra high intensity and ultra high laser energy regimes

With the recent rapid progresses in high power laser technologies, new laser systems are under construction and will allow to explore the ultra high intensity regime (Apollon in France and then ELI in Europe) and the ultra high energy regime (OmegaEP at Rochester, PETAL in France, ARC at LLNL and FIREX in Japan). This section presents the limits of the laser ion acceleration mechanisms described in the previous sections in these extreme regimes. These studies aim at obtaining a better understanding of how the predictions given by the existing laser ion acceleration models will be affected in these regimes and to estimate more realistically their potential.

5.1. Laser ion acceleration in the ultra high intensity (UHI) regime: Effects of radiation losses

Several PIC codes have recently been modified to take into account radiation losses by charged particles [Sokolov et al., 2009a; Tamburini et al., 2010; Capdessus et al., 2012] These models were then applied to the study of laser ion acceleration in the ultra high intensity regime (> 10^{22} W/cm^2) [Naumova et al., 2009; Tamburini et al., 2010; Capdessus et al., 2012; d'Humières et al., 2012] when radiation losses become important and can not be neglected anymore. This section discusses these results and shows how, in this regime, radiation losses strongly depend on the target density, thickness and on the laser pulse polarization. Even if radiation losses always lead to electron cooling, their effect on the ion distribution depends on the interaction conditions. For thin and moderate density foils, radiation losses can even enhance laser ion acceleration while decreasing electron heating [Capdessus et al., 2012]. These results have been confirmed using 2D PIC simulations [d'Humières et al., 2012].

A new generation of laser systems such as planned in the Extreme Light Infrastructure (ELI) project [http://www.extreme-light-infrastructure.eu/] will produce laser intensities as high as 10^{24}W/cm^2. New physical processes are expected under these conditions such as emission of high energy photons, the radiation reaction force acting on electrons, electron-positron pair production, acceleration of ions to relativistic energies, etc. [Bulanov, 2009]. One of the important applications of ultra intense laser pulses is acceleration of charged particles to extremely high energies. Recent numerical simulations and theoretical analysis show that at laser intensities exceeding 10^{22} W/cm^2 the ions can be accelerated to relativistic energies under the laser radiation pressure. While the radiation reaction force has been applied to the motion of a single particle for a long time, it has only recently been considered in plasma physics. First kinetic simulations of laser plasma interaction with a particle-in-cell (PIC) code accounting for the radiation force were reported in Refs. [Naumova et al., 2009; Schlegel et al., 2009; Tamburini et al., 2010, Chen et al., 2011]. They demonstrated the role that the radiation reaction plays in the radiation pressure acceleration by high intensity laser pulses.

The effect of radiation losses on the process of ion acceleration by ultra intense laser pulse has also been studied in 1D in [Capdessus et al., 2012]. This effect becomes important for laser intensities exceeding 10^{22} W/cm^2, where the radiation friction force slows down electrons and affects the ion dynamics through the self-consistent electrostatic field. The effect of radiation losses depends strongly on the target density and thickness and also on the laser polarization. It is less important in the case of strongly overdense targets and for a circular polarization, where the relative density n_e/n_c is larger than the laser amplitude, a_L. This is explained by clear spatial separation of the particles and fields. On the contrary, the radiation losses are important in the induced transparency regime where $n_e/n_c < a_L$. Although radiation losses are always leading to cooling of electrons, their effect on the ion distribution depends on the target thickness. In the case of thin targets, where the areal density is small, radiation losses may improve ion acceleration. On the contrary, in the piston regime, radiation losses lead to a reduction of the piston velocity and less efficient ion acceleration. These simulation results were limited to 1D simulations. However, the particle momentum has three components and arbitrary angles of electron propagation with respect to the laser wave are accounted for. No qualitative changes in 2D or 3D simulations are therefore expected. Recent results published in [Tamburini et al., 2012] confirm that statement. They demonstrated that results obtained for lower dimensionality remain valid qualitatively, although the maximum energy of ions in 3D is found to be higher than in corresponding simulations in 1D and 2D. The self-generated magnetic fields are also expected to be responsible for stronger radiation emission.

2D simulations were performed in [d'Humières et al., 2012]. The goal was to study how laser ion acceleration using thin overdense targets evolves in the ultra high intensity regime (> 10^{22} W/cm^2) and to assess the importance of radiation losses in this regime. The radiation losses model used is based on the so-called Sokolov model [Sokolov et al., 2009a]. In a first set of simulations, the wavelength of the incident pulse is 0.8 μm, its pulse duration is 21 fs and its irradiance is 1.6×10^{22} W/cm^2. The FWHM of the focal spot is 5 μm. The p-polarized Gaussian pulse interacts with the target in normal incidence. The plasma is composed of protons and electrons with a constant 400 n_c density. The plasma slab thickness was varied from 0.2 to 0.8 μm. Figure 10.a shows the simulated proton spectra with and without radiation losses for the 0.5 μm case. In this range of thicknesses, radiation losses have a measurable effect on ion acceleration but this effect remains small. In a second set of simulations, the wavelength of the incident pulse is 0.8 μm, its pulse duration is 48 fs and its irradiance is 1.5×10^{23} W/cm^2. The FWHM of the focal spot is 5 μm. The circularly-polarized trapezoidal pulse interacts with the target in normal incidence. The plasma is composed of deuterons and electrons with a constant 10 n_c density. The plasma slab thickness was varied from 0.4 to 5 μm. Figure 10.b shows the simulated electron spectra with and without radiation losses for the 5 μm case. For such a high intensity the radiation losses effects become important and the energetic electron population is strongly affected. Very high ion energy are nevertheless obtained (a maximum energy of 1.8 GeV for Deuterons in the 5 μm case) [d'Humières et al., 2012].

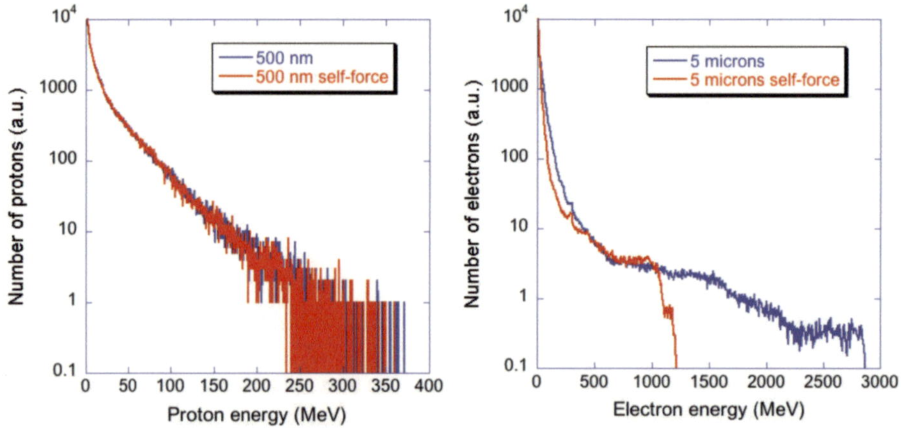

Figure 10. (a) Proton spectra for the 1.6×10^{22} W/cm^2 pulse interacting on a 500 nm high density foil. (b) Electron spectra for the 1.5×10^{23} W/cm^2 pulse interacting on a 5 μm overdense foil. Blue curve: without radiation losses. Red curve: with radiation losses.

The interaction of a high intensity short pulse with overdense targets is now actively studied in the UHI regimes to estimate ion acceleration possibilities on future laser facilities. This study is performed using Particle-In-Cell codes in which charged particles energy losses through radiation have been implemented. In the ultra high intensity regime, radiation losses will start affecting laser ion acceleration using thin overdense targets for intensities higher than 10^{22} W/cm^2, but maximum proton energies of a few hundreds MeV can still be reached.

5.2. Laser ion acceleration in the ultra high energy (UHE) regime

The interaction of a ultra high energy (up to 3.5 kJ) and high intensity laser with various types of targets has recently been modelled using PIC simulations to prepare laser ion acceleration experiments on high energy installations [d'Humières et al., 2012]. Laser ion acceleration in the ultra high energy regime (> 1 kJ) has also recently been studied experimentally [Flippo et al., 2010]. This section presents a study of this regime to analyze the changes brought by such high laser energies on the main accelerating mechanisms presented in the previous sections. Energies of more than 100 MeV can be obtained in a robust manner using solid foils with thicknesses of a few tens of μm. These beams could then be used to radiograph the implosion of a DT target when coupling PETAL with the LMJ for instance, or for other experiments (isochoric heating, laboratory astrophysics...).

A new era of plasma science started with the first experiments on the National Ignition Facility (NIF) at the Lawrence Livermore National Laboratory (LLNL) in the USA. The Laser MégaJoule (LMJ) under construction near Bordeaux in France is following the trail opened by the NIF with its planned 160 laser beams for more than 1 MJ to reach ignition of a deuterium - tritium target using the indirect drive method. Besides the physics of ICF (plasma physics, shock / fast ignition), NIF & LMJ will be essential for basic science, exploring fields such as plasma astrophysics (e.g. study of shocks to simulate violent events in the Universe such as supernovæ, accretion disks), planetary physics (highly compressed and warm matter), stellar interiors with large coupling between radiation field and matter & nuclear physics. A petawatt short pulse laser will be added to the ns pulse beams of the LMJ. This is the PETAL system (PETawatt Aquitaine Laser) [Blanchot et al., 2008], under construction on the LMJ site near Bordeaux (France) and funded by the Région Aquitaine. The ultimate goal is to reach 7 PW (3.5 kJ with 0.5 ps pulses). For the beginning of operation, the PETAL energy will be at the 1 kJ level, corresponding to an intensity on target of ~ 10^{20} W/cm^2.

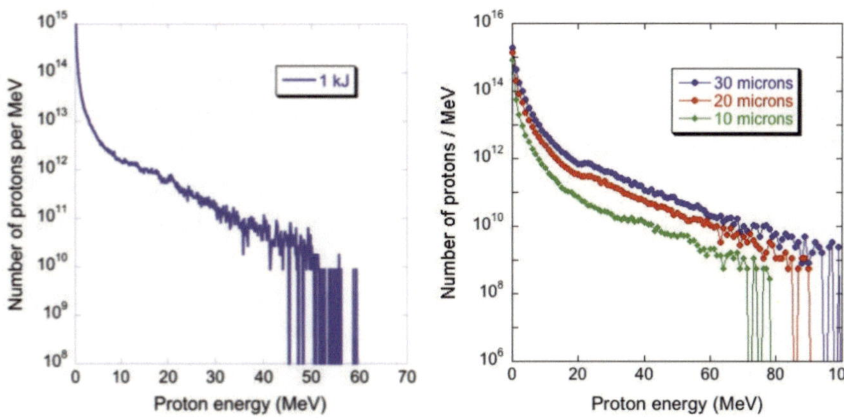

Figure 11. (a) Computed proton energy spectrum emitted from the PETAL target with 1kJ energy. (b) Variation of the proton energy spectra for three laser FWHM: 10 μm (green line), 20 μm (red line) and 30 μm (blue line) with 1.75 kJ at 2ω.

In the experiments proposed for LMJ - PETAL, the petawatt laser may be focused on a secondary target, where a short (~ 20 ps length) bunch of particles (electrons, protons, ions) is produced and may be used to probe the plasma generated by the ns LMJ pulses. The energetic ions are produced using the TNSA mechanism detailed in Section 2. First calculations were performed using the 2D particle-in-cell code PICLS to compute the energy spectra and angular divergences of the protons produced with PETAL (Fig. 11.a) for 1 kJ of energy in the PETAL beam and for 3.5 kJ. Protons with a maximum energy higher than 100 MeV are measured in the 3.5 kJ case. Within these conditions, these calculations can be considered as majoring the intensities and electron energy spectra end-points. After the accurate characterization of the particle emission from the PETAL target, the diagnostics will be used for plasma experiments, which, for example, requires proton radiography to determine the magnetic field or the electric field structure at the plasma scale or to measure the density of the LMJ plasmas.

Multi-kJ Petawatt-class laser systems therefore open new and exciting opportunities for laser ion acceleration. The interaction of a high intensity short pulse with underdense, near-critical and overdense targets has been studied using 2D Particle-In-Cell simulations in these regimes [d'Humières et al., 2012]. The goal was to study laser ion acceleration in the ultra high energy regime (> 1 kJ). 2D Particle-In-Cell simulations were performed to study the interaction of the expected PETAL laser pulse with a thin target in the transparency regime [Dong et al., 2003; d'Humières et al., 2005]. The wavelength of the incident pulse is 1 μm, its pulse duration is 500 fs and its irradiance is 10^{21} W/cm^2. The FWHM of the focal spot is 30 μm. The p-polarized Gaussian pulse interacts with the target in normal incidence. The plasma is composed of protons and electrons with a constant 400 n_c density. The plasma slab thickness was varied from 0.2 to 1 μm.

The maximum proton energy measured in the simulation depends on target thickness and reaches several hundreds of MeV at 1ω. The results in [d'Humières et al., 2012] indicate that the most energetic protons come from the rear surface of the target. It will not be straightforward to accelerate protons with such thin targets without controlling the laser contrast. A plasma mirror or doubling the frequency could improve this contrast. With a 250 nm foil and at 2ω (limiting the total energy to 1.75 kJ), the maximum proton energy is lowered to 160 MeV and 3.1×10^{10} protons/MeV at 100 MeV are expected. Figure 11.b shows the variation of the proton energy spectra for three laser FWHM: 10 μm (green line), 20 μm (red line) and 30 μm (blue line). As expected, the number of protons at high energies increase with the laser FWHM. The maximum proton energy also increases but at a slower rate. A simulation using a 10 microns CH target at solid density with a preplasma and similar laser parameters predicts a maximum proton energy of around 110 MeV.

Another promising way to accelerate ions to high energies is to use underdense or near-critical density targets (see Section 3). The interaction of a laser pulse with a near-critical density target and an underdense target was simulated using the above laser parameters in 2D. In both cases the plasma is composed of protons and electrons. For the first case, the target density is constant and its thickness is 100 μm. For the underdense case, the target density profile is a cosine square with a full-width at half maximum of 100 μm and a maximum density of 0.4 n_c. A high laser absorption of 75.9% is measured in the 2 n_c case and 18.9% is measured in the 0.4 n_c case as the laser propagates through the plasma leaving a large channel behind. The generated hot electron population produces strong electrostatic fields at the back of the target.

Figure 12 shows the proton spectra in both cases (a) and the proton phase space at the middle of the simulation box on the laser propagation axis 0.99 ps after the beginning of the simulation (b). For both cases the most energetic protons come from a region close to the back surface of the target. Wave breaking is clearly visible in the underdense case as expected from previous theoretical studies. The maximum proton energy in the near-critical case is more than 400 MeV, which is comparable with the solid density target simulation, and 7×10^9 protons/MeV at 200 MeV are expected. It reaches a little more than 200 MeV in the underdense case with 2×10^{10} protons/MeV at 100 MeV.

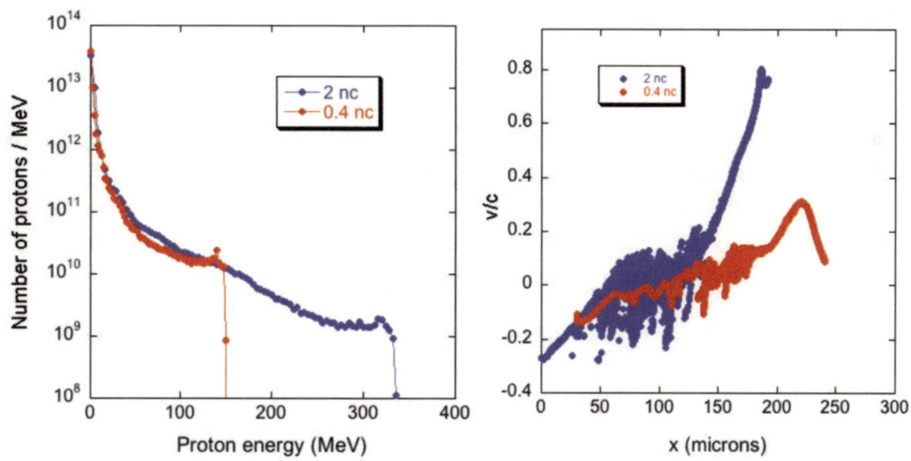

Figure 12. (a) Proton spectra for the 2 n_c case (blue line) and for the 0.4 n_c case (red line). (b) Proton phase space at the middle of the simulation box on the laser propagation axis 0.99 ps after the beginning of the simulation.

Interaction conditions are not optimized. In this low density regime it is possible to find an optimum thickness and a key parameter controlling the acceleration process is the product of the density and the thickness. This optimum thickness also depends on laser parameters. The interaction of a PETAL laser pulse with a low density target could therefore lead to even higher maximum proton energies than solid targets. The near-critical target could be achieved experimentally using a foam, an aerogel or a cryogenic target, the underdense target using a gas jet or a foil exploded by a nanosecond pulse (see Section 3).

The interaction of a high intensity short pulse with underdense, near-critical and overdense targets has therefore been studied with 2D PIC simulations in the UHE regime to estimate ion acceleration possibilities on future laser facilities. In the ultra high energy regime, proton beams with maximum energies of several hundreds of MeV and a high number of high energy protons could be accelerated using thin solid foils or underdense and near-critical targets.

6. Applications of laser ion acceleration

As described in the previous sections, laser ion acceleration has already been applied to diagnose electromagnetic fields in plasmas and to produce warm dense matter. Other important applications are envisioned in laboratory astrophysics, as injectors for conventional accelerators or for the production of radioisotopes and hadrontherapy. Several of these promising applications of laser accelerated ions are presented in this section. This list, which is not exhaustive, aims at illustrating the important potential of compact high intensity ion accelerators.

6.1. Laser accelerated ions as a diagnostic of hot electron transport in dense targets

The possibility to control the shape of MA electron currents in solids by high intensity laser pulses has been studied using different target materials [Sentoku et al., 2011]. This Section details how these effects can be diagnosed using laser accelerated ion beams as the accelerating electron sheath is affected by the transport of the hot electron beam. These results are beneficial to applications like the production of secondary sources and the ion fast ignition concept for inertial confinement fusion.

Ultra-high currents (MA) of suprathermal (MeV) electrons, that are driven through solids using relativistic laser pulses (with intensity $I > 10^{18}$ W/cm^2), lie at the heart of numerous applications, either present-day like the generation of ultra-short secondary sources of particle and radiation (ions [Malka, 2008], X-rays [Mancic et al., 2010b], positrons [Chen et al., 2010], or neutrons [Disdier et al., 1999]), or potential like fast ignition of inertial confinement targets [Tabak et al., 1994] or laser-driven hadrontherapy [Bulanov et al., 2002a; Fourkal et al., 2002; Malka et al., 2004]. Having the possibility to spatially shape these currents is crucial in order to optimize the efficiency of each process. Sentoku et al. [2011] showed by combining experiments and modelling, how dynamical shaping of the currents can be achieved in various conductor materials. By tuning the target ionization dynamics that depends both on the target material properties and on the input electron beam characteristics, it is possible to control the growth of resistive magnetic fields that feedback on the current transport. As a result, collimating, hollowing or filamenting the electron beam can all be obtained and observed on the laser accelerated ions characteristics.

For all these applications, metals are preferable as target materials as they can provide enough cold electron return current to neutralize the forward laser-generated current and allow its propagation at current levels (MA) exceeding the Alfven critical current [Alfven, 1939]. At solid density, such targets are little prone to the Weibel relativistic electromagnetic two-stream instability [Weibel, 1959; Sentoku et al., 2003b] that is collisionally damped [Sentoku and Kemp, 2008]. Instead, resistive magnetic fields have been suggested to play more of a role [Kar et al., 2009; Robinson et al., 2007; Storm et al., 2009; Solodov et al., 2010]. These fields, that could collimate or focus the otherwise divergent electron beams [Debayle et al, 2010; Perez et al., 2010b], are driven by the Ohmic fields with the resistivity dynamically changing during target heating due to the hot electrons flow. The fields evolve according to Faraday's law. Weighting the role of magnetic fields has been relying up to now on hybrid simulations. These however present significant limitations due to their inability to self-consistently model the laser-generated hot electron source evolving at the dynamically ionizing interface. This has thus limited so far our understanding of the magnetic fields influence and impacted the possibility to adequately and quantitatively plan future progress, e.g. in designing fast ignition targets [A.A. Solodov et al., 2009] for present large-scale international programs [M. Schirber et al., 2005]. It is now possible to use Particle-In-Cell codes treating relativistic collisions and ionization processes to solve this issue.

Modulations of the sheath potential near the target surface will be imprinted in the protons angular distributions as they are detaching from the rear surface during the acceleration [H. Ruhl et al., 2004]. In [Sentoku et al., 2011], the 10 μm thick Au target exhibits a single peak distribution with the tightest profile, consistent with the experimental observation. The 15 μm thick Cu target on the other hand has a twin peak distribution, also consistent with the doughnut pattern of the experimental measure. Finally, the 40 μm thick Al target has a lower potential, since it is the thickest target, with a wider and modulated distribution, again consistent with the experiments. The trends observed in the experiment when increasing the target thickness or reducing the laser energy, i.e. a disappearance of the modulation in Cu or Al, are also observed in simulations performed with PICLS using collisions and ionization. For a 40 μm Cu target, strong ionization ($Z>15$) proceeds in a distance $x \sim 15$ μm, which is consistent with the heated region seen in the electron energy density map. Thus, strong resistive magnetic fields only grow in this region before breaking into weak filaments, ~ 5 MG, that cannot modulate strongly MeV electrons. As a result, the electrons spray out and form a smooth sheath potential at the target rear surface. Similar simulation results from reduced laser energy cases were also obtained.

The technique of manipulating the target resistivity developed here allows, even using a monolithic material, to control MA current flows in solids, e.g. excite the pinching, hollowing or filamenting of the currents. For this, the target thickness should be equal to or a bit thinner than the propagation distance of ionization waves during the laser pulse duration. Beyond the demonstration in [Sentoku et al., 2011], an important practical point is that keeping the same laser energy but changing the laser intensity (by e.g. defocusing the laser or increasing the pulse length), one can engineer transporting the same electron charge through different spatial form, e.g. pinched or hollowed out. Diagnosed using laser accelerated protons, it could help get higher maximum proton energies and smoother proton spatial distributions.

6.2. Magnetic lens for laser accelerated ions

High resolution proton deflectometry has been applied to measure magnetic fields in dense plasmas in order to quantitatively compare with simulations of the evolution of the magnetic field produced by intense electron currents generated by high intensity short laser pulses [Albertazzi et al., 2012]. Indeed, as seen in Section 6.1, such magnetic fields have the potential to limit the divergence of fast electron beams, a crucial factor in the context of secondary sources production or for achieving fast ignition of fusion targets. The capacity to measure these fields allows to have a better understanding of the implied physical mechanisms and to optimize them. In Section 6.1, these fields were diagnosed directly by ions leaving the target. In this section, two targets are used: one to accelerate protons and a second one irradiated by a secondary laser and radiographed by the proton beam coming from the first target.

Preliminary studies [Sentoku et al., 2003b] have shown that one of the main mechanisms to achieve collimation of of an energetic electron beam could rely on self-generated magnetic fields. These fields grow in the solid since the fast electron beam is not neutralized everywhere by the cold return current. The magnetic field arises from the resistive electric field $\mathbf{E} = \eta \mathbf{j}$ which depends on the resistivity η and on the current $\mathbf{j}$. Taking into account Maxwell-Faraday equation, two terms for the temporal derivative of the variation of the magnetic field can be obtained [Sentoku et al., 2011]. One term is a current term and the other one is a resistive term. This last term depends on the collision time, atomic number of the target, temperature and time since the resistivity will evolve in time. In order to study the influence of the self-generated magnetic fields on the transport of fast electrons, the self-generated magnetic fields were measured using face-on proton radiography and the influence of target material was also studied [Albertazzi et al., 2012]. Using a plasma mirror and thin targets, it was possible to probe self-generated magnetic fields in dense target. The experiment gives the dynamics of this self-generated magnetic field over a very large time: ~100 ps. The temporal dependence of the magnetic field patterns shows a field inversion which appears at different times for different materials. 2D PIC simulations including ionization and collisions performed with the code PICLS and coupled to test particles simulations using the fields obtained with the PIC simulations have allowed to reproduce some important features of the experiments and thus to better understand the physical mechanisms at play. This setup can also be used to constitute a magnetic lens to improve the characteristics of laser accelerated ions (divergence and energy spread).

6.3. Collision of laser accelerated plasmas to study collisionless shocks for astrophysics

Ion beam instabilities play an essential role in the formation of collisionless shocks in laser-plasma interaction or in astrophysics. Weakly relativistic shocks are considered as potential sources of high energy cosmic rays. Experiments performed in the laboratory would allow to have a better understanding of the microphysics involved in this process. As demonstrated in the previous sections, laser systems delivering short high intensity laser pulses can accelerate ions to high energies. Low energy electrons neutralize these ion beams. It is therefore possible to produce counter-propagating beams to study the collision of fast plasmas and the formation of collisionless shocks (Figure 13) [Kato et al., 2008; Davis et al., 2010; Davis et al., 2012].

Collisionless shocks are an ubiquitous phenomena in the Universe. Weakly relativistic shocks are created in supernova explosions. It is believed that relativistic shocks are responsible for strong electron heating, magnetic field generation and subsequent emission of hard X and gamma rays in Gamma Ray Bursts (GRB). Collisionless shocks are considered to be a source of energy redistribution in Nature and high energy cosmic rays. However, a detailed mechanism of energy transformation of fast plasma flows into relativistic electrons and large amplitude magnetic fields is not known. The hypothesis of energy equipartition in a collisionless shock between the electrons, ions and magnetic fields has been shown in recent astrophysical models using large scale numerical simulations [A. Spitkovsky, 2008]. Laboratory experiments may provide further understanding of this obscure process. However, acceleration of sufficiently large volumes of matter to relativistic energies requires concentration of high densities of energy in a short time scale. Only high energy short pulse laser systems could be suitable for such experiments. The Target Normal Sheath Acceleration (TNSA, see Section 2) [Snavely et al., 2000; Wilks et al., 2001] can provide protons with energies of several tens of MeV corresponding to the streaming velocities of 20 – 30% of the velocity of light. The number of fast ions produced in one shot exceeds 10^{12}, which should be sufficient for the collective processes to become significant. The range of such protons is of the order of 1 g/cm^2, so the collisional effects should not be significant for distances of many ion inertia length, which define the characteristic thickness of the shock front.

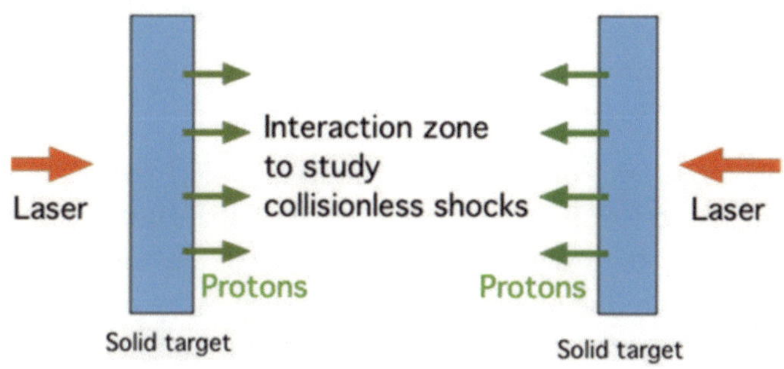

Figure 13. Schematic of the experimental setup required to perform the experiments discussed in this Section.

It was theoretically predicted that collisionless shocks may occur at mildly relativistic streaming velocities with a fraction ~ 10^{-4} of the total energy converted into magnetic field unless the electrons are heated significantly [Kato et al., 2008]. Experiments with non-relativistic laser pulses of sub-nanosecond duration and energy of a few hundred joules demonstrate a possibility of plasma streams formation with high Mach numbers, but the plasma temperature is relatively low and the ion mean free path remains shorter or comparable to the shock front thickness [Kuramitsu et al., 2011; Davis et al., 2010]. In a previous paper [S. Davis et al., 2010] it was proposed to collide the TNSA accelerated ion bunch with appropriately chosen low density plasma so the collective effects will dominate the ion stopping. [S. Davis et al. 2012] is dedicated to a more detailed study of electron heating and magnetic field generation in the overlapping region of the colliding plasmas: the first stage of formation of a collisionless shock. The characteristics of the high energy proton bunch were optimized in large scale PIC simulations. The ions were accelerated from an overdense target to sub-relativistic velocities by the TNSA mechanism. The plasma instabilities are excited by the streaming ions. The overall interaction time of about one ion plasma period, $\omega_{pi}t_{max} \sim 6$, is not sufficient for excitation of ion instabilities and full shock formation. In this time interval one can see the development of electron instabilities, magnetic field generation and electron heating.

This study has allowed to show that during the beginning of the plasma collision, a strong electron heating is generated by a Weibel type filamentation and each ions beam is then slowed. The energy partition between ions, electrons and electromagnetic fields, at the foundation of astrophysical models of gamma-ray bursts was therefore studied and has allowed to have a better understanding of the mechanisms at play. The better knowledge of laser ion acceleration described in the other sections of this chapter allow to envision collisions with even higher velocities with more energetic lasers (see Section 5.2). This would allow to get closer to relativistic shocks and study other types of astrophysical shocks (pulsar winds, other types of shocks involved in gamma-ray bursts…). Using an external magnetic field, it will be possible to study shocks with different magnetizations (ratio of the magnetic energy density and of the kinetic energy density) and thus to study in the lab for the first time collisionless shocks with parameters similar to the ones in supernovae remnants (SNRs).

7. Conclusions and perspectives

Laser acceleration of intense, collimated and multi-MeV ion beams is a recent research domain, rising fast. Its advent has been made possible thanks to short pulse lasers with extremely high intensities and progresses at the fast pace of the development of these lasers. Because of its pulsed nature (the duration of the source is of the order of a few ps), its high beam quality, of the high number of produced ions (10^{11}-10^{13}), of the possibility to modulate the spectrum as well as the divergence of the beam, the ion source produced by laser appears useful and promising for a number of applications. One can quote high temporal and spatial resolution radiography, fast ignition, production of warm dense matter and, later on, high intensity injectors for accelerators and sources for protontherapy or radio-isotope production.

This chapter presents several important contributions to the domain of laser ion acceleration and its applications. These contributions, experimental, theoretical and numerical, cover the possibilities to obtain in various regimes compact ion accelerators both with existing high intensity lasers and with future laser systems. Several promising applications are detailed, as well as some of the necessary developments to obtain these results. It is shown that the TNSA mechanism is robust and used for existing applications and that the transparency regime and low density laser ion acceleration could be used for applications requiring high ion energies.

The applicability perspectives offered by these beams are interesting and will be pushed further, in particular thanks to the progresses that can be anticipated for the lasers and the close opening of new installations like ILE, Gemini (at RAL), and further in time Apollon, ELI and diode pumped lasers with very high repetition rate and high energy. Thanks to these installations, it will be possible to increase the maximum energy and the dose of the ions (i) either by extrapolating simply using the mechanisms known nowadays, (ii) or by using other mechanisms (predicted or to be discovered) that will be more efficient.

To do so, the understanding of the accelerating parameters and the optimization of the laser parameters need to improved, by coupling theory and experiments, in order to obtain proton beams of a few hundred MeV, reproducible, controllable and predictable. This will require in particular to explore the scaling laws of existing mechanisms and to test the various mechanisms supposed to take place at ultra-high intensities 10^{21}-10^{22} W/cm^2 or higher (acceleration at the back surface, direct acceleration by the ponderomotive force, acceleration by electrostatic shock, acceleration by radiation pressure).

Future installations will possess various « lines », coupling laser pulses with similar intensities but with very variable durations ultra-short (10 fs) or moderately short (500 fs or more), which will offer the possibility to couple interaction regimes that are different, or even separated, and to take advantage of this coupling. For instance, the adjunction of bright X-ray sources and high energy proton sources will allow to probe dense plasmas heated by the protons. The proton sources will also be used to probe electron acceleration mechanisms. On these installations, it will be important to be able to change the parameters (pulse duration, energy, focalization) because if some applications require an excellent emittance, for others on the other hand the priority will go to ultra high energies (GeV and beyond), or to a maximum number of well collimated ions at low energy, different characteristics that require different ranges of laser parameters.

The progresses mentioned, the increase of energy in particular, should also open the way, as already mentioned, to new applicability perspectives, like injectors for high energy accelerators, the physics of ion beams (injection of multi-charged ion beams in matter), pulsed neutron sources, or even medical applications as well as material spallation and transmutation. Ultra high laser intensities (> 10^{24} W/cm^2) will even allow to test laser wakefield acceleration concepts, that are nowadays limited to electrons [Tajima and Dawson, 1979], with ions.

An intense international competition is underway both on the modeling of new acceleration regimes and on the experimental validation of these regimes. The recent results obtained at Los Alamos National Laboratory using some of the results presented in this chapter are an important illustration of this.

Author details

Emmanuel d'Humières
Université de Bordeaux – CEA - CNRS - CELIA, France

Acknowledgement

The author would like to thank Prof. V. T. Tikhonchuk, Prof. Y. Sentoku and Dr. J. Fuchs for fruitful discussions.

8. References

Adam J.C. et al. (2006), Phys. Rev. Lett. 97, 205006.
Albertazzi B. et al. (2012), submitted to Eur. Phys. J. Web of Conf.
Albright B.J. et al. (2007), Phys. Plasmas 14, 094502.
Allen M. et al. (2004), Phys. Rev. Lett. 93, 265004.
Alfvèn, H. (1939). On the Motion of Cosmic Rays in Interstellar Space, Phys. Rev. 55, 425.
Andresen G. B. et al. (2007), Phys. Rev. Lett. 98, 023402.
Antici P. et al. (2007), Phys. Plasmas 14, 030701.
Antici P. et al. (2008a), Phys. Rev. Lett. 101, 105004.
Antici P. et al. (2008b), J. Appl. Phys. 104, 124901.
Antici P. et al. (2009), New J. Phys. 11, 023038.
Atzeni S. et al. (2002), Nucl. Fusion 42, L1.
Begay F. et al. (1982), Phys. Fluids 25, 1675.
Bin J. H. et al. (2009), Phys Plasmas, 16, 043109.
Blanchot N. et al. (2008), Plasma Phys. Control. Fusion 50, 124045.
Borghesi M. *et al. (2002)*, Phys. Plasmas 9, 2214.
Borghesi M. et al. (2004), Phys. Rev. Lett. 92, 055003.
Borghesi M. et al. (2006), Fusion Science and Technology 49, 412.
Brunel F. (1987), Phys. Rev. Lett. 59, 52.
Buffechoux S., *et al. (2010)*, Phys. Rev. Lett. 105, 015005.
Bulanov S. V. et al. (2002a), Plasma Physics Reports 28, 453.
Bulanov S. V. et. al. (2002b), Physics Letters A 299, 240.
Bulanov S. V. et al. (2007), Phys. Rev. Lett. 98, 049503.
Bulanov S. V. (2009), Plasma Phys. Control. Fusion 48, B29.
Capdessus R. et al. (2012), submitted to Phys. Rev. E.
Carrié M. et al. (2011), High Energy Density Physics 7, 353.
Ceccotti T. et al. (2007), Phys. Rev. Lett. 99, 185002.
Chen Z. L. et al. (2005), Phys. Rev. E 71, 036403.
Chen H. et al. (2009), Phys. Rev. Lett. 102, 105001.
Chen H. *et al. (2010)*, Phys. Rev. Lett. 105, 015003.
Chen M. et al. (2011), Plasma Phys. Control. Fusion 53, 014004.
Chen S. et al. (2012), Phys. Rev. Lett. 108, 055001.
Clark E. et al. (2000a), Phys. Rev. Lett. 84, 670.
Clark E. et al. (2000b), Phys. Rev. Lett. 85, 1654.
Cowan T. et al. (2002), AIP Conference Proceedings 647, 135.
Cowan T. et al. (2004), Phys. Rev. Lett. 92, 204801.
d'Humières E. et al. (2005), Phys. Plasmas 12, 062704.

d'Humieres E. et al (2006), AIP Conference Proceedings 877, pp. 41-50.
d'Humières E. et al. (2007), Chinese Optics Letters, Vol. 5, Supplement, S136.
d'Humières E. et al. (2010a), J. Phys.: Conf. Ser. 244, 042023.
d'Humières E. et al. (2010b), AIP Conf. Proc. 1299, 704.
d'Humières E. et al. (2011a), MOP153, PAC11 proceedings.
d'Humières E. et al. (2011b), Eur. Conf. Abstracts 35G, P5.005.
d'Humières E. et al. (2012), submitted to Eur. Phys. J. Web of Conf.
Davis S. et al. (2010), Journal of Physics: Conference Series 244, 042006.
Davis S. et al. (2012), submitted to Eur. Phys. J. Web of Conf.
Debayle A. et al. (2010), Phys. Rev. E 82, 036405.
Denavit J. (1992), Phys. Rev. Lett. 69, 3052.
Disdier L. et al. (1999), Phys. Rev. Lett. 82, 1454.
Dong Q. et al. (2003), Phys. Rev. E 68, 026408.
Doumy G. et al. (2004), Phys. Rev. E 69, 026402.
Esirkepov T. Zh. et al. (1999), JETP Lett. 70, 82.
Esirkepov T. et al. (2006), Phys. Rev. Lett. 96, 105001.
Estabrook K. et al. (1978), Phys. Rev. Lett. 40, 42.
Fews A. et al. (1994), Phys. Rev. Lett. 73, 1801.
Flippo K. et al. (2008), Physics of Plasmas 15, 056709.
Flippo K. et al. (2010), J. Phys. Conf. Ser. 244, 022033.
Fourkal E. et al. (2002), Medical Physics 29, 2788.
Fritzler S. *et al. (2003)*, Appl. Phys. Lett. 83, 3039.
Fuchs J. *et al. (1999)*, Phys. Plasmas 6, 2569.
Fuchs J. et al. (2003), Phys. Rev. Lett. 91, 255002.
Fuchs J. et al. (2005), Phys. Rev. Lett. 94, 045004.
Fuchs J. et al. (2006), Nature Physics 2, 48-54.
Fuchs J. et al. (2007a), Phys. Plasmas 14, 053105.
Fuchs J. *et al.* (2007b), Phys. Rev. Lett. 99, 015002.
Fuchs J. et al. (2009), C. R. Physique 10, 176–187.
Gauthier M. et al. (2012), 39th EPS Conference on Plasma Physics proceedings – Stockholm, Sweden.
Gaillard S. et al. (2011), Phys. Plasmas 18, 056710.
Gitomer S. et al. (1986), Phys. Fluids 29, 2679.
Grech M. et al. (2011), *New J. Phys.* 13 123003.
Grismayer T. et al. (2006), Phys. Plasmas 13, 032103.
Gurevich A. V. et al. (1966), Sov. Phys. JETP 22, 449.
Habara H. *et al. (2004)*, Phys. Rev. E 70, 046414.
Haberberger D. et al. (2011), Nature Physics 7, 2130.
Hatchett S. *et al. (2000)*, Phys. Plasmas 7, 2076.
Hegelich B. M. et al. (2006), Nature 439, 441.
Henig A. et al. (2009), Phys. Rev. Lett. 103, 245003.
Jung D. (2012), Ph.D. dissertation – Ludwig Maximilians Universität München, Ion acceleration from relativistic laser nano-target interaction.

Kaluza M. et al. (2004), Phys. Rev. Lett. 93, 045003.
Kar S. et al. (2009), Phys. Rev. Lett. 102, 055001.
Kar S. et al. (2011), Phys. Rev. Lett 106, 225003.
Kato T. et al. (2008), Astrophys. J. 681, L93.
Key M. et al. (2006), Fusion Science and Technology 49, 440.
Klimo O. et al. (2008), Phys. Rev. Special Topics-Accelerators And Beams, 11(3):031301.
Klimo O. et al. (2011), New J. Phys. 13, 053028.
Kodama R. *et al*. (2004), Nature 432, 1005.
Krushelnick K. et al. (1999), Phys. Rev. Lett. 83,737.
Krushelnick K. et. al. (2000), IEEE Trans. Plasma Sci 28, 1184.
Kuramitsu Y. et al. (2011), Phys. Rev. Lett. 106, 175002.
Lancaster K. et al. (2007), Phys. Rev. Lett. 98, 125002.
Ledingham K.W. D. et al. (2004), J. Phys. D: Appl. Phys. 37 2341.
Leemans W. et al. (2006) Nat. Phys. 2, 696.
Lefebvre E. et al. (1997), Phys. Rev. E 55, 1011.
Lefebvre E. et al. (2006), J. of App. Phys. 100, 113308.
Limpouch J. et al. (2008), Laser & Part. Beams 26, 225.
Macchi A. et al. (2009), Phys. Rev. Lett. 103, 085003.
Mackinnon A.J. *et al. (2002)*, Phys. Rev. Lett. 88, 215006.
Mackinnon A.J. et al. (2004), Rev. Sci. Inst. 75, 3531.
Malka G. *et al. (1996)*, Phys. Rev. Lett. 77, 75.
Malka V. et. al. (2004), Medical Physics 31, 1587.
Malka V. (2008), Principles and applications of compact laser–plasma accelerators, Nature Physics 4, 447-453.
Maksimchuk A. et al. (2000), Phys. Rev. Lett. 84, 4108.
Mancic A. *et al. (2010a)*, High Energy Density Physics 6, 21.
Mancic A. et al. (2010b), Phys. Rev. Lett. 104, 035002.
Mora P. (2003), Phys. Rev. Lett. 90, 185002.
Mora P. (2005), Phys. Rev. E 72, 056401.
Mourou G. et al. (2006), Rev. Modern Phys. 78, 309.
Murakami Y. et al. (2001), Phys. Plasmas 8, 4138.
Nakamura T. et al. (2004), Phys. Rev. Lett. 93, 265002.
Nakamura T. et al. (2007), Phys. Plasma 14, 103105.
Nakatsutsumi M. et al. (2007), Phys. Plasma 14, 050701.
Nakatsutsumi M. et al. (2008), J. of Physics: Conf. Series 112, 022063.
Naumova N. et al. (2009), Phys. Rev. Lett. 102, 025002.
Neely D. et al. (2006), App. Phys. Lett 89, 021502.
Nemoto K. et al. (2001), Appl. Phys. Lett. 78, 595.
Offermann D.T. et al. (2011), Phys. Plasmas 18, 056713.
Palmer C.A. et al. (2011), Phys. Rev. Lett. 106, 014801.
Patel P. et al. (2003), Phys. Rev. Lett. 91, 125004.
Peano F. et al. (2007), Phys. Plasmas 14, 056704.
Pelka A. et al. (2010), Phys. Rev. Lett. 105, 265701.

Perez F.et al. (2010a), Phys. Rev. Lett. 104, 085001.
Perez F. et al. (2010b), Phys. Plasmas 17, 113106.
Pommier L. et al. (2003), Laser Part. Beams 21, 573.
Popescu H. *et al. (2005)*, Phys. Plasmas 12, 063106.
Psikal J. et al. (2008), Phys. Plasmas 15, 053102.
Psikal J. et al. (2010), Phys. Plasmas 17, 013102.
Pukhov A. (2001), Phys. Rev. Lett. 86, 3562.
Quéré F. et al. (2006), Phys. Rev. Lett. 96, 125004.
Renard-Le Galloudec N. et al. (2008), Rev. Sci. Inst. 79, 083506.
Renard-Le Galloudec N. et al. (2009), Phys. Rev. Lett 102, 205003.
Renard-Le Galloudec N. et al. (2010), Laser and Particle Beams 28, 513.
Robinson A. P. L. et al. (2007), Phys. Plasmas 14, 083105.
Robinson A. P. L. et al. (2008), New Journal Of Physics, 10:013021.
Robson L. et al. (2007), Nat. Phys. 3, 58.
Romagnani L. *et al.* (2005), Phys. Rev. Lett. 95, 195001.
Roth M. et al. (2001), Phys. Rev. Lett. 86, 436.
Roth M. *et al.* (2002), Phys Rev ST-AB 5, 061002.
Ruhl H. et al. (2001), Plasma Phys. Report, 27, 5, 363-371.
Ruhl H. et al. (2004), Phys. Plasmas 11, L17.
Sack C. et al. (1987), Physics Reports 156, 311.
Santala M. et al. (2000), Phys. Rev. Lett. 84, 1459.
Santala M. et al. (2001), Appl. Phys. Lett. 78, 19.
Sarkisov G. S. et al. (1999), Phys. Rev. E 59, 7042.
Schirber M. et al. (2005), Science 310, 1610-1611.
Schlegel T. et al (2009)., Phys. Plasmas 16, 081303.
Schollmeier M. et al. (2011), Bull. Am. Phys. Soc. http://meetings.aps.org/link/BAPS.2011.DPP.NO7.1.
Schwoerer H. et al. (2006), Nature 439, 445.
Sentoku Y. (2000) et al., Phys. Rev. E 62, 7271.
Sentoku Y. et al. (2003a), Phys. Plasmas 10, 2009.
Sentoku Y. et al. (2003b), Phys. Rev. Lett. 90, 155001.
Sentoku Y. et al. (2004), Phys. Plasmas, 11, 3083.
Sentoku Y. et al. (2008), J. Comput. Phys. 227, 6846.
Sentoku Y. et al. (2011), Phys. Rev. Lett. 107, 135005.
Silva L. et al. (2004), *Phys. Rev. Lett.* 92, 015002.
Snavely R. et al. (2000), Phys. Rev. Lett. 85, 2945.
Snavely R. et al. (2007), Phys. Plasmas, 14, 092703.
Sokolov I. V. (2009a), Journ. Exp. Theor. Phys. 109, 207.
Sokolov I.V. et al. (2009b), Phys. Plasmas 16, 093115.
Sokolov I.V. et al. (2010), Phys. Rev. E 81, 036412.
Sokolov I. V. et al. (2011), Phys. Plasmas 18, 093109.
Solodov A.A. et al. (2009), Phys. Plasmas 16, 056309.
Solodov A. A. et al. (2010), J. Phys. Conf. Series 244, 022063.

Spitkovsky A. (2008), Astrophys. J. 673, L39.
Stephens R. et al. (2004), Phys. Rev. E 69, 066414.
Storm M. et al. (2009), Phys. Rev. Lett. 102, 235004.
Tabak M. et al. (1994), Phys. Plasmas 1, 1626.
Tabak M. et al. (2006), Fusion Sci. and Tech. 49, 254.
Tajima T. and Dawson J. (1979), Phys. Rev. Lett. 43, 267–270.
Tamburini, M. et al. (2010), New Journ. Physics 12 123005.
Tamburini M. et al. (2012), Phys. Rev. E 85, 016407.
Temporal M. et al. (2002), Phys. Plasmas 9, 3098.
Temporal M. (2006), Phys. Plasmas 13, 122704.
Tikhonchuk V.T. et al. (2005), Plasma Phys. and Control. Fusion 47, 869.
Toncian T. et al. (2006), Science 312, 410.
Weibel E.S. (1959), Phys. Rev. Lett. 2, 83.
Wilks S. et al. (1992), Phys. Rev. Lett. 69, 1383.
Wilks S.C. (1993), *Simulations of ultraintense laser-plasma interactions,* Phys Fluids B 5, 2603.
Wilks S. C. et al. (2001), Phys. Plasmas 8, 542.
Willingale L. et al. (2006), Phys. Rev. Lett. 96, 245002.
Willingale L. et al. (2007), Phys. Rev. Lett. 98, 049504.
Willingale L. et al. (2009), Phys. Rev. Lett. 102, 125002.
Wittmann T. et al. (2006), Rev. Sci. Instrum. 77, 083109.
Yamagiwa M. et al. (1999), Phys. Rev. E 60, 5987.
Yang J. M. et al. (2004), J. Appl. Phys. 96, 6912.
Yin L. et al. (2006), Laser and Particle Beams 24, 291.
Yogo A. et al. (2008), Phys. Rev. E 77, 016401.
Youssef A. et al. (2006), Phys. Plasmas 13, 030702.

Chapter 12

Progress in High Average Power, Short Pulse Solid State Laser Technology for Compton X-Ray Sources

Akira Endo

Additional information is available at the end of the chapter

http://dx.doi.org/10.5772/46236

1. Introduction

Laser Compton X-ray source has been developing in more than decade as an accelerator-laser hybrid technology to realize a compact, high brightness short wavelength source. The basic principle is similar to an undulator emission, in which a high intensity laser field plays as the modulating electromagnetic field. Basic principle of the laser Compton X-ray source is explained in this chapter with recent examples of phase contrast imaging of bio samples. Single shot imaging is critical for many practical applications, and the required specification is explained as the laser pulse must exceeds some threshold parameters. It is already well studied on the optimization of the laser-Compton hard X-ray source by single shot base (John, 1998, Endo, 2001). Experimental results agreed well with theoretical predictions. Highest peak brightness is obtained in the case of counter propagation of laser pulse and electron beam bunch with minimum focusing area before nonlinear threshold (Babzien et.al, 2006: Kumita, et.al, 2008). The new short wavelength light source is well matured to demonstrate a single-shot phase contrast bio imaging in hard X-ray region (Oliva, et.al, 2010). The employed laser is a ps CO_2 laser of 3J pulse energy (Pogorelsky, et.al, 2006), but the laser system is not an easy and compact one for further broad applications in various laboratories and hospitals.

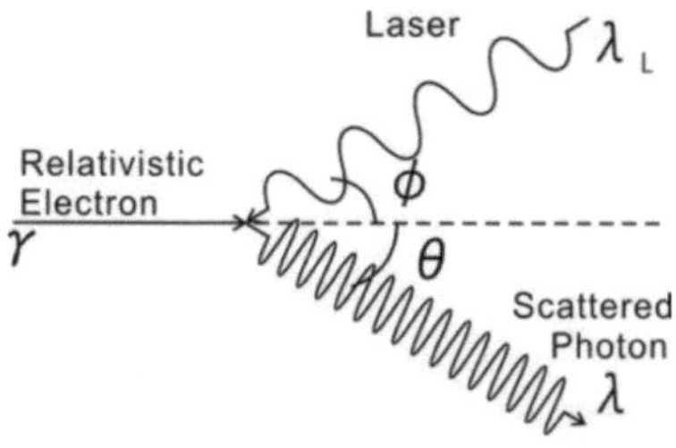

Figure 1. Schematic of laser-Compton scattering process

The major challenge of the laser Compton source for single shot imaging is the generation of threshold X-ray brightness, which in turn results in a clear sample imaging. Figure 1 describes the schematic of the laser-Compton interaction between electron beam and laser.

Laser-Compton scattering photon spectrum has a peak in the forward direction at a wavelength;

$$\lambda_p = \frac{\lambda_L(1+\frac{K^2}{2})}{2\gamma^2(1+\beta cos\phi)} \tag{1}$$

where γ and β are Lorentz factors, λ_L is the laser undulation period (laser wavelength), K is the K parameter of the undulator, which is equivalent to the laser intensity parameter, and Φ is the angle between electrons and laser propagation direction. The spectrum depends on the angular distribution; the wavelength λ is emitted at

$$\theta = \frac{1}{\gamma}\sqrt{\frac{\lambda-\lambda_p}{\lambda_p}} \tag{2}$$

It is seen that higher γ electron beam produces higher brightness of generated X-ray beams. The general formula of obtainable X-ray photon flux N_0 is calculated in the normal collision by the following expression,

$$N_0 \propto \frac{\sigma_c N_e N_p}{4\pi r^2} \tag{3}$$

where σ_c is the Compton cross section (6.7×10^{-25} cm^2), N_e is the total electron number, N_p is the total laser photon number, and r is the interaction area radius. Longer wavelength laser like ps CO_2 laser is advantageous to generate higher brightness X-rays at a fixed wavelength due to higher γ factor of employed electron beam, namely higher energy accelerator. Same energy laser pulse contains 10 times photons compared to solid state laser ones. Disadvantage is that the total system size becomes larger compared to the case of solid state laser based Compton source.

The approach to increase the photon flux is equivalent to increase N_e, N_p and decrease r, but there are instrumental limitations to realize these simultaneously. The practical limitation is the maximum electron number N_e and minimum interaction area diameter r. These are determined by emittance of the accelerated electron bunch and Coulomb repulsion. We would like to suppose it as 1nC, 3ps and focusable down to 10μm diameter at 38MeV acceleration energy. Another limitation is the onset of the nonlinear threshold of the higher harmonics generation, which is evident over 10^{17}W/cm^2 CO_2 laser irradiation intensity (Kumita, et.al. 2008). Laser pulses with 1ps pulse width focused down to 10μm, reaches at this threshold with 100mJ pulse energy. The nonlinear Compton threshold is characterized by the laser field strength

$$a_0 = eE/m\omega_L c \quad (4)$$

where E is the amplitude of laser electric field, ω_L is the laser frequency and c is the speed of light. The laser field strength is linearly depending on the laser wavelength. The laser energy for the nonlinear threshold of $a_0 \sim 0.6$ corresponds to 1J with 1ps at 10μm focusing in case of solid state laser. Single shot imaging was already realized by a 3J, 5ps CO_2 laser pulse focused onto 0.5nC, 32μm electron bunch (Oliva, et.al 2010). The focused laser intensity is over the nonlinear threshold as $a_0 > 1$. The X-ray spectrum was evidently overlapped with higher harmonics of X-rays. We can then estimate as it is also possible to expect a single shot imaging with equivalent solid state laser pulse, once it is possible to focus down to 10μm diameter to overcome the magnitude lower laser photon number. Table 1 summarizes the design laser parameters optimized for single shot imaging. It is clear from the table that a one pulse configuration is not possible to realize a single shot imaging because of the nonlinear threshold.

Nonlinear threshold	1J
Single shot imaging	4J
Pulse width	1ps
Focus diameter	10μm

Table 1. Solid state laser parameters for single shot imaging by Compton X-ray source

Usual approach is to increase the repetition rate of the event, and the obtainable X-ray photon average flux is expressed as;

$$N = f \times N_0 \quad (5)$$

where f is the repetition frequency. Fundamental characterization of the laser-Compton X-ray source has been undertaken with f typically as 1-10 Hz. High flux mode requires f in 100MHz range in burst mode for an equivalent single shot imaging.

The first approach is the pulsed laser storage in an optical enhancement cavity for laser-Compton X-ray sources (Sakaue, et.al. 2010, 2011). The enhancement factor P inside the optical cavity was 600 (circulating laser power was 42kW), in which the Finess was more than 2000, and the laser beam waist of 30μm (2σ) was stably achieved using a 1μm wavelength Nd:Vanadium mode-locked laser with repetition rate 357MHz, pulse width 7ps, and average power 7W. The schematic of the employed super-cavity is shown in Figure 2.

Short laser pulse *input* is injected through mirror 1 with transmittance T_1 and reflectance R_1. The mirror curvature is given as ϱ. The beam waist is given as W_0 and the cavity length is given as L_{cav}. The injected pulses overlap with the following pulses inside the cavity indicated as *Stored*. The loss is caused due to transmissions T_1 and T_2 of both mirrors.

An enhancement cavity requires high reflectivity and low transmittance mirror i.e. ultra-low loss mirror as an input and high reflectivity mirror as an output for high enhancement. The enhancement P is expressed by using cavity finesse F as (Hodgson, et.al, 2005);

$$P = \frac{F}{\pi} \quad (6)$$

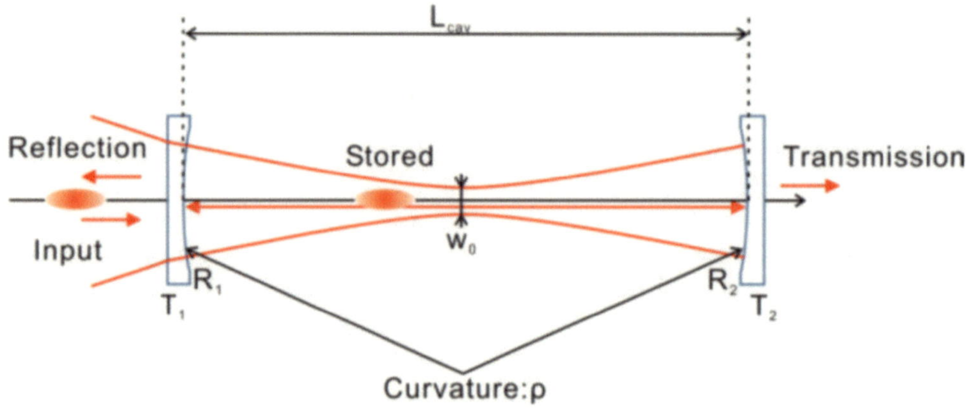

Figure 2. Schematic of laser storage enhancement cavity of Sakaue

It is noted that the assumed cavity length is perfectly matched with the repetition rate of input laser pulses. Finesse *F* is given by;

$$F = \frac{\pi\sqrt{R_{eff}}}{1 - R_{eff}} \tag{7}$$

where R_{eff} is $\sqrt{R_1R_2}$. As is described above, higher reflectivity provides a higher enhancement cavity. Particularly the loss, which includes both absorption and scattering on the reflection coating, is the critical issue for storing high power laser beam. The beam waist of an enhancement cavity is described as;

$$w_0^2 = \frac{\lambda}{\pi}\frac{\sqrt{L_{cav}(2\rho - L_{cav})}}{2} \tag{8}$$

where λ is the wavelength of the laser, L_{cav} is the cavity length, ϱ is the curvature of the cavity mirror. While high enhancement is relatively easier, smaller waist cavity down to 10μm is difficult as described in Eq. (8). Another work reported an enhancement of *P*~1400 with a 22 μm beam waist and 72kW storage power (Pupeza, et.al. 2010). The scaling limit is given by optics damage, which is around 100kW with ps pulse in this research stage. It was reported by Sakaue on an imaging demonstration by using the enhancement cavity approach of Fig.2, in which the stored pulse energy was 200μJ level in a burst mode of 100 pulses. The equivalent macro pulse energy was 20mJ. The larger focusing spot decreased available X-ray photons in each collision event, and the required time for imaging was much longer than equivalent single shot imaging (Sakaue, et.al, 2012). The repetition rate was 3Hz, and imaging of a fish bone was taken in 30 min with total laser energy of 108J. Figure 3 shows an imaging example in this experiment. Once the laser is focused to 10 μm diameter, and electron beam is focused to 30μm diameter, then the required total laser energy deceases to 3J, which indicates the design parameter of Table 1 as a good measure.

Grating based X-ray phase contrast imaging is now developing as a more sensitive imaging technology (Momose, et al. 2012), and a high repetition rate X-ray source, based on an enhancement cavity combined with a compact synchrotron, was recently introduced in preclinical demonstration of biological samples (M.Bech, et.al 2009). The X-ray peak energy was 13.5keV with 3% band width. The source size was relatively large as 165μm due to the focusing limit of circulating electron bunch in the compact ring. The repetition rate was typically in continuous 100MHz region, but the unit imaging time period was around 100 seconds (~2 minutes) due to lower X-ray photon flux per each event.

Classical low repetition rate laser Compton X-ray source demonstrated earlier a successful in-line phase contrast imaging of biological samples (Ikekura-Sekiguchi,et.al 2008). The repetition rate was at 10Hz with 40μm diameter source size. The imaging was undertaken by 3ps pulse width X-ray beam of 30keV energy. The required shot number for imaging was 18000 (30 minutes). It was indicated by this experiment that a solid state laser must have higher pulse energy more than 1J, and a better beam quality for 10μm focusing, for single shot imaging. We evaluate a possible solid state laser technology in the following sections on this subject, by reviewing practical instrumental limitations and propose the most promising approach for a compact single shot laser-Compton X-ray imaging.

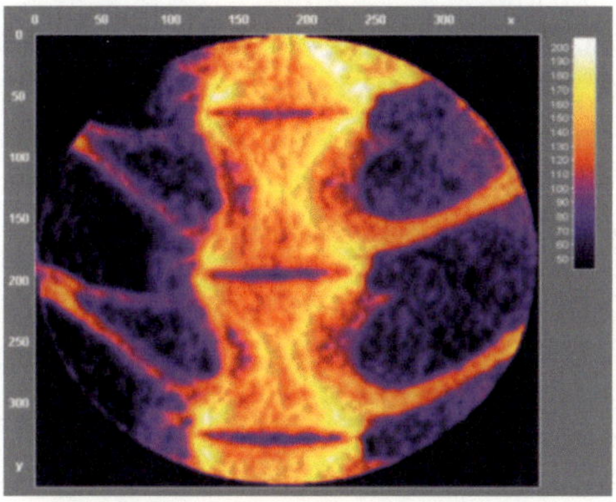

Figure 3. Refraction contrast imaging of bio sample (fish bone) by a laser-Compton X-ray source (Sakaue, et.al. 2012)

2. Temporal and spatial synchronization between electron beam and laser pulses

The essential technology for the laser-Compton X-ray source has been well studied in the Femtosecond Technology Project in Japan, and the achieved performance of the X-ray beam was also well characterized. Mathematical formula was obtained on its fluctuation depending on the temporal and spatial jitters (Yorozu, et.al 2002). Synchronization and stabilization technology was developed to the stage that the resulting pulse–pulse X-ray fluctuation almost reflects the laser pulse energy fluctuation (Yanagida, et.al 2003). The achieved overall performance was reported by T.Yanagida in a SPIE conference (Yanagida, et.al 2005). Figure 4 and table 2 show the system configuration and the summary of the specification of the laser-Compton X-ray source, studied and developed in the FESTA program. A phase contract imaging was also demonstrated by this light source of bubbles in solidified adhesives.

The electron beam is generated from a photo cathode RF gun driven by a synchronized picosecond UV laser, and accelerated to 38MeV energy by a S-band Linac. The achieved normalized emittance was 3 πmm-mrad, and resulted in the focused beam size as 30μm. It was demonstrated as further reduction of emittance was possible by spatial and temporal shaping of irradiation laser pulse for electron beam from photo cathode (Yang, et.al.2002). The employed laser for X-ray generation was a 4TW Ti:Sapphire laser with 800nm wavelength. The laser pulse was focused down to 10μm diameter and the peak intensity was around 10^{18} w/cm^2. The number of generated X-rays was measured with Micro Channel Plate located 2.6m downstream from the interaction point (source point). The MCP gain was calibrated using a standard ^{55}F X-ray source with known strength. The pulse width was estimated from measured electron beam and laser pulse width. The X-ray pulse width is almost determined by longer electron beam pulse width in case of normal incidence (165°interaction angle) and the cross section of the focused electron beam in case of 90°interaction angle. The long term fluctuation of the generated X-ray pulses is shown in Figure 5 in case of normal incidence arrangement. The repetition rate was 10Hz and the X-ray fluctuation was 6%, which is almost equivalent to the fluctuation of incident laser pulse energy. The laser focused intensity is around the nonlinear laser-Compton threshold as a_0~0.6. This was confirmed by a calculation by CAIN code in Figure 6. It is observed in the calculation of a nonlinear effect in the higher component of the generated X-ray energy distribution by blue dots (calculation by K.Sakaue).

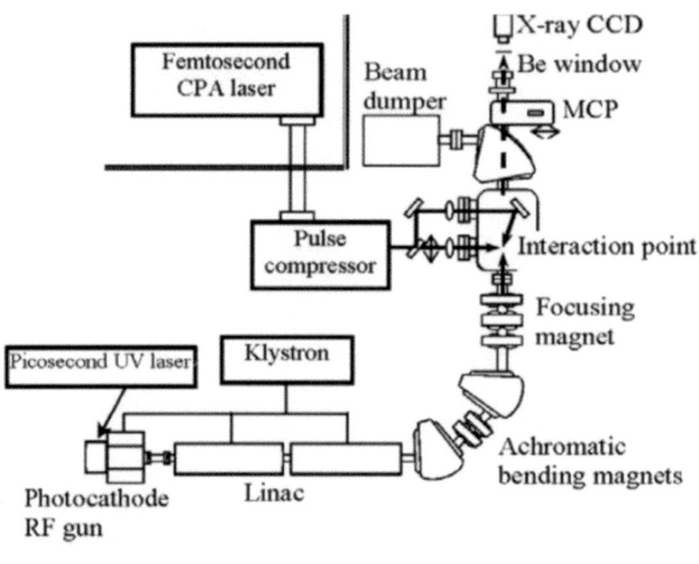

Figure 4. System configuration of laser-Compton X-ray source

Electron parameters		X-ray parameters		
Electron energy	38 MeV	Interaction angle	165°	90°
Bunch charge	0.8 nC	Maximum energy	33.7 keV	17.1 keV
Bunch width	3 ps (rms)	Total number of photons (photons/pulse)	2×10^6	5×10^5
Beam size	30 μm (rms)			
Normalized emittance	3 π mm·mrad	Pulse width	3 ps (rms)	150 fs (rms)
		Intensity fluctuation	6%	11%
Laser parameters		Repetition rate	10 Hz	10 Hz
Pulse energy	200 mJ			
Pulse width	50 fs (FWHM)			
Wavelength	800 nm			
Beam size	10 μm (rms)			

Table 2. Summary of the electron beam and laser parameters and obtained X-ray parameters

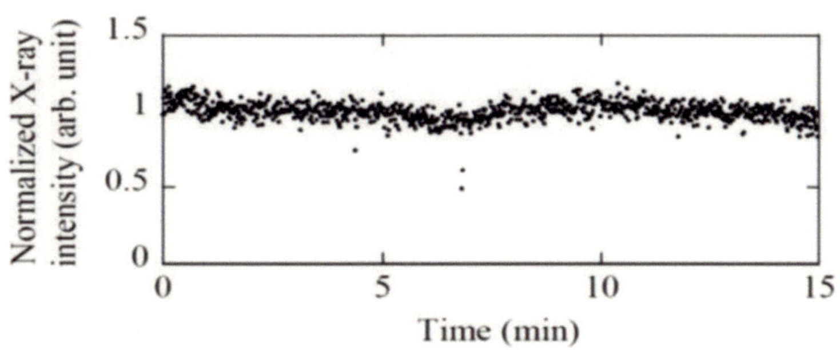

Figure 5. Fluctuation of X-ray intensity in the normal incidence laser-Compton X-ray generation

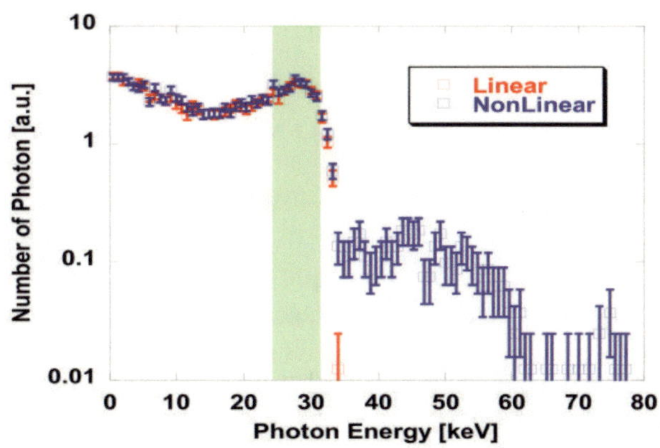

Figure 6. Calculated results of the X-ray energy distribution by CAIN code in log plot for linear (red) and nonlinear (blue) laser Compton scattering. Higher energy components are accompanying the linear ones. Light green region indicates the x-ray spectrum employed for imaging.

It is noticed that the component technologies for a single shot imaging by laser-Compton X-ray is well matured. There are but still several concerns necessary to design an optimized multi pulse method to realize the threshold (effective) laser energy of 4J in 10μm focus spot overlapped with electron bunch. The spatial stability of the laser-Compton X-ray source is essentially guaranteed in the order of the focus spot, because laser and electron beam must synchronize spatially (also temporally) each other to generate X-ray beam. Stable multi pulse electron beam generation is needed for efficient and stable laser-Compton X-ray source, to avoid higher harmonics noise of X-rays by limiting laser pulse intensity in each interaction. The RF photocathode gun is irradiated by synchronized ps laser pulses to generate flat top electron beam pulse train. An earlier experiment was reported by T.Nakajyo in 2003 of 60 micro pulses generation with a flat-top shape (Nakajyo, et.al 2003). The essential technology is temporal modulation of the seed laser pulse trains by Pockels Cell, to compensate the amplification saturation of the seeded pulse trains in the power amplifiers. Figure 7 shows the example of the pulse train amplification without and with intensity modulation. The obtained flatness of the 60 bunch electron beam was equal to that of the incident laser train (<7%) during 0.5μsec duration. The time duration is regarded for bio imaging enough short for effective single shot imaging.

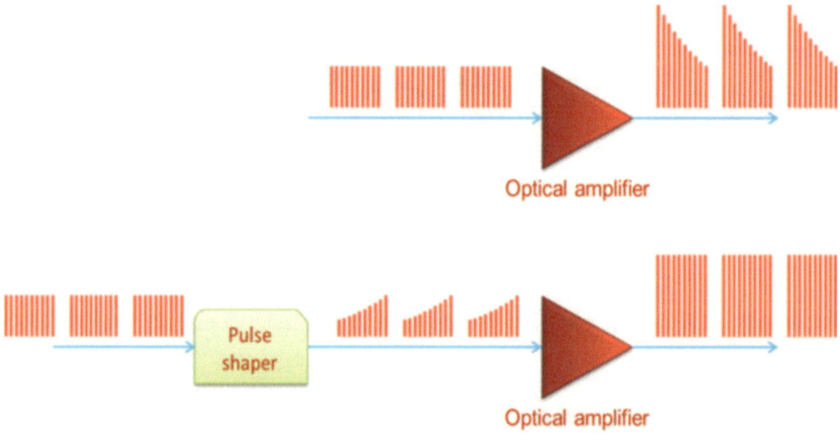

Figure 7. Pulse train amplification without (upper) and with (lower) intensity modulation

The other consideration is the selection of the amplifier module. It is required to focus 1J, 1ps laser pulses onto 10μm spot in spatial multiplexing in a near normal incidence arrangement. The requirement for the beam quality is expressed by M^2 parameter of the laser beam. Figure 8 shows the relationship between M^2 and focused beam spot size with the beam diameter as the working parameter. It is clear that M^2 is required to be less than 2 to realize a 10μm focal spot diameter with 20mm original beam diameter.

3. Thermal distortion in solid state amplifier

The basic requirement for a laser driver in a single shot laser-Compton X-ray imaging is summarized as Table 3. The M^2 of the laser beam is less than 2 from the discussions in the section 2. Detailed design work is required on the spatial configuration of 8 focusing optics to satisfy the 10 μm focus spot by avoiding radiation damage to the optics from scattered X-ray and electrons. The main consideration is an evaluation of the innovative solid state laser technology in the last decade like fiber, thin slab and thin disc lasers.

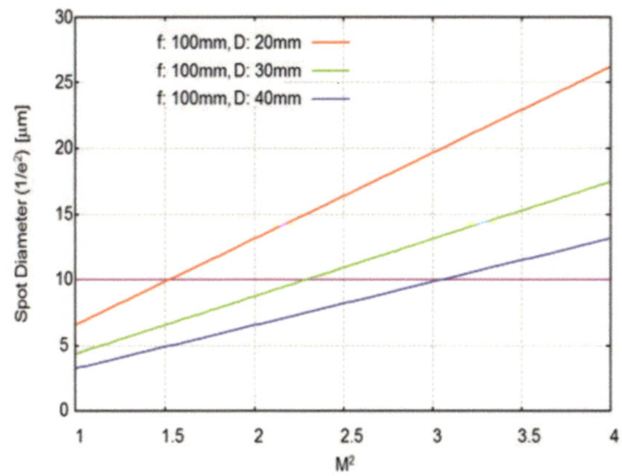

Figure 8. M^2 and focused beam size with beam diameter as parameters

Module pulse energy	>500mJ/ps
Module number	8 units
Multiplexed energy	4 J
Micro pulse time interval	8.4ns (119MHz)
Macro pulse width	~60ns

Table 3. Laser parameters for a laser driver in a single shot laser-Compton X-ray imaging around 30keV

Laser diode pumped rod type laser was regarded as the most suitable laser to meet the simultaneous requirement of high pulse energy, high average power together with high beam quality, before the fundamental solid state laser innovation. It was well known that flush lamp pumped solid state laser suffered from high thermal distortion of the laser medium due to low optical-optical conversion efficiency. Laser diode pumping was expected to solve the thermal distortion problem by improved energy conversion efficiency in the same configuration. Figure 9 is an example of a LD pumped rod Nd:YAG amplifier of 9mm diameter. Maximum LD pump power was 2.1kW, and optical-optical conversion efficiency was 41% (Endo et.al, 2004).

The fundamental difficulty of the LD pumped large rod amplifier comes from slow cooling speed of the laser material from the water jacked located around the rod. The resulting temperature gradient causes thermal lensing, which is expressed analytically by the following expression (Koechner, 1999).

$$f = \frac{KA}{P_a}\left(\frac{1}{2}\frac{dn}{dT} + aC_{r,\phi}n_0^3 + \frac{ar_0(n_0-1)}{L}\right)^{-1} \tag{9}$$

Temperature profile becomes radially parabolic. The first term corresponds to the temperature depending refraction index change of 70% contribution to f, the second term is stress induced refraction index change of 20% contribution to f, and the last term is temperature depending surface effect of 10% contribution. The cumulative effect of beam amplification and propagation of spatially non uniform beam results, combined with slight non uniform initial gain distribution, in a chaotic wave front with higher M^2. Figure 10 shows an example of beam cross section after booster amplifier of Fig.9 with 1.1kW average power at 10 kHz repetitive amplification of 6ns pulses. The beam was focused by f=10cm lens to 350 μm diameter with 10mm initial beam diameter. The resulting M^2 was nearly 35. The fundamental problem of rod amplifiers comes from temperature gradient. This was the main motivation of the enthusiastic search for a new architecture of low *dn*/*dT* solid state laser technology in the last two decades (Injeyan. et.al, 2011).

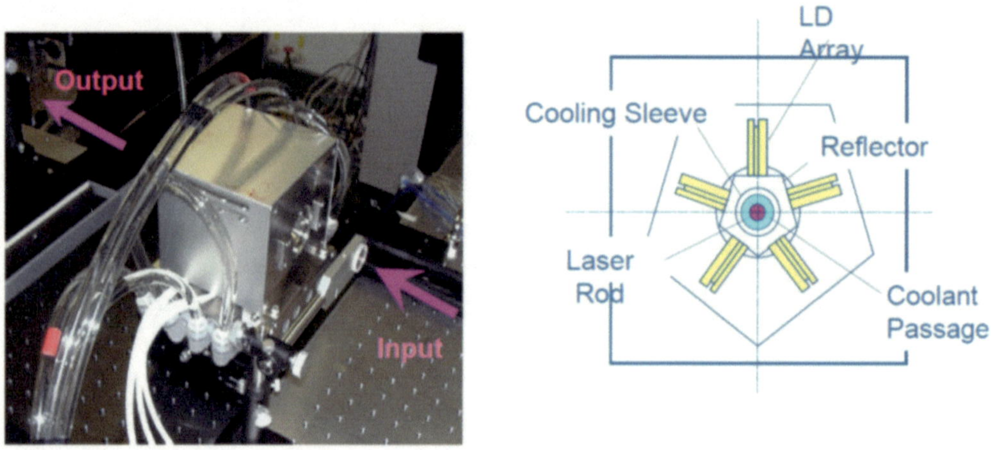

Rod amplifier

Cross section of rod amplifier

Figure 9. Outlook and cross section configuration of 9mm diameter Nd:YAG rod amplifier

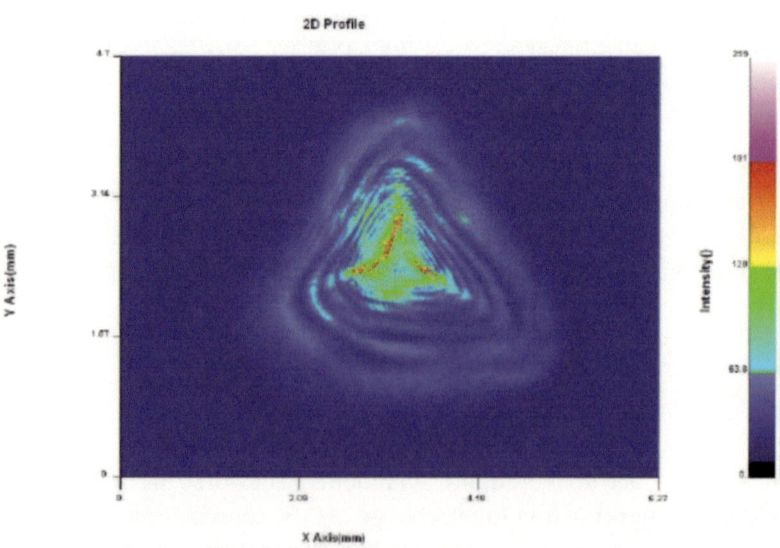

Figure 10. Beam shape after rod amplifier of 9mm diameter at 1.1kW average power

4. Thin disc laser as a high beam quality, short pulse solid state amplifier

Cryogenic cooling was considered to solve the temperature gradient problem in rod type LD pumped laser. MIT laser scientists are working following on this concept with recent unprecedented results of $M^2<1.05$ from cryogenically cooled (77k) bulk Yb:YAG laser in Q-switched mode of 20mJ/16ns at 5kHz. The average power was modest 100W in this experiment. The pointing stability was reported as 20 μ radian as mean deviation (Manni, et.al, 2010). One disadvantage of the cryogenic cooled Yb:YAG is the gain bandwidth narrowing, and compression to 1ps pulse width is not appropriate due to this effect (Hong, et.al. 2008). Fiber laser technology is progressing significantly with various laser specifications in CW and pulsed mode due to its efficient cooling characteristics owning to larger surface area/volume ratio. One drawback of fiber laser is its limited short pulse energy due to smaller medium diameter. There are still significant progresses in this field from its early work by a cladding-pumped, Yb doped large core fiber amplifier with specifications of 50W average power by 80MHz repetition rate of 10ps pulses with $M^2<1.3$ (Limpert, et.al,2001). Recent experiments achieved high pulse energy of 26mJ with 60ns pulse width at 5kHz repetition rate in Q-switched mode by a large-pitch fiber with a core diameter of 135 μ m (Stutzki,et.al. 2012). The average power is approaching to kW level with femtosecond pulse at high repetition rate. The reported performance was $M^2=1.3$ with 0.9ps pulse width with average power 830W at 80MHz repetition rate (Limpert, et.al. 2011). Sandwiched thin slab geometry is also promising to realize low temperature gradient inside laser medium by efficient cooling from both sides of thin slab. Multi-pass amplification is successfully employed inside the medium with expanding beam shape to keep the laser intensity constant during amplification. 1.1kW average power was reported with Yb:YAG as laser medium. The repetition rate was 80MHz of 615fs pulses, with M_x=1.43 and M_y=1.35 (Russbueldt, et.al. 2010). All these approaches are remarkable, especially regarding the beam quality M^2, but the achievable pulse width, and energy is limited due to cryogenic temperature or limited beam diameter in each technology. It is to be noticed that kW level, 80MHz femtosecond source could improve the average stored laser power in an enhancement cavity, once present limitation of optics damage is eliminated.

Thin disc laser is characterized with its larger diameter, and fundamentally suited for high pulse energy amplification. The schematic of a thin disc laser is shown in Fig.11. Thin disc of laser active medium like Yb:YAG of typical diameter 25mm is molded on a high reflectivity mirror (both wavelength of multi-pass LD; 940nm and laser wavelength; 1030nm). Water cooling from the backside of thin disc keeps the medium temperature around 15 degree. Mechanical distortion of the surface and ASE gain depletion is the main subject to be considered for high beam quality, short pulse high energy amplification. There are several activities to realize one J pulse energies with $M^2<1.3$ in ps pulse length at high repetition rate by thin disc laser technology. Conceptual design of a spatially multiplexed laser driver for single shot laser Compton imaging is presented in the next section, by showing several research examples.

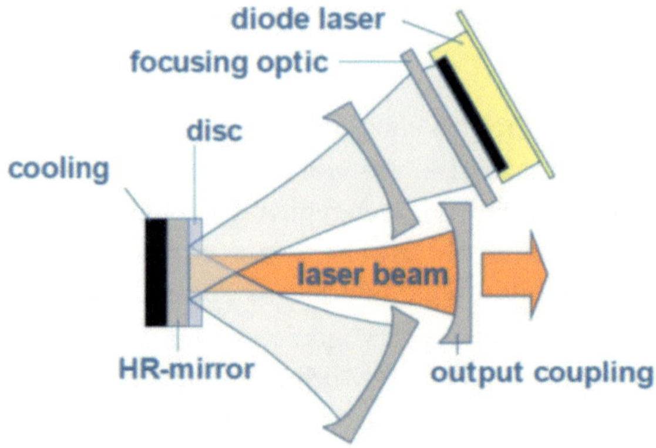

Figure 11. Schematic of thin disc laser configuration

5. Laser driver for single shot laser-Compton imaging

Candidate materials are considered for this particular application as Yb:S-FAP (Yb:Sr5(PO4)3F) or Yb:YAG. Comparison of both material characteristics are shown in table 4 (Payne,et.al, 1994). Both are characterized with higher quantum efficiency (Stokes factor), which is advantageous to less thermal stress after pulse energy depletion. Crystal growth to a larger diameter is important to avoid laser induced damage on the laser medium surface for 1J, ps pulse amplification at high repetition rate. It was tried to select Yb:S-FAP as the laser material by an end pumped square bar configuration, for the development as the future laser driver for high brightness laser Compton X-ray source (Ito et.al, 2006). The oscillator was a Yb:glass mode locked laser with 200fs, 170mW average power at 79.33MHz repetition rate, tuned at 1043nm wavelength. The oscillator pulse was stretched by a grating pair, and seeded into a cavity of a regenerative Yb:S-FAP laser by a Pockels Cell. Stacked laser diode array irradiated the Yb:S-FAP square rod (3.5 x 3.5 x 21 mm^3) with 900 nm wavelength, for 1.3ms duration of 1J pulse energy, through a lens duct and aspheric lens. The regenerative amplifier delivered 24mJ and the pulse was compressed down to 2ps in an initial experiment. Pre-amplifiers and main amplifiers were designed on the same architecture. Main amplifier employed square rods of geometrical size as 8 x 8 x 24 mm^3. Heat removal at higher repetition rate was not efficient from these amplifiers and the amplification was not perfect due to thermally induced birefringence. It was recently reported that "Mercury Laser Program" has achieved 100J in ns pulse length at 10Hz repetition rate from a side pumped thin slab Yb:S-FAP module of 3cm x 5cm aperture with a powerful cooling by He gas flow (Ebbers, et.al 2009). It is essentially proved from these experiments that Yb:S-FAP is usable as a laser material for specific ps application with higher pulse energy, once a large gas flow system is allowed in the whole system.

Another candidate is Yb:YAG for short pulse, high repetition rate operation for various applications. It is discussed that there is an obstacle to obtain large pulse energy in J level, from a bulk structure Yb:YAG material like a rod due to thermal population of the lower laser level (Ostermeyer, et.al. 2007). Solution might be found in a new configuration

	Yb:S-FAP	Yb:YAG
Pump wavelength (nm)	900	940
Laser wavelength (nm)	1047	1030
Fluorescence lifetime (ms)	1.26	1.0
Emission cross section ($10^{-20}cm^2$)	7.3	2.3
Saturation fluence (J/cm^2)	3.2	9.6
Pump saturation intensity (kW/cm^2)	2.3	32
Spectral bandwidth (nm)	3.5	9.5
Thermal conductivity (W/mK)	2	10

Table 4. Specific characteristics of Yb:S-FAP and Yb:YAG materilas

optimized for efficient cooling. Thin disc configuration is advantageous for the sake of efficient heat removal from gain media. It was tried to develop a pulsed thin disc laser with 1kW average power at 10 kHz repetition rate (Miura, et.al. 2005). Cavity optimization was performed for a regenerative amplifier, composed of two Yb:YAG thin disc modules, by compensating the deformation of optical components inside the cavity, with high beam quality at 500W CW operation. The extinction rate of linear polarization was more than 1:140. The developed regenerative amplifier module was connected with a seeder, which was a Yb:glass mode locked oscillator with 325fs pulse width and a fiber pulse stretcher. The extended pulse was injected into the regenerative amplifier cavity at 10 kHz repetition rate. The experimental configuration is shown in Fig12. Figure 13 is the pulse build up inside the regenerative amplifier cavity. Output average power was 33W in single mode, and 73W in multi mode with 50-100 ps pulse length (Miura,et.al. 2006). It was reported that an average power of 75W was achieved at 3kHz repetition rate with pulse energies exceeding 25mJ, a pulse-pulse stability of <0.7% (rms), a pulse duration of 1.6ps from an improved single thin disc module configurated in a regenerative amplifier with high beam quality as $M^2<1.1$ (Metzger, et.al 2009).

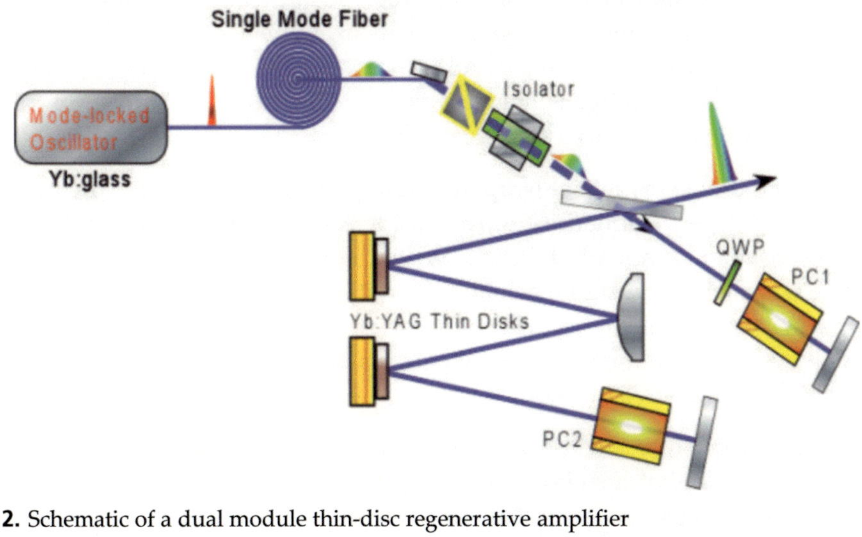

Figure 12. Schematic of a dual module thin-disc regenerative amplifier

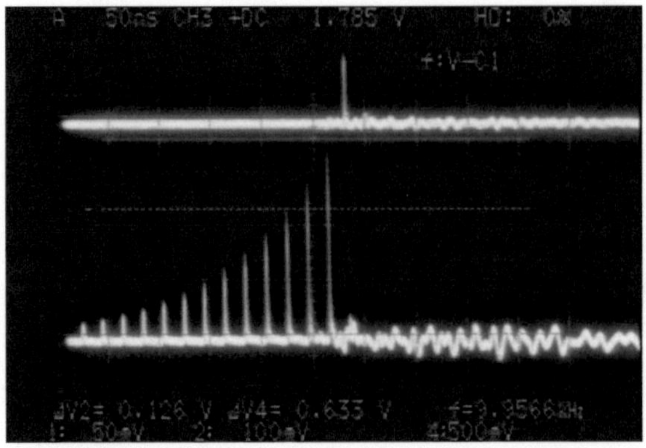

Figure 13. Operation of thin disc regenerative amplifier. Upper trace shows sliced out pulse, and lower trace shows building up inside cavity

Pulse energy increase to J level needs a multi pass amplifier without intra cavity Pockels Cell. The thickness of a thin disc medium is less than mm length for efficient water cooling from back side, and the single pass gain is lower than that of a rod medium in general. Multi pass optical cavity is required for this purpose, without any beam distortion during the amplification. A study was tried to design an optimized multi pass mechanical structure (Neuhaus, et.al.2008). A progress was recently reported from a group of Max Born Institute, Berlin, Germany on a development of a diode pumped chirped pulse amplification (CPA) laser system based on Yb:YAG thin disk technology, with a repetition rate of 100 Hz and output pulse energy aiming in the joule range (Tuemmler, et.al, 2009). Regenerative amplifier pulse energy was more than 165 mJ at a repetition rate of 100 Hz with a stability of 0.8% over a period of more than 45 min. The optical to optical conversion efficiency was 14%. The following main amplifier increased pulse energy to more than 300 mJ by a multi pass configuration. A nearly bandwidth limited recompression to less than 2 ps was also demonstrated. Further scaling of this technology is possible by enlargement of the thin disc diameter by careful optimization of the mitigation of surface deformation and ASE gain depletion. The latter phenomenon is well known in a small aspect ratio laser medium (Lowental, 1986). Numerical modeling of ASE gain depletion is useful to optimize working parameters, and HiLASE project is engaged in this effort to achieve 1J level picosecond pulses with high beam quality from thin disc amplifiers (Smrz, et.al.2012).

It is possible to design a spatial-temporal multiplexing of 0.5J, 1ps pulses onto the interaction point with low emittance electron bunch as is shown in Fig.14. Multiplexing of 8 pulses in polarisation combined 4 beams is the natural configuration. Timing jitter is possible in fs range which causes no actual X-ray output fluctuation. Spatial overlapping on 10μm diameter spot is challenging with pointing stability in the 10μrad range. Figure 14 indicates the multiplexing scheme to realize the laser specification of Table 2, based on 0.5J, ps thin disc laser modules of 8 units.

The generated forward directed X-ray beam has an effective pulse width <70ns, which is enough short for single shot imaging of bio samples. It is noted that the relative interaction angle between electron bunch and laser beams are fixed as 165 degree each other, in axial symmetry. It is proposed in a white book published by ELI Nuclear Physics working group, as the first stage of gamma ray program based on laser-Compton scheme, assumes 20 micro pulses with 0.15J, ps laser pulses, which is 3J effectively (Barty,C. et.al. 2011). The macro pulse repetition rate is expected as 120Hz. The average laser power is 360W. This is a manageable specification by usable laser technology described in this article.

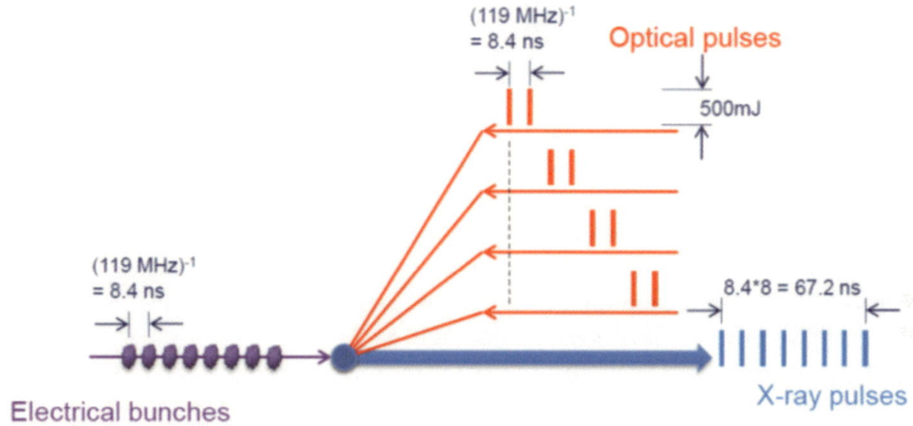

Figure 14. Spatial-temporal multiplexing of 0.5J,ps laser pulses

6. Conclusion

This chapter described the laser-Compton X-ray generator. The compact, high brightness X-ray source has been designed, fabricated and tested. This technology provides successful single-shot imaging of bio samples with multi J solid state laser pulses of ps pulsewidth. Advanced laser technologies were evaluated to realize a high beam quality, 1J level pulses. Thin disc laser was shown to be the best candidate for this application with Yb:YAG as the active medium. Spatial-temporal laser multiplexing was proposed to avoid nonlinear Compton effect. Some further research effort may bring us the realization of this technology.

The author deeply appreciates to his former colleagues in the Femtosecond Technology Project (FESTA), Extreme Ultraviolet Lithography Project (EUVA), supported by New Energy and Industrial Technology Development Organization (NEDO) in Japan to their productive and advanced works. Dr.Sakaue of Waseda University in Tokyo, Japan and Dr.Miura of HiLASE Project in Prague, Czech Republic, are especially appreciated to prepare various materials in this article.

This work was partially supported by the Czech Republic`s Ministry of Education, Youth and Sports to the HiLASE project (reg.No. CZ.1.05/2.1.00/01.0027).

Author details

Akira Endo
Research Institute for Science and Engineering, Waseda University, Tokyo, Japan
HiLASE Project, Institute of Physics AS CR, Prague, Czech Republic

7. References

Babzien, M. Ben-Zvi, I. Kusche, K. Pavlishin, I.V. Pogorelsky, I.V. Siddons, D.P. Yakimenko, V. (2006); "Observation of the second harmonic in Thomson scattering from relativistic electrons", Phys. Rev.Lett. 96 (2006) 054802

Barty, C. Wormser, G. Hajima, R. (2011); "Infrastructure producing high intensity gamma-rays for ELI nuclear physics Bucharest-Magurele, Romania", *The White Book of ELI Nuclear Physics Bucharest-Magurele, Romania*

Bech, M. Bunk, O. David, C. Ruth, R. Rifkin, J. Loewen, R. Feidenhaus, R. Pfeiffer, F. (2009); "Hard X-ray phase contrast imaging with the Compact Light Source based on inverse Compton X-rays", J.Synchrotron Rad. 16, pp43-47

Ebbers, C. Caird, J. Moses, E. (2009); "High-Power Solid State Lasers: The Mercury laser moves toward practical laser fusion", *www.laserfocusworld.com/article/355406*

Endo, A. Yang, J. Okada, Y Yanagida, T. Yorozu, M. Sakai, F. (2001); "Characterization of the monochromatic laser Compton X-ray beam with picosecond and femtosecond pulsewidths", *Proceedings SPIE 4502, pp100-108*

Endo, A (2004); „Performance of a 10 kHz laser produced plasma light source for EUV lithography", *Proceedings SPIE 5374, pp160-167*

Hodgson, N & Weber, H (2005). *Laser resonators and beam propagation: Fundamentals, Advanced Concepts and Applications 2nd edition*, ISBN-10: 0387400788, Springer, Berlin

Hong, K-H. Siddiqui, A. Moses, J. Gopinath, J. Hybl, J. Ilday, F.O. Fan, T.Y. Kaerntner, F.X. (2008); "Generation of 287W, 5.5ps pulses at 78MHz repetition rate from a cryogenically cooled Yb:YAG amplifier seeded by a fiber chirped-pulse amplification system", Opt.Lett. 33, pp2473-2474

Ikekura-Sekiguchi, H. Kuroda, R. Yasumoto, M. Toyokawa, H. Koike, M. (2008); "In-line phase contrast imaging of a biological specimen using a compact laser-Compton scattering-based x-ray source", Appl.Phys.Lett. 92, 131107(2008)

Injeyan, H. Goodno, G (2011); *High power laser handbook*, ISBN-978-0-07-160901-2, McGrow-Hill, New York

Ito, S. Nakajyo, T. Yanagida, T. Sakai, F. Endo, A. Torizuka, K. (2006); "Diode-pumped chirped pulse Yb:S-FAP regenerative amplifier for laser-Compton X-ray generation" Opt.Commun. 259, pp812-815

John, R.W (1998); "Brilliance of X rays and gamma rays produced by Compton back scattering of laser light from high energy-electrons", Laser and Particle Beams, 16 (1998) 115-127

Koechner, W. (1999); *Solid-State Laser Engineering*, Springer, Berlin

Kumita, T Kamiya, Y Babzien, M Ben-Zvi, I Kusche, K Pavlishin, I.V. Pogorelsky, I.V. Siddons, D.P. Yakimenko, V. Hirose, T Omori, T Urakawa, J Yokoya, K Cline, D. and Zhou, F (2008); "Observation of the Nonlinear Effect in Relativistic Thomson

Scattering of Electron and Laser Beams", Laser Phys. 16 pp267-271

Lowenthal, D.D. Egglestone, J.M. (1986); "ASE effects in small aspect ratio laser oscillators and amplifiers with nonsaturable absorption", IEEE J.Quantum Electron. QE-22, pp1165-1173

Limpert, J. Liem, A. Gabler, T. Zellmer, H. Tuennermann, A. Unger, S. Jetschke, S. Mueller, H.R. (2001); "High average power picosecond Yb-doped fiber amplifier", Opt.Lett.26, pp1849-1851

Limpert, J. Haedrich, S. Rothhardt, J. Krebs, M. Eidon, T. Schreiber, T. Tuennermann, T. (2011) " Ultrafast fiber lasers for strong-field physics experiments" Laser Photonics Rev. 5 pp634-646, DOI 10.1002/lpor.201000041

Manni, J.G. Hybl, J.D. Rand, D. Rippin, D.J. Ochoa, J.R. Fan, T.Y. (2010); "100-W Q-switched cryogenically cooled Yb:YAG laser", IEEE J.Quantum Electron. QE-46, pp95-98

Metzger, T. Schwarz, A. Teisset, C.Y. Sutter, D. Killi, A. Kienberger, R. Krausz, F. (2009); "High-repetition-rate picosecond pump laser based on a Yb:YAG disc amplifier for optical parametric amplification, Opt.Lett. 34, pp2123-2125

Miura, T. Suganuma, T. Endo, A. (2005) ; "High average power pulsed Yb:YAG thin-disc laser", *Proceedings SPIE, 5707, pp91-98*

Miura, T. Endo, A. (2006); "High power Yb:YAG thin-disc laser for EUV lithography", *Proceedings of 2nd EPS-QEOD 2006 Europhoton Conference, ThD5, 10-15 September, 2006, Pisa, Italy*

Momose, A (2012); International workshop on X-ray and neutron phase imaging with gratings, March 5-7, 2012 Tokyo, Japan

Nakajyo, T. Yang, J. Okada, Y. Yanagida, T. Yorozu, M. Sakai, F. Aoki, Y. (2003); "Multi-bunch electron beam source with a magnesium-photocathode radio-frequency gun", *Proceedings of RadTech ASIA 2003, pp.234-237 (December 9-12, 2003, Yokohama, Japan)*

Neuhaus, J. Kleinbauer, J. Killi, A. Weiler, S. Sutter, D. Decorsy, T. (2008) ; "Passively mode-locked Yb:YAG thin-disc laser with pulse energies exceeding 13μJ by use of an active multipass geometry", Opt.Lett. 33, pp726-728

Oliva, P. Carpinelli, M. Golosio, B. Delogu, P. Endrizzi, M. Park, J. Pogorelsky, I. Yakimenko, V. Williams, O. Rosenzweig, J (2010); "Quantitative evaluation of single-shot inline phase contrast imaging using an inverse Compton x-ray source", Appl.Phys.Lett. 97, 134104

Ostermeyer, M. Straesser, A. (2007); "Theoretical investigation of feasibility of Yb:YAG as laser material for nanosecond pulse emission with large energies in the joule range", Opt.Commun. 274, pp422-428

Payne, S.A. Smith, L.K. Deloach, L.D. Kway, W.L. Tassano, J.B. Krupke, W.F. (1994); "Laser, optical and thermomechanical properties of Yb-doped fluorapatite", IEEE J. Quantum Electron. QE30, pp170-179

Pogorelsky, I.V. Babzien, M. Pavlishin, I. Stolyarov, P. Yakimenko, V. Shkolnikov, P. Pukhov, A. Zhidkov, A. Platonenko, V.T. (2006); "Terwatt CO_2 laser; a new tool for strong field research", *Proceedings of SPIE, vol 6261, 626118*

Pupeza, I. Eidam, T. Bernhardt, B. Ozawa, A. Raushenberger, J. Fill, E. Apolonski, A. Udem, T. Limpert, J. Alahmed, Z.A. Azzeer, A.M. Tuennermann, A. Haensch, T.W. Krausz, F. (2010); "Power scaling of a high repetition rate enhancement cavity", Opt.Lett. 35, pp2052-2054

Sakaue, K. Washio, M. Araki, S. Fukuda, M. Higashi, Y. Honda, Y. Omori, T. Taniguchi, T. Terunuma, N. Urakawa, J. Sasao, N (2009); "Observation of pulsed x-ray trains produced by laser-electron Compton scatterings", Rev.Sci.Instrum. 80 123304 1-7

Sakaue, K. Araki, S. Fukuda, M. Higashi, Y. Honda, Y. Sasao, N. Shimizu, H. Taniguchi, T. Urakawa, J. Washio, M. (2011); "Development of a laser pulse storage technique in an optical super-cavity for a compact X-ray source based on laser-Compton scattering", Nucl.Instrum.Meth. A637 S107-S111

Sakaue, K. Aoki, T. Araki, S. Fukuda, M. Honda, Y. Terunuma, N. Urakawa, J. Washio, M (2012);"First refraction contrast imaging via laser-Compton scattering X-ray at KEK", *Proceedings of International workshop on X-ray and neutron imaging with gratings, P36, March 5-7, 2012, Tokyo, Japan*

Smrz, M. Severova, P. Mocek, T (2012); "Design and modelong of kW-class thin disc lasers", *Proceedings to be published, SPIE Photonics West, Jan 2012, San Francisco*

Stutzki, F. Jansen, F. Liem, A. Jauregui, C. Limpert, J. Tunnermann, A. (2012) "26mJ, 130W Q-switched fiber-laser system with near-diffraction-limited beam quality", Opt.Lett. 37, pp1073-1075

Tuemmler, J. Jung, R. Stiel, H. Nickles, P.V. Sandner, W. (2009); "High-repetition rate chirped-pulse-amplification thin disc laser system with joule level pulse energy", Opt.Lett. 34, pp1378-1380

Yang, J. Sakai, F. Yanagida, T. Yorozu, M. Okada, Y. Takasago, K. Endo, A. Yada, A. Washio, M. (2002); "Low emittance electron beam generation with laser pulse shaping in photocathode radio-frequency gun", J.App.Phys. 92, pp1608-1612

Yanagida, T. Kobayashi, Y. Maeda, K. Ito, S. Sakai, F. Torizuka, K. Endo, A.(2003); "Synchronization of two different repetition rate mode-locked laser oscillation for Laser-Compton X-ray generation", *Proceeding of SPIE, 5914 pp149-156*

Yanagida, T. Nakajyo, T. Ito, S. Sakai, F (2005); "Development of high brightness hard X-ray source by Laser-Compton scattering", *Proceedings of SPIE 5918, 59180V*

Yorozu, M. Yang, J. Okada, Y. Yanagida, T. Sakai, F.Takasago, K. Ito, S and Endo, A. (2002); "Fluctuation of femtosecond X-ray pulses generated by a laser-Compton scheme", Appl.Phys.B74, pp327-331

Yorozu, M Yang, J Okada, Y Yanagida, T Sakai, F Ito, S and Endo, A (2003); "Spatial beam profile of the femtosecond X-ray pulses generated by a laser-Compton scheme", Appl.Phys. B76 pp293-297

Chapter 13

GeO_2 Films with Ge-Nanoclusters in Layered Compositions: Structural Modifications with Laser Pulses

Evgenii Gorokhov, Kseniya Astankova, Alexander Komonov
and Arseniy Kuznetsov

Additional information is available at the end of the chapter

http://dx.doi.org/10.5772/53481

1. Introduction

In this chapter, we will discuss issues related to the development of new materials and device technology for micro-, nanoelectronics and optics. More specifically, our research activities were devoted to the study of objects in which 3D quantum size effect was revealed. These are a group of materials consisting of indirect-band-gap semiconductor nanoclusters embedded in insulator (Knoss, 2008; Molinari et al., 2003; Takeoka et al., 1998).

Changes in the optical properties of Ge-nanoclusters due to 3D quantum size effect in metastable germanium monooxide (GeO(solid)) layers after decomposition of such layers into Ge and GeO_2 (GeO_2<Ge-NCs>) were observed by us as early as in the late seventies, i.e. a few years before this effect was for the first time reported in the literature (Ekimov & Onuschenko, 1981). It should be emphasised here that in our experiments this effect was observed in a thin-film heterosystem rather than in bulk SiO_2 glasses with CuCl or CdS precipitates. The latter circumstance is important for using such materials in modern film technology of micro-, nano- and optoelectronics.

Apart from the detection of the photoluminescence coming from GeO_2<Ge-NCs> heterolayers, we showed that the 3D quantum size effect, radically changing the properties of the electron subsystem of the solid, could also be used to achieve a dramatic modification of the lattice subsystem of the solid matter. So, a new material, nanofoam, a solid similar to aerogels (Hrubesh & Poco, 1995), was obtained from germanium dioxide in GeO_2<Ge-NCs> heterolayers (Gorokhov et al., 2011). Simultaneously, we have developed a new technique allowing easy production of Ge-quantum dots (Ge-QDs) of a very small size. The dispersion of QDs sizes in obtained GeO_2<Ge-NCs> layers could be significantly reduced in comparison with the dispersion of Ge-QD sizes in GeO_2<Ge-NCs> heterolayers.

In our work, thin films from the following germanium oxides were investigated:

1. GeO(solid) films, which were layers of (usually amorphous) stoichiometric germanium monooxide (also, there are reported data about the existence of crystalline modifications of GeO(solid) (Martynenko et al., 1973));
2. GeO_x films, which were nonstoichiometric GeO(solid) layers with x, standing to indicate the chemical composition of the material, ranging in $0 < x < 2$ (I) or $1 < x < 2$ (II). In case I, the composition of GeO_x varies from pure germanium to GeO_2, and in case II, it varies between the stoichiometric compositions of GeO(solid) and GeO_2;
3. GeO_2 films, which can be amorphous or have crystalline modifications;
4. heterogeneous material GeO_2<Ge-NCs>, obtained from metastable GeO(solid) during its chemical decomposition into two components: an amorphous GeO_2 matrix and Ge-nanoclusters dispersed throughout the matrix.

These film materials, belonging to dielectrics, were recognised inappropriate for use in planar Si-based IC technology, and they were therefore forgotten. The main reason was that germanium monooxide was a metastable material readily undergoing decomposition even without any additional heating (Marin, 2010). Second, the layers of amorphous germanium dioxide rapidly dissolve in water (Kamata, 2008). Therefore, layers of amorphous GeO(solid) and GeO_2 were recognised inferrior to thermal silicon dioxide films, as well as to high-temperature silicon nitride films, which also proved to be chemically resistant and mechanically strong. In addition, it was absolutely unclear how patterning or selective etching of germanium oxide layers could be achieved. Other layers of germanium oxides, such as hexagonal GeO_2 and GeO_2<Ge-NCs> heterolayers, were unknown at that time when the possibility of using germanium oxide films in semiconductor industry was under evaluation. Also, there were no reported data on the interaction of layers of amorphous GeO_2 with SiO_2 or Si_3N_4 and on subsequent crystallisation of such binary compositions (Gorokhov et al., 1987, 1998).

However, the study of properties and specific features of all modifications of film compounds based on germanium oxides have led us to a revision of the appropriateness of such compounds for use in modern technology of solid-state devices in micro-, nano- and optoelectronics. The main advantage of germanium oxide layers over other dielectric layers consists in the capability of such layers (both atomic and electronic subsystem) to be easily modified during treatments given to the layers, and the modifications lead to a significant change of the initial properties of the material. Therefore, a deposited film of germanium oxides should now not be considered as a fully complete and final result of some process sequence; instead, it should be considered as a material suitable for subsequent modifications. In particular, based on such layers, one can create light-emitting diodes, photodiodes, optoelectronic couples, optical fibers, interference filters, mirrors, lenses, diffraction gratings and holograms, as well as single electron transistors, memory elements, resists for laser and probe nanolithography, low-k and high-k dielectrics, and a component of colloidal solutions of Ge-nanoparticles to fight against cancer in medicine (Tyurnina et al., 2011). To date, not all device applications of germanium oxide layers have been identified.

From our point of view, the specific properties of germanium oxide compounds outlined above show great promise in the development of the nanotechnology of layered systems in nanoelectronics. A pressing problem in this field is developing methods for local modification of film coatings aimed at imparting desired properties to a small area of the film only. Especially, there is a problem of formation of elements having extremely small sizes in one or two dimensions in a thin continuous planar layer.

Studying the potential offered by narrow linear and dot scanning laser treatments as applied to thin-film materials or surfaces exhibiting photosensitivity or susceptible of high-intensity focused laser radiation, i.e. laser micro- and nanolithography, was one of the promising research lines at this topic. A big volume of research has been done by us in this field, and that has brought forth many interesting effects and successful developments. Achievement of modification effects in thin near-surface layers of substance while leaving the material intact at larger depths is another research area in using laser treatments in micro- and nanoelectronics. Local modifications of material properties could also be achieved in the volume of bulk samples using a laser beam focused at a point inside the sample, and then the material remains non-modified on the surface and in subsurface layers.

Obviously, for a stronger display of necessary effects, special materials should be used in such treatments. Easy initiation of processes leading to structural modification of materials is the main property of required layers. The lower is the energy threshold of such modifications in a substance, the easier is the initiation of the process, and the stronger is this effect induced by the treatment. There exist substances in which processes leading to a change of their structure and chemical composition can be activated with an increase of the level of specific energy introduced in unit volume of the substance. Each of such processes is characterised by its own activation energy, and the activation energies of different processes normally differ in value quite widely so that the processes proceed separately. Of course, the more such potential processes can proceed in a material and the lower are the activation energies of the processes, the wider is the diversity of forms and transformations that can be induced in such an initial material.

Metastable germanium monoxide layers can serve such a material. The capability of such layers to transformation into a chemically and structurally stable germanium dioxide allows us to include modifications typical of GeO_2 in a number of possible modifications of initial GeO(solid) layers. During the decomposition of GeO(solid) layers, atomic germanium forms quantum-sized Ge-nanoparticles; this process also adds to the potential of possible transformations.

The two unique properties of film systems based on germanium oxides, their capability to easy modification of their electron and lattice subsystems, are well complemented with a third unique capability – easy transformation of material properties under pulsed laser irradiation. Local laser pulse treatments of the samples were found to be a technique enabling easy modification of germanium oxide layers. This technique offers us a unique tool for realising the potential inherent to germanium oxide layers. Thus, the content of this chapter aims to acquaint the reader with the main results of our investigations in the indicated field.

2. Experimental methods

2.1. Methods of synthesis of GeO(solid) layers and GeO_2<Ge-NCs> heterolayers

The studied films were obtained using three film deposition methods. First, heterolayers of GeO_2<Ge-NCs> were deposited from supersaturated GeO vapour in a low-pressure chemical vapour deposition (LP CVD) process (Knoss, 2008) onto substrates located in a quartz flow reactor (Fig. 1). The implemented LP CVD process involves two stages. The first stage (Fig. 1, zone A) is the formation of germanium monoxide molecules (GeO(gas)) according to the reactions:

$$2Ge(solid) + O_2(gas) \rightarrow 2GeO(gas) \quad (1)$$

$$Ge(solid) + H_2O(gas) \rightarrow GeO(gas) + H_2(gas) \quad (2)$$

The second stage (Fig. 1, zone B) involves two subsequent processes (a and b) proceeding during deposition of supersaturated GeO(gas) onto substrates. Process a is the condensation of vapour molecules (GeO(gas) → GeO(solid)) proceeding with the formation of a homogeneous metastable solid layer of germanium monoxide (GeO(solid)) on the substrate. The solid GeO film is metastable, and it readily decomposes into Ge and GeO_2 (process b) in several minutes at relatively low temperatures about 300°C and over:

$$GeO(solid) \rightarrow ½Ge + ½GeO_2 \quad (3)$$

In reaction (3), the germanium dioxide GeO_2 forms a glassy matrix, with excess germanium atoms being segregated as Ge-nanoparticles. The film growth rate depends on GeO vapour pressure and substrate temperature. In this way, using reactions (1) – (3) we were able to obtain either GeO_2 films with Ge-nanocrystals with sizes ranging from ~ 2 nm to ~ 10 nm (higher deposition temperatures, see Fig. 1 b, area III), or GeO_2 films with amorphous Ge-nanoclusters (lower deposition temperatures, Fig. 1 b, area II), or GeO(solid) films (deposition at room temperature, Fig. 1 b, area I). It should be emphasised here that the heterostructures used in our experiments had the following remarkable property: their molar ratio between Ge and GeO_2 was always fixed at exactly 1:1 independently on particular implemented growth conditions. The surface density of Ge-nanoclusters in a single-layer coating could range from $\sim 10^{10}$ to $\sim 10^{14}$ NC/cm^2, the average distance between Ge-nanoclusters being 1/2 of their diameter.

To obtain thin, stoichiometric, nondecomposed and homogenous GeO films on various substrates, a second method was employed. GeO(solid) films were additionally deposited onto substrates using thermal re-evaporation in a high-vacuum (10^{-7} Pa) flow reactor of thick GeO_2<Ge-NC> heterolayers (400–500 nm) grown by the first method (Sheglov, 2008). In such a process, a sample with a GeO_2<Ge-NCs> heterolayer was heated to a temperature of 550–600 °C by an ohmic heater. The heated GeO_2<Ge-NCs> film evaporated in accordance with the reverse of the deposition reaction:

$$\tfrac{1}{2}GeO_2(solid) + \tfrac{1}{2}Ge(solid) \rightarrow GeO(gas) \tag{4}$$

The resulting GeO vapour condensed onto a cold substrate (T ~ 25 °C), yielding a layer of metastable solid germanium monoxide GeO(solid): GeO(gas) → GeO(solid) (Fig.2). At temperatures $T > 250$ °C the GeO(solid) films decomposed, according to reaction (3), into Ge and GeO_2 with the formation of a new GeO_2<Ge-NCs> heterolayer.

In the third method (Ardyanian et al., 2006), using evaporation of GeO_2 with an electron beam in a high-vacuum chamber followed by deposition of evaporated species onto substrates at a low temperature (100 °C), we were able to obtain non-decomposed GeO_x films with $x = 1,2$. Those films, of thicknesses about 300 nm, were obtained in Laboratoire de Physique des Matériaux (LPM), Nancy-Université, CNRS, France. In both cases, with the help of the decomposition of GeO(solid) proceeding according to reaction (3), we could obtain layers of GeO_2 with embedded Ge-nanoclusters.

All studied films were deposited onto either Si (100) substrates or glass. To avoid evaporation of the GeO(solid) films during laser treatments, the samples were covered with protective SiO_2 or SiN_xO_y layers. Protective SiN_xO_y cap layers about 25 nm thick were deposited at 100 °C using the plasma enhanced chemical vapour deposition method (PE CVD), in Institute of Semiconductor Physics of the Siberian Brunch of Russian Academy of Science (ISP SB RAS). Protective SiO_2 cap layers were prepared using two methods. The first method was the evaporation of fused silica glass with an electron beam at 100 °C in a high vacuum (10^{-8} Torr); this method was implemented at Nancy-Université, France (Jambois et al., 2006). The second method was chemical vapour deposition of SiO_2 layers with simultaneous monosilane oxidation in the pressure range 0,5-1,2 Torr at 150°C; this method was developed and implemented at ISP SB RAS.

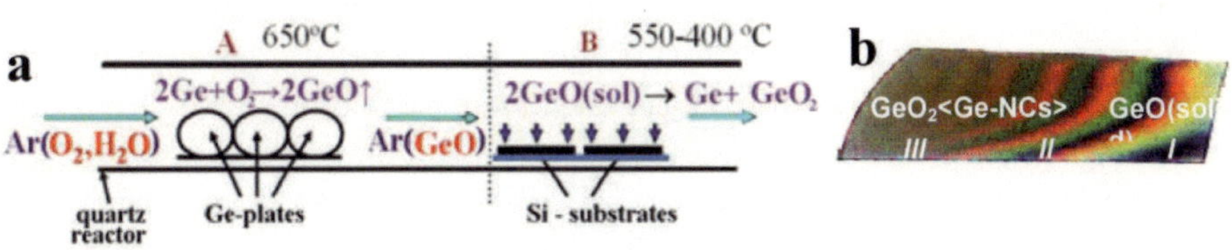

Figure 1. The scheme of LP CVD process of GeO_2<Ge-NCs> heterolayer growth (a); and photo of the GeO_2<Ge-NCs> film on Si-substrate of gradient thickness, structure and features (b).

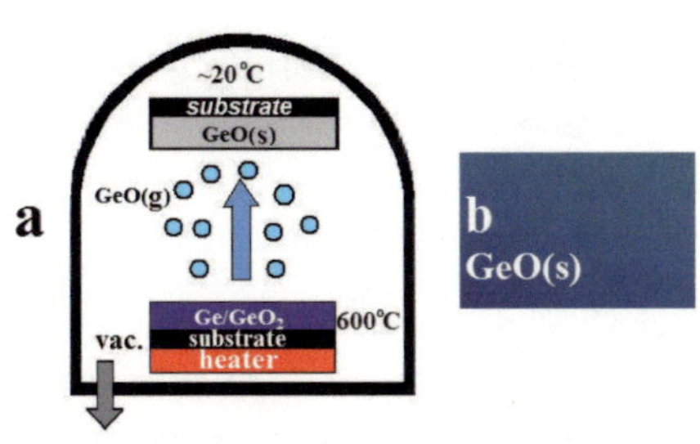

Figure 2. Scheme of vacuum GeO(solid) formation (a); and photo of the GeO(solid) film of uniform properties on Si-substrate (b).

2.2. Methods of structural analysis

Investigation into the structural properties of thin layers of amorphous dielectrics, and also investigation into the impact of various growth factors and technological treatments on those properties, present most challenging problems in modern material science for micro- and nanoelectronics. This usually requires a complex of specific instruments and methods. We examined our films using Raman scattering spectroscopy, IR spectroscopy, ellipsometry, scanning electron microscopy (SEM), transmittance electron microscopy (TEM), and atomic force microscopy (AFM).

The optical constants of our films were studied with the help of the scanning ellipsometry method, implemented with step δl=0,5 mm, using a He-Ne laser (633 nm). The results were interpreted using a specially developed algorithm described elsewhere (Marin et al., 2009). The spectral dependences of absorption and refraction indexes, k(D) and n(D), versus the diameter of Ge-nanoclusters (D) contained in the films were studied using many-angle, multiple-thickness spectral (250-800 nm) scanning ellipsometry. With the combination of spectral many-angle, multiple-thickness measurements, we were able to significantly improve the accuracy of ellipsometric data on the thickness and optical constants of studied films in the visible range.

Atomic force microscopy (Solver P-47H, NT-MDT, Russia) and scanning electron microscopy (Hitachi S4800) were used to examine the surface morphology of our films before and after applied modifications. Optical microscopy was used to register changes in the optical constants of the films in laser-modified areas. The direct observation of Ge-nanoclusters in GeO_2 films was carried out with the help of transmission electron microscopy (TEM) on specially prepared thin SiO_2 membranes (Gorokhov et al., 2006).

Raman spectroscopy was used to reveal and identify the structure (amorphous or crystalline) of Ge-nanoclusters in the films. Raman spectra were registered in quasi back-scattering geometry with a 514,5-nm Ar^+ laser used as an excitation source. A double DFS-52 spectrometer and a triple T64000 Horiba Jobin Yvon spectrometer equipped with a micro-Raman setup were employed. In the latter case, slightly unfocused laser beam was used, producing a spot of about 3-4 micrometer diameter on the surface of the sample, with the laser power reaching the sample being diminished to 10-20 mW in order to avoid overheating of the films. All Raman spectra were measured at room temperature.

IR spectroscopy (FT-801 Fourier spectrometer, ISP SB RAS, Russia, equipped with a micro-setup) was used to study the chemical composition and structure of dielectric layers after different treatments given to them in small areas of the sample. IR absorption measurements carried out at normal incidence were performed at a resolution of 4 cm^{-1}.

Treatments of GeO(solid) layers were performed using a nanosecond KrF excimer laser (wavelength 248 nm, pulse duration 25 ns). Laser fluences ranging from 130 to 170 mJ/cm^2 were applied. A Ti-Sapphire laser (FemtoPower Compact Pro, Femtolasers Produktions GmbH) with 800 nm central wavelength and pulse duration < 30 fs was used for laser treatments. The energy distribution in the laser spot was assumed to have a Gaussian form.

The laser fluence was changed by varying the laser pulse energy E_{pulse}: $E_0 = 2E_{pulse}/\pi r_0^2$, where E_0 is the maximum energy and r_0 is the laser spot radius. The treatments were done in scanning mode with the laser spot diameter being 70 or 1 μm.

3. Peculiarities of GeO(solid) films and GeO_2<Ge-NCs> heterolayers

3.1. The structure and the decomposition process of the metastable GeO(solid) layers

Because of the weak interest of researchers to the films of germanium oxides, nowadays in the scientific literature there are no unambiguous data about the structure of GeO(solid) layers, about the mechanisms of formation of such layers under different conditions, and about their decomposition according to the reaction (3) or evaporation of decomposed GeO(solid) films under heating according to the reaction (4) (Kamata, 2008). However, for controllable variations of the GeO(solid) layers structure and properties, adequate to reality considerations on these aspects are required. This urged us to look for answers to these questions. The qualitative considerations described below are different in that its main statements do not contradict to any fact from known experimental data on GeO(solid) layers. By analogy, one can extend this consideration to the description of the behaviour and properties of SiO(solid) films, since Ge and Si are both group IV elements, this circumstance makes compounds of germanium and silicon with other elements quite similar.

The technology of the films consisting of Si- or Ge-QDs introduced in a dielectric SiO_2 or GeO_2 matrix, respectively, demands understanding of the role and properties of SiO and GeO monoxides. They can be solid and gaseous. However, the physical chemistry of Ge and Si lower oxides is studied insufficient. The actuality of their studies originates from the fact that GeO(gas) condensation in the form of a thin GeO(solid) film is the simplest way of obtaining Ge-QDs dispersed in the glassy GeO_2 film. As the GeO(solid) structure is thermodynamically unstable, the film quickly decomposes due to reaction GeO(solid) → ½Ge + ½GeO_2. To obtain vapour from GeO molecules, one can easy use inverse reaction ½Ge + ½GeO_2 → GeO(gas) at T > 500 ^{0}C (Knoss, 2008). The produced heterolayers with a high Ge-QDs concentration in GeO_2 matrix have the peculiarities which are interesting from both scientific and practical standpoints (Knoss, 2008). For controllable variations of the GeO_2<Ge-NCs> heterolayers structure (in particular, of Ge-nanoparticles characteristics), reliable control of the decomposition process of GeO(solid) layers and a good understanding of the nature of its thermodynamic metastability are required. So we had to look for an explanation to one of the unsolved properties of solid germanium monoxide – thermodynamic metastability of GeO(solid).

In studying germanium monoxide, there arises a question why GeO molecules in gaseous state are stable (dissociation energy ΔH^o ~159 kcal/mole (Jolly & Latimer, 1952)), whereas in solid GeO germanium atoms turn out to be capable of easily breaking their bonds with oxygen atoms without any activation at room temperature. The difference in the stability of the two, gaseous and solid, aggregate states of GeO can be attributed to the fact that the Ge- and O-atoms in a GeO vapour molecule have no immediate interaction with other atoms.

Obviously, sp^3-hybridization (tetrahedral orientation of the four orbitals of Ge-atom) in this molecule is not possible. Therefore in a GeO(gas) molecule the Ge-atom is bonded with the O-atom with one σ-bond and one π-bond (Tananaev & Shpirt et. all, 1967). Transformation of Ge-atom orbitals from sp^3-hybridized state (inherent to solid Ge and GeO_2) into the structure of valence bonds inherent to GeO(gas) molecules requires energy ΔH_{298}~54kcal/mole (Jolly & Latimer, 1952) and heating.

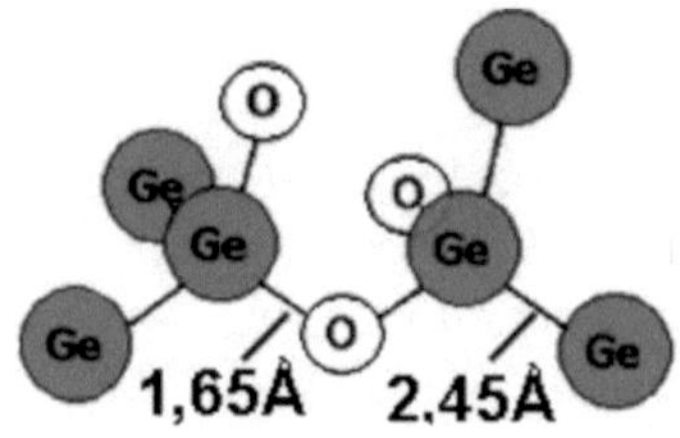

Figure 3. Assumed structure of the atomic network of GeO(solid) satisfying the following two conditions: the valence orbitals of Ge-atoms are sp^3-hybridized and the chemical composition of the GeO(solid) film complies with the Ge-O stoichiometry.

When diatomic GeO(gas) molecules condense, the distances between individual GeO molecules become the order of the atomic size. As a result, Ge- and O-atoms become able to be bonded into 3D network from chains of Ge- and O-atoms. The latter allows sp^3-hybridization with tetrahedral orientation of the valence orbitals of Ge-atoms to form the base for construction of an atomic network. The driving force for such a transformation can be the release of free energy spent on the formation of a new valence bond structure of GeO(gas) molecules (reaction (4)). Thus, as the lattices of germanium dioxide and germanium, the lattice GeO(solid) should consist of linked tetrahedra. According to the stoichiometry of GeO each Ge-atom at the vertex of a tetrahedron, formed by its sp^3-hybridized bonds, should be bonded to two other Ge-atoms and two O-atoms (see Fig. 3).

However, there are differences between the lengths of Ge-Ge (2,45 Å) and Ge-O (1,73 Å) bonds in the elemental tetrahedrons of the GeO(solid) lattice. Therefore, this lattice is heavily deformed and has strongly distorted bond angles. So, the real structure of GeO(solid) is not compact, its lattice energy is not minimised to ensure the stability of the atomic network. The latter situation is typical of materials and their basic lattice elements have a low degree of symmetry. When such elements become arranged in a continuous atomic network by successive translations, large cavities should appear in the film. To make such cavities filled with the material, strong deformations of the translation elements, tetrahedrons, become necessary. In any case, the formed atomic network possess a high level of internal strain energy. This will force GeO(solid) to decompose into components, the structure of such components is constructed from (a) basic elements with a higher degree of symmetry and based on sp^3-hybridization of Ge-atom orbitals (b). I.e. these basic elements must provide a layering of unstable material Ge(solid) into the components made without the high internal strain energy. Such components are Ge and GeO_2 which have stable lattices based on elements with high degree of symmetry – $Ge(Ge^{3-})_4$ and $Ge(O^{4-})_4$ tetrahedra, respectively.

A model for the atomic structure of GeO(solid) were developed on data received by IR and Raman spectroscopy, scanning probe microscopy and new methods in high-resolution TEM and ellipsometry. The model allows us to consciously control the structure of heterolayers GeO_2<Ge-NCs> during their growth and further treatment. An analysis of structural transformations of germanium monoxide based on this model allows gaining a better insight into some important details of the formation of the metastable GeO(solid) atomic network during condensation of GeO(gas) molecules and of decomposition under activating treatments. According to the above model, the formation of stable GeO_2<Ge-NCs> heterolayers proceeds in two steps. The first step is the formation of a homogeneous metastable GeO(solid) structure at T< 100-200°C, when GeO(gas) molecules condense from vapour onto the surface of the growing GeO(solid) film. It is a very rapid process (femtoseconds) proceeding when the atomic bonds of Ge- and O-atoms in absorbed GeO(gas) molecules get rearranged into tetrahedral configuration of sp^3-hybridized bonds; the process has a very high rate and the duration of this process is defined by the switching time of valence bonds between chemically interacting atoms.

Note that if during the condensation of GeO(gas) molecules its diatomic molecules cannot approach each other at atomic distances, the structure of the valence bonds of those molecules remains unaltered. It follows from the data by Ogden and Ricks (Ogden & Ricks, 1970), who found that during condensation GeO(gas) molecules readily become incorporated into the solid matrix of frozen N_2 or Ar at T ~ 20 K. Under such conditions, the vibrational spectra of GeO(gas) molecules (Fig. 4 a) show no changes during freezing in the solid state, and their IR absorption spectra are no difference from the spectra of GeO vapour molecules. In contrast, in the case of condensation of GeO(gas) molecules proceeding with the formation of a GeO(solid) film, vibrational characteristics of the atomic network change considerably (see Fig. 4 b). IR spectrum of frozen GeO molecules has no band, which is typical for GeO(solid) during its thermal decomposition in the range of 770 - 870 cm^{-1}.

The second step is a process of rearrangement of the homogeneous but metastable lattice of GeO(solid) layer into a complicated heterogeneous film structure such as a solid glassy GeO_2 matrix with incorporated Ge-nanoparticles. The latter process is more difficult and by order of magnitude slower than the former process. At this step, two stages can be marked out. The first stage involves decomposition of the lattice of the GeO(solid) layer, with half of the Ge-atoms of this lattice becoming bonded to the bridge O-atoms with all the four sp^3-hybridized bonds of each Ge-atom. Here, a solid GeO_2 network made by $Ge(O^{2-})_4$-type tetrahedrons forms. This network fills up the volume that was previously fully occupied by the 3D GeO(solid) network. On the contrary, the second half of the Ge-atoms in the decaying GeO(solid) network tends to fully disrupt its sp^3-hybridized valence bonds with O atoms to get out of the atomic network of solid germanium oxide. However, because of the impossibility to do that immediately, these atoms are forced to form various defects in glassy GeO_2. The variety of atomic configurations in GeO_2 layers could be rather large. But the numerous atomic configurations can be subdivided into two fundamentally different types: type 1 is Ge-atoms at interstitial sites not bonded to the lattice; and type 2 is Ge-atoms partially preserving their bonds with the lattice. In the latter case, such atoms are elements of sp^3-hybridized distorted tetrahedra involved in continuous chains formed by undistorted, successively aligned tetrahedra. The distorted tetrahedra are tetrahedra lacking some bridge O-atoms on their tops; i.e. the central Ge-atom of such a tetrahedron has one to three dangling sp^3-hybridized valence bonds or those bonds link the central Ge-atom to other Ge-atoms (here, different combinations of the two variants are also possible). Note that during the formation of the GeO_2 network from the lattice of GeO(solid), there is no need in a transfer of large masses of substance between remote sites in the volume of the material. Structural rearrangements of the 3D matrix proceed simultaneously in a huge number of centres, and the rearrangement processes proceed in a local vicinity of each centre.

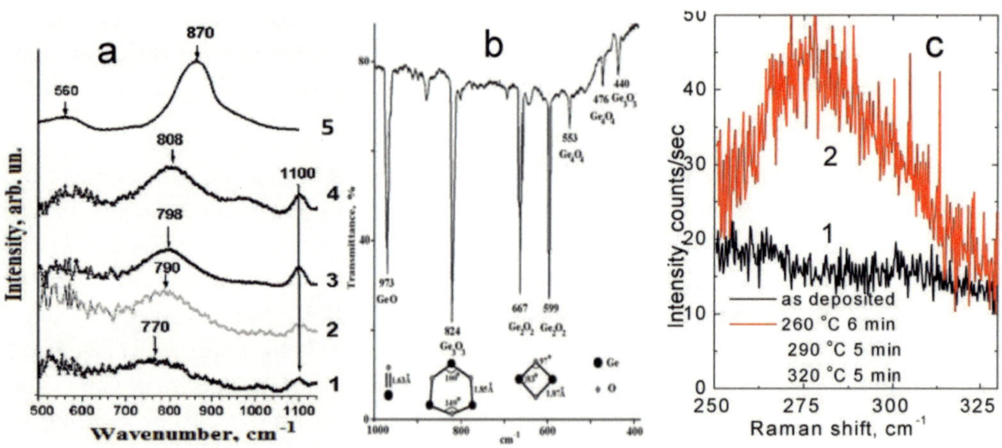

Figure 4. a - IR transmission spectra of gaseous GeO molecules (monomer, dimer, trimer, tetramer), embedded in matrix of frozen N_2 or Ar at T = 20 K (Ogden & Ricks, 1970); b - IR absorption spectrum of GeO(solid), prepared by condensation of GeO vapour: **1** - as-deposited; **2, 3, 4, 5** - annealed at 260ºC/6 min; 290ºC/5 min; 320ºC/5 min; 600ºC/10 min, respectively; c - Raman spectra of GeO(solid): **1**- as deposited; **2** – after series of annealings at 260ºC/6 min; 290ºC/5 min; 320ºC/5 min.

That is why migrations of many atoms turn out to be quite localised processes. However, simultaneous decay of one lattice and formation of the other lattice lead to the formation of a multitude of various defects in the forming atomic network. The density of such defects is several orders higher than the values typical of the equilibrium lattice states. As a result, simultaneously with the intensive formation of defects in the GeO_2 network, there may be no less intensive annihilation of such defects and complementary defects. Hence, during structural rearrangement of a GeO(solid) film into a heterostructure, density and viscosity of the material should considerably decrease. Of course, the characteristic times of such rearrangements must be large and depend on the temperature and other external factors. The structural modification process of germanium oxide lattice in the initial GeO(solid) layer subjected to successive anneals is shown in IR spectroscopy data (Fig. 4 b). Film of 55 nm thickness was deposited at T~ 25 °C. After two annealings (260 and 290 °C) shrinkage of layer thickness occurred by ~ 6-8% according to ellisometric data, as a result the structure of discussed heterosystem seeks to optimise its stable parameters: the size, the number of defects in the atomic net, the excess of internal energy, etc.

The above specific features of the rearranging lattice of the oxide layer facilitate the nucleation of Ge-nanoclusters in the volume of the GeO_2 matrix, the second component of the structural arrangement process of GeO_2<Ge-NCs> heterolayer. Very likely, this process is the slowest one among the above-discussed processes, as Ge-atoms need to lose their continuous bonds with the oxide lattice and, then, migrate over sufficiently large distances (much larger than the size of the "unit cell" in the oxide network) to enter the structure of growing Ge-clusters. The segregation kinetics of amorphous germanium in GeO(solid) layers annealed at different temperatures is reflected in the Raman data shown in Fig. 4 c. Combined Raman spectroscopy and HRTEM data show that germanium initially forms small amorphous Ge-particles sized several nanometers; on increasing the temperature and the duration of the film synthesis process and subsequent anneals, those particles grow in size to form larger Ge-particles sized several ten nanometers. According to Raman data, at T ~ 470-490 ^{0}C, amorphous Ge-particles undergo crystallisation. At T > 600^{0}C, in a GeO_2<Ge-NCs> heterostructure the GeO_2 matrix may also transform into hexagonal phase, which is isomorphous to α-quartz (Gorokhov, 2005).

3.2. Elementary GeO_2 lattice defects in modifications of GeO_2<Ge-NCs> heterolayers

Analysis of rearrangement processes proceeding during decay of metastable germanium monoxide layers should be performed considering the fact that such processes are controlled by regularities typical of glasses rather than crystals, as microscopic mechanisms underlying lattice transformation and relaxation processes in glasses and crystals are radically different. For gaining a better insight into substance modification processes proceeding in GeO_2<Ge-NCs> heterosystems under pulsed laser treatments, we have to first consider the role of elementary defects in glass atomic network during its formation and subsequent treatments.

In glasses transitions during melting and freezing do not have a sharp boundary. Such processes proceed gradually as the glass viscosity (η) continuously decreases (upon heating) or increases (upon cooling). Vitrification temperature (T_g), at which glass viscosity $\eta=10^{13}$ P, is accepted as the phase transition point. At $T < T_g$ glassy materials are solid, with the majority of atomic bonds in their bulk being not broken; at $T >> T_g$ the materials are melted, with their atomic bonds undergoing rupture, and that provides the material's easy flow ability. Silicon and germanium oxides are basic natural glass-forming materials, but the vitrification temperatures of pure SiO_2 and GeO_2 are strongly different ($T_g(GeO_2)$ =570^{0}C, $T_g(SiO_2)$ =1170^{0}C), causing a considerable difference in material properties of silicate and germanate glasses. However, the physicochemical mechanisms underlying the behavior and properties of silicate and germanate glasses and the mechanisms controlling structural and chemical modifications of the two groups of glasses, do not have vital differences, although quantitative characteristics defining material properties and processes in glasses are usually considerably different.

The viscous glass flow is a thermally activated process: $\eta(T)=A\cdot exp(Q/RT)$, where Q is the viscosity activation energy and A is a constant. In amorphous materials, the viscous flow clearly deviates from the Arrhenius law: in such materials, the viscosity activation energy Q changes its magnitude from a higher value Q_H at low temperatures (glass state) down to a smaller value Q_L at high temperatures (liquid state). In the theory of glass viscosity (SiO_2, GeO_2 and others), much attention is normally paid to defect migration in the glass atomic network. Presently available models of this phenomenon are based on the Mueller hypothesis (Mueller, 1955, 1960) about switch-over of oxygen bridge bonds in material. According to this hypothesis, during viscous melt flow, covalent Si-O (or Ge-O in GeO_2 glass) bonds do not break (Appen, 1974), but they switch over (translate, reorient). That is why the mechanism of viscous flow in mechanically loaded materials consists of two different stages, preliminary local re-grouping of valent bonds due to thermal fluctuations and switch-over of oxygen bridge bonds. At first stage, low-activated glass network extension proceed. Due to thermal fluctuations, basic structural elements (SiO^4 tetrahedrons) undergo deformation and regrouping of their valent Si-O bonds, i.e. the bond angles slightly change their values and atoms in Si-O-Si bridges become displaced from their previous positions. As the interaction between covalent atoms is short-range and oriented, the pair interaction energy sharply increases at large changes of interatomic distance. The activation energy of atom regrouping proceeding without chemical bond rupture is close to the mean thermal motion energy of atoms in glass melt; and, according to estimates, 100-150 atoms should participate in formation of microvoids of the required size (Nemilov, 1978). Such local fluctuations of valent bond configurations are necessary for subsequent elementary kinetic acts, namely switch-over of oxygen bond in Si-O-Si bridges (Filipovich, 1978; Mueller, 1955, 1960). Here, one of the oxygen bonds with silicon atoms undergoes disruption, and it switches over to another Si-atom having a vacant bond (Fig. 5). As a result of the rearrangement process, the O-atom shifts at one interatomic distance, and that is possible due to the presence of a microvoid earlier formed in the vicinity of the O-atom. Such bonds switches over occur in crystals during dislocations glides in the time of plastic flow.

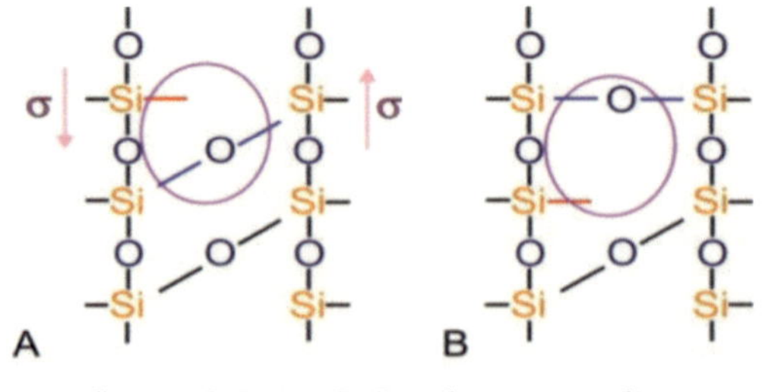

Figure 5. Mechanism of viscous flow in SiO_2 (or GeO_2) glass proceeding as a result of silicon vacant bond switching in Si-O-Si bridges: A - initial state; B - after Si vacant bond switching

The activation energy of the bridge oxygen switch-over is a sum of two components. The first component is the relatively low microvoid formation energy, or fluctuating local silicon-oxygen network deformation energy ~4 - 5 kcal/mol (Sanditov, 1976). Herein, the energy barrier of the second process, direct switch-over of bridge oxygen bonds, decreases to some minimal level of ~20 - 30 kcal/mol, which is different for various glasses. However, if a small number of atoms (3 - 4) participate in the bridge oxygen bond switch-over process, up to ~100-150 atoms are involved in the local deformation of the silicon-oxygen bulk. The latter circumstance provides for a substantial contribution of the entropy term to the effective glass viscosity activation energy, H^{*}_{η}. Vacant bonds switch-over activation energy in the glass viscosity theory is connected with formation mechanism of bonds (Filipovich, 1978; Nemilov, 1978; Sanditov, 1976; Zakis, 1981; Mueller, 1955, 1960; Nemilov, 1978), i.e. with a predominant way of vacant bonds formation and switch-over type. However, rather low activation energies of ~25-35 kcal/mol and large pre-exponent multipliers are typical of the temperature dependencies of bond switch-over rates, providing for most defects migration in the glass network and for the viscous flow of glass at $T \sim T_g$. Therefore, the processes conditioned by directed bonds switch-over (viscous flow, crystallisation, chemical interaction, and diffusion) proceed, as a rule, at relatively low rates, which is typical of the chemical properties of polymeric glass-forming substances (Appen, 1974).

All the above regularities suggest that, in GeO_2<Ge-NCs> heterolayers, oxygen vacancies generated in glassy GeO_2 matrix at T > 500°C (i.e. at temperatures close to $T_g(GeO_2)$) at the boundary of Ge-nanoparticles can easy move within matrix volume. Translation of bridge oxygen bonds to vacant bonds of Ge-atoms underlies migration of oxygen vacancies. Maximum values of equilibrium vacancies concentrations in GeO_2 glass are high; as a result, this mechanism becomes capable of providing mass transfer of Ge-atoms through oxide in the heterolayer towards the boundaries of Ge-nanoclusters. In this way this mechanism contributes to Ge-nanoclusters growth. During decay of GeO(solid) atomic network at temperatures over ~250 °C, the decomposition process of this network, leading to releasing of half the total amount of Ge-atoms initially contained in GeO(solid), ensures excessive supply of vacant oxygen bonds towards the forming GeO_2 matrix. The rearrangement of valence bonds of Ge-atoms in the sp^3-hybridization form is accompanied by the release of considerable energy. During rearrangement, the switch-over of bridge oxygen bonds to vacant bonds requires the least energy in comparison with all other potential barriers. For a number of glasses, the activation energy of the switch-over falls in the range from 23 to 37 kcal/mol, these values are close to the activation energy of oxygen atoms hops in alkaline-silicate glasses (Gorokhov, 2005).

3.3. Densification of GeO_2<Ge-NCs> heterolayers and them moisture absorption

The majority of physical and chemical processes in the modification of lattices in GeO(solid) and GeO_2 layers should be analysed in terms of structural rearrangements with active participation of oxygen vacancies. First of all, such processes include densification of heat-treated layers and adsorption/desorption of moisture from atmosphere in the material.

The densification process is due to the fact that the high rate of GeO molecules condensation from vapour at low temperatures (< 200-250 °C) forms low-density GeO(solid) layers with high concentration of lattice defects. Concentrations of various point defects and their ensembles, pores and other microvoids in the film lattice are much higher than the values typical for the quasi-equilibrium state of atomic network. Such a structure is typical of CVD SiO_2 layers, especially deposited at T < 500 °C. This structure has a zeolite-like behaviour, which disappears completely after annealing at T > 700 °C and is manifested in the ability of CVD SiO_2 films during the storing to absorb in their volume water molecules and other atmospheric vapours. The absorbed moisture fills all kinds of defects in the volume of the film atomic network. Besides, part of H_2O molecules absorbed in the film dissociates into hydroxyl groups OH^- and hydrogen ions H^+ fix at vacant bonds of atoms in the oxide lattice.

Annealing of CVD SiO_2, GeO(solid) and GeO_2 layers at temperatures over 250-300 °C activates processes leading to a decrease of nonequilibrium defects concentration in the atomic lattices of the oxides. Usually this leads to an increase of film density with simultaneous decrease of film volume (shrinkage), as well as to increased optical constants, decreased conductivity, increased tensile stress and viscosity of the material. The intensity of the process and the degree of film properties change depend on the temperature, anneal duration and atmosphere in which anneals were held, as well as on the defects density (or imperfection) of the initial atomic network of the film. Typically, the stronger the thermal treatment and more defective the films, the higher the degree of change of basic film characteristics. For instance, the shrinkage degree in thickness of CVD SiO_2 and Si_xN_y(H) films varies within the range from 3-5 to 25-30% (Gorokhov et al., 1982; Pliskin et al., 1965). The presence of absorbed moisture in the film increases its ability to shrink. Therefore, protective (cap) layers are used for protection of GeO(solid) and GeO_2 layers from the degrading effects of atmosphere. Thin (tens of nm) low-temperature PE Si_xN_y(H) films are the best encapsulating coatings. The protection ability of CVD SiO_2 layers is worse, so one should increase their thickness. Note that during anneals moisture leaves the GeO(solid) and GeO_2 layers faster than CVD SiO_2 layers. Anneals at 150-180 °C during ~ 30-40 min in vacuum or in a flow of a pure and dry inert gas are sufficient for effective moisture desorption from germanium oxide layer 150-200 nm thickness.

3.4. Other methods of structure modification of GeO_2<Ge-NCs> heterolayers

The composition and structure of GeO_2<Ge-NCs> heterolayers can be easily modified chemically. Removal of the glassy GeO_2 matrix from GeO_2<Ge-NCs> heterolayer is the simplest way. Glassy GeO_2 is readily soluble in water and, especially, in aqueous solutions of HF, whereas crystalline or amorphous Ge-nanoparticles remain unaffected by such solutions. On dissolution of GeO_2 matrix, Ge-nanoclusters initially contained in the volume of a GeO_2<Ge-NCs> heterolayer agglomerate with each other due to very weak electrostatic forces and settle on the substrate forming a weakly coupled, very loose and porous Ge-layer. So, very thin GeO_2<Ge-NCs> heterolayers can be used to produce, for example, single-layer coatings formed by Ge-nanoparticles on substrates of different materials. An annealed agglomerative layer of Ge-nanoparticles becomes sintered, and a more durable Ge-coating with a porous structure forms. Such a layer was formed inside a multilayer structure containing SiO_2 films in order to examine the effect of femtosecond (fs) laser treatments on this layer (see Fig. 6 a). Under ablation, after the explosive action of fs laser pulses a region of porous germanium was detached from the thick SiO_2 layer, turned over and thrown away onto the undestroyed part of the multilayer system. As a result, we became able to obtain a SEM image of the upper and lower surfaces of the layer under study (see Fig. 6 b).

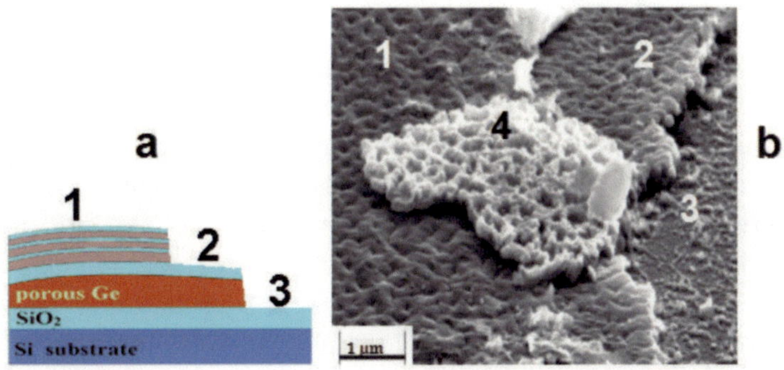

Figure 6. Surface morphology of a multilayer coating after fs laser treatment : **a** - scheme of coating structure consisting of three tiers: the bottom tier is Si-substrate with the first SiO_2 film (of 100 nm thick), middle tier is a Ge-porous layer (of ~ 70-230 nm thick) coated by second SiO_2 film (of 100 nm thick), and upper level is three bilayer structures, each of them consists of thin heterolayer GeO_2<Ge-NCs> (of ~ 15 -25 nm thick) coated by the SiO_2 cap-layer (of 10 nm thick) (CVD SiO_2 is shown as blue, Ge-porous and GeO_2<Ge-NCs> layers – as brown); **b** - SEM image of the multilayer coating surface: **1** – unmodified by radiation, **2** – after liftoff of the top tier; **3** - surface of the lower SiO_2 layer after liftoff of the Ge-porous layer; **4** - surface morphology of the back side of the Ge-porous layer fragment produced by laser damage.

Processes of oxidation and shrinkage of GeO(solid) layers and GeO_2<Ge-NCs> heterolayers are similar. We found that oxidation of GeO_2<Ge-NCs> heterolayers in an oxygen flow at normal pressure differs little from the oxidation of pure germanium wafers under the same conditions. The process can be well monitored ellipsometrically at temperatures over ~500^{0}C. Under such conditions, GeO(solid) layers disproportionate into Ge and GeO_2 according to reaction (3) for a time not longer than 2-3 minutes, and then they undergo oxidation as GeO_2<Ge-NCs> heterolayers. Ge-nanoparticles oxidation front propagates from the surface of the film into depth. The kinetics is close to kinetics of pure germanium oxidation even quantitatively. The mechanism of the process is that the Ge/GeO_2 interface in the system generates vacant bonds ("oxygen semi-vacancies") on Ge-atoms incorporated in the formed oxide network. Further on, these vacant bonds tend to maximally fill the GeO_2 lattice by their translation mechanism. O-atoms from the molecules of ambient O_2 get captured by vacancies on the exposed external surface of the oxide layer. After this captured O-atom becomes attached with one of its bonds to the GeO_2 network while the other bond usually remains free. In the oxide this free bond presents one quarter of a Ge-atom vacancy and is excessive (non-equilibrium). Thus, the interdiffusion of oxygen semi-vacancies and quarters of germanium vacancies in the oxide layer network, accompanied by gradual annihilation of these complementary defects, is the oxidation process of both germanium wafers and Ge-nanoparticles in the glassy GeO_2 layer.

Crystallisation of the glassy GeO_2 matrix in GeO_2<Ge-NCs> heterolayers is an exceptionally structural (not chemical) process. We have studied this process using layers of pure thermal germanium oxide on Ge(111) substrates. The most important is that GeO_2 layers begin to rearrange into the low-quartz structure already at 600-620 °C; it is worth noting that there are conditions for forming large-area continuous block-epitaxial coatings from crystallised hex-GeO_2 over quite a large area. Interestingly, an analogous process also proceeds in thermal silicon oxide with formation of block-epitaxial β-crystabalite SiO_2 layers. But the latter process proceeds very slowly even at temperatures $T > 1200^0C$. It makes the material practically unpromising in the silicon integral planar technology. Unlike SiO_2, low-quartz GeO_2 layers are more suitable for application, as crystallisation of glassy GeO_2 layers considerably increases the mechanical strength and chemical stability of the films. For removing crystallised layers, etching solutions with considerable HF acid proportion are necessary. It is described in the monography (Knoss, 2008) where reactions of glassy GeO_2 layers with CVD SiO_2 and Si_3N_4 films deposited onto such layers were also described.

Interaction of thermal GeO_2 layers with ammonia at ~ 700-750^{0}C according to reaction $3GeO_2 + 4NH_3 \rightarrow Ge_3N_4 + 6H_2O\uparrow$ proceeds in the oxide lattice in compliance with the micromechanisms similar to those involved in germanium oxidation (Gorokhov, 2005). At these temperatures the solid reaction product, Ge_3N_4, dissolves in the practically liquid oxide layer and forms a glassy GeO_2:Ge_3N_4 (m:n) layer. Under the described conditions, germanium (both the wafer material and Ge-nanoparticles) also reacts with NH_3 and forms Ge_3N_4. Therefore, relatively thin GeO_2<Ge-NCs> heterolayers can be used to increase the Ge_3N_4 content of the formed GeO_2:Ge_3N_4 glassy layers.

The above reactions are interesting because they allow a drastic modification of the structure and composition of the GeO_2 matrix in GeO_2<Ge-NCs> heterolayers up to ~100 nm thickness. Low viscosity of glassy GeO_2 at temperatures T>Tg provides for rapid mutual solution of SiO_2 layers with germanium dioxide, practically to a homogeneous composition at thickness amounting to hundreds of nanometers; this process transforms the layers into germanium-silicate glass. Varying the thickness ratio of GeO_2/SiO_2 double-layers we can change the composition of composite GeO_2:SiO_2 glass in a wide range of values.

Glassy GeO_2 reacts with CVD Si_3N_4 films at the interface at $T > 700^0C$ according to the formula $3GeO_2 + 2Si_3N_4 \rightarrow 2Ge_3N_4 + 3SiO_2$. Like SiO_2, the low viscosity of glassy GeO_2 alleviates the interdiffusion of initial and final products of this reaction across the interface of the reacting layers. As a result, a layer of glassy GeO_2:Ge_3N_4:SiO_2-type compound several tens-hundreds of nm thickness forms at the interface. The composition of this layer depends on the thickness ratio of initial layers, and on the temperature and duration of the reaction. Under heating held at temperature T ~750^{0}C for an hour, in the obtained germanium-silicate glass and multi-component GeO_2:Ge_3N_4:SiO_2 → k:p:q layers crystallisation begins, proceeding similar to that in pure glassy GeO_2. The crystal lattices of the glasses are isomorphic to the low-quartz structure, and they excel hex-GeO_2 layers in chemical stability.

Interestingly, the dramatic increase of density and chemical stability in composite glassy GeO_2:SiO_2 and GeO_2:SiO_2:Ge_3N_4 – k:p:q layers begins already at their formation stage. With the example of GeO_2:SiO_2, it is seen that this effect is much more pronounced in comparison with the effect due to mere linear growth of the more chemically stable SiO_2 fraction in the two-component composition. In other words, the etching rate of the glassy GeO_2:SiO_2 layer in a structurally sensitive etchant turns out to be a few times lower than the individual etching rates of glassy GeO_2 and SiO_2 layers (even when the fraction of SiO_2 in the composition is not high). We suppose this effect been due to the fact that the formation of glassy GeO_2:SiO_2 and GeO_2:Ge_3N_4:SiO_2 – k:p:q films occurs in the vicinity of the crystallisation temperatures of these substances. Therefore, during annealing such glasses should undergo the stage preceding crystallisation. At this stage, the glass atomic network is getting ready to transform into a high-ordered crystal lattice so that the density of quasi-equilibrium defects rapidly decreases. Crystallised glass is known to have lattice defect concentrations orders lower than the initial concentrations of the same defects at the initial stage. Hence, introduction of GeO_2 into glassy SiO_2 provides for a sharp decrease of vitrification and crystallisation temperatures of the composition, together with a sharp decrease of vacancy concentration. Thus, CVD SiO_2 layers with reduced density, due to a high concentration of non-equilibrium defects in rather unstable atomic network, can be etched in etchant sensitive to their structure tens of times faster than quasi-equilibrium thermal SiO_2 layers formed at temperatures ~ 1000 ^{0}C (Pliskin and Lehman, 1965).

Thus, the described structural-chemical modification processes of glassy oxide GeO(solid), GeO_2 layers and GeO_2<Ge-NCs> heterolayers have a high modification ability, which allows us to obtain chemically stable coatings with an original stable structure and chemical composition, high mechanical strength and easily controlled physical, electrical and optical properties from chemically unstable, structurally imperfect, physically and mechanically soft, and electrically unstable layered films.

4. Laser treatments of GeO_2<Ge-NCs> heterolayers

Thermal anneals, traditionally used in microelectronics, were the tool for the described modifications of the composition and structure of the films under study, in which different Ge oxide layers played the main role. As it was noted in Introduction, pulsed laser irradiations offer a considerable resource for enlarging the list of possible chemical and structural transformations can be implemented in the materials of interest. Irradiation of film systems with electromagnetic radiation can be considered as a basic kind of possible treatments. Unlike common thermal treatments, irradiation of samples with light pulses allows a radical widening of the range of conditions that can be realized in the high-energy treatment of layered materials. First, here we mean the value of the energy input in irradiated materials, which can be orders higher than the limits that can be reached in ordinary heat treatments of samples. Second, the range of dynamic parameters of high-energy effects in irradiated films can also be widened by many orders, together with the spatial localization parameters of such effects. Due to these factors, irradiation of samples with short laser pulses considerably extends the spectrum of possible modifications of thin films in micro- and nanoelectronic structures.

Over more than thirty years, Pulsed laser Annealing (PLA) has been successfully used to achieve crystallisation and re-crystallisation of semiconductor materials in thin-film structures (Dvurechenskiy et al., 1982). Earlier, PLA treatments of thin-film structures with light quanta exceeding in energy the bandgap width of semiconductor material were recognized as a useful means for rapid (typically, during some tens of nanoseconds or less) restoration of the crystalline structure in disordered or even completely amorphized near-surface layers of Si (Ahmanov et al., 1985). At a proper choice of laser parameters, almost all laser radiation can be absorbed within the film; hence, this radiation does not reach the wafer and does not heat it. As a result, short laser pulses allow one to avoid overheating of the substrate during film cooling due to diffusion of heat into the substrate (tens of nanoseconds). That is why PLAs can be used to crystallize amorphous silicon films on substrates not withstanding high temperatures.

Powerful laser impacts lead to fast, high-quality re-crystallisation of amorphized near-surface semiconductor layers. Despite the fact that PLA has already become a well mastered technique, laser annealing experiments raised a number of still unsolved fundamental physical questions (Chong et. al., 2010). Recent contributions have shown that fast laser-induced phase transformations in near-surface semiconductor layers, such as melting-hardening, phase transitions 'amorphous solid – crystal' and 'crystal – amorphous solid', etc., proceed over nano-, pico-, and even subpicosecond time scale. To explain the various and, in many respects, unexpected phenomena observed during PLA treatments of semiconductor structures, it is necessary to give an answer to a number of fundamental questions regarding the behavior of semiconductors in strong laser fields.

PLA is widely used to eliminate structural imperfections and radiation defects introduced during implantation of ions in the near-surface crystal layers (Kachurin et al., 1975). PLAs provide an important possibility for the technology, obtaining perfect crystal structures in subsurface layers with impurity concentrations unreachable in the common thermal annealing (to 10^{21} cm^{-3} and higher). Diffusion of impurities is highly suppressed under PLA conditions (Kachurin et al., 1975). In addition, PLA allows realization of processes that do not occur during common thermal treatments. Such processes, which are of physical concern, are conditioned by the presence of dense plasma in the samples. The range of observed PLA-induced processes includes softening of phonon modes with increasing plasma density, change of bandgap width with increasing n_c (non-equilibrium carrier concentration), registration of optical and electrical phenomena in semiconductors depending on n_c (Ahmanov et al., 1985), etc. Spatial coherence of laser radiation used for pulse laser treatments of semiconductor materials allows creation on the surface of annealed samples of periodic spatial structures, e.g. lattices formed by alternating crystalline and amorphous regions (the so-called interference laser annealing).

Later, nanosecond pulsed treatments using excimer XeCl (λ = 308 nm) (Volodin et al., 1998) and ArF laser radiation (λ = 193 nm) were used to obtain silicon nanoclusters and achieve crystallisation of a-Si inclusions in SiN_x and SiO_x films (Rochet et al., 1988). Some of the unusual properties of such heterolayers were manifested in samples treated with ultrashort pulsed fs laser radiation (Korchagina et al., 2012). Reports on laser treatments performed to modify the structure, chemical composition, and properties of GeO(solid) and GeO_2 films, and also to achieve formation of c-Ge nanoparticles and crystallisation of a-Ge inclusions in dielectric matrices are encountered much more rarely (Gorokhov et al., 2011).

4.1. Peculiarities of conventional nanosecond PLA

Considering the complexities of PLA impact mechanisms on thin-film semiconductor structures and a complex nature of physicochemistry and structure of glassy GeO(solid), GeO_x, and GeO_2 films, and GeO_2<Ge-NCs> heterolayers, unusual effects could be expected in our films during PLA treatments. Two lasers with different photon energies and durations of light pulses, a KrF excimer laser (wavelength λ = 248 nm, pulse duration 25 ns) and a Ti-Sapphire laser (central wavelength λ = 800 nm, pulse duration < 30 fs), were used in this investigation. As expected, the impacts of short-pulse treatments on our samples for the two lasers turned out different.

The effect of ns radiation was studied for two bilayer systems, GeOx/SiO_2 and Si/GeO_2<Ge-NCs>/SiN_xO_y, both prepared on Si(100) substrates (SiO_2 films of thickness 100 nm and SiN_xO_y films of thickness 25 nm were the cap layers). A sketch of the second system is presented in Fig. 7 a, b. The laser beam was focused, through a square-section diaphragm (200 x 200 μm), on the sample surface. Laser pulses followed at a frequency of 100 Hz, and the beam moved along one of the sides of the square on the sample in ~180-μm increments. After exposure of one band on the sample surface to the beam with a set laser pulse energy (see Fig. 7 d) the beam was shifted a step further to treat the next band under the first one. An exposed film section was formed out of several bands. Because of overlapping of laser spots, the film was exposed to the laser beam two times on the edges of the square and four times in the corners (Fig. 7 c, d and e). Such neighborhood of film regions exposed to different numbers of identical laser pulses (1, 2 and 4) allowed us to trace the pulse-by-pulse changes induced in the regions where the film system acquired different thicknesses.

In the absence of cap layers, the used PLA completely "evaporated" the germanium oxide layers on our samples transforming them into GeO(gas) according to reaction (4). The SiO_2 and SiN_xO_y cap layers, impeding the removal of GeO(gas) molecules, substantially slowed down the formation reaction of those molecules in the protected films. Nonetheless, even in the presence of cap layers the irradiated regions exhibited a notable change in interference colors (Fig. 7 b, c, d, e).

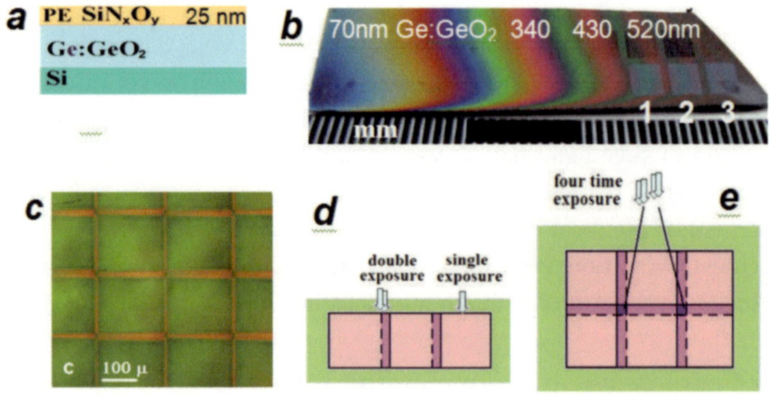

Figure 7. **a** – schematic representation and **b** - photo of the Si/GeO_2<Ge-NCs>/SiN_xO_y bilayer system on Si substrate irradiated with laser radiation to fluences 170, 150 and 130 mJ/cm^2 in regions 1, 2 and 3 of the heterolayer (thicknesses ~390, ~480 and ~520 nm, respectively); **c** - micrograph and **d, e** – schematic illustrating the formation of single-, double- and four-time exposed areas of the film at stepped irradiation.

4.1.1. Shrinkage of GeO_2<Ge-NCs>/SiN_xO_y bilayer coatings under ns PLA

Data obtained with a set of experimentaltechniques in irradiated regions 1, 2 and 3 of the sample with ~390-, ~480- and ~520-nm thick heterolayers provided a comparison of the type and degree of changes in the structure and material properties of examined films after their exposure to different laser fluences, 130, 150, and 170 mJ/cm^2. Changes of interference colors in the bilayer system occurred due to strong shrinkage of the materials forming in the system individual layers, which was usually accompanied by increase of film optical constants. The effect here is similar to the one which was studied by ellipsometry and which proceeded with a change of optical constants of GeO(solid) layers during thermal anneals. In the latter case, GeO(solid) did not evaporate because of low annealing temperature. Results of ellipsometric measurements of our samples performed following their thermal treatments in vacuum are shown in Fig. 8. The measurements were carried out using a special method that combined spectral and scanning ellipsometry with multi-thickness measurements (Marin et al., 2009) and that allowed us to obtain sufficiently precise data on optical constants of annealed GeO_2<Ge-NCs> heterolayers throughout the whole visible range of the spectrum.

The data for $k(\lambda)$ (curves 1 - 4 in Fig. 8) show the changes in the bandgap of as-deposited GeO(solid) films and GeO(solid) films given traditional anneals at different temperatures. These changes are the result of the chemical decomposition of film material connected with the appearance in the film and subsequent growth of Ge nanoparticles in which, due 3D confinement effects, the efficient band gap value decreases with an increase of nanoparticle size, tending at large nanoparticle sizes to the bandgap value of bulk α-Ge. The $k(\lambda)$ data for all examined films were plotted in the Tauc coordinates, with the square root of absorbance coefficient α being plotted along the vertical axis. This coefficient was obtained from the extinction coefficient by formula $\alpha_j(E)=2\pi k_j(E)/\lambda$; here, the parameter j is the number of a studied film in Fig. 8. Using the well-known formula for the absorption coefficient of indirect-band semiconductors, $\alpha^{1/2} \sim (E - E_g)$, and interpolation of the linear dependences of $\alpha^{1/2} (E- E_g)$ to $(E- E_g) \rightarrow 0$, we were able to evaluate the effective optical gap $E^j{}_g{}^{eff}$ in the GeO(solid) films and in the GeO_2<Ge-NCs> heterolayers formed upon decomposition of the films. The variation of the effective optical gap in GeO(solid) films with the growth of annealing temperature T_{an} is shown in Fig. 9. For comparative analysis of obtained $E^j{}_g{}^{eff}$– values and for analysis of Ge-nanocluster sizes, the values of $E^j{}_g{}^{eff}$ for amorphous bulk germanium and for GeO_2<Ge-NCs> heterolayers are also shown in Fig. 9, curves 5 and 6. For the effective optical *gap* $E^5{}_g{}^{eff}$ in GeO_2<Ge-NCs> heterolayers, a value of 1,15 eV was found, and the mean size of Ge NCs in the heterolayers proved to be ~5 -6 nm (according to TEM data). Since the value of $E^j{}_g{}^{eff}$ in annealed GeO(solid) films always remained appreciably higher than 1,15 eV, it can be concluded that a-Ge nanoparticles in those films were smaller than nanocrystal sizes in GeO_2<Ge-NCs> heterolayers. These data have also allowed us to evaluate the sizes of a-Ge nanoparticles in other examined films.

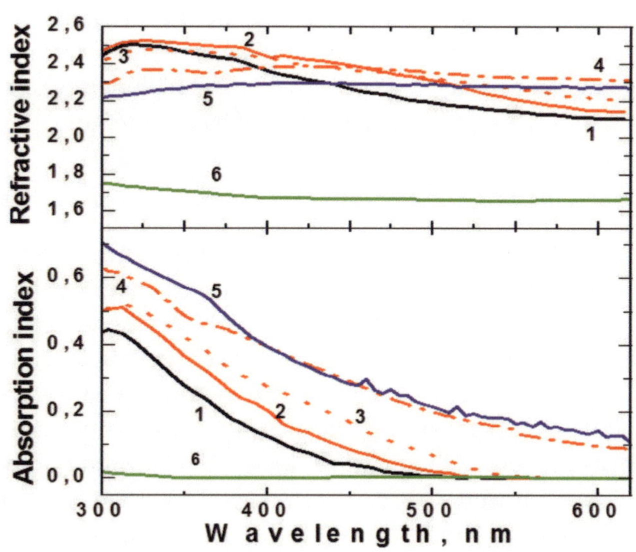

Figure 8. Spectral dependences of optical constant (n(λ) and k(λ)) in the various films: 1, 2, 3, 4 – as-deposited GeO(solid) film and films annealed at 260, 290, and 320 °C, 5 – LP CVD GeO_2<Ge-NC> heterolayer, 6 - thermal GeO_2 film grown on a single-crystal Ge(111) substrate.

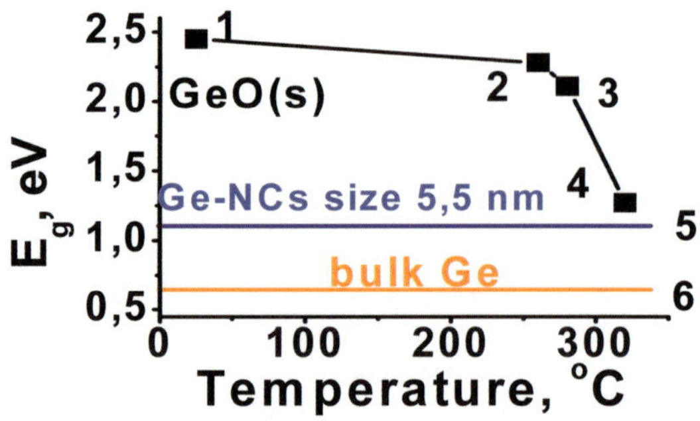

Figure 9. The effective optical bandgap E^{eff}_g versus annealing temperature: **1** – in as-deposited GeO(solid) films (with Ge-NCs sized 0,2 - 0,4 nm) and after subsequent anneals (**2** - 260 °C, 6 min; **3** - 290°C, 4 min, **4** – 320°C, 4 min); **5** – E^{eff}_g for a GeO_2<Ge-NCs> heterolayer with 5,5-nm Ge NCs; **6** - E^{eff}_g for bulk Ge.

According to AFM data, on a sample area irradiated with one laser pulse the shrinkage degree of the film was ~5 - 9%; this value increased up to ~23 - 34% under four laser pulses (Figs. 10 and 11). The cause for the film shrinkage was the low initial density of the film conditioned by the presence of many defects in the atomic network of GeO(solid). Similar data were also obtained for GeO_x layers covered by a cap SiO_2 layer (on a silicon substrate) in which the predominant part of the GeO(solid) component has already decayed during the growth of those layers. Notable film shrinkage is usually observed on anneals in loose films grown by CVD at low temperatures. Normally, thermal treatments modify the structure of CVD films and improve their quality. Anneals of multilayer structures activate both the shrinkage processes in individual layers of multilayer films and the chemical reactions proceeding among neighboring layers (Knoss, 2008). The high defect content alleviates mass transfer processes in the layers and rearrangement of their structure.

4.1.2. Changes in the structure and properties of thin bilayer coatings under ns PLA

Raman spectroscopy and microprobe IR-spectroscopy were used to analyze the structure of germanium inclusions in examined films. The non-destructive Raman method combined with calculations is a very informative tool for nano-object studies. The presence of Raman peaks at 301,5 and 520 cm^{-1} in the spectra is an indication of c-Si and c-Ge micro-inclusions present in the layers in which light is scattered by long-wave lattice phonons. It is in this way that micro- and nanoparticles are usually detected in the films of these materials. By the way, if the crystals are small-sized (~nm), then the spectral position of related Raman peaks exhibits a red shift in comparison with the spectral position of the peaks due to bulk c-Si and c-Ge. According to the peak shift value, one can determine the mean size of semiconductor nanocrystals in the films of interest using the phonon localization method (Volodin et al., 2005). The greater is the Raman shift of the peaks toward smaller wavenumbers in comparison with the position of the peaks in bulk c-Ge (301,5 cm^{-1}), the smaller is the typical size of Ge nanocrystals in the films (Fig. 12). Broad scattering bands peaking at 275 - 280 cm^{-1}, which are due to the presence of amorphous germanium, are also clearly seen in the Raman spectra of examined GeO(solid) films and GeO_2<Ge-NCs> heterolayers since it is into this spectral region where the maximum density-of-state value of optical vibrations in a-Ge falls. In the films under study, amorphous germanium is normally contained in the form of nanoparticles dispersed in glassy GeO_2 matrix (Knoss, 2008).

The decrease of film thickness and increase of film material density after PLA as manifested in the Raman spectra of Si/GeO_2<Ge-NCs>/SiN_xO_y films is accompanied with the growth of the intensity of the peak at 520 cm^{-1} due to light scattering by long-wave optical phonons

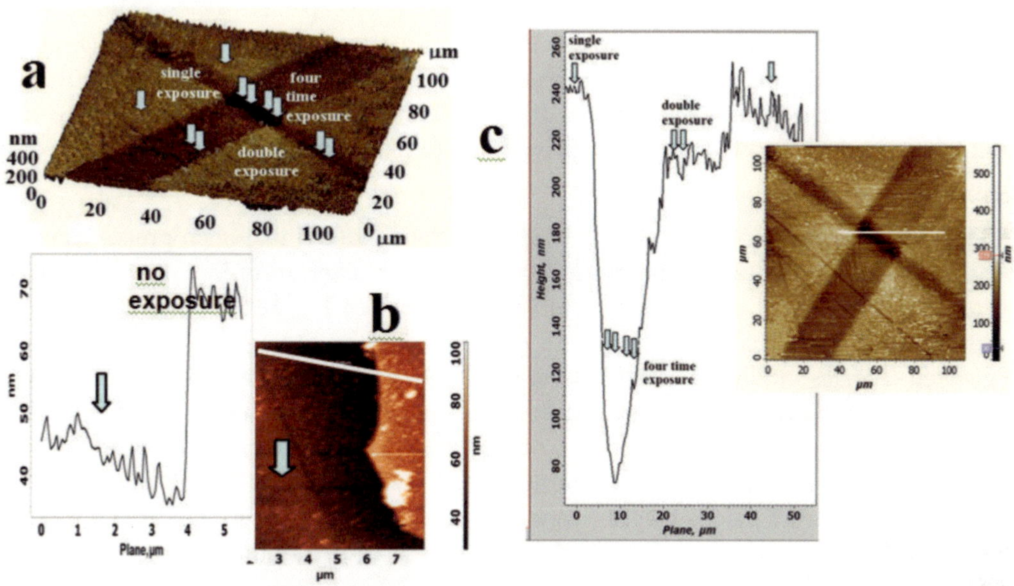

Figure 10. **a** - AFM image illustrating the shrinkage of the GeO_2<Ge-NCs>/SiN_xO_y bilayer system in region 2 (see Fig. 7 b) after PLA of the sample with ns laser pulses (fluence E_0=150 mJ/cm²); **b** - measurements of the shrinkage of the bilayer film in the region treated with one laser pulse, **c** – the same for two and four laser pulses. The arrows ↓, ↓↓ and ↓↓↓↓ indicate regions subjected to the single, double and four-time exposures.

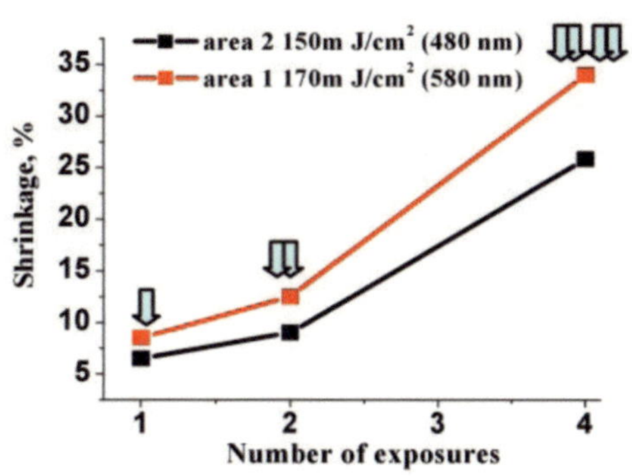

Figure 11. Effect of the number of laser pulses, laser fluence, and thickness of GeO_2<Ge-NCs> heterolayer on the shrinkage value of the Si/GeO_2<Ge-NCs>/SiN_xO_y bilayer system.

in the Si substrate (Fig. 12). This effect is manifested as an increase of the transparency of the bilayer film for the excitation laser radiation in Raman measurements (514,5 nm Ar+ laser) falling onto, and reflected by, the substrate. The cause for the shrinkage effect are PLA-induced changes of optical constants of the films in the Si/GeO_2<Ge-NCs>/SiN_xO_y bilayer system, as optical constants of Si substrate remained essentially unaltered in the spectral region around the excitation wavelength. Let us consider now possible reasons for such changes. Some minor increase in the refractive index of a very thin SiN_xO_y film due to its shrinkage and compaction cannot be the factor appreciably affecting the optical constants of the bilayer system. The absorption index of the film in this spectral region is initially very small, and the irradiation of the sample with laser radiation could not considerably increase its value. Hence, the increase in transparency of the bilayer system was most probably due to the change in the optical constants of the GeO_2<Ge-NCs> heterolayer or, alternatively, it could be a result of the chemical interaction of the two layers in the interfacial region of the structure.

Indeed, micro-Raman and micro-IR-spectroscopy data showed that such processes indeed proceed in the film system under study. Raman spectra taken from local areas of the film system irradiated with one, two, or four laser pulses (see Fig. 12 a) were measured with excitation laser beam focused to a ~1-μm diameter spot on the sample surface. From the registered changes in Raman spectra, we were able to judge the rate of release of Ge atoms during decomposition of the GeO(solid) structure, and also the formation of amorphous (α-Ge) and, then, crystalline (c-Ge) phases of Ge. The formation kinetics of α-Ge and c-Ge phases within the heterolayer is characterized by initially the highest amorphous germanium content of the layer system and by almost complete absence of c-Ge phase from it (see the initial spectrum in Fig. 12 a). After one laser pulse, the fraction due to α-Ge component sharply decreased in the system, and c-Ge phase emerged in quite large quantities. Two and four laser pulses gave rise to an increased amount of c-Ge phase at a comparatively slow regeneration of α-Ge component. During annealing in these

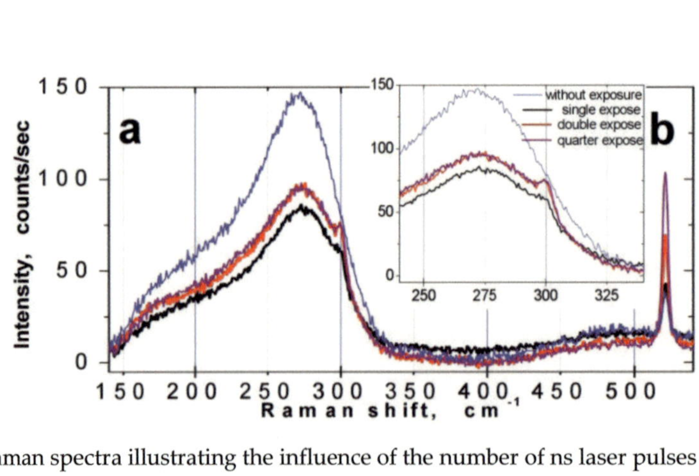

Figure 12. **a** - Raman spectra illustrating the influence of the number of ns laser pulses given to a local region on the sample (region 2 in Fig.7 b, E_0 =150 mJ/cm²) on the structure of the GeO_2<Ge-NCs> heterolayer in the Si/GeO_2<Ge-NCs>/SiN_xO_y bilayer system; **b** – magnified fragment of the same spectra.

heterolayers, amorphous germanium phase usually forms a big amount of amorphous Ge nanoparticles growing in size during high-energy treatments of the system, and then such nanoparticles undergo crystallisation without interrupting their continuous growth. The growth of Ge nanoparticles proceeds primarily due to decay of nanocluster fractions with minimal sizes. The formation processes of α-Ge and c-Ge nanoparticles are also maintained by the decay of the atomic network of metastable GeO(solid), which supplies the nanoparticles with atomic germanium.

According to micro-IR-spectroscopy data (Fig. 13), the atomic network of the insulator matrix also undergoes modifications that proceed simultaneously with the transformation processes of germanium inclusions in the GeO_2<Ge-NCs> heterostructure. Yet, the process turned out different than the expected formation of pure glassy GeO_2 by reaction (3) during anneals of GeO(solid) films (Fig. 4 b). In the heterolayer region treated with one laser pulse the IR absorption band starts moving towards longer wavelengths instead of showing the expected trend to the maximum at 860-870 cm^{-1}. On the other hand, the maximum IR absorption intensity in the region treated with two laser pulses decreases markedly, and the whole band notably widens. In the bilayer structure under study, in addition to the formation of GeO_2 during decomposition of metastable GeO(solid), the GeO_2<Ge-NCs> heterolayer and the SiN_xO_y film can react with each other in the interfacial region. The GeO_2 matrix of the heterolayer, and the Si_3N_4 and SiO_2 materials forming the cap SiN_xO_y layer, are involved in this reaction. In particular, the products of the reaction between GeO_2 and Si_3N_4 proceeding at T>650^{0}C ($3GeO_2 + Si_3N_4 \rightarrow 3SiO_2 + Ge_3N_4$), SiO_2 and Ge_3N_4, can be dissolved by glassy GeO_2 matrix. As described above, these processes finally yield a GeO_2:SiO_2:$Ge_3N_4 \rightarrow$ k:p:q glassy compound. Since pure Ge_3N_4 and glassy SiO_2 exhibit IR absorption peaks respectively at about 770 cm^{-1} and 1070 cm^{-1}, it can be expected that the peaks due to the Ge_3N_4:GeO_2 and GeO_2:SiO_2 double glasses will tend to shift in opposite directions from the peak due to pure GeO_2; i.e. the first peak will shift towards longer wavelengths and the second peak, towards shorter wavelengths. On the whole, the material of the GeO_2 matrix is expended on the formation of both SiO_2 and Ge_3N_4, and also on the formation of the three-component GeO_2:SiO_2:$Ge_3N_4 \rightarrow$ k:p:q compound. As a result, a predominant part of the cap SiN_xO_y layer may turn into Ge_3N_4 and SiO_2 that will subsequently undergo dissolution in the upper part of the GeO_2<Ge-NCs> heterolayer. Such transformation will be capable of reaching depths comparable with the thickness of the initial cap layer. This may become the reason for the observed modification of the IR absorption band in the irradiated film, including reduced absorption at the band maximum, shift of the band, and its strong widening (Fig. 13).

Of course, the properties of the upper layer in the initial bilayer structure will undergo dramatic changes. In particular, the optical constants and the optical bandgap will strongly change in this layer. Dissolution of wide-bandgap glassy SiO_2 ($E_g^{SIO2} \sim 9$ eV) in the matrix of the GeO_2<Ge-NCs> heterolayer will radically change the energy-band characteristics of the heterosystem. A similar behavior is also demonstrated by glass-like Ge_3N_4, which is close to GeO_2 in terms of optical gap, $E_g^{Ge3N4} \sim$ 5-5,5 eV, yet differs from GeO_2 in its high refractive index (1,90-2,00) and permittivity (~9-11). On the contrary, the refractive index of SiO_2 (1,45) is lower than that of glassy GeO_2 (1,61-1,63) (Gorokhov, 2005).

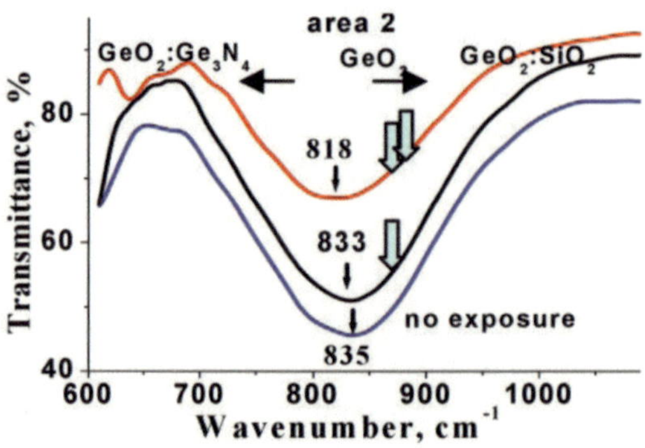

Figure 13. Micro IR transmittance spectra illustrating the influence of the number of ns laser pulses given to a local region on the sample (region 2 in Fig.7b, E_0 =150 mJ/cm^2) on the structure of the dielectric in the Si/GeO_2<Ge-NCs>/SiN_xO_y bilayer system.

The degree of compositional changes in the upper part of the layer will depend on material viscosity. In the described process, this degree will grow from one to the other irradiation pulse as regions with altered chemical composition will spread away from the interface across both films of the bilayer system. In turn, the compaction of the material in each layer will more and more impede the diffusion process in the intermixing region of the bifilm. Formation of fine cracks at the periphery of the squares irradiated with laser pulses serves an indication for the growth of viscosity in the upper part of the bilayer structure. The formation process of such cracks is defined by two factors: (i) growth of the internal tensile stress in the film due to material compaction (the shrinkage of the film in the direction normal to the film plane does not completely eliminate the lateral internal tensile stress in the film) and (ii) growth of viscosity in the compacted layer due to increase of strength characteristics of the material proceeding with simultaneous vanishing of defects from the atomic network and with modification of the chemical composition of the layer (Gorokhov et al., 1998). As a result, quite a thick intermediate layer (up to ~50-80 nm), in which the insulator matrix has a variable chemical composition and altered structural, mechanical, physical and other material properties, forms during the reactions at the interface between the GeO_2<Ge-NCs> heterolayer and the SiN_xO_y film. There are no Ge nanoparticles in the upper part of this layer, but the concentration of such nanoparticles gradually increases as we approach the pure GeO_2<Ge-NCs> heterolayer. The optical constants of the upper layer, as well as its optical gap value, also vary in a complex manner over the thickness of the layer. But, since glassy SiO_2, a very broad-band material, presents one of the composite-glass components, the optical gap in the mixture will be everywhere wider than the optical gap in the regions where only narrow-band glassy components, having an optical gap not wider than 5 - 5,5 eV, are present. Hence, in the course of PLA treatments of the samples with KrF excimer laser pulses the upper multi-component layer will become increasingly more transparent to the excitation beam of the Ar+ laser (514,5 nm) in comparison with the bulk of the lower GeO_2<Ge-NCs> heterolayer.

Note that all the discussed processes can be realized using traditional heat treatments at $T \geq 400^0C$. Yet, the duration of such heat treatments is typically greater than one minute, whereas, in our case, many atoms simultaneously participated in several structural-chemical rearrangement processes for only several tens of ns. To gain an insight into the whole picture of involved processes, let us focus on their most important features. First, AFM data on the shrinkage of the layered structure and IR-spectroscopy data on the modifications of its chemical composition reveal a gradual increase in the rate of structural-chemical transformations in the system from the first to last laser pulse (Fig. 11). Second, the dynamics of irradiation-induced changes in the optical properties of the bilayer film shows a distinct saturation. Those changes first occur rapidly and, then, their rate sharply decreases. The latter is evident from changes in the Raman spectra of the Si/GeO_2<Ge-NCs>/SiN_xO_y system observed 275 to 300 cm^{-1} (Fig. 12 a) at the beginning and at the end ot its irradiation with a sequence of ns laser pulses. During the first pulse, the input energy was absorbed more readily by the bilayer film in comparison with subsequent pulses since the changes in the Raman spectra due to the first pulse were manifested more distinctly in comparison with the changes induced by the second pulse and, the more so, by the two subsequent pulse. The successive growth of the intensity of the Raman peak (520 cm^{-1}) due light scattering by the lattice of Si substrate for Ar+ laser excitation (Fig. 12) also agrees with the general tendency in pulse-by-pulse modification of film-structure properties although, here, a more steady growth of peak intensity is observed. The effect is explained by an increase of bilayer film transparency, or reduced absorption of laser radiation by the film in this spectral region). A common feature for all pulses was that a predominant part of their energy dissipating in GeO_2<Ge-NCs> was spent on rearrangement of its lattice structure that proceeded mainly during the time intervals between the pulses.

An unusual thing here is that, with each subsequent radiation pulse, the Si/GeO_2<Ge-NCs>/SiN_xO_y hetrosystem becomes more and more transparent to the probing Raman-spectrometer radiation. Initially, we expected quite a contrary thing, namely, an increase in the absorption of Ar^+ laser beam radiation by the GeO_2<Ge-NCs> heterolayer. Such an increase was expected to be a result of the considerable amount of free Ge atoms released in the heterolayer bulk. Ge atoms become the material for nucleation and growth of many Ge-nanoparticles, which was to be manifested as an increased absorption coefficient of the heterolayer in the visible spectral region. For instance, such a growth of the absorption coefficient was observed during successive anneals of GeO_2<Ge-NCs> heterolayers and GeO(solid) and GeO_x films (Fig. 8). Similarly, during growth of silicon nanoparticles in the bulk of non-stochiometric SiO_x and Si_xN_y(H) films under chemical and laser treatments no enhanced transparency of such films was observed in the visible spectral region (Korchagina et al., 2012; Rinnert et al., 2001; Volodin et al., 2010). However, it proved to be a very difficult task to reveal the true nature of such an unusual effect in the experiments with the bilayer system under study, as in PLA-treated samples several different chemical and structural processes proceed simultaneously in the bulk of the GeO_2<Ge-NCs> heterolayer and at its boundaries. It was required to simplify the complex of such structural and compositional modifications in the films. Therefore, we first undertook an analysis of material transformations proceeding during traditional heat treatments in a GeO(solid) film on Si substrate without cap layer (ISP SB RAS) and in a GeO_x film on Si substrate protected by a cap layer formed from sputtered SiO_2 (~100 nm) (Nancy University, France).

A GeO(solid) layer of uniform thickness 61 nm was characterized with the follows values of optical constants: n=1,86 and k=0 (obtained from many-angle ellipsometric measurements using an He-Ne laser with λ=632,8 nm at beam incidence angles 45^0, 50^0, 55^0, 60^0, 65^0, and 70^0). A broad band peaking at 280 cm^{-1} due to Ge nanoparticles was observed in the Raman spectra of the as-deposited film (Fig. 14). The film was thin and its decomposition just began; therefore the amplitude of this Raman band was very low, especially at the sample edges (spectrum 3) where the film was less heated by the evaporator during the film deposition process. The highest degree of decomposition of the GeO(solid) film was observed at the centre of the sample (spectrum 2). Note that 4-minute exposition of a local sample area to the excitation laser beam focused to a ~1-μm diameter spot on the film surface at the centre of the sample caused a strong reduction in the beam scattering by a-Ge nanoparticles (spectrum 4). This effect was due to photo-stimulated oxidation of the particles by air oxygen in the locally heated surface area. Here, the concentration of a-Ge nanoparticles decreased in the film; yet, the film thickness grew in value due to the formation of an additional amount of germanium dioxide in the film. A comparison of the Raman spectrum of Si substrate without a film (spectrum 1) with the other spectra in Fig. 14 shows that deposition of a GeO(solid) layer onto the film surface (spectrum 3), growth of the fraction of a-Ge in the film in the form of nanoparticles (spectrum 2), and transformation of part of a-Ge nanoparticles into glassy GeO_2, i.e. increase in the volume of heterolayer matrix (spectrum 4) lead to multiply decreased amplitude of the peak at 520 cm^{-1} in the Raman spectrum of the Si/GeO(solid) system, this peak being due to light scattering by long-wave optical phonons in the Si substrate. The latter finding can be attributed to enhanced absorption of Ar^+ laser radiation in the GeO(solid) film for all these cases.

The cap layer prevents evaporation of the GeO_x layer in $Si/GeO_x/SiO_2$ structures (with layer thicknesses 100 nm/100 nm) annealed at temperatures T>450^{0}C. As a result, the initial metastable atomic network of GeO(solid) in such structures undergoes decomposition with the emission of all the excessive germanium and its accumulation in a-Ga nanoparticles. On increasing the annealing temperature, such a-Ga nanoparticles transform in Ge nanocrystals. These processes can be easily monitored using Raman scattering measurements (see Fig. 15). The temperatures used to anneal the $Si/GeO_x/SiO_2$ sample were insufficient for the onset of a reaction between the GeO_2 matrix in the decomposed GeO_x film and the SiO_2 cap layer; this sample was found to also not absorb the Ar+ laser radiation and undergo no heating due to the wide optical gap of the material. Hence, all the changes proceeding with the Raman scattering peak due to light scattering by phonons in Si substrate bear no relation with modification processes in the structure of the GeO_x layer. On the other hand, it is seen that the process of germanium release in the GeO_x film proceeding with the formation of a-Ge nanoparticles and their subsequent crystallisation successively reduces the amplitude of the peak due to Raman scattering in Si substrate. The same result was also observed in deposition of the initial GeO_x layer, like in the case of GeO(solid) layer synthesis (see Fig.14).

Note that the spectral position of the Raman peak due to Ge-nanocrystals in the GeO_x film is no difference from the spectral position of the peak in bulk Ge crystals (301 cm^{-1}). The latter finding is indicative of rather large mean sizes of Ge nanocrystals (~8-10 nm and greater) formed in this film during anneals.

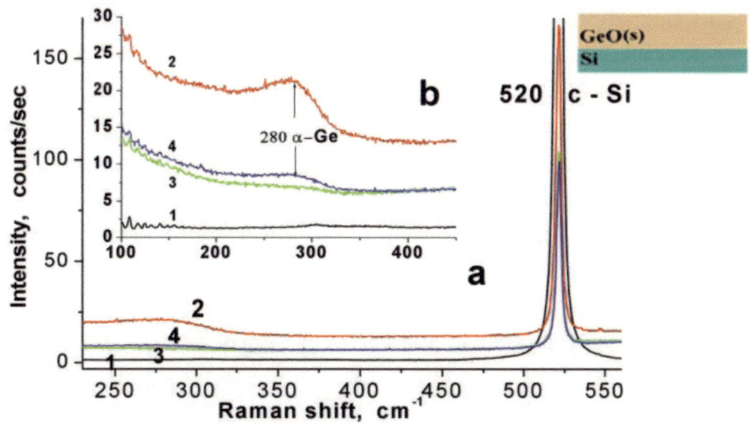

Figure 14. **a** –Raman spectra of GeO(solid) films of different thicknesses after decomposition of GeO(solid) into GeO_2 and Ge nanoclusters and oxidation of the nanoclusters: **1** – Si substrate, **2** and **3** - thick and thin areas of the GeO(solid) film, **4** - thick area of the GeO(solid) film where Ge nanoclusters were oxidized by the Ar + laser; **b** – magnified fragments of the same spectra.

Thus, the experiments with GeO(solid) and GeO_x films showed that the modification processes of their structures proceeding during traditional anneals are qualitatively similar to PLA-induced processes. All such processes (film shrinkage and densification, all decomposition stages of the metastable GeO(solid) atomic network, formation of a-Ge nanoparticles and their oxidation and crystallisation) do not make the layers more transparent for Ar^+ laser excitation radiation. Among all the above-considered physicochemical processes in the analyzed one- and bilayer compositions, only reactions of the glassy GeO_2 matrix in GeO_2<Ge-NCs> heterolayers with Si_3N_4 cap layers could lead to such an effect. Besides, the decrease of the Raman-spectrometer beam absorption in the Si/GeO_2<Ge-NCs>/SiN_xO_y system with increasing the number of ns KrF excimer laser pulses given to the sample (Fig. 12) could be explained assuming that a-Ge nanoclusters formed by short laser pulses in the GeO_2<Ge-NCs> heterolayer were substantially smaller than a-Ge nanoclusters formed during traditional thermal treatments. Thus, the whole mass of free Ge atoms leaving the decaying atomic lattice of GeO(solid) during irradiations formed a multitude of very small c-Ge nanoparticles.

The latter is quite natural a situation under conditions of abrupt heating of the heterolayer not only with laser impulses, but also with the energy released during decay of the unstable atomic network of GeO(solid) undergoing decomposition during the discussed treeatments. However, due to the short width of laser pulses, the length of possible diffusion of free Ge toward Ge nanoclusters does not exceed 2-3 cluster diameters, as nanoclusters are separated with distances close to the mean radius of nanoclusters in studied heterolayers. Hence, in small-sized a-Ge nanoparticles, due to the 3D quantum confinement effect, the effective optical gap will be substantially wider than that of larger nanocrystals formed during traditional anneals and, hence, their radiation absorption effect will not be so high as that of the larger Ge nanoparticles.

The sensitivity of our Raman spectrometer has allowed us to reveal in our experiments specific features directly defined by the sizes of Si and Ge nanocrystals in thin-film materials (Volodin et al., 2005). In large Ge nanocrystals, the spectral position of the Raman peak is at 301,5 cm^{-1}, with its blue shift becoming notable at nanocrystal sizes < 6-7 nm. Data on the spectral position of the Raman peaks due to Ge nanocrystals hosted in a GeO_2 matrix as calculated from two theoretical models (data of (Nelin & Nilsson, 1972) and data calculated from the model of effective density of folded vibration states (Volodin et al., 2005)) are shown in Fig. 16. Raman spectra of irradiated GeO_2<Ge-NCs> heterostructures measured in the spectral region around the wavenumber 301,5 cm^{-1} are shown in Fig. 12 b. In these spectra, a red shift of Raman peaks, typical of Ge NCs, is clearly observed. According to model calculations, the sizes of Ge NCs increase from 1,4 – 1,8 nm to 1,8 – 2,3 nm (Fig. 16)

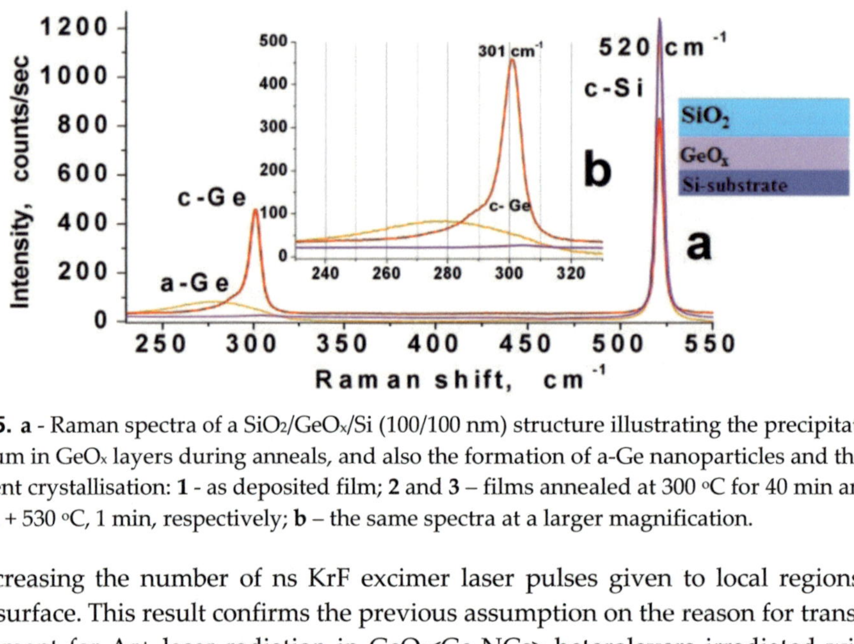

Figure 15. a - Raman spectra of a $SiO_2/GeO_x/Si$ (100/100 nm) structure illustrating the precipitation of germanium in GeO_x layers during anneals, and also the formation of a-Ge nanoparticles and their subsequent crystallisation: **1** - as deposited film; **2** and **3** – films annealed at 300 °C for 40 min and at 480 °C, 3 min + 530 °C, 1 min, respectively; **b** – the same spectra at a larger magnification.

with increasing the number of ns KrF excimer laser pulses given to local regions on the sample surface. This result confirms the previous assumption on the reason for transparency enhancement for Ar+ laser radiation in GeO_2<Ge-NCs> heterolayers irradiated with 25-ns KrF excimer laser pulses (λ=248 nm).

Note that analogous treatments of non-stochiometric silicon-rich SiO_x and $Si_xN_y(H)$ glassy films with high-energy KrF excimer laser pulses (λ=248 nm, t_p=25 ns) activate a complex structure rearrangement process which begins with the emission of excessive Si atoms from the atomic network of the material due to breakage of Si-O bonds (in SiOx films) or Si-H and Si-N bonds (in $Si_xN_y(H)$ films). The process ends with the formation of many a-Si nanoparticles in the film bulk from released Si atoms followed with subsequent crystallization of the nanoparticles (Gallas et al., 2002). In such cases, the silica films undergo shrinkage. Here, all processes proceed like in GeO_2<Ge-NCs> heterolayers. However, unlike in the latter case, in PLA-treated GeO_2<Ge-NCs> heterolayers there is no growth of transparency for Ar^+ laser radiation; on the contrary, according to (Korchagina et al., 2012), the absorption of 514,5-nm radiation increases. Today, the reason for the different behaviors demonstrated by the GeO_2<Ge-NCs> and SiO_2<Si-NCs> systems under PLA treatments remains unclear. In this effect, which depends on many factors, the contribution due to each factor changes under PLA differently in Ge- and Si-based heterosystems.

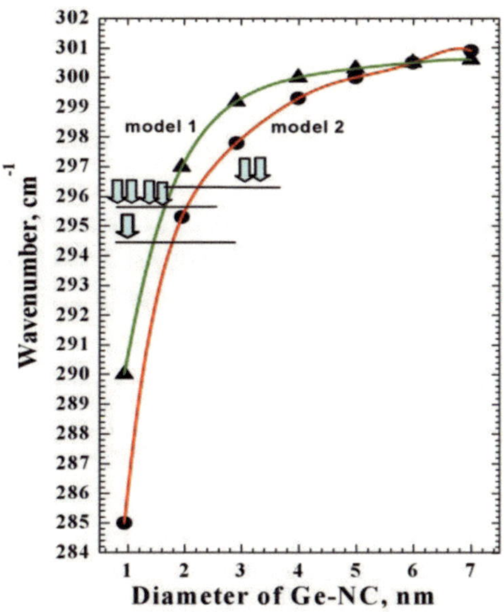

Figure 16. The spectral position of the Raman peak due to Ge nanocrystals versus nanocrystal size in GeO_2<Ge-NCs> heterolayers: • – data of (Nelin & Nilsson, 1972), ▲ - data calculated by the model of effective density of folded vibration states (Volodin et al., 2005). Using the Raman spectra in Fig. 12, one can determine the sizes of Ge NCs size formed in local areas of the GeO_2<Ge-NCs>/ SiN_xO_y bilayer system treated with one, two and four ns laser pulses.

Finally, we would like to emphasize here that irradiation of samples with 25-ns KrF excimer laser pulses (λ=248 nm) produces a very important effect on GeO_2<Ge-NCs> heterolayers, namely, its allows one to decrease both the sizes and size dispersion of Ge nanoparticles formed in metastable GeO(solid) and GeO_x layers. This process proceeds in the strongly densifying GeO_2<Ge-NCs> heterolayer that consists of 70% vol. of GeO_2 matrix, which, during the same irradiation treatment, is capable of controllably changing its chemical composition, structure, and physical, optical and electrophysical properties.

4.2. Femtosecond treatments of samples with Ti-Sapphire laser

Differences in laser radiation parameters, and also differences in other irradiation conditions, strongly affect the material modification process and its ultimate results. The idea of to which extent the final effect due to PLA might differ in various irradiated systems can be grasped through making a comparison of PLA data for film structures involving GeO, GeO_x layers, and GeO_2<Ge-NCs> heterolayers obtained in treatments performed using two different lasers, a KrF excimer laser and a Ti-Sapphire laser.

Phase transition mechanisms in solid materials under femtosecond laser treatments are radically different from analogous PLAs implemented using nanosecond pulses. These differences are most clearly manifested in the case of ultra-high-power femtosecond laser treatments, when nonlinear effects play an important role in the light absorption processes. There are many contributions devoted to the study of the interaction of femtosecond laser pulses with matter (Juodkazis et al., 2009). Results of an experimental study of the dynamics of electron-hole plasma formation and recombination of photo-induced charge carriers in Si under fs PLA at laser energy lower the threshold value for surface melting and ablation were reported by Ashitkov et al. (Ashitkov et al., 2004). During the action of a fs laser pulse, electron-hole plasma with electron concentration in the conduction band up to 10^{22} cm^{-3} forms, both thermally activated and athermal surface-layer melting processes being observed along with a transition of material to metallic state. In (Ashitkov et al., 2004), time-dependent optical reflection was measured for heating-pulse energies.

In phase transitions under such conditions, athermal effects may emerge. Indeed, the duration of fs laser pulses is much shorter then the duration of the electron- phonon interaction in semiconductor (approx. 1-2 ps). That is why in silicon, during laser pulse the "hot" electron-hole plasma does not excite vibrational modes and relaxes insignificantly. The temperatures of the electron and atomic sub-systems are much different. According to theoretical calculations, the material becomes unstable when the electron concentration excited from the valence to conduction band reaches 9-20% of the atom concentration in Si lattice (Bok, 1981). This metastable state may relax to a more stable crystalline phase without melting yet with release of latent crystallisation heat. The process is similar to "explosive" crystallisation. In several picoseconds after the pulse, the phonon and electron temperatures should equalize. Post-pulse processes, e.g. post-pulse atomic diffusion in heated films, start manifesting. Very probably, diffusion can be stimulated by laser radiation not only via film heating, but also due to the breaking of valence bonds that occurs when electrons undergo excitation to the conduction band. Under nanosecond laser treatments, the time of Si film cooling to a temperature below the melting point is known to amount to about 100 nanoseconds (Sameshima & Usui, 1991). For fs laser treatments, the cooling time can be expected to be the same. This time can turn out sufficient for diffusion of excess silicon and for formation of nanocrystals only in the case of melt.

Zabotnov et al. (Zabotnov et al., 2006) showed that femtosecond laser annealing of Si surfaces leads to their nanostructuring, two types of formed nanostructures being observed. In one case, these are ordered lattices with a period somewhat shorter than laser radiation wavelength. The formation of such structures can be explained by the interference between the fs radiation waves and the non-equilibrium plasma excited by radiation in the surface layer. Here, ablation of Si atoms is accompanied with their rapid oxidation in air, and also with the formation of surface Si nanocrystals sized 3-5 nm.

Martsinovsky et al. (Martsinovsky et al., 2009) showed that, under the action of femtosecond laser pulses, intense photo-excitation of semiconductor surface, which dramatically modifies its optical response and provides conditions for generation of surface electromagnetic waves of various types, becomes possible. Martsinovsky et al. have also discussed the interrelation of electronic processes optically induced in the near-surface layer with the formation of periodical surface microstructures that were observed in experiments on irradiation of silicon targets.

Femtosecond PLAs were also used to crystallise silicon films. In PLA treatments, a critical point for thin films is proper choice of irradiation conditions since at high pulse energies such films may readily suffer ablation. At present, the issue of Si-nanocrystal formation in silicon nitride and silicon oxide films under nanosecond PLAs is a well-studied matter. A process for fabricating nanocrystals in dielectric films with the the help of nanosecond laser pulses was patented as involved in the production technology of non-volatile flash memory devices. Femtosecond treatments were used for modification of Si nanoclusters in SiO_2 films (Korchagina et al., 2012). Yet, for femtosecond treatments structural changes in silicon clusters in SiN_x and SiO_x films have not been studied. As for germanium, Ge films, and Ge nanoparticles embedded in dielectric layers, to the best of our knowledge, only one reported communication on fs-laser-initiated crystallisation in thin layers of amorphous Ge is presently available (Salihoglu et al., 2011).

On the whole, the physics underlying processes in film systems subjected to to femtosecond anneals still remains scantily understood. That is why it would be interesting to perform a study of such processes in samples treated with nanosecond and femtosecond laser pulses with revealing all involved characteristic features, also including possible athermal effect on Ge nanoparticles hosted in an insulator matrix, both with the aim of detecting new manifestations of PLA-induced effects in germanium-based materials and with the aim of comparison of similar processes in systems based on two chemically akin semiconductors, Si and Ge. Assumes that the results of these studies will prove also useful for applications.

4.2.1. Femtosecond PLA of GeO_2<Ge-NCs> heterolayers

Our study of fs PLA-treated thin-film systems based on germanium oxide layers was the first one devoted to the subject matter of interest. Therefore, the first task in such investigations was to develop a general concept of the complex effects and phenomena occurring in the film systems of interest subjected to PLAs under different conditions. In particular, we examined GeO_2<Ge-NCs> heterofilms of different thicknesses grown by the LP CVD method on (bare and CVD SiO_2-covered) silicon or quartz glass substrates. The films either had one layer and no cap layer or they were protected by a thin cap layer of PE CVD SiN_xO_y or CVD SiO_2. In another set of samples, ~ 20 – 30 nm and ~200-250 nm thick, GeO_2<Ge-NCs> layers were part of multilayer compositions where they were separated with thin (~7-10 nm) and thick (~100 nm) CVD SiO_2 layers as shown in Fig. 17 a. Scanning was made with a focused fs-laser beam at a velocity of 100 μm/s. The frequency of laser pulses was 10 pulses per 1 μm. The laser fluence (E_{pulse}) varied from 15 to 1 mJ in ~70-μm diameter spot and from 0.028 to 0.0018 mJ in 1-μm diameter spot. In the experiments, samples with GeO_2<Ge-NCs> heterolayers were used. In addition, in some cases optical bench vibrations of ~1,5-μm amplitude in two dimensions, in the plane of the optical bench and in the direction normal to it, were implemented. A typical result for layered systems with GeO_2<Ge-NCs> heterolayers on Si substrate scanned with Ti-Sapphire laser beam focused to a 70-μm diameter spot is shown in Fig. 17 b. From line to line, the laser fluence was decreased in 1-mJ steps within 15-mJ wide intervals of E_{pulse}. Thus, the adopted femtosecond PLA regimes and conditions for examined films were different from their treatments with the nanosecond laser in terms of the majority of basic physical parameters (laser radiation wavelength, pulse duration, intervals between pulses, arrangement of the irradiation process). Besides, in the case of fs PLAs the adopted irradiation technique did not allow us to trace changes in the film structure following successive irradiations of the sample with a series of laser pulses.

The first result was that optical microscopy was quite a sensitive technique allowing us to reveal both changes in the surface relief and the variations in the optical parameters of the material induced by structural and chemical modification processes proceeding in heterolayers under laser beam treatments, whereas the SEM method was efficient only when studying surface morphology modifications (cp. Figs. 17 a and b). The main fundamental difference in the PLA mechanisms for the two lasers is manifested when the

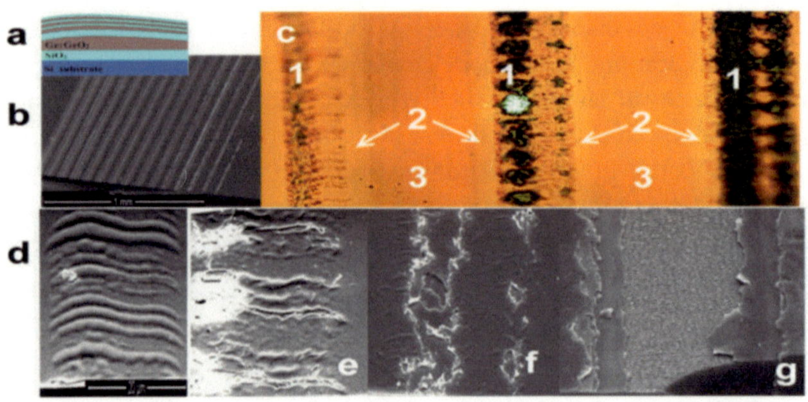

Figure 17. Surfaces of multilayer coatings involving alternating GeO_2<Ge-NCs> heterolayers and SiO_2 layers after fs PLA (beam diameter 70 μm): **a** – schematic of the multilayer coating (similar to that in Fig. 6); **b** – SEM image of the film coating treated with laser pulses of various energies; **c** – optical microscopy image of the same sample: (1) evolution of surface morphology at the centerline of laser spot path; (2) along coating edges there is no relief, although light stripes defined by the changes of optical properties of some layers inside the coating are clearly seen; (3) laser treated paths are separated with wide stripes where the film coating structure has remained unchanged (E_{pulse} was not high enough); **d, e, f** and **g** – SEM images obtained at the various stages of the coating laser damage process implemented with laser energy increase: **d** – development of a wavy surface relief as a result of transformation of 3 thin GeO_2<Ge-NCs> heterolayers in the top tier into nanofoam (E_{pulse} = 8 μJ); **e** - beginning of layer destruction in the top tier (E_{pulse} = 9 μJ); **f** – simultaneous damage of upper and middle tiers (E_{pulse} = 10 μJ); **g** – simultaneous damage of all layers down to the Si substrate and the beginning of its damage by formed ripples (E_{pulse} =13 μJ).

laser are used to modify GeO_2<Ge-NCs> heterolayers. Only Ge-nanoclusters, whose optical gap E_g^{eff}(Ge-NC) < 2 eV, absorbed femtosecond laser radiation with λ=800 µs in GeO_2<Ge-NCs> heterolayers (see Figs. 8 and 9). Nanosecond KrF laser radiation with λ=248 nm is additionally absorbed by the GeO_2 matrix, which constitutes 70% of the heterolayer volume, its optical gap being $E_g^{eff}(GeO_2)$ ~4,5-5 eV. The actions of the light emitted by the two lasers on the two material components in the heterolayers are radically different: while KrF laser radiation heats both the Ge-nanoclusters and the dielectric matrix, Ti-Sapphire lase radiation heats only Ge-nanoparticles while leaving the matrix cold. It is this difference that makes the effects due to PLA processes in GeO_2<Ge-NCs> heterolayers under irradiation with the two lasers cardinally opposite. Namely, in GeO_2<Ge-NCs> heterolayers treated with the fs laser the heated Ge-nanoclusters begin to react with the surrounding glassy GeO_2 and, during this process, the initial heterostructure transforms into a totally different kind of solid with fundamentally different physical properties.

4.2.2. Formation of GeO_2 nanofoam using fs PLA of GeO_2<Ge-NCs> heterolayers

Indeed, Ge-nanoclusters surrounded by GeO_2 matrix were rapidly heated by fs laser radiation, to which the oxide was transparent, unless the chemical reaction (4) between both components in GeO_2<Ge-NCs> heterostructure was initiated. The emitted germanium monooxide molecules (GeO(gas)) are initially localised in the vicinity of hot Ge-nanoparticles surrounded by a cooler insulator (Fig. 18). Affected by the vapor temperature and pressure, the glassy GeO_2 matrix gets heated and it extends when softening. On cooling, the GeO_2<Ge-NCs> heterostructure turns into a swelled mass of nano-dimensional glass cells. Each of the initial Ge-nanoclusters, as its grows smaller and lighter, is to remain inside a formed nano-cells (Fig. 18 c). The most surprising thing here is that, in the outcome of nano-foam formation process, all Ge-nanoparticles should have a uniform minimal size independently of their size in the initial heterolayer. Due to the 3D quantum-dimensional effect, Ge-nanoclusters, as they grew smaller during their reaction with the GeO_2 matrix, should stop absorbing the heating light according to the condition $h\nu = E_g^{eff}(NCs)$. This technique allows us to obtain ensembles of isolated indirect gap semiconductors of very small dimensions that also have a low size dispersion.

The process is completed as follows. While the nano-cell sizes in the glass softened under the internal gas pressure of the gas inside grow in value, their glass walls start cooling. The latter is due to the fact that, first, the gas in the nano-cells volumetrically expands under near-adiabatic conditions and, hence, the gas temperature will decrease. In addition, the mechanical expansion of the nano-cells in the soft GeO_2 nanofoam consumes the energy from the gas in cells. The gas gives off heat and its pressure decreases. Simultaneously, the nanocell-hosted Ge-nanoclusters, as they grow smaller, start absorbing less fs laser radiation (due to the 3D quantum-confinement effect). This reduces the efficiency of the reaction yielding new portions of GeO(gas) in the nano-cells. Thus, in the nanofoam the nano-cell walls start cooling at a sufficiently high internal pressure due to GeO(gas) vapor. Cooling of the GeO_2 glass rapidly increases its viscosity, which prevents the nano-cells from being compressed with decrease of the internal GeO(gas) pressure. Thus, the individual nano-cells, as well as all of the nanofoam, will preserve their dimensions in the cooling process.

PLA results of multilayer film systems consisting of a conventional dielectric layers and GeO_2<Ge-NCs> heterolayers illustrate in Fig.17 the complex effect of fs laser radiation on the system. The system consisted of nine layers belonging to two types: thick and thin SiO_2 films and GeO_2<Ge-NCs> heterolayers that alternated among each other (according to the diagram in Fig. 17 a). With increase of laser pulse energy, distinct indications of film structure modification started showing up, becoming manifested more and more clearly. The sample holder vibration accompanying the scanning process proved useful in revealing important details of the mechanisms underlying the impact of laser pulse radiation on the structure of irradiated films. Changes in film properties began to evolve in the volume of the films, and then they continued in surface morphological changes. To describe the observed processes, it was necessary to employ the whole complex of available experimental methods, each method, as a rule, allowing characterisation of a limited number (one or two) of the changing properties of the objects under study.

For instance, optical microscopy (Fig. 17 c) was sensitive to variations of optical parameters of the heterolayers under laser beam, manifested in color changes. This was an indication of the beginning of modification of material properties in some layers inside the coating. Those processes preceded the changes of the thickness and surface relief of the coating. SEM was effective only when studying surface morphology modifications (cp. Figs. 17 b, d, e, f and g). Besides, it has limited capabilities in height measurements of many relief details. However, when combined, the two methods were capable of revealing a complete picture of the dynamics of fs PLA induced changes in investigated multilayer structures on increasing the laser pulse energy.

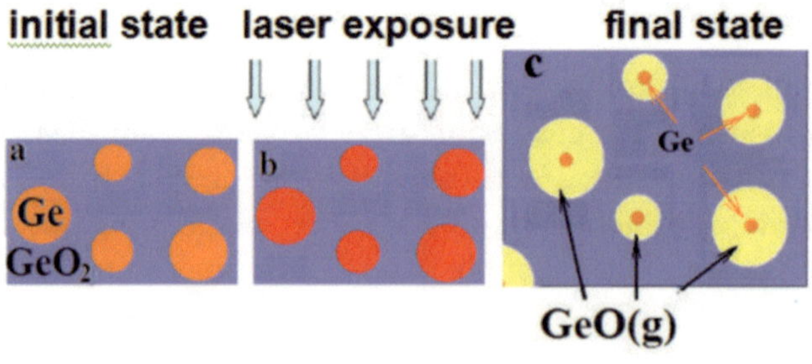

Figure 18. Schematic illustrating the process of transformation of a GeO_2<Ge-NCs> heterolayer in nanofoam-like glassy GeO_2 in a film structure subjected to fs laser annealing: **a** - initial heterolayer structure; **b** - heating of Ge-nanoclusters by fs laser pulses, which initiates the formation of GeO(gas) around these clusters; **c** - final structure of nanofoam-like glassy GeO_2.

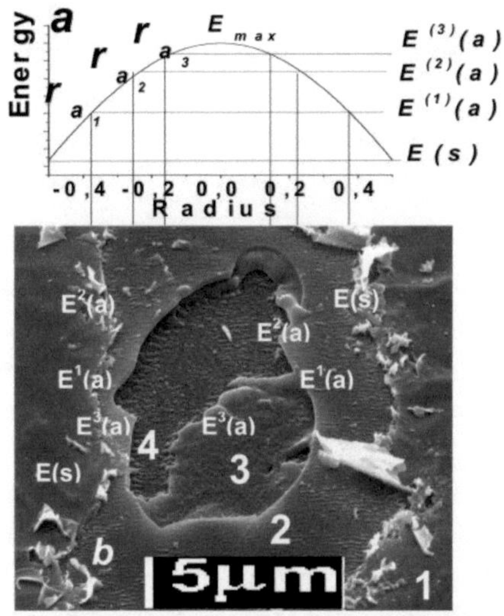

Figure 19. The simultaneous laser damage of two tiers of the multilayer system in Fig. 17 g during fs PLA (SEM image): **a** - radial Gaussian distribution of energy inside the laser beam; **b** - $E^{(1)}(a)$, $E^{(2)}(a)$, $E^{(3)}(a)$ - threshold energies for laser damage: (1) - in the top tier of the multilayer film system, (2) and (3) - in a thick SiO_2 layer and in a thick GeO_2<Ge-NCs> heterolayer of the middle top, respectively; $2r_{a1}$, $2r_{a2}$, $2r_{a3}$ - width of damaged areas in the respective layers of the system; *E(s)* - threshold energy for the onset of the foaming process in thin GeO_2<Ge-NCs> heterolayers of the top tier of the multi-layer film system.

The experiment revealed some interesting features of the nanofoam formation process. First of all, the process was indeed implemented. The latter was indirectly proved by several methods of structural analysis. For instance, at an early stage of the process a wavy relief not violating the continuity of the multilayer coating emerged on the coating surface. The formation of this relief was registered by optical microscopy and SEM (Figs. 17 c and d). With increasing the laser pulse energy, the top tier in the coating starts destructing, and the smooth thick SiO_2 layer of the middle tier stands out. Then, the top tier of the coating detaches from the lower tier, tears and rolles, with all its thin layers becoming laminated (see Figs. 17 f and 19 b). This is the first stage of the laser damage of the whole multilayer system. Further on, the process laser damage develops involving more and more layers (Fig. 17 f and g and Fig 19 b), the boundaries of destructed zones in each layer spreading from the centerline of the laser treated path on the sample surface with growth of laser pulse energy. The ablated zone (band) in each layer is always wider than that in the lower layer. Therefore, SEM images allow observation of the structure and morphology of the uncovered surface of any exposed coating layer down to the Si-substrate surface.

Ultra-short laser pulses of the investigated heterostructures cause structural rearrangements and chemical reactions as: films shrinkage, GeO(solid) layers decay on reaction (3), growth of Ge-nanoparticles in the heterolayer and their crystallisation, chemical reactions of GeO_2 matrix of heterolayers with cap layers on their interface, possible crystallisation of GeO_2 glass around Ge-nanoparticles and reaction of GeO vapour formation from GeO_2<Ge-NCs>

heterolayer components, also ablation — explosive damage of films and substrates (Korte et al., 1999). Many of these processes are also expressed in radiation of GeO_2<Ge-NCs> heterolayers by fs Ti-Sapphire laser, just as in case of Kr excimer laser. But, as is seen from Fig. 17, the interaction character of fs laser radiation with the studied films and substrate was considerably different.

Results of using PLA and fs lasers for the studied film coatings showed that a number of the above-mentioned structural rearrangement processes and chemical reactions excite one by one with laser fluence growth. In general, to activate a new (i)-process, it is necessary to have laser fluence, absorbed by a film coatings square unit, exceeding the energy threshold of ($E^{(i)}$). Energy intervals among thresholds of different processes successively excited in films are usually so large that they can be definitely separated from each other. Therefore, the processes observed in multi-layer systems could be divided into two main groups. The first one consists of almost all the processes from the presented list, which proceed in the volume and interfaces of each of a multi-system's layers independently from processes in its other layers. Ablation processes belong to the second group as they, proceeding in one of a multi-system's layers, as a rule, determine all other processes in its lower layers.

4.2.3. Laser damage in multilayer coatings

In investigated heterostructures, ultra-short laser pulses induce various structural rearrangements and chemical reactions such as film shrinkage, decomposition of GeO(solid) layers according to reaction (3), growth of Ge-nanoparticles followed with their crystallisation in the heterolayer, chemical reactions of the GeO_2 matrix with the cap layers in the interfacial region, possible crystallisation of GeO_2 glass around Ge-nanoparticles and reaction (4) of GeO vapor formation from GeO_2<Ge-NCs> heterolayer components, and also ablation and explosive laser damage of the films and substrates (Korte et al., 1999). Many of these processes are manifested in GeO_2<Ge-NCs> heterolayers irradiated with fs Ti-Sapphire laser in a similar way like in the case of Kr excimer laser. However, as it is seen from Fig. 17, the interaction of fs laser radiation with examined films and Si substrate was largely different.

Results of using ns and fs PLA for treating examined film coatings showed that a sequence of the above-mentioned structural rearrangement processes and chemical reactions gets initiated on increasing the laser fluence. In general, for activation of a new (i-th) process, it is required to reach a laser fluence absorbed by a film coatings in unit square in excess of the energy threshold for the process ($E^{(i)}$). The energy intervals between the thresholds of the various processes successively initiated in the films are usually so large that they can be definitely separated from each other. Therefore, the processes observed in our multilayer systems could be divided into two main groups. The first group consists of almost all the processes from the above process list, which proceed in the bulk and interfacial regions of each of the involved layers independently of the processes proceeding in other layers. Ablation and laser damage processes belong to the second group as they, proceeding in one of the layers, as a rule, govern the processes proceeding in lower layers.

Successive onsets of structural-chemical modification processes of group I was observed in bilayer coatings given ns PLAs. In the case of fs PLA, successive (layer-by-layer) initiation of structural-chemical modification processes becomes distinctly pronounced during ablative destruction of multilayer coatings (see Fig. 19 b that shows a fragment of Fig. 17 f). It is seen that the laser pulse fluence absorbed by the multilayer coating is sufficiently high to exceed four energy threshold values, $E(s)$, $E^{(1)}(a)$, $E^{(2)}(a)$, and $E^{(3)}(a)$ (see Fig. 19 a). Three of the threshold energies are the thresholds for laser damage initiation in the two top tiers of the irradiated system (in the top tier and in the two layers of the middle tier, a thick SiO_2 layer and a GeO_2<Ge-NCs> heterolayer). The very first threshold energy is the non-ablative threshold $E(s)$ of the foaming process of three thin GeO_2<Ge-NCs> heterolayers in the top coating tier. The radial distribution of light energy in the laser beam obeys the Gauss law $E(r) = E_0 \exp(-2r^2 / r_0^2)$ (Korte et al., 1999) as shown in Fig. 19 a; therefore, destruction of the layered structures begins at the centerline of the laser treated path on the film surface when the pulse energy exceeds the necessary energy threshold of the top tier. $E^{(1)}(a)$ (see Fig. 19 b). If the pulse energy is sufficient to exceed the damage thresholds of the first, second and subsequent film, the laser damage will be initiated in all these layers simultaneously. Yet, ablative removal of the lower- coating material is impossible without a preliminary removal of the upper films. That is why, in most cases multilayer coatings are damaged successively, layer by layer. In this case the positions of the boundaries in destructed areas of each of the layers are determined by the radial energy distribution in the laser-beam cross section. Qualitatively, the theoretical expectation for this regularity is illustrated by Fig. 19 a. The experimental data of Fig. 17 f, g, and Fig. 19 b, in principle, do not disagree with the stated concept, although a multi-mode laser beam structure in real cases often distorts the strictly circular pattern of radial energy distribution in laser spots on irradiated surfaces. Thus, the processes activated in PLA-treated films experimentally revealed the real energy distribution round the laser-beam center.

The particular case under consideration is interesting in that respect that, apart from the propagation of the laser damage process into depth in a multilayer coating involving two types of films of different origin, here there was also a process that can be classed to intra-layer processes that can also proceed in layers covered by other films. This was the transformation process of fs PLA treated GeO_2<Ge-NCs> heterolayers into a nanofoam. The formation of nanofoam in germanate heterolayers was not hindered by thin SiO_2 layers that covered such heterolayers. Such SiO_2 layers were weakly heated by the laser beam, and they do not change their properties due to chemical and structural modifications. Simultaneously, this process progressed along the centerline of the fs laser treated path when the energy threshold of nanofoam formation E(s) was overcome during the fs PLA. If the laser fluence many times exceeded the threshold energy $E(s)$, then the top tier was immediately torn off from the middle tier as a homogeneous film according to the common explosive machanism of film damage. Here, the chemical nanofoam formation process did not have sufficient time to evolve in the GeO_2<Ge-NCs> heterolayers of this tier. Here, under fs laser irradiation the system of upper layers behaves as a solid film of homogeneous material. The thickness of this layer system at its edges along the line of damage rupture was almost no difference from the thickness of the whole layer on the substrate, and the edges of the layer were like those of glassy films under brittle fracture. In other words, the edges mostly looked like broken line segments (Fig. 20 b, area 1), which were similar to the edges of thick SiO_2 layers after damage ruptures in the same multilayer coating (Fig. 20 b, areas 2 and 4).

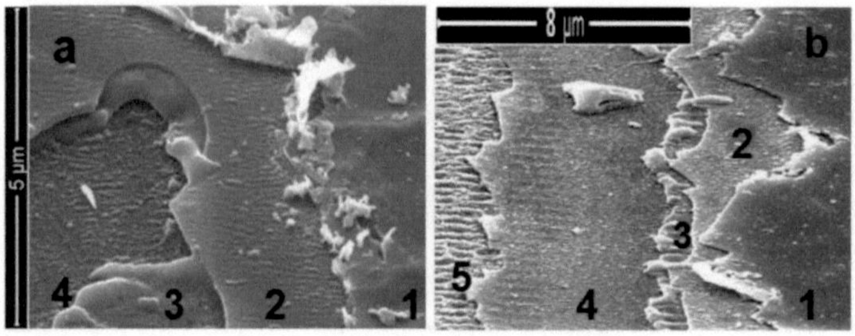

Figure 20. SEM image of layer-by-layer destruction processes in a multilayer structure exposed to laser radiation (**a** – E_{pulse} = 10 mJ , **b** – E_{pulse} = 13 mJ) : **1** - the top tier of thin layers, **2, 3** – the middle tier of thick layers (SiO_2 and nanofoamed GeO_2<Ge-NCs> heterolayer, respectively), **4, 5** – the bottom tier (lower SiO_2 layer and Si substrate, respectively.

However, laser damage took a different course if, under fs PLA, the laser fluence was only a little in excess of the threshold E(s) and the process of GeO_2<Ge-NCs> heterolayer foaming had sufficient time to develop. The thicknesses of those layers increased appreciably together with the thickness of the whole top tier in the multilayer coating; the top tier swelled and began to gradually tear off from the underlying SiO_2 layer. Then the areas of the top tier showing a most pronounced nanofoam formation suffered destruction accompanied by tearing off of its fragments (see Figs. 19 b and 20 a). Here, there are many thickened monolayer fragments of the top tier, exfoliated from the substrate incompletely and left in many areas over the edges of the layer along the centerline of the damage rupture zone. Those fragments roll in microtubes and laminate into very thin separate layers that constitute this tier (Figs. 19 b and 20 a).

During fs PLA, the nanofoam formation process of the thick GeO_2<Ge-NCs> heterolayer in the middle tier of the film coating proceeded in a different way than that in thin GeO_2<Ge-NCs> heterolayers of the top tier. The reason for this was that the thickness of GeO_2<Ge-NCs> heterolayer and its mass allowed it to accumulate more laser energy per unit volume. Moreover, a longer time was required for the damage process to cause removal of the thin film layers and the upper thick SiO_2 layer. Both factors allow the heterolayer to accumulate in its volume a considerable energy and considerably prolong the action of this energy transforming the heterolayer into nanofoam. Therefore, in all SEM images (Figs. 17 g, 19 b, 20 a and b) one can observe under the upper thick SiO_2 layer removed by laser damage a substance that looks like a liquid light foamy mass.

4.2.4. *Some characteristic parameters of the nanofoam formation process*

The thicknesses of the layers in the bottom and middle tiers as measured in SEM images were close to the values obtained in as-synthesized films. We could also conveniently measure in the SEM images the thicknesses of the foam-like layers, which were found to 1,5-3 times exceed the initial thicknesses of the GeO_2<Ge-NCs> heterolayer. In some areal parts of the foam-like layers (where the layers became uncoated following the laser damage of the SiO_2 layer, Fig. 20), the foamy mass often increased, approximately by 2-4 times, the thickness of the GeO_2<Ge-NCs> heterolayer. During of the most foamed layer, some individual fragments of that layer were often explosively thrown up onto the SiO_2 layer. The thicknesses of those fragments were bigger than the three- or four-fold thickness of the SiO_2 layer.

As a rule, AFM data on the sizes of important relief features of the film coatings (Fig. 21) complied with the data obtained by other measuring techniques. Thus, AFM measurements proved that the top tier tears off from the underlying SiO_2 layer during laser damage ablation (Fig. 22). Besides, it is possible to determine the initial thickness (h) of the top tier (h ~50-60 nm) and the degree of its changes during fs PLA using the surface relief profile (Fig. 22 b) measured along the $\boldsymbol{p}$-line (Fig. 22 c). The obtained value for the thickness of the top tier (H) in the immediate vicinity of the damage rupture (Fig. 22 a, area 2) is five times greater than the initial layer thickness (h). But the value H is not only the result of transformation of the three GeO_2<Ge-NCs> heterolayers in the top tier into nanofoam. The laminating effect of the edges of the top-tier layers from the lower films in this area also contributed to the found value of H since after the tearing the free edges bent upwards.

The energy threshold of the foam formation process $E(s)$ in the thin GeO_2<Ge-NCs> heterolayers of the top tier during fs PLA was exceeded over the whole irradiated film surface (Figs. 21 and 22). Moreover, the rate of irradiation along the laser treated path (close to the line $\boldsymbol{q}$ in Fig. 22 a) varied periodically and at the points of maximum it reached the lower threshold energy for laser damage in this tier $E^{(1)}(a)$. In accordance with these periods, crests in swollen and buckled regions of the top tier formed. On the average, the separation between crests is 2,4 - 2,5 μm (line $\boldsymbol{q}$ in Fig. 22 a). There are zones in between the crests, where there was a minimal foaming of GeO_2<Ge-NCs> heterolayers. From the relief profile measured along the cutting line $\boldsymbol{q}$ (Fig. 22 d), we could determine the modulation frequency

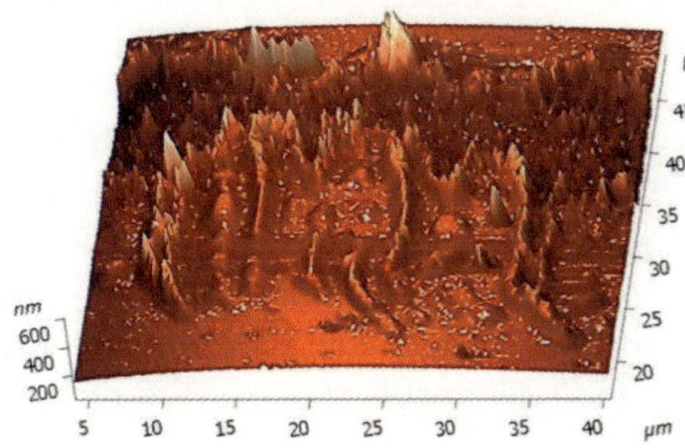

Figure 21. 3D AFM image of fs laser treated path on the multilayer surface shown in Fig. 17 e.

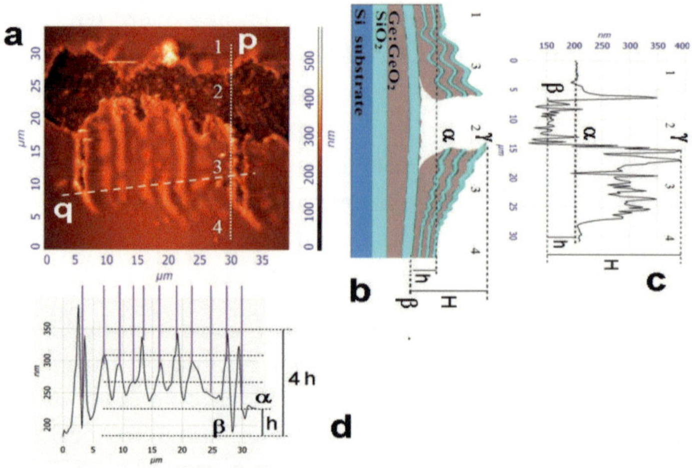

Figure 22. **a** - AFM image of the surface of the multilayer coating shown in Fig. 21: ***p*** and ***q*** - cutting lines of the sample surface across and along the fs laser treated path, respectively; **b** – schematic of the sample cross section along the line ***p*** drawn along the rupture edge of the the top tier of the coating with swelled films (**1** and **4** - areas with unchanged relief, **2** - laser damage zone of the top tier, **3** - zone of the maximum change in relief at the edge of rupturing film during its swelling; **c** - relief of the film coating along the line ***p***: α - the initial level of coating surface, β - surface of the upper thick SiO_2 layer, γ - the highest level of coating relief changes, **d** - relief of the film coating along the line ***q***.

of laser beam energy connected with the sample vibration. An analysis of this relief in the described fs PLA conditions showed that, at this modulation frequency of laser irradiation, its energy changes within the interval between the threshold energies of the two processes $E^{(1)}(a)$ and E(s) during the period of ~ 0,025 sec. The thickness of GeO_2<Ge-NCs> heterolayers may increase 3-4 times for the time equal to half the period.

As a result, the vibration experiment revealed: (i) a low threshold energy *E(s)* of the process; (*ii*) a high capability of GeO_2<Ge-NCs> heterolayers to increase of their volume (by 3-5 times) during transformation into nanofoam material; (*iii*) high dynamic characteristics of the nanofoam formation process during increase of fs laser pulse energy.

Micro-Raman spectra of fs laser treated GeO_2<Ge-NCs>heterolayers (Fig. 23) produce another evidence of their transformation into a nanofoam-like material. For simplification of the analysis, the spectra were registered in GeO_2<Ge-NCs> heterolayers deposited onto a Si substrate protected by a 25-nm thick SiN_xO_y cap layer (the effect due to ns PLA was also studied on this sample). The effect due to laser fluence value exciting the nanofoam formation process in local regions of the GeO_2<Ge-NCs>/SiN_xO_y bilayer system is shown in Fig. 23. In Fig. 23 a, in spectrum 1, which corresponds to non-irradiated film at point 1 (Fig. 23 e), only a broad Raman band due to light scattering by transversal optical phonons of α-Ge is shown, like in the case of ns laser treatment of this film system (Fig. 12).

As it can be judged from the weak scattering peak of the beam of the exciting Ar^+ laser of Raman spectrometer due to the Si-substrate lattice (scattering band in the region of 520 cm^{-1}), the film system was little transparent to this beam. Raman spectra 2 and 3 in Fig.

23 a were obtained at two closely located points (2 and 3) on one of the lines treated by the fs laser beam (Fig. 23 d and e, respectively). Yet, because of heterogeneity of the radial energy distribution in the beam, intensive formation of nanofoam mass from the GeO_2<Ge-NCs> heterolayer proceeded at both points of this line during PLA. At point 3, the film coating was at an early stage of destruction, still remaining almost intact, whereas at point 2, damage was well activated. The latter is evident from the intensity of the Raman peaks of Ar^+ laser beam due to Si-substrate lattice (in the corresponding spectra); at point 2 the film practically does not hinder the Ar^+ laser beam to reach the substrate. In other words, at points 2 and 3 we observe an early and the final stage of the destruction process of the foamed layer. Nonetheless, despite the possibility of mixing up film fragments belonging to different areal parts of the film, Ge-nanocrystal sizes were not large, ~2,3 – 3,1 nm as calculated by model (1) and ~3,1 - 4,1 as calculated by model (2). At point 4 in Fig. 23 c the fs laser energy was slightly lower than that at points 2 and 3. Therefore, the film foamed without and violation of its continuity. Ge-NCs of the least diameter ~1,0 (model (1)) and 1,4 nm (model (2)) were found here (spectrum 4 in Fig. 23 a).

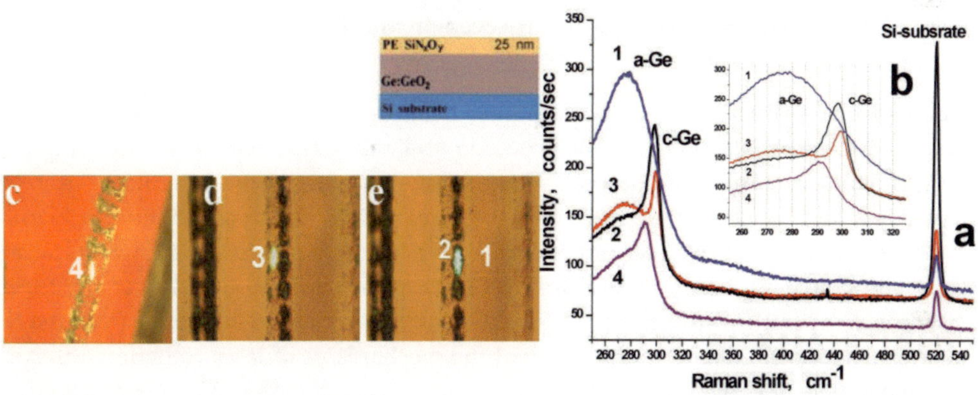

Figure 23. **a, b** - Raman spectra and **c, d, e** - photos of GeO_2<Ge-NCs> heterolayers transformed to nanofoam under a SiN_xO_y cap layer on the Si substrate: **1** – non-irradiated films; **2** - strong laser damage of the films; **3** - laser damage at an early stage; **4** - the formation of foam without destruction of the cap layer.

Thus, the both main features which are to accompany the formation of nanofoam from GeO_2<Ge-NCs> heterolayers during their fs laser pulse treatments were observed in the experiment. Direct images of the studied material are supposed to be obtained with HR TEM, but additional information about the properties of the nanofoam was also obtained with the SEM method. Characterizing the most impressive effects produced in GeO(solid) films and GeO_2<Ge-NCs> heterolayers by fs PLA, we would like to note that, here, the same structural and chemical modification processes like those in ns PLA treated samples proceed both consequently and in parallel in the material; those process include: decomposition of the metastable solid germanium monoxide atomic network with its transformation into the atomic network of glassy GeO_2 oversaturated with Ge atoms; clusterization of released germanium into amorphous nanoparticles; and growth and crystallisation of α-Ge nanoparticles. The effect of film transparency increase is also common for both kinds of laser treatments during crystallisation of α-Ge nanoparticles in the film bulk. On the other hand, the main difference between the effects due to UV ns and IR fs PLAs on GeO_2<Ge-NCs> heterolayers consists in that UV ns laser radiation leads to a considerable shrinkage of the material, whereas IR fs radiation leads to transformation of the whole irradiated heterolayer material into a very loose foam containing nanocavities separated with ultra-thin glassy GeO_2 barriers. An SEM image of the nanofoam obtained by fs PLA of GeO_2<Ge-NCs> heterolayers unprotected by cap layers (Fig. 24 d) confirms such a microstructure of the obtained material.

Another remarkable feature is that the ablation and laser damage of the multilayer system proceeding together with the formation of nanofoam in GeO_2<Ge-NCs> heterolayers (Fig. 17 and 23) are processes highly sensitive to even small variations in the fs laser fluence distributed over the surface of the multilayer. On the contrary, the shrinkage process of the same multilayer system (illustrated by Figs. 7 c and 10) under ns pulsed laser radiation, even at the highest energies, demonstrates high uniformity of the irradiated coating surface. In GeO_2<Ge-NCs>/Si_xN_yO bilayer films consisting predominantly of GeO_2<Ge-NCs>, no ablation or laser damage effects were observed in the examined ranges of heterolayer thicknesses and laser fluences.

The causes for this difference may be somehow related with the fact that the shrinkage of heterolayer thickness is a process that proceeds at a rate ten times slower in comparison with the nanofoam formation process. The volume of a foaming heterolayer normally increases by a few hundred per cent, whereas the maximum shrinkage of film thickness (in the case of ns laser treatment) does not exceed 40 per cent. It is possible that during the latter relatively slow process local temperature non-uniformities get smoothed over the film. In other words, temperature and stress gradients have enough time to disappear from the film. Another cause may be related to the fact that the mechanisms of laser radiation absorption by the material in the two cases are very different. In GeO_2<Ge-NCs> heterolayers, the ultraviolet radiation emitted by ns pulsed laser is readily absorbed not only in Ge NCs, but also in the glassy GeO_2 matrix. As a result, the matrix, absorbing the main part of the incident light, undergoes quick heating. Ge NCs absorb less energy of light pulses in the heterolayer as they occupy only about 30% of its volume. Microscopic regions that do not absorb laser radiation are absent from heterolayer, making the distribution of absorbed heat in the film volume less inhomogeneous. If the average temperature in heterolayer does not exceed the energy threshold for GeO(gas) formation due to reaction (4) at the interface between the GeO_2 matrix and Ge-nanoclusters, the heterolayer shrinkage process will mainly depend on the matrix viscosity. The temperature of the film defines the film material viscosity. Then, it can be anticipated that the shrinkage process in the film will most likely proceed similarly to the shrinkage during traditional anneals of ordinary CVD SiO_2 and SiO_3N_4 films (as described in Section 3.3). Due to this shrinkage mechanism, much of the internal lateral stresses turn out to be reduced by viscous flow of the material. The lower the viscosity the more readily the film shrinking process proceeds.

In the case of fs laser treatments of GeO_2<Ge-NCs> heterolayer, radiation will be absorbed only in Ge-nanoclusters. As a result, the distribution of the heat absorbed in the heterolayer should be much less homogeneous in comparison with the case of ns laser pulses. Temperature gradients in heterolayer, which are created by fs laser pulses between the cold glass and the heated Ge-nanoclusters can be expected quite high because all the energy of the light absorbed in the heterolayer is transferred into the heterolayer bulk only through the boundaries of Ge-NCs with the glass matrix. Thus, in this system only a small part of the glass directly bordering on heated Ge-nanoclusters can be quickly heated to high temperatures, while the rest part of the glass matrix will undergo heating at a much slower rate. Such a state of the heterogeneous material will be characterised by abrupt spatial variations of temperature and compressive stress over the volume of the heterolayer occurring over distances comparable with Ge-NC sizes. In particular, effects associated with thermal expansion coefficients of GeO_2 glass and Ge-NCs, as well as with the formation of high-pressure GeO(gas) bubbles around Ge-clusters, will facilitate smoothening of internal stress inhomogeneity. The impact action of fs laser pulses stimulates rapid evolution of the latter state of the material in heterolayer.

4.3. Prospects for use of GeO_2<Ge-NCs> heterolayers in medicine and laser lithography

High sensitivity to laser irradiation readily activating the evaporation reaction of GeO(solid) films and GeO_2<Ge-NCs> heterolayers proceeding with the formation of volatile germanium monoxide is an interesting prerequisite for using such films as a nanoresist in laser lithography. Therefore, some experiments were carried out to evaluate the potential offered by the examined films in this field. In our study, GeO(solid) films and GeO_2<Ge-NCs> heterolayers not covered with cap-layers were used. Such films and heterolayers were deposited onto pure Si substrates or onto SiO_2 layers grown by CVD on Si. Then, the films were subjected to fs Ti-Sapphire laser PLA in the same way like in previously described experiments with multilayer coatings, except for the laser beam was focused in ~1-μm diameter spot and the pulse energy falling onto the sample was decreased with the help of optical filters. The maximum laser-beam energy (E_{max}) was chosen to exceed the threshold energy ($E^{(1)}$(nfoam)) of the foam formation process in the heterolayer yet it was lower than the threshold energy $E^{(2)}(evap)$ of the evaporation reaction of the film GeO (Fig. 24 a). Such a beam was used to scan the surface of a heterolayer on a Si substrate covered with a SiO_2 film. The sample was mounted on an optical bench vibrating in the lateral plane and vertically within 1 - 1,5 μm. In the sample vertical vibration, the laser beam focus was on the film at one turning point 1 of the oscillation period, and the focus raised over the film surface during the reverse motion of the optical bench. In the first case, the beam had a minimal diameter on the sample ($2r_{o1} \sim \lambda$) and a maximal energy value (E'_{max}) which was higher than the evaporation threshold of the film $E^{(2)}$(evap) and its foaming threshold $E^{(1)}$(nfoam) (i.e. $E'_{max} > E^{(2)}(evap) > E^{(1)}(nfoam)$). The defocused beam in point 2 had the biggest diameter $2r_{o2} > 2r_{o1}$ and a minimal value of the energy maximum (i.e. $E''_{max} < E^{(1)}(nfoam)$ and $E^{(2)}(evap)$).

The results of such scanning of the used GeO_2<Ge-NCs> heterolayer ~ 60-80 nm thick with a vibrating beam spot are shown in Fig. 24 b and c. The track width periodically varies due to pulse energy modulation by the sample vertical vibration. No changes in film material properties were observed within the laser treated path at point 2 (Fig. 24 b) as the laser pulse energy on the layer surface was minimal here. The foam formation process is initiated in the heterolayer when the laser pulse energy on its surface grows in magnitude along the laser treated path due to light beam narrowing. The reaction zone eventually broadens up, and the film thickness at the beam centre increases by ~2 - 3 times. At the moment when the rate of the process is maximal, the heterolayer evaporation energy becomes exceeded (point 1 in Fig. 24 b) as the energy E'_{max} in the experiment was a little higher than $E^{(2)}(evap)$. Evaporation begins at the beam centre, analogously to the reaction of foam formation. The stretch of laser treated path over which this reaction proceeds is not long as the distance between the focusing lens and the sample changed due to vibrations simultaneously with the lateral shift of the beam over its trajectory. So, the beam on the sample surface again becomes defocused, and the nanofoam evaporation process ceases. In this case, the width of the cavities (~11-50 nm) formed by the beam in the foamed layer is quite appreciable (Fig. 24 b and c). This result attracts attention, as in geometric optics a spot having a diameter less than $\sim\lambda/2$ is considered to be the diffraction limit of beam focusing, and the traces of its effect for photoresist should be close to this size considered as the physical limit for laser lithography line width. Through-holes shaped as round dots or prolonged windows, whose transversal sizes were ten times larger than the expected limit were obtained in our experiments with laser radiation wavelength λ=800 nm.

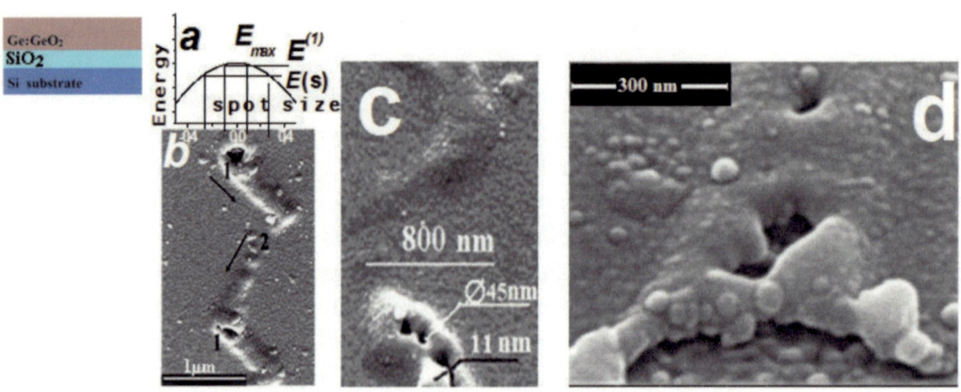

Figure 24. SEM images of the local formation of nanofoam in a GeO_2<Ge-NCs> heterolayer without a cap layer after fs laser irradiation (beam diameter ~ 1 µm;): **a** - radial Gaussian distribution of light energy in the laser beam; *E(s)* and $E^{(1)}(evap)$ are the threshold energies for foaming and evaporation of the heterolayer material, respectively; **b** and **c** - focused beam treated paths on the heterolayer surface during horizontal and vertical microvibrations of the optical bench; **d** – nanofoam.

The mechanism behind the laser beam effect on the GeO_2<Ge-NCs> heterolayer has not been adequately understood. Possible explanations of the sharp narrowing of the high-flux zone in the laser beam section on the film surface were considered, including intermode interference of light rays inside the laser beam and the effect of two-photon absorption of light by the film in areas with high degree of radiation power localization on the film surface.

Note that in this experiment no additional measures were taken to stabilize the irradiation conditions of GeO(solid) films and GeO_2<Ge-NCs> heterolayers used as a resist in laser lithography. We, however, believe that further considerable progress can be achieved along this line of research. However, in this way, even at the technological level available to us it was possible to fabricate some simple optical devices. In particular, the evaporation process of GeO_2<Ge-NCs> heterolayers by fs laser beam focused to a 1-μm diameter spot was used to fabricate prototypes of diffraction gratings (see Fig. 25). GeO_2<Ge-NCs> heterolayers 25 – 300 nm thick were scanned with laser beam at a speed 100 μm/s and frequency 1 kHz in the air. For the chosen heterolayer thickness, laser pulse energy was matched to slightly exceed the film evaporation energy threshold. The results showed that, under such conditions, it was possible to prepare diffraction gratings with line density up to 1000 lines per 1 mm.

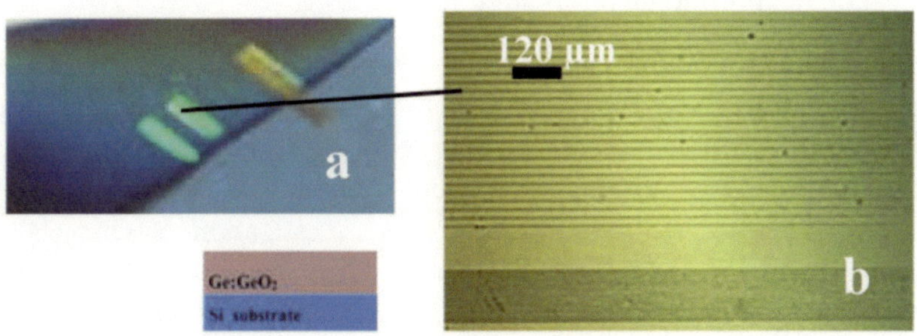

Figure 25. Diffraction of white light (**a**) on a set of lines in GeO_2<Ge-NCs> heterolayer on Si-substrate formed by a dry laser lithography (**b**) (optical microscopy photos).

As is seen, the main idea behind this method is that laser irradiation can be used to heat Ge-nanoparticles incorporated in a continuous medium. This idea differs little from the idea of using such Ge-nanoparticles in nanofoam formation from GeO_2<Ge-NCs> heterolayers. Therefore, we tried to employ Ge-nanoparticles in destructing malignant tumors. First, it is necessary to become able to prepare germanium particle based colloid solutions. It may be not so difficult a problem. A GeO_2 matrix with embedded Ge-nanoclusters is readily soluble in water or aqueous solutions of various active organic or inorganic chemicals. It is easy to find among the various chemicals those inert to germanium. Using such solutions, we can transfer Ge-nanoclusters from GeO_2<Ge-NCs> heterolayers into a suitable colloidal solution. This is one possible way towards solving the problem. Another strategy here is based on the ability of germanium to relatively slowly dissolve not only in liquid aggressive media, both acid and alkaline, but also in water. The solving proceeds in two stages: (*i*) oxidation of Ge-nanoparticle surface to GeO_2; (*ii*) dissolution of the oxide in a solvent. In some time, the colloidal germanium particles introduced in human organism will become completely dissolved to be removed from it in a natural way.

Another possible application of the results described above may be the use of Ge-NCs in medicine to suppress malevolent tumour formation. A method of tissue destruction in malignant tumours is widely known; it employs injection of colloidal solutions containing Ag-NCs with average diameter ~ 5-7 nm into tumor (Tyurnina et al., 2011). Ag-NCs spread over the tumour body and penetrate some part of its cells. Then the tumour is to be exposed to a laser beam which burns out the soft tumour tissues to a big depth due to heating of Ag-NCs.

5. Conclusions

The main aim of this paper was to describe the technological potential of pulsed laser treatments in modification of properties of film heterosystems involving germanium oxide layers with embedded Ge-nanoparticles. The goal of the study was the development of a general concept of physical-chemical mechanisms underlying such processes, i.e. suggestion 1 of a physical picture, or materials-science concept with analysis of reasons and outcomes, rather than the search for particular processes ensuring required modifications together with identification of appropriate technological conditions.

The general scientific aspect of the research involved two tasks. One of the tasks concerned fundamental problems in studying effects and processes connected with the input of large amounts of energy into small volumes of a solid, i.e. with the behavior of substance absorbing ultra-short laser pulses and subsequent relaxation of excitations in the solid body and surrounding medium. The second task was investigation of changes of structura chemical, physical and electrical properties of thin films under such treatments. New was the study of structural changes under pulsed laser treatments in multi-layer insulating coatings containing the films with a high concentration of QDs in the form of semiconductor nanoparticles. More specifically, we examined only one kind of such films, formed by germanium oxide layers (GeO and GeO_2) with embedded Ge-QDs.

A characteristic feature of both kinds of the effects due to PLA on studied multilayer coatings was that ultra-short laser pulses locally absorbed in small volumes of some cubic microns led to relaxation of this energy in volumes many times bigger than the sizes of the absorption zone while the duration of the dissipation process of absorbed energy lasted for a time that exceeded the laser pulse duration by many orders.

If it were not so, then we could not detect in the irradiated films indications of such slow and very large-scaled processes as films densification (by some ten percents), diffusion (up to 30% of the film material mass at distances of some hundred nm), rearrangement of the atomic network of the material from one kind into another, and ablation. In other words, all these modification processes of the material structure show that part of the material absorbed laser pulse energy from subsystems with superfast processes (subsystems of electromagnetic radiation, electron and phonon subsystems) spread into the subsystem of lattice processes. The latter subsystem has longest excitation relaxation times during chemical and structural rearrangement processes. Those long relaxation times are due to the fact that these rearrangements are collective, involving many atoms. Within such ideas, it can be speculated that, increasing the laser power, we will enter the conditions under which part of the energy absorbed by the solid will become sufficient for excitation of shock waves in the sample. These waves will begin to mechanically destroy the sample into fragments scattering around. This effect is called ablation.

An analysis of experimental data showed that, under PLA, layered films capable of good absorption of laser radiation play a big role in laser pulse energy accumulation. In the present case, those were GeO(solid) layers, GeO_2<Ge-NCs> heterolayers, and also the silicon substrate. All of the layers conveyed the absorbed energy either in heat or in structural chemical modification processes. The latter is clearly seen in Fig. 17 e and f, where the ablation process in the treated materials begins as the formation of ripples well before the thick SiO_2 layers covering the materials are exploded. In cases where thick GeO_2<Ge-NCs> heterolayers were absent from the multilayer coatings or when the silicon substrate absorbing radiation was replaced with transparent glass, increased threshold energies for all the processes activated in the multilayer system under PLA were observed.

From these standpoints, one can suggest an explanation to the opposite results obtained for two kinds of fs and ns PLA of GeO_2<Ge-NCs> heterostructures when, in one case, we had the formation process of super-loose nanofoam-type structure and, in the other case, a layer of a material having a very high density. In fs laser treated GeO_2<Ge-NCs> heterolayers, the laser pulses, 30 fs long, follow at 1-ms intervals, with the predominant part of their energy being absorbed by Ge-NCs. Suppose that this energy has enough time to completely dissipate for less than 1 msec. Then, strictly speaking, an act of energy absorption by this Ge-nanoparticle is no different from a similar act for the next pulse. This means that, in each of the pulses, identical portions of nanocluster-absorbed energy will produce identical amounts of gaseous monoxide at the boundary of the Ge-nanocluster with the surrounding GeO_2 glassy matrix. But to return the system to its strictly initial state till the coming of the next pulse, gas GeO formed around the Ge-nanoparticle should again transforms into initial solid components of Ge and glassy GeO_2. Otherwise, the absorption act of the second laser pulse by Ge-nanoparticle will not be strictly identical to the first one. It means that the whole nanofoam formation process will be confined to the process with a periodical alternation of two phases – one during which minimal setting nanofoam quantities form for less than 1 ms and the phase during which they disappear so quickly.

A different result was obtained in the experiment: nanofoam formation turned a stable and irreversible process. Logically, henceforth it follows that, in a series of fs laser pulses, each is absorbed by the GeO_2<Ge-NCs> heterolayer in the conditions different from the absorption of the previous ones. So, the effect of their impact should be considered not as a set of discreet and disconnected flashes, but as some continuous process proceeding during the duration of a whole pulse series. In fact, it shows that the complete time of energy relaxation by GeO_2<Ge-NCs> heterolayer absorbed from a separate pulse is longer than the intervals between them. Therefore, accumulation of some energy part of each pulse proceeds in the heterolayer. This accumulation integrates the whole series of discreet flashes in a continuous action that consists of three stages – the initial stage, the stage of stationary mode and the final stage that proceeds upon the completion of radiation. During PLA of GeO_2<Ge-NCs> heterolayer with ns laser, the time spans between laser pulses are so big that one cannot say about the thing that their effect is connected with part of their energy accumulation by GeO_2<Ge-NCs> heterolayer from pulse to pulse. Besides, GeO_2 matrix considerably absorbs ns laser pulse energy in the

GeO_2<Ge-NCs> heterolayer. Accordingly, the part of the pulse energy absorbed by Ge-nanoparticles considerably decreases and, along with it, their heating rate also decreases. If the speed of Ge-NCs heating in the heterolayer is lower than that of GeO_2 matrix, then the matrix will be cool down close to the border between these two heterolayer components, as it spends part of its heat for heating colder Ge-NCs. In the rest part of the glassy GeO_2 matrix, with the growing temperature, the decay processes of atomic net remaining parts of metastable GeO(solid) and its shrinkage by lowering atomic GeO_2 net defectiveness proceed. The latter of these processes is usually accompanied by a viscosity increase in glasses and that is to cause threshold energy growth of a beginning of foam formation in the GeO_2<Ge-NCs> heterolayer. Thus, the totality of all the described factors impairs, in this case, GeO_2<Ge-NCs> heterolayers transformation into a nanofoam-like matter under their ns laser pulse radiation under the conditions used in our experiments.

At the end of this research work, we emphasise one more time that it is difficult so far to find out of a multitude of film coatings and materials this or that way involved in microelectronics such their kinds that would characterised by so high capability to properties and structure modification and absolutely all the properties as germanium oxide layers. Although, as we believe, silicon oxide-based films are the most chemically close to them. These materials also have the capability to different modification forms and radical changes of many physicochemical properties. Particularly, in our viewpoint, SiO_2 layers transformation into nanofoam-like material may be one of the most interesting out of modifications similar to those of germanium oxides. It is not excluded that it will be realised by the way analogous to the used one for nanofoam formation from glassy GeO_2.

The authors hope that the possibilities demonstrated in our paper for germanium oxide layers modification assisted with laser treatments will attract attention of material scientists in the field of film coatings used in nano- and optoelectronics and also of researchers engaged in applied trends of nanotechnology.

Author details

Evgenii Gorokhov, Kseniya Astankova, Alexander Komonov and Arseniy Kuznetsov
Institute of Semiconductor Physics of SB RAS, Russia
Laser Zentrum Hannover, Germany

Acknowledgement

This investigation was supported by the Russian Fundation for Basic Research (projects Nos. 07-08-00438 and 10-07-00537). E.B. Gorokhov is grateful to Universite' de Nancy (France) and Laser Zentrum Hannover (Germany) for a visit grant. Also, the authors thank their colleagues for assistance in work: Dr. D.V. Marin, Dr. T.A. Gavrilova, Dr. V.A. Volodin, M. Slabuka, Prof. Boris Chichkov from Laser Zentrum Hannover (Germany), Prof. M. Vergnat from Institute Jean Lamaur and l'Universite' de Lorraine for their help in studying GeO(solid) films and Prof. A.V. Latyshev for financial support.

6. References

Ahmanov, S.A.; Yemel'yanov, V.I.; Koroteev, N.I. Seminogov, V.N. (1985). The Impact of High-Power Laser Irradiation on the Surface of Semiconductors and Metals: Nonlinear Optical Effects and Nonlinear Optical Diagnostics. *Advances in Physical Sciences,* Vol. 147, No. 4, pp. 675-745

Appen, A.A. (1974). Chemistry of Glass. 2nd Edition. *Publ.: Khimiya, Leningrad (In Russian).*

Ardyanian, M.; Rinnert, H.; Devaux, X. & Vergnat, M. (2006). Structure and Photoluminescence Properties of Evaporated GeO_x Thin Films. *Appl. Phys. Lett.,* Vol.89, pp. 011902-1-3

Ashitkov, S.I.; Ovchinnikov, A.V. & Agranat, M.B. (2004). Recombination of an Electron-hole Plasma in Silicon Under the Action of Femtosecond Laser Pulses. *JETP Letters,* Vol.79, No 11, pp. 529-531

Bok, J. (1981). Effect of electron-hole pairs on the melting of silicon. *Phys. Lett.,* Vol. 84 A., No. 8., pp. 448-450

Chong, T.C.; Hong, M.H. & Shi, L.P. (2010) Laser Precision Engineering: From Microfabrication to Nanoprocessing. *Laser&Photon. Rev.,* Vol.4, No 1, pp. 123-143

Dvurechenskiy, A.V.; Kachurin, G.A.; Nidaev, N.V.; Smirnov, L.S. (1982). Pulsed Annealing of Semiconductor Materials. *Publ.: Nauka, Novosibirsk (In Russian).*

Ekimov, A.I. & Onushchenko, A.A. (1981). Quantum-Size Effect in Three-dimensional Microscopic Semiconductor Crystals. *JETP Lett.,* Vol.34, No.6, pp. 345-348

Filipovich, V.N. (1978). The Theory of Self-Diffusion of Oxygen in Glassy SiO_2, GeO_2. *Glass Physics and Chemistry.* Vol.4, No.1, pp. 22-30

Gallas, B.; Kao, C.-C.; Fisson, S.; Vuye, G.; Rivory, J.; Bernard, Y. & Belouet, C. (2002). Laser Annealing of SiO_x Thin Films. *Appl. Surf. Science,*Vol. 185, pp. 317–320

Gorokhov, E.B.; Kosulina, I.G.; Pokrovskaya, S.V. & Neizvestny, I.G. (1987). Mechanical and Electrical Properties of the Double-layer Film System GeO_2-Si_3N_4 on Ge. *Phys. Stat. Sol. (a),* Vol.101, pp. 451-462

Gorokhov, E.B.; Noskov, A.G.; Sokolova, G.A.; Stenin, S.I.; Trukhanov, E.M (1982). Mechanical Stability of Pyrolytic Silicon Dioxide Films. *Physics, Chemistry and Mechanics of Surfaces (In Russian – Surface),* No.2, pp.25-33

Gorokhov, E.B.; Prinz, V.Y.; Noskov, A.G. & Gavrilova, T.A. (1998). A Novel Nanolithographic Concept Using Crack-assisted Patterning and Self Alignment Technology. *J. Electrochem. Soc.,* Vol.145, No.6, pp. 2120-2131

Gorokhov, E.B. (2005). *Evaporation and Crystallisation Processes in Germanium Oxide Films on Germanium.* PhD thesis, SB RAS, Novosibirsk, Russia

Gorokhov, E.B.; Volodin, V.A.; Kuznetsov A.I.; Chichkov B.N.; Astankova, K.N. & Azarov, I.A. (2011). Laser Treatment of the Heterolayers "GeO_2:Ge-QDs". *Proc. of SPIE,* Vol.7994, pp. 79940W-10

Hrubesh, L.W. & Poco, J.F. (1995). Thin Aerogel Films for Optical, Thermal, Acoustic and Electronic Applications. *Journal of Non-Crystalline Solids,* Vol. 188, pp. 46-53

Jambois, O.; Rinnert, H.; Devaux, X. & Vergnat M. (2006). Influence of the Annealing Treatments on the Luminescence Properties of SiO/SiO_2 Multilayers. *J. Appl. Phys.*, Vol.100, pp. 123504-6

Jolly, M. & Latimer, W.M. (1952). The Equilibrium Ge(s) + GeO_2(s) = 2GeO(gas). The Heat of Formation of Germanic Oxide. *J. Amer. Chem. Soc.*, Vol.74, No20, pp. 5757-5758

Juodkazis, S.; Mizeikis, V. & Misawa, H. (2009). Three-dimensional Microfabrication of Materials by Femtosecond Lasers for Photonics Applications. *J. Appl. Phys.*, Vol. 106. pp. 051101-1-14 Kachurin, G.A.; Pridachin, N.B.; Smirnov, L.S. (1975). Annealing of Radiation Defects by Pulsed Laser Irradiation. *Semiconductors,* Vol. 9, No. 7, pp. 1428-1429

Kamata, Y. (2008). High-k/Ge MOSFETs for Future Nanoelectronics. *Materials Today,* Vol.11, No.1-2, pp. 30-38, ISSN 1369-7021

Knoss, R.W. (2008). *Quantum Dots: Research, Technology and Applications,* Nova Science Publishers Inc., New York

Korchagina, T. T.; Gutakovsky, A. K.; Fedina, L. I.; Neklyudova, M. A. & Volodin, V. A. (2012). Crystallisation of Amorphous Si Nanoclusters in SiO_x Films Using Femtosecond Laser Pulse Annealings. *Journal of Nanoscience and Nanotechnology* (in print)

Korte, F.; Nolte, S.; Chichkov, B.N.; Bauer, T.; Kamlage, G.; Wagner, T.; Fallnich, C. & Welling H. (1999). Far-field and near-field Material Processing with Femtosecond Laser Pulses. *Appl. Phys. A.*, Vol. 69, pp. S7-S11

Marin, D.V.; Gorokhov, E.B.; Borisov, A.G. & Volodin, V.A. (2009). Ellipsometry of GeO_2 Films with Ge Nanoclusters: Influence of the Quantum-size Effect on Refractive Index. *Optics and Spectroscopy,* Vol.106, No3, pp. 436-440, ISSN 0030-400X

Marin, D.V.; Volodin, V.A.; Gorokhov, E.B.; Shcheglov, D.V.; Latyshev, A.V.; Vergnat, M.; Koch, J. & Chichkov, B.N. (2010). Modification of Germanium Nanoclusters in GeOx Films During Isochronous Furnace and Pulse Laser Annealing. *Technical Physics Letters,* Vol.36, No5, pp. 439-442

Martynenko, A.P.; Krikorov, V.S.; Strizhkov, B.V. & Marin, K.G. (1973). Physicochemical Properties of Silicon and Germanium Monoxide. *Inorganic materials,* Vol.9, No.9, pp. 1394-1399

Martsinovsky, G.A.; Shandybina, G.D.; Dement'eva, Yu.S.; Dyukin, R.V.; Zabotnov, S.V.; Golovan', L.A. & Kashkarov, P.K. (2009). Generation of Surface Electromagnetic Waves in Semiconductors Under the Action of Femtosecond Laser Pulses. *Semiconductors,* Vol.43, No10, pp. 1298-1304

Molinari, M.; Rinnert, H. & Vergnat, M. (2003). Visible Photoluminescence in Amorphous SiOx Thin Films Prepared by Silicon Evaporation Under a Molecular Oxygen Atmosphere. *Appl. Phys. Lett.*, Vol.82, No.22, pp. 3877-3879

Mueller, R.L. (1960). Chemical Characteristics of Polymer Glass-Forming Substances and the Nature of Glass Formation. Book: Proceedings of the Third All-Union Conference on the Glassy State. *Publ.: The USSR Academy of Sciences, Moscow and Leningrad.* Pp.61-71

Mueller, R.L. (1955). A Valence Theory of Viscosity and Fluidity for High-Melting Glass-Forming Materials in the Critical Temperature Range. *Journal of Applied Chemistry (Russian Journal of Applied Chemistry).* Vol.28, No.10, pp. 1077-1087

Nelin, G. & Nilsson, G. (1972). Phonon Density of States in Germanium at 80 K Measured by Neutron Spectrometry. *Phys. Rev. B*, Vol.5, pp. 3151-3160

Nemilov, S.V. (1978). The Nature of the Viscous Flow of Glasses with Frozen Structure and Some of the Consequences of Valence-Configuration Theory of Viscous Flow. *Glass Physics and Chemistry*. Vol.4, No.6, pp. 662-674

Nemilov, S.V. (1978). The Valence-Configuration Theory of Viscous Flow of Supercooled Glass-Forming Liquids and its Experimental Validation. *Glass Physics and Chemistry*. Vol.4, No.2, pp. 129-148

Ogden, J.S. & Ricks, M.J. (1970). Matrix Isolation Studies of Group IV Oxides. II. Infrared Spectra and Structures of GeO, Ge_2O_2, Ge_3O_3, and Ge_4O_4. *J. Chem. Phys.*, Vol.52, No1, pp. 352-357

Pliskin, W.A.; Lehman, H.S. (1965). Structural Evaluation of Silicon Oxide Films. *J. Electrochem. Soc.*, Vol.122, pp.1013-1019

Rinnert, H.; Vergnat, M.; Burneau, A. (2001). Evidence of Light-Emitting Amorphous Silicon Clusters Confined in a Silicon Oxide Matrix. *J. Appl. Phys.* Vol. 89, No. 1, pp. 237-243

Rochet, F.; Dufour, G.; Roulet, H.; Pelloie, B.; Perrier, J.; Fogarassy, E.; Slaoui, A. & Froment M. (1988). Modification of SiO Through Room-temperature Plasma Treatments, Rapid Thermal Annealing, and Laser Irradiation in a Nonoxidizing Atmosphere. *Phys. Rev. B.*, Vol.37, No 11, pp. 6468-6477

Salihoglu, O.; Kürüm, U.; Yaglioglu, H.G., Elmali, A.; Aydinli, A. (2011). Femtosecond laser crystallization of amorphous Ge. *J. Appl. Rhys.* Vol.109, pp. 123108-1

Sameshima, T. & Usui, S. (1991). Pulsed Laser-induced Amorphization of Silicon Films. *J. Appl. Phys.*, Vol. 70. No. 3, pp. 1281-1289

Sanditov, D.S. (1976). On the Mechanism of Viscous Flow of Glasses. *Glass Physics and Chemistry*. Vol.2, No.6, pp. 515-518

Sheglov, D.V.; Gorokhov, E.B.; Volodin, V.A.; Astankova, K.N. & Latyshev, A.V. (2008). A Novel Tip-induced Local Electrical Decomposition Method for Thin GeO Films Nanostructuring. *Nanotechnology*, Vol.19, pp. 245302-1-4

Takeoka, S.; Fujii, M.; Hayashi, S. & Yamamoto, K. (1998). Size-dependent Near-infrared Photoluminescence From Ge Nanocrystals Embedded in SiO_2 Matrices. *Phys. Rev. B.*, Vol.58, pp. 7921-7925

Tananaev, I. V.; Shpirt, M. Ya. (1967). The Chemistry of Germanium. Khimiya, Moscow (in Russian).

Tyurnina, A.E.; Shur, V.Ya.; Kuznetsov, D.K.; Mingaliev, E.A. & Kozin, R.V. (2011). Synthesis of Silver Nanoparticles by Laser Ablation in Liquid, *Proceedings of 19th International Symposium "Nanostructures: Physics and Technology"*, pp. 111-112, ISBN 978-5-93634-042-0, Ekaterinburg, Russia, June 20-25, 2011

Volodin, V.A.; Efremov, M.D; Gritsenko, V.A. & Kochubei S. A. (1998). Raman Study of Silicon Nanocrystals Formed in SiN_x Films by Excimer Laser or Thermal Annealing. *J. Appl. Phys. Lett.*, Vol.73, pp. 1212–1214

Volodin, V.A.; Gorokhov, E.B.; Marin, D.V.; Cherkov, A.G.; Gutakovskii, A.K. & Efremov, M.D. (2005). Ge Nanoclusters in GeO_2: Synthesis and Optical Properties. *Solid State Phenomena*, Vol.108-109, pp. 83-90

Volodin, V.A.; Korchagina, T.T.; Kamaev, G.N.; Antonenko, A.Kh.; Koch, J.; Chichkov, B.N. (2010). Femtosecond and Nanosecond Laser Assistant Formation of Si Nanoclusters in Silicon-Rich Nitride Films. *Proceedings of International Conference "Micro- and Nanoelectronics"*, pp. 75210X1-(X8), SPIE, Vol. 7521, Zvenigorod, Moscow region, Russia, October 5-9, 2009.

Zabotnov, S.V.; Golovan', L.A.; Ostapenko, I.A.; Ryabchikov, Yu.V.; Chervyakov, A.V.; Timoshenko, V.Yu.; Kashkarov, P.K. & Yakovlev, V.V. (2006). Femtosecond Nanostructuring of Silicon Surfaces. *JETP Letters,* Vol. 83, No. 2, pp. 69-71

Zakis, Yu.R. (1981). Applicability of Ideas about *Quasi-Particles and Defects* to *Glass*. *Glass Physics and Chemistry.* Vol.7, No.4, pp. 385-390

Chapter 14

Jet Engine Based Mobile Gas Dynamic CO_2 Laser for Water Surface Cleaning

V. V. Apollonov

Additional information is available at the end of the chapter

http://dx.doi.org/10.5772/39264

1. Introduction

What is the best to start our paper with? "Deep water Horizon" (BP's operation) case in the Mexican gulf is the best example of man made natural disasters. Today, the notion that "offshore drilling is safe" seems absurd. The Gulf spill harks back to drilling disasters from few decades past — including one off the coast of Santa Barbara, Calif. in 1969 that dumped three million gallons into coastal waters and led to the current moratorium. The "Deep water Horizon" disaster is a classic "low probability, high impact event" — the kind we've seen more than our share of recently, including space shuttle disasters, 9/11, Hurricane Katrina and earth quake in Japan. And if there's a single lesson from those disparate catastrophes, it's that pre-disaster assumptions tend to be dramatically off-base, and the worst-case scenarios downplayed or ignored. The Gulf spill is no exception. Fire boats battle the fire on the oil rig "Deep water Horizon" after the April 21 terrible explosion.

The post-mortem is only the beginning, so the precise causes of the initial explosion on the drilling platform and the failure of a "blowout preventer" to deploy on the sea floor probably won't be established for weeks or months. But the outlines of serious systemic problems have already emerged, indicating just how illusory the notion of risk-free drilling really was, while pointing to some possible areas for reform. A "blowout" on an oil rig occurs when some combination of pressurized natural gas, oil, mud, and water escapes from a well, shoots up the drill pipe to the surface, expands and ignites. Wells are equipped with structures called blowout preventers that sit on the wellhead and are supposed to shut off that flow and tamp the well. "Deep water Horizon"'s blowout preventer failed. Two switches — one manual and an automatic backup — failed to start it. When such catastrophic mechanical failures happen, they're almost always traced to flaws in the broader system: the workers on the platform, the corporate hierarchies they work for, and the government bureaucracies that oversee what they do. According to the study 600 major equipment failures in offshore drilling structures 80 percent were due to "human and organizational factors", and 50 percent of those due to flaws in the engineering design of equipment or processes. With near-shore and shallow reserves of fossil fuels largely depleted, drilling has moved farther off shore, into deeper waters and deeper underground. The technology for locating oil and gas reserves and for drilling has improved, but the conditions are extreme and the challenges more formidable. This is a pretty frigging complex system. You've got equipment and steel strung out over a long piece of geography starting at surface and terminating at 18,000 feet below the sea floor. So it has many potential weak points. Just as Katrina's storm surge found weaknesses in those piles of dirt — the levees — gas likes to find weakness in anything we connect to that source. It must be questioned, whether energy companies and government agencies have fully adapted to the new realities. The danger has escalated exponentially. We've pushed it to the bloody edge in this very, very unforgiving environment, and we don't have a lot of experience. Disaster has several possible insights for the oil spill: one was that BP and other corporations sometimes marginalize their health, safety, and environmental departments. BP and other companies tend to measure safety and environmental compliance on a day-to-day, checklist basis, to the point of basing executive bonuses on those metrics. But even if worker accident rates fall to zero, that may reveal nothing about the risk of a major disaster. These things we are talking about are risks that won't show up this year, next year — it may be 10 years down the road before you see one of these big blowouts or refinery accidents.

That assumption — that catastrophic risks were so unlikely they were unworthy of serious attention — appears to have driven a lot of the government decision-making on drilling as well. One, published in 2007, estimated the "most likely size" of an offshore spill at 4,6 K barrels. Current, conservative estimate of the Gulf spill put its total at more than 80 K barrels, increasing at a rate of 5 K per day. Why we have paid in our paper such a detailed attention to the policy of BP`s "Deep water Horizon"? The contamination of large water areas (oceans, seas, lakes, and rivers) with petroleum products as a result of accidents and not so smart industrial activity of similar companies and officials around them is one of major problems of protection of environment. Any other contaminant can not be compared with petroleum on a basis of universal utilization, number of contamination sources and degree of effect on all components of environment. Through penetration of petroleum products into water, there are deep, frequently irreversible changes of its chemical, physical and microbiological and even global (redirection of the ocean streams) properties.

The following methods of disposal of petroleum contamination of water: natural, mechanical, physical-chemical, chemical, biochemicals are widely used nowadays.

Natural method demands to leave the oil alone so that it breaks down by natural means. If there is no possibility of the oil polluting coastal regions or marine industries, the best method is to leave it to disperse by natural means. A combination of wind, sun, current, and wave action will rapidly disperse and evaporate most oils. Light oils will disperse more quickly than heavy oils.

The mechanical methods are such: collecting of petroleum from a surface manually or with different installations. All these methods are effective during limited time (from several hours to several days) only – time period during which the thickness of a petroleum film is great enough. Contain the spill with booms and collect it from the water surface using skimmer equipment. Spilt oil floats on water and initially forms a slick that is a few millimeters thick. There are various types of booms that can be used either to surround and isolate a slick, or to block the passage of a slick to vulnerable areas such as the intake of a desalination plant or fish-farm pens or other sensitive locations. Boom types vary from inflatable neoprene tubes to solid, but buoyant material. Most rise up about a meter above the water line. Some are designed to sit flush on tidal flats while others are applicable to deeper water and have skirts which hang down about a meter below the waterline. Skimmers float across the top of the slick contained within the boom and suck or scoop the oil into storage tanks on nearby vessels or on the shore. However, booms and skimmers are less effective when deployed in high winds and high seas.

Use dispersants to break up the oil and speed its natural biodegradation. Dispersants act by reducing the surface tension that stops oil and water from mixing. Small droplets of oil are then formed, which helps promote rapid dilution of the oil by water movements. The formation of droplets also increases the oil surface area, thus increasing the exposure to natural evaporation and bacterial action. Dispersants are most effective when used within an hour or two of the initial spill. However, they are not appropriate for all oils and all locations. Successful dispersion of oil through the water column can affect marine organisms like deep-water corals and sea grass. It can also cause oil to be temporarily accumulated by sub-tidal seafood. Decisions on whether or not to use dispersants to combat an oil spill must be made in each individual case. The decision will take into account the time since the spill, the weather conditions, the particular environment involved, and the type of oil that has been spilt and many other parameters involved into that hard consideration.

Introduce biological agents to the spill to hasten biodegradation. Most of the components of oil washed up along a shoreline can be broken down by bacteria and other microorganisms into harmless substances such as fatty acids and carbon dioxide. This action is called biodegradation. The natural process can be speeded up by the addition of fertilizing nutrients like nitrogen and phosphorous, which stimulate growth of the microorganisms concerned. However the effectiveness of this technique depends on factors such as whether the ground treated has sand or pebbles and whether the fertilizer is water soluble or applied in pellet or liquid form.

Besides of that the biological activity of microorganisms strongly depends on the temperature of water. As for physical-chemical methods it is necessary to mention, first of all, application of various adsorbing materials (polyurethane foam, coal dust, sawdust etc.), however all these methods are labor-consuming and low efficient. Besides they require secondary reprocessing of adsorbents.

Chemical method is the method of petroleum removal with the help of chemical substances. The basic disadvantages of this method are high price and fact, that detergents frequently are more toxic for water microorganisms, than petroleum.

The laser method of cleaning of water surface from a thin petroleum film is one of physical-chemical methods. Some time later after penetration into water, the petroleum spreads on a surface of a water and forms very thin film (thickness several microns). This film cannot be forced to burn, since because of a good thermal contact to a surface of water the film cannot be heated to temperature ensuring steady combustion.

The principle idea of a laser method consists in following. The laser beam passes through a petroleum film and then is absorbed in thin layer of water. The water heats very fast up to the boiling temperature, and the forming vapor destroys a film, bursting it in small-sized fragments, which, mixing up with the hot vapor, are decomposed quickly with formation of simple un-toxic substances.

The main advantages of a laser method consist of the following.

a. This method is "fast-response" one, since does not require any special preparation; in emergency situations the time from the moment of obtaining of the alarm signal to the beginning the laser installation operation is determined only by time necessary for arrival of the ship or the helicopter with the installation on board in given area.
b. The method is contactless, i.e. does not require realization of preparatory or other activities in the oil spillage.

For realization of the proposed technology, it is possible to utilize different types of lasers, as continuous (power up to 250 kW), and high repetition rate pulse-periodical (average power of the same level with duration of pulses 100 ns and repetition rate up to 100kHz).

Theoretical estimations and the experiments have shown, that as a result of a laser beam action on a surface of a water and land, covered with film of hydrocarbon contamination, following effects can be observed:

Evaporation and burning, and in a continuous mode the consumption of energy for 1 gram of vaporized liquid on the order of value surpasses energy necessary for heating up to boiling temperature and evaporation of a film, that is explained by a heat consumption for heating of water. In a pulse mode, the consumption of energy it is per unit of vaporized mass of a liquid (film) approximately in 5 times less, thus the process of film ignition start more easily.

Knocking out the particles of polluting substance above a surface of water under action of pulse or powerful scanning laser radiation should be considered [1]. The physical mechanism of this effect is explained by sharp evaporation (boiling-up) of a thin layer of water under a layer of polluting substance. This process takes place where as hydrocarbon polluting thin film has an absorption coefficient less, than water. Knocked out particles of petroleum at a height up to 50 cm is possible. For this case, the energy consumption per unit mass of a raised liquid is sufficiently less than energy necessary for evaporation and ignition (mode number one) or for sucking up and saving for future efficient usage (mode number two).

The character of a task to be solved superimposes certain conditions on operating characteristics of the laser device concerning both parameters of laser source, and concerning auxiliary systems.

Technological lasers of 10 - 15 kW power range, that widely are used in production, will have the output completely insufficient for liquidation of large scale contamination presenting the greatest danger according to the above-stated estimations. Such lasers can be utilized, at the best, for improvement of a process engineering of cleaning under modeling conditions.

Besides, the character of this task dictates impossibility or extreme undesirability of use of the stationary civil engineering service line (electric power network, water pipe, main gas line etc.), in view of the requirement of a high self-sufficiency and mobility. The power unit should not limit mobility and thus provide totally energy needed of all installation, the necessary reserve of expendables and fuel on board a complex. A capability of a fast redeployment from one type of a vehicle on another is also desirable.

In the present technical paper, the basic design concept of the mobile laser installation on basis of gas dynamic CO_2-laser is developed. The activities were executed in accordance with the working plan of OOO "Energomashtechnika"

In the first paragraph, the possibility of using of various types of lasers for solution of required tasks is considered and the selection of GDL is justified.

The second paragraph is devoted to selection and substantiation of basic performances of the laser installation and principal schematics of GDL.

In the third paragraph, the possible schemes of organization of air supply in the laser installation are considered. The selection of jet engines for high quality power unit of GDL installation has been provided. The necessary volume of the selected jet engine adaptation/modification experimental works is presented.

The fourth paragraph of the paper is devoted to the description of the design concept of GDL, and also laser installation as a whole. The pneumohydraulic schematics are described, the structure of the equipment, fuel tanks and operation control units are determined. The description of jet engine based CO2 GDL and detailed analysis of the laser installation components is also presented.

2. The substantiation of selection of laser type according to specific parameters, operation autonomy, and mobility

There are few the most effective and scalable modern high power continuous/pulse-periodical operated gas lasers should be considered for realization of clamed in the title of that paper tasks: electro discharge laser (EDL), gas dynamic laser (GDL), chemical laser

HF/DF (CL) and chemical oxygen - iodine laser (COIL).We are not going here to details of high power high repetition rate P-P laser systems operation.Main results of oil films elimination detailed consideration taken in the past can be summarized and reduced to the paper format and presented here as following:

2.1. Analysis of applicability of various types of lasers to the task

The laser with output power up to 250 kW of continuous operation (CW) or high repetition rate pulse-periodical operation (P-P) during few hours (minimum requirement) is required for realization of the mentioned above task. Besides of that, according to the physics of the process of destruction of a petroleum film, explained in Introduction, laser radiation should be weakly absorbed by petroleum film and should be absorbed effectively by water. In the table 1, the computational data for depth of radiation penetration for 4 types of mentioned above lasers in petroleum and in water are presented.

The laser type	Depth of penetration, micron	
	Petroleum	Water
GDL, EDL	100 - 260	10
CL	26 - 200	0,8 - 40
COIL	50	2000

Table 1. Laser radiation penetration depth for petroleum and water.

The range of values of depth of penetration for petroleum is the sequence of the fact, that absorption coefficient of various grades of petroleum and water (for the CL case) in relation to wavelength in CL radiation spectral band is considered.

From the Table 1, it is obvious, that the COIL radiation is rather weakly absorbed by water and is rather strongly absorbed by petroleum. However to heat a petroleum film by thickness ~100 microns through the absorption mechanism of COIL radiation practically is impossible because of strong heat transfer from a film into water through the heat conductivity. The conclusion is: the use of COIL to solve this task is impossible.

As the lengths of waves of radiation GDL and EDL are identical, the comparison of these two types of lasers is carried out on the basis of mass - dimension characteristics and other parameters that are listed in Table 2. The comparison is carried out for two real installations, which were developed and tested in the past to sufficient degree, precisely - GDL of rated power up to 250 kW, and EDL with 20kW output power.

The analysis of the data, presented in the Table 2 shows, that GDL has decisive advantages in comparison with EDL. It is necessary to be mentioned here, that EDL specific fuel consumption is 2 times less, than that of GDL. However, influence of this factor on the total complex weight will demonstrate a negative influence only for large operation time (5 hours or more). At the arrangement of a laser complex on the board of helicopter, the operating time will not exceed 1 hour, therefore in this case smaller EDL specific fuel consumption will not take decisive value.

#1	The name	GDL	EDL
1.	Power laser (max), kW	Up to 250	20
2.	Overall dimensions of the laser installation, m	4x2.4x2.4	2.5x3x3.5 + 2 x 0,8 x 0,8
3.	"Dry weight " (free of fuel), metric ton	5	18.6
4.	Specific, 1 kW of laser power dry weight, t/kW	Up to 0,02	0,93
5.	Specific, 1 kW of laser power, dry volume, cubic m/kW	0,1	1,5
6	Specific volume fuel consumption for 1 kW of laser power, cubical meter/ hour / kW	0,02	0,01

Table 2. The comparison of the parameters for GDL (250 kW) and EDL (20 kW).

Thus, GDL and CL are remaining the only competing systems. The comparative analysis of these two types of lasers with reference to the considered task is stated below.

2.1.1. Specific power of laser generation

GDL specific power, Wsp is about 20-35 J/g (C2H4+toluene); this value for CL is much higher and reaches 150 J/g. From this point of view, CL has big advantage in comparison with GDL. But taking into account few other parameters one can say it is not the final conclusion for this particular story. Very important parameters for our task to be solved should be considered as well: wavelength, scalability of the system, technical maturity of the technology, safety, life time of hardware and so on. Up to now GDL looks like the best system for the task under consideration - cleaning of water surface from petroleum films.

2.1.2. Wavelength of radiation

GDL radiation wavelength, λ = 10,6 microns and CL wavelength, λ= 2,7 - 4,5 microns. From this point of view, CL has one more potential advantage over GDL, as the diffraction limited angle of divergence of CL beam is 3 - 4 times less in comparison with GDL (at identical aperture). However this problem requires the further presize consideration.

a. Radiation extraction from the resonator.

In high-power lasers, extraction of laser beam from the resonator, where the pressure is much lower than atmospheric pressure, in external space through a rigid window is practically impossible for the reason of thermal destruction of such a window. For a extraction of such a beam out of installation, the aerodynamic window (AW) usually can be utilized. The supersonic gas stream cross-sectional to the beam direction fulfills the task of transparent boundary. As there are areas of various density gas sheaths limited by curvilinear surfaces in such flow, the beam transmission through this gas inuniformity introduces the distortions of a wave front resulting in increase of an angular divergence of a beam. These distortions do not depend on a wavelength, as the dispersion of an index of refraction of gas is extremely small and therefore these distortions have the same absolute value for GDL and CL. However relative values of these distortions of a wave front set (i.e. reduced to wavelength) are for CL 3-4 times higher, than for GDL, therefore from the point of view of an angular divergence of radiation the advantage CL in comparison with GDL can be not so essential.

b. External optical system.

The optical systems GDL and CL for delivery beam energy to remote objects include necessarily external (in relation to the resonator) optical elements and systems, intended for expansion of a beam with the purpose of reduction of its divergence. A principal component of a telescope is the main mirror, which diameter can reach the value up to several meters (depending on required power density on the target). The one piece glass or metal mirror will be too heavy. For this reason the main mirror is usually considered as consisting from large number of small mirrors (facets), assembled in one unit. Every facet has the own adjusting device. The accuracy of relative positioning these facets should be not worse $\lambda/4$, otherwise distortions of a wave front after reflection of a laser beam from this mirror will cause the essential increase of an angular divergence in comparison with it diffraction limit. This is $\lambda/4 \cong 2{,}5$ microns for GDL , $\lambda/4 \cong 0{,}75$ microns for CL. Therefore in practice required accuracy is much more difficult to ensure for CL.

This brief discussion shows, that the efficiency of CL external optical system can be much below, than that of GDL and, hence, reduces advantage of shorter wavelength of CL radiation.

Besides of that, for small delivery length of the beam energy down to the water surface, the angular divergence has no great importance, since density of power on a water surface appears sufficiently high without application of an additional external optical system.

2.1.3. Length of radiation generation zone

In GDL working section (the resonator region) the transfer of oscillation energy from N_2 to CO_2 has taken place. The characteristic time of this process is relatively large. Besides the relaxation time of oscillatory - exited molecules CO_2 at their collisions with other particles is also large. Therefore for full extraction of N_2 accumulated oscillatory energy, and transformation it in laser radiation, length of a working zone in flow down direction should be 10 - 15 cm. Such large length provides certain freedom of selection of the resonator scheme (one pass or multipass, symmetrical or asymmetrical, etc.). Thus the mirror operating surfaces are large enough, that results in a reduction of its thermal loading and distortion of a surface due to inhomogeneous heating.

In CL, the length of beam generation section is small, less then 2 cm. This circumstance causes difficulties in optimization of the laser resonator and leads to increase of thermal loading of mirrors, which results to significant thermal distortion of mirrors and distortion of a wave front, i.e. to significant increase of an angular divergence of laser radiation.

2.1.4. Direct exhaust of used gas to atmosphere

If GDL or CL are placed on ground or on the board of flying vehicle (plane or helicopter), there is a problem of exhaust of spent gas to atmosphere. In the case of GDL this problem can be solved rather simply by use of the diffuser, as the operating pressure in a zone of the resonator is rather high. For reaching a maximum degree of pressure recovery in the diffuser it should have a very special geometry.

In the CL case the operating pressure in the resonator is much lower in comparison with GDL, therefore the direct exhaust to atmosphere with the help of the diffuser is impossible, it is necessary to utilize additionally gas or water vapor ejector to ensure necessary pressure in output section of the diffuser. Such ejector requires large additional gas flow rate, that results in fundamental complexity of the laser installation design and significantly decrease specific power of generation, i.e. in this case one of main CL advantages is lost in comparison with GDL.

2.1.5. Toxic characteristics

The degree of danger CL from the point of view of toxic characteristics of working components and exhaust gases is much higher in comparison with GDL. In CL, for creation of a working mixture the extremely toxic substances containing fluorine should be utilized. Besides the exhaust CL contains a significant quantity of fluorine hydride HF, which is supertoxicant. In this connection with use CL in a system of laser cleaning of water areas it will be necessary to ensure special safety measures, that will cause the significant complication of the installation design and rise of its price.

In the case of GDL exhaust gas is ecologically clean ($N_2 + CO_2 + H_2O$). However, CO, carbon monoxide which is toxic gas can be used as a fuel for GDL. Hence, system of storage and supply of CO should ensure absence of its leakage. On the other hand, CO is lighter-than-air, therefore it floats in atmosphere and is fast blown away by wind. If as fuel in GDL will be used liquid hydrocarbon fuel (benzole, toluene), the system of its storage and supply does not require special measures to secure the ecological safety.

2.1.6. Absorption of laser radiation by atmosphere

The wavelengths of GDL and CL radiation are within of windows of atmosphere transparency. These windows are usually determined for vertical transmission of beam through atmosphere, and for this case power dissipation on track is ~20 % for GDL and 25 % for CL. For vertical tracks of length 50 - 150 m, which are of interest with reference to a considered problem, the radiation power dissipation can be neglected.

2.2. Substantiation of selection gas dynamic CO_2-laser

Proceeding from the comparison of various type lasers, explained in the previous paragraphs, it is possible to make a conclusion, that the most suitable laser for development of the installation for disposal of petroleum films is GDL:

2.2.1. Simplicity of design

From design point of view GDL components are the most simple in comparison with the considered lasers. The combustion chamber operates on usual components, for example, kerosene + air. The design of such chambers is practically very much mature, there is a wide experience of their operation in various technical devices. The nozzle unit of GDL is made from heat resisting steel and can work long time without forced cooling. Wide zone of the media population inversion behind the nozzle unit (10-15 cm) stipulates simplicity of selection of the scheme of the resonator for obtaining laser generation with maximum efficiency. Thus the characteristic size of a mirror of the resonator is approximately equal to length of a zone of generation, i.e. 10-15 cm, therefore manufacturing of these mirrors does not call serious technological difficulties.

The capability of GDL activity at high pressure of gas in the resonator in front of the nozzle unit (up to 3 MPa and more) provides with a straight line an exhaust of spent gas in atmosphere with the help of the supersonic diffuser of a special design. Such capability all remaining considered lasers are deprived. The aft ejector is necessary for maintenance of a direct exhaust of these lasers which requires additional gas flow rate, that complicates the installation and reduces specific energy of radiation.

It is necessary to note also, that GDL output power rather weak varies at change over a wide range of parameters of gas (pressure, temperature, chemistry) in the combustion chamber, i.e. GDL is not critical to accuracy of a task and maintenance of an operational mode. One more virtue GDL is the small time (some seconds), necessary for start.

Thus, GDL is the rather simple, reliable and flexible tool ensuring high output power of laser radiation.

2.2.2. Mobility

From all considered types of lasers GDL for today has, apparently, least "weight" and "volume" of 1 kW of laser power. It allows rather simple to place GDL on any vehicle, whether it will be airplane or helicopter, ship or railway platform, etc. Such GDL mobility allows with its help to solve many tasks, including connected with disposal of petroleum films.

2.2.3. Selection of propellant components

One of specific singularities of working process of the CO_2-laser is, that the working mixture of gases should contain in a receiver rather small quantity of water vapor (no more than 5-6 volume %). On the other hand, the operation temperature of gas in a receiver should be enough high (1300-1600 K). It superimposes limitations on selection of propellant components (fuel and oxidizer). From the point of view of maximum specific energy of radiation (J/kg) optimum components are gaseous at standard conditions with damp (CO) and liquid nitrous oxide (N_2O), thus additional nitrogen (or air), necessary for working process, moves in a receiver from a separate source, and the water vapor in quantity 1,5-2 % is formed by burning small quantity of hydrogen or alcohol in air.

2.2.4. Wavelength of radiation

The analysis of the data presented in tab. 1 show, that for the solution of a problem of disposal of petroleum films, the optimal wavelength of radiation is that of the CO_2-laser, as this radiation is weakly absorbed by the film and is strongly absorbed by water. From the point of view of a diffraction limited angular divergence of radiation, the CO_2-lasers (EDL and GDL) lose for CL. However with reference to a considered problem, the value of a divergence is not critically important. The estimations show, that at length of a beam from the laser source down to a surface of water ~ 50-100 m and power of radiation up to 250 kW, the precision focussing of laser beam is not required, for the reason of obtaining required density of power in a spot on water surface. It is possible to utilize unfocussed or partially focussed beam. Therefore with reference to a considered problem, the relatively large wavelength of GDL radiation is not the factor of insufficiency.

The explained above reasons allow to make a unequivocal conclusion for the selection GDL as the laser radiation generator for the mobile installation intended for disposal of a petroleum film on a surface of water. At the estimated power of radiation, GDL is preferable in weight and volume factors in comparison with EDL, HCL and COIL: it does not require the electric power, use to utilize a low toxic fuels, provides the direct exhaust of utilized gas to atmosphere. GDL is very simple in control, it is not critical to changes of working parameters in a sufficiently broad range, and it is convenient in operation. The important circumstance is that we had in our hands a very reliably and operable GDL with output power 100 kW, which in our days is a very effective tool for intensive research program to be carried out. In particular, the design of the aerodynamic window permitting to extract a laser beam from the resonator zone to atmosphere without application of transparent for the working wavelength materials has been effectively developed.

3. Selection and substantiation of basis performances of the laser installation and GDL principal diagram

In the previous chapter the comparison of various types of high power lasers was carried out from the point of view of their application for the solution of this task, where was shown, that the most appropriate type of the laser is CO_2 - GDL. However, there are different types of GDL therefore it is necessary to choose among them the optimal version, and also to develop the general concept of the power installation for optimal supply of working components in GDL.

In the present chapter the analysis of the various GDL schemes, their advantages and disadvantages are given from the point of view of application in the considered installation, the selection of propellant components for GDL is justified from the point of view of their power efficiency, production and toxic characteristics, and ecological safety.

Last part of the chapter is devoted to the substantiation of the GDL characteristics working on the chosen components. For this purpose the calculation results which have been carried out according to [2], are permitting to define specific power of laser radiation at the given initial parameters, such as fuel chemistry, temperature and pressure in front of the nozzle block, expansion ratio of the nozzle etc are used. The calculation results and computer analysis are shown.

For definition of the main dimensional characteristics of the laser and its flow organizing parts it is necessary to define the basic level of laser radiation power, which, in turn determines the total fuel consumption and remaining geometrical characteristics of the laser.

3.1. Selection of optimal radiation power of GDL

3.1.1. Theoretical substantiation of power level

The basic capability of disposal of petroleum contaminations from water surface by means of irradiation of a petroleum film by a high flux of laser radiation has been proved experimentally and presented in our publications. However, the published data give only the general representation about basic possibility of using of such method. It is obvious, that the combustion is complex multiparameter process and, if we talk about operational use of the mentioned principles in the particular technical device intended for practical use, the optimization of parameters of the installation is necessary. The important value in this case is the efficiency, which it is possible to understand as the area of cleaned water surface divided to mean energy, used on clearing, (m_2/kW). For optimization of parameters of the installation it is necessary to conduct a cycle of research works on study of evaporation, ignition and combustion processes of petroleum used for the above mentioned method. The experimental research of the given problem even without qualitative understanding of involving processes represents a very difficult task in connection with vast quantity of optimization parameters and requires heavy material costs and time. In this connection the preliminary development of theoretical model representing process of disposal of petroleum film is desirable.

The second phase [3] of the process is presented, the various interaction mechanisms of a laser beam with a petroleum film on a water surface are theoretically considered. The following mechanisms are included:

- laser heating active absorbing, translucent and transparent films;
- an explosive boiling-up and conditions of process efficiency;
- evaporation and vaporization.

The basic conclusions of the technological approach are the following.

1. For reaching temperatures of evaporation of the film the necessary time of radiation action is about 100 nanoseconds at laser power 100 kW and diameter of an irradiated zone is about 30 cm. Temperature of water layer under the surface of film can essentially exceed boiling point. Such situation arises due to superior velocities of laser heating (up to several millions degrees of Celsius per one second) and inertial character of heat transfer process. At such temperatures the water passes in a metastable state with active outgassing. Under a surface of a film there will be micro explosions of air - steam bubbles. As result, there will be a separation of a film from the surface of water, after it`s breaking the outflow of drops of petroleum take place to the open air.
2. The scanning velocity at power of the laser of 100 kW and diameter of a spot 30 cm ought to be in range of 10-12 m/s. Thus the velocity of cleaning of a surface 18000-24000 m^2/hour can be provided.

It is necessary to note, that all calculations are fulfilled for thickness of a petroleum film about 100 microns, which is rather large value and is observed only in the initial moments of formation of oil spillages. For much thinner films, the velocity of cleaning will be increased proportionally.

3.1.2. Experimental researches

For experimental confirmation of basic physical principles included in the basis of the Project, the participants carried out preliminary experimental research of interaction of laser radiation with films of various petroleum types on a surface of water. The research works were conducted within the framework of financing of activities.

During our experimental research works the GDL developed by our team, and prepared for this particular task to be solved had been used. The optical scheme of realization of experiments is presented in Fig. 1.

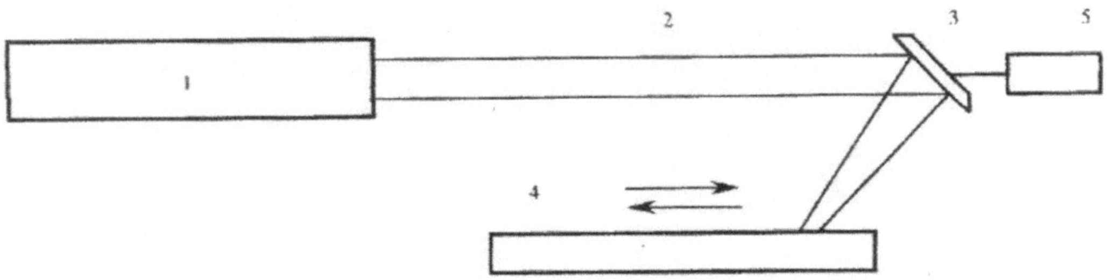

Figure 1. Optical scheme of oil film interaction with laser beam.

The GDL 1 generates radiation 2 with a wavelength 10,6 microns. The radiation is directed on a concave focusing mirror 3 and, being reflected, impacts on a surface of test heterogeneous structure in a cuveta 4. The mirror realizes periodic oscillations with the help of the mechanical device 5, thus the laser beam scans on a surface of a dish. Diameter of laser beam cross section on a surface of a dish changed from 15 cm to 30 cm. The laser radiation power was within the limits of 100 kW.

In Figs. 2, 3 and 4 the results of experimental research works are presented, on which it is possible to make following preliminary conclusions.

Figure 2. Intensive combustion of petroleum contamination under laser radiation.

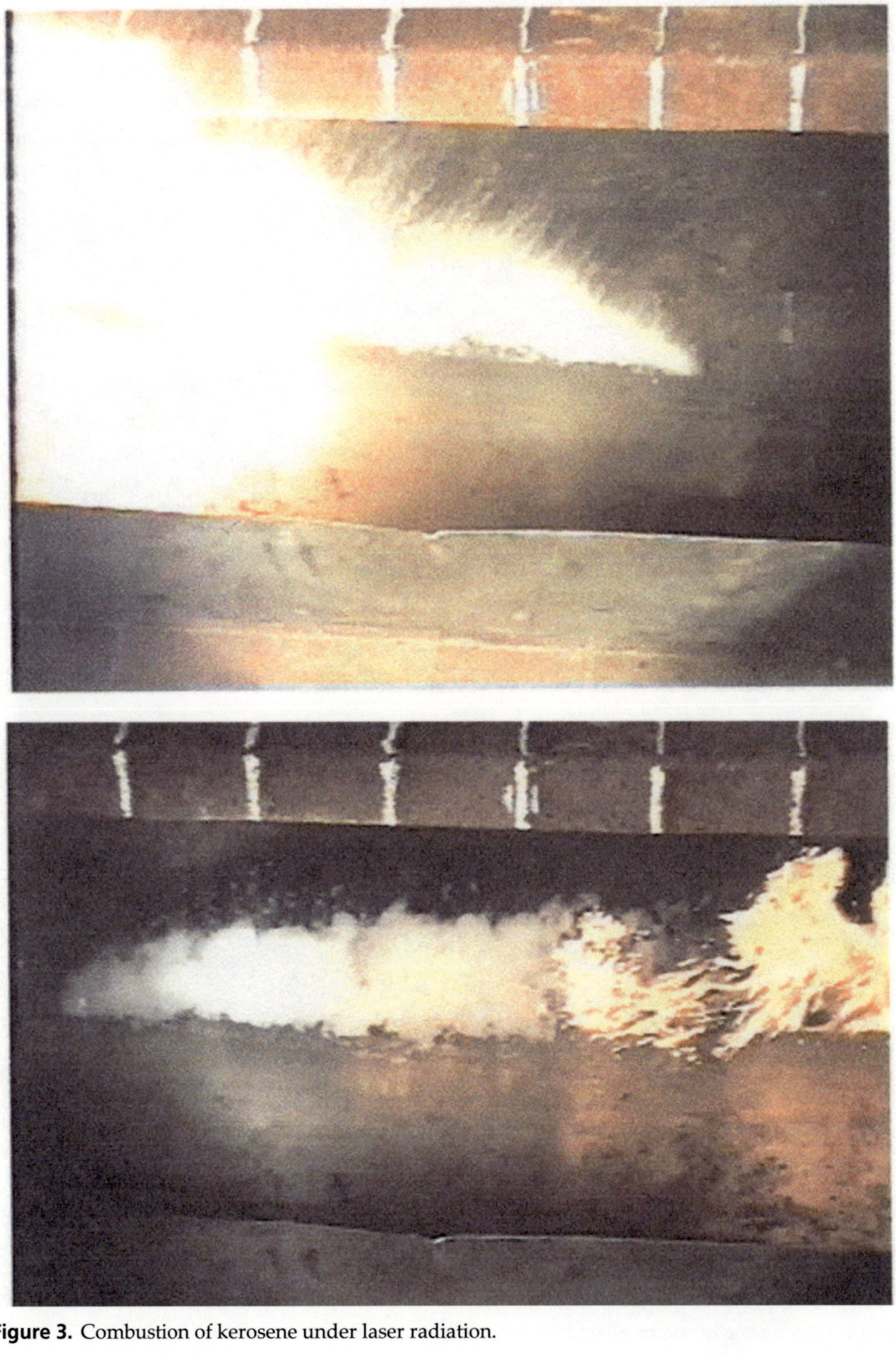

Figure 3. Combustion of kerosene under laser radiation.

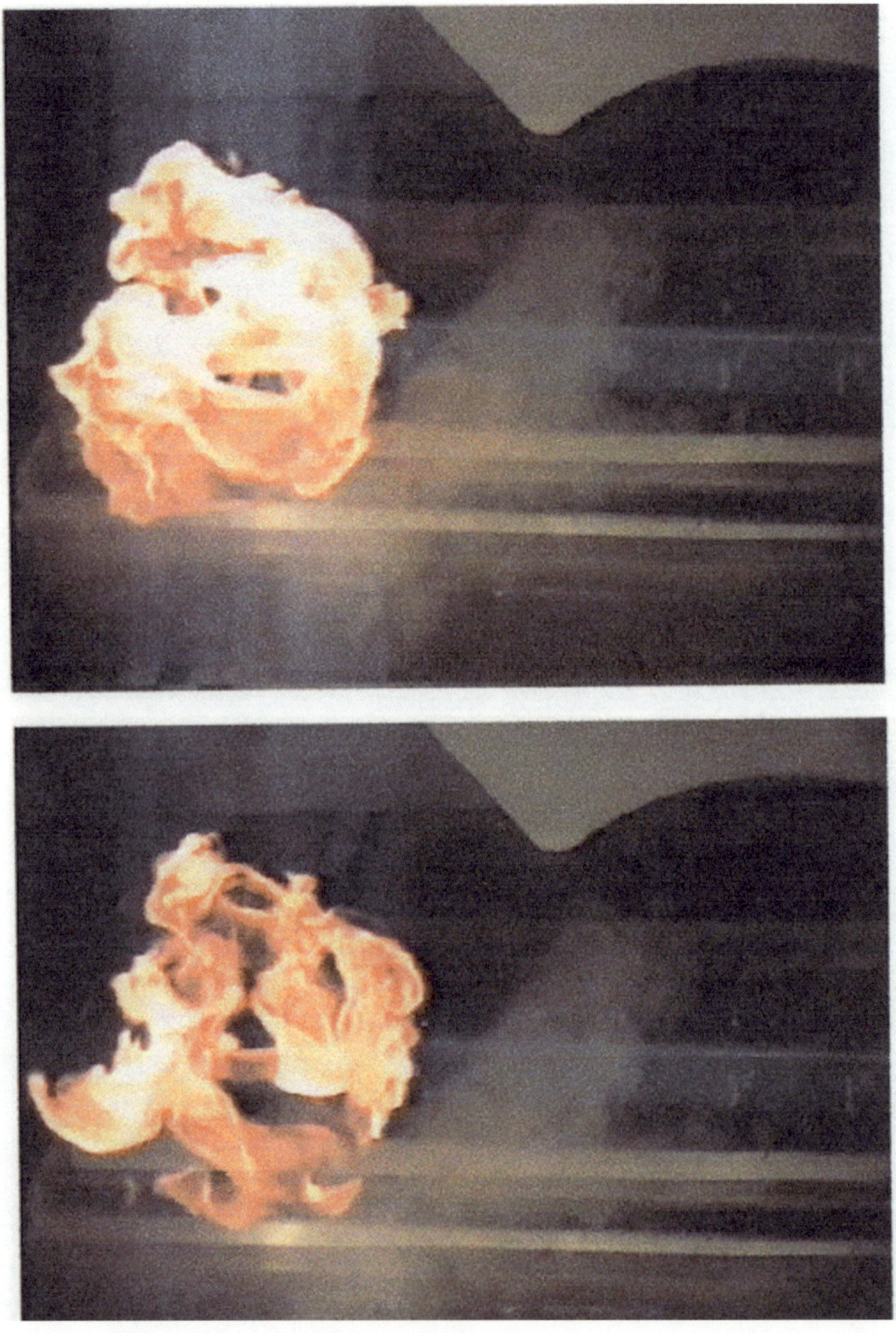

Figure 4. Combustion of engine oil under laser radiation.

1. The carried out experiments testify the possibility of using of the laser installations for cleaning of an aqueous medium from contaminations of petroleum and of it processing products.
2. For definition of optimal parameters of the laser installation and operation modes in practical conditions it is necessary to select experimental conditions of the working regime according to the special program approved by a potential customer.
3. For necessary reserve, the radiation power of GDL of 100 kW for the installation to be developed according to the present consideration is chosen.

3.2. Selection of the GDL scheme and its substantiation

3.2.1. Conventional scheme of GDL

The principle of GDL operation [4] is based on non-equilibrium fast expansion of the heated up working gas in the supersonic nozzle, at which there is a partial freezing population of oscillatory levels of molecules CO_2 and N_2 and formation of population inversion of upper and low laser levels. The heating of gas to an operation temperature 1300 –1700K at pressure 2-3 MPa is conducted in the combustion chamber through of combustion of chosen fuel – oxidizer components. The transformation of the accumulated oscillatory energy of gas to laser radiation has taken place in the optical resonator located down the nozzle. After the resonator the working gas passes through the supersonic diffuser and will be exhausted in atmosphere.

The most fast gas expansion take place near the critical zone of the nozzle at the length equal to the value of one order of magnitude of the critical cross-section size. Therefore for fast freezing of oscillatory energy it is necessary to use the nozzles with a small size of critical cross-section (< 1 mm). On the other hand, the pressure of gas after the nozzle (i.e. in the resonator region) should be small enough (< 100 mm Hg) to reduce to a minimum the losses of oscillatory energy due to collision relaxation. It has been solved by use of nozzles with large expansion ratio (area ratio of output to critical cross-sections $A / A^* \geq 20$, that corresponds to the value of a Mach number M 4).

For effective transformation of the accumulated oscillatory energy to laser radiation the amplification of a small signal during the pass the length of an active zone should be large enough to compensate optical losses (removal of radiation, absorption in mirrors, dissipation in medium) and to ensure a high enough radiation power in the resonator. In GDL of very low powers (tens W) is possible to use the so-called mononozzle, which represents the flat (rectangular) nozzle with critical cross-section as a narrow slot (Fig. 1,a). However for high-power GDL implementation of the mononozzle is practically impossible because of its large length (1 m and more); arising thermal deformations result in essential to nonuniform width of critical cross-section, h *, down to full blocking of critical cross-section in some zones. Besides the mononozzle does not allow to receive enough large transversal size of an active zone, that causes certain problems at resonator selection. Therefore in high power GDL the so-called nozzle unit representing set of a many relatively short (~100 mm) slot-type nozzles (Fig. 1.1, a) is used. These nozzles were formed by contoured nozzle blades, made from thermal resisting metal. Such design of the nozzle unit allows changing in a wide range width and height of an active zone. Besides the nozzle unit provides a uniform enough density field of gas in an active zone, that allows realizing a small angular divergence of GDL radiation.

In view of such advantages of nozzle blocks, this technology now is widely used for GDL, and therefore it is possible now to name such GDL systems as the "GDL of traditional scheme". The radiation power of such a GDL is in our days up to 250 kW and more for continuous and P-P radiation modes of operation [5, 6].

3.2.2. GDL with a set of axisymmetrical nozzles

The nozzle unit of the conventional scheme has the disadvantage that it is complex in manufacturing, and also is subject to influence of thermal deformations. There is an alternative approach to the development of nozzle units, which consists of the following. Instead of two-dimensional nozzles the axisymmetrical nozzles with small diameter of critical cross-section are used. The supersonic part of such nozzle can be made as a conical or contoured. In this case nozzle unit looks like a plate, where the numerous small-sized nozzles located close as possible to each other (Fig. 1.1, c) are made. The manufacturing of such a nozzles is possible, they are not subject of thermal deformations, by the appropriate arrangement of these nozzles on a holding plate (quantity of rows and quantity of nozzles in a row) it is possible to give output cross-section of the nozzle unit with required configuration.

However such nozzle unit has at least two disadvantages caused by presence of "empty" zones between nozzles, since complete filling of output cross-section by round nozzles basically is impossible.

The first disadvantage consists of the empty zones, which are the sources of shock waves in a supersonic flow. That wave results in density inhomogeneous structure of gas in a zone of the resonator and, hence, leads to increase of an angular divergence of laser radiation.

The second disadvantage consists of the effect, which creates in empty zones the viscous vortices flows, which results in reduction of total pressure of a main flow. In turn, the reduction of total pressure has negative consequences on the supersonic diffuser operation, which can be appreciated as reduction a pressure recovery coefficient. In this case for realization of direct exhaust of working gas to atmosphere it is necessary to raise combustion-chamber pressure, which results in the decrease of output power of laser radiation.

For these two reasons GDL with the nozzle unit as a set of axisymmetrical nozzles is not widely used at the noticeable production scale.

3.2.3. GDL with mixing of working gases in a supersonic flow

In the case of traditional GDL the simultaneous heating of all components of a working mixture - N_2, CO_2, H_2O has taken place in the combustion chamber. However initial reserve of laser energy contains, primarily in oscillatory - exited molecules N_2, which part in gas mixture makes ~ 90 %. The role CO_2 and H_2O consists only of transforming accumulated in N_2 oscillatory energy to laser radiation.

The pure nitrogen has extremely large time of an oscillatory relaxation stipulated by collisions of molecules N_2 among themselves. As the result, the oscillatory levels N_2 are easily frozen at fast expansion of the working mixture in the nozzle. However other components of a mixture (CO_2 and H_2O) considerably accelerate a collision relaxation N_2 and by this reduce a reserve of oscillatory energy.

For increase of a reserve of oscillatory energy it is desirable to raise temperature in the combustion chamber. However in the traditional GDL temperature is limited by value about 2300K; the dissociation CO_2 begins at higher temperatures, and the efficiency of conversion of the accumulated oscillatory energy in laser radiation decreases. In contrast to CO_2 the molecular dissociation N_2 begins at temperature ~ 4000 K.

The presented ideas have resulted in the new GDL scheme, which concept consists in following: the pure nitrogen, heated up in a prechamber, expands in the supersonic nozzle, and at the exit of the nozzle to it are mixed cold CO_2 and H_2O. Such scheme has three basic advantages for the usual scheme GDL: a) temperature of nitrogen in a prechamber can be essentially increased, b) more effective freezing of oscillatory energy of nitrogen is reached, c) energy loss for CO_2 and H_2O heating is not necessary.

In view of the indicated advantages mixing GDL can basically ensure the increase of specific power of radiation many times in comparison with the GDL conventional scheme. A main problem arising at realization mixing GDL - necessity to ensure fast intermixing hot N_2 with cold CO_2, but the disturbances of a supersonic flow N_2 should be kept at minimum.

Some solutions of this problem (Fig. 1.2) are possible. Technically it is much simply to organize mixing of CO_2 and N_2 in wakes (Fig. 1.2, a). In this case flow CO_2 moves in parallel to flow N_2, in this case two sets of supersonic nozzles - one set for N2, second for CO_2 are used. A advantage of such scheme of supply is that arising at interaction of wakes disturbances are minimum in comparison with other schemes of intermixing.

The basic disadvantage of this scheme is that the mixing of wakes is rather slowly, therefore for effective work of the laser the working section (from nozzle exit down to the diffuser inlet) should have sufficient length, which results in large relaxation losses and reduction of a pressure recovery coefficient of the diffuser. For these reasons the scheme GDL with intermixing of supersonic wakes, apparently, is not effective.

The theoretical analysis and the experimental data show, that the GDL schemes are more effective, in which the injection of cold CO_2 is made near to critical cross-section of the nozzle in it transonic part (Fig. 1.2,b,c). In experiments with heating N_2 in electric arc plasma generating device and supply of CO_2 under the scheme 1.2,c, was received record for GDL value of a gain equal 3 % cm^{-1}, that exceeds maximum value of a gain for usual GDL almost to an order of magnitude. However specific power of radiation was insignificant, that is connected to a small scale of the installation.

Despite of basic promising character of mixing GDL, there is a lot of problems connected with their practical realization.

One of the most important problems is obtaining hot nitrogen at temperature 3000-4000K. Application of electric arc plasma generating device for high-power mixing GDL is unreal, since at the consumption of nitrogen 10 kg\sec and more required electrical power exceeds 100 MW. It is possible to use for obtaining hot nitrogen special fuels, however all of them are strong toxicants.

As a source of hot nitrogen, the devices can also be considered, in which a combustion of metal in air has taken place. As a result of this strong exothermal reaction hot nitrogen and oxide of metal as particles are formed, which are extracted from the formed flow with the help of a cyclone separator. However, necessity of a very high degree of a nitrogen stream cleaning from particles makes this way problematic.

Other important technical problem is the development of the nozzle (selection of a material, development of a robust design, which would maintain without destruction large heat flows in the region of critical cross-section.

The supersonic diffuser causes one more important problem. For an effective work of the laser translation temperature of working gas in an active zone should be sufficiently low, whereas temperature in a prechamber should be relatively high. For the coordination of these two conditions it is necessary to use supersonic nozzle with large Mach numbers (M=5 and more). However is known, the higher Mach number is, the lower is pressure recovery coefficient for the diffuser. Therefore for a direct exhaust of gas in atmosphere it is necessary to develop special diffusers, thus can appear, that such task basically is impracticable without use of a fodder ejector. However consumption of inducing gas in some times exceeds the consumption of induced gas, and in this case specific laser power designed on summarized gas flow rate, considerably decreases.

From explained above it is possible to make a unequivocal conclusion: taking into account problems connected to development of the high power mixing laser, and also absence of the prototypes of such lasers, for the solution of a task of disposal of petroleum film on water surface it is necessary to choose GDL of conventional scheme.

3.3. Selection of fuel for mobile GDL

In the present part the analysis of possibility of using various fuels for support of operation of the mobile CO_2-GDL, intended for disposal of petroleum film on a surface of water is carried out.

CO_2-GDL provides the laser radiation generation in a continuous/P-P modes of operation with radiation wavelength 10.6 microns. A basic physical principle of operation gas dynamic of the CO_2-laser is the fast expansion in the supersonic nozzle of a mixture of gases (CO_2, N_2, H_2O) preheated up to temperature 1500K- 2000K. In a supersonic flow with Mach number M = 4 - 5 temperature of gases down to values 300K – 350K, necessary for population inversion of laser levels obtaining of a molecule CO_2. For nozzle operation at direct exhaust in atmosphere after pressure recovery in the diffuser (the static pressure in a zone of generation is equal 5 10^3-1 10^4 Pa) the nozzle inlet pressure (stagnation pressure of a flow), equal 2-3 MPa is necessary. At small transversal and longitudinal sizes of the supersonic nozzle (the height of critical cross-section is equal 0.3 - 0.5 mm, and length of the nozzle 20 - 40 mm) characteristic time of the gas flowing through the nozzle appears small and comparable with time of oscillating relaxation of molecules of a mixture of gases, i.e. oscillating - oscillating exchange (*V-V* process) and oscillating - translation relaxation (*V-T* process) has taken place in non-equilibrium mode. As a result of non-equilibrium processes in oscillating exited molecules at certain gas mixture ratio and flow parameters in the supersonic nozzle the inversion population of the upper laser level of molecules CO_2 (0001) is formed. The molecules CO_2, are light generating molecules (laser radiation is realized on transition 00^01 -10^00), N_2 molecules are donors molecules transmitting oscillating energy to the upper laser level of molecules CO_2 (00^01), the molecules H_2O are intended for population of the lower laser level of molecules CO_2, (10^00). The optimal ratio of mixture of gases for an effective work of gas dynamic CO_2-laser should contain CO_2 = 0.1, N_2 = 0.89, H_2O = 0.01 (volumetric fraction). It is necessary to note, that the increase of a volumetric vapor fraction of (H_2O) results in decrease of the laser radiation power with other things being equal.

For mobile GDL, laser radiation, ensuring high power in CW/P-P modes, a basic method of thermal excitation is the combustion of such fuel components, which as final products give necessary components of a laser mixture at relatively high temperature. The optimal mixture ratio of gases at high temperature of mixture can be obtained, for example, at burning of carbon oxide (CO) and hydrogen (H_2) in air or other oxidizer (N_2O, N_2O_4) with subsequent use of ballast nitrogen.

The specific requirements to the analyzed mobile installation impose some limitations on use this or that fuel, oxidizer and ballast gas. The main difference of this installation from existing lasers is the requirement of large duration of operation.

For processing of the greatest possible area of the polluted basin, the laser installation should have an operating time appropriate to operation capabilities of a vehicle, on which it is secured. For the helicopter this time makes t ~ 30 min. The carried estimations of interaction of a beam with a water surface covered with thin petroleum film, have shown, that the power losses by heating water can be reduces through fast scanning of a surface by a beam of power W ~100 kW. These requirements allow estimating total of fuel and ballasting gas necessary for mobile GDL operation. The specific power (w_{out}) existing homogeneous CO_2-GDL, as a rule, does not exceed 10 kJ/kg. Thus weight of fuel and ballast gas for such GDL operation should be not less than $m = Wt / w_{out}$ = 18000 kg. This value essentially will increase at the account of weight of the equipment for storage of working components. Apparently, that such large weight is unacceptable for laser arrangement on a flight vehicle.

At the same time, the main part (80 95 %) of weight of design components used for creation of a propulsive mass GDL, is the weight of an oxidizer and ballast gas. Thus, use of free air as an oxidizer and a ballast gas is a unique opportunity to guaranty this installation operation. As CO_2-GDL fuel usually carbon monoxide is used with the hydrogen addition.

Carbon monoxide allows receiving high values of specific power of radiation in comparison with liquid fuels, such as toluene, benzole, kerosene, but weight this fuel together with a system of storage is rather high. Weight CO is 13 - 15 % from total mass of spent components, i.e. in our case of 2000 - 3000 kg. Weight of a storage system of such quantity gaseous carbon monoxide is ≈7000 kg, and summarized weight fuel and system of storage m_Σ = 9000 - 10000 kg. In case of use cryogenic state carbon monoxide summarized weight m_Σ can be reduced approximately twofold, but nevertheless it considerably surpasses appropriate weight for liquid hydrocarbon. Weight of liquid hydrocarbon is 4 5 % from total mass of spent components, and the tanks for storage are much lighter than for carbon monoxide, as the hydrocarbon are stored at atmospheric pressure. In our case weight of liquid hydrocarbon is 700 - 900 kg, and summarized weight fuel and system of storage m_Σ = 1000 - 1500 kg.

Taking into account the mentioned above reasons, in the present operation as possible fuel are considered benzole, toluene and kerosene, and as an oxidizer and a ballast gas (instead of nitrogen) free air compressed by the compressor.

As it was mentioned above, for normal operation of supersonic nozzles of the laser it is necessary to ensure high total pressure of a stream (P_0 = 2 - 3 MPa). The aviation compressors can ensure a compression ratio up to $P_0/Pa = 20$. Here P_0 stagnation pressure of a working mixture in a GDL channel, Pa atmospheric pressure. In all further calculations the value of stagnation pressure of a flow P_0 = 2 MPa and maximum values, appropriate Mach number of the nozzle M = 4.7 with the area ratio of an exit of the nozzle to the critical cross-section $F/F^* \approx 25$ was accepted.

At comparison various fuels, basic parameter is the power efficiency, namely specific power of laser radiation w_{out} (power of laser radiation referred to summarized mass flow of laser components), achievable at use of the given fuel. In this connection the analytical-theoretical research of influence of various parameters on value w_{out} was carried out.

Temperature and chemistry of a working mixture formed at combustion various fuels in air, were determined by thermodynamic calculation. The necessary for calculation enthalpy of initial components was taken from [4]. For the reason of uncertainty of efficiency of the compressor (approximately it is 0.75-0.83) in a main part of executed calculations temperature of inlet air in the combustion chamber was assumed to equal appropriate temperature at adiabatic compression up to value

$P_0 / Pa = 20$. This temperature calculated according to the following formula

$$T = T_a \cdot \left(\frac{P_0}{P_a} \right)^{(\gamma-1)/\gamma},$$

where T_a - temperature of free air, γ - isentropic exponent. At adopted values T_a = 298K, γ= 1.4 temperatures of inlet air in the combustion chamber is T ~700K. The enthalpy, appropriate to this temperature of air is equal I = 400 kJ/kg.

At thermodynamic calculation the thermal losses which depend on a design of the combustion chamber and beforehand are not known were not taken into account. In this connection in the subsequent calculations of flow of a working mixture in the nozzle block inlet temperature *(K)* was taken equal to 0.95 of temperature obtained in thermodynamic calculation.

The thermodynamic calculations have shown, that the working mixture contains in the investigated temperature range (or oxidizer-to-fuel ratios) only necessary for operation CO_2 - GDL components (CO_2, N_2, H_2O) and oxygen (O_2), as the content of remaining components is insignificant. Therefore in the further researches it was supposed, that the working mixture consists only of carbon dioxide, nitrogen, water and oxygen.In the paper the calculations of non-equilibrium flow of a mixture CO_2 - N_2 - H_2O - O_2 in the supersonic nozzle are carried out as well. The kinetic model describing population change of separate oscillating levels of molecules CO_2 and N_2 at processes of *VT*- and *VV*- of exchange was used. All oscillating levels of a molecule CO_2 with characteristic temperatures $\theta < 5000K$ were taken into account. It was supposed also, that the velocity of exchange of oscillating energy to a Fermi-resonance considerably surpasses a velocity of all remaining *VT*- and *VV*-processes. Thereof, the modeling system of levels was introduced, some of which have a large degree of degeneration and correspond to several levels in a full system of levels of a molecule CO_2. As a result of calculations the population inversion and gain coefficient of the nozzle unit outlet was determined. The calculations of specific output power of laser radiation were conducted at following input data.

1. The working mixture of gas dynamic laser is formed at combustion hydrocarbon fuel in air. The stagnation pressure of a flow is equal 2 MPa, total temperature is equal 1600K.
2. The nozzle unit of the installation consists of flat contoured nozzles with the height of critical cross-section 0.8 mm and expansion ratio 25.
3. The sizes of exit cross-section of the nozzle unit: length - 1600 mm, height - 130 mm.
4. The size of a cavity of the resonator along the flow direction - 200 mm.
5. The three pass unstable resonator with a multiplication factor

$M = 1.4$. A reflection coefficient of mirrors $r = 0.98$, the losses on dissipation make ◎= 0.03 m^{-1}.

The contour of a supersonic part of the nozzle was obtained as follows. The flow of gas with an isentropic exponent equal 1.35 in the contoured nozzle with an angular point was calculated and as a contour of a supersonic parts of the nozzle the form of a streamline ◎= 0.9 was used.

Fuel	$k_◎$, m^{-1}	w_{max}, kJ/kg	w_{out}, kJ/kg
Benzole	0.41	23.6	9.6
Toluene	0.39	23.1	9.2
Kerosene	0.32	20.8	7.4

Table 3. Values of specific output power of laser, based on different fuels.

Thus, value of specific output power of laser radiation, which it is possible to expect at use as fuel benzole and toluene, and as an oxidizer and ballast gas - dry air, is ≈ 9 9.5 kJ/kg. are presented in Table 3. At use of kerosene this value is approximately 2 kJ/ kg less.

By selection of fuel, except for power efficiency, it is necessary to take into account production and toxic characteristics, and cost of components.

The application of benzole is inconvenient at low environment temperature. The freezing temperature is 5.5°C. For toluene this value is equal -80°C, and for kerosene -38°C.

The limit for a room vapor concentration of researched fuel has following values: 5 mg/m^3 for benzole, 50 mg/m^3 for toluene, 300 mg/m^3 for kerosene.

From said it is possible to conclude, that the application of benzole is unwise, as it a little surpasses toluene according to power efficiency, but is significantly worse for toxic than toluene and is less convenient in operation.

As to selection between toluene and kerosene, the application of toluene provides specific power increase approximately 25 %. However use of kerosene has a number of advantages. At first, kerosene is approximately 3 times cheaper than toluene. Secondly, kerosene is fuel for airplane and engines. Therefore at the arrangement of the laser installation working on kerosene on a flight vehicle the additional system for fuel storage is not required.

As conclusion for this chapter it is possible to note the following.

The homogeneous laser working on combustion hydrocarbon fuel products is most reasonable for development of the mobile installation.

The specific requirements to the mobile laser installation,impose the limitations on use this or that fuel, oxidizer and ballast gas. The main limiting factor is the required long duration of continuous operation. The carried out estimations have shown, that summarized weight of fuel, oxidizer and ballast gas necessary for GDL operation, is very hight and is unacceptable for arrangement on a flight vehicle. At the same time, the main part of weight of fuel components used for GDL operation mass, is weight of an oxidizer and a ballast gas. In this connection, use of free air compressed by the aviation compressor, as an oxidizer and a ballast gas is the only way to organized operation of this installation.

The carried out comparative analysis of various fuel has shown, that most reasonable among them are toluene and kerosene. The application of toluene provides specific power of the installation approximately 25 % higher than that of kerosene. However, use of kerosene has a number of the advantages. At first, kerosene is approximately 3 times cheaper than toluene. Secondly, kerosene is the fuel for airplane and helicopter. Therefore at arrangement of the kerosene operating installation, on a flight vehicle the additional system for storage fuel is not required.

4. Selection of the power installation for GDL

4.1. Selection of a schematic of the power installation

4.1.1. Special conditions

In the previous chapter devoted to the computational substantiation of the basic power characteristics, the capability of development of the mobile CO_2 - GDL with use of kerosene as fuel, and as an oxidizer and a ballast gas - atmospheric air supplied by the compressor is shown.

Besides, some GDL parameters, permitting to determine the required characteristics of the compressor are added.

The calculations have shown, that in case of GDL the condition Po /Pn = 20 should be realized

Where: Po - stagnation pressure of a working mix(mixture) in GDL channel;
Pn - atmospheric pressure.

In view of channel losses from the compressor to GDL inlet of the chosen compressor should have the pressure ratio not less than 22 - 23.

Temperature of inlet air in GDL combustion chamber chosen at calculations is 720K, which corresponds to the temperature range of compressed air for compressors with $P_K > 22 \div 23$.

For an estimation of necessary value of airflow rate through GDL it is rational to set specific energy extraction at known accepted power range of 100 kW. As was shown in the previous chapter, the GDL specific power with combustion of kerosene in air is 7,5 kJ/kg. However this value can be reached at dry air only. At increase of humidity of atmospheric air the specific power begins to decrease and at humidity value 100 % (T = 25°C) it is 5,4 kJ/kg.

To achieve the higher specific power it is possible in principle to include air dryer in an air flow path, however it large overall dimensions and power consumption can result in technical and operational problems and, besides the automatic control of dryer operation is necessary if air humidity and fuel have changed. It will be more conveniently to have certain losses of specific power at humid air, having compensated losses by increased air flow.

Therefore, having accepted average specific power of 6 kJ/kg, instead of 7,5 kJ/kg we shall receive an estimated value of an air flow - 16,25 kg / s.

Thus, for a operation of the laser installation in power range of 100 kW it is necessary to have the compressor with an air flow rate ≈ 16 kg / s and $P_K > 22 \div 23$.

In essence task of air compression at flow rate of 16,25 kg /s and pressure in access of 20 atm does not contain unsolved technological problems, since compressor engineering is widely applied in industry. The task consists to find the optimal solution ensuring the reasonable cost of the installation, reliability of its activity and convenience in operation. Let's consider three possible schematics of power installation.

4.1.2. Development special power installation for the above mentioned parameters of the compressor

In this case the installation will consist (Fig.5) of the compressor itself, turbine for driving, gas generator for rotation the turbine, and also fuel supply system, control and regulation units.

As there is no installation for the mentioned parameters in industry, it is necessary to develop such installation from very beginning, to conduct preproduction activity and to find a producer. For all reason, the price of the produced installation will be high, and development - time consuming.

Besides, short order book for such installations (single production) will lead initially to unprofitable operation for any producer.

4.1.3. *The installation of air compressor on aviation engine shaft*

In general approach, the task can be decided rather simple. On the shaft of the aviation gas-turbine engine the compressor with the required characteristics is installed instead of the propeller.

The calculations show, that the shaft horsepower about 5,5-8 MW is necessary for compressor driving with the flow rate of 16 kg / s and Pк 23 kg / cm^2. It is power of the aviation gas-turbine engine of middle power class, a lot of which is in series production. However, as engineering realization the task has a number of serious problems.

The nominal speed of the propeller shaft of existing aviation engines is in the range of 5-8 thousand rpm, and all high efficiency centrifugal compressors require operation speed of the order above. For example, industry compressor TB7-117 being far from the best in efficiency already requires 30000 rpm.

The installation between the gas-turbine engine and compressor of the high-speed gear reducer (multiplicator) because of complexity in operation and large overall dimensions excludes such solution as acceptable.

Technically task is solved by the installation on the shaft of the engine of an axial multistage compressor (Fig.6). But even the principal diagram general view demonstrates the technical irrationality of such solution, it practically two aero-engines, one of which a production one, and the second requires large design and manufacturing costs connected refurbishment for realization of rigidity of the shaft, organization of turbine side exhaust etc. Besides of that, the overall dimensions and accordingly the weight of the installation are considerably increased.

4.1.4. *Air bleed from aero-engine compressor final stage*

As it is visible from the scheme Fig.7, if it is decided to take an air bleed for GDL after a final stage of the engine compressor, practically there is no necessity redesign the power installation with installing new aggregates. Moreover, the dismantling from the engine of the free running turbine operating for driving the propeller shaft for the reason of propeller nonuse, considerably simplifies a design of the installation as a whole.

This solution bribes by the simplicity, since the engines not only will be used with minimum modification, but also all control systems are kept, including the use of the control panel which is taken off from a flight vehicle.

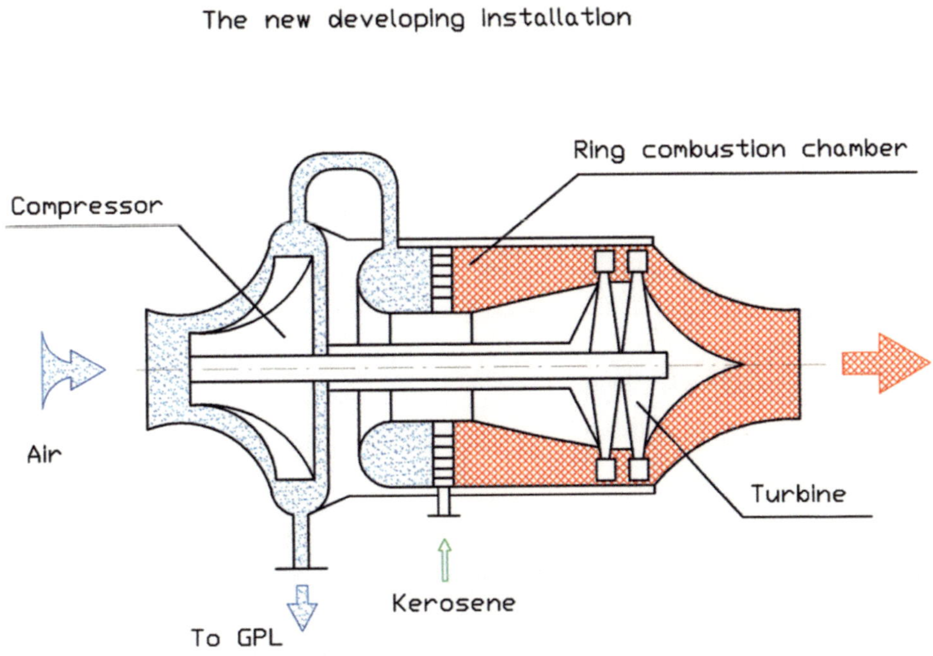

Figure 5.

Besides, from point of view of the power, dismantling of the free running turbine together with nozzle block allows to lower required power of the turbine of system of the compressor - free running turbine.

The decrease of required power of the system offers the possibility to reduce quantity of required air and kerosene in the combustion chamber of the gas-turbine engine. With constant power rate of the turbine for the drive of the compressor and of the compressor flow rate, probably, it exist the possibility to take off a part of the air and to direct it to the combustion chamber of gas dynamic laser. At the same time, utilizing the characteristics of the combustion products down the main turbine (ρ, m, T), it is possible to organize an ejection of GDL gases down the supersonic diffuser for improvement of the gas dynamic laser overall performance, especially at start.

4.2. Selection of the aviation gas-turbine engine

With the purpose of optimal selection of an aviation-engine for this task it is necessary to formulate the initial requirements.

1. For obtaining air pressure in GDL inlet not less than 20 atm and in view of losses in a flow channel the compressor of the engine P_K should be in the range of 22-25, not less.
2. The GDL inlet air flow rate is 16,25 kg /s.

As for elimination of extreme engine power reduction the air bleeding from the compressor should not exceed 15-18 %, the total air flow rate through the engine should be not less than 100 kg /s.

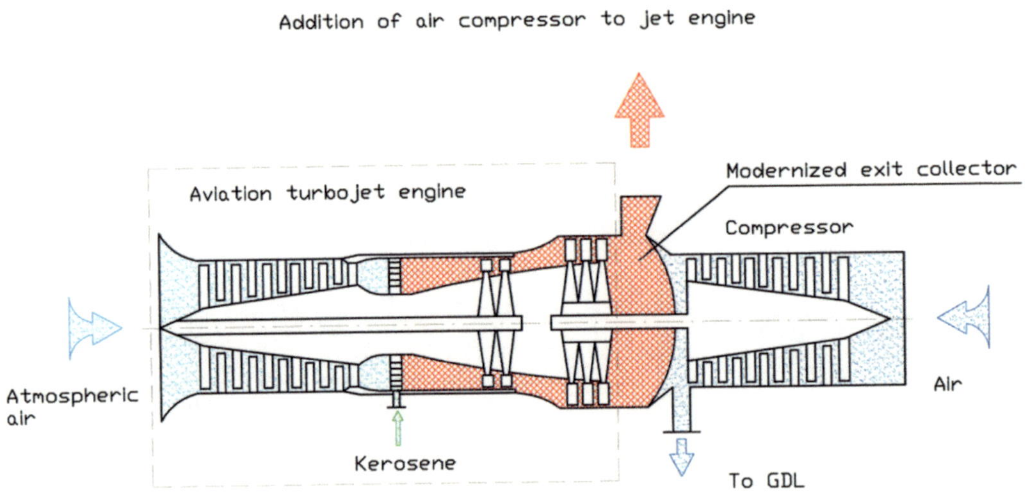

Figure 6.

3. The chosen aviation engine should widely be used by standard aviation operation, that enables to utilize engines which are decommissioned from flight vehicles, having passed prescribed operational life limit, but still are operable for ground installation. To buy them is much cheaper, than to buy new engines.
4. It is desirable to make a choice of domestic (Russian made) engines that enables in case of realization of the project to connect to activities, both engine developing company and manufacturer.

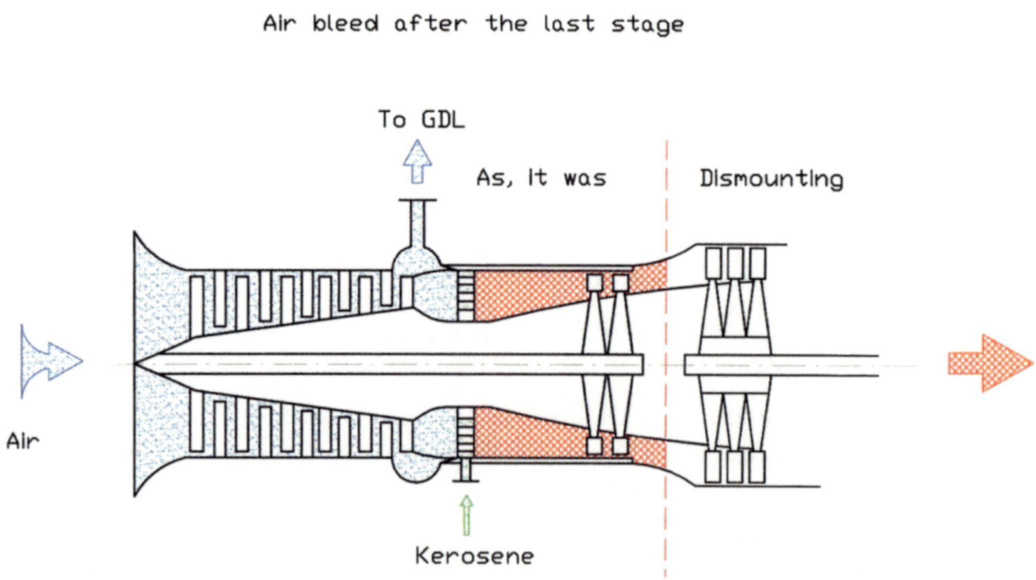

Figure 7.

The table data of some Russian engines characteristics: Пк and G of airflow rate (kg / s) are given below.

Engine	НК-12СТ	НК-14СТ	НК-14Э	НК-16СТ	НК-17	НК-36СТ	ПС-90А	РД-33
$P_к$	8,8	9,5	9,5	9,68	9,68	23,12	19,6	22,0
G (kg/s)	56,0	37 .1	39,0	102,0	102,0	101,4	56,0	80,0

Table 4.

The analysis of the table results in a conclusion, that most reasonable for the laser installation could become the engine НК-36СТ serially produced by Samara plant «Motorostroitel», which satisfies to all requirements presented above. It is obvious, this is not the single engine, which can be put to use for this task, but in advantage of it would be desirable to note the following.

The high performance engine НК-36СТ of the Samara Technological complex «Motorostroitel»," is developed in 1990 on the basis of an aero-engine НК-321 and is designed for the drive of the centrifugal supercharger in a structure gas pipeline pumping aggregate.The modular design of the engine facilitates transport and assembly. The engine has the remote control panel and record of long life ground operation. The general view of the engine is shown in Fig.8. In Fig.9 the design concept is shown.

Figure 8. Gas-turbine engine NK-368T

4.3. Gas dynamic assessment of the chosen aero-engine as the power unit for mobile GDL

In the previous chapter the task of GDL supply with oxidizer (air) by use of the HK-36CT engine compressor was considered.

However, one more function - use of the engine exhaust gas as working medium for ejection of a GDL exhaust system is planed to the chosen engine.

The practice of development GDL with direct exhaust of combustion products in atmosphere has shown, that the aft diffuser of the laser produced the pressure recovery of gases up to atmospheric pressure requires for start significant the combustion chamber pressure increase (more than 30 atm), or decrease of exit pressure below atmospheric one. It is possible to ensure the last requirement, through installation of an ejector, for which the exhaust gases down the engine turbine are the working medium.

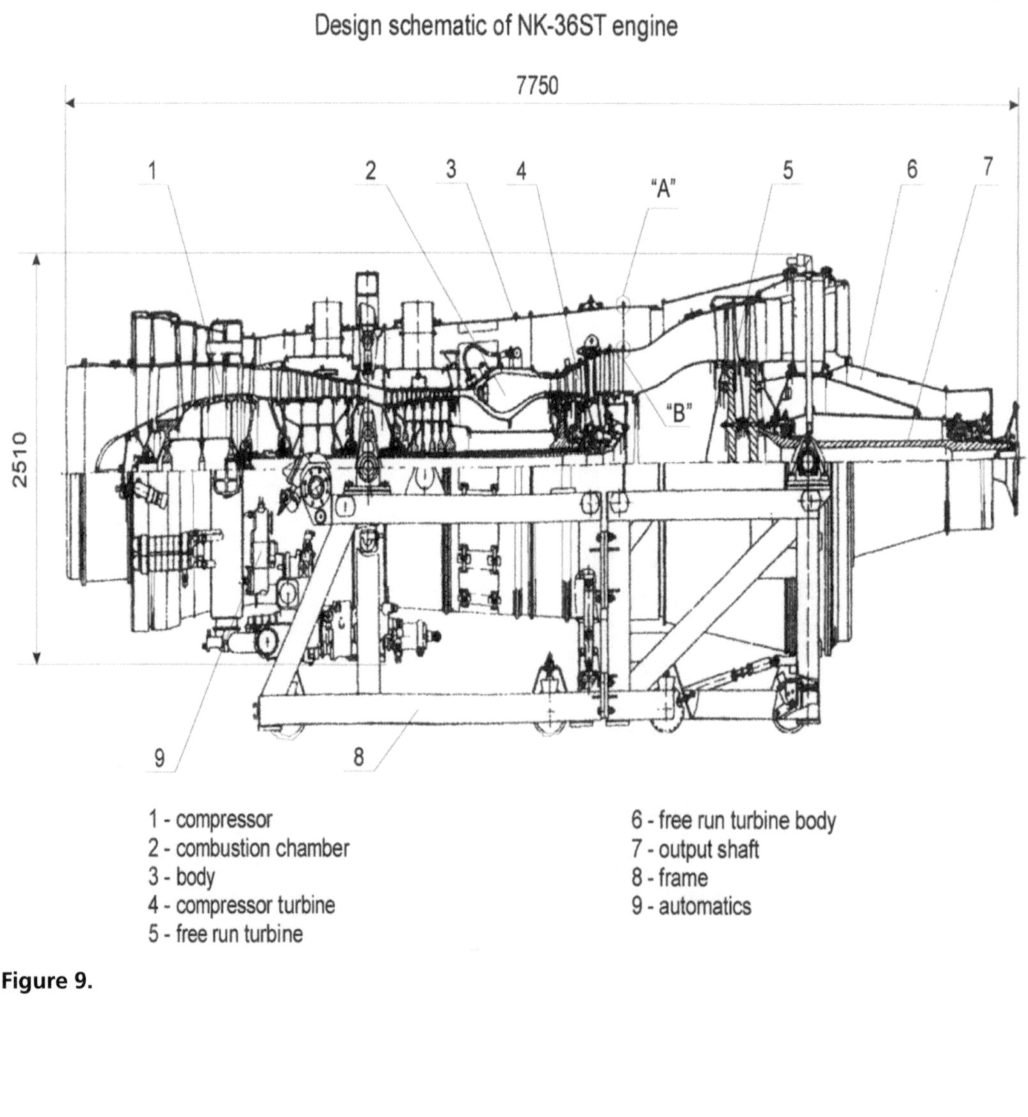

Figure 9.

Gas dynamic estimation of the capability of the HK-36CT engine of GDL diffuser start support is shown below.

As it was already indicated, the basic adaptation of the HK-36CT gas-turbine engine is the elimination of design structure of the free running turbine, that allows to improve characteristics from the point of view of possible application in the laser installation, for which it is necessary to bleed air with the consumption m_B = 16,25 kg\sec by pressure P ≈22 kg / cm² and temperature T = 720K.

Reasonable temperature for a GDL gas flow channel is T = 1600K at combustion-chamber pressure of 20 kg / cm².

The given temperature is realized in the combustion chamber for fuel kerosene - air at mass ratio K_K = 38,5, that is at kerosene flow rate m_n = 0,422 kg / s.

In GDL working section it is necessary to ensure static temperature of combustion products Ta = 370 K, that is realized velocity exhaust of combustion products through supersonic nozzles with velocity Wà = 1700 m\s at a Mach number Ma = 4,46, isoentropy index κ = 1,333, pressure Pa = 0,057 kg / cm².

For realization in GDL flat working section shock wave free flow of the of combustion products in atmosphere is necessary to utilize the flat aft supersonic diffuser. One of major parameters of functioning of the diffuser in GDL operation conditions is the start pressure.

The start pressure (Pst) is determined on intensity of a direct shock wave relation:

$$\frac{Pst.}{P_H} = \frac{g(\frac{1}{\lambda a})}{g(\lambda a)};$$

Where: Pn - pressure behind the diffuser;

$\lambda a = \frac{Wa}{a_*}$ - velocity factor equal to the ratio of exhaust velocity in working section of GDL (Wa) to the critical velocity of sound (a_*).

The value of a velocity factor λa is connected to a Mach number by a ratio

$$\lambda a = \sqrt{\frac{X+1}{2} \cdot \frac{Ma}{\sqrt{1+\frac{X-1}{2}Ma^2}}}$$

and then to a Mach number Ma = 4,46 corresponds $\lambda a = 2,32$.

The function $g = \frac{Fkr.}{Fa}$ determines the ratio of the geometrical area of critical cross-section of the diffuser to the area of gases exhaust and is according to expression

$$g(\lambda a) = (\frac{X+1}{2})\frac{1}{X-1} \cdot \lambda a(1 - \frac{X-1}{X+1}\lambda a)\frac{1}{X-1}$$

For direct shock wave the following ratio is valid

$$\lambda a \cdot \lambda_{n.c.} = 1;$$

Where: $\lambda_{n.c.}$ - dimensionless velocity factor after front of a direct shock wave

Thus, the ratio of pressure of start to pressure behind the diffuser is equal

$$\frac{Pstor.}{Pn} = 13{,}7.$$

The wind tunnel pressure of start is usually above their steady state operating pressure up to 10-30 %.

At the same time, in GDL laser conditions, when there is a gas volume involved in a working section (region of resonator mirrors) pressure of start of the diffuser can is still increasing to the value, at which the diffuser could not started.

Hence, it is necessary to lower exhaust pressure from the diffuser, that is possible to realized through an ejection of a gas jet, flowing out from the diffuser, by a gas jet flowing out after the turbine of the engine (Fig.10).

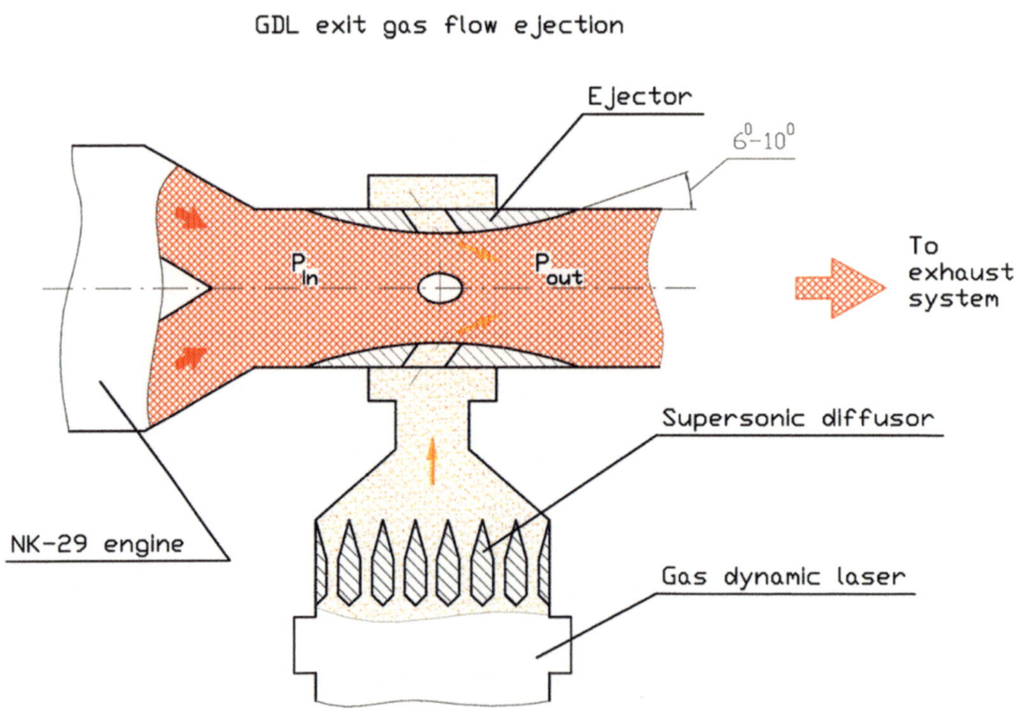

Figure 10.

The designs of similar ejectors with maximum pressure recovery are known at high speeds of flow of gas in minimum cross-section. Such ejectors designs are characterized by smooth profile of an inlet part and with small cone angles (L) on an exit, which make value L = 6 – 10 degree.

The pressure recovery coefficient for such ejectors can reach value $\frac{Pexit}{Pinlet} = 0{,}955 \div 0{,}965$ and even higher.

Thus the velocity factor value in minimum cross-section should be

$$\lambda = 0{,}75 - 0{,}85.$$

Thus, such ejector inlet pressure should be Pinlet. = 1,076 kg / cm^2, which can be provided with the gas-turbine engine without free running turbine. (Pexit. = 1,033 kg / cm^2).

At a velocity factor $\lambda = 0{,}8$ in minimum cross-section and ejector inlet pressure Pinlet. = 1,076kg / cm^2 the static pressure in minimum cross-section becomes equal Pmin =0,73kg / cm^2, that allows to organize introduction of flowing out from the GDL diffuser gas in area of minimum cross-section of the exhaust device of the gas-turbine engine and by that to lower pressure behind the diffuser GDL at least down to Pexit = 0,8 instead of 1,033.

Such reduction of pressure will allow to lower start pressure of the GDL down, to a reasonable level of 18-20 atm.

4.4. Design solution for GDL power unit on the basis of the gas-turbine engine

In view of the technical plan generated by the authors of the present technical proposal, experts of the Samara complex «Motorostroitel», has developed on the basis of the HK-36CT gas-turbine engine the engine for a laser complex under index HK-29.

The HK-29 engine (Fig.11) represents the gas-turbine engine obtained as a result of adaptation of the engine HK-36CT, which consists of the following:

4.4.1. Dismantling of the free running turbine with the output shaft

The dismantling is made through the joint "A" flange of a engine body (see Fig. 4), through the joint "B" of the hot gas pipe down to the turbine of the compressor and finally the free running turbine together with a turbine casing is removed from the engine. As the frame, on which the engine is mounted, consists of two not connected permanently sections, the removal of the second turbine side section does not represent any complexity.

For organization of the smooth gas flow from the modified engine, and also for a fixing of the exhaust system to the engine, two conical adapters (casing) and internal fairing made of a stainless steel are added.

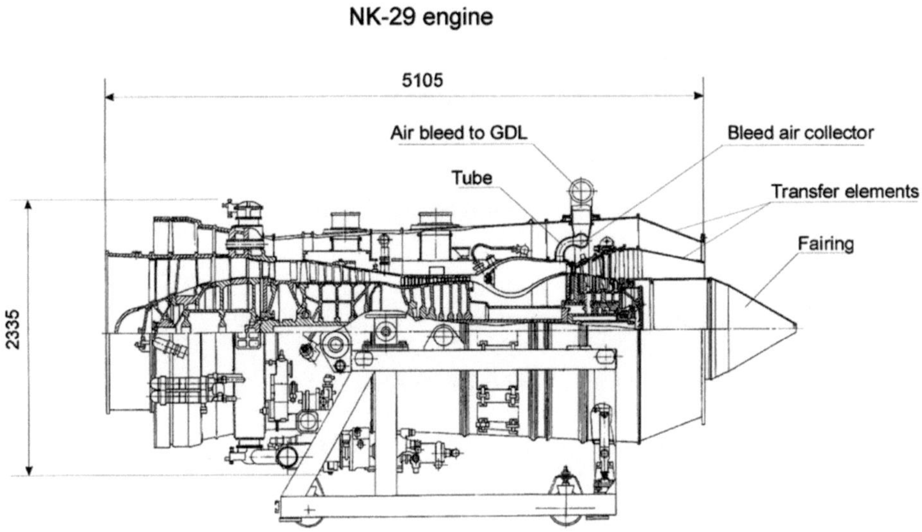

Figure 11.

The external adapter is terminated by a flange to be fixed with the engine ejector. The adapters and fairing were designed by «Motorostroitel», supplier in accordance to the aerodynamic characteristics of a gas stream and the strength requirements.

Ejector, which scheme is shown in Fig.9, is not the engine item. The ejector design is calculated and finally developed on the basis of engine, and laser complex exhaust system parameters and GDL exit characteristics (down the supersonic diffuser).Ejector design represents a pipe from high-temperature steel having symmetrical narrowing in the middle part ahead of critical cross-section. The pipe can be manufactured from three parts with flange or welded connections.

The middle narrowing part can be turned from a thick-walled billet or from a rolled ring and has mating collars for different side collector for convenience of assembly for welding. The orifices connecting cavities of exhausts of the engine and GDL are located around the perimeter of critical cross-section. The collector is manufactured by stamping of a sheet, has the welded branch pipe and serves for uniform distribution of gas after GDL on all cross-section of an ejector. The adaptation of the engine with removal of the free running turbine considerably reduces overall dimensions and weight of the installation. For clarity of estimation of advantage in overall dimension reduction of the modified engine the presentations of Figs. 8 and 10 are made in the same scale.

4.4.2. Air bleeding to GDL

The next considered problem is the air bleeding after the compressor to GDL.

During our long searching of the engineering solution for the engine adaptation the aviation specialists have offered to realize the air bleeding from the exit of the combustion chamber second contour, not from the compressor final stage exit, i.e. from a channel of a cooling flow of the combustion chamber. This solution is connected to the fact, that combustion chamber of the gas-turbine engine is one of the most thermal stressed items of the engine, and the cooling parameters are results of long development stage. Therefore the air bleed behind the compressor, reducing the air flow of the combustion chamber cooling flow, will change steady state thermal status, which can result in undesirable consequences for engine operation.

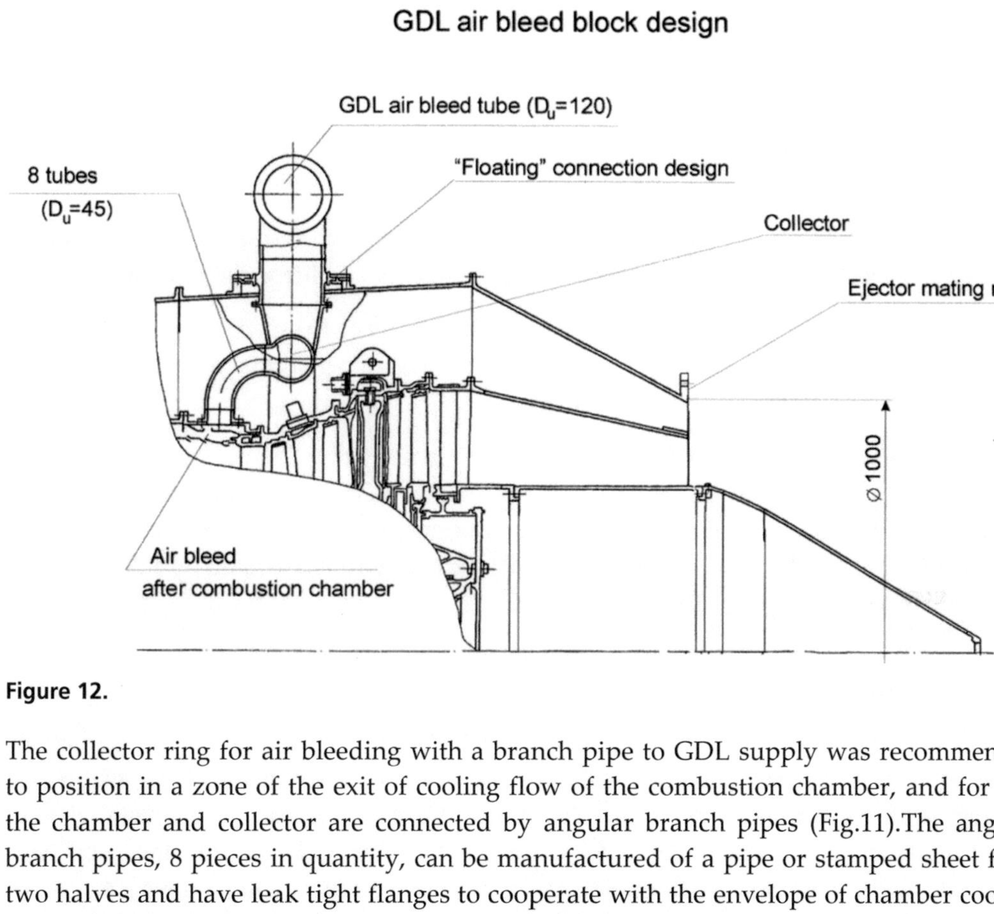

Figure 12.

The collector ring for air bleeding with a branch pipe to GDL supply was recommended to position in a zone of the exit of cooling flow of the combustion chamber, and for that the chamber and collector are connected by angular branch pipes (Fig.11).The angular branch pipes, 8 pieces in quantity, can be manufactured of a pipe or stamped sheet from two halves and have leak tight flanges to cooperate with the envelope of chamber cooling flow channel. The angular branch pipes are fixed by welding to the collector ring, which has the terminal flanged tap. To this tap the branch pipe to supply air to GDL is fixed. This branch pipe passes through an outer shell of a engine body and has "floating" attachment to it excluding deformation of designs at a thermal expansion of contacting elements.

It exists from the experience of such bleed scheme, however particular design should pass a number of development tests with the purpose of experimental selection of channel pressure reduction to ensure calculated bleed air flow rate. All other elements of the engine remain without change, including all start and cutoff automatics and operation control.

Thus engine HK-29, obtained as a result of adaptation of the engine HK-36CT will require minimum development, and manufacturing costs.

The basic performances of the engine HK-29. All parameters of the engine are indicated for a nominal mode.

Fuel	- Kerosene TC-1
Power, MW	- 25
Air flow, kg / s	- 107,2
Compressor pressure ratio, Pk	- 23,12
CC gas temperature, K	- 1420
Gas exit temperature, K	- 698
Propellant consumption, kg / s	- 1,434
Engine weight, with frames, kg	~- 5000

The engine HK-29 is single mode, as by activity of the laser installation is required neither decrease, nor increase of the power, that provides reliability and life value growth.

The engine HK-29 should be manufactured by aviation plant irrespective of, whether it is refurbished through adaptation of the engine HK-36CT or again assembled with use of furnished engine parts from production engines.

Thus, as a result of the carried out research work not only the optimistic results for possibility of adaptation of the gas-turbine engine for GDL gas dynamic complex were obtained, but also design activities by definition of general realization of such engine are executed in practice. And, what is very valuable, this activity is executed by the experts of aero-engines developing - producing company.

5. The mobile laser installation

In view of all computational investigations, design studies and engineering estimations within the present paper the provisional structure of the autonomous laser installation for clearing of surfaces of reservoirs of petroleum recordings is produced. The general view of laser installation is shown in Fig.13.As it can be seen from the figure, the installation represents modular assembly. The engine has an own fixing frame, autonomous subsystems and producer - supplier. GDL also represents the complete aggregate, the developer and manufacturer of which is not connected to other suppliers. The same concerns are valid for the fuel storage system (is not shown in figure).Task of the installation developer as a whole is the coordination of these units both on arrangement on a vehicle, and organization of interaction during operation. For example, relative arrangement of the engine and laser can be arbitrarily, depending on overall dimensions and device of a vehicle, however in any case the fulfillment of following conditions is necessary:

- The engine and GDL should not have of rigid connection on the frame, the GDL frame should have vibration absorption plates.

- The engine and GDL should be placed as close as possible, as the unreasonable elongation of gas lines connecting them, will result in to additional losses because of pressure reduction.
- On fire protection control requirement fuel storage container should be placed on distance from the engine and GDL, and is separated from them by a fire-prevention wall.
- For protection of the attendants the remote control panel should be born for limits of the installation, and input part of the engine and line of a beam are protected by special fences. The control of a scanning mirror also is carried out for limits of the installation.
- The direction of exhaust gases should not intersect structural items of the installation.

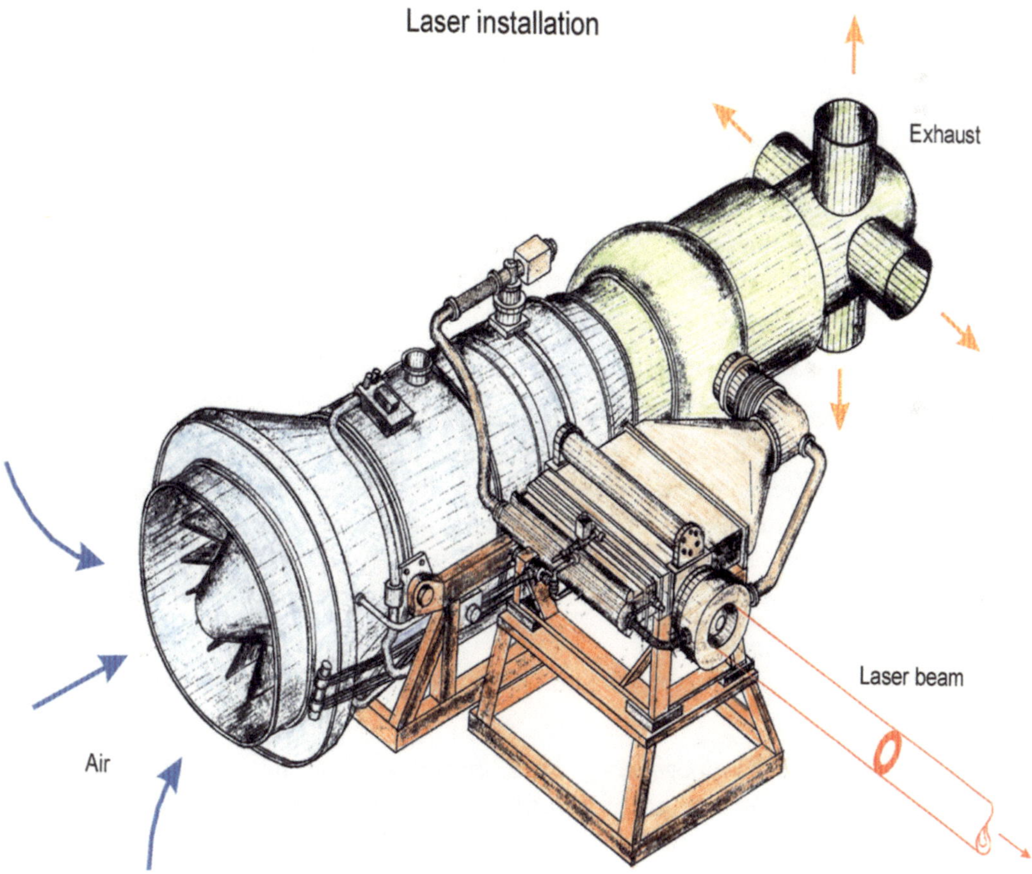

Figure 13. Laser installation.

As the development of the laser installation does not assume development of testing site for development testing, prior to deliveries the development installation should be assembled, on which all designer and operational problem solutions will be found.

6. Conclusion

The Purpose of presented in the paper project is to develop a laser technology of cleaning of large areas of seas and other water surfaces from oil film contamination. This paper also aims development of composition scheme, technological requirements and engineering design solution of aggregates and units of vehicle, vessel or aircraft compatible mobile GDL (100 - 250 kW) intended to solve this important problem of environment protection. This method proposed for development is expected to complement other traditional methods, which usually more successfully treat bulk layer oil pollution but do not match to eliminate up to 100 μ oil films, the latter usually being spread over numerous square kilometers of sea and kilometers of seacoasts. Note about one million tons of such films drift now in the World Ocean, as consequence of such disasters.

Thus, this promising approach has a fundamental science basis, from one hand, and skill and experience of missile jet specialists and their laser prototypes, from another hand. The investigations are planned to be accomplished in three basic directions:

1. theoretical and experimental research of the laser radiation action on thin oil and it's derivatives films over water surface, evaporation in three basic directions:
2. experimental optimization of laser optical scheme and operation modes in order to achieve the maximum optical and gas - dynamic efficiency.
3. engineering design, composition scheme and accompanied problems solution in order to adapt the laser module for vehicle, ship or aircraft transportation.

Results expected are as following:

- Certain recommendations on oil films elimination from large water areas, including conditions of real tanker and pipeline disaster, by means of laser radiation; development of technologies of oil and it's derivatives films burning out (mode number one) and oil films sucking up and saving of oil products for future efficient usage (mode number two) by means of high power mobile CO_2- GDL.
- Development of self-contained mobile CO_2 GDL tailored for these environmental problems solution.

These investigations may result in significant progress in the following branches of science:

a. physics of liquid inhomogeneous films, phenomena on the boundary of water-carbohydrate composition, phase transitions and chemical processes in two-phase or multi-phase compositions under high power CO_2 laser radiation and related phenomena;
b. problem of active resonators for high power GDL, different methods of laser efficiency calculation and simulation, control of temporal radiation modulation and optimization of performance, with respect to the problem of beam quality.

The author has completed the conceptual analysis of GDL installation development for disposal of petroleum films from a water surface. The basic capability of development of such installation is shown. The basic characteristic calculations of the system are made, the schematic and design solutions of basic installation are presented.

Besides of that, the GDL bench has been restituted by the author of the paper and his colleagues on their own costs and some demonstration operations with scanning the beam over water surface covered with petroleum film were carried out. The experimental confirmation of film effective burning mode and mode of oil film gathering was obtained. The next part of realization phase can be the development of the installation preliminary design, in case of the customer objective and specific technical tasks, formulated by him. The preliminary design will include complete study of a design of all elements of the installation, technological works, will define all suppliers of the developing systems and sub suppliers from obtaining from them of the consent on terms and conditions. In the preliminary design, schedules of all phases of installation development will be worked out: issue of the working documentation, technology modification, manufacturing of a prototype, prototype development and beginning of installations deliveries.

Thus, according to the estimation the experimental mobile CO2 GDL installation for water surface cleaning can be designed, manufactured and tested for not longer than two - two and half years.

Author details

V. V. Apollonov
Prokhorov General Physics Institute of RAS, Moscow, Russia

Acknowledgement

The author would like to acknowledge the valuable contributions made to realization of this work **"JET ENGINE BASED MOBILE GAS DYNAMIC CO2 LASER FOR water SURFACE CLEANING"** by V.V. Kijko, Yu.S. Vagin, and A.G. Suzdal'tsev.

7. References

[1] V.V. Apollonov, A.M. Prokhorov, "Universal laser for industrial, scientific and ecological use", Proceedings of GCL/HPL Conference, St. Petersburg, p.140, 1998.
[2] V.V. Apollonov, "Ecologically safe High power Lasers", Proceedings of Lasers-2001 Conference, p.3, 2001;
[3] V.V. Apollonov, "High power autonomous CO2 GDL (100kW,CW/P-P modes) for new technologies development and environment protection"; Russia –NATO Int. Seminar, Moscow, 2005, Proceedings of GCL/HPL Conference, Lisbon, p.35,2008.
[4] V.V. Apollonov "New application for high power high repetition rate pulse-periodic lasers", "Laser pulse Phenomena and Applications" Intech,p.19,2010.

[5] V.V. Apollonov, "Oil films elimination by laser", Oboronzakaz, №17, December, p. 33, 2007.

[6] V.V. Apollonov, Yu, S. Vagin, V.V. Kijko, "High rep.rate P-P lasers", Patent RF № 2175159

Section 5

Seeing Invisible

Chapter 15

Single-Molecule Recognition and Dynamics with Pulsed Laser Excitation

Guofeng Zhang, Ruiyun Chen, Yan Gao, Liantuan Xiao and Suotang Jia

Additional information is available at the end of the chapter

http://dx.doi.org/10.5772/48688

1. Introduction

Single-molecule spectroscopy has evolved from a specialized variety of optical spectroscopy into a versatile tool used to address a broad range of questions in physics, chemistry, biology, and materials science. Due to the ultra short time duration, the pulsed laser can be applied widely in the research of single-molecule dynamics. In the chapter, we will discuss the laser pulse application in two aspects of single-molecule detection and spectroscopy: fast recognition of single molecules, and manipulation of interfacial electron transfer dynamics.

The chapter is organized as follows. In the part of fast recognition of single molecules, we will discuss the Mandel's Q-parameter of single-event photon statistics for single-molecule fluorescence, recognition of single molecules using Q parameter, and the influence of signal-to-background ratio and the error estimates for fast recognition of single molecules. In the other part of the chapter, we will discuss the manipulation of interfacial electron transfer dynamics. First, we will introduce the principle of fluorescence lifetime measurement. Next, we will show the experiment results of the single-molecule and the ensemble under the external electric currents. Last, we will present our analysis and discussion for the results.

2. Fast recognition of single molecules

Although people most often think about and model molecule systems in terms of individuals, experimental science has been dominated by measurements that result in ensemble averages. This has traditionally hidden much of the rich variety present at microscopic scales. Although detecting single molecules optically was an old dream, the first convincing detection of a single molecule was achieved in 1989 by Moerner and Kador in an absorption measurement (Moerner et al., 1989). Optical spectroscopy offers a wealth of information on the structure, interaction, and dynamics of molecule species. Soon here after, fluorescence was shown to provide a much better signal-to-noise ratio, in cryogenic condition (Orrit et al., 1990) as well as at room temperature (Shera et al., 1990; Basché et al., 1992). The microscopy of single molecules at room temperature took off in 1993 with Betzig and Chichester's detection of immobilized molecules on a solid surface by means of excitation with a near-field optical source (Betzig et al., 1993). The scope of the method expanded suddenly when several groups (Nie et al., 1994; Trautman et al., 1994; Funatsu et al., 1995; Dickson, et al., 1996) showed that single molecules could be detected at ambient conditions with a simple confocal microscope. Single-molecule microscopy by fluorescence at room temperature has now become a versatile and general technique, opening investigations of the nanoworld (Orrit, 2002; Bartko et al., 2002). In above experiments it is critical to ascertain that the observed signal actually comes from a single molecule, but not the random mixture of emissions from nearby molecules. Usually low concentration and equivalently small excitation volumes are the most common experimental strategies (Deniz et al., 1999).

Once experimental conditions favoring single-molecule detection are satisfied, a number of criteria have to be met to ascertain that the observed signal actually comes from a single emitter (Michalet et al., 2002 ; Nie et al., 1994). The most common criteria is to detect an antibunching curve by two-time correlation measurements: based on quantum properties of single photon states, the absence of coincidence at zero delay gives clear evidence of single photon emission (Brunel et al., 1999). For several molecules, coincident emission of photons by the different molecules are likely and will result in an autocorrelation function that does not cancel out for zero time-delay.

Although the phenomenon of photon antibunching is demonstrated most clearly by two-time correlation measurements, it is, in principle, also exhibited by the probability $P(n)$ that n photons are emitted (or detected) in a given time interval T. Antibunching implies sub-Poissonian statistics, in the sense that the probability distribution is narrower than a Poisson distribution with the same $\langle n \rangle$ (Mandel, 1979). Traditional photon number counting statistics (Huang et al., 2006) is seriously affected by the blinking in the fluorescence, due to the molecular triplet state. Then single event photon statistics (Huang et al., 2007; Treussart et al., 2002) is suggested to character the single-molecule fluorescence. And based on moment analysis, Mandel's Q-parameter, which is defined in terms of the first two photon count moments, is an attractive alternative to two-time correlation measurements. It is quite robust with respect to molecular triplet state effects (Sanchez-Andres et al., 2005). In this work we suggest a novel approach to distinguish single-molecule system based on single event photon statistics characterization of single-molecule fluorescence. Using Hanbury Brown and Twiss (HBT) configuration (Hanbury Brown et al., 1956), by analyzing and comparing the Mandel parameters of actual single molecule fluorescence and ideal double molecule fluorescence, we present a new criterion based on single event photon statistics measurement.

2.1. Detection system of single-molecule fluorescence

Standard confocal microscopy technique (Michalet & Weiss, 2002) which is the commonly used experimental setup for single-molecule fluorescence detection is summarized in Fig.

1(a) , the emitted light from molecules is focused on a pinhole in order to reject out-of-focus background light and then recollimated onto two single photon counting modules (SPCM) after partition by a 50/50 beamsplitter (BS), which is a standard HBT configuration. For each detected photon, the SPCM would generate a TTL pulse.

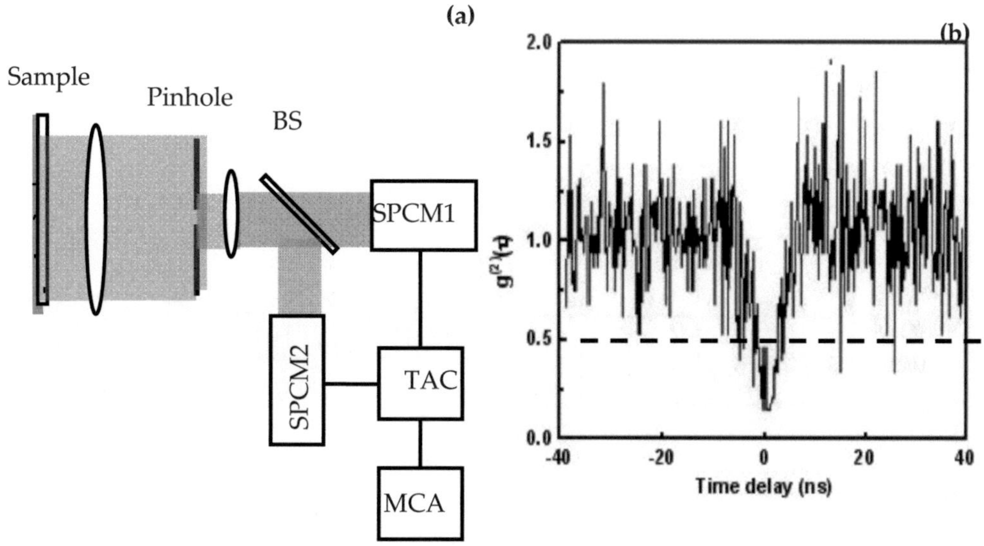

Figure 1. (a) Schematic of a confocal setup used for single-molecule fluorescence detection. (b) The second-order autocorrelation function measured, which indicates that fluorescence of a single molecule was being detected.

2.2. Detection of second-order coherence function

The signals from SPCMs are inputed to a time-to-amplitude converter (TAC). The start-stop technique with allows us to build a coincidences histogram as a function of time delay between two consecutive photodetections on each side of the beamsplitter. Then the pulses from TAC, whose amplitude is proportional to the time delay, are discriminated according to their heights and accumulated by a multichannel pulse height analyzer (MCA). The second-order degree of coherence at zero delay $g^{(2)}(0)$ can be directly gained and displayed on the screen, as shown in Fig. 1(b).

In experiment the effect of background signals makes it impossible for the perfect absence of coincidence at zero delay. Commonly as long as the second-order degree of coherence $g^{(2)}(\tau)$ at zero delay (τ=0) is less than 0.5, which corresponds with two ideal molecules detected, the fluorescence can be considered from a single emitter. Obviously if $g^{(2)}(0)$ is bigger than 0.5, it indicates that the fluorescence detected is emitted from more than one molecule. In contrast, in Fig. 1(b), $g^{(2)}(0)$ is smaller than 0.5, which indicates that one single molecule was being detected. This method is widely used to distinguish single-molecule in experiments. However, for a small number of blinking molecules, the total number of coincident emission might be too low to reject the single-molecule hypothesis with the antibunching curve. Furthermore the criteria need long time (typically several minutes to tens minutes, depending on the mean photon number $\langle n \rangle$ and the dead-time of detection system) to detect enough photons to accumulate and display a visible antibunching curve. It is severely affected by the molecule's photostability. And the existence of irreversible photobleaching compelled the long time for single molecules recognition to be insufferable in experiments. Using this method, the fluorescent photons detected should be very weak ($\langle n \rangle$ is less than 0.1), otherwise the start-stop technique will bring an error that can not be ignored.

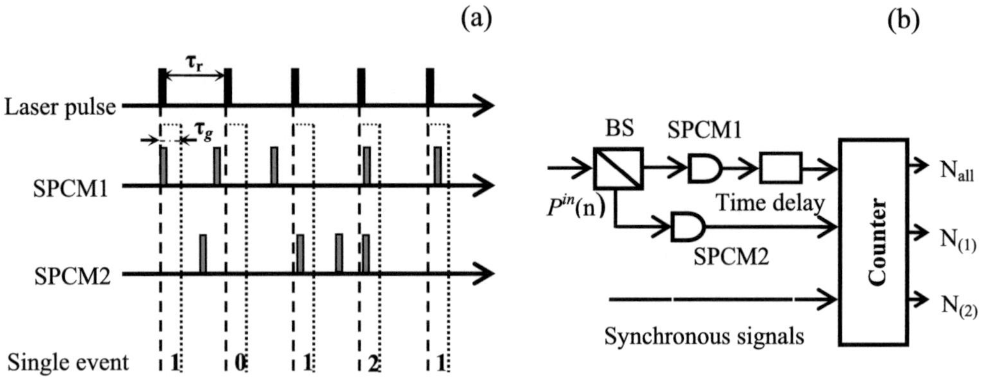

Figure 2. (a) The single event photon statistics measurement. The sample gate time τ_g, dead time for SPCM τ_d and laser pulse period τ_r fulfill $\tau_g<\tau_d<\tau_r$. (b) The schematic of detection setup used for single event photon statistics measurement. The synchronous signals provide a counting time-gate.

2.3. Single event photon statistics characterization of single-molecule fluorescence

The single event photon statistics measurement is described in Fig. 2(a). In order to eliminate the background signals as much as possible, time-gated technique (Shera, et al., 1990) is used after the excitation pulse. The records within the time gates are considered to be the detected signals while all records outside the time gates are rejected. This time-filtering procedure can filter out the real photodetection events from the most of non-synchronous background photocounts not rejected by optical filters, and is often an efficient way to improve the signal-to-noise ratio. The time gate duration must be shorter than the laser period and much longer than the molecule excited-state lifetime so that the probability of discarding a fluorescence signal is negligible. In the above experiments the gate durations are usually ten times the radiative lifetime of the molecule (Alléaume et al., 2004). Furthermore most after-pulse which will show up within 3 ns after the detector dead time, can be eliminated by time-gated technique, when time gate is shorter than the dead time. For the influence of detectors' dead time, each SPCM gives only one count (supposedly with 100% detection efficiency) for one or more than one incident photon within the dead time. For every excitation pulse cycle, the number of detected photons cannot exceed two if we using two identical SPCMs operating in the photon counting regime compulsory for our system. The existence of detectors' dead time in each detection channels will results in a non-linear relation-ship between detected photon statistics and source photon statistics. Thus the measured photon probabilities should be corrected.

Fig. 2(b) shows the schematic of our detection setup used for single event photon statistics measurement. The synchronous signals provide a counting time-gate and a nanosecond time delay box is used to compensate the different transport distance between two detection channels. The numbers of pulse cycles which are detected N_{all}, in which only one photon is detected $N_{(1)}$, and in which two photons are detected $N_{(2)}$ can be directly gained from a counter. Then the ratio of $N_{(1)}$ to N_{all} and $N_{(2)}$ to N_{all} respectively is the probability of detected one-photon and of detected two-photon in every pulse cycle.

Denoting by $P^{in}(n)$ the photon number probability distribution of incoming light on HBT detection set-up, the non-linear transformation relating this probability to the detected photon probability $P(n = 0, 1, 2)$ is simply calculated for 'ideal' detectors. The 'ideal' means that each SPCM clicks with 100% efficiency immediately upon receiving a photon, but that no more than one click can occur in a given repetition period. The joint probability of detecting i photons on SPCM1 and j photons on SPCM2 can be written as $P(i, j)$, $i, j = 0$ or 1. Actually, there are total four measured photon probabilities: $P(0, 0)$, $P(0, 1)$, $P(1, 0)$and $P(1, 1)$.

With our experimental detection scheme, random splitting of photons on two sides of 50/50 beamsplitter gives

$$P(0) = P(0,0) = P^{in}(0),$$

$$P(1) = P(0,1) + P(1,0) = \sum_{n\geq 1}^{\infty} P^{in}(n)\frac{1}{2^{n-1}}, \tag{1}$$

$$P(2) = P(1,1) = \sum_{n\geq 2}^{\infty} P^{in}(n)\left(1 - \frac{1}{2^{n-1}}\right).$$

The mean photon number per excitation pulse period is

$$\langle n \rangle = P(1) + 2P(2). \tag{2}$$

2.4. Mandel's parameter for single event photon statistics

In order to quantify the fluctuations of the number of photons detected per pulse, an important figure of merit is the Mandel parameter $Q = (\langle(\Delta n)^2\rangle - \langle n\rangle)/ \langle n\rangle$, where $\langle n\rangle$ is the average number of photons detected within a time interval T and $\langle(\Delta n)^2\rangle$ is the mean variance (Short et al., 1983). The parameter Q is a natural measure of the departure of the variance of the photon number $\langle n\rangle$ from the variance of a Poisson process, for which $Q = 0$. Negative and positive Q-values indicate sub- and super-Poissonian behavior, respectively.

From single event photon statistics probability the Mandel parameter can be computed directly

$$Q = \frac{2P(2)}{\langle n \rangle} - \langle n \rangle. \tag{3}$$

Note that an ideal single molecule should produce photons containing exactly one photon per pulse, P(2)=0, would yield $Q_i = -\langle n \rangle =, -\eta$, which means Q is only limited by the detection efficiency. Negative Q confirms that single-molecule fluorescence indeed exhibits sub-Poissonian photon statistics which is an explicit feature of a quantum field.

2.5. Distinguishing single-molecule using Q-parameter

The Mandel's Q-parameter provides an alternative for differentiating single-molecule fluorescence system. The Q-parameter of ideal double molecules (which is used to model two molecules that can not be separated by diluting) fluorescence can be used as a boundary between that of an actual single molecule fluorescence and actual double molecules fluorescence. From this boundary we can deduce an explicit criterion of single molecules based on photon statistics.

Now suppose that in the excitation volume there has been more than one molecule, the number is s, and all of them are ideal single photon emitters. We call η the overall detection efficiency, which includes the optical collection efficiency, all linear propagation losses and quantum efficiency of photon detectors. Then the photon number probability distribution of incoming light on detection set-up is

$$P_I^{in}(n) = \frac{s!}{n!(s-n)!}(1-\eta)^{s-n}\eta^n. \tag{4}$$

Applying Eq. (4) in Eq. (1) one can show

$$P\ (0) = (1-\eta)^s,$$

$$P(1) = \sum_{n\geq 1}^{s} \frac{s!}{n!(s-n)!}(1-\eta)^{s-n}\eta^n \frac{1}{2^{n-1}}, \tag{5}$$

$$P(2) = \sum_{n\geq 2}^{s} \frac{s!}{n!(s-n)!}(1-\eta)^{s-n}\eta^n \left(1 - \frac{1}{2^{n-1}}\right).$$

Then

$$\langle n \rangle = s\eta(1-\eta)^{s-1} + \sum_{n\geq 2}^{s} \frac{2^n - 1}{2^{n-1}} \frac{s!}{n!(s-n)!}(1-\eta)^{s-n}\eta^n. \tag{6}$$

Now suppose that in the excitation volume there are two ideal molecules excited and the fluorescent photons emitted are detected without background noise at all. Denote by Q_D and $P_D(n)$ the Mandel parameter and single event photon statistics probability of the fluorescence signals. From Eq. (4) one obtains

$$
\begin{aligned}
P_D(0) &= (1-\eta)^2, \\
P_D(1) &= 2\eta - \frac{3}{2}\eta^2, \\
P_D(2) &= \frac{1}{2}\eta^2, \\
\langle n \rangle &= 2\eta - \frac{1}{2}\eta^2.
\end{aligned}
\tag{7}
$$

In an experiment a problem we have to face is the background signals, which includes radiation from the environment and the dark counts of the SPCM. Because the scattering background light from surroundings can be thought as a thermal field with very large bandwidth and very short coherent time, the usual photon counting time (nano-seconds) discussed here is much longer than the coherent time and the photocounts of such time-average stationary background show a Poisson distribution. In a dark environment, the SPCM also generates random dark counts that follow a Poisson distribution. Both of these two random counts appear in the Poisson distribution, and thus we can use a weak coherent field with a Poisson photon distribution $P_B^{in}(n) = e^{-\eta\gamma}(\eta\gamma)^n / n!$ with γ to simulate the backgrounds. Actual single-molecule fluorescence can be modeled as the superposition of an ideal single molecule and a background emission that can be modeled as a Poisson distribution. Denote by Q_A and $P_A(n)$ the Mandel parameter and single event photon probability statistics of the actual single molecules fluorescence. Using Eq. (1) and Eq. (4), the photon statistics probability can be written as

$$
\begin{aligned}
P_A(0) &= e^{-\eta\gamma}(1-\eta), \\
P_A(1) &= 2\left(e^{-\eta\gamma/2} - e^{-\eta\gamma}\right) + \eta\left(2e^{-\eta\gamma} - e^{-\eta\gamma/2}\right), \\
P_A(2) &= \left(1 - e^{-\eta\gamma/2}\right)^2 + \eta\left(e^{-\eta\gamma/2} - e^{-\eta\gamma}\right).
\end{aligned}
\tag{8}
$$

Then

$$\langle n \rangle = 2(1 - e^{-\eta\gamma/2}) + \eta e^{-\eta\gamma/2}.$$

The measured mean photon number of fluorescence signals is $S = \eta$, and the measured mean photon number of background signals is $B = 2\left(1 - e^{-\eta\gamma/2}\right)$. According to Eq. (7), we get

$$P_A(0) = (1-S)\left(1-\frac{B}{2}\right)^2,$$

$$P_A(1) = (S+B-SB)\left(1-\frac{B}{2}\right), \tag{9}$$

$$P_A(2) = \frac{BS}{2} + \frac{B^2}{4} - \frac{B^2 S}{4},$$

$$\langle n \rangle = B + S(1-\frac{B}{2}).$$

When

$$P_A(1) \geq 2\sqrt{P_A(2)} - 3P_A(2),$$

the signal-to- background ratio (SBR) can be expressed as

$$SBR = \frac{S}{B} = \frac{P_A^{\,2}(1)}{2P_A(2)}. \tag{10}$$

This equation can be directly applied for the measurement of SBR in experiment.

Using Eq. (6) one can show that with the same average photon number $\langle n \rangle$, if $P_A(2)<P_D(2)$, then $Q_A<Q_D$. From Eq. (10) one can obtains

$$P_A(2) < \frac{1}{2}\left(2-\sqrt{4-2\langle n \rangle}\right)^2. \tag{11}$$

At the same time,

$$P_A(1) > \langle n \rangle - \left(2-\sqrt{4-2\langle n \rangle}\right)^2. \tag{12}$$

It is not possible to have fluorescence from more than one fluorophore with it's Q-parameter smaller than Q_D. So when Eq. (11) or (12) is satisfied, $Q_A<Q_D$， the fluorescence can be deduced origin from a single molecule system.

2.6. Signal-to-background ratio effect on the criterion

However low background signal is an important precondition of the criterion of Eqs. (11) and (12). With the high background signals, $Q_A > Q_D$ will come into existence for a single molecule system. This criterion will not be applicable. So it is necessary to make certain the range of SBR when Eqs. (11) and (12) is satisfied.

Corresponding to Eqs. (11) and (12) using Eqs. (8) and (9) one can obtains

$$SBR_A > SBR_0 = \frac{\sqrt{\langle n\rangle^2 - 2\left(2-\sqrt{4-2\langle n\rangle}\right)^2}}{\langle n\rangle - \sqrt{\langle n\rangle^2 - 2\left(2-\sqrt{4-2\langle n\rangle}\right)^2} - \frac{1}{2}\left(\langle n\rangle - \sqrt{\langle n\rangle^2 - 2\left(2-\sqrt{4-2\langle n\rangle}\right)^2}\right)^2}. \quad (13)$$

It is well known that SBR of the actual single molecule system varies with average photon number $\langle n\rangle$. Based on Eq. (13), the SBR curve versus $\langle n\rangle$ is shown in Fig. 3. When $\langle n\rangle$ is between 0 and 1, the variation range of SBR is within 1.63 to 2.41. So with the same $\langle n\rangle$ when $SBR_A > SBR_0$, for actual single molecule fluorescence and ideal double molecules fluorescence, it has $Q_A < Q_D$. Especially when SBR > 2.41, the Eqs. (11) and (12) can be the criterion used to distinguish single molecule system. Contrarily when $SBR_A < SBR_0$, even an actual single molecule system can not satisfy the Eqs. (11) and (12). It is obvious that SBR is also an important parameter of distinguishing single molecule system. Nevertheless most of the SBR in single molecule fluorescence experiments is large than 2.41.

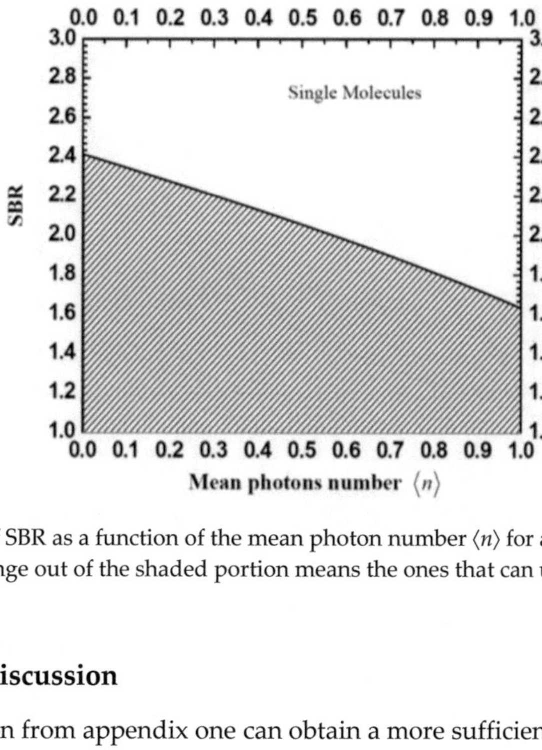

Figure 3. The curve of SBR as a function of the mean photon number $\langle n\rangle$ for actual single molecular photon source. The range out of the shaded portion means the ones that can use the criterion of Eqs. (11) and (12).

2.7. Results and discussion

After error correction from appendix one can obtain a more sufficient criterion,

$$P_A(1) > \langle n\rangle - \left(2-\sqrt{4-2\langle n\rangle}\right)^2 - \delta P_A(1) + \langle \Delta P_A(1)\rangle = P_1, \quad (14)$$

$$P_A(2) < \frac{1}{2}\left(2-\sqrt{4-2\langle n\rangle}\right)^2 - \delta P_A(2) - \langle \Delta P_A(2)\rangle = P_2, \quad (15)$$

In the above inequalities the critical values P_1, P_2 are impacted by three parameters, the mean photon number γ of Poissonian background, the mean overall detection efficiency η, and factor Δ which represents the unbalance of two channels of imperfect detection system.

Now let us assume that the size of sample cycles M was 10^4. In this way the statistical fluctuation of the $\langle \Delta P_A \rangle$ induced by finite sample pulse cycles detected can be negligible, which is less than 1/100 of the statistics probability itself. The error δP_A caused by imperfect detection system is the main correction factor of the critical value.

Fig. 4 shows the critical values P_1, P_2 as the function of mean overall detection efficiency η and factor $\Delta = 0.3$, $\gamma = 0.2$, which corresponds to SBR = 5. It is found that the critical values P_1, P_2 all increase with increasing efficiency η as respected, however the P_1 increases slowly while P_2 increases fast. Furthermore because in single molecule experiments $P_A(2)$ is less than $P_A(1)$, the effect of the error on P_2 is more evident than on P_1. So using $P_A(1)$ as the criterion is more feasible.

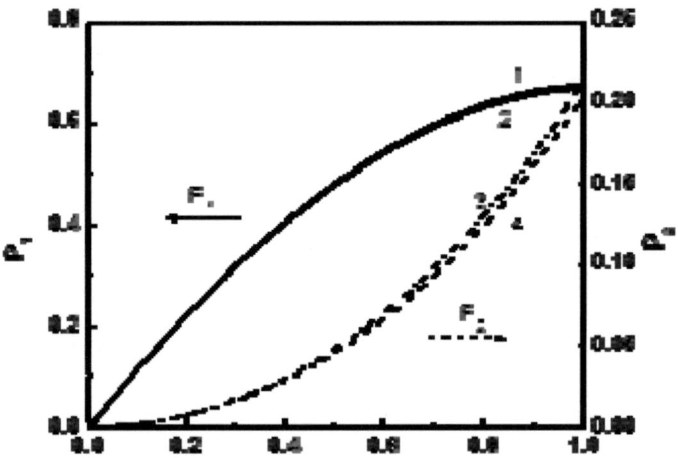

Figure 4. The critical values P_1 (solid line), P_2 (dashed line) as a function of mean overall detection efficiency η. And factor $\Delta = 0.3$, $\gamma = 0.2$, which corresponds to SBR = 5. P_1 (solid line) using left coordinate, P_2 (dashed line) using right coordinate. Curves 2 and 3 are the critical values P_1, P_2 without error estimate. Curves 1 and 4 are the critical values P_1, P_2 after error correction.

The experiment data, shown in Table I, is obtained by using typical single molecule experiment setup (Basché et al., 1992). The sample were dye molecules Cy5 (5×10^{-10} M, Molecular Probes) doped in Polymethyl Methacrylate (PMMA, Sigma-Aldrich) polymer films. Molecules in the sample were excited by 50 ps pulses at a wavelength of 635 nm and repetition rate of 2 MHz, generated by a ps pulsed diode laser (PicoQuant, PDL808), at the focus of a confocal inverted microscope (Nikon). Fluorescence photons were detected in two detection channels, using two identical single photon counting modules (SPCMs, PerkinElmer SPCM-AQR-15).

For example, in a typical experiment to sample 1, during 299613 periods (about 149 ms) there are 13917 recorded photons including 13902 single photon events, 15 two-photons events. These data allow us to extract the photon probabilities P(0) = 0.9535, P(1) = 0.0464, and P(2) = 5 × 10^{-5} and the mean number of detected photon per pulse $\langle n \rangle$ = 0.0465, then yield Mandel parameter Q = - 0.04435.

We observed fluorescence photons for three sample molecules in one polymer film.

As shown in Table 1, for sample 1 and sample 2, both P(1) are much more than critical values P_1 which indicates that a single molecule was being detected; for sample 3, P(1) is less than critical values P_1 which indicates that more than one molecule was being detected. As a reference, the interrelated results for coherent light (pulse laser) are attached both from theoretical analysis and experimental measurement. The typical time for single event photon statistics measurement is about 150 ms.

	n	P(1)	P_1	$g^{(2)}$ (0)	SM?	SBR
Sample 1	0.0465	0.0464	0.04590	0.19	Yes	21
Sample 2	0.0372	0.0370	0.03685	0.30	Yes	7
Sample 3	0.0521	0.0508	0.05150	0.65	No	--
Coherent light (theory)	0.1046	0.0991	0.0991	1.0	--	--
Coherent light (experiment)	0.1046	0.0992	0.0995	0.98	--	--

Table 1. Experimental results of single event photon statistics for Cy5 molecules and coherent light. The mean number $\langle n \rangle$ of detected photons per pulse is calculated by the measured values P(1) and P(2) from Eq. (2). The critical value P_1 is estimated from Eq. (14), the relative error is less than 1%. And the $g^{(2)}$ (0) is measured by two-time correlation measurements. Whether the sample detected is a single molecule is determined by that P(1) is more than critical value P_1 or not. And the determination is confirmed by the measured value $g^{(2)}$ (0). To a single molecule, the SBR is acquired from Eq. (9).

3. Manipulation of interfacial electron transfer dynamics

Interfacial electron transfer (IET) dynamics play an important role in many chemical and biological processes (Weiss, 1676; Wang et al., 2009). However, IET processes are usually very complex due to high dependence on its local environments. Single-molecule spectroscopy has led to many surprises and has now become a standard technique to study complex structures (Kulzer et al., 2010) or dynamics (Orrit et al., 2002; Uji-i et al., 2006; Zhang et al., 2010) including photoinduced, excited-state intramolecular, and IET dynamics. The optical signals of single molecules provide information about dynamics of their nanoscale environment, free from space and time averaging. Single-molecule studies of photoinduced electron transfers in the enzyme flavin reductase have revealed formational fluctuation at multiple time scales (Yang et al., 2003). Single-molecule studies of photosensitized electron transfers on dye containing nanoparticles also showed fluorescence fluctuations and blinking, and the fluctuation dynamics were found to be inhomogeneous from molecule to molecule and from time to time (Funatsu et al., 1995). Recently developed single-molecule spectroelectrochemistry extends single-molecule approaches to ground state IET by simultaneously modulating the electrochemical potential while detecting single molecule fluorescence (Palacios et al., 2006; Lei et al., 2009).

Indium tin oxide (ITO) films are the most widely used material as a transparent electrode of organic light emitting diode and also in other devices like solar cells (Friend et al., 1999; Tak et al., 2002; Hanson et al., 2005). It is interesting to study electron transfer of ITO to suitably modify interactions at the interfaces of dissimilar materials so that desired electronic properties of devices incorporating them can be realized. Single dye molecules dispersed on the semiconductor surface of ITO were used to measure IET from excited cresyl violet molecules to the conduction band of ITO or energetically accessible surface electronic states under ambient conditions by using a far-field fluorescence microscope, and single-molecule exhibited a single-exponential electron transfer kinetics (Lu et al., 1997). Here we apply an external electric current (EEC) to manipulate the IET rate between single dye molecules and its neighboring ITO nanoparticles by probing the fluorescence intensity change of individually immobilized single dye molecules dispersed in ITO film.

3.1. Experimental section

Cover glass substrates were cleaned by acetone, soap solution, milliQ water sonication and irradiation with ultraviolet lamp. The single molecules samples were prepared by spin coating (3000 rpm) a solution of 1,1′- dioctadecyl - 3, 3, 3′, 3′-tetramethylindodi-carbocyanine (DiD, 10^{-9}M, Molecular Probes) in chlorobenzene onto the cover glass substrate. The chemical structure of DiD molecule is shown in Fig. 5(a). The indium tin oxide (ITO) was purchased from Sigma-Aldrich (Product Number: 700460, dispersion, <100 nm particle size (DLS), 30 wt. % in isopropanol, composition: In_2O_3 90%, SnO_2 10%). ITO film in hundreds of nanometer thicknesses was spin-coated onto the dye molecules. Two aluminum leads were fixed to the ITO film, and the interval between the two leads is about 4 mm. After vacuum-dried, the samples were further covered with a poly-(methyl methacrylate) (PMMA) (Mw = 15,000, Tg = 82 °C, Aldrich) film in order to insulate oxygen. The samples were subsequently annealed in vacuum at 350 K for 5 h to remove residual solvent, oxygen and to relax influences of the spin coating technique on the polymer conformations. The method is depicted schematically in Fig. 5(b).

A 70 picosecond pulse diode laser of λ = 635 nm (PicoQuant, PDL808) with a repetition rate of 40 MHz was used to excite single dye molecules. The output of the pulse laser passed through a polarizing beamsplitter cubes (New Focus 5811) to obtain linear polarization light. A 1/4 wave-plate was used to change the polarized laser into circular polarization light. The laser beam was sent into a conventional inverted fluorescence microscope (Nikon ECLIPSE TE2000-U) from its back side, reflected by a dichroic mirror (BrightLine, Semrock, Di01-R635-25x36), and focused by an oil immersion objective lens (Nikon, 100 ×, 1.3 NA) onto the upper sample surface of the cover glass substrate. The fluorescence was collected by the same objective lens and then passed through the dichroic mirror, an emission filter (BrightLine, Semrock, FF01-642/LP-25-D), and a notch filter (BrightLine, Semrock, NF03-633E-25), is focused onto a 100 μm pinhole for spatial filtering to reject out-of-focus photons.

Fluorescence photons were subsequently focused through a lens and collected by a single-photon detector (PerkinElmer, SPCM-AQR-15). A piezo-scan stage (Piezosystem jena, Tritor 200/20 SG) with an active x-y-z feedback loop mounted on the inversion microscope was used to scan the sample over the focus of the excitation spot, producing a two-dimensional fluorescence imaging. All measurements were conducted in a dark compartment at room temperature. We use an alternative power source to supply EEC for the DID/ITO film system. Here, the applied electric current is proportional to the amplitude of applied external bias voltage, with the relation being I = ZU with Z = 0.87ohm^{-1}. The dye molecules embedded in the ITO film are distinct from the molecules in electric field experiments (Hania et al 2006). In the electric field experiments, a large electric field intensity of about 100 V/μm is needed, and dyes must be insulated from the electrode. In our experiment, dye molecules directly contact with ITO nanoparticles for IET controlling with much low voltage needed (less than 0.02 V/μm). In order to distinguish with the normal electric field experiments, we discuss the results ensuing from EECs here.

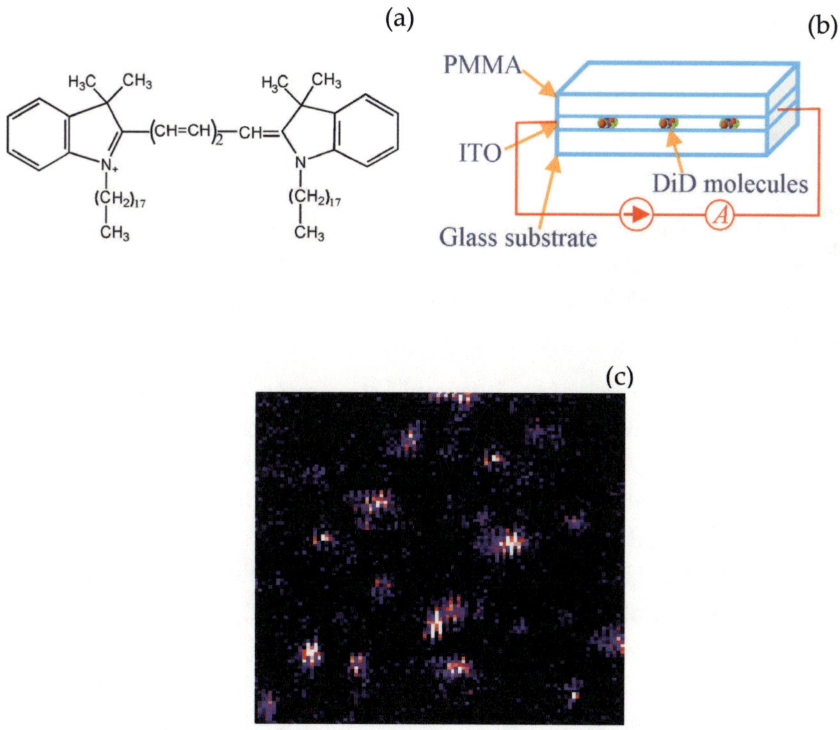

Figure 5. Fig. 5 (a) Structure of the 1, 1′-dioctadecyl-3, 3, 3′, 3′-tetramethylindodicarbocyanine (DiD) dyes molecule. (b) Schematic structure of the sample preparation. (c) Confocally scanned fluorescence image (10 μm × 10 μm) of DiD molecules dispersed in an indium tin oxide (ITO) film. Each bright feature may be attributed to a single DiD molecule.

3.2. Results and discussion

3.2.1. Fluorescence imaging of single dye molecules in ITO film

Fluorescence imaging using a confocal arrangement has superior sensitivity for spectroscopic measurements and is most suitable for studying single-molecule behavior in dilute sample. Fig. 5(c) displays the confocal fluorescence image of the single DiD molecules within a 10 μm × 10 μm area, which is obtained by scanning a sample containing randomly placed isolated single fluorescent molecules dispersed in an ITO film. The imaging is taken in 225 s (150 pixels × 150 pixels), with a pixel integration time of 10 ms. The single molecules are excited with laser excitation intensity of 1.5 kW/cm^2 at 40 MHz. The concentration of the DiD molecules was kept at such a low level that either one molecule or no molecules were in the focus. The bright features in the image then represent the fluorescence from individual molecules. The full width at half maximum of typical spots is about 300 nm. Meanwhile, it is also found that the spots have different intensities due to the molecular orientations and local environments. We have found that the average intensities of the molecules were about an order of magnitude weaker than that of DiD molecules in PMMA without ITO under the same experimental conditions. The weaker signal intensities result from the molecules undergoing the IET with surrounding ITO nanoparticles (Lu, et al., 1997).

3.2.2. Electric current response of the ensemble fluorescence

To study the IET dynamics between dye molecules and ITO nanoparticles, we placed dyes embedded in ITO film. An EEC was added to the DiD/ITO film to manipulate the IET rate. The molecules fluorescence intensity was measured by centering ensemble or one molecule in the laser focus and recording the transient fluorescence intensity, while a time-dependent EEC was applied.

We measured the EEC response of the ensemble at different dyes concentration, and the ensemble averaged study would establish how the average fluorescence intensity and lifetime are affected by the electrical current. The Fig. 6 shows three typical ensemble average results. Fig. 6(a) shows that the fluorescence intensity of an ensemble decreases rapidly while EEC applyed. Fig. 6(b) shows a fluorescence quenching trajectory of an ensemble under the EEC. However, fluorescence can be enhanced sometimes for some ensembles at relative small EECs as shown in Fig. 6(c). The similar fluorescence enhancement was also observed(Palacios et al., 2009), which was explained that potential-induced modulation of the excited state reduction processes (i.e., electron transfer from ITO to the molecules) dominate the low-potential fluorescence-modulation effect. We can find from the Fig. 6 that the ensembles do not show complete fluorescence quenching at the relative large EEC. This may be due to that those dye molecules in the ensemble are not in good contact with the ITO nanoparticles, which shows poor IET.

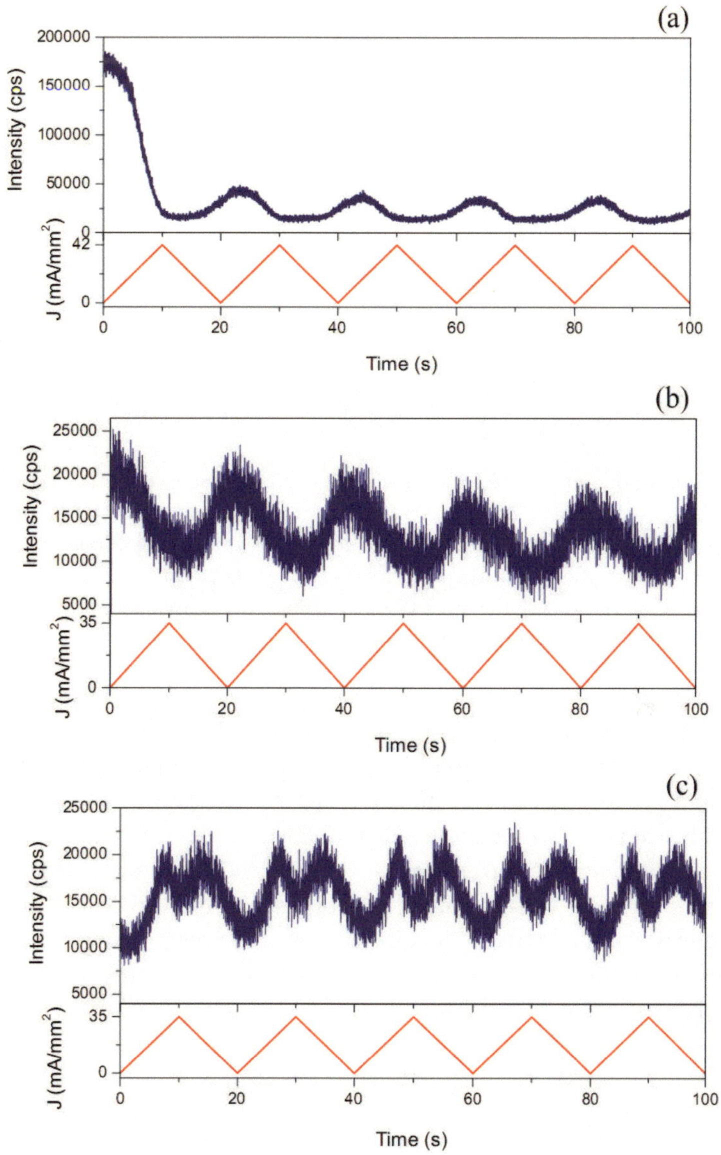

Figure 6. Three typical patterns of ensemble fluorescence intensity trajectories that were obtained while repeatedly applying a triangle wave EEC sequence to ITO shown by the bottom red curves. (a) The fluorescence intensity shows a fast quenching while applying the EEC to the ITO. (b) The fluorescence intensity shows a decrease while applying the EEC to the ITO. (c) The trajectory shows that fluorescence increases at smaller EEC and then decreases at larger EEC.

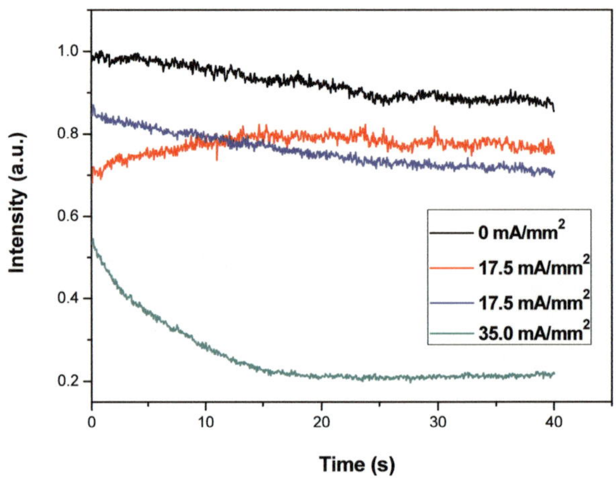

Figure 7. Normalized fluorescence intensity trajectories of 25 ensembles obtained at different EECs. The black curve shows the average of 25 ensemble trajectories obtained at zero current; the blue curve shows the average of 16 ensemble trajectories with fluorescence quenching obtained at the currentF of 17.5 mA/mm^2; the red curve shows the average of 9 ensemble trajectories with fluorescence enhancement obtained at the current density of 17.5 mA/mm^2; the green curve shows the average of 25 ensemble trajectories obtained at the current density of 35.0 mA/mm^2.

Normalized fluorescence intensity trajectories of 25 ensembles with different dyes concentration obtained under different EECs is shown in Fig. 7. Under relative small EECs, some ensembles show fluorescence quenching and the other ensembles show fluorescence enhancement, and the red curve and blue curve are constructed by sorting trajectories with fluorescence enhancement and fluorescence quenching at the current density of 17.5 mA/mm^2. In our experiment, for approximately 30% of the ensemble data show fluorescence enhancement effects at relative small current. The green curve is the average of 25 ensemble trajectories obtained at the current density of 35.0 mA/mm^2, and it shows fluorescence quenching of the ensemble at a large current. The various responses to EECs arise from the heterogeneity of site-specific molecules.

3.2.3. Electric current response of single-molecule fluorescence

We have detected several hundreds of single DiD molecules and all of the molecules sensitively responded to EEC. Fig. 8 shows the typical fluorescence emission for different individual DiD molecules in dependence of the EEC as a function of time. Fig. 8(a) is the fluorescence intensity time trace of one DiD molecule as an EEC of 32.0 mA/mm^2 periodically applied to the ITO film, which shows that the EEC can effectively quench the fluorescence emission of single-molecule. Fluorescence blinking observed in the trace shows a single DiD molecule emission and the blinking is related to the triplet state and/or charge transfer between single molecule and its local environment. While the electric current

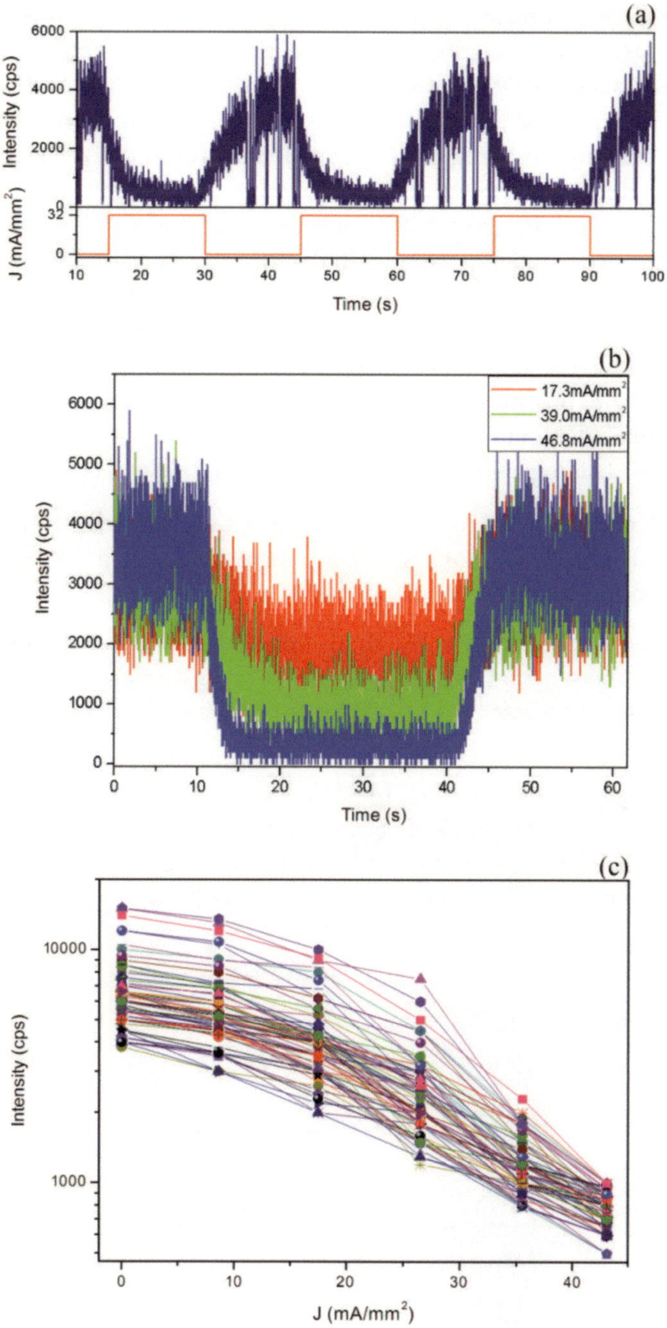

Figure 8. EEC response of DiD single-molecule emission intensity. (a) One single molecule fluorescence emission patterns as electric current applied periodically to the electrodes. (b) Another single-molecule emission intensities under different electric currents applied between 10 s and 40 s. (c) Single-molecule fluorescence intensity trajectories (data points) for 75 DiD single molecules that were obtained while applying different EECs.

density of 32.0 mA/mm² is applied to the electrodes, the fluorescence intensities of the single molecule exhibit an exponential decay with the time constant about 2.24 ± 0.23 s. The fluorescence emission gradually recovers to the initial value which needs about 10 s after switching off the EEC. We find that both of the decay time and the recovering time depend on the molecular neighboring environment. The emission intensities of another single molecule at different electric currents are showed in Fig. 8(b), note that the rates of intensity decay are different, when electric currents are applied between 10 s and 40 s. These intensity decay traces were fitted by single-exponential function with the time constants of 5.80 ± 1.20 s, 3.20 ± 0.54 s and 1.38 ± 0.09 s under the electric current density of 17.3 mA/mm², 39.0 mA/mm², and 46.8 mA/cm² respectively. Note that the bigger current, the faster response. The intensities recover to the initial values which need about 7.5 s after switching off the EEC. It is also found from the Fig. 8(b) that the 46.8 mA/mm² current density (and even bigger) can quench almost completely the fluorescence emission. The two molecules have the different recovering times, which arises from the heterogeneity of site-specific molecular interactions.

In addition, we record the average fluorescence intensities of each single DiD molecule at different currents. The statistical data of 75 DiD single molecules are shown in Fig. 8(c). All the DiD single molecules exhibit intensity decrease with the large EEC.

We present a model, as shown in the Fig. 9, to explain the results of fluorescence quenching of single-molecule. The schematic representation of energy levels, basic photoinduced and electric current driving processes, and multiple electron transfer cycles in DiD / ITO system are shown.

When an EEC is applied to the ITO film, the Fermi level of the ITO is tuned by the potential. With a positive potential, the Fermi level of ITO is decreased and there is more driving force for the forward electric transfer (FET) and the electron transfer of ground state (GET), but backward electron transfer (BET) is suppressed simultaneously.

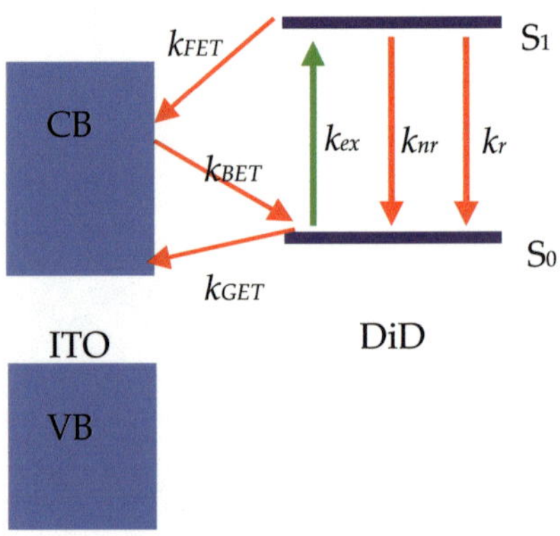

Figure 9. Schematic representation of energy levels, basic photoinduced and electric current driving process, and multiple electron transfer cycles in DiD / ITO system. k_r, radiative decay rate; k_{nr}, intrinsic nonradiative decay rate; k_{FET}, the rate of forward interfacial electron transfer from excited molecule to semiconductor; k_{BET}, the rate of backward electron transfer; k_{GET}, the rate of electron transfer from ground state of molecule to ITO; CB, the conduction band of ITO; VB, the valence band of ITO.

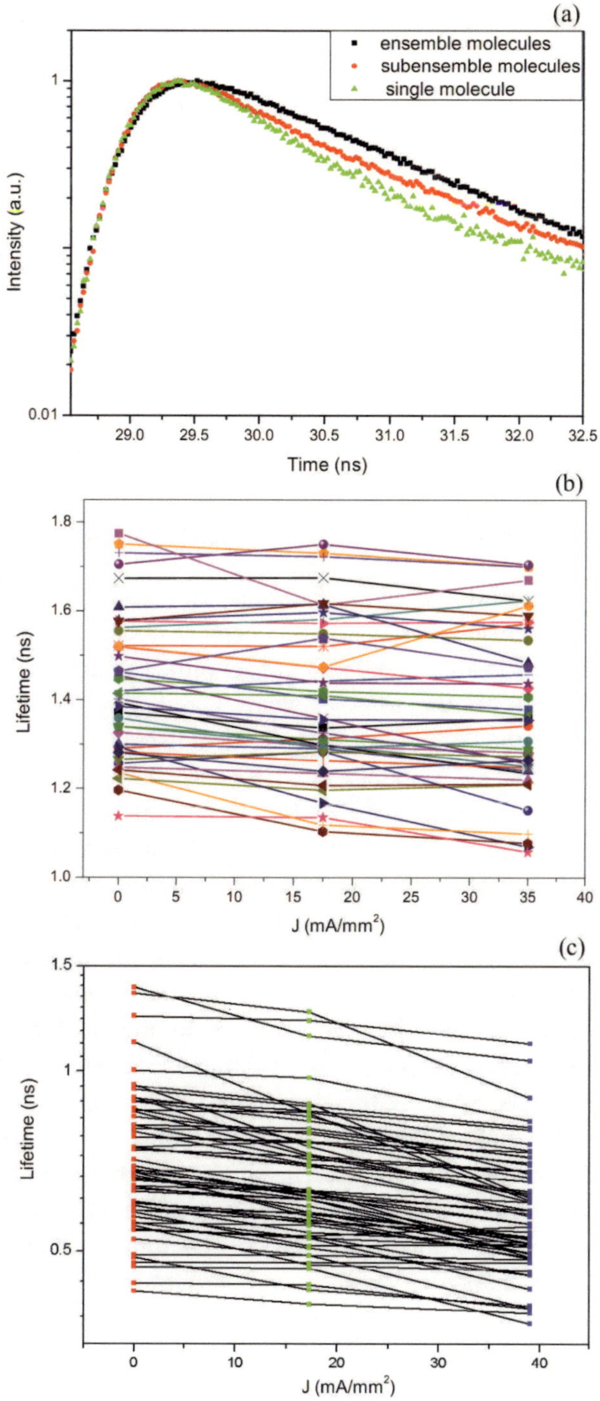

Figure 10. (a) Fluorescence decays for an ensemble molecules (black curve), a subensemble molecules (red curve) and a single molecule (green curve) in ITO film, measured by time correlated photon counting. The decay curves are fitted with single-exponential decay with the time constant about 1.41

ns, 1.16 ns and 0.90 ns respectively. (b) The fluorescence lifetimes of ensemble average for 45 ensembles at three different electric currents. The average fluorescence lifetime is 1.43 ns at 0 mA/mm^2, 1.40 ns at 17.5 mA/mm^2, and 1.38 ns at 35.0 mA/mm^2 respectively. (c) The fluorescence lifetimes of 65 single molecules at three different electric currents. The average fluorescence lifetime is 0.73 ns at 0 mA/mm^2, 0.67 ns at 17.3 mA/mm^2, and 0.59 ns at 39.0 mA/mm^2 respectively.

While the BET is suppressed completely at a large EEC, the fluorescence is completely quenched. When EEC is switched off, the Fermi level of ITO will gradually recover to its original value and the driving force for the FET and GET will decrease, simultaneously the BET will increase, thus the fluorescence will recover gradually.

3.2.4. Electric current dependence of fluorescence lifetime

Lu and Xie have presented that each single molecule exhibits a single-exponential IET dynamics in dye molecules / ITO film system, the rate of FET, k_{FET}, can be measured by the fluorescence decay of excited dye molecules. And the changes of fluorescence lifetimes were attributed to the FET (Guo et al., 2007; Wang et al., 2009; Jin, 2010).

Fig. 10(a) shows three typical fluorescence decay curves for an ensemble (average intensity is about 800 k cps), a subensemble (average intensity is about 40k cps) and a single DiD molecule (average intensity is about 8 k cps) in ITO film. The decay curves are fitted with single-exponential decay with the time constant about 1.41 ns, 1.16 ns and 0.90 ns respectively. Accordingly, the lifetimes of DiD single-molecule are shorter than the 2.5 ns ~ 3.0 ns lifetime measured in the polymer (Vallée et al., 2003). Fig. 10(b) shows that the fluorescence lifetimes of ensemble average for 45 ensembles at three different electric currents, ranging from 1.1 ns to 1.8 ns. The average fluorescence lifetime is 1.43 ns at 0 mA/mm^2, 1.40 ns at 17.5 mA/mm^2, and 1.38 ns at 35.0 mA/mm^2 respectively. Fig. 10(c) shows fluorescence lifetimes of 65 DiD single molecules at three different electric currents, ranging from ~300 ps to 1.4 ns. The average fluorescence lifetime is 0.73 ns at 0 mA/mm^2, 0.67 ns at 17.3 mA/mm^2, and 0.59 ns at 39.0 mA/mm^2 respectively. Unfortunately, we cannot measure a single-molecule lifetime shorter than 300 ps due to the limited sensitivity for time resolution of the instrumental response. As mentioned above, the FET rate is highly sensitive to the interactions between the dyes and ITO nanoparticles. When some dye molecules in the ensemble that are not in good contact with the ITO nanoparticles there will induce the poor IET dynamics. Thus, fluorescence lifetime of the ensemble is longer than the lifetime of single-molecule. In the typical dye-photosensitization system, FET can shorten the fluorescence lifetime and reduce fluorescence quantum yield of the dye molecules. Under the EEC, the small change of fluorescence lifetimes may indicate the change of the FET rate, and the fluorescence quenching may be mainly dominated by the BET and GET.

4. Conclusion

We present a fast and robust method to recognise single molecules based on single event photon statistics. Mandel's Q-parameter provides an attractive approach to two-time correlation measurements, because it is easy to implement, requires little time, and is immune with respect to the effects of molecular triplet state. Compared with common two-time correlation measurements, our approach has some advantages: (1) the effect of molecular triplet state can be ignored, whereas its effect can only be contained in the non-perfect detection efficiency analysis; (2) ~ ms level measurement time is needed as only ~10^4 fluorescence photons are needed for photon statistic; (3) it is not limited only to weak photons emitted, which means it is independent with the fluorophores photon intensity. The method can also be applied for the other single emitters recognition, such as single atoms, quantum dots and color centers.

Individual DiD dye molecules were dispersed in ITO semiconductor films as probes of IET dynamics, which should help to understand the intrinsic properties of electron transfer at interface between organic molecules and transparent semiconductor materials. While the EEC was used to drive the FET and GET, and suppress the BET, the change of k_{FET} induces the change of fluorescence lifetime and the increasing k_{GET} and decreasing k_{BET} would quench the fluorescence. Due to the inhomogeneous nature of the interactions from molecule to molecule, the lifetime under EEC is inhomogeneous. The EEC dependence of lifetime distribution clearly demonstrates a manipulated IET dynamics, which can be revealed by the single-molecule experiments instead of by the ensemble-averaged measurements. The results could open up a new path to manipulate single-molecule electron transfer dynamics by using EEC while measuring single-molecule fluorescence intensity and lifetime simultaneously.

Author details

Guofeng Zhang, Ruiyun Chen, Yan Gao, Liantuan Xiao and Suotang Jia
State Key Laboratory of Quantum Optics and Quantum Optics Devices,
Laser Spectroscopy Laboratory, Shanxi University, Taiyuan, China

Acknowledgement

The project sponsored by 973 Program (Nos. 2012CB921603, 2010CB923103), 863program (No. 2011AA010801),the Natural Science Foundation of China (Nos. 11174187，11184145, 60978018 and 10934004), NSFC Project for Excellent Research Team (No. 61121064), the Natural Science Foundation of Shanxi province, China (No. 2011091016), TSTIT and TYMIT of Shanxi, and Shanxi Province Foundation for Returned scholars.

5. References

Alléaume, R.; Treussart, F.; Courty, J. M. & Roch, J. F. (2004). Experimental open-air quantum key distribution with a single-photon source, *New J. Phys.*, Vol. 6, No. 92, (March 2004) page number (1-12), ISSN: 1367-2630

Bartko, A. P.; Xu, K. & Dickson, R. M. (2002). Three-dimensional single molecule rotational diffusion in glassy state polymer films, *Phys. Rev. Lett.*, Vol. 89, No. 2, (July 2001) page numbers (026101-1-4), ISSN: 0031-9007

Basché, T. & Moerner, W. E. (1992). Energetics of negatively curved graphitic carbon, *Nature*, Vol. 355, No. 6358, (January 1992) page numbers (333-335), ISSN: 0028-0836

Betzig, E. & Chichester, R. J. (1993). Single molecules observed by near-field scanning optical microscopy, *Science*, Vol. 262, No. 5138, (September 1993) page number (1422-1425), ISSN: 0036-8075

Brunel, C.; Lounis, B.; Tamarat P. & Orrit, M. (1999). Triggered source of single photons based on controlled single molecule fluorescence, *Phys. Rev. Lett.*, Vol. 83, No. 14, (February 1999) page numbers (2722-2725), ISSN: 0031-9007

Deniz, A. A.; Dahan, M.; Grunwell, J. R.; Ha, T.; Faulhaber, A. E.; Chemla, D. S.; Weiss, S. & Schultz, P. G. (1999). Single-pair fluorescence resonance energy transfer on freely diffusing molecules: Observation of Förster distance dependence and subpopulations, *Proc. Nat. Acad. Sci.USA*, Vol. 96, (March 1999) page number (3670-3675), ISSN: 0027-8424

Dickson, R. M.; Norris, D. J.; Tzeng, Y. L. & Moerner, W. E. (1996). Three-dimensional imaging of single molecules solvated in pores of poly(acrylamide) gels, *Science*, Vol. 274, No. 5289, (November 1996) page number (966-968), ISSN: 0036-8075

Friend, R. H.; Gymer, R. W.; Holmes, A. B.; Burroughes, J. H.; Marks, R. N.; Taliani, C.; Bradely, D. D. C.; Dos Santos, D. A.; Bredas, J. L.; Logdlund, M. & Salaneck, W. R. (1999). Electroluminescence in conjugated polymers, *Nature*, Vol. 397, (January 1999) page numbers (121-128), ISSN: 0028-0836

Funatsu, T.; Harada, Y.; Tokunaga, M.; Saito, K. & Yanagida, T. (1995). Imaging of single fluorescent molecules and individual ATP turnovers by single myosin molecules in aqueous solution, *Nature*, Vol. 374, (April 1995) page numbers (555-559), ISSN: 0028-0836

Guo, L.; Wang, Y. & Lu, P. H. (2010). Combined single-molecule photon-stamping spectroscopy and femtosecond transient absorption spectroscopy studies of interfacial electron transfer dynamics, *J. Am. Chem. Soc.*, Vol. 132, No. 6, (January 2010) page numbers (1999-2004), ISSN: 0002-7863

Hanbury, B. R. & Twiss, R. Q. (1956). Correlation between photons in two coherent beams of light, *Nature*, Vol. 177, (January 1956) page numbers (27-29), ISSN: 0028-0836

Hania, P. R., Thomsson, D. & Scheblykin, I. G. (2006). Host matrix dependent fluorescence intensity modulation by an electric field in single conjugated polymer chains, *J. Phys. Chem. B*, Vol. 110, No. 51, (December 2006) page numbers (25895-25900), ISSN: 1520-6106

Hanson, E. L.; Guo, J.; Koch, N.; Schwartz, J. & Bernasek, S. L. (2005). Advanced surface modification of indium tin oxide for improved charge injection in organic devices, *J. Am. Chem. Soc.*, Vol. 127, No. 28, (June 2005) page numbers (10058-10062), ISSN: 0002-7863

Huang, T.; Dong, S. L.; Guo, X. j.; Xiao, L. T. & Jia, S. T. (2006). Signal-to-noise ratio improvement of photon counting using wavelength modulation spectroscopy, *Appl.*

Phys. Lett., Vol. 89, No. 6, (November 2005) page number (061102-061102-3), ISSN: 0003-6951

Huang, T.; Wang, X. B.; Shao, J. H.; Guo, X. J.; Xiao, L. T. & Jia, S. T. (2007). Single event photon statistics characterization of a single photon source in an imperfect detection system, *J. Lumin.*, Vol. 124, No. 2 (August 2005) page number (286-290), ISSN: 0022-2313

Jin, S.; Snoeberger III, R. C.; Issac, A.; Stockwell, D.; Batista, V. S. & Lian, T. (2010). Single-molecule interfacial electron transfer in donor-bridge-nanoparticle acceptor complexes, *J. Phys. Chem. B,* Vol. 114, No. 45, (March 2010) page numbers (14309-14319), ISSN: 1520-6106

Kulzer, F.; Xia, T. & Orrit, M. (2010). Single molecules as optical nanoprobes for soft and complex matter, *Angew. Chem. Int. Ed.,* Vol. 49, No. 5, (January 2010) page numbers (854-866), ISSN: 1433-7851

Lei, C. H.; Hu, D. H. & Ackerman, E. (2009). Clay nanoparticle supported single-molecule fluorescence spectroelectrochemistry, *Nano Lett.,* Vol. 9, No. 2, (January 2009) page numbers (655-658), ISSN: 1530-6954

Lu, H. P. & Xie, X. S. (1997). Single-molecule kinetics of interfacial electron transfer, *J. Phys. Chem. B,* Vol. 101, No. 15, (February 1997) page numbers (2753-2757), ISSN: 1520-6106

Mandel, L. (1979). Sub-poissonian photon statistics in resonance fluorescence, *Opt. Lett.*, Vol. 4, No. 7, (March 1979) page number (205-207), ISSN: 0146-9592

Michalet, X. & Weiss, S. (2002). Single-molecule spectroscopy and microscopy, *C. R. Physique*, Vol. 3, No. 5, (March 2002) page number(619-644), ISSN: 1631-0705

Moerner, W. E. & Kador, L. (1989). Optical detection and spectroscopy of single molecules in a solid, *Phys. Rev. Lett.*, Vol. 62, No. 21, (March 1989) page numbers (2535-2538), ISSN: 0031-9007

Nie, S.; Chiu, D. T. & Zare, R. N. (1994). Probing individual molecules with confocal fluorescence microscopy, *Science,* Vol. 266, No. 5187, (November 1994) page number (1018-1021), ISSN: 0036-8075

Orrit, M. (2002). Single-molecule spectroscopy: The road ahead, *J. Chem. Phys.,* Vol. 119, No. 24, (December 2002) page numbers (10938-10946), ISSN: 0021-9606

Orrit, M. & Bernard, J. (1990). Single pentacene molecules detected by fluorescence excitation in a p-terphenyl crystal. *Phys. Rev. Lett.*, Vol. 65, No. 21, (July 1990) page numbers (2716-2719), ISSN: 0031-9007

Palacios, R. E.; Fan, F. R. F.; Bard, A. J. & Barbara, P. F. (2006). Single-molecule spectroelectrochemistry, *J. Am. Chem. Soc.,* Vol. 128, No. 28, (June 2006) page numbers (9028-9029), ISSN: 0002-7863

Palacios, R. E.; Chang, W. S.; Grey, J. K.; Chang, Y. L.; Miller, W. L.; Lu, C. Y.; Henkelman, G.; Zepeda, D.; Ferraris, J. & Barbara, P. F. (2009). Detailed single-molecule spectroelectrochemical studies of the oxidation of conjugated polymers, *J. Phys. Chem. B,* Vol. 113, No. 44, (October 2009) page numbers (14619-14628), ISSN: 1520-6106

Sanchez-Andres, A.; Chen, Y. & Müller, J. D. (2005). Molecular brightness determined from a generalized form of mandel's Q-parameter, *Biophys. J.*, Vol. 89, No. 5, (May 2005) page number (3531-3547), ISSN: 0006-3495

Shera, E. B.; Seizinger, N. K.; Davis L. M.; Keller, R. A. & Soper, S. A. (1990). Detection of single fluorescent molecules, *Chem. Phys. Lett.*, Vol. 174, No.6, (July 1990) page numbers (553-557), ISSN: 0009-2614

Short, R. & Mandel, L. (1983). Observation of sub-poissonian photon statistics, *Phys. Rev. Lett.*, Vol. 51, No. 5, (April 1983) page numbers (384-387), ISSN: 0031-9007

Tak, Y. H.; Kim, K. B.; Park, H. G.; Lee, K. H. & Lee, J. R. (2002). Criteria for ITO (indium–tin-oxide) thin film as the bottom electrode of an organic light emitting diode, *Thin Solid Films*, Vol. 411, No. 1, (May 2002) page numbers (12-16), ISSN: 0040-6090

Trautman, J. K.; Macklin, J. J.; Brus, L. E. & Betzig, E. (1994). Imaging and time-resolved spectroscopy of single molecules at an interface, *Science*, Vol. 272, No. 5259, (April 1996) page number (255-258), ISSN: 0036-8075

Treussart, F. ; Alléaume, R. ; Le Floc'h, V., Xiao, L. T.; Courty, J. M. ; & Roch, J. F. (2002). Direct measurement of the photon statistics of a triggered single photon source, *Phys. Rev. Lett.*, Vol. 89, No. 9, (February 2002) page numbers (093601-1-4), ISSN: 0031-9007

Uji-i, H.; Melnikov, S. M.; Deres, A.; Bergamini, G.; De Schryver, F.; Herrmann, A.; Müllen, K.; Enderlein, J. & Hofkens, J. (2006). Visualizing spatial and temporal heterogeneity of single molecule rotational diffusion in a glassy polymer by defocused wide-field imaging, *Polymer*, Vol. 47, No.7, (March 2006) page numbers (2511-2518), ISSN: 0032-3861

Vallée, R. A. L.; Tomczak, N.; Kuipers, L.; Vancso, G. J. & van Hulst, N. F. (2003). Single molecule lifetime fluctuations reveal segmental dynamics in polymers, *Phys. Rev. Lett.*, Vol. 91, No. 3, (July 2003) page numbers (1038301-1--1038301-4), ISSN: 0031-9007

Wang, Y. M.; Wang, X. F.; Ghosh, S. K. & Lu, H. P. (2009). Probing single-molecule interfacial electron transfer dynamics of porphyrin on TiO2 nanoparticles. *J. Am. Chem. Soc.*, Vol. 131, No. 4, (January 2009) page numbers (1479-1487), ISSN: 0002-7863

Wang, Y. M.; Wang, X. F. & Lu, H. P. (2009). Probing single-molecule interfacial geminate electron–cation recombination dynamics, *J. Am. Chem. Soc.*, Vol. 131, No. 25, (June 2009) page numbers (9020-9025), ISSN: 0002-7863

Weiss, S. (1999). Fluorescence spectroscopy of single biomolecules, *Science*, Vol. 283, No. 5408, (March 1999) page numbers (1676-1683), ISSN: 0036-8075

Yang, H.; Luo, G. B.; Karnchanaphanurach, P.; Louie, T. M.; Rech, I.; Cova, S.; Xun, L. Y. & Xie, X. S. (2003). Protein conformational dynamics probed by single-molecule electron transfer, *Science*, Vol. 302, No. 5643, (October 2003) page numbers (262-266), ISSN: 0036-8075

Zhang, G. F.; Xiao, L. T.; Zhang, F.; Wang, X. B. & Jia, S. T. (2010). Single molecules reorientation reveals the dynamics of polymer glasses surface, *Phys. Chem. Chem. Phys.*, Vol. 12, (December 2009) page numbers (2308-2312), ISSN: 1463-9076

Chapter 16

Ultrashort Laser Pulses for Frequency Upconversion

Kun Huang, E Wu, Xiaorong Gu, Haifeng Pan and Heping Zeng

Additional information is available at the end of the chapter

http://dx.doi.org/10.5772/48541

1. Introduction

Single-photon frequency upconversion is a nonlinear process where the frequency of the signal photon is translated to the higher frequency with the complete preservation of all the quantum characteristics of the "flying" qubits [1,2]. One promising application of the single-photon frequency upconversion is converting the infrared photons to the desired spectral regime (usually visible regime) where the high performance detectors are available for sensitive detections [3-6]. The infrared single-photon frequency upconversion detection technique has been successfully used in a variety of applications, including infrared imaging [7] and infrared ultra-sensitive spectroscopy [8,9]. Additionally, such upconversion detector has greatly benefited the applications stringently requiring efficient photon detection in optical quantum computation and communication [10-12]. Furthermore, due to the phase-matching requirement in the frequency upconversion process, the upconversion detectors have some unique features, such as narrow-band wavelength acceptance [13] and polarization sensitivity [14], both of which render them very useful for fiber based quantum systems [15]. In the recent decade, experiments bear that the photon correlation, entanglement, and photon statistics are all well preserved in the coherent frequency upconversion [16-18].

Although the demonstration of frequency upconversion technology for strong light could be dated to the late 70s of last century [19], the single-photon frequency upconversion was just achieved at the beginning of this century [20]. The advent and rapid evolution of fabricating the periodically reversed nonlinear media led to the widespread use of quasi-phase matching (QPM), which has opened up new operating regimes for nonlinear interactions [21]. Rapidly growing interest has been focused in recent years on proposing novel schemes for achieving single-photon frequency upconversion with high efficiency and low noise and expanding its applications in all-optical nonlinear signal processing and in quantum state manipulation. With the help of extracavity enhancement, highly efficient up-conversion at single-photon level has been demonstrated by using bulk periodically poled lithium niobate (PPLN) crystals

[3]. By means of intracavity enhancement, a stable and efficient single-photon counting at 1.55 μm was achieved without the requirement of sensitive cavity servo feed-back [4,22]. Recently, frequency upconversion based on the PPLN waveguide attracts more and more attention [9,23]. All of the above single-photon frequency upconversion systems were pumped by the strong continuous wave (CW) lasers which would inevitably bring about severe background noise due to parasitic nonlinear interactions. Frequency upconversion based on pulsed pump arises to provide an effective solution [24]. To make sure that every signal photon can be interact with the pulsed pump field, synchronous pumping frequency upconversion system is thus introduced, which greatly improves the total detection efficiency. VanDevender *et al.* reported a frequency upconversion system based on electronic synchronization of the pump source and signal source with a repetition rate of just 40 kHz [5]. For satisfying applications of high-speed single-photon detection, fast and efficient single-photon frequency upconversion detection system operating at tens of MHz was realized based on the all-optical synchronized fiber lasers [25-27].

Thanks to the ultrashort optical pulses, intense peak power to obtain unit conversion efficiency could be achieved with a modest average pump power, thus loosening the restriction for the available output power from the pump laser [3,4]. Additionally, the background noise induced by the strong pump can be effectively reduced due to the pulsed excitation. Thus ultrashort pulses constitute ideal means as the pump source of frequency upconversion. The pulsed pumping system could not only permit efficient photon counting of infrared photons, but also quantum manipulation of single photons. One can realize a quantum state router by having a control on the multimode pump source [28]. Current researches also demonstrate simultaneous wavelength translation and amplitude modulation of single photons from a quantum dot by pulsed frequency upconversion [29]. Additionally, the concept of a quantum pulse gate is presented and an implementation is proposed based on spectrally engineered frequency upconversion [30].

In this chapter, we review the recent experimental progress in single-photon frequency upconversion with synchronous pulse pumping laser. The signal photons were tightly located within the synchronous pump pulses [25]. For improving the conversion efficiency, the specific control of the synchronized pulses was required, which led to the development of temporally and spectrally controlled single-photon frequency upconversion for pulsed radiation [26,27]. The compact fiber-laser synchronization system for fast and efficient single-photon frequency upconversion detection is of critical potential to stimulate promising applications, such as infrared photon-number-resolving detector (PNRD) [18,31] and ultrasensitive infrared imaging at few-photon level. This chapter is organized as follows. After this brief introduction, we present the basic theory for frequency upconversion in quantum frame in Section 2. Section 3 presents the experimental realization of synchronous pulsed pumping. Applications based on the synchronous pumping frequency upconversion system will be highlighted in Section 4 by the examples of infrared photon-number-resolving detection and several-photon-level infrared imaging. We conclude the chapter in Section 5 by emphasizing that the synchronous pumping frequency upconversion system will benefit numerous applications not only in infrared photon counting and also in manipulation of the quantum state of single photons.

2. Quantum description of frequency upconversion

Frequency upconversion based on sum frequency generation (SFG) is an optical process by which two optical fields combine in a quadratic nonlinear medium to generate a third field at a frequency equal to the sum of the two inputs as shown in Fig. 1. The theory of the SFG is well established for a long time and applied in a variety of classical and quantum optical applications. In this section, we go through the fundamental theory of SFG and derive the important results in quantum architecture. We investigate the quantum features according to different pumping methods for frequency upconversion, respectively. Specially, we will concentrate on the detailed discussion of the quantum characteristics in the multimode pumping scheme.

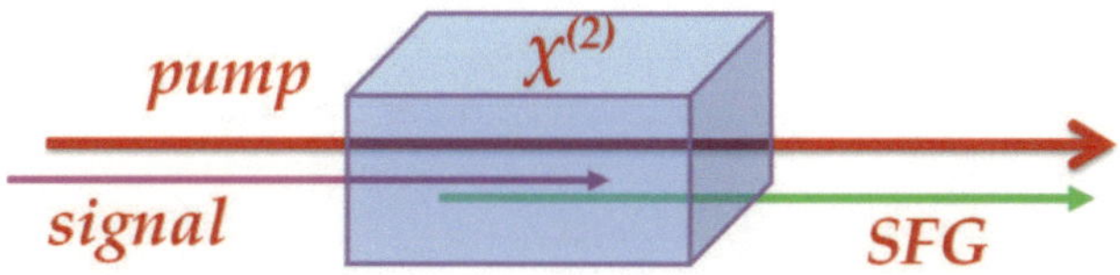

Figure 1. Schematic illustration of the frequency upconversion process.

2.1. Single-mode frequency upconversion

In quantum optics, the frequency upconversion is a nonlinear optical process, in which a photon at one frequency is annihilated and another photon at a higher frequency is created. We start with a Hamiltonian for three-wave mixing:

$$\hat{H}_{3W} = i\hbar g(\hat{a}_p \hat{a}_1 \hat{a}_2^{\dagger} - H.c.), \tag{1}$$

where $\hat{a}_1$ and $\hat{a}_p$ is the annihilation operators corresponding to the signal photon at ω_1 and pump photon ω_p, respectively. $\hat{a}_2^{\dagger}$ is the creation operator corresponding to the upconverted photon ω_2, g is the coupling constant which is determined by the second-order susceptibility of the nonlinear medium, and *H.c.* denotes a Hermitian conjugate. If the pump field is very strong with negligible depletion, as in the case here, we can classically treat it as a constant *Ep*. So the Hamiltonian in Eq. (1) becomes

$$\hat{H} = i\hbar g E_p(\hat{a}_1 \hat{a}_2^{\dagger} - H.c.). \tag{2}$$

Since for most cases, the upconverted frequency wave was in vacuum state at the input of the nonlinear medium. So the initial condition at the input facet of the nonlinear medium could be written as

$$|\Phi\rangle = |\Psi_1, 0_2\rangle, \tag{3}$$

where $|\Psi_1\rangle$ represents the input signal state and $|0_2\rangle$ represents the vacuum input of the SFG state. The dynamics of the input and output quantum fields in phase-matched SFG process can be described by the coupled-mode equations as

$$\hat{a}_1(L) = \hat{a}_1(0)\cos(|gE_p|L) - \hat{a}_2(0)\sin(|gE_p|L), \quad (4)$$

$$\hat{a}_2(L) = \hat{a}_2(0)\cos(|gE_p|L) + \hat{a}_1(0)\sin(|gE_p|L), \quad (5)$$

where L is the interaction length in the nonlinear crystal. The corresponding creation operators can be found by taking the Hermitian conjugates of these equations.

Eq. (5) indicates that frequency translation of any quantum state at ω_1 to the same quantum state at ω_2 with unity efficiency is possible, even at the single-photon level, if $|gE_p|L = \pi/2$. Note that the coherent quantum transduction from one frequency to a higher frequency would preserve complete coherence properties including entanglement, quantum correlation, and photon statistics.

At the output of the nonlinear medium, the average photon number of infrared signal can be calculated from expected value of the photon number operator:

$$\begin{aligned} N_1(L) &= \left\langle \hat{a}_1^\dagger(L)\hat{a}_1(L) \right\rangle \\ &= \langle \psi_1 | \hat{n}_0 | \psi_1 \rangle \cos^2(|gE_p|L) \\ &= N_1^0 \cos^2(|gE_p|L), \end{aligned} \quad (6)$$

where $\hat{n}_0$ denotes the photon number operator of the input infrared signal; N_1^0 is the average photon number of the input signal.

Meanwhile, the average number of the SFG photons can be likewise given by

$$N_2(L) = N_1^0 \sin^2(|gE_p|L). \quad (7)$$

From Eq. (6) and Eq. (7), the upconversion efficiency can be given as

$$\eta = \frac{N_2(L)}{N_1(0)} = \sin^2(|gE_p|L). \quad (8)$$

Additionally, we can get $N_2(L) + N_1(L) = N_1^0$, indicating the energy conservation during the conversion process. The correlation between the upconverted photons and unconverted photons is shown in Fig. 2.

The joint probability P_{12} of simultaneously detecting a photon at both ω_1 and ω_2 at the output of the frequency upconverter is proportional to $\langle \hat{n}_1\hat{n}_2 \rangle$. With the help of Eq. (4,5) we readily obtain that

$$\begin{aligned} \langle \hat{n}_1\hat{n}_2 \rangle &= \left\langle \hat{a}_1^\dagger(L)\hat{a}_2^\dagger(L)\hat{a}_2(L)\hat{a}_1(L) \right\rangle \\ &= [\langle \hat{n}_0^2 \rangle - \langle \hat{n}_0 \rangle]\cos^2(|gE_p|L)\sin^2(|gE_p|L). \end{aligned} \quad (9)$$

Inserting Eq. (8) into the above equation, we have

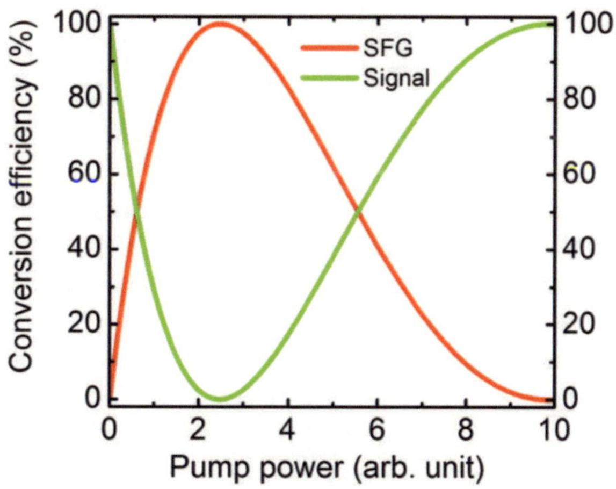

Figure 2. Correlation between the upconverted SFG photons and unconverted signal photons.

$$\langle \hat{n}_1 \hat{n}_2 \rangle = [\langle \hat{n}_0^2 \rangle - \langle \hat{n}_0 \rangle]\eta(1-\eta), \tag{10}$$

which is dependent on the conversion efficiency of the frequency upconversion process.

Moreover, the intensity cross-correlation function $g^{(2)}(\tau)$ at $\tau = 0$ is then obtained from

$$g^{(2)}(0) = \frac{\langle \hat{n}_1 \hat{n}_2 \rangle}{\langle \hat{n}_1 \rangle \langle \hat{n}_2 \rangle} = \frac{\langle \hat{n}_0^2 \rangle - \langle \hat{n}_0 \rangle}{\langle \hat{n}_0 \rangle^2}. \tag{11}$$

When input signal photons are in single photon state, meaning $\langle \hat{n}_0^2 \rangle = \langle \hat{n}_0 \rangle$, the probability of detecting a photon both in frequency ω_1 and ω_2 is then zero. The unconverted infrared photons and SFG photons are anti-correlated. When the incident photons are in a coherent state, meaning $\langle \hat{n}_0^2 \rangle - \langle \hat{n}_0 \rangle = \langle \hat{n}_0 \rangle^2$, the intensity cross-correlation equals to 1, which means no anti-correlation will be observed. It indicates that the frequency upconversion process is a random event for the individual incident signal photons.

All above results are obtained on the basis of perfect phase matching. The phase matching condition is sensitive in the SFG interaction due to the limit spectral bandwidth of the nonlinear medium. Taking into account the phase mismatching [32], the quantum state of the SFG photons can be expressed as

$$\begin{aligned} \hat{a}_1(L,\Delta\omega) = {} & \cos(qL)\exp(-i\Delta\omega\Delta kL/2)\hat{a}_1(0,\Delta\omega) \\ & -\frac{igE_p}{q}\sin(qL)\exp(-i\Delta\omega\Delta kL/2) \\ & \times[\frac{\Delta\omega\Delta k}{2gE_p}\hat{a}_1(0,\Delta\omega)+\hat{a}_2(0,\Delta\omega)], \end{aligned} \tag{12}$$

$$\begin{aligned}\hat{a}_2(L,\Delta\omega) &= \cos(qL)\exp(-i\Delta\omega\Delta kL/2)\hat{a}_2(0,\Delta\omega) \\ &+\frac{igE_p}{q}\sin(qL)\exp(-i\Delta\omega\Delta kL/2) \\ &\times[\frac{\Delta\omega\Delta k}{2gE_p}\hat{a}_2(0,\Delta\omega)+\hat{a}_1(0,\Delta\omega)],\end{aligned} \tag{13}$$

where Δk represents the phase-mismatch at frequency detuning $\Delta\omega$, and $q=[(gE_p)^2+(\Delta\omega\Delta k/2)^2]^{1/2}$. At zero detuning $(\Delta\omega=0)$, Eqs. (12) and (13) can be deduced to Eqs. (4) and (5), respectively. The unity single-photon frequency conversion can be achieved when $|gE_p|L=\pi/2$ is fulfilled. At nonzero detuning $(\Delta\omega\neq 0)$, although a similar condition $qL=\pi/2$ can be fulfilled, the perfect quantum conversion is impossible due to the disturbance of the vacuum state $\hat{a}_2(0,\Delta\omega)$. Therefore, for practical experiments, the spectral requirement of the pump field for frequency upconversion should be well concerned.

2.2. Synchronous pumping frequency upconversion

Could the complete quantum state transduction be feasible when the signal and the pump source were both in the multi-longitudinal modes? This is the instance of the synchronous pumping frequency upconversion system. Obviously, the frequency conversion process becomes more complicated due to the commutative nonlinear coupling of the pump modes and signal modes.

The Hamiltonian can be rewritten as

$$\hat{H}=i\hbar g\sum_{ij}E_{pi}(\hat{a}_{1j}\hat{a}_{2ij}^{\dagger}-H.c.), \tag{14}$$

where the E_{pi} is the pump electric field related to each longitudinal mode numbered by *i*, and the $\hat{a}_{1j}$ is the annihilation operator of the infrared signal photons related to each longitudinal mode numbered by *j*, the $\hat{a}_{2ij}^{\dagger}$ is the creation operator of the SFG photons corresponding to the longitudinal mode of both pump field and infrared signal photons. Then the dynamics of the input and output quantum fields in phase-matched SFG processes can be described by the coupled-mode equations as

$$\begin{aligned}\frac{d\hat{a}_{1j}}{dt} &= \frac{1}{ih}[\hat{a}_{1j},\hat{H}]=-g\sum_i E_{pi}\hat{a}_{2ij} \\ \frac{d\hat{a}_{2ij}}{dt} &= \frac{1}{ih}[\hat{a}_{2ij},\hat{H}]=gE_{pi}\hat{a}_{1j}.\end{aligned} \tag{15}$$

The superposition state of the infrared signal photons and SFG photons are represented by the operators $\hat{a}_1$, $\hat{a}_2$, which are defined as

$$\begin{aligned}\hat{a}_1 &= \sum_i \hat{a}_{1j} \\ \hat{a}_2 &= \sum_j\sum_i C_i\hat{a}_{2ij},\end{aligned} \tag{16}$$

where $E_p^2 = \sum_i E_{pi}^2$ and $C_i = E_{pi} / E_p$ denotes the probability amplitude of each longitudinal mode. Then the coupled-mode equation can be trivially solved by using the initial condition at the input facet of the nonlinear medium to yield

$$\hat{a}_1(L) = \hat{a}_1(0)\cos(|gE_p|L) - \hat{a}_2(0)\sin(|gE_p|L), \tag{17}$$

$$\hat{a}_2(L) = \hat{a}_2(0)\cos(|gE_p|L) + \hat{a}_1(0)\sin(|gE_p|L). \tag{18}$$

We get the same results as that in a single-longitudinal mode situation (Eqs. (4, 5)). When $|gE_p|L = \pi/2$ is satisfied, the coherence properties of the incident signal photons could be maintained during the upconversion process in this multi-mode system.

Here we can clearly see that either in single-mode regime or in multi-mode regime, the upconversion can, in principle, be used to transduce one photon at a given wavelength to another wavelength in preservation of all the quantum characteristics. Such quantum upconverter would not only be useful in the fields related to efficient infrared photon counting, but also helpful to implement novel quantum functions such as quantum interface and quantum gate or quantum shaper. In next section, experimental realization of the single-photon frequency upconversion with synchronously pumping will be demonstrated at length.

3. Experimental realization of single-photon frequency upconversion system

To achieve efficient and low-noise single-photon frequency upconversion system, various schemes have been proposed by the researchers for different applications in recent years. The upconversion technique typically requires a sufficiently strong pump to achieve unity nonlinear frequency conversion efficiency in a quadratic nonlinear crystal. The requisite strong pump can be achieved by using an external cavity or intracavity enhancement [3,4] or a waveguide confinement [9,23]. With such high intensity of pump, frequency up-conversion was implemented with almost 100% internal conversion efficiency. Nevertheless, a strong pump field inevitably brings about severe background noise because of parasitic nonlinear interactions, such as spontaneous parametric downconversion and Raman scattering. Pulsed pumping technique is given rise to circumscribe induced noise within the narrow temporal window of the pump pulses. Ultralow background counts of 150/s was reported with the help of a long-wavelength pump scheme [24]. In order to include every signal photon within the pump pulse for improving the total detection efficiency, researchers recently develop a coincidence single-photon frequency upconversion system [5,25-27]. In this section, we will focus on the synchronous pulsed pumping technique and its experimental results.

Fig. 3 shows the experimental setup of the synchronous single-photon frequency upconversion detection. The system was composed of a passive master-slave synchronization fiber-laser system and a single-photon frequency upconversion counting system. In the synchronization fiber-laser system, both two fiber lasers were passively mode-locked by the nonlinear polarization rotation in the fiber cavity, operating at the repetition rate of 17.6 MHz to

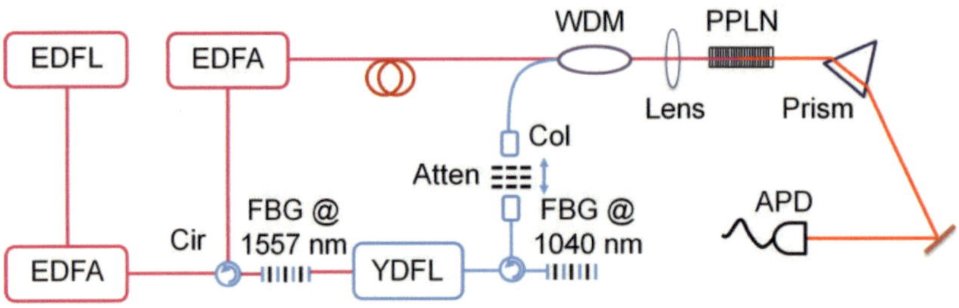

Figure 3. Experimental setup of the synchronous single-photon frequency upconversion detection. YDFL, ytterbium-doped fiber laser; EDFL, erbium-doped fiber laser; EDFA, erbium-doped fiber amplifier; Cir, circulator; Col, collimator; Atten, attenuator; WDM, wavelength-division multiplexer; FBG, fiber Bragg grating; PPLN, periodically poled lithium niobate crystal; APD, avalanche photodiodes.

satisfy the high-speed detection. The master laser was an Er-doped fiber ring laser. The ring cavity consisted with 1.5-m Er-doped fiber, standard single-mode fiber and dispersion-shifted fiber. By optimizing the lengths of the single-mode and dispersion-shifted fibers while maintaining the total cavity length, the dispersion in the cavity was well controlled. The spectrum of the output laser centered at 1563.8 nm with a full width at half maximum (FWHM) of 6.2 nm. A narrow spectral portion of the master laser was extracted by the FBG reflection for approaching the QPM bandwidth of the PPLN. This part was further amplified by an Er-doped fiber amplifier (EDFA) as the pump source. The maximum output power was about 60 mW and the output spectrum from the amplifier centered at 1557.6 nm with an FWHM of 0.56 nm [Fig. 4(a)]. The transmission from the FBG was injected as a seed to the slave laser cavity via a 1040/1557-nm wavelength-division multiplexer (WDM) to trigger the mode-locking of the Yb-doped fiber laser. An FBG was used as well at the slave laser output to control the spectral matching of the signal photons. The filtered spectrum was centered at 1040.0 nm with the bandwidth about 0.35 nm (FWHM) [Fig. 4(b)]. Injection-locking of the slave laser ensured the synchronization between the master and slave laser due to the fact that the transmission from the FBG shared exactly the same spectral mode separation with the reflection.

With this master-slave configuration, the two synchronized lasers could be isolated from each other, and thus free from mutual perturbation. This all-optical synchronization technique is superior to other passive synchronization methods due to its simplicity and robustness [33-35]. If self-started mode-locking existed in the slave laser, synchronized injection-locking could be obtained as well, but only within a sensitive slave cavity length match and could not last long, which would not be implemented in the frequency upconversion system. The self-started mode-locking would compete with the injection-triggered mode-locking, leading to the instability. To reduce the instability, self-started mode-locking was avoided in the slave fiber laser by properly adjusting the laser polarization state in the cavity. In such a critical position, the slave laser could only be mode-locked with the presence of the seed injection. The cavity length of the slave laser was adjusted to be the same as the master laser by carefully moving one of the fiber collimators on the stage. Eventually, the slave laser was synchronized with the master laser at the same repetition rate. And the stability of synchronization was enhanced by optimizing the master laser injection polarization. With such setup, we achieved the maximum cavity mismatch tolerance of 25 μm with the long-term stability of several hours.

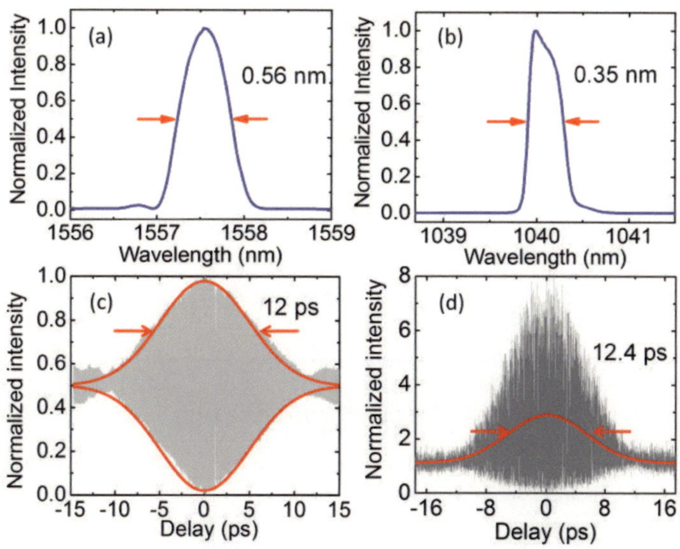

Figure 4. Spectrum of the pump source from the EDFL (a) and the signal source from the YDFL (b), respectively. Autocorrelation pulse profiles of the pump source (c) and the signal source from (d), respectively. Solid symbols are the experimental data and solid curves are the Gaussian fits to the data.

The signal pulse duration was measured by an autocorrelator with the FWHM bandwidth of 12 ps [Fig. 4(c)], corresponding to the actual pulse duration of 6 ps by taking the denominator of 2. In order to optimize the upconversion efficiency, the pump pulse duration relative to the signal pulse needs to be considered carefully [5]. The total conversion efficiency dependent on the pulse overlapping is given by

$$P_{overlap} = \int_{-\infty}^{+\infty} P_0(I_p(t))I_s(t)dt, \tag{19}$$

where P_0 is the probability of upconversion dependent on pump intensity I_P as $\sin^2(I_p^{1/2})$, and $I_s(t)$ is the normalized input pulse profile

$$\int_{-\infty}^{+\infty} I_s(t)dt = 1. \tag{20}$$

The simulated conversion efficiency shown in Fig. 5 indicates that the total efficiency increases as the pump pulse duration goes longer by assuming a constant FWHM duration (6 ps) for the signal pulse. When the pump pulse duration is much shorter than the signal pulse, the conversion efficiency is quite low as most of the photons distribute outside of the pump pulse temporal window. On the other hand, too long pulse duration will lead to reduction of the energy utilization efficiency and increase of the background counts. Therefore, to achieve a conversion efficiency over 90% with experimentally available pump intensity, it is theoretically necessary to make pulse duration ratio between the pump and signal slightly larger than 1.2.

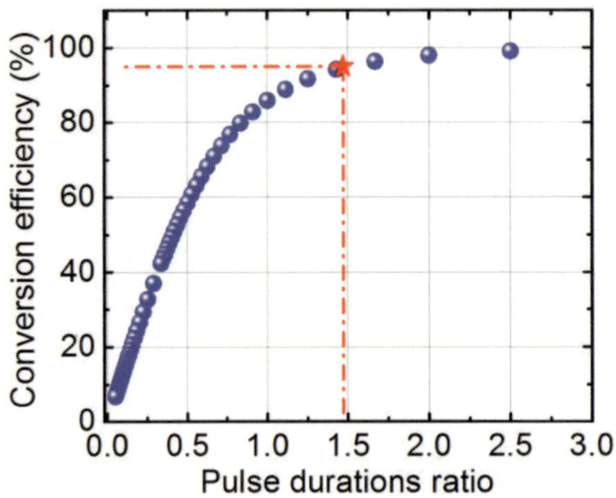

Figure 5. Simulated conversion efficiency as a function of the pulse duration ratio between the pump and signal. The red star indicates the experimental situation.

Thanks to the spectral filtering by the FBGs in the experiment, the pulse durations of the master and slave lasers were stretched according to the time-frequency Fourier transform. With the help of cavity dispersion management in the master laser cavity, the desired temporal match was achieved. The pump pulse duration was measured to be 8.8 ps [Fig. 4 (d)]. In this way, the signal photons and pump pulses were well matched temporally, guaranteeing an efficient upconversion with a relatively low background noise. According to the theoretical simulation, the conversion efficiency can reach 95% as shown by the red star in Fig. 5.

In the experiment, the timing jitter between the signal and pump pulses was measured to be as low as 45 fs shown in Fig. 6. Such a low timing jitter would impose a negligible influence on the temporal distribution of the signal photons within the pump pulse window. So the single-photon signal could be synchronously gated with a sufficiently high stability for coincidence upconversion detection.

As shown in Fig. 3, the signal beam and pump beam were then combine by a 1040/1557-nm WDM before being focused at the center of the 50-mm-long PPLN crystal. The temperature of the crystal was optimized at 130.4 °C for the grating period of 11.0 μm. The temperature was high enough to avoid photorefractive effects [36]. The infrared object beam interacted with the pump beam, and was upconverted through SFG to generate the visible photon at 624 nm.

For a single-photon frequency upconversion system, besides efficiency and stability, background noise was also a very important metrics for the detector performance. To analyze the noise, the spectrum before the filtering system was recorded. No observable peaks around the wavelength of the sum-frequency photons appeared when no signal photons were incident, revealing that the major noise from pump-induced parametric fluorescence was almost eliminated in our experiment. Considering the pulsed pump mode, the noise was localized within a very short pump time window which was much shorter than any electronic gates applied on the APD, leading to a very low noise counts on the APD.

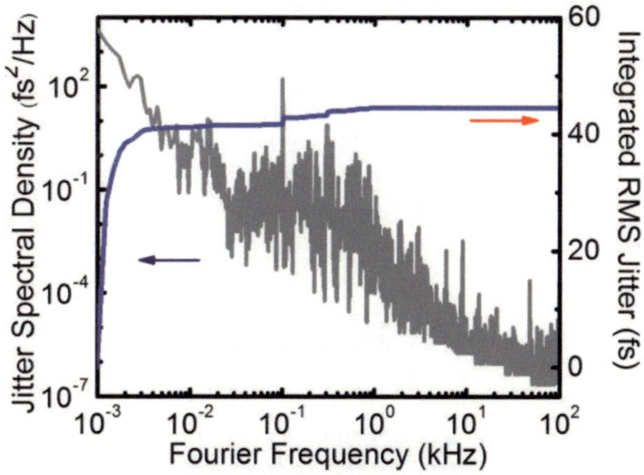

Figure 6. Timing jitter power spectral density and the integrated timing jitter in Fourier domain.

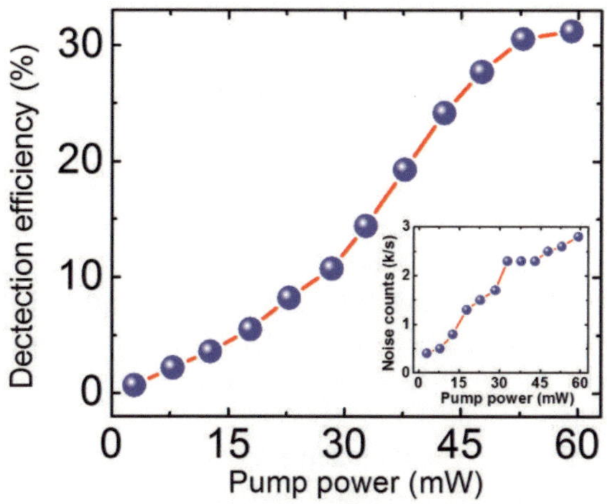

Figure 7. Conversion efficiency as a function of the pump power. Inset: background counts vs the pump power.

As a result, the maximum detection efficiency of 31.2% was achieved at the pump power of 59.1 mW, with the corresponding background counts of 2.8×10^3 s^{-1} shown in Fig. 7. The maximum conversion efficiency of the system was calculated to be 91.8% after taking into account the transmittance of the filters and the quantum efficiency of the Si-APD SPCM. Since the signal and pump pulses were focusing on PPLN crystal, the imperfection of the conversion efficiency might be mainly caused by the spatial mode mismatching of the pump and signal. Compared with the background counts of CW pumping (1×10^5 s^{-1}) [3,4], it was about two orders smaller in synchronously pulsed pumping scheme. As a figure of merit, we calculated the noise equivalent power ($NEP = h\upsilon(2R_{BC})^{1/2}/\eta$) divided by the operation rate f, where $h\upsilon$, R_{BC} and η are the energy of a signal photon, the background noise, and the detection efficiency, respectively. NEP/f was an important parameter of the sensitivity of an optical detector, especially referring to the determinant of the data acquisition time in general and the key generation rate in quantum key distribution (QKD) systems [23]. At the peak detection efficiency, the NEP/f was as low as 2.6×10^{-24} W/Hz$^{3/2}$, thus showing that such scheme is suitable for the fast and efficient infrared single-photon detection.

The synchronous pumping single-photon frequency upconversion system is not only applied in the infrared photon detection with high efficiency and low background noise. More importantly, due to the unique quantum features in coherence preservation, it may stimulate promising applications with compact all-fiber devices in temporal and spectral control of single-photon nonlinear photonics. In Section 4, some applications are presented.

4. Applications in infrared PNRD and imaging

We believe that the compact fiber-laser synchronization system for fast and efficient single-photon frequency upconversion is of critical potential to stimulate other promising applications, such as high-speed QKD and quantum interface. The synchronous frequency upconversion system will also benefit infrared photon number resolving detection or few-photon-level infrared imaging.

4.1. Infrared photon-number-resolving detection

PNRD supports promising and important applications in few-photon detection, nonclassical photon statistics measurements, fundamental quantum optics experiments, and practical quantum information processing [37–39]. In the visible regime, PNRD with high quantum efficiency and low dark counts could be realized by employing silicon-based multipixel photon counter (Si-MPPC) [40]. Meanwhile at telecom wavelengths, PNRD could be achieved with InGaAs-based avalanche photodiodes (APD) operating in non-saturated mode [41]. However, PNRD for the wavelengths around 1 μm is still a bottleneck because both Si- and InGaAs-detectors are insensitive at those wavelengths. Recent advances in infrared photon detection technology show that coherent upconverison of quantum states is feasible, where the photon statistics would be conserved consequently [10,11]. Therefore, it is possible to count the visible replicas via frequency upconversion to realize PNRDs around 1 μm.

Based on coincidence frequency upconversion presented in Section 3, photon-number-resolving detection at 1.04 μm has been realized. The experimental setup was schematically illustrated in Fig. 8, consisting of two parts: synchronization fiber laser system for the pump pulses and signal photons, and photon-number-resolved detection system based on coincidence frequency upconversion and the Si-MPPC. The signal photons at 1.04 μm were then upconverted by the synchronized pump pulses at 1.55 μm in the PPLN crystal. Then the SFG photons passed through a group of filters before impinging on the Si-MPPC with a multimode fiber pigtail. The Si-MPPC (Hamamatsu Photonics S10362-11-100U) was composed of 10×10 APDs which were arranged on an effective active area of 1 mm^2 with a photon detection efficiency of 16.0% (including the fiber coupling efficiency). When the incident photons were injected onto different pixels, the output voltage of the superposition from all APD pixels was proportional to the number of incident photons [40]. The histogram of the peak voltage from the output of Si-MPPC was showed in Fig. 9(a). The pulse area spectrum featured a series of peaks representing the different photon number states. By fitting each peak to Gaussian function, the probability distribution was obtained as shown in Fig. 9(b). The area under each Gaussian curve gave the number of events presenting that photon number state. The area of each peak could be normalized by the total area to give the probability distribution. As the input light was in coherent state, the upconverted photons statistics obeyed the Poissionian law. By fitting the experimental data according to Poisson distribution, we got the detected photon number of 4.65. With different incident photon flux, the photon numbers detected by SPCM were shown with circles in Fig. 9(c). Since SPCM could not discriminate more than one photon per shot, it was obviously saturated with large incident photon numbers. In contrast, as shown with squares in Fig. 9(c), the detected photon numbers of Si-MPPC linearly increased with the incident photon numbers. It showed that the Si-MPPC could correctly identify the photon numbers per pulse with a large dynamic range, which was promising in few-photon-level detection.

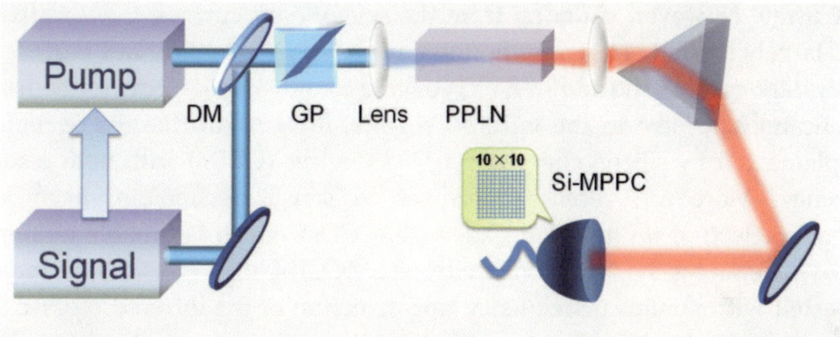

Figure 8. Experimental setup for the 1.04 μm photon-number-resolved detection. DM, dichroic mirror; GP, Glan prism; Si-MPPC, silicon multipixel photon counter.

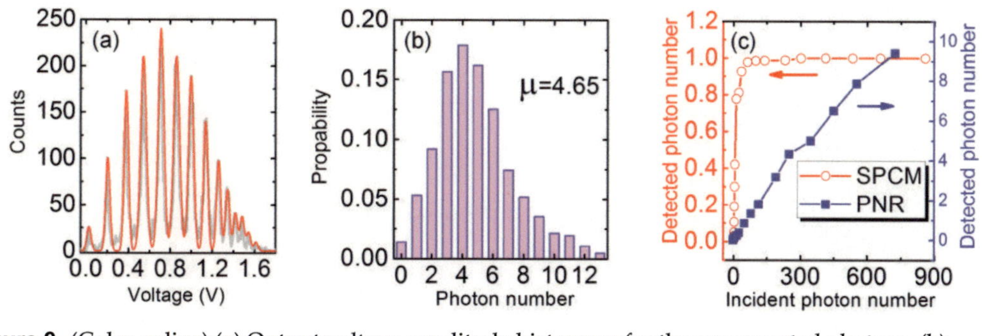

Figure 9. (Color online) (a) Output voltage amplitude histogram for the upconverted photons. (b) Photon number distribution. (c) Detected photon numbers by SPCM (circles) and Si-MPPC (squares) as a function of incident photon numbers.

The photon-number-resolving performance was improved by reducing the background counts with a synchronous pump as the coincidence gate and reducing the intrinsic parametric fluorescence influence with long-wavelength pumping. As a result, a total detection efficiency of 3.7% was achieved with a quite low noise probability per pulse of 0.0002. Such a low background noise probability could remarkably improve the sensitivity of the frequency-upconversion PNRD. The remarkable decrease in background noise would optimize applications such as quantum entanglement and quantum teleportation [42] and improve the signal-to-noise ratio of widely used light detection and ranging system, since the background counts would randomly couple into the modes of the quantum states, which may significantly affect the original photon number distribution [38]. The approach may find promising applications in various quantum optical experiments using nonclassical light sources to demonstrate the features of quantum states around 1 μm [43].

4.2. Ultrasensitive infrared imaging

Realization of ultra-sensitive infrared imaging has critical importance for applications such as astronomy, medical diagnosis, night-vision technology and chemical sensing. Currently, the infrared imaging detectors are available, like the commonly used linear InGaAs photodiode array. However, suffering from the severe dark current, the sensitivity of such detectors is largely limited [44]. Even though liquid-nitrogen-cooling can provide a solution for a smaller dark current, the additional cryogenic cooling device reduces the feasibility for lots of applications. Unlike in the infrared regime, imaging in the visible regime can be readily implemented by silicon charged coupled devices (CCDs) with high resolution and high efficiency. Moreover, recent progresses in sensor technology have led to the development of electron multiplying CCDs (EMCCDs) which is capable of single-photon detection. By leveraging the high sensitivity of EMCCDs, ultra-sensitive infrared imaging can be expected with nonlinear frequency upconversion of the infrared electric field to the visible spectral region [10,11]. Here we will introduce a few-photon-level two-dimensional infrared imaging detector by coincidence frequency upconversion. The upconversion imaging apparatus is shown in Fig. 10.

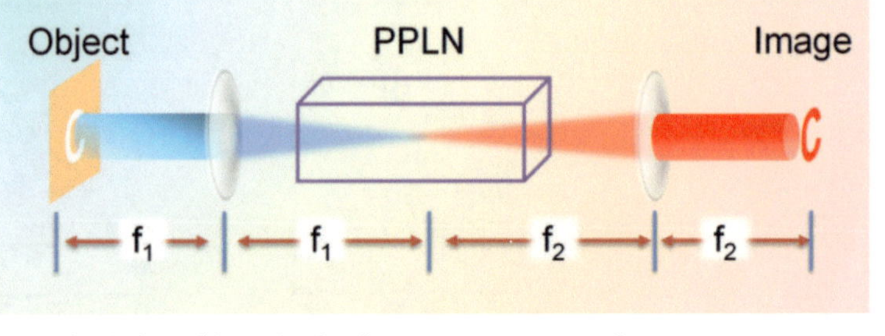

Figure 10. Schematic for infrared imaging by frequency upconversion.

In our experiment, the pump and signal sources were taken from two fiber lasers mode-locked at 19.1 MHz [33]. The signal beam illuminated a transmission mask with a character "C" to form the object beam shown in Fig. 11 (a). By a dichroic mirror, the object beam and the pump beam were then combined into a 4-f imaging system with lens L1 (f_1=250 mm) and L2 (f_2=300 mm). Fourier plane was arranged right at the middle of 50-mm-long PPLN crysta. In order to facilitate the type I quasi-phase matching of the PPLN crystal, a Glan prism was employed before the crystal for enforcing the polarization [45]. The temperature of the crystal was optimized at 104.3 °C for the grating period of 11.0 μm. The upconverted image at 622 nm was captured by a silicon EMCCD (Andor iXon3 897) bearing 512×512 pixels. The pixel size was 16×16 μm, which was very suitable for high spatial resolution imaging. To improve the signal-to-noise ratio, the EMCCD was thermoelectrically cooled to -85 °C.

With the pump power of 40.0 mW, corresponding to a peak power of about 210 W, the internal conversion efficiency of 27% could be inferred by correcting for the filtering transmittance. The object beam was attenuated to 2.0 photons per pulse. The upconverted imaging photons were then registered by the EMCCD with integration time of 30 s and accumulated for 50 times. The upconverted image was shown in Fig. 11 (c), from which the character "C" could be identified. Thanks to the pulsed pump field together with the long-wavelength pumping, the background noise was remarkably reduced. The theoretical calculation based on coherent imaging theory was given in Fig. 11 (b), which was in good agreement with the experimental result. The image blurring was attributed to the spatial filtering relating to the point spread function in the upconversion imaging system [7]. To improve the resolution, large pump beam profile at the Fourier plane would be better to enwrap the transformed object field as much as possible. However, increasing the diameter of the pump beam would decrease the intensity, thus reduce the conversion efficiency. Thus optimization of the trade-off should be in consideration.

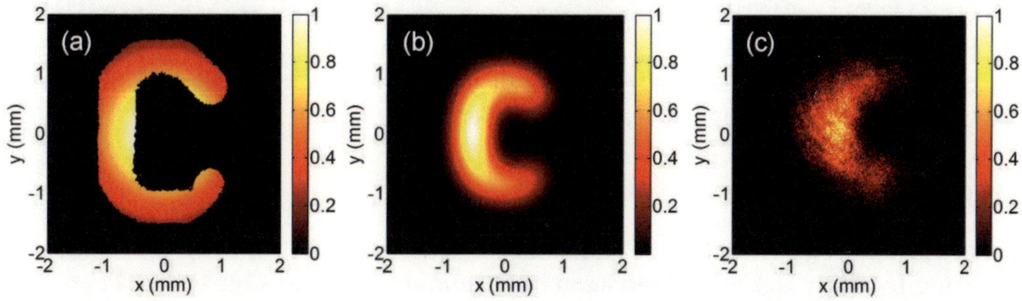

Figure 11. (a) The coherently illuminated mask. The theoretical (b) and measured (c) upconverted images at the image plane.

In summary, we demonstrated full 2D infrared image upconversion at few-photon level with a high conversion efficiency of 31%. The infrared image at 1040 nm was upconverted to the visible regime where the imaging photons were registered by the silicon EMCCD with high sensitivity and resolution. The imaging performance was remarkably improved with the reduction of the background noise by the synchronized pulsed excitation of the 1549 nm pump source, as well as the long-wavelength pumping [46]. Such all-optical upconversion imaging technique can offer an attractive method for ultra-sensitive infrared imaging.

5. Conclusion

In this review chapter, the quantum theory of the frequency upconversion is introduced and indicates that either in single-mode regime and multi-mode regime, complete quantum transduction can be realized in principle with two necessary requirement, sufficiently intensive pump field and perfect phase matching. Several upconversion systems with high conversion efficiency are presented and discussed in detail. Synchronous pumping frequency upconversion system shows superior performance with high conversion efficiency and low background noise. Thanks to the short time window of the synchronized pump pulse together with long-wavelength pumping scheme, the detection sensitivity was improved remarkably by reducing the background noise. This technique facilitates not only many traditional applications, such as classical optical communication, imaging, photobiology, and astronomy, but also novel quantum optics applications, such as quantum interface to transfer quantum entanglement, linear optical quantum gates, single photon polarization switches, and nonlinear control of single photons.

Author details

Kun Huang, E Wu, Xiaorong Gu, Haifeng Pan and Heping Zeng
State Key Laboratory of Precision Spectroscopy, East China Normal University, China

Acknowledgement

This work was funded in part by National Natural Science Fund of China (10990101, 60907043, 61127014 & 91021014), International Cooperation Projects from Ministry of Science and Technology (2010DFA04410), Key project sponsored by the National Education Ministry of China (109069), Research Fund for the Doctoral Program of Higher Education of China (20090076120024), ECNU Reward for Excellent Doctors in Academics.

6. References

[1] Boyd R (2008) Nonlinear Optics, Academic Press, ISBN 978-0123694706, New York.
[2] Huang J, Kumar P (1992) Observation of Quantum Frequency Conversion. Phys. Rev. Lett. 68: 2153–2156.
[3] Albota M, Wong F (2004) Efficient Single-Photon Counting at 1.55 μm by Means of Frequency Upconversion. Opt. Lett. 29: 1449–1451.
[4] Pan H, Dong H, Zeng H, Lu W (2006) Efficient Single-Photon Counting at 1.55 μm by Intracavity Frequency Upconversion in a Unidirectional Ring Laser. Appl. Phys. Lett. 89: 191108.
[5] Vandevender A, Kwiat P (2004) High Efficiency Single Photon Detection via Frequency Up-Conversion. J. Mod. Opt. 51: 1433–1445.
[6] VanDevender A, Kwiat P (2007) Quantum Transduction via Frequency Upconversion. J. Opt. Soc. Am. B, 24: 295-299.
[7] Pedersen C, Karamehmedović E, Dam J, Tidemand-Lichtenberg P (2009) Enhanced 2D-Image Upconversion Using Solid-State Lasers. Opt. Express 17: 20885-20890.
[8] Kuzucu O, Wong F, Kurimura S, Tovstonog S (2008) Time-Resolved Single-Photon Detection by Femtosecond Upconversion. Opt. Lett. 33: 2257-2259.

[9] Zhang Q, Langrock C, Fejer M, Yamamoto Y (2008) Waveguide-Based Single-Pixel Up-Conversion Infrared Spectrometer. Opt. Express 16: 19557–19561.

[10] Tanzilli S, Tittel W, Halder M, Alibart O, Baldi P, Gisin N, Zbinden H (2005) A Photonic Quantum Information Interface. Nature 437: 116–120.

[11] Rakher M, Ma L, Slattery O, Tang X, Srinivasan K (2010) Quantum Transduction of Telecommunications-Band Single Photons from a Quantum Dot by Frequency Upconversion. Nature Photonics 4: 786–791.

[12] Xu H, Ma L, Mink A, Hershman B, Tang X (2007) 1310-nm Quantum Key Distribution System with Up-Conversion Pump Wavelength at 1550 nm. Opt. Express 15: 7247–7260.

[13] Fejer M, Magel G, Jundt D, Byer R (1992) Quasi-Phase-Matched Second Harmonic Generation: Tuning and Tolerances. IEEE J. Quantum Electron. 28: 2631-2654.

[14] Takesue H, Diamanti E, Langrock C, Fejer M, Yamamoto Y (2006) 1.5 μm Single Photon Counting Using Polarization-Independent Up-conversion Detector. Opt. Express 26: 13067-13072.

[15] Ma L, Slattery O, Tang X (2010) NIR Single Photon Detectors with Up-Conversion Technology and Its Applications in Quantum Communication Systems. InTech. pp. 315–336.

[16] Ramelow S, Fedrizzi A, Poppe A, Langford N, Zeilinger A (2012) Polarization-Entanglement-Conserving Frequency Conversion of Photons. Phys. Rev. A 85: 013845.

[17] Gu X, Huang K, Pan H, Wu E, Zeng H (2012) Photon Correlation in Single-Photon Frequency Upconversion. Opt. Express 20: 2399-2407.

[18] Huang K, Gu X, Ren M, Jian Y, Pan H, Wu G, Wu E, Zeng H (2011) Photon-Number-Resolving Detection at 1.04 μm via Coincidence Frequency Upconversion. Opt. Lett. 36: 1722-1724.

[19] Midwinter J, Warner J (1967) Up-conversion of Near Infrared to Visible Radiation in Lithium-Meta-Niobate, J. Appl. Phys. 38: 519-523.

[20] Kim Y, Kulik S, Shih Y (2001) Quantum Teleportation of A Polarization State with a Complete Bell State Measurement. Phys. Rev. Lett. 86: 1370-1373.

[21] Hum D, Fejer M (2007) Quasi-Phasematching. C. R. Physique 8: 180–198.

[22] Pan H, Zeng H (2006) Efficient and Stable Single-Photon Counting at 1.55 μm by Intracavity Frequency Upconversion. Opt. Lett. 31: 793-795.

[23] Langrock C, Diamanti E, Roussev R, Yamamoto Y, Fejer M, Takesue H (2005) Highly Efficient Single-Photon Detection at Communication Wavelengths by Use of Upconversion in Reverse-Proton-Exchanged Periodically Poled $LiNbO_3$ Waveguides. Opt. Lett. 30: 1725-1727.

[24] Dong H, Pan H, Li Y, Wu E, Zeng H (2008) Efficient Single-Photon Frequency Upconversion at 1.06 μm with Ultralow Background Counts. Appl. Phys. Lett. 93: 071101.

[25] Gu X, Li Y, Pan H, Wu E, Zeng H (2009) High-Speed Single-Photon Frequency Upconversion with Synchronous Pump Pulses. IEEE J. Sel. Top. Quantum Electron. 15: 1748-1752.

[26] Gu X, Huang K, Li Y, Pan H, Wu E, Zeng H (2010) Temporal and Spectral Control of Single-Photon Frequency Upconversion for Pulsed Radiation. Appl. Phys. Lett. 96: 131111.

[27] Huang K, Gu X, Pan H, Wu E, Zeng H (2012) Synchronized Fiber Lasers for Efficient Coincidence Single-Photon Frequency Upconversion. IEEE J. Sel. Top. Quantum Electron. 18: 562-566.

[28] Pan H, Wu E, Dong H, Zeng H (2008) Single-Photon Frequency Up-Conversion with Multimode Pumping. Phys. Rev. A 77: 033815.

[29] Rakher M, Ma L, Davanco M, Slattery O, Tang X, Srinivasan K (2011) Simultaneous Wavelength Translation and Amplitude Modulation of Single Photons from a Quantum Dot. Phys. Rev. Lett. 107: 083602.

[30] Eckstein A, Brecht B, Silberhorn C (2011) A Quantum Pulse Gate Based on Spectrally Engineered Sum Frequency Generation. Opt. Express 19: 13770-13778.

[31] Pomarico E, Sanguinetti B, Thew R, Zbinden H (2010) Room Temperature Photon Number Resolving Detector for Infrared Wavelengths. Opt. Express 18: 10750-10759.

[32] Albota M, Wong F, Shapiro J (2006) Polarization-Independent Frequency Conversion for Quantum Optical Communication. J. Opt. Soc. Am. B 23: 918-924.

[33] Rusu M, Herda R, Okhotnikov O (2004) Passively Synchronized Two-Color Mode-Locked Fiber System Based on Master-Slave Lasers Geometry. Opt. Express 12: 4719-4724.

[34] Hao Q, Li W, Zeng H (2009) High-Power Yb-Doped Fiber Amplification Synchronized with a Few-Cycle Ti: Sapphire Laser. Opt. Express 17: 5815-5821.

[35] Li Y, Gu X, Yan M, Wu E, Zeng H (2009) Square Nanosecond Mode-Locked Er-Fiber Laser Synchronized to a Picosecond Yb-Fiber Laser. Opt. Express 17: 4526-4532.

[36] Xu P, Ji S, Zhu S, Yu X, Sun J, Wang H, He J, Zhu Y, Ming N (2004) Conical Second Harmonic Generation in a Two-Dimensional $\chi^{(2)}$ Photonic Crystal: A Hexagonally Poled $LiTaO_3$ Crystal. Phys. Rev. Lett. 93: 133904.

[37] Allevi A, Bondani M, Andreoni A (2010) Photon-Number Correlations by Photon-Number Resolving Detectors. Opt. Lett. 35: 1707-1709.

[38] Waks E, Diamanti E, Sanders B, Bartlett S, Yamamoto Y (2004) Direct Observation of Nonclassical Photon Statistics in Parametric Down-Conversion. Phys. Rev. Lett. 92: 113602.

[39] Afec I, Natan A, Ambar O, Silberberg Y (2009) Quantum State Measurements Using Multipixel Photon Detectors. Phys. Rev. A 79: 043830.

[40] Eraerds P, Legré M, Rochas A, Zbinden H, Gisin N (2007) SiPM for Fast Photon-Counting and Multiphoton Detection. Opt. Express 15: 14539-14549.

[41] Wu G, Jian Y, Wu E, Zeng H (2009) Photon-Number-Resolving Detection Based on InGaAs/InP Avalanche Photodiode in the Sub-Saturated Mode. Opt. Express 17: 18782-18787.

[42] Honjo T, Takesue H, Kamada H, Nishida Y, Tadanaga O, Asobe M, Inoue K (2007) Long-Distance Distribution of Time-Bin Entangled Photon Pairs over 100 km Using Frequency Up-Conversion Detectors. Opt. Express 15: 13957-13964.

[43] Vasilyev M, Choi S, Kumar P, D'Ariano G (2000) Tomographic Measurement of Joint Photon Statistics of the Twin-Beam Quantum State. Phys. Rev. Lett. 84: 2354–2357.

[44] Liang Y, Jian Y, Chen X, Wu G, Wu E, Zeng H (2011) Room-Temperature Single-Photon Detector Based on InGaAs/InP Avalanche Photodiode with Multichannel Counting Ability. IEEE Photon. Tech. Lett. 23: 115–117.

[45] Thew R, Zbinden H, Gisin N (2008) Tunable Upconversion Photon Detector. Appl. Phys. Lett. 93: 071104.

[46] Pelc J, Zhang Q, Phillips C, Yu L, Yamamoto Y, Fejer M (2012) Cascaded Frequency Upconversion for High-Speed Single-Photon Detection at 1550 nm. Opt. Lett. 37: 476-478.

Chapter 17

Generation of Tunable THz Pulses

J. Degert, S. Vidal, M. Tondusson, C. D'Amico, J. Oberlé and É. Freysz

Additional information is available at the end of the chapter

http://dx.doi.org/10.5772/46234

1. Introduction

Many phenomena in materials science, physics, chemistry, biology and medicine involve fundamental processes with a spectral signature in the THz range [20]. However, for many years, these processes have been little studied due to the lack of high quality THz sources. Fortunately, the recent developments in THz technology have filled this so-called THz gap [28]. With the emergence of reliable THz sources, the need for arbitrarily shaped THz pulses dedicated to specific applications like communications, signal processing, spectroscopy or coherent/optimal control is more and more felt. Concerning coherent control, selective excitations in the THz spectral region of phonon modes in molecular crystal [41], charge oscillations in semiconductor heterostructures [5, 26] or phonon-polaritons in ferroelectric crystal [9, 40] have already been achieved by means of temporally shaped near-infrared laser pulses. From this point of view, the possibilities of control would be increased if one makes it possible to excite directly the system with shaped THz pulses [27]. To this end, several methods have been developed to extend the generation of arbitrary pulse shapes from the visible and mid-infrared spectral region up to the THz range [1, 8, 10, 16–19, 21, 30, 31, 43]. All are based on the generation of THz pulses by excitation of different types of emitters with spatially and/or temporally shaped optical pulses.

Here, we will focus our attention on the generation, tuning and shaping of narrow-band THz pulses. Such pulses have already been implemented by means of various pulse shaping techniques which differ from each other by their respective spectral width, their ease of use and their versatility. Some of them use traditional pulse shaping setup: for example, Sohn et al. [30] used a specially designed mask placed in the Fourier plane of a zero-dispersion line to produce modulated optical pulses subsequently photomixed to generate THz pulses tunable from 0.5 to 3 THz, with a spectral width of $\sim$ 500 GHz. Despite its tunability, this method is limited by the use of a non programmable mask. Likewise, using optical pulse shaping with a liquid crystal spatial light Fourier filter combined with optical rectification in ZnTe, Ahn et al. were able to generate THz waves tunable from 0.5 to 2 THz, with a spectral bandwidth of $\sim$ 200 GHz [1]. To do this, they have implemented a Gerchberg-Saxton algorithm to find the best spectral phase required for the optical pulse in order to generate the desired THz pulse. However, this technique, characterized by its great ability to generate various pulse shapes, has one drawback: it relies on the ability of the algorithm to converge towards the expected solution. On the other hand, interesting examples of tunable narrow-band THz pulses with possibilities of shaping the waveform have also been reported in lithium niobate crystals. For example, Lee et al. [16, 18, 19] employed a single pulse or a pair of temporally separated optical pulses rectified in PPLN crystals with a specially engineered domain structure, generating very narrow THz pulses (bandwidth $\sim$ 25 GHz) tunable between 0.5 and 2.5 THz. Nevertheless, it requires a new crystal with a specific domain structure for each temporal shape or an adjustment of the crystal position to tune the frequency. At last, using a transient polarization grating induced by two femtosecond laser pulses propagating in a $LiNbO_3$ crystal, Stepanov et al. [31] were able to generate and to shape THz pulses tunable from 0.5 to 3 THz with a bandwidth of 100 GHz. In this experiment, the tunability is controlled by the angle between the pump beams and the shaping is accomplished by filtering with a slit or a shield the spatial intensity distribution of the pump beams. While the spatial filtering can be done with a liquid crystal modulator, enabling thus to do optimal control experiments, the control of the tunability is less flexible.

In the present chapter, we present two different techniques that make it possible to generate tunable THz pulses. The first one, that can be regarded as analytical, is based on the spectral tailoring of an ultrashort femtosecond laser pulse, whereas the second one is relying on the spatial shaping of the transverse profile of the femtosecond laser beam. In both techniques, liquid crystal devices are used to modulate in the spatial or spectral domain the femtosecond pulses that are incident on a nonlinear crystal in which optical rectification is taking place. Hence, it is flexible and needs no moving mechanical parts or Mach-Zender like interferometers. We demonstrate that the first one makes it possible to generate THz pulses tunable between 0.5 and 2.5 THz with spectral bandwidth as narrow as 140 GHz. The second one, based on a geometrically-assisted optical rectification technique, enables us to generate THz spatiotemporal interferences in the intermediate field zone beyond the rectifying crystal. We will show that the spatiotemporal properties of the generated THz field in the intermediate zone are tightly tied to the geometry of the transverse profile of the laser beam attaining the rectifying crystal. Therefore by shaping the transverse beam profile one can change the temporal and the transverse profile of the emitted THz pulse.

2. Model and preliminary analysis of THz generation by optical rectification in zinc blende crystals

Whatever the way used to obtain shaped THz waves, it is strongly correlated to the physical mechanisms involved in the THz wave generation. Here, since our attention will be focused on THz generation by optical rectification in Zinc Blende crystals, we are going to remind, at first, the main theoretical results concerning this process. Then, based upon these latter, we will show how one can obtain shaped THz pulses.

Let us consider an ultrashort near-infrared (NIR) laser pulse exciting, at normal incidence, a ZnTe emitter cut along the $\langle 110 \rangle$ plane, with a thickness L_e. Assuming a propagation in the z-direction, we can write its electric field in the following form:

$$E(t,x,y,z) = u(x,y,z)\mathcal{E}(t) \exp\left[i(\omega_0 t - k_0 z)\right], \tag{1}$$

where $u(x,y,z)$ and $\mathcal{E}(t)$ are, respectively, the spatial and temporal amplitudes of the laser beam, ω_0 its carrier frequency (corresponding to a central wavelength $\lambda_0 = 2\pi c/\omega_0$, c is the velocity of light in vaccum), and k_0 the mean wave vector ($k_0 = \omega_0 n(\omega_0)/c$, where n is the refractive index of ZnTe). Here, we assume that this pulse is not modified during its propagation in the nonlinear crystal, so that:

$$u(x,y,z) = \mathcal{F}(x,y) \exp\left(i\frac{\omega z}{v_g}\right), \tag{2}$$

v_g being the group velocity of the laser pulse: $v_g = c/n_g(\omega_0)$, where $n_g(\omega_0) = n(\omega_0) + \omega_0 \left(dn/d\omega\right)_{\omega_0}$ is the group index of the NIR pulse.

At each point of the medium, this pulse induces via the frequency difference between its different spectral components a second order nonlinear polarization whose spectral components Ω lie in the terahertz range. More precisely, the nonlinear polarization induced at frequency Ω is given by:

$$P^{(2)}(\Omega,x,y,z) = \varepsilon_0 \chi^{(2)}(\Omega) \int \frac{d\omega}{2\pi} E(\omega,x,y,z) E^*(\omega-\Omega,x,y,z), \tag{3}$$

where ε_0 is the vacuum permittivity, $\chi^{(2)}$ the second order nonlinear susceptibility of ZnTe, and $E(\omega,x,y,z) = \int E(t,x,y,z)\exp(i\omega t)dt$. Note that, in Eq. (3), we have assumed that the spectral dependence of $\chi^{(2)}$ in the NIR range is negligible. Taking into account Eqs. (1) and (2), and introducing the spectral amplitude of the optical field $\mathcal{E}(\omega) = A(\omega)\exp\left[i\phi(\omega)\right] = \int \mathcal{E}(t)\exp(i\omega t)dt$, one obtains:

$$P^{(2)}(\Omega,x,y,z) = \varepsilon_0 |\mathcal{F}(x,y)|^2 e^{i\frac{\Omega z}{v_g}} \chi^{(2)}(\Omega)\, C(\Omega), \tag{4}$$

where $C(\Omega)$ is the power spectrum of the rectified NIR pulse, given by:

$$\begin{aligned} C(\Omega) &= \int |\mathcal{E}(t)|^2 \exp(i\Omega t)dt, \\ &= \int \frac{d\omega}{2\pi} \mathcal{E}(\omega)\mathcal{E}^*(\omega-\Omega), \\ &= \int \frac{d\omega}{2\pi} A(\omega)A(\omega-\Omega) e^{i[\phi(\omega)-\phi(\omega-\Omega)]}. \end{aligned} \tag{5}$$

From a quantum mechanical point of view, $C(\Omega)$ is proportional to the probability amplitude that two photons of the NIR pulse with frequencies ω and $\omega - \Omega$ interact in the nonlinear crystal to generate a THz photon at frequency Ω.

So, this second order polarization emits THz radiations at different frequencies Ω, and, consequently, acts as a source term in the wave equation for the THz wave generated by optical rectification. This latter equation, written in the spatial and spectral Fourier domains, is given by [2, 6]

$$\left[\frac{\partial^2}{\partial z^2} + k_\parallel^2\right] E_{\mathrm{THz}}(\Omega,k_x,k_y,z) = -\frac{\Omega^2}{c^2}\chi^{(2)}(\Omega)C(\Omega)\mathcal{G}(k_x,k_y)e^{i\Omega z/v_g}, \tag{6}$$

where

$$\mathcal{G}(k_x,k_y) = \frac{1}{(2\pi)^2}\iint |\mathcal{F}(x,y)|^2 e^{-i(k_x x + k_y y)} dxdy, \tag{7}$$

and $k_\| = \sqrt{k^2(\Omega) - k_\perp^2}$, with $k_\perp = \sqrt{k_x^2 + k_y^2}$ and $k(\Omega) = n(\Omega)\Omega/c$ (wave vector of the THz wave).

At the exit of the ZnTe crystal ($z = L_e$), the amplitude of each spectral component of the THz pulse generated by optical rectification is given by [6, 7]

$$E_{\text{THz}}(\Omega, k_x, k_y, L_e) = E_{\text{THz}}(\Omega, k_\perp, L_e) = iL_e \frac{\Omega^2}{c^2} \frac{\chi^{(2)}(\Omega)}{k_\|(\Omega) + \Omega/v_g} e^{i[k_\|(\Omega)+\Omega/v_g]\frac{L_e}{2}} \times C(\Omega)\mathcal{G}(k_\perp) \operatorname{sinc}\left[\frac{L_e \Delta k}{2}\right], \tag{8}$$

where $\mathcal{G}(k_\perp) = \mathcal{G}(k_x, k_y)$, and $\Delta k = k_\|(\Omega) - \Omega/v_g$ is the wavevector mismatch between the generated THz wave and the incident optical wave.

In Eq. (8), three terms play a key role concerning the shape of the emitted THz spectrum and the way it is related to the NIR pulse shape: the phase mismatch factor sinc $[L_e\Delta k/2]$, $C(\Omega)$ and $\mathcal{G}(k_\perp)$. Here, we will discuss only the influence of the sinc term, the discussion concerning $C(\Omega)$ and $\mathcal{G}(k_\perp)$ being postponed to sections 3 and 4 respectively. The phase mismatch factor gives its main contribution to the THz spectrum for the frequency Ω such that $\Delta k(\Omega) = 0$, the so-called phase matching condition. Assuming that the NIR pulse is almost a plane wave across the crystal, *i.e.* $k_\perp \simeq 0$, this latter condition may be written:

$$\Delta k = k(\Omega) - \frac{\Omega}{v_g} = \Omega\left(\frac{1}{v_\varphi} - \frac{1}{v_g}\right) = 0, \tag{9}$$

where $v_\varphi = c/n(\Omega)$ is the phase velocity of the THz wave. Thus, the phase matching condition sets that:

$$v_\varphi(\Omega) = v_g(\omega_0) \Longleftrightarrow n(\Omega) = n_g(\omega_0). \tag{10}$$

Dispersion of ZnTe in the NIR range is well describe by the following Sellmeier equation [25]:

$$n^2(\lambda) = 4.27 + \frac{3.01\lambda^2}{\lambda^2 - 0.142}, \tag{11}$$

where λ is in μm, and in the THz range by the dielectric function given in [12]:

$$\tilde{\varepsilon}(\Omega) = (n + i\kappa)^2 = \varepsilon_{\text{el}} + \frac{\varepsilon_{\text{st}}\Omega_{\text{TO}}^2}{\Omega_{\text{TO}}^2 - \Omega^2 - i2\gamma\Omega}, \tag{12}$$

with $\varepsilon_{\text{el}} = 7.44$, $\varepsilon_{\text{st}} = 2.58$, $\Omega_{\text{TO}}/2\pi = 5.32$ THz and $\gamma/2\pi = 0.022$ THz. So, a NIR pulse with a spectrum centered at $\lambda_0 \sim 800$ nm leads to a THz wave centered around 2.5 THz (see, for example, the black curve of figure 2b). Actually, the way the phase mismatch factor affects the shape of the THz spectrum depends on two factors: λ_0 and L_e [1]. As a consequence, it offers little possibilities to generate shaped THz spectra with an easy tunability. However, we will see that $C(\Omega)$ and $\mathcal{G}(k_\perp)$ are more promising candidates for this purpose.

Finally, in order to compare the experimental results presented below with the theoretical predictions of Eq. (8), we have to take into account the detection of the THz electric field via electro-optic (EO) sampling in a second ⟨110⟩ ZnTe crystal, with a thickness L_d, using a weak probe NIR ultrashort pulse. To this end, we use the expression of the EO signal given in [11]:

$$S(\tau) \propto \int_{-\infty}^{+\infty} E_{\mathrm{THz}}(\Omega, k_\perp, L_e) f(\Omega) e^{-i\Omega\tau} d\Omega = \int_{-\infty}^{+\infty} S(\Omega) e^{-i\Omega\tau} d\Omega, \tag{13}$$

where τ is the time delay between the THz and the NIR pulses, $S(\Omega)$ is the spectral amplitude of the EO signal, and

$$f(\Omega) = C_{\mathrm{probe}}(\Omega) \chi^{(2)}(\Omega) \frac{e^{iL_d\Delta k} - 1}{i\Delta k}, \tag{14}$$

$C_{\mathrm{probe}}(\Omega)$ being the power spectrum of the probe pulse.

Furthermore, in Eqs. (8) and (14), we will assume that the frequency dependence of $\chi^{(2)}$ is given by [39]:

$$\chi^{(2)}(\Omega) = 2d_{14} \left[1 + C \left(1 - \frac{\Omega}{\Omega_{\mathrm{TO}}} \right)^{-1} \right], \tag{15}$$

where $d_{14} = 90\,\mathrm{pm/V}$ and $C = -0.07$.

3. Generation of tunable THz pulses by optical pulse shaping in the spectral domain

3.1. Further analysis of THz generation by optical rectification I

In section 2, we have shown the influence of the phase mismatch factor on the THz spectrum. Here, we are going to study in which way one can control the shape of the THz pulse by acting on the power spectrum $C(\Omega)$. To this end, we assume that the NIR pulse is a plane wave in the crystal, so that $k_\perp = 0$ and $\mathcal{G}(k_\perp) = 1$ in Eq. (8). With these assumptions, it is clear that, apart from the sinc term, this is the power spectrum, *i.e.* the *temporal* pulse shape of the incident NIR pulse, since $C(\Omega) = \int |\mathcal{E}(t)|^2 \exp(i\Omega t) dt$, that controls the shape of $E_{\mathrm{THz}}(\Omega, L_e)$. Actually, Eq. (5) clearly shows that $C(\Omega)$ strongly depends on the NIR pulse spectrum and more especially on the relative phase $\phi(\omega) - \phi(\omega - \Omega)$ between all frequency pairs within the pulse separated by Ω. So, by acting on the spectral phase $\phi(\omega)$ of the NIR pulse in the appropriate way it is possible to finely control $C(\Omega)$ and, consequently, the THz spectrum shape.

In what follows, we will show some examples of tunable narrow-band THz spectra, the tunability being achieved through a sinusoidal spectral phase modulation and the spectral narrowing through an additional triangular phase.

3.2. Experimental setup

The experimental setup is depicted on Fig. 1: a Ti:sapphire chirped pulse amplifier (Femtopower Compact Pro) delivering laser pulses of duration $\tau_p = 35$ fs at 790 nm with a 1 kHz repetition rate is used to generate and detect THz waves. About 90% of the output of the CPA is tailored by a pulse shaper and is used for the generation of THz pulses by optical rectification in a ⟨110⟩ ZnTe crystal with a thickness L_e ($L_e = 300$ µm for the experiments of section 3.3 and 1 mm for those of sections 3.4 and 3.5).

The pulse shaper consists in a half-zero-dispersion line made of a 600 g/mm grating and a $f = 600$ mm cylindrical mirror that spatially disperses and focuses all the spectral components

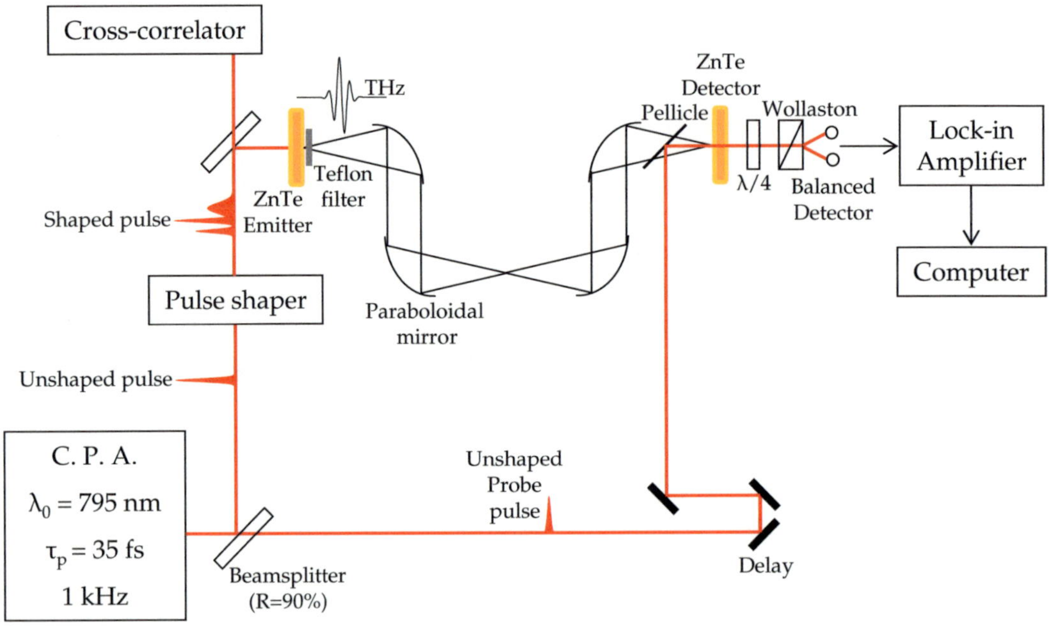

Figure 1. Experimental setup for the generation of tunable THz pulses by optical pulse shaping in the spectral domain.

of the pulse in the Fourier plane where a programmable single liquid crystal spatial light modulator (LC SLM) with 640 pixels, manufactured by Jenoptik AG, acts as a spectral filter [22, 23, 33, 42]. A plane mirror positioned just after the SLM folds the line to perform an inverse Fourier transform and reassemble all the spectral components of the shaped pulse. The SLM is such that it is only possible to change the spectral phase of the incoming pulse. The intensity profile of the shaped laser pulse is measured by cross-correlation, in a type I BBO crystal, of this pulse with an unshaped reference pulse.

The beam exiting from the pulse shaper then excites the ZnTe emitter at normal incidence. After removal of the residual NIR pulse by a Teflon filter, the generated THz wave is imaged into a second $\langle 110 \rangle$ ZnTe crystal with a thickness L_d by means of four off-axis paraboloidal mirrors ($L_d = 300$ μm for the experiments of section 3.3 and 500 μm for those of sections 3.4 and 3.5). There, it is measured via electro-optic effect using a weak unshaped NIR pulse coming directly from the exit of the Ti:sapphire amplifier [11]. By varying the delay τ between the THz pulse and the ultrashort NIR pulse, one gets an EO signal, $S(\tau)$, corresponding to a cross-correlation between the laser pulse and the THz wave integrated over the EO crystal length. This signal is then recorded by ellipsometry (the ellipsometer consists in a $\lambda/4$ wave plate, a Wollaston polarizer, and two balanced detectors).

3.3. Sinusoidal phase modulation: generation of THz pulse trains

It is well known that selectivity in the THz spectral range can be achieved by means of terahertz-rate sequences of femtosecond pulses [1, 21, 41]. Experimentally, such a pulse train with a repetition rate F_{rep} is generated by applying a periodic phase modulation with a period F_{rep}, the most used being $\phi(\omega) = a \sin(\omega / F_{\text{rep}})$. In the absence of any phase modulation, *i.e.*

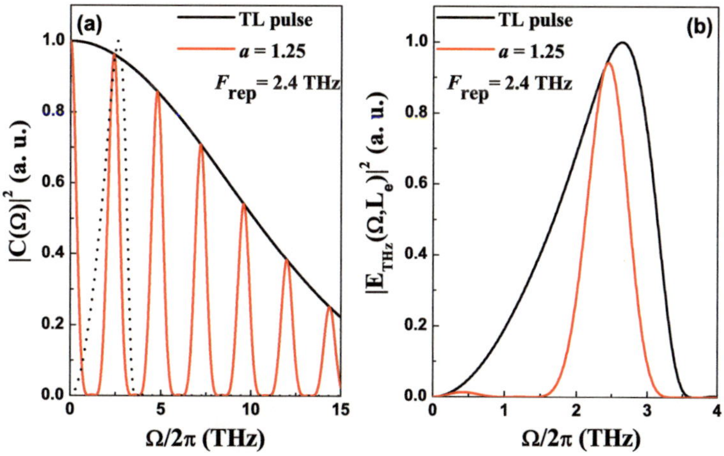

Figure 2. (a) Power spectrum of an optical pulse with $\lambda_0 = 790$ nm, and a pulse duration $\tau_p = 35$ fs: transform-limited (TL) pulse (black curve), shaped pulse with a sinusoidal phase modulation $\phi(\omega) = a\sin(\omega/F_{\text{rep}})$ (red curve). The dotted curve superimposed to $|C(\Omega)|^2$ corresponds to the THz spectrum generated by a TL pulse. (b) Terahertz spectra generated by optical rectification in a $\langle 110\rangle$ ZnTe of a: TL pulse (black curve), sinusoidally phase modulated pulse (red curve).

for a transform-limited pulse corresponding to $\phi(\omega) = 0$, the power spectrum is a bell-shaped curve centered on $\Omega = 0$, with a half-width at maximum equal to the pulse bandwidth (black curve, Fig. 2a). However, if a sinusoidal phase modulation is applied, then $C(\Omega)$ exhibits spectral interference fringes with maxima centered at the frequencies $\Omega_n = 2\pi n F_{\text{rep}}$ (where n is an integer) (red curve, Fig. 2a, where $a = 1.25$ and $F_{\text{rep}} = 2.4$ THz). Now, if one takes into account the spectral filtering of the powerspectrum induced by the phase matching condition, corresponding to a multiplication of $|C(\Omega)|^2$ by the dotted curve of Fig. 2a, it results in a narrower THz spectrum centered at the frequencies Ω_n within the THz band authorized by the phase matching condition (Fig. 2b). This spectral narrowing can also be understood by a reasoning in the time domain. Actually, each pulse of the NIR pulse train, two consecutive pulses being separated by $T_{\text{rep}} = 1/F_{\text{rep}}$, leads to its own THz waveform. All these waveforms add coherently to generate a THz train with a repetition rate F_{rep} whose spectrum precisely consists in peaks centered at frequencies $\Omega_n = 2\pi n/T_{\text{rep}}$, the bandwidth of these peaks decreasing as the number of pulses in the THz train increases.

Fig. 3a displays the EO signals, $S(\tau)$, associated to THz trains generated with our setup for different values of F_{rep} and $a = 1.25$, whereas the corresponding spectra, defined as $|S(\Omega)|^2$, are represented in Fig. 3c. In agreement with the former analysis, we observe a THz pulse train with a spectrum that can be tuned within the spectral range obtained for an unshaped pulse by adjusting the repetition rate of the NIR pulse train. There is a very good agreement between our results and the spectra calculated with the model described in section 2 (Fig. 3c and d). These two last figures clearly show that we can tune the carrier frequency of THz pulse simply by changing the repetition rate of the NIR pulse train.

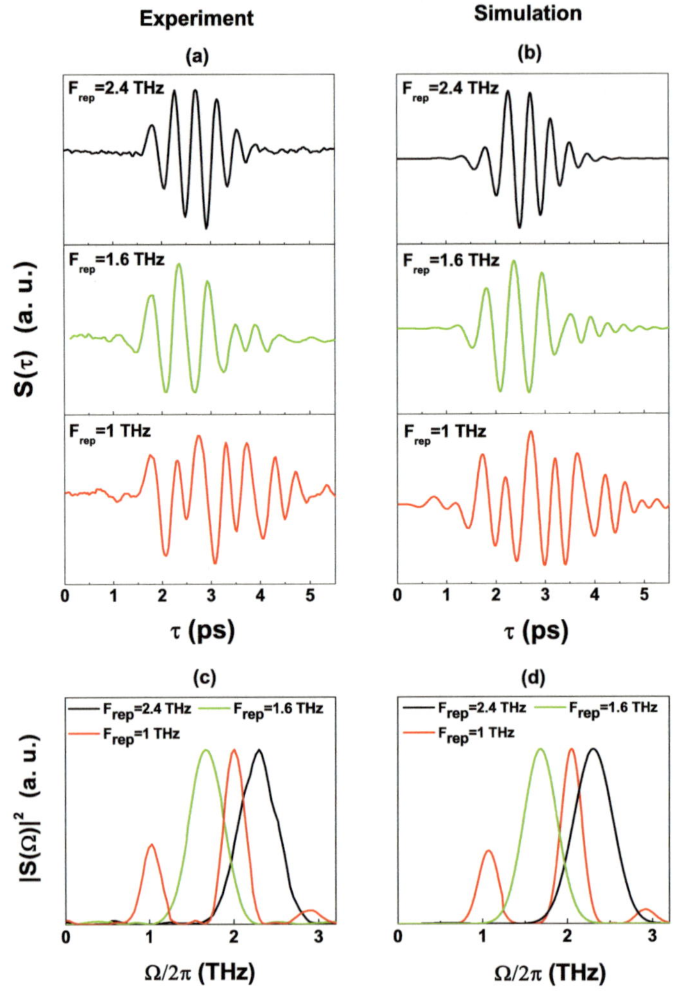

Figure 3. THz pulse train with an adjustable repetition rate F_{rep}, the paramater a being equal to 1.25. Experiment: (a) waveforms and (c) corresponding spectra. Simulation: (b) waveforms and (d) corresponding spectra (taken from [37]).

However, for $F_{rep} = 1$ THz, one can note that the spectrum also displays second and third-harmonic peaks centered at 2 and 3 THz respectively. They are inherent to the use of a sinusoidal phase modulation: for $F_{rep} = 1$ THz, the phase matching condition is less selective with respect to the peaks of the power spectrum, leading to these harmonics. From a "temporal" point of view, these harmonics arise because, for this repetition rate, the THz waveform generated by each NIR pulse of the train has a duration shorter or comparable to T_{rep}. Thus, when all these partial waveforms add coherently to generate the total THz electric field $E_{THz}(t)$, they leads to an overall shape that is more complex and damps more slowly than the sinusoidal waveform obtained for $F_{rep} = 1.6$ or 2.4 THz. Thanks to the higher

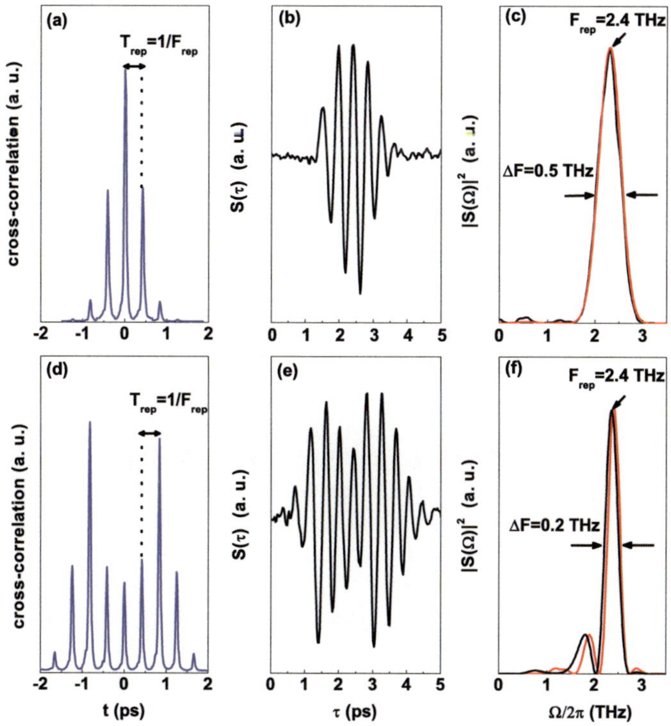

Figure 4. Experimental illustration of the spectral narrowing by a sinusoidal phase modulation $\phi(\omega) = a \sin(\omega/F_{\text{rep}})$: (a) cross-correlation of the optical pulse train generated with $a = 1.25$ and $F_{\text{rep}} = 2.4$ THz. (b) Associated THz train and (c) corresponding spectrum: experiment in black and simulation in red. (d) cross-correlation of the NIR pulse train generated with $a = 2.65$ and $F_{\text{rep}} = 2.4$ THz. (e) Associated THz train and (f) corresponding spectrum (taken from [37]).

number of oscillations, that is to the longer THz pulse duration, the width of the harmonics for $F_{\text{rep}} = 1$ THz is also reduced (~ 280 GHz) compared to the spectra for $F_{\text{rep}} = 1.6$ or 2.4 THz (bandwidth ~ 500 GHz).

To see if it is possible to further decrease the bandwidth of the THz spectrum when a sinusoidal phase modulation is applied to the rectified NIR pulse, let us consider a transform-limited laser pulse with a duration τ_{p}, and a gaussian amplitude

$$A(\omega) = E_0 T_p \sqrt{\pi} \exp\left(-\frac{\omega^2 T_p^2}{4}\right), T_p = \tau_p / \sqrt{2 \ln 2}. \tag{16}$$

When a sinusoidal phase $\phi(\omega)$ is applied to this pulse, it results in a pulse train with a temporal dependence of the electric field given by [29]:

$$\mathcal{E}(t) \propto \sum_{m=-\infty}^{+\infty} J_m(a) \exp\left(\frac{im\pi}{2}\right) \exp\left[-\left(\frac{t - mT_{\text{rep}}}{T_{\text{p}}}\right)^2\right], \tag{17}$$

where the J_m are Bessel functions of the first kind. This equation shows that, *whatever* the phase modulation period F_{rep}, the amplitude of the mth pulse of the train is given by $|J_m(a)|$. So, by playing on the modulation strength a one can control, to some extend, the weight of the different pulses of the train. More precisely, the evolution of $|J_m(a)|$ is the following [29]: except for $|J_0|$, it increases when a is varied from 1.25 to 2.65, leading to an enhancement of the number of pulses in the train. Hence, increasing a will lengthen the THz pulse duration and lead to a narrower spectrum, this regardless of the repetition rate F_{rep}. Fig. 4f displays the THz spectrum obtained by applying a sinusoidal phase modulation with $a = 2.65$ and $F_{rep} = 2.4$ THz. With this modulation strength, there are nine pulses in the NIR pulse sequence (Fig. 4d) instead of five for $a = 1.25$ (Fig. 4a), giving a spectrum with a bandwidth of 0.2 THz. Here again, experiment and simulation are in good agreement (Fig. 4f).

Nevertheless, one also observes some ripples on the spectrum of Fig. 4f. They can be attributed to the difference in amplitude between the different pulses of the NIR pulse train, and especially to the two main peaks centered around ±800 fs in Fig. 4d, resulting from the fact that, for $a = 2.65$, $|J_{\pm 2}(a)| > |J_{\pm n}(a)|$ ($|n| \neq 2$). Unfortunately, their amplitude cannot be modified easily without affecting the amplitude of the other peaks. Indeed, as shown in reference [29], it is impossible to produce pulse trains with more than three pulses of equal amplitude with a simple sinusoidal phase. On the other hand, one can solve this problem by adding higher harmonic orders to the applied phase modulation, *i.e.* by means of the following spectral phase: $\phi(\omega) = \sum_n F_n \sin\left(n\omega/F_{rep} + \phi_n\right)$. However, this phase is harder to implement since it requires the use of an optimization algorithm to find the best value for the weight F_n of each harmonic. In the same spirit, Ahn et al. [1] have used a Gerchberg-Saxton algorithm to find the best spectral phase in order to generate many nearly equally spaced identical optical pulses. But, here again, this method relies on the ability of the algorithm to converge towards the expected solution. Finally, we will show in section 3.5 that it is possible to shrink the THz pulse bandwidth simply by adding a soundly chosen triangular phase modulation to the sinusoidal one. Before that, we will consider the effect of a triangular phase modulation on the THz generation.

3.4. Triangular phase modulation: generation of phase-locked THz pulse pairs

Let us consider a triangular phase modulation defined by

$$\phi(\omega) = -\Delta\tau|\omega - \delta\omega|, \tag{18}$$

where $\Delta\tau$ is the slope of the spectral phase, and $\delta\omega$ is the detuning with respect to the center of the spectral amplitude $\mathcal{E}(\omega)$. Such a phase, when it is applied to a transform-limited pulse with a duration τ_p and a gaussian shape

$$\mathcal{E}(t) = E_0 \exp\left[-\left(\frac{t}{T_p}\right)^2\right], T_p = \tau_p/\sqrt{2\ln 2}, \tag{19}$$

leads to a pair of pulses whose temporal amplitude is given by [29]:

$$\mathcal{E}(t) = \mathcal{E}_+(t + \Delta\tau) + \mathcal{E}_-(t - \Delta\tau), \tag{20}$$

with

$$\mathcal{E}_\pm(t) = \frac{E_0}{2} e^{-(t/T_p)^2} \left[1 - \text{erf}\left(i\frac{t}{T_p} \pm \frac{T_p\delta\omega}{2}\right)\right], \tag{21}$$

with the error function, erf, defined by

$$\operatorname{erf}(x) = \frac{2}{\sqrt{\pi}} \int_0^x \exp(-y^2) dy. \tag{22}$$

So it leads to two pulses separated by a time delay $t_p = 2\Delta\tau$, with a ratio of their amplitudes $(\mathcal{E}_+(0)/\mathcal{E}_-(0))$ controlled by the detuning $\delta\omega$, this ratio being equal to 1 for $\delta\omega = 0$. This kind of sequence of ultrashort phase-locked pulses has been used extensively to perform temporal coherent control experiments with optical pulses [3]. It has also been used successfully in the terahertz range to control various processes [15, 16, 44] (see also section 3.5 for a brief presentation). In these latter experiments, the THz pulse pair was generated from a NIR pulse pair created by a Michelson or a Mach-Zender interferometer. Here, we will demonstrate that it is possible to generate a THz pulse pair by applying a triangular spectral phase to the rectified NIR pulse. Since temporal coherent control experiments require a balanced pair of pulses to achieve the best efficiency, we will only consider, in what follows, the case where $\delta\omega = 0$, so that $\mathcal{E}_+(t) = \mathcal{E}_-(t)$.

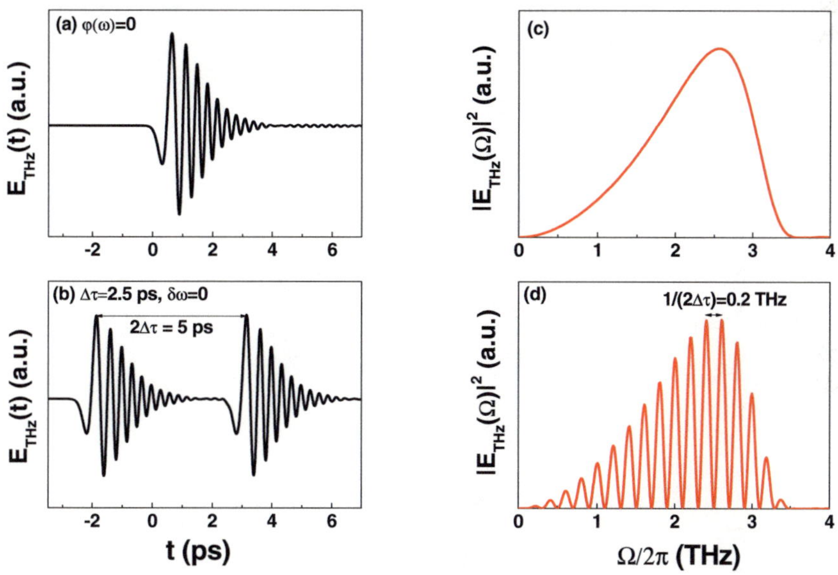

Figure 5. THz waveforms generated by optical rectification in a 300 μm-thick ZnTe crystal of: (a) a transform-limited NIR pulse, (b) a NIR pulse having a triangular phase modulation with $\Delta\tau = 2.5$ ps and $\delta\omega = 0$. (c) and (d) respectively show the spectra of (a) and (b). In (d), the fringe spacing is $1/t_p = 0.2\,\text{THz}$.

By optical rectification of a pair of balanced NIR pulses in a ZnTe crystal, one obtains a THz pulse pair with the following electric field:

$$E_{\text{THz}}(t) = E_{\text{THz}_+}(t + \Delta\tau) + E_{\text{THz}_+}(t - \Delta\tau), \tag{23}$$

where $E_{\text{THz}_+}(t)$ is the electric field of the THz pulse generated by optical rectification of a NIR pulse whose temporal amplitude is $\mathcal{E}_+(t)$. Its amplitude spectrum is given by

$$\begin{aligned} E_{\text{THz}}(\Omega) &= E_{\text{THz}_+}(\Omega) e^{-i\Omega\Delta\tau} + E_{\text{THz}_+}(\Omega) e^{i\Omega\Delta\tau} \\ &= E_{\text{THz}_+}(\Omega) e^{-i\Omega\Delta\tau} \left(1 + e^{i\Omega t_p}\right), \end{aligned} \tag{24}$$

so that the expected spectrum of the EO signal is

$$|S(\Omega)|^2 \propto |E_{\mathrm{THz}}(\Omega)|^2 = 2|E_{\mathrm{THz}_+}(\Omega)|^2 \left[1 + \cos\left(\Omega t_p\right)\right]. \quad (25)$$

It has the same enveloppe as the spectrum $|E_{\mathrm{THz}_+}(\Omega)|^2$ of a single THz pulse (shown on Fig. 5c), but with modulations due to the spectral interference of the two THz pulses, the fringes spacing being $1/t_p$ (see Fig. 5d).

Fig. 6 shows the THz pulse pair generated experimentally by applying a triangular phase modulation with $\Delta\tau = 2.5$ ps and $\delta\omega = 0$ to the NIR pulse (the same parameters as in the simulation of Fig. 5). As mentioned previously, for $\delta\omega = 0$, both NIR pulses are balanced (Fig. 6d), and give rise to a pair of THz pulses with the same carrier frequency and separated by $t_p = 5$ ps (Fig. 6e). The corresponding EO signal spectrum (Fig. 6f) has the same enveloppe as the EO signal spectrum of a single THz pulse (Fig. 6c), but exhibits oscillations with a period $1/t_p = 0.2$ THz.

However, in addition to the NIR pulse pair, the cross-correlation trace of Fig. 6d also has an unwanted TL pulse centered at $t = 0$. This distortion is due to the combination of two limitations of the LC SLMs [35]: first, since LC SLMs have a phase operating range slightly above 2π, the applied phase modulation is "wrapped" to be within the $[0, 2\pi]$ interval

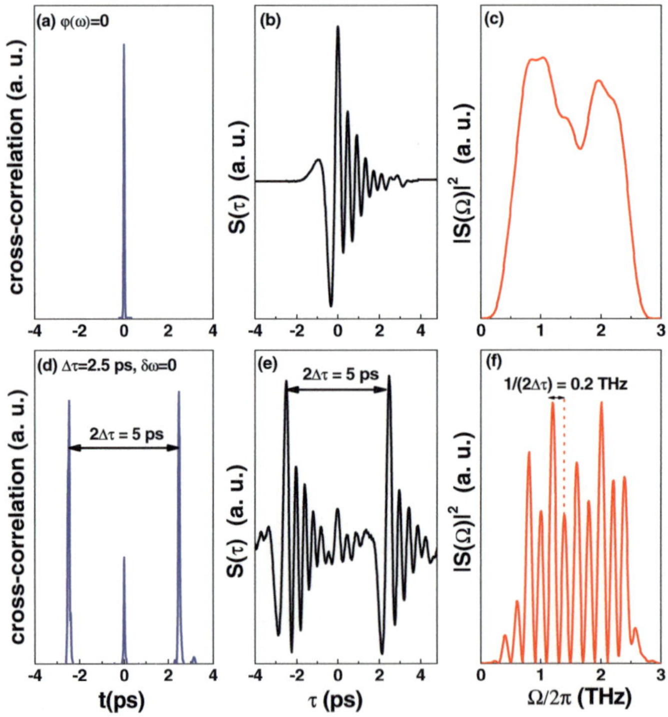

Figure 6. Sequence of THz phase-locked pulses experimentally generated by applying a triangular phase modulation with $\Delta\tau = 2.5$ ps and $\delta\omega = 0$ to the rectified NIR pulse: (a) cross-correlation of the initial TL NIR pulse, (b) associated THz waveform and (c) corresponding spectrum. (d) cross-correlation of the NIR pulse pair, (e) associated THz waveform and (f) corresponding spectrum (taken from [37]).

[33]. Second, the pixels, and consequently the gaps, of the LC SLM are not sharply defined, leading to some smoothing of the pixelated phase. Combined together these limitations lead to the apparition of "modulator replica" pulses in addition to the desired pulse shape. In our experiment, the phase varies from -223π to 0 over the first half of the spectrum and, symetrically, from 0 to -223π over the second half. This leads to numerous phase-wraps and, in combination with the smoothed spatial response of the pixels, gives the pulse centered at $t = 0$ in Fig. 6d. This unwanted effect can be avoided by the use of a diffraction-based pulse shaper [36]. Nevertheless, as we will see in the next section, it does not affect significantly the THz spectrum shape.

3.5. Sinusoidal + triangular phase: tunable narrow band THz pulses.

We now consider the action of both phases introduced previously, *i.e.*

$$\phi(\omega) = a\sin(\omega/F_{\text{rep}}) - \Delta\tau|\omega - \delta\omega|. \tag{26}$$

According to sections 3.3 and 3.4, we expect a sequence of pulse trains with a repetition rate F_{rep} separated by a delay $t_p = 2\Delta\tau$. This is confirmed by Fig. 7a obtained for $a = 1.25$, $F_{\text{rep}} = 2$ THz, $\Delta\tau = 2.5$ ps and $\delta\omega = 0$. As in section 3.4, the applied triangular phase modulation leads to numerous phase-wraps and, in combination with the limitations of the LC SLM, gives an additional TL pulse centered at $t = 0$. This undesired pulse and the pulse train sequence generate their own THz waveforms that add coherently (Fig. 7b). The THz spectrum is displayed on Fig. 7c. Due to the triangular phase modulation, it still exhibits interference fringes with a spacing equal to $1/t_p$, but, thanks to the sinusoidal phase, it is narrower than the one of Fig. 6f and is centered in F_{rep}. Note that it is not affected significantly by the THz pulse generated by the TL pulse due to the limitations of the SLM as can be seen from the simulation of Fig. 7f which does not include these latter.

These THz pulse trains sequences are of great importance for temporal coherent control and the generation of tunable narrow-band THz pulses. The goal of temporal coherent control is to excite selectively some levels of a quantum system by means of a pair of ultrashort phase-locked pulses in order to steer the evolution of this system towards a desirable state through its interaction with light (for example to control the outcome of a photo-induced chemical reaction). Let us consider, for example, a quantum system with a ground state $|g\rangle$ and N excited states $|e_k\rangle$ (with energies $\hbar\Omega_k$), which is excited by a sequence of two identical ultrashort pulses separated by a delay t_p. We suppose that the bandwidth of the pulses is broad enough to excite the N states $|e_k\rangle$. In the weak field regime, the state vector $|\Psi(t)\rangle$ of the system after interaction with this sequence of pulses is given by [3]:

$$|\Psi(t)\rangle = |g\rangle + \sum_{k=1}^{N} b_k e^{-i\Omega_k t}|e_k\rangle, \tag{27}$$

with

$$b_k = \frac{1}{i\hbar}\mu_{kg}E_{tot}(\Omega_k), \tag{28}$$

where μ_{kg} is the transition dipole moment from state $|g\rangle$ to state $|e_k\rangle$, and $E_{tot}(\Omega_k)$ is the spectral amplitude of the total electric field taken at frequency Ω_k, given by

$$E_{tot}(\Omega_k) = E(\Omega_k)\left(1 + e^{i\Omega_k t_p}\right), \tag{29}$$

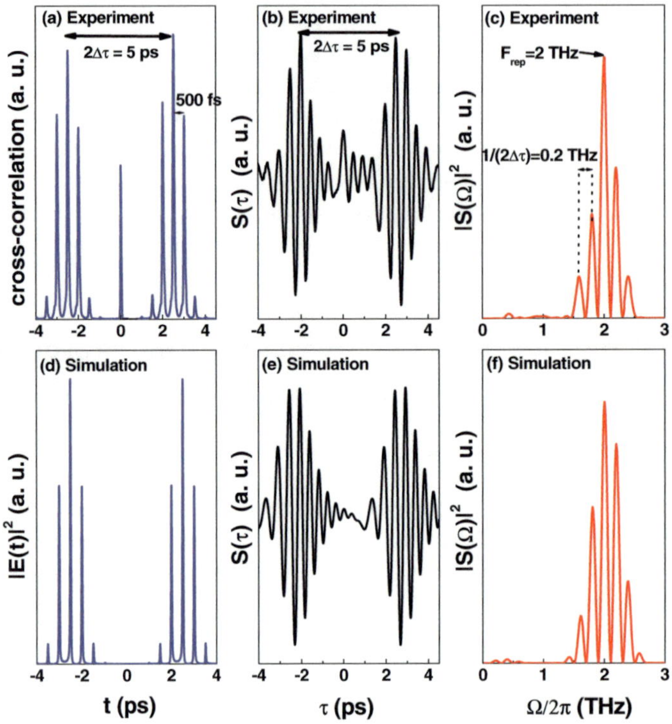

Figure 7. Sequence of THz pulse trains generated by applying simultaneously a triangular and a sinusoidal phase with $a = 1.25$, $F_{\text{rep}} = 2$ THz, $\Delta\tau = 2.5$ ps and $\delta\omega = 0$. (a) cross-correlation of the NIR pulse, (b) associated THz waveform and (c) corresponding spectum. Panels (d), (e) and (f) correspond respectively to the simulations of panels (a), (b) and (c) (taken from [37]).

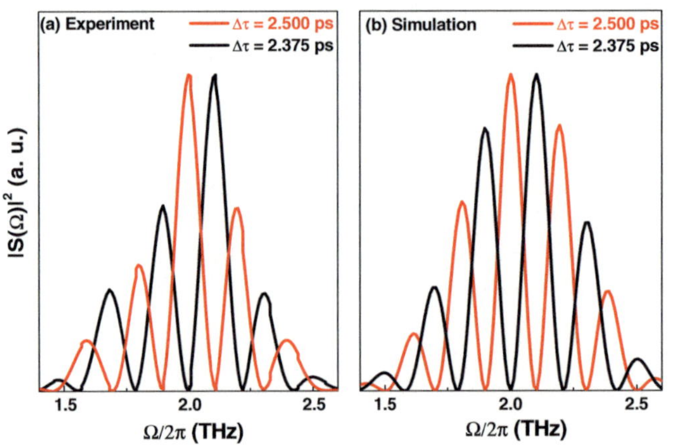

Figure 8. "In-phase" and "out-of-phase" sequences of pulse trains at $\Omega_k/2\pi = 2$ THz: (a) Experimental THz spectra obtained for $a = 1.25$, $F_{\text{rep}} = 2$ THz, $\delta\omega = 0$, $\Delta\tau = 2.5$ ps (red curve, corresponding to $\Delta\varphi = 2\pi p$) and $\Delta\tau = 2.375$ ps (black curve, corresponding to $\Delta\varphi = 2\pi(p + 1/2)$). (b) Corresponding simulations (taken from [37]).

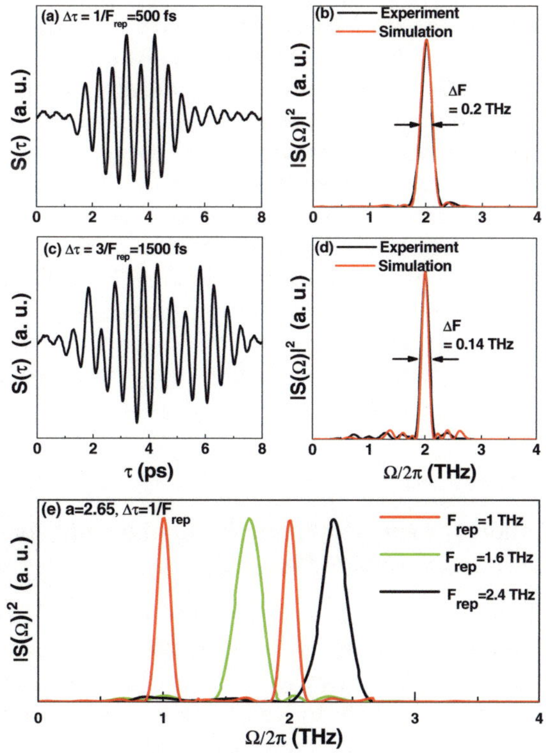

Figure 9. Spectral smoothing ((a) and (b)) and spectral narrowing ((c) and (d)) with a sequence of THz pulse trains. In both cases $a = 2.65$, $F_{\text{rep}} = 2$ THz and $\delta\omega = 0$. (e) Experimental illustration of the tunability of the spectral smoothing shown in (a): for all the curves $a = 2.65$, $\Delta\tau = 1/F_{\text{rep}}$ and $\delta\omega = 0$ (taken from [37]).

$E(\Omega)$ being the amplitude spectrum of a single pulse. Thus, the probability to excite the state $|e_k\rangle$ is

$$|\langle e_k|\Psi(t)\rangle|^2 = |b_k|^2 \propto |E(\Omega_k)|^2 \left[1 + \cos\left(\Omega_k t_p\right)\right]. \tag{30}$$

This probability strongly depends on the relative phase $\Delta\varphi(\Omega_k) = \Omega_k t_p$ between the two pulses of the sequence: the excitation is "constructive" for $\Delta\varphi = 2\pi p$ (p is an integer) and "destructive" for $\Delta\varphi = 2\pi(p + 1/2)$. The fine tuning of $\Delta\varphi$ is achieved via the tuning of the delay t_p, and strongly depends on its (interferometric) stability. In the THz range, our setup makes it possible to generate easily these "in-phase" and "out-of-phase" pulse pairs, simply by changing the slope $\Delta\tau$ of the triangular phase. For instance, Fig. 8 shows two THz pulse sequences leading respectively to a constructive excitation (red curve) and a destructive excitation (black curve) at $\Omega_k/2\pi = 2$ THz.

One can also take advantage of the combined triangular and sinusoidal phases to generate less structured THz spectra than the one of Fig. 4f, simply by choosing a delay between the THz pulses shorter than their duration. Then the THz waveforms are superposed and, for a suitably choosen value of $\Delta\tau$, give rise to a single THz train with pulses having almost the same amplitude, resulting in a spectrum with a well defined peak, centered at the pulse train repetition rate F_{rep}. This remarkable value of $\Delta\tau$ is $1/F_{\text{rep}}$, $\forall$ F_{rep}. Figures 9a and 9b display an example of such a "smooth" narrow-band THz spectrum for $F_{rep} = 2$ THz, whereas Fig. 9e shows the tunability of this method. On this figure, one can see that, contrary to Fig. 3c, the peaks at 1 and 2 THz for $F_{rep} = 1$ THz have the same amplitude. It results from the fact that this spectrum was obtained with a thicker crystal than the one used in section 6, leading to an unshaped THz spectrum with almost the same energy around 1 and 2 THz (see Fig. 6c) unlike the unshaped spectrum corresponding to Fig. 3. Now, for a given value of F_{rep}, if one tunes $\Delta\tau$ around $1/F_{rep}$, it is then possible to generate relatively clean THz spectra with a narrower bandwidth. An example is given in figures 9c and 9d: for $\Delta\tau = 3/F_{rep} = 1.5$ ps, the THz waveform exhibits more oscillations than in Fig. 9a, leading to a spectrum with a bandwidth of 140 GHz. However, for repetition rates like $F_{rep} = 1$ THz, one should note that although this method makes it possible to narrow and smooth the peak at 1 THz and its harmonics, it cannot suppress these latter since they are inherent to the use of a sinusoidal phase modulation.

3.6. Limitations of this method

In this section, we will address some points concerning the limitations of our approach and the possibility to generate even narrower THz spectra.

First, we will consider the limitations imposed by the pulse shaper. We have already met such limitations in sections 3.4 and 3.5, leading to the apparition of "modulator replica" pulses in addition to the desired pulse shape. Now, we will see that the pixelated nature of the LC SLM imposes upper limits to T_{rep} and $\Delta\tau$. Indeed, to be correctly sampled by the modulator's pixels, and thus avoid temporal aliasing, the phase modulation must satisfy the Nyquist's criterion: the increase of phase between two adjacent pixels of the SLM must not exceed π [22, 33, 42]. For the sinusoidal and triangular phase modulations considered separately, this criterion, applied to our setup, leads respectively to $T_{rep} \leq 4$ ps (*i.e.* $F_{rep} \geq 250$ GHz) and $\Delta\tau \leq 4$ ps (*i.e.* $t_p \leq 8$ ps). If these conditions are not satisfied, then sampling replica pulses alter the desired output waveform [22, 35]. So, the characteristics of the pulse shaper limit the ability to generate smooth and tunable THz spectra at low frequencies, that is below 500 GHz. Nevertheless, these frequencies are not of great concern here since the THz spectrum generated by optical rectification in this spectral range is almost zero.

Actually, the main limiting factor of our approach, especially if one intends to generate THz spectra with a bandwidth below 100 GHz, is the use of a simple sinusoidal phase modulation. As already mentioned in section 3.3, one can circumvent this limitation by adding harmonics to the applied phase, the number of harmonics increasing inversely with the required THz bandwidth. Indeed, an enhancement of the number of harmonics makes it possible to increase the number of pulses with the same amplitude in the NIR pulse train and, consequently, to reduce the bandwidth of the generated THz spectrum. However, one must not forget that the determination of the coefficients F_n and ϕ_n requires the use of optimization methods like the one of reference [1]. Moreover, it is important to have in mind that in LC SLM pulse shapers, the tailored output pulse reflects accurately the response of the SLM over a restricted temporal window. The duration T of this window is governed either by the finite spectral resolution of the zero-dispersion line or by the pixelation of the SLM (*i.e.* the Nyquist's criterion)[23, 35, 42]. In our setup, T is limited by the pixelation of the SLM, and $T = 8$ ps, so the whole NIR pulse train must not exceed this duration, imposing an upper limit to the number m of pulses in the train. Indeed, a train of m pulses, with a repetition rate F_{rep}, has a duration Δt given by $(m-1)/F_{\text{rep}}$. If $\Delta t \leq T$, then

$$m \leq m_{\max} = 1 + T \times F_{\text{rep}}. \tag{31}$$

For $F_{\text{rep}} = 0.5$, 1 and 2 THz, $m_{\max}$ is respectively equal to 5, 9 and 17 pulses. If we consider a pulse train with a repetition rate F_{rep} and a number of identical (same amplitude and same duration) pulses equaling the maximum number of pulses $m_{\max}$ authorized by the temporal window $T = 8$ ps for this repetition rate, a numerical calculation of the corresponding power spectrum $C(\Omega)$ shows that the spectral interference fringes induced by this pulse train (centered at $\Omega_n = 2\pi n F_{\text{rep}}$) have a bandwidth $\sim$ 100 GHz, whatever the value of F_{rep}. It means that, with our setup, we can't generate narrow-band THz pulse with a bandwidth below 100 GHz. To go below this limit, it is necessary to work with transform-limited NIR pulses having a longer duration, the number of pixels of the SLM being unchanged. Indeed, the temporal window T increases as the bandwidth of the input NIR pulse decreases [35]. For example, in reference [22], the input pulse duration is 130 fs and the LC SLM pulse shaper has a time window $T = 28$ ps (the number of pixels is 640 like in our pulse shaper). With these values, it is possible to enhance $m_{\max}$, and, consequently, to generate narrow-band THz spectra with bandwidths down to $\sim$ 30 GHz for all repetition rates as shown by numerical simulations. However, the tuning range is reduced, since working with longer NIR pulses decreases the width of the power spectrum, and consequently, the support of the THz spectrum.

4. Generation of tunable THz pulses by optical pulse shaping in the spatial domain

4.1. Further analysis of THz generation by optical rectification II

Our problem is to propose a way to spectrally tune the THz field generated by optical rectification in a zinc blende crystal by adjusting the transverse beam profile of femtosecond pump laser pulse. To solve this problem we will exploit the intermediate field properties of the emitted THz pulse. In order to describe the THz radiation properties in the intermediate field, we will use the Fresnel diffraction theory. To further simplify our analysis, we will consider that the beam profile has a radial symmetry and is only a function of r. Assuming a propagation in the z-direction, we will slightly modify Eq. (1) and write the electric field in the following form:

$$E(t,x,y,z) = E(t,r,z) = u(r,z)\mathcal{E}(t)\exp\left[i(\omega_0 t - k_0 z)\right], \tag{32}$$

where $r = \sqrt{x^2+y^2}$ and $u(r,z)$ and $\mathcal{E}(t)$ are, respectively, the spatial and temporal amplitudes of the laser beam. Here again, we assume that this pulse is not modified during its propagation in the nonlinear crystal. After its generation in a very thin nonlinear crystal located at $z = z_0$, the THz electric field $E(\Omega,r,z)$ at the transverse plane z, is the convolution of the THz field $E(\Omega,r,z_0)$ leaving the crystal with the Fresnel propagator

$$\frac{\exp\left(ik(\Omega)(z-z_0)\left[1+\frac{r^2}{2(z-z_0)^2}\right]\right)}{i\lambda(z-z_0)} \tag{33}$$

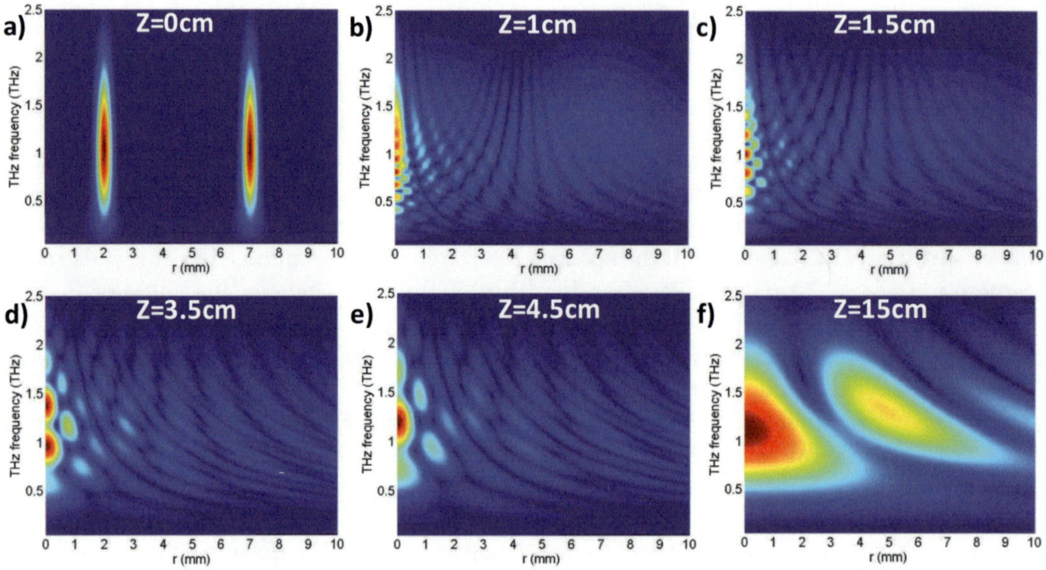

Figure 10. Numerical spatiotemporal behavior of a THz pulse emitted by a ZnTe crystal at different position along the propagation direction. The near I.R. femtosecond pump pattern is represented by two concentric circles that have a radius of 2 mm and 7 mm and a thickness of 200 μm (taken from [7])

where Ω and λ are respectively the THz frequency and the associated wavelength. In the spatial frequency domain it reads:

$$E(\Omega, k_\perp, z) = E(\Omega, k_\perp, z_0) \cdot \mathbb{F}^2 \left(\frac{\exp\left(ik(\Omega)(z-z_0)\left[1 + \frac{r^2}{2(z-z_0)^2}\right]\right)}{i\lambda(z-z_0)} \right). \tag{34}$$

The expression of $E(\Omega, k_\perp, z_0)$ is given by Eq. (8). If the pump beam has a radial symmetry, the function $G(k_\perp)$ is given by:

$$G(k_\perp) = 2\pi \int_0^\infty F(r) J_0(k_\perp r) r dr. \tag{35}$$

It represents the Hankel transform of the pump laser fluence $F(r)$, $J_0(k_\perp r)$ being the 0th order Bessel function. The latter equation makes it possible to compute the spatial and spectral profiles of a THz pulse propagating in a linear medium after its generation in a nonlinear crystal.

To appreciate the possibilities offered by our approach of the near field, the spatiotemporal evolution of the THz beam generated by a spatially patterned femtosecond laser pulse is shown in Fig. 10. The spatial profile of the pump beam used to generate the THz pulse in a 1 mm-thick ZnTe crystal consists in two concentric circles having a radius of 2 mm and 7 mm. At the exit of the crystal, the THz and pump beam pattern matches. However as shown in Fig. 10, along short distances, as the THz pulse propagates in free space most of its energy is confined around the z-axis ($r = 0$). In fact, in the r-THz frequency plane, one can observe along $r = 0$ the modulation of the THz spectrum. To account for this behavior, we have proposed the following phenomenological interpretation. At the exit of the crystal, each NIR circle simultaneously generates a THz pulse. The THz pulses are emitted as Cherenkov-radiation-like conical beams. This is partly due to the thickness of each circle which is about the THz central wavelength (or much smaller for lower THz frequencies) [4, 32]. Under these conditions, emitted THz pulses can cross themselves in the intermediate field, generating a THz pulse train in a given point along the longitudinal axis. As shown in Fig. 10, under favorable geometries, spectral interference are expected along the propagation axis.

4.2. Comparison between experimental results and theory.

In order to check the validity of our approach we have used the setup shown in Fig. 11. The pump pulse (about 100 μJ, 800 nm, 50 fs at 1 kHz) generates THz pulses in the first ZnTe crystal. The THz spectrum generated in the latter crystal is then detected in a second ZnTe crystal. The measurements are performed along the axis for different distance d between the emitting and detecting ZnTe crystals. Both crystals were 1 mm-thick. The NIR pump and probe beams are separated by mean of a thin undoped silicon wafer. The latter is almost transparent to the THz pulse. As shown in Fig. 11 to pattern the pump beam profile we used a liquid crystal Spatial Light Modulator (SLM). The pump beam reflected by the SLM that carries the spatial phase added by the SLM is focussed by means of a 1 m focal length lens. At the focal point of the lens, where one records the spatial Fourier transform of the pump beam, we place the first ZnTe crystal. This device makes it possible to impinge patterned pump beam profiles in the ZnTe crystal that generates the THz pulse.

In the first set of experiments, THz pulse spectra are recorded along the propagation axis, under a two concentric NIR circles cross-section geometry, at three different distances from the generating crystal ($z = 1.5$ cm, 2.5 cm and 4.5 cm). The two NIR circles have radii of 2 mm and 4 mm. The result is shown in Fig. 12. Note that the spectrum shape changes completely as the position of the detecting crystal changes, in accordance with the theoretical considerations above. This behavior can be explained by the theoretical model. In fact, the optical path difference between the two THz pulses, simultaneously emitted by the two NIR circles, decreases as the detection distance increases along the z-axis. As a consequence the number of interference fringes in the spectrum decreases and the spectral peak moves at the

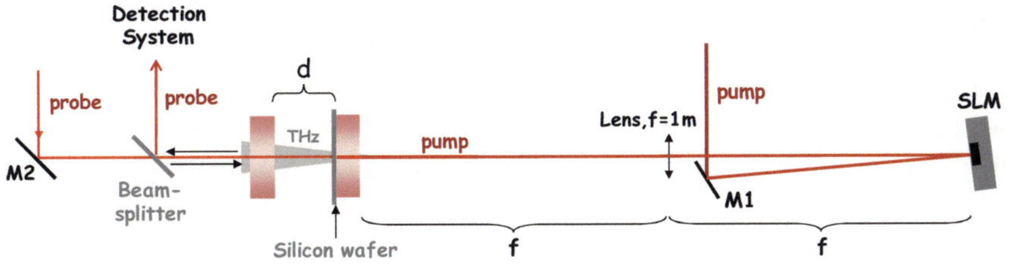

Figure 11. Experimental setup. Two 1 mm-thick ZnTe crystals are used to generate and detect the THz pulse. The pump and probe pulses are delivered by a regenerative amplifier (Coherent Legend) at 800 nm, with 50 fs duration, working at a 1 kHz repetition rate. The pump beam is spatial patterned by a SLM working in reflection. The THz pulse is measured along the propagation direction for different distances d between the generating and detecting crystal. The pump beam is focussed in the generating crystal by means of a 1 m focal lens. The latter also realizes the spatial Fourier transform (taken from [7]).

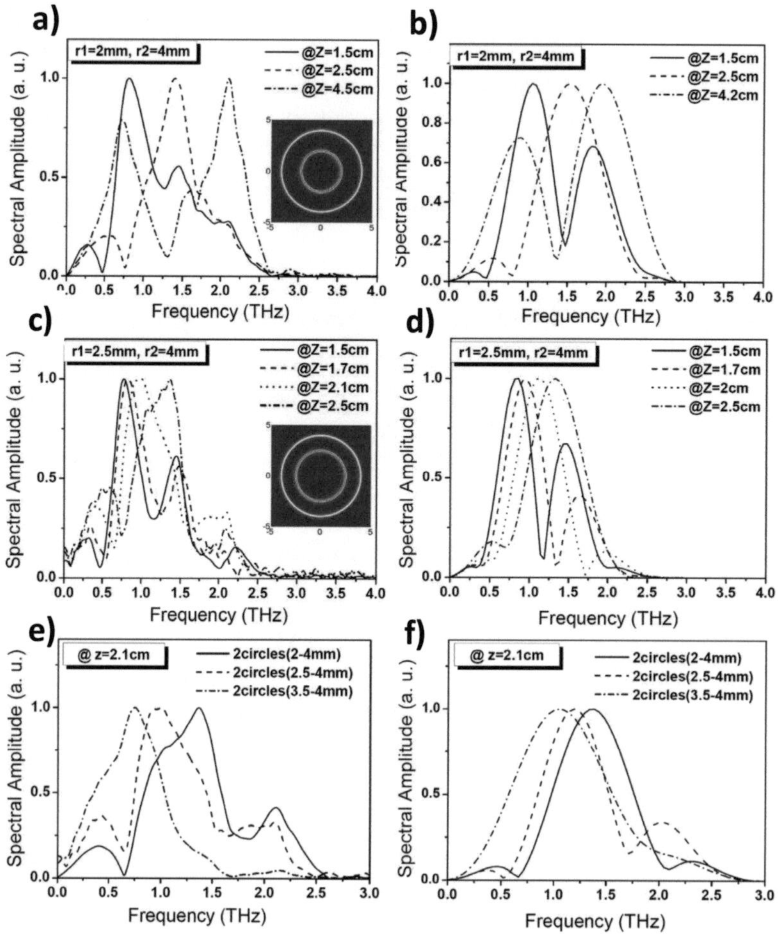

Figure 12. Experimental results (figures 3a, 3c, 3e) and corresponding numerical simulations (figures 3b, 3d, 3f) for THz spectra obtained by optical rectification in ZnTe under two concentric circles configurations of the pump laser beam profile. The insets show input beam rings. Each time, spectral amplitudes are normalized (taken from [7]).

same time toward the higher frequencies. A similar result is obtained for two concentric NIR circles with radii of 2.5 mm and 4 mm, and it is shown in Fig. 12c. Very good agreement is found between experiment (Fig. 12a and c) and numerical simulations (Fig. 12b and d). In the second set of experiments the detecting-ZnTe is placed at a fixed distance of 2.1 cm from the source-ZnTe. We still use a two concentric circles configuration for the pump geometry fixing the radius of the external circle at 4 mm but changing the radius of the internal circle from 2 mm to 3.5 mm. Spectra recorded for three different diameter values are shown in Fig. 12e. Note that the spectral peak moves toward the lower frequencies as the diameter of the internal circle increases. Here again, the numerical simulations (Fig. 12f) are in very good agreement with the experimental results (Fig. 12e).

Another important characteristic of this source is that most of the emitted THz radiation is spatially confined around the z-axis, during propagation, as shown by the numerical

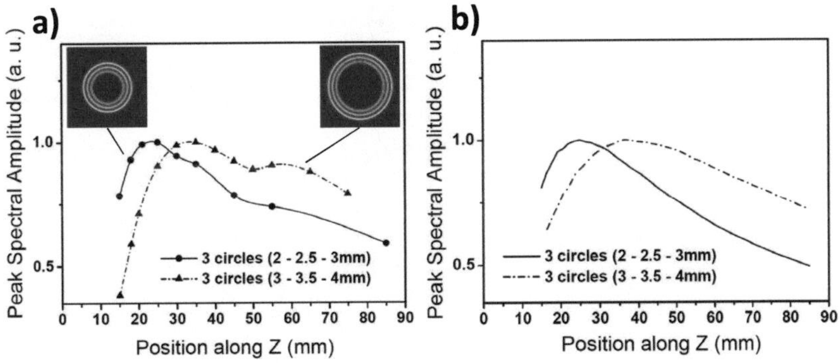

Figure 13. THz peak spectral amplitude against the position of the detecting ZnTe crystal along z. THz radiation shows interesting "self-focusing" properties. The "focal" position changes with the radius values of the circles. Fig. 4a Experiment (the insets show representative ring configuration, lines are guides for eyes); Fig. 4b simulations (taken from [7]).

simulation of Fig. 10. This behavior can be explained by the following simple argumentation. The z-axis is the zone of space which minimizes the optical path difference between the pulses coming from each NIR circle. Therefore constructive spatial interference between the spectral components of the THz pulses generated by each NIR circle occur mainly along the propagation axis. Furthermore, for a given cross-section geometry, the peak spectral amplitude exhibits a maximum value at a given point z (the focal point). The device acts as a tunable chromatic lens. This behavior is shown in the experimental curves of Fig. 13a, obtained using a three concentric circles geometry for the pump fluence. We study two configurations. The first one is represented by three circles with radii of 2 mm, 2.5 mm and 3 mm respectively (black solid circles); in the second configuration the three circles have radii of 3 mm, 3.5 mm and 4 mm respectively (black solid triangles). In the first case the peak spectral amplitude exhibits a maximum around $z = 20$ mm, in the second case the maximum is found around $z = 35$ mm. Corresponding numerical simulations, in Fig. 13b, are in very good agreement with the experimental results.

4.3. Generation of spectrally shaped THz pulses: theoretical approach.

In the previous sections, we have shown that the use of spatially patterned pump pulses strongly impacts the spectral shape of the near field generated THz pulses. In fact numerical simulations based on Eq. (34) indicate that very interesting spectrally shaped THz pulses can be achieved under optimized experimental conditions. In Fig. 14 are shown results predicted by the model under two "opposite" geometrical conditions: geometries with few concentric circles widely separated between them (Fig. 14a) and geometries with many close concentric circles (Fig. 14b). In the first case (two circles) the spectrum exhibits a large number of modes, which can be translated along the frequency axis by changing the detection position along z and/or the circle diameters. In the second case (five circles) a very narrow-band THz spectrum can be obtained, and its peak can be translated along the frequency axis by changing the detection position along z and/or the circle diameters. From the experimental point of view, the ZnTe crystal size is a major issue for fine tuning the spectral and spatial shapes of the THz pulses. In our experiments, the maximum diameter of the circle diameter was limited by the crystal diameter (10 mm). For instance, the result in Fig. 14a and b can only be achieved

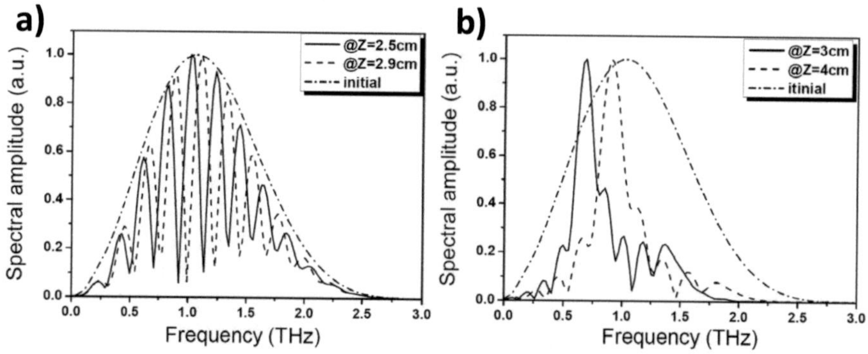

Figure 14. (a) Multi-mode theoretical THz spectra obtained with a two circles configuration (radii of 3 mm and 9 mm), at 2.5 cm (solid line) and 2.9 cm (dashed line) from the generating ZnTe. (b) Narrow-band THz theoretical spectra obtained with a five circles configuration (radii of 11, 12, 13, 14 and 15 mm) at 3 cm (solid line) and 4 cm (dashed line) from the generating ZnTe. Each time the initial THz spectrum is shown (dash-dotted line) (taken from [7]).

with 20 mm-large and 30 mm-large ZnTe crystals respectively. Furthermore, due to the initial Gaussian spatial distribution of the pump beam impinging the SLM, NIR circles in the ZnTe crystal have not the same intensity, as a consequence, they generate THz pulses with unequal amplitude. The good compromises in the present experiment were achieved in Fig. 12 when the circle separation was in between 0.5 mm and 2 mm, and in between 0.5 mm and 1 mm in Fig. 13.

5. Conclusion

We have presented two new approaches to synthesize THz pulse sequences and narrow-band tunable THz pulses via optical rectification of shaped NIR pulses in ZnTe. In the first one, the NIR pulses are shaped through a simple sinusoidal and/or triangular spectral phase modulation. Our experimental results show a tunability spanning the bandwidth permitted by the phase matching condition in ZnTe, *i.e.* ranging from 0.5 to 2.5 THz, and the possibility to generate spectra with bandwidth as narrow as 140 GHz. These results are in good agreement with a theoretical model taking into account THz generation by optical rectification, the shape of the optical pulses, dispersion in the THz range and the electro-optic detection. The second one has been less explored. It relies on the spatial shaping of the NIR pulse beam profile. Here again, we have demonstrated our ability to generate tunable THz pulses that agreed with our numerical simulations. One should note that this second technique is potentially as versatile as the first one, even if it is not relying on the same kind of pulse shaping method.

One important point is related to the intensity of the THz pulses we can generate using these techniques. A critical issue is the peak intensity that the used crystal can stand. Indeed, since we are dealing with second order nonlinear effects, one could want to increase as much as possible the peak intensity of the NIR femtosecond pulses. Unfortunately, as one increases the peak intensity new nonlinear phenomena are usually taking place. For example, we have recently shown that increasing the peak intensity of the pump beam in ZnTe induces two-photon absorption that impacts quite strongly the optical rectification efficiency as well as the available THz bandwidth [38]. Hence there is a compromise to find between peak power, crystal thickness, THz conversion efficiency. Such an effect becomes negligible if one uses for example $LiNbO_3$ crystal, thanks to its higher optical gap. However, this latter crystal has a lower nonlinear $\chi^{(2)}$, a higher intrinsic THz absorption above 3 THz and a lower THz spectral acceptance. But the latter limitation can be circumvented using the so-called tilted pulsefront phase matching technique [14], leading to THz fields with an amplitude up to $400\,\mathrm{kV}\cdot\mathrm{cm}^{-1}$ [24]. The combination of tunable THz pulse generation and nonlinear optics makes it possible to envision new developments in THz remote sensing, hyperspectral imaging, study and characterization of THz materials as well as high resolution spectroscopy largely emphasized by recent review articles [13, 14, 34].

Acknowledgments:

The Conseil Régional d'Aquitaine and the French National Research Agency (ANR 09-BLAN-0212) are acknowledged for financial supports.

Author details

J. Degert, M. Tondusson, J. Oberlé and É. Freysz
Université de Bordeaux, Laboratoire Ondes et Matiï£¡re d'Aquitaine, France

S. Vidal
Université de Bordeaux, Laboratoire Ondes et Matiï£¡re d'Aquitaine, France
Present address: Université de Bordeaux, Centre Lasers Intenses et Applications, France

C. D'Amico
Université de Bordeaux, Laboratoire Ondes et Matiï£¡re d'Aquitaine, France
Present address: Université Jean Monet, Laboratoire Hubert Curien, France

6. References

[1] Ahn, J.; Efimov, A. V.; Averitt, R. D. & Taylor, A. J. (2003). Terahertz waveform synthesis via optical rectification of shaped ultrafast laser pulses. *Opt. Express*, Vol. 11, 2486-2496, ISSN 1094-4087

[2] Akhmanov, S. A.; Vysloukh, V. A.; Chirkin, A. S. *Optics of Femtosecond Laser Pulses*, AIP, ISBN 0883188511, New York

[3] Amand, T.; Blanchet, V.; Girard, B. & Marie, X. (2005). Coherent Control in Atoms, Molecules and Solids, In: *Femtosecond Laser Pulses*, C. Rullière, (Ed.), 333-394, Springer-Verlag, ISBN 1441918507, New-York

[4] Auston, D. H. ; Cheung, K. P.; Valdmanis, J. A. & D. A. Kleinman, D. A. (1984). Cherenkov Radiation from Femtosecond Optical Pulses in Electro-Optic Media. *Phys. Rev. Lett.*, Vol. 53, 1555-1558, ISSN 0031-9007

[5] Brener, I.; Planken, P. C. M.; Nuss, M. C.; Pfeiffer, L.; Leaird, D. E. & Weiner, A. M. (1993). Repetitive excitation of charge oscillations in semiconductor heterostructures. *Appl. Phys. Lett.*, Vol. 63, 2213-2215, ISSN 0003-6951

[6] Caumes, J.-P.; Videau, L.; Rouyer, C. & Freysz, E. (2002). Kerr-Like Nonlinearity Induced via Terahertz Generation and the Electro-Optical Effect in Zinc Blende Crystals. *Phys. Rev. Lett.*, Vol. 89, 047401, ISSN 0031-9007

[7] D'Amico, C.; Tondusson, M.; Degert, J. & Freysz, E. (2009). Tuning and focusing THz pulses by shaping the pump laser beam profile in a nonlinear crystal. *Opt. Express*, Vol. 17, 592-597, ISSN 1094-4087

[8] Danielson, J. R.; Jameson, A. D.; Tomaino, J. L.; Hui, H.; Wetzel, J. D.; Lee, Y.-S. & Vodopyanov, K. L. (2008). Intense narrow band terahertz generation via type-II difference-frequency generation in ZnTe using chirped optical pulses. *J. Appl. Phys.*, Vol. 104, 03311, ISSN 0021-8979

[9] Feurer, T.; Vaughan, J. C. & Nelson, K. A. (2003). Spatiotemporal Coherent Control of Lattice Vibrational Waves. *Science*, Vol. 299, 374-377, ISSN 1095-9203

[10] Feurer, T.; Vaughan, J. C.; Hornung, T. & Nelson, K. A. (2004). Typesetting of terahertz waveforms. *Opt. Lett.*, Vol. 29, 1802-1804, ISSN 0146-9592

[11] Gallot, G. & Grischkowsky, D. (1999). Electro-optic detection of terahertz radiation. *J. Opt. Soc. Am. B*, Vol. 16, 1204-1212, ISSN 0740-3224

[12] Gallot, G.; Zhang, J.; McGowan, R. W.; Jeon, T. & Grischkowsky, D. (1999). Measurements of the THz absorption and dispersion of ZnTe and their relevance to the electro-optic detection of THz radiation. *Appl. Phys. Lett.*, Vol. 74, 3450-3452, ISSN 0003-6951

[13] Hebling, J. ; Yeh, K.-L.; Hoffmann, M. C. & Nelson, K. A. (2008). High Power THz Generation, THz Nonlinear Optics, and THz Nonlinear Spectroscopy. *IEEE J. Selected Topics in Quantum Electronics*, Vol. 14, 345-353, ISSN 1077-260X

[14] Hoffmann, M.C. & Fülöp, J.A. (2001). Intense ultrashort terahertz pulses: Generation and applications. *J. Phys. D: Appl. Phys.*, Vol. 44, 083001, ISSN 0022-3727

[15] Huggard, P. G.; Cluff, J. A.; Shaw, C. J.; Andrews, S. R.; Linfield, E. H. & Ritchie, D. A. (1997). Coherent control of cyclotron emission from a semiconductor using sub-picosecond electric field transients. *Appl. Phys. Lett.*, Vol. 71, 2647-2649, ISSN 0003-6951

[16] Hurlbut, W. C.; Norton, B. J.; Amer, N. & Lee, Y.-S. (2006). Manipulation of terahertz waveforms in nonlinear optical crystals by shaped optical pulses. *J. Opt. Soc. B*, Vol. 23, 90-93, ISSN 0740-3224

[17] Kohli, K. K.; Vaupel, A.; Chatterjee, S. & Rühle, W. W. (2009). Adaptive shaping of THz-pulses generated in ⟨100⟩ ZnTe crystals. *J. Opt. Soc. B*, Vol. 26, 74-78, ISSN 0740-3224

[18] Lee, Y.-S.; Meade, T.; Norris, T. B. & Galvanauskas, A. (2001). Tunable narrow-band terahertz generation from periodically poled lithium niobate. *Appl. Phys. Lett.*, Vol. 78, 3583-3585, ISSN 0003-6951

[19] Lee, Y.-S.; Amer, N. & Hurlbut, W. C. (2003). Terahertz pulse shaping via optical rectification in poled lithium niobate. *Appl. Phys. Lett.*, Vol. 82, 170-172, ISSN 0003-6951

[20] Lee, Y.-S. (2009). *Principles of Terahertz Science and Technology*, Springer-Verlag, ISBN 038709539X, New-York.

[21] Liu, Y.; Park, S.-G. & Weiner, A. M. (1996). Terahertz Waveform Synthesis via Optical Pulse Shaping. *IEEE J. Sel. Top. Quantum Electron.*, Vol. 2, 709-719, ISSN 1077-260X

[22] Monmayrant, A. & Chatel, B. (2004). New phase and amplitude high resolution pulse shaper. *Rev. Sci. Instrum.*, Vol. 75, 2668-2671, ISSN 0034-6748

[23] Monmayrant, A.; Weber, S. & Chatel, B. (2010). A newcomer's guide to ultrashort pulse shaping and characterization. *J. Phys. B: At. Mol. Opt. Phys.*, Vol. 43, 103001, ISSN 0953-4075

[24] Nagai, M.; Matsubara, E. & Ashida, M. (2012). High-efficiency terahertz pulse generation via optical rectification by suppressing stimulated Raman scattering process. *Opt. Express*, Vol. 20, 6509-6514, ISSN 1094-4087

[25] Nahata, A.; Weling, A. S. & Heinz, T. F. (1996). A wideband coherent terahertz spectroscopy system using optical rectification and electro-optic sampling. *Appl. Phys. Lett.*, Vol. 69, 2321-2323, ISSN 0003-6951

[26] Planken, P. C. M.; Brener, I.; Nuss, M. C.; Luo, M. S. C. & Chuang, S. L. (1993). Coherent control of terahertz charge oscillations in a coupled quantum well using phase-locked optical pulses. *Phys. Rev. B*, Vol. 48, 4903-4906, ISSN 1098-0121

[27] Qi, T.; Shin, Y.-H.; Yeh, K.-L.; Nelson, K. A. & Rappe, A. M. (2009). Collective Coherent Control: Synchronization of Polarization in Ferroelectric $PbTiO_3$ by Shaped THz Fields. *Phys. Rev. Lett.*, Vol. 102, 247603, ISSN 0031-9007

[28] Reimann, K. (2007). Table-top sources of ultrashort THz pulses. *Rep. Prog. Phys.*, Vol. 70, 1597-1632, ISSN 0034-4885.

[29] Renard, M.; Chaux, R.; Lavorel, B. & Faucher, O. (2004). Pulse trains produced by phase-modulation of ultrashort optical pulses: tailoring and characterization. *Opt. Express*, Vol. 12, 473-482, ISSN 1094-4087

[30] Sohn, J. Y.; Ahn, Y. H.; Park, D. J.; Oh, E. & Kim, D. S. (2002). Tunable terahertz generation using femtosecond pulse shaping. *Appl. Phys. Lett.*, Vol. 81, 13-15, ISSN 0003-6951

[31] Stepanov, A. G.; Hebling, J. & Kuhl, J. (2004). Generation, tuning, and shaping of narrow-band, picosecond THz pulses by two-beam excitation. *Opt. Express*, Vol. 12, 4650-4658, ISSN 1094-4087

[32] Stepanov, A. G.; Hebling, J. & Kuhl, J. (2005). THz generation via optical rectification with ultrashort laser pulse focused to a line. *Appl. Phys. B*, Vol. 81, 23-26, ISSN 0946-2171

[33] Stobrawa, G.; Hacker, M.; Feurer, T.; Zeidler, D.; Motzkus, M. & Reichel, F. (2001). A new high-resolution femtosecond pulse shaper. *Appl. Phys. B*, Vol. 72, 627-630, ISSN 0946-2171

[34] Tanaka, M.; Hirori, H. & Nagai, M. (2011). THz Nonlinear Spectroscopy of Solids. *IEEE Transactions on Terahertz Science and Technology*, Vol. 1, 301-312, ISSN 2156-342X

[35] Vaughan, J. C.; Feurer, T.; Stone, K. W. & Nelson, K. A. (2006). Analysis of replica pulses in femtosecond pulse shaping with pixelated devices. *Opt. Express*, Vol. 14, 1314-1328, ISSN 1094-4087

[36] Vaughan, J. C.; Hornung, T.; Feurer, T. & Nelson, K. A. (2005). Diffraction-based femtosecond pulse shaping with a two-dimensional spatial light modulator. *Opt. Lett.*, Vol. 30, 323-325, ISSN 0146-9592

[37] Vidal, S.; Degert, J.; Oberlï£¡, J. & Freysz, E. (2010). Femtosecond optical pulse shaping for tunable terahertz pulses generation. *J. Opt. Soc. Am. B.*, Vol. 27, 1044-1050, ISSN 0740-3224

[38] Vidal, S.; Degert, J.; Tondusson, M.; Oberlï£¡, J. & Freysz, E. (2011). Impact of dispersion, free carriers and two-photon absorption on the generation of intense THz pulses in ZnTe crystals. *Appl. Phys. Lett.*, Vol. 98, 191103, ISSN 0003-6951